SECOND INTERNATIONAL SYMPOSIUM ON NEGATIVE IONS, BEAMS AND SOURCES

To learn more about the AIP Conference Proceedings Series,
please visit http://proceedings.aip.org

SECOND INTERNATIONAL SYMPOSIUM ON NEGATIVE IONS, BEAMS AND SOURCES

Takayama City, Japan   16 – 19 November 2010

EDITORS

Yasuhiko Takeiri
Katsuyoshi Tsumori
National Institute for Fusion Science
Japan

All papers have been peer reviewed.

SPONSORING ORGANIZATION
National Institute for Fusion Science

Melville, New York, 2011
AIP I CONFERENCE PROCEEDINGS ■ 1390

Editors

Yasuhiko Takeiri
Katsuyoshi Tsumori

National Institute for Fusion Science
322-6 Oroshi-cho, Toki
509-5292 Japan

Email: takeiri@nifs.ac.jp
tsumori@nifs.ac.jp

Authorization to photocopy items for internal or personal use, beyond the free copying permitted under the 1978 U.S. Copyright Law (see statement below), is granted by the American Institute of Physics for users registered with the Copyright Clearance Center (CCC) Transactional Reporting Service, provided that the base fee of $30.00 per copy is paid directly to CCC, 222 Rosewood Drive, Danvers, MA 01923, USA. For those organizations that have been granted a photocopy license by CCC, a separate system of payment has been arranged. The fee code for users of the Transactional Reporting Services is: 978-0-7354-0955-2 /11/$30.00

© 2011 American Institute of Physics

No claim is made to original U.S. Government works.

Permission is granted to quote from the AIP Conference Proceedings with the customary acknowledgment of the source. Republication of an article or portions thereof (e.g., extensive excerpts, figures, tables, etc.) in original form or in translation, as well as other types of reuse (e.g., in course packs) require formal permission from AIP and may be subject to fees. As a courtesy, the author of the original proceedings article should be informed of any request for republication/reuse. Permission may be obtained online using Rightslink. Locate the article online at http://proceedings.aip.org, then simply click on the Rightslink icon/"Permission for Reuse" link found in the article abstract. You may also address requests to: AIP Office of Rights and Permissions, Suite 1NO1, 2 Huntington Quadrangle, Melville, NY 11747-4502, USA; Fax: 516-576-2450; Tel.: 516-576-2268; E-mail: rights@aip.org.

L.C. Catalog Card No. 2011935869
ISBN 978-0-7354-0955-2
ISSN 0094-243X
Printed in the United States of America

AIP Conference Proceedings, Volume 1390
Second International Symposium on Negative Ions, Beams and Sources

Table of Contents

Preface: Second International Symposium on Negative Ions, Beams and Sources — 1
Yasuhiko Takeiri

Symposium Organization — 3

List of Participants — 5

FUNDAMENTAL PROCESSES AND MODELING

The negative ion mean free path and its possible implications — 13
M. Bacal, R. McAdams, and B. Lepetit

3D Monte Carlo modeling of the EEDF in negative hydrogen ion sources — 22
R. Terasaki, A. Hatayama, T. Shibata, and T. Inoue

Origin of extracted negative ions by 3D PIC-MCC modeling. Surface vs volume comparison — 30
S. Mochalskyy, A. F. Lifschitz, and T. Minea

Modeling of neutrals and H⁻ transport in a large negative ion source — 39
N. Kameyama, D. Matsushita, S. Koga, R. Terasaki, and A. Hatayama

Fluid model analysis of the distribution of the negative ion density before the extraction from a tandem type of a plasma source — 48
Stiliyan St. Lishev, Antonia P. Shivarova, and Khristo Ts. Tarnev

Numerical analysis of the extraction of volume produced negative hydrogen ions in the extraction region of negative ion source — 58
S. Wada, S. Kuppel, T. Fukuyama, K. Miyamoto, A. Hatayama, and M. Bacal

Electric potential near the extraction region in negative ion sources with surface produced negative ions — 68
A. Fukano and A. Hatayama

A sheath model for negative ion sources including the formation of a virtual cathode — 78
R. McAdams, D. B. King, A. J. T. Holmes, and E. Surrey

About the extraction of surface produced ions in negative ion sources — 88
Francesco Taccogna, Pierpaolo Minelli, Savino Longo, and Mario Capitelli

Fluid and kinetic models of negative ion sheaths — 97
M. Cavenago

Monitoring surface condition of plasma grid of a negative hydrogen ion source — 107
M. Wada, T. Kasuya, T. Kenmotsu, and S. Tokushige

Effect of ion escape velocity and conversion surface material on H⁻ production — 113
O. Tarvainen, T. Kalvas, J. Komppula, H. Koivisto, E. Geros, J. Stelzer, G. Rouleau, K. F. Johnson, and J. Carmichael

Towards understanding the cesium cycle of the persistent H⁻ beams at SNS — 123
Martin P. Stockli, B. X. Han, S. N. Murray, T. R. Pennisi, M. Santana, and R. F. Welton

Simulation studies of hydrogen ion reflection from tungsten for the surface production of negative hydrogen ions — 134
Takahiro Kenmotsu and Motoi Wada

A discharge with a magnetic X-point as a negative hydrogen ion source — 140
Tsanko Tsankov and Uwe Czarnetzki

Cusp loss width in multicusp negative ion source: A rigorous mathematical treatment — 150
Ajeet Kumar and V. K. Senecha

Low-pressure small-radius hydrogen discharge as a volume-production based source of negative ions — 165
Tsvetelina V. Paunska, Antonia P. Shivarova, and Khristo Ts. Tarnev

Vibrational states of hydrogen molecules in one third LHD negative ion source — 175
M. Nishiura, Y. Matsumoto, K. Tsumori, M. Wada, and T. Inoue

Factors affecting VUV emission spectrum near Lyman-α from a hydrogen plasma source — 186
K. Ogino, T. Kasuya, Y. Kimura, M. Nishiura, S. Shimamoto, and M. Wada

Laser-photodetachment and Faraday-cup measurements in the expansion region of a tandem-type plasma source — 192
Stiliyan St. Lishev and Antonia P. Shivarova

ION SOURCES FOR ACCELERATORS

Latest results from the front end test stand high performance H⁻ ion source at RAL — 205
D. C. Faircloth, S. R. Lawrie, A. P. Letchford, C. Gabor, M. Whitehead, T. Wood, and M. Perkins

Performance of the H⁻ ion source supporting 1-MW beam operations at SNS — 216
B. X. Han, T. Hardek, Y. Kang, S. N. Murray Jr., T. R. Pennisi, C. Piller, M. Santana, R. F. Welton, and M. P. Stockli

Ion source development at the SNS — 226
R. F. Welton, N. J. Desai, B. X. Han, E. A. Kenik, S. N. Murray, T. R. Pennisi, K. G. Potter, B. R. Lang, M. Santana, and M. P. Stockli

Operation status of the J-PARC negative hydrogen ion source
    H. Oguri, K. Ikegami, K. Ohkoshi, Y. Namekawa, and A. Ueno — 235

Measurement of optical emission from the hydrogen plasma of the Linac4 ion source and the SPL plasma generator
    J. Lettry, S. Bertolo, A. Castel, E. Chaudet, J.-F. Ecarnot, G. Favre, F. Fayet, J.-M. Geisser, M. Haase, A. Habert, J. Hansen, S. Joffe, M. Kronberger, D. Lombard, A. Marmillon, J. Marques Balula, S. Mathot, O. Midttun, P. Moyret, D. Nisbet, M. O'Neil, M. Paoluzzi, L. Prever-Loiri, J. Sanchez Arias, C. Schmitzer, R. Scrivens, D. Steyaert, H. Vestergard, and M. Wilhelmsson — 245

Magnetic cusp configuration of the SPL plasma generator
    Matthias Kronberger, Elodie Chaudet, Gilles Favre, Jacques Lettry, Detlef Küchler, Pierre Moyret, Mauro Paoluzzi, Laurent Prever-Loiri, Claus Schmitzer, Richard Scrivens, and Didier Steyaert — 255

CERN LINAC4 H⁻ source and SPL plasma generator RF systems, RF power coupling and impedance measurements
    M. M. Paoluzzi, M. Haase, J. Marques Balula, and D. Nisbet — 265

Operation of negative ion sources at the cooler synchrotron COSY/Jülich
    R. Gebel, O. Felden, and R. Maier — 272

Ion source development for the proposed FNAL 750keV injector upgrade
    D. S. Bollinger — 284

Impedance measurement of an antenna with hydrogen plasma driven by 13.56MHz-rf for J-PARC H⁻ ion source
    A. Ueno, Y. Namekawa, K. Ohkoshi, K. Ikegami, and H. Oguri — 292

ION SOURCES FOR FUSION

Dependence of the performance of the long pulse RF driven negative ion source on the magnetic filter field
    W. Kraus, U. Fantz, D. Wünderlich, and NNBI Team — 303

Beam homogeneity dependence on the magnetic filter field at the IPP test facility MANITU
    P. Franzen, U. Fantz, and NNBI Team — 310

Characteristics of hydrogen negative ion source with FET based RF system
    A. Ando, T. Matsuno, T. Funaoi, N. Tanaka, K. Tsumori, and Y. Takeiri — 322

Improvement of plasma production for large area multi-antenna RF ion source
    Y. Oka, T. Shoji, and NIFS-NBI Group — 329

Analysis of discharge initiation in a RF hydrogen negative ion source
    T. Hayami, S. Yoshinari, R. Terasaki, A. Hatayama, and A. Fukano — 339

Quantification of cesium in negative hydrogen ion sources by laser absorption spectroscopy
    U. Fantz, Ch. Wimmer, and NNBI Team — 348

Cavity ring-down system for density measurement of negative hydrogen ion on negative ion source
Haruhisa Nakano, Katsuyoshi Tsumori, Kenichi Nagaoka, Masayuki Shibuya, Ursel Fantz, Masashi Kisaki, Katsunori Ikeda, Masaki Osakabe, Osamu Kaneko, Eiji Asano, Tomoki Kondo, Mamoru Sato, Seiji Komada, Haruo Sekiguchi, and Yasuhiko Takeiri ... 359

Comparison of optical emission spectroscopy and cavity ring-down spectroscopy in large-scaled negative-ion source
K. Ikeda, H. Nakano, K. Tsumori, U. Fantz, O. Kaneko, M. Kisaki, K. Nagaoka, M. Osakabe, and Y. Takeiri ... 367

Measurement of electron density near plasma grid of large-scaled negative ion source by means of millimeter-wave interferometer
K. Nagaoka, T. Tokuzawa, K. Tsumori, H. Nakano, Y. Ito, M. Osakabe, K. Ikeda, M. Kisaki, M. Shibuya, M. Sato, S. Komada, T. Kondo, H. Hayashi, E. Asano, Y. Takeiri, O. Kaneko, and NBI Group ... 374

Experimental mapping and benchmarking of magnetic field codes on the LHD ion accelerator
G. Chitarin, P. Agostinetti, A. Gallo, N. Marconato, H. Nakano, G. Serianni, Y. Takeiri, and K. Tsumori ... 381

OTHER ION SOURCES

Production of a high brightness H$^-$ beam by charge exchange of a hydrogen atom beam in a sodium jet
V. Davydenko, A. Ivanov, A. Kolmogorov, and A. Zelenski ... 393

Study of fluctuations in the CW Penning surface-plasma source of negative ions
Yuri Belchenko, Andrey Sanin, and Valery Savkin ... 401

Surface plasma source electrode activation by surface impurities
Vadim Dudnikov, B. Han, Rolland P. Johnson, S. N. Murray, T. R. Pennisi, M. Santana, Martin P. Stockli, and R. F. Welton ... 411

Negative hydrogen ion beam extracted from a Bernas-type ion source
N. Miyamoto and M. Wada ... 422

Development of hydrogen pair-ion source on the basis of catalytic ionization
W. Oohara, T. Maeda, and T. Higuchi ... 430

BEAM FORMATION, ACCELERATION, NEUTRALIZATION AND TRANSPORT

Application of 3D code IBSimu for designing an H$^-$/D$^-$ extraction system for the Texas A&M facility upgrade
T. Kalvas, O. Tarvainen, H. Clark, J. Brinkley, and J. Ärje ... 439

Improvement of voltage holding and high current beam acceleration by MeV accelerator for ITER NB
M. Taniguchi, M. Kashiwagi, T. Inoue, N. Umeda, K. Watanabe, H. Tobari, M. Dairaku, H. Yamanaka, K. Tsuchida, A. Kojima, M. Hanada, and K. Sakamoto ... 449

Study of beamlet deflection and its compensations in a MeV accelerator
Mieko Kashiwagi, Takashi Inoue, Masaki Taniguchi, Naotaka Umeda, Larry R. Grisham, Masayuki Dairaku, Jumpei Takemoto, Hiroyuki Tobari, Kazuki Tsuchida, Kazuhiro Watanabe, Haruhiko Yamanaka, and Keishi Sakamoto ... 457

Acceleration of 500 keV negative ion beams by tuning vacuum insulation distance on JT-60 negative ion source
A. Kojima, M. Hanada, Y. Tanaka, M. Taniguchi, M. Kashiwagi, T. Inoue, N. Umeda, K. Watanabe, H. Tobari, S. Kobayashi, Y. Yamano, L. R. Grisham, and JT-60 NBI Group ... 466

Magnetic insulation for electrostatic accelerators
L. R. Grisham ... 476

Space charge neutralization of DEMO relevant negative ion beams at low gas density
Elizabeth Surrey and Michael Porton ... 482

SIPHORE: Conceptual study of a high efficiency neutral beam injector based on photo-detachment for future fusion reactors
A. Simonin, L. Christin, H. de Esch, P. Garibaldi, C. Grand, F. Villecroze, C. Blondel, C. Delsart, C. Drag, M. Vandevraye, A. Brillet, and W. Chaibi ... 494

Improving negative ion beam quality and purity with a RF quadrupole cooler
Y. Liu ... 505

BEAMLINES AND FACILITIES

Stability of high power beam injection in negative-ion-based LHD-NBI
K. Tsumori, O. Kaneko, Y. Takeiri, M. Osakabe, K. Ikeda, K. Nagaoka, H. Nakano, M. Shibuya, E. Asano, T. Kondo, M. Sato, S. Komada, and H. Sekiguchi ... 517

Modeling activities on the negative-ion-based neutral beam injectors of the large helical device
P. Agostinetti, V. Antoni, M. Cavenago, G. Chitarin, H. Nakano, N. Pilan, G. Serianni, P. Veltri, Y. Takeiri, and K. Tsumori ... 526

Development of the JT-60SA neutral beam injectors
M. Hanada, A. Kojima, T. Inoue, K. Watanabe, M. Taniguchi, M. Kashiwagi, H. Tobari, N. Umeda, N. Akino, M. Kazawa, K. Oasa, M. Komata, K. Usui, K. Mogaki, S. Sasaki, K. Kikuchi, S. Nemoto, K. Ohshima, Y. Endo, T. Simizu, N. Kubo, M. Kawai, and L. R. Grisham ... 536

Status of the 1 MeV accelerator design for ITER NBI
M. Kuriyama, D. Boilson, R. Hemsworth, L. Svensson, J. Graceffa, B. Schunke, H. Decamps, M. Tanaka, T. Bonicelli, A. Masiello, M. Bigi, G. Chitarin, A. Luchetta, D. Marcuzzi, R. Pasqualotto, N. Pomaro, G. Serianni, P. Sonato, V. Toigo, P. Zaccaria, W. Kraus, P. Franzen, B. Heinemann, T. Inoue, K. Watanabe, M. Kashiwagi, M. Taniguchi, H. Tobari, and H. De Esch ... 545

Design and overview of 100 kV bushing for the DNB injector of ITER
Sejal Shah, S. Rajesh, S. Nishad, B. Srusti, M. Bandyopadhyay, C. Rotti, M. J. Singh, G. Roopesh, A. K. Chakraborty, B. Schunke, R. Hemsworth, J. Chareyre, and L. Svensson 555

RAMI analyses of heating neutral beam and diagnostic neutral beam systems for ITER
D. H. Chang, S. Lee, R. Hemsworth, D. van Houtte, K. Okayama, F. Sagot, B. Schunke, and L. Svensson 567

Physics and engineering studies on the MITICA accelerator: Comparison among possible design solutions
P. Agostinetti, V. Antoni, M. Cavenago, G. Chitarin, N. Pilan, D. Marcuzzi, G. Serianni, and P. Veltri 574

Numerical assessment of the diagnostic capabilities of the instrumented calorimeter for SPIDER (STRIKE)
M. Dalla Palma, M. De Muri, R. Pasqualotto, A. Rizzolo, G. Serianni, and P. Veltri 584

Sensitivity analysis of the off-normal conditions of the SPIDER accelerator
P. Veltri, P. Agostinetti, V. Antoni, M. Cavenago, G. Chitarin, N. Marconato, N. Pilan, E. Sartori, and G. Serianni 594

RF - plasma source commissioning in Indian negative ion facility
M. J. Singh, M. Bandyopadhyay, G. Bansal, A. Gahlaut, J. Soni, Sunil Kumar, K. Pandya, K. G. Parmar, J. Sonara, Ratnakar Yadava, A. K. Chakraborty, W. Kraus, B. Heinemann, R. Riedl, S. Obermayer, C. Martens, P. Franzen, and U. Fantz 604

Cesium delivery system for negative ion source at IPR
G. Bansal, K. Pandya, M. Bandyopadhyay, A. Chakraborty, M. J. Singh, J. Soni, A. Gahlaut, and K. G. Parmar 614

Conceptual design, implementation and commissioning of data acquisition and control system for negative ion source at IPR
Jignesh Soni, Ratnakar Yadav, A. Gahlaut, G. Bansal, M. J. Singh, M. Bandyopadhyay, K. G. Parmar, K. Pandya, and A. Chakraborty 624

Computer system for unattended control of negative ion source
P. V. Zubarev, A. D. Khilchenko, A. N. Kvashnin, D. V. Moiseev, E. A. Puriga, A. L. Sanin, and V. Ya. Savkin 634

Status of NIO1 construction
M. Cavenago, E. Fagotti, M. Poggi, G. Serianni, V. Antoni, M. Bigi, F. Fellin, E. Gazza, M. Recchia, P. Veltri, S. Petrenko, and T. Kulevoy 640

Simulation and design of a reflection magnet for the EAST neutral beam system
Liang Li Zhen and Hu Chun Dong 650

NEW APPROACHES AND APPLICATIONS

Negative ions for emerging interdisciplinary applications
Samar K. Guharay 661

Extraction and acceleration of ions from an ion-ion plasma   668
   Lara Popelier, Ane Aanesland, and Pascal Chabert

Development of a negative hydrogen ion source for spatial beam profile measurement of a high intensity positive ion beam   675
   Katsuhiro Shinto, Motoi Wada, Tomoaki Nishida, Yasuhiro Demura, Daichi Sasaki, Katsuyoshi Tsumori, Masaki Nishiura, Osamu Kaneko, Masashi Kisaki, and Mamiko Sasao

Electron strippers for compact neutron generators   684
   K. Terai, N. Tanaka, M. Kisaki, K. Tsugawa, A. Okamoto, S. Kitajima, M. Sasao, T. Takeno, A. J. Antolak, K. N. Leung, and M. Wada

Extraction of low-energy negative oxygen ions for thin film formation   692
   M. Vasquez, Jr., D. Sasaki, T. Kasuya, S. Maeno, and M. Wada

Author Index   701

Preface: Second International Symposium on Negative Ions, Beams and Sources

The Second International Symposium on Negative Ions, Beams and Sources — NIBS2010 — was held in Takayama, Japan, from November 16 until 19 in 2010. NIBS2010 covered all areas of science and technology related to negative ions. The symposium originally started in 1977 at Brookhaven National Laboratory with the name of "Symposium on Production and Neutralization of Negative Ions and Beams (PNNIB)". Since then, the symposia had provided a discussion forum mainly for research and development related to hydrogen negative ions and beams in the 1980s and 1990s. During these periods remarkable progress was made especially for the development of negative ion sources used in fusion and accelerator fields. After entering the new century, the symposium has evolved including other application fields, and for further development the name of the symposium was changed at the previous symposium in 2008. Thus, NIBS2010 was the second symposium after the renewal and the thirteenth one in total. Presently, the symposium is the only international forum dedicated to all aspects of negative ions in physics and technology from formation to application.

NIBS2010 was organized by the National Institute for Fusion Science, Toki, Japan, and this was held in Asia for the first time. The symposium venue, Takayama, is located in a mountain area which preserves the Japanese traditional atmosphere. The registered participants were 94 from Europe, US, India, Korea, Russia, China, and Japan. The presentations numbered 94, and 73 articles are contained as scientific papers in these proceedings.

The main subjects in the symposium are negative ion sources and the related beam acceleration and transport, and the fundamental physics and modeling for negative ion production and acceleration are discussed. The modeling and related physics are being remarkably developed with 2D- and 3D-simulations, which treat the negative ion formation in the plasma or on the cesiated surface. The negative ion transport to the extraction boundary and the sheath formation in the negative-ion-dominated plasma are discussed as a currently noteworthy topic. The reliable measurement of the negative ions is reported, which is expected to accelerate the model development. RF-driven negative ion sources are also highlighted with a view of long-period operation for both fusion and accelerator applications. The usage of cesium for negative ion production is an old and new subject. While cesium behavior in the plasma is investigated, efficient negative ion production without cesium is explored. The facilities for high-current and high-energy DC acceleration to 1 MeV are currently designed and constructed for the ITER NBI system, and the related activities are reported. New approaches of negative ion application are introduced, such as negative heavy ions used for material analyses and biological application, space propulsion, and material syntheses.

Here, it is a great pleasure to express my gratitude to all the people who made NIBS2010 a successful symposium. All members of the International Program Committee, who played a significant role in organizing the symposium in addition to developing the program, are greatly acknowledged. I also appreciate the support of Takayama City. For the Proceedings, Kenichi Nagaoka greatly supported the editorial works, and we would like to thank him as well as every referee for making these proceedings valuable. Also, I would like to thank Yohko Nishiyama for arranging various affairs related to collecting the proceedings.

I am pleased to announce that the next symposium will be held in 2012 at Jyvaskyla in Finland, hosted by the University of Jyvaskyla, Accelerator Laboratory which is internationally recognized as a center of nuclear and accelerator based physics. I send my best wishes to Olli Tarvainen, the Chair of the symposium for a successful NIBS2012.

Last but not least, I would like to express my sincere gratitude to all participants for making this symposium a great success.

Yasuhiko Takeiri
Proceedings Editor
Chair, NIBS2010

National Institute for Fusion Science
Toki, Japan

Symposium Organization

INTERNATIONAL PROGRAM COMMITTEE

M. Bacal (Ecole Polytechnique, France)
Y. Belchenko (INP, Russia)
D. Faircloth (STFC, UK)
R. Hemsworth (ITER Organization)
Y. Hwang (SNU, Korea)
T. Inoue (JAEA, Japan)
W. Kraus (IPP, Germany)
J. Lettry (CERN, Switzerland)
K.N. Leung (LBNL, USA)
Y. Mori (Kyoto University, Japan)
M. Stockli (ORNL, USA)
E. Surrey (CCFE, UK)
Y. Takeiri (NIFS, Japan)
O. Tarvainen (University of Jyvaskyla, Finland)

LOCAL ORGANIZING COMMITTEE

Y. Takeiri (Chairman)
K. Tsumori (Scientific Secretary)
M. Osakabe
K. Ikeda
K. Nagaoka
H. Nakano
M. Kisaki
Y. Nishiyama

SYMPOSIUM LOCATION

Hida Earth Wisdom Center, Takayama City, Gifu, Japan

SPONSORED BY

Japan Society for the Promotion of Science (JSPS)
Takayama city
HITACHI, Ltd.
TOSHIBA Corporation

LIST OF PARTICIPANTS

Aanesland	Ane	Ecole polytechnique	ane.aanesland@lpp.polytechnique.fr
An	YoungHwa	Seoul National University	ayh1800@snu.ac.kr
Ando	Akira	Tohoku University	akira@ecei.tohoku.ac.jp
Bacal	Marthe	Ecole Polytechnique	marthebacal@free.fr
Bansal	Gourab	Institute for Plasma Research	bansal@ipr.res.in
Bollinger	Daniel S.	Fermilab	bollinger@fnal.gov
Cavenago	Marco	INFN-LNL	cavenago@lnl.infn.it
Chakraborty	Arun	ITER-India, Institute for Plasma research	arunkc@iter-india.org
Chang	Doo Hee	Korea Atomic Energy Research Institute	doochang@kaeri.re.kr
Chitarin	Giuseppe	Univ. of Padova and Consorzio RFX	chitarin@igi.cnr.it
Cho	Won Hwi	Seoul National University	whjo05@snu.ac.kr
Davydenko	Vladimir I.	Budker Institue of Nuclear Physics	V.I.Davydenko@inp.nsk.su
Demura	Yasuhiro	Doshisha University	dtj0101@mail4.doshisha.ac.jp
Dudnikov	Vadim	Muons, Inc.	dvg43@yahoo.com
Duré	Franck	Paris-Sud XI unversity	franck.dure@u-psud.fr
Engeln	Richard	Eindhoven University of Technology	r.engeln@tue.nl

Faircloth	Dan	ISIS, RAL, STFC	dan.faircloth@stfc.ac.uk
Fantz	Ursel	Max-Planck-Institut für Plasmaphysik	ursel.fantz@ipp.mpg.de
Franzen	Peter	Max-Planck-Institut für Plasmaphysik	peter.franzen@ipp.mpg.de
Fubiani	Gwenael	CNRS	gwenael.fubiani@laplace.univ-tlse.fr
Fukano	Azusa	Tokyo Metropolitan College of Industrial Technology	fukano@s.metro-cit.ac.jp
Fukumasa	Osamu	Ube National College of Technology	fukumasa@ube-k.ac.jp
Gebel	Ralf	Forschungszentrum Juelich	r.gebel@fz-juelich.de
Grisham	Larry Richard	Princeton University	lgrisham@pppl.gov
Guharay	Samar Kumar	The MITRE Corporation	sguharay@mitre.org
Han	Baoxi	Oak Ridge National Laboratory	hanb@ornl.gov
Hanada	Masaya	JAEA	hanada.masaya@jaea.go.jp
Hatayama	Akiyoshi	Keio University	akh@ppl.appi.keio.ac.jp
Hayami	Takahiro	Keio University	taka0803@ppl.appi.keio.ac.jp
Heinemann	Bernhard	Max-Planck-Institut fuer Plasmaphysik	bernd.heinemann@ipp.mpg.de
Hillion	François	Cameca	hillion@cameca.fr
Ikeda	Katsunori	NIFS	ikeda.katsunori@lhd.nifs.ac.jp

Inoue	Takashi	JAEA	inoue.takashi52@jaea.go.jp
Ji	Qing	Lawrence Berkeley National Laboratory	qji@lbl.gov
Kalvas	Taneli	University of Jyvaskyla	taneli.kalvas@jyu.fi
Kameyama	Nobuhiko	Keio University	kameyama@ppl.appi.keio.ac.jp
Kaneko	Osamu	NIFS	kaneko.osamu@LHD.nifs.ac.jp
Kenmotsu	Takahiro	Doshisha University	tkenmots@mail.doshisha.ac.jp
Kisaki	Masashi	NIFS	kisaki.masashi@LHD.nifs.ac.jp
Kojima	Atsushi	JAEA	kojima.atsushi@jaea.go.jp
Kraus	Werner	Max-Planck-Institut für Plasmaphysik	kraus@ipp.mpg.de
Kronberger	Matthias	CERN	matthias.kronberger@cern.ch
Kuriyama	Masaaki	ITER	Masaaki.Kuriyama@iter.org
Lettry	Jacques Alexandre	CERN-ABP	jacques.lettry@cern.ch
Liang	Li Zhen	Institute of Plasma Physics, Chinese Academy of Sciences	lzliang@ipp.ac.cn
Lishev	Stiliyan Stiliyanov	Faculty of Physics, Sofia University	lishev@phys.uni-sofia.bg
Liu	Yuan	Oak Ridge National Laboratory	liuy@ornl.gov
McAdams	Roy	CCFE	Roy.McAdams@ccfe.ac.uk
Mochalskyy	Serhiy	Université Paris Sud 11	serhiy.mochalskyy@u-psud.fr

Mori	Yoshiharu	Kyoto University	mori@rri.kyoto-u.ac.jp
Nagaoka	Kenichi	NIFS	nagaoka@nifs.ac.jp
Nakano	Haruhisa	NIFS	nakano@nifs.ac.jp
Nakano	Ryotaro	Kyoto University	rnakano@rri.kyoto-u.ac.jp
Nishiura	Masaki	NIFS	nishiura.masaki@LHD.nifs.ac.jp
Niwa	Yusuke	Fukui University	niwa@rri.kyoto-u.ac.jp
Ogino	Ken	Doshisha University	dtj0139@mail4.doshisha.ac.jp
Oguri	Hidetomo	JAEA	oguri.hidetomo@jaea.go.jp
Oka	Yoshihide	NIFS	oka@LHD.nifs.ac.jp
Okabe	Kota	Fukui University	kota@rri.kyoto-u.ac.jp
Oohara	Wataru	Yamaguchi University	oohara@yamaguchi-u.ac.jp
Osakabe	Masaki	NIFS	osa@nifs.ac.jp
Paoluzzi	Mauro Mario	CERN	mauro.paoluzzi@cern.ch
Paunska	Tsvetelina Venelinova	Faculty of Physics, Sofia University	cwalchew@phys.uni-sofia.bg
Popelier	Lara	Ecole polytechnique	lara.popelier@lpp.polytechnique.fr
Robert	Welton Frederick	ORNL-SNS	welton@ornl.gov

Rotti	Chandramouli	ITER-India, Institute for Plasma Research	crotti@iter-india.org
Sakakita	Hajime	National Institute of Advanced Industrial Science and Technology	h.sakakita@aist.go.jp
Sanin	Andrey	Budker Institute of Nuclear Physics	sanin@inp.nsk.su
Sasao	Mamiko	Tohoku University	mamiko.sasao@qse.tohoku.ac.jp
Senecha	Vinod Kumar	Raja Ramanna Centre for Advanced Technology	senecha@rrcat.gov.in
Serianni	Gianluigi	Consorzio RFX	gianluigi.serianni@igi.cnr.it
Shah	Sejal P.	Institute for Plasma Research, Gandhinagar	sshah@iter-india.org
Shinto	Katsuhiro	JAEA	shinto.katsuhiro@jaea.go.jp
Simonin	Alain	CEA	alain.simonin@cea.fr
Singh	Mahendrajit	ITER-India	mahendrajit@iter-india.org
Soni	Jignesh Jayendrabhai	Institute for plasma resesarch	jsoni2k@yahoo.com
Stockli	Martin	Oak Ridge National Laboratory	stockli@ornl.gov
Surrey	Elizabeth	CCFE	elizabeth.surrey@ccfe.ac.uk
Taccogna	Francesco	IMIP-CNR	francesco.taccogna@ba.imip.cnr.it
Takagi	Akira	High Energy Accelerator Research Organization	akira.takagi@kek.jp
Takeiri	Yasuhiko	NIFS	takeiri@nifs.ac.jp
Tanaka	Nozomi	Tohoku University	nozomi@ecei.tohoku.ac.jp

Taniguchi	Masaki	JAEA	taniguchi.masaki@jaea.go.jp
Tarvainen	Olli	University of Jyvaskyla	olli.tarvainen@jyu.fi
Terai	Kensuke	Tohoku University	kensuke.terai@ppl2.qse.tohoku.ac.jp
Tsankov	Tsanko Vaskov	Ruhr-University Bochum	Tsanko.Tsankov@rub.de
Tsuji	Hiroshi	Kyoto University	tsuji@kuee.kyoto-u.ac.jp
Tsumori	Katsuyoshi	NIFS	tsumori@nifs.ac.jp
Ueno	Akira	J-PARC	akira.ueno@j-parc.jp
Vasquez	Magdaleno Jr Rigodon	Doshisha University	eti1106@mail4.doshisha.ac.jp
Veltri	Pierluigi	Consorzio RFX	pierluigi.veltri@igi.cnr.it
Wada	Motoi	Doshisha University	mwada@mail.doshisha.ac.jp
Wada	Shohei	Keio University	wady328@ppl.appi.keio.ac.jp
Welton	Robert Frederick	ORNL-SNS	welton@ornl.gov
Yamashita	Yasuo	Hitaci	yamashita-yasuo@k-kec.co.jp

FUNDAMENTAL PROCESSES AND MODELING

The Negative Ion Mean Free Path And Its Possible Implications

M. Bacal[a], R. McAdams[b] and B. Lepetit[c,d]

[a] LPP, Ecole Polytechnique, Palaiseau, UPMC, Université PARIS-SUD 11,
UMR CNRS 7648, France
[b] EURATOM/CCFE Association, Culham Science Center, Abingdon, Oxfordshire,
OX14 3DB, UK
[c] Université de Toulouse, UPS, Laboratoire Collisions, Agrégats, Réactivité, IRSAMC,
F-31062 Toulouse, France
[d] CNRS, UMR 5589, F-31062 Toulouse, France

Abstract. The knowledge of the mean free path (mfp) of the negative ions is important for many purposes, *e.g.* evaluating the distance from which the ions can be extracted, or understanding the significance of experimental results. We will present the mfp for similar conditions in volume and caesium seeded sources. It appears that the mfp is longer in the caesium seeded source than in the volume source under identical conditions.

Keywords: negative ion, hydrogen, mean free path
PACS: 41.75.Cn, 52.75.-d, 32.80.Gc

INTRODUCTION

Several types of collisions occurring in negative ion sources destroy the negative ions. These are collisions with positive ions (mutual neutralisation), collisions with electrons (electron detachment) and collisions with hydrogen atoms (associative and non-associative detachment.

The knowledge of the mean free path (mfp) is useful in order to evaluate the distance from which negative ions can be extracted and to understand the relative importance of various collisions. In the case of collisions of negative ions with another heavy particle (a positive ion or an atom) the mfp, λ, is defined as

$$\lambda = \frac{1}{n\sigma} \qquad (1)$$

where n is the density of the positive ions or atoms, colliding with the H⁻ ion, and σ is the cross section of the corresponding collision process. In the case of negative ion destruction in collisions with electrons leading to electron detachment (ED), the mfp, λ_{ED}, is :

$$\lambda_{ED} = \frac{v^-}{n_e <\sigma v_e>}. \qquad (2)$$

Here v^- is the negative ion velocity, n_e is the electron density and $<\sigma v_e>$ is the reaction rate coefficient. The latter can be found expressed by the electron temperature e.g. in [1].

Since the mfp is inversely proportional to the density of the particles colliding with the H⁻ ion, and to the cross section of the corresponding collision process (or in the case of electron collisions to its reaction rate coefficient), the shortest mfp corresponds to the dominating destruction process. One is most interested in the value of the mfp near the extraction opening, in order to evaluate the distance from which the negative ions can be extracted. We will estimate the mfp for the background plasma. No estimate will be made for the mfp in the sheath, since its size is small compared to the mfp.

THE MEAN FREE PATH IN VOLUME AND CAESIUM SEEDED SOURCES

The mfp for the same collision process and the same density of the collision partner may be very different in the volume source compared to the caesium seeded source, due to the different characteristics of the sheath in these two cases. In the volume source the plasma grid is usually biased at a potential close to the plasma potential, in order to optimize the extracted H⁻ current and reduce the extracted electron current. In this case the sheath is accelerating the plasma electrons and negative ions towards the plasma grid, but does not affect the energy of the volume produced negative ions in the presheath and background plasma. In the caesium seeded source, the plasma grid is biased negative with respect to the plasma potential and a sheath is formed accelerating the incident positive ions towards the plasma grid, and also accelerating the negative ions formed on the surface of the plasma grid (by atoms and positive ions) towards the plasma.

In the case of hydrogen, the cross sections of most processes can be found in Data Bases such as, e.g. EIRENE [2]. Some cross sections are presented versus the collision energy in eV / atomic mass unit (e.g. the resonant charge exchange cross section of H⁻ ion with a hydrogen atom). However most of the cross sections are presented versus the relative energy in the frame of the center of mass (E_{CM}). The calculation of E_{CM} in simple cases is exposed in textbooks of Classical Mechanics, e.g. [3]. In a first approach we will discuss the mfp in the background plasma, where positive ions and atoms have an isotropic distribution of the velocity v_1. We are interested in the following two cases :

Case 1. In the volume source negative ions have an isotropic distribution of the velocity v_2 in the complete space. The square of average relative velocity between two particles $<v^2>$ is then given by

$$<v^2> = <v_1^2> + <v_2^2>$$

or in terms of energy :

$$\frac{E_{CM}}{\mu} = \frac{E_1}{m_1} + \frac{E_2}{m_2} \qquad (3)$$

where $\mu = \frac{m_1 m_2}{m_1 + m_2}$ is the reduced mass. If the two particles have equal masses then this becomes:

$$E_{CM} = \frac{E_1}{2} + \frac{E_2}{2} \qquad (4)$$

Case 2. In the caesium seeded source the negative ions are emitted from the caesiated plasma grid surface with velocities having a cosine angular distribution (as suggested by Seidl et al [4]) and also acquire in the sheath the energy $e\Delta V$ (ΔV is the sheath voltage). Since the distribution of the atoms and ions in the plasma is isotropic, Eq. 4 is also relevant.

In order to estimate the emission energy of the negative ions it will be assumed, following Seidl *et al* [4] that the atoms are reflected with slightly reduced energy E':

$$E' = R_E E \qquad (5)$$

where E is the incident atom kinetic energy However, according to [4] only atoms with energy greater than the threshold energy E_{thr},

$$E_{thr} = \phi - E_A \qquad (6)$$

can form negative ions, where ϕ is the caesiated surface work function assumed to be 1.5 eV, and E_A = 0.75 eV is the electron affinity of the hydrogen atom. Based on an initial estimate a value of 1.2 eV has been chosen for the emission energy of the negative ions for an atom temperature of 0.8 eV, but this could be an upper value.

An important feature of the cross section for mutual neutralisation of H⁻ ions with positive ions and associative detachment of H⁻ with H atoms is that their highest values correspond to the lowest H⁻ ion energy. However, a second, non-associative detachment process exists along with the associative detachment in the collisions with H atoms, which becomes dominating at $E_{CM} >$ 4 eV.

In the background plasma of volume H⁻ ion sources, assuming an isotropic distribution of both atoms and negative ions, E_{CM} for (associative and non-associative) detachment by H atoms (denoted H_{det}) is calculated using Eq. 4. If the temperatures of the H⁻ ions and hydrogen neutral atoms are known, their energy can be calculated as $E = (3/2) kT$. A similar calculation can be carried out for the H⁻ destruction by mutual neutralisation with positive ions.

We will consider the caesium seeded rf source studied by Wünderlich *et al* [5] using a PIC code for the plasma and sheath. The plasma parameters used as input for the PIC code are the electron temperature T_e = 2 eV, H⁺ density $n(H^+)$ = 4x10^{17} m^{-3}, H⁺ temperature $T(H^+)$ = 0.8 eV, H density $n(H)$ = 10^{19} m^{-3}, H temperature $T(H)$ = 0.8 eV,

$n(Cs^+) = 10^{16}$ m^{-3}. The electron density is found as a result of the simulation to be 2.74 x10^{17} m^{-3}. The negative ion temperature is not mentioned. We will calculate the mfp for a volume source and a caesium seeded source having these parameters.

Associative and non-associative detachment

Case 1: the volume source. We consider the negative ion temperature T(H$^-$) = 0.1 eV. This value was found in [6] from laser photodetachment experiments performed in the ion source extraction region.

With T(H) = 0.8 eV, we find from Eq. 4 E_{CM} = 0.67 eV and the corresponding σ_{Hdet} is 1.42 x10^{-19} m^2. With n(H) = 1 x 10^{-19} m^{-3} the mfp is λ_{Hdet} = 0.70 m.

Case 2: the caesium seeded source. The energy of the atomic population is E_1 = 1.2 eV while the negative ions emitted with an average energy 1.2 eV by the plasma grid, acquire an energy of ΔV = 4.2 eV in the sheath [5] and leave the sheath edge with an energy E_2 = 5.4 eV. Thus using Eq. 4 leads to E_{CM} = 3.31 eV. The cross section for detachment by atoms is in this case 1.46 x 10^{-19} m^2. With the same value n(H) = 1 x 10^{-19} m^{-3} as in the case of the volume source, λ_{Hdet} = 0.69 m.

Mutual Neutralisation

Case 1: the volume source. With E_{CM} = 0.67 eV relevant to the volume operation of the source, and σ_{MN} = 4.1x10^{-18} m^2, λ_{MN} = 0.61 m.

Case 2: the caesium seeded source. With the E_{CM} = 3.31 eV and σ_{MN} = 1.35 x10^{-18} m^2, the mfp is λ_{MN} = 1.85 m, three times longer than in volume operation. This is due to the higher negative ion energy and E_{CM} leading to a lower cross section for mutual neutralisation.

Detachment by Electrons

The reaction rate coefficients for electron detachment, calculated for a Maxwellian electron energy distribution, are from [1], where they are expressed versus the electron temperature. In this case for T_e = 2 eV $<\sigma v_e>$ = 8x10^{-8} cm^3/s.

Case 1: the volume source. With T(H$^-$) = 0.1 eV, the negative ion velocity in Eq. 2 is calculated from the H$^-$ ion energy E_2 = 0.15 eV. Thus one finds, using Eq. 2, with n_e = 2.74 x10^{17} m^{-3}, λ_{ED} = 0.23 m.

Case 2: the caesium seeded source. In this case, the negative ion velocity in Eq. 2 is calculated from the energy at the sheath edge E_2 = 4.2 + 1.2 = 5.4 eV. One finds λ_{ED} = 1.47 m, seven times longer than in volume operation. This is due to the higher negative ion velocity in caesium seeded operation.

These calculations are summarised in Table 1. Note that these estimates were done for a hydrogen plasma and it was assumed that all the positive ions were atomic ones, H$^+$. Note that in the volume source the dominant destruction process is electron detachment, while in the caesium seeded source the H$_{det}$ is dominant.

The mfp of the three destruction processes considered above are taken into account in the combined mfp $\lambda_C = (1/\lambda_{Hdet} + 1/\lambda_{MN} + 1/\lambda_{ED})^{-1}$. In the case of the caesium seeded source, the combined mfp $\lambda_C = 0.375$ m is almost two times shorter than that of the dominating destruction process due to atoms. In the volume source $\lambda_C = 0.135$ m is also almost two times shorter than that for electron detachment, which is the dominating destruction process in this case.

In the caesium seeded source the presence of caesium ions enhances the destruction by mutual neutralization. The cross section for this process is reported in [7]. With a caesium ion density $n(Cs^+) = 3 \times 10^{15}$ m^{-3}, reported by Fantz et al [8], the mfp for mutual neutralisation with caesium ions is 166 m. The change of the combined mfp indicated in Table I due to mutual neutralisation with Cs^+ ions was neglected.

Resonant Charge Exchange and Elastic collisions

The MFP for momentum transfer to H_2 molecules [9] has been estimated for a hydrogen pressure of 0.45 Pa to be much longer than that for resonant charge exchange of H^- with hydrogen atoms.

The resonant charge exchange can, under certain conditions, interfere with the detachment process, since this collision, very efficient at low energy, can replace a fast H^- ion by a slow one, having the energy of the atom with which the ion exchanged its charge. Thus the detachment processes having large cross sections at low energy will become dominant and shorten the calculated mfp.

The cross section of resonant charge exchange is reported in [2, 10] versus the collision energy eV/amu.

The resonant charge exchange cross section relevant to the starting negative ion energy in the cesium seeded source (5.4 eV) is 1.08×10^{-18} m^2 and the mfp $\lambda_{CX} = 0.092$ m. In this case λ_{CX} is four times shorter than the combined mfp for detachment processes (0.375 m). Thus the effect of detachment processes will be enhanced in caesium seeded sources due to resonant charge exchange which will reduce the negative ion energy and thus increase the cross sections of H_{det} and mutual neutralisation, relevant to this case, and also reduce λ_{ED} (by reducing v^- in the equation for λ_{ED}). The situation is different in the volume source, since we consider a negative ion temperature much lower than the atomic temperature ($T(H^-) = 0.1$ eV, $T(H) = 0.8$ eV). Here the effect of resonant charge exchange will be opposite to its effect in caesium seeded sources, namely the heating of the negative ions and the increase of their mfp.

In the preceeding discussion, the collisions of negative ions with excited atoms $H(n>2)$ have not been considered. These cross sections are not known. However it is suggested in [2] that in these collisions the detachment (associative and non-associative) will dominate, and thus suppress the charge exchange channel.

In conclusion in caesium seeded sources resonant charge exchange in the collisions with atoms, in ground or excited states, is shortening the negative ion mfp. In volume sources if the atomic temperature is higher than that of negative ions, the resonant charge exchange may have the opposite effect, *i.e.* enhance the negative ion mfp.

TABLE I. Mean Free Path and combined mfp (λ_c) for three destruction processes in a volume and a caesium seeded negative ion source. In the case of the caesium seeded source the sheath voltage is 4.2 V. The mfp for resonant charge exchange of H⁻ and H, and the mfp due to mutual neutralisation of H⁻ with Cs+ are also shown.

Type of source	Mean free paths (m)					
	λ_{Hdet}	λ_{MN}	λ_{ED}	λ_C	λ_{CX}	$\lambda_{MN(Cs^+ + H^-)}$
Volume	0.70	0.61	0.23	**0.135**	0.077	
Caesium seeded	0.69	1.85	1.47	**0.375**	0.092	166

ESTIMATE OF THE MFP IN A CAESIUM SEEDED DEUTERIUM ION SOURCE

The purpose of this section is to estimate the mfp in the caesium seeded source experiment using deuterium as discussed by Christ-Koch *et al* [11]. In this experiment the negative ion density and extracted beam current are reported versus the plasma grid bias which varies in the range 14 V to 24 V. The sheath voltage ΔV is reported to be 2.2 V at 14 V and 0 at 21 V (the plasma potential). The strong decrease of the negative ion density (measured at 0.022 m from the plasma grid) with increasing plasma grid bias is attributed by the authors to the decrease of the mfp of the H⁻ ions. Therefore the evaluation of the mfp at plasma grid bias values of 14 and 21 V is important.

Since no other information is provided in [11] on the plasma parameters in this experiment we used the data obtained for the same ion source by Langmuir probe measurements by McNeely *et al.* [12] The measurements of McNeely *et al* were done in hydrogen. However, it was reported by Fantz *et al* [13] that the electron temperatures and densities in the plane parallel to the plasma grid were the same in hydrogen and deuterium. We assumed that the gas temperature in this plasma was 1200 K as indicated by Fantz *et al* [9]. We used the indication that the deuterium discharges show a factor 1.5 higher atomic density in the extraction region than in hydrogen [9].

The experiment [11] was carried out at a deuterium pressure of 0.45 Pa and a power of 53 kW. We estimated the atomic density in deuterium to be 7.6×10^{18} m⁻³ (30% of the neutral particle density). The electron and positive ion density and electron temperature from [12] are as follows: electron temperature 0.75 eV, electron density 2×10^{17} m⁻³, positive ion density 3.2×10^{17} m⁻³. The decrease of the positive ion density with plasma grid bias was taken from Fig. 20 of [12]: 10^{17} m⁻³ at plasma grid potential 14 V, and 3×10^{16} m⁻³ at 20 V. The electron density was also decreased, accordingly.

Electron impact detachment and mutual neutralisation cross sections for deuterium can be scaled from hydrogen cross sections by scaling the collision energy (at equal relative velocity) [14]. The associative and non-associative detachment cross sections in H + H⁻, and D + D⁻ collisions contain a genuine isotope effect [14], and the corresponding cross sections for deuterium are not available. However the cross section of these reactions can be derived from those in hydrogen, just by scaling the energy, as in the case of mutual neutralisation [14]. This may not be a quite accurate approach at very low energies, but it should be still appropriate even in the 5-20 eV energy range [14]. We will assume the same scaling for the associative and non associative detachment (denoted D_{det}), as for mutual neutralization. Note that the energy scaling factor is 2 when going from H to D, if the H data are given versus center of mass energy. For detachment by electrons the energy scaling factor is 1.

In the case of the resonant charge exchange of D⁻ with D atoms the cross sections can be easily scaled from those for hydrogen, since in [2, 10] the energy is reported in eV/amu.

The energy in the frame of the center-of-mass was calculated using Eq. 4. This energy has the same value for D_{det} and mutual neutralisation because the atom and positive ion temperatures are the same. E_{CM} are reported in Table 2 along with the mfp of D⁻ ions for the three detachment processes considered and for two values of the plasma grid bias: 14 V and 21 V. This corresponds to the sheath voltage values 2.2 V and 0 respectively. In both cases the combined mfp is more than ten times longer than the distance from the plasma grid to the probe (0.022 m). The mfp for resonant charge exchange of D⁻ with D atoms, also shown in Table 2, is in both cases at least two times shorter than the combined mean free path for detachment processes, but still much longer than the distance from the plasma grid to the probe.

TABLE 2. The mean free path for three detachment processes and the combined mfp (λ_c) in the experiment in deuterium reported in [11]. The mfp was calculated for two values of the plasma grid bias (PG bias). The corresponding sheath voltage ΔV and center of mass energy E_{CM} used for calculating the mutual neutralization and associative detachment (D_{det}) mfp are also indicated. The electron temperature used for calculating the mfp for electron detachment is 0.75 eV.

PG bias (V)	ΔV (V)	E_{CM} (eV)	λ_{Ddet}	λ_{MN}	λ_{ED}	λ_C	λ_{CX}
14	2.2	2.3	1.05	3.91	0.36	**0.25**	0.11
21	0	1.2	0.94	6.59	1.08	**0.47**	0.1

Thus the lower starting energy of the negative ions in the case of plasma grid bias 21 V does not explain the reduced negative ion density indicated by the photodetachment measurement at 0.022 m from the plasma grid, as suggested by the authors of [11].

ISOTOPE EFFECT IN EXTRACTED CURRENT

The comparison of the mfp of negative ions of hydrogen and that of its isotope deuterium is important since it can help understanding the difference in extracted currents. Table 3 presents the values of the mfp in hydrogen and deuterium for the same plasma conditions in a volume source: $n(H) = 10^{19}$ m^{-3}, $n(H^+) = 4\times10^{17}$ m^{-3}, $n_e = 2.74\times10^{17}$ m^{-3}, $T_e = 2$ eV. It can be noted that the combined mfp is shorter in deuterium than in hydrogen:

$$\frac{\lambda(D^-)}{\lambda(H^-)} = 0.73$$

This can explain in part the higher extracted hydrogen negative ion beam current.

TABLE 3. Comparison of the Mean Free paths in Hydrogen and Deuterium volume sources

Gas	Mean free paths (m)			
	λ_{det}	λ_{MN}	λ_{ED}	λ_C
Hydrogen	0.70	0.61	0.23	0.14
Deuterium	0.64	0.35	0.17	0.10

CONCLUSION

Several results have been obtained in this work.
1. The combined mfp in typical operation conditions of the rf caesium seeded source is several tens of cm. This is much longer than usually quoted.
2. The mfp in hydrogen is longer than in deuterium. This can explain in part the higher extracted hydrogen beam current under the same plasma conditions.
3. The mfp relevant to the combined extraction-photodetachment experiment at IPP Garching [11] is obtained. It is shown that the reduction of the negative ion density with increasing plasma grid bias is not due to the reduction of the mfp to values comparable to the distance between the plasma grid and the probe (0.022 m), as proposed by the authors.

ACKNOWLEDGMENTS

This work was funded partly by the United Kingdom Engineering and Physical Sciences Research Council under grant EP/G003955 and the European Communities under the contract of Association between EURATOM and CCFE. The views and opinions expressed herein do not necessarily reflect those of the European Commission. This work was carried out within the framework of the European Fusion Development Agreement.

REFERENCES

1. R.K Janev, W.D. Langer, K. Evans, D.R. Post, *Elementary Processes in Hydrogen-Helium Plasmas* Springer-Verrlag, Berlin-Heidelberg (1987).
2. R.K Janev, D. Reiter, U. Samm, *Collision Processes in Low Temperature Hydrogen Plasmas*; Jülich Forschungszentrum., Report 4105 (2003).
3. H.C. Corben, P. Stehle, *Classical Mechanics*, 2nd Edition, New York, Wiley (1977).
4. M. Seidl, H.L. Cui, J.D. Isenberg, H.J. Kwon, B.S. Lee, S.T. Melnychuk, *J. Appl. Phys.*, **79**, 2896 (1996).

5. D.Wünderlich, R. Gutser, U. Fantz, *Plasma Sources Sci.& Technol.*, **18**, 045031 (2009).
6. A.A. Ivanov, A.B. Sionov, F. El Balghiti-Sube, M. Bacal, M., *Phys. Rev.* E, **55**, 956 (1977).
7. R.K. Janev, Z.M. Radulovic, *Phys. Rev.* A **17**, 889 (1978).
8. U. Fantz. et al, *Nucl. Fusion*, **46**, S297-S306 (2006).
9. A.V. Phelps, J. Phys. Chem. Ref. Data, **19**, 653 (1990).
10. D. Reiter, D., private communication.
11. S. Christ-Koch, U. Fantz and M. Berger, *Plasma Sources Sci. Technol.* **18** 025003 (2009).
12. P. McNeely, S.V. Dudin, S. Christ-Koch, U. Fantz, U. and NNBI Team, *Plasma Sources Sci.&Technol.*, **18**, 014011 (2009).
13. U. Fantz. et al., CCNB meeting (2004), unpublished.
14. R.K. Janev, private communication.

3D Monte Carlo Modeling of the EEDF in Negative Hydrogen Ion Sources

R. Terasaki[a], A. Hatayama[a], T. Shibata[a] and T. Inoue[b]

[a]Graduate School of Science and Technology, Keio University,
3-14-1 Hiyoshi, Yokohama 223-8522, Japan.
[b]Japan Atomic Energy Agency, 801-1 Mukouyama, Naka 311-0193, Japan

Abstract. For optimization and accurate prediction of the amount of H$^-$ ion production in negative ion sources, analysis of electron energy distribution function (EEDF) is necessary. We are developing a numerical code which analyzes EEDF in tandem-type arc-discharge sources. It is a three-dimensional Monte Carlo simulation code with the realistic geometry and magnetic configuration. Coulomb collision between electrons is treated with "Binary Collision" model and collisions with hydrogen species are treated with "Null-collision (NC)" method. We have applied this code to the analysis of the JAEA 10 ampere negative ion source. The numerical result shows that the obtained EEDFs reasonably agree with experimental results.

Keywords: Electron, EEDF, Monte Carlo method
PACS: 29.25Ni, 52.65Pp

INTRODUCTION

Negative hydrogen ion (H$^-$) source is a key component of neutral-particle beam injection (NBI) system which is an efficient method for heating fusion core plasmas. The understanding of the electron energy distribution function (EEDF) inside the H$^-$ sources and its control is very important to optimize the H$^-$ source performances, e.g., the total amount of H$^-$ volume/surface production and its spatial uniformity.

In most of the previous modeling of arc-discharge H$^-$ sources, the EEDF has been calculated by the spatially 0D (zero dimension) model, e.g., in Ref [1]. These modelings are useful for the basic understanding the H$^-$ production process. In these 0D modelings, however, realistic geometry and the magnetic configuration of the H$^-$ sources were not taken into account. Instead, a simple model for the electron confinement time has been assumed for the electron spatial transport to fit the experiment results.

Recently, it has been shown experimentally in JAEA 10A negative ion source [2] that the non-uniformity of the electron distribution function (EEDF) inside the source and the resultant non-uniformity of the H$^-$ production strongly affect the H$^-$ beam optics, which in turn increases the heat load of the acceleration grid. Therefore, modeling of the EEDF and analysis of the spatial non-uniformity of the EEDF is necessary to optimize H- ion source and the beam optics.

In order to solve these problems, we are developing the 3D3V (three dimension in real space and three dimension in velocity space) Monte Carlo modeling of the EEDF including realistic 3D geometry with realistic multi-cusp and filter magnetic field configuration of the tandem-type arc-discharge sources. In the previous paper [3,4], we have applied the code to the study of the EEDF in JAEA 10A negative ion source. The focus was placed on the spatial non-uniformity of the EEDF observed in JAEA 10A negative ion source mentioned above. The calculated results of the EEDF in the driver region have been compared with the experimental results. The good agreement has been obtained, i.e., the EEDF in the driver region shows the spatial non-uniformity. In addition, the EEDF in driver region consists of two temperature components, i.e., the low energy component and the high energy component. Their temperatures (T_e~3eV for the low energy component, while T_e~20eV for the high energy component) well agree with those obtained by the two temperature fit of the Langmuir probe measurement in the experiments[2].

In the present paper, we focus our attention on the spatial profile of the EEDF in the direction along the H⁻ extraction axis inside the JAEA 10A negative ion source from the driver region towards the extraction region. In the experiments, the electron temperature decreases towards the Plasma Grid (PG) in the extraction region and becomes smaller (T_e~1eV) than in the driver region (T_e~3eV). The numerical results of the EEDF obtained by our 3D Monte-Carlo kinetic modeling will be compared with these experimental results.

BASIC EQUATIONS AND NUMERICAL TECHNIQUES

In order to calculate the EEDF in the source, we directly solve equations of motion for electrons,

$$m_e \frac{d\mathbf{v}}{dt} = -e(\mathbf{E} + \mathbf{v} \times \mathbf{B}) + \text{(collision term)} \tag{1}$$

where m_e, $\mathbf{v}$, e, $\mathbf{B}$ and $\mathbf{E}$ are the mass, velocity, electric charge of electron, the magnetic field and the electric field. The equation of motion is numerically solved by the Boris-Buneman version of the leap-frog method [5]. Being based on the real magnet dimension and location (see Fig.1), the external magnetic field at each local point is calculated in advance with the analytic solution based on the magnetic charge model [6]. On the other hand, we assume the quasi-neutrality condition $n_e \approx n_i$ (n_e: electron density, n_e: ion density) in the bulk source-region. The electric field $\mathbf{E}$ has not been taken into account in the present analysis, since the Debye length becomes significantly small ($\lambda_D \approx 2-5 \times 10^{-2}$ mm) for typical plasma parameters of large H⁻ sources such as JAEA 10A source ($T_e = 1\text{eV} - 4\text{eV}$, $n_e = 1 \times 10^{17} \text{m}^{-3}$) compared with the characteristic scale length/volume of the system ($\approx 240\text{mm} \times 480\text{mm} \times 200\text{mm}$: see Fig.1). The potential variation in the bulk plasma volume is considered to be relatively small except for the thin sheath layer at the plasma boundary.

Various collision processes are taken into account in Eq. (1) by Monte Carlo techniques. They are classified mainly into two categories. The first one is the collision processes of electrons with hydrogen particles (H atoms, H_2 molecules, and ions). Totally about 540 collision processes are included in the analysis. Among these processes, main collision species are summarized in Table 1. The second category is the electron-electron collision, i.e., Coulomb collision. The collision with hydrogen particles is modeled by the "Null-Collision (NC)" method [7] to speed up the calculation, while the "Binary Collision (BC)" model" [8] is applied to the Coulomb collision.

TABLE 1. Main collision processes included in the model

Collision species	Reaction process
Vibrational excitation (eV-process)	$H_2(v)+e \rightarrow H_2(v \pm 1)+e$
Vibrational excitation (EV-process)	$H_2(X^1\Sigma_g^+, v=0,14)+e \rightarrow H_2(X^1\Sigma_g^+, v''=0,14)+e$
Electronic excitation (H_2)	$e+H_2(X^1\Sigma_g^+) \rightarrow e+H_2^*(B^1\Sigma_u^+ 2p\sigma)$
Electronic excitation (H)	$e+H(1s) \rightarrow e+H^*(2p)$
Ionization (H_2)	$e_1+H_2(X^1\Sigma_g^+) \rightarrow e_1+H_2^+(v)+e_2$
Ionization (H)	$e_1+H(1s) \rightarrow e_1+H^++e_2$
Dissociative attachment	$e+H_2(X^1\Sigma_g^+;v) \rightarrow H^-+H(1s)$
Recombination	$H_2(v>5)+H^*(n=2)$
Elastic collision with H_2	$H_2+e \rightarrow H_2+e$
Elastic collision with H	$H+e \rightarrow H+e$

The main procedure of NC method[7] is summarized as follows. (1) First, whether collision occurs or not is determined based on the maximum collision frequency. (2) Next, when collision occurs, the kind of collisions is determined based on the ratio of collision frequency of each kind to the total collision frequency.

The main procedure of BC method[8] is summarized as follows. (1) First, a pair of electrons is randomly chosen from the system and their relative velocity u is calculated. (2)Next, the scattering angles, χ, and Φ, are calculated in the following manner. The scattering angle χ is given by $\chi = 2\tan^{-1}\delta$, where δ is a random number sampled from the Gaussian distribution with the mean $<\delta>=0$ and the variance $<\delta^2>=n_e e^4 \ln\Lambda \Delta t_c / 2\pi\varepsilon_0^2 m_e^2 u^3$ (n_e: electron density in the numerical cell, $\ln\Lambda$: Coulomb logarithm, Δt_{BC}: numerical time step for the Coulomb collision, ε_0: permittivity of free space, u: relative speed of the electron pair). The angle Φ is given by $\Phi = 2\pi U$, where U is a uniform random number. (3) From these scattering angles, the relative velocity change Δu due to the collision is obtained. (4) Finally, the velocity change of each electron involved in the collision is calculated.

NUMERICAL SIMULATION OF JAEA 10A SOURCE

We have applied our simulation code described in Sec. II to the analysis of the EEDF in JAEA 10A source [8]. Figure 1 shows the schematic diagram of the source. In the simulation, the realistic source dimension, geometry, magnetic configuration and filament locations are taken into account. We have adopted the (X,Y,Z) coordinate system shown in Fig.1. The model dimensions are summarized in Table II. In the experiments, sixty-six magnets are installed surrounding the vacuum chamber, as schematically shown in Fig.1. The number, dimension and location of these cusp and filter magnets in our simulation are exactly the same those in the experiments. The resultant magnetic field inside the source is calculated by the surface magnetic charge model [5] as mentioned above.

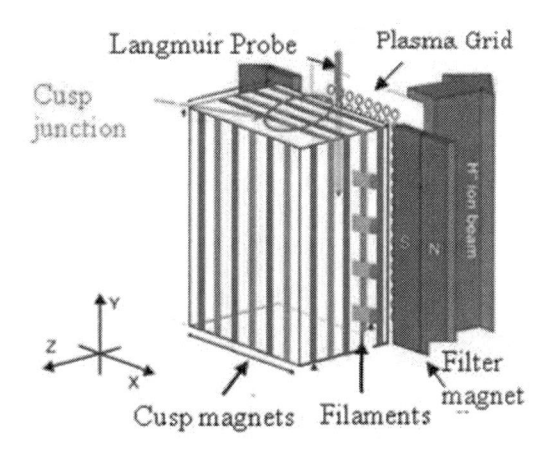

FIGURE 1. The schematic diagram of the JAEA 10 ampere negative ion source [8].

TABLE 2. Dimensions of the sources in the present model

Direction of the axes in the source	Length (mm)
Transversal direction (X)	$-120 < X < 120$
Longitudinal direction (Y)	$-240 < Y < 240$
Extracting direction (Z)	$0 < Z < 203$

TABLE 3. Typical operation condition used in the simulation

Operation parameters	Parameters' values
Arc power	10kW
Arc voltage	60V
Arc current	166.7A
Gas pressure	0.3Pa

Being based on a typical operation condition of the JAEA 10A source, the operational parameters in our simulation are calculated and summarized in Table 3. The initial energy of the primary electrons emitted from the filaments is set to be 60eV with the assumption that these electrons are accelerated by the arc-voltage in the thin sheath region surrounding the filaments immediately after their thermal emission. The emitted angle is chosen from the isotropic distribution for simplicity. The number of electron test-particles emitted from the filaments at each time-step and the statistical weight of each test-electron are determined from the arc-discharge current in Table 3.

With above initial conditions, trajectories of test electros are followed by Eq. (1) with the numerical time-step $\Delta t = 10^{-10}$ s, which is much smaller than the Larmor period of in the magnetic field inside the source. The time step Δt_{NC} and Δt_{BC} for the NC method and the BC model are chosen, respectively, as $\Delta t_{NC} = 10^{-8}$ s and $\Delta t_{BC} = 10^{-8}$ s. The former is much smaller than the minimum flight time of the NC method and the latter is chosen so as to simulate the important feature of the small angle scattering of the Coulomb collision. The main collision processes included in the simulation has already been shown in Table 1. The main simulation parameters related to the collision processes with hydrogen particles, which are necessary to calculate the collision frequency used in the NC method, are summarized in Table 4.

TABLE 4. Parameters for the hydrogen species

Simulation parameters		Values
Hydrogen molecule temperature T_{H_2}		300K(initial)
		2000K(during discharge)
Hydrogen molecule density	n_{H_2}	2.8×10^{19} m^{-3}
Hydrogen atom temperature	T_H	5802K
Hydrogen atom density	n_H	2.8×10^{18} m^{-3}
Hydrogen ion density	n_{H^+}	$H^+ : 4 \times 10^{17}$ m^{-3}
	$n_{H_2^+}$	$H_2^+ : 1 \times 10^{17}$ m^{-3}
	$n_{H_3^+}$	$H_3^+ : 5 \times 10^{17}$ m^{-3}

In order to take into account the sheath potential drop near the walls and PG, here, we have used the following simple model. If these test electrons reach the upper/side/bottom walls, those electrons with the energy E_e larger than the sheath potential drop ($E_e \geq e|V_{sh}|$, where V_{sh} sheath-potential drop) are absorbed at the wall, while those with the low energy ($E_e \leq e|V_{sh}|$) are reflected. The sheath-potential drop V_{sh} can be estimated from the sheath theory [10],

$$V_{sh} \approx (kT_e/2e)\ln[(1/2\pi)(m_i/m_e)], \qquad (2)$$

where T_e and m_i are the electron temperature and ion mass, respectively. In the simulation, T_e in Eq.(2) is given by the average energy of the test electrons in the cell closest to the wall.

In order to discuss the spatial variation of the EEDF, we divide the discharge volume into 11 spatial cells as shown in Fig.2 (a) and (b). Figure 2(a) shows a bird-eye's view of the numerical cells, while Fig.2(b) shows the cross-sectional view in the X-Z plane of the numerical cells used in the simulation. The lower part of the discharge volume (0mm $\leq Z <$ 100mm, $|X| \leq$ 120mm, $|Y| \leq$ 220mm) is divided into the 10 cells with the same volume from the bottom (the Plasma Grid surface at Z=0mm) towards the top along the Z direction. The remaining part of the discharge volume, i.e., the upper region with 100mm ≤ Z < 200mm is treated as one large cell with the center position $(X, Y, Z) = (0, 0, 150\text{mm})$. This last cell almost covers the upper driver region of JAEA 10A source as shown in Fig.2.

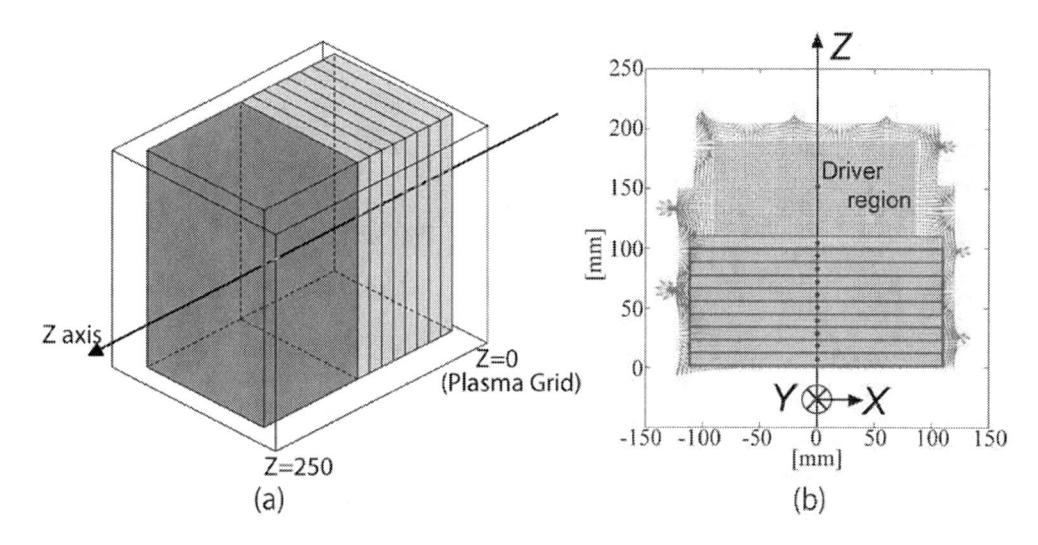

FIGURE 2. Schematic diagram of the numerical cells used in the simulation.

RESULTS AND DISCUSSION

Figure 3 shows the calculated EEDF in JAEA 10A source. The EEDF in the bottom cell (open square) just above the PG surface (the cell center : Z=0.5mm) is compared with the EEDF in the cell (closed square) in the driver region. In the driver region, the EEDF consists of two energy component as in the similar manner in the previous result in Ref. [4]. Not only the low energy component, but also the high energy tail of the EEDF is clearly seen in Fig.3 in the driver region. On the other hand, high energy electrons are effectively filtered out and hardly seen in the EEDF at the bottom cell where the transverse filter-magnetic field is strong (~60-70G: see Fig.4).

From the EEDF obtained in each cell, the electron temperature T_e profile along the Z-axis can be estimated. The results are shown in Fig.4. These electron temperatures are calculated by the linear fitting of the $f(E)/E^{1/2}$ vs. E plot [$f(E)$: EEDF] with the least squares method. In the driver region, the EEDF consists of two energy components with the different temperature as mentioned above. The electron temperatures in this region shown in Fig.4 are those for the low energy bulk component of the electrons in the range $E \leq 20$ eV.

The strength of the magnetic field along the Z axis is also shown in Fig.4 with the broken line. The electron temperature T_e gradually decreases towards the PG surface at first. Then, it starts rapidly decreasing around $Z \approx 30$ mm, where $B \approx 60$G and the magnetic field by the filter magnet is effective. The experimental results of T_e by the Langmuir probe measurement[2] are also shown with the open square in Fig.4.

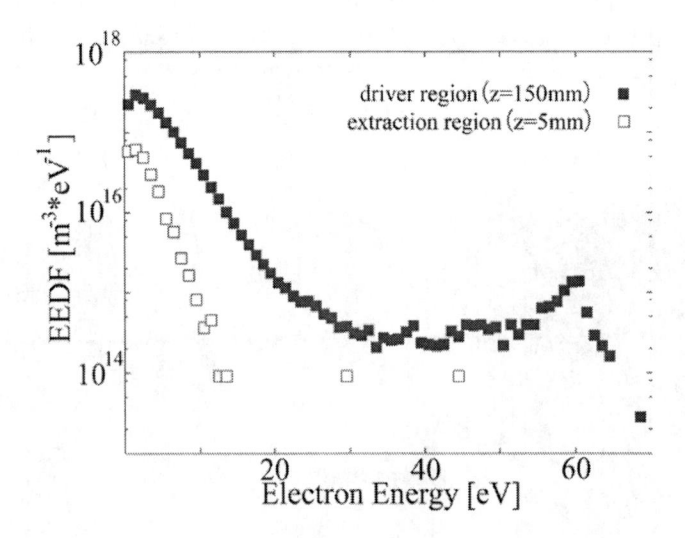

FIGURE 3. The calculated EEDFs in the driver region (Z=150mm) and in the magnetic filter region (Z=5mm).

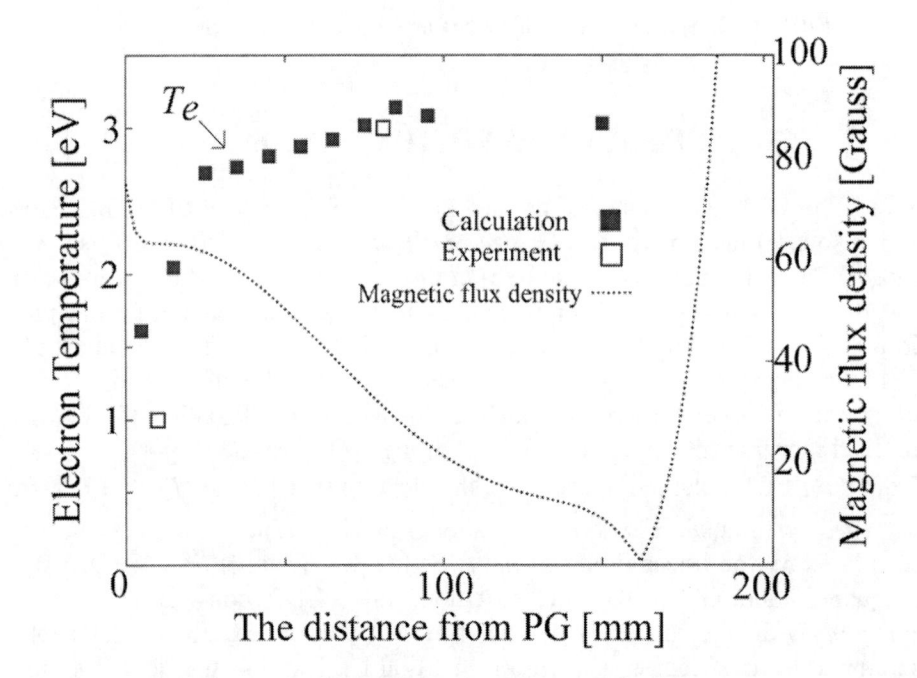

FIGURE 4. Electron temperature profile along the Z-axis (closed square: simulation, open square: experiment[2]). The strength of the magnetic field is also shown by the broken line.

Although only the data for two locations (at Z=14mm and 84mm from the PG surface) are available along the Z-axis. The numerical result reproduces the characteristic feature of T_e observed in the experiments. In addition, the numerical results of T_e reasonably agree with the experimental results both in the driver and the extraction region.

We have saved the data of the kinetic energy of each electron which is lost at the side wall along the field line in the magnetic filter region. From the data, the average kinetic energy of the electrons lost at the side walls in the magnetic filter region can be estimated as ~ 20eV, which is larger than the average kinetic energy of the electrons lost at the PG. This result suggests that the filter effect of the high energy electrons, i.e., the magnetic filter effect can be explained by the parallel loss of fast electrons along the field line. Their confinement time of fast electrons along the field line is possibly shorter than their diffusion time across the magnetic field. Coulomb collision time of fast electrons is larger than that of slow electrons. More detailed analysis of the transport process of slow and fast electrons in the real space parallel and perpendicular to the magnetic field is now underway.

SUMMARY

The 3D3V Monte Carlo kinetic model of the EEDF including realistic 3D geometry with realistic multi-cusp and filter magnetic field configuration of the tandem-type arc-discharge sources has been developed. The code has been validated by the comparison with the experiments in JAEA 10A negative ion source.

As in the previous study of the non-uniformity of the EEDF in JAEA 10A source, the 3D3V Monte-Carlo kinetic code of the EEDF reproduces the most of the important characteristic features of the magnetic filter effects observed in the experiments. These results shows that the 3D3V Monte Carlo kinetic code developed is useful not only for the basic physics understanding of the energy relaxation process of the electrons and its optimization in arc-discharge type negative H⁻ source, but also the design study of the new large H⁻ ion sources.

REFERENCES

1. C. Gorse, M. Capittlli, M. Bacal, J. Bretagne, A. Lagana, *Chem. Phys.* **117**, 177 (1987).
2. N. Takado, H. Tobari, T. Inoue, J. Hanatani, A. Hatayama, M. Hanada, M. Kashiwagi*et al*, *J. Appl. Phys.* **103**, 5, 053302 (2008).
3. I. Fujino, A. Hatayama, N. Takado, and T. Inoue, *Rev. Sci. Instrum.* **79**, 02A510 (2008).
4. R. Terasaki, I. Fujino, A. Hatayama, T. Mizuno, T. Inoue, *Rev. Sci. Instrum.* **81** (2010) 02A728/1-3.
5. C. K. Birdsal, A. B. Langdon, *Plasma Physics via Computer Simulation* (New York: McGraw-Hill) p.13-15,58-63 (1985).
6. Y. Ohara, M. Akiba, H. Horiike, H. Inami, Y. Okumura, and S. Tanaka, *J. Appl. Phys.* **61**, 15 (1987).
7. K. Nanbu, *IEEE Trans. Plasma Sci.* **28**, 971 (2000).
8. T. Takizuka and H. Abe, *J. Comput. Phys.* **25**, 205 (1977).
9. M. Hanada, T. Seki, N. Takado, T. Inoue, H. Tobari, T. Mizuno, A. Hatayama *et al*, *Rev. Sci. Instrum.* **77** 03A515 (2006).
10. G. A. Emmert, *Phys. Fluids* **23**, 4 (1980).

Origin of extracted negative ions by 3D PIC-MCC modeling. Surface vs Volume comparison

S. Mochalskyy*,[†], A.F. Lifschitz* and T. Minea*

*LPGP, UMR 8578: CNRS, Universite Paris Sud, Orsay, France
[†]CEA/IRFM, Cadarache, France

Abstract. The development of a high performance negative ion (NI) source constitutes a crucial step in the construction of Neutral Beam Injector (NBI) of the future fusion reactor ITER. NI source should deliver 40 A of H^- (or D^-), which is a technical and scientific challenge, and requires a deeper understanding of the underlying physics of the source and its magnetic filter. The present knowledge of the ion extraction mechanism from the negative ion source is limited and concerns magnetized plasma sheaths used to avoid electrons being co-extracted from the plasma together with the NI. Moreover, due to the asymmetry induced by the ITER crossed magnetic configuration used to filter the electrons, any realistic study of this problem must consider the three spatial dimensions. To address this problem, a 3D Particles-in-Cell electrostatic collisional code was developed, specifically designed for this system. Binary collisions between the particles are introduced using Monte Carlo Collision scheme. The complex orthogonal magnetic field that is applied to deflect electrons is also taken into account. This code, called ONIX (Orsay Negative Ion eXtraction), was used to investigate the plasma properties and the transport of the charged particles close to a typical extraction aperture [1].This contribution focuses on the limits for the extracted NI current from both, plasma volume and aperture wall. Results of production, destruction, and transport of H^- in the extraction region are presented. The extraction efficiency of H^- from the volume is compared to the one of H^- coming from the wall.

Keywords: PIC MCC modeling, negative ion extraction, NBI plasma source
PACS: 52.65.Rr,52.40.Kh,52.65.-y,52.50.Dg

INTRODUCTION

The experimental fusion reactor ITER will include two 17 MW Neutral Beam Injector (NBI) systems to heat the plasma ions to thermonuclear temperatures ($T \sim 10$ keV). In the injector, negative ions coming from a negative ion source are electrostatically accelerated to 1 MeV, and then stripped of their extra electron before entering the Tokamak chamber [2]. In the source hydrogen plasma is created by radio-frequency inductively coupled coils [3, 4]. It diffuses into the expansion chamber where it becomes relatively homogeneous. The extraction region of the source is composed of two grids: the plasma grid (PG) and the extraction grid (EG). The positive potential applied between them drains out negative ions and repeal positive ones. Negative ions are mainly produced by electron dissociative attachment of vibrationally exited $H_2(v)$ molecules. Due to a number of destruction processes, the survival length of the negative ions is just few centimeters [5]. Therefore, only NI created close to the plasma grid can be extracted. To enhance the negative ion yield production in the vicinity of the extraction apertures,

Cs atoms (with a low work function $\sim$ 2 eV) are injected into the plasma source and deposited over the plasma grid [6, 7]. The interaction of gas molecules and plasma positive ions with the cessiated walls will produce a significant amount of negative ions. The main problem of the source is the co-extraction of plasma electrons. This electron current has a deleterious effect over the injector. On one hand, electrons can destroy NI after the extraction. On the other hand, the acceleration of these electrons will increase the power consumption of the accelerator, that will mainly deposited over the accelerator grids and walls. To limit the undesirable electron current, two orthogonal magnetic fields are applied close to the appertures, the filter field and the deflecting field. The first one is created in the chamber close to the plasma grid, whereas the second field is created by permanent magnets embedded in the extraction grid.

Several numerical models have been developed during last years to describe the physics inside the plasma source and in particular plasma behavior close to the extraction system, including Monte Carlo transport calculations [8, 9, 10, 11] and 1D and 2D Particle-in-Cell simulations [12, 13, 14, 15, 16]. The Monte Carlo codes do not include the plasma screening effect in a self-consistent manner, a critical issue in the modeling of the extraction. Low-dimensional PIC simulations, on the other hand, are not able to bring a quantitative description of the extraction problem. This is because the presence of the magnetic fields used to filter and deflect electrons breaks up the cylindrical symmetry of the system close to an extraction aperture. As consequence of this lack of symmetry, a realistic description of this problem requires to model the plasma screening of the external field and the particle transport in three-dimensions.

In order to get a self-consistent plasma description in the vicinity to the extraction system, we developed 3D electrostatic Particle-in-Cell code coupled with a Monte Carlo module for binary particle collisions [17]. The present work is focused on the contribution of volume and of surface NI produced ions to the extracted current. The self-consistent meniscus formation will be shown together with screened potential distribution in the volume. The description of the simulation model will be given in the next section.

NUMERICAL MODEL

A specifically designed electrostatic 3D Particle-in-Cell Monte Carlo collisions code was developed to study extraction from the negative ion source [17]. The code is parallelized for distributed memory multi-CPU computers using spatial domain decomposition for fields and particles, via the Message Passing Interface. The potential distribution is calculated by solving Poisson equation using a preconditioned Conjugate Gradient algorithm [18]. To get a realistic potential in the vicinity of the extraction hole, which has a circular cross section, the Poisson solver implements a technique to deal with boundaries not lying over the grid points [17, 19].

The simulation domain includes only a single extraction aperture of the plasma grid (Fig. 1). Its dimensions are $25mm \times 20mm \times 20mm$ in x, y, z directions. The plasma grid is centered at $x = 19$ mm with thickness of 2 mm. The shape of the extraction aperture was improved compared to the previous model [17]. The apperture has a conical shape, that corresponds to the design of the real machine [1, 2]. Periodical boundary conditions

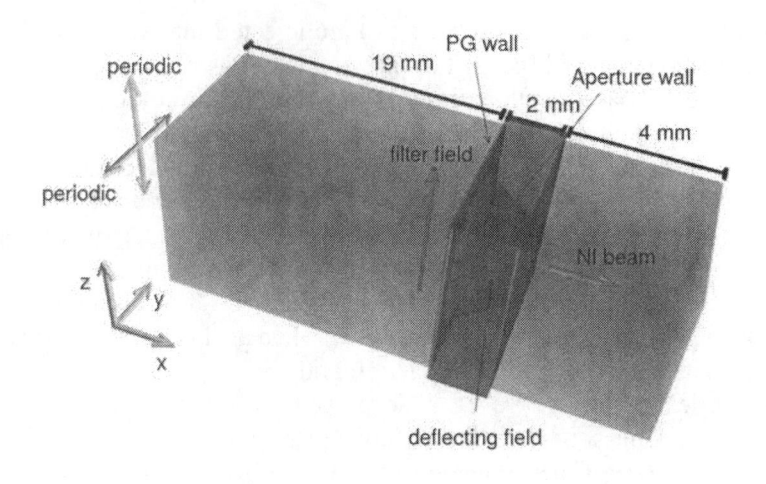

FIGURE 1. Simulation domain used in ONIX code.

are used in the y and z direction to model an infinite 2D array of apertures. The transverse size of the simulation box is 20 mm x 20 mm, that corresponds to the distance between the extraction apertures in each direction.

The filter magnetic field is in the z direction, with a spatial profile along x given by:

$$B_z(x) = B_{max} \exp\left(-\frac{(x-x_{max})^2}{2\sigma^2}\right) \qquad (1)$$

with $B_{max} = 7$ mT, $\sigma = 35$ mm, $x_{max} = 0$ mm. The deflecting magnetic field is in the y direction. Its profile is given by equation 1 with parameters $B_{max} = 80$ mT, $\sigma = 6$ mm and $x_{max} = 30$ mm.

The plasma density and temperature of each species far from the apperture are uniform, and they were taken from the experimental data [1, 3, 21]. The initial plasma is assumed to be composed of H^+, H_2^+ and electrons. We set the density to $n_0 = 10^{17} m^{-3}$ with source ratio $S_e/S_{H^+}/S_{H_2^+} = 1/0.6/0.4$, and temperatures $T_e/T_{H^+}/T_{H_2^+} = 3/1/1 (eV)$. Our code does not follow the neutrals dynamics but treats them as a background with constant density $n_H = 4 \times 10^{19} m^{-3}$ and $n_{H_2} = 10^{19} m^{-3}$. The simulation is started without negative ions, which are created during the simulation via three main processes [20]: 1) electron dissociative attachment to the vibrationally exited molecules $H_2(v)$ in the volume, 2) interaction of the positive ions (H^+ and H_2^+) with the aperture wall, and 3) collisions of the neutral gas (H, H_2) with the aperture wall.

The volume production of the negative ions results from the collision of low energy electrons (about 1eV) to the vibrationally exited molecules [reaction 1, Table 1] [20]. For level of H_2 lower than v=5, the cross section of the reaction is very negligible. For vibration levels with v$\geq$ 5, the cross section increases by five order of magnitude [24]. Our code does not calculate the population of the vibrational states. We assume a total effective density of vibrational excited $H_2(v)$ (in vibrational states $5 \leq v \leq 9$) taken from calculations reported in ref.[14].

The second mechanism of negative ion formation is interaction of positive ions (H^+ and H_2^+) with the Cs covered surface [25, 26] of the plasma grid. The negative ions are produced in backscattering of the incident positive ions from the plasma grid wall. NI yield is calculated as the function of the incident energy by the formula [26]:

$$Y(E_{in}) = R_N \eta_0 \left(1 - \frac{E_{th}/R_E}{E_{in}}\right), E_{in} \geq E_{th}/R_E, \qquad (2)$$

where R_N and R_E are particle and energy reflection coefficients, η_0 is height and E_{th} - the threshold energy. We take $R_N \eta_0 = 0.3$ and $E_{th}/R_E = 2eV$ [26] for both H^+ and H_2^+ ions, because all molecular ions are dissociated before colliding with the surface. New negative ions emitted from the PG surface have an initial energy of 1 eV.

The most important mechanism of the negative ion production is the collision of the neutral gas (H, H_2) with the Cs covered surface [27, 28]. The external potential applied between the plasma and the extraction grid penetrates deeply inside the expansion chamber, thus it repels most of the positive particles close to the extraction aperture. Eventually, the production of the negative ions via impact of H^+ and H_2^+ with the PG wall (described above) becomes inefficient. However external positive potential does not have any influence on the neutrals flux onto the plasma grid surface. The neutral H, H_2 particles collide the Cs covered wall could return as atoms or negative ions. The probability to emit negative ions from the surface is given by Langmuir-Saha relation [29], that depends on the electron affinity and the surface work function. In our model we do not follow trajectories of the neutrals, but treat them as the background gas with a density and temperature. Therefore we installed in the code a given flux of the negative ions from the PG surface ($\sim 2000 A/m^2$), that assumes the NI created via neutrals impact. The produced particles are launched with energy of 1eV in the normal direction.

The three volume processes are mainly responsible of the destruction of the negative ions [20]: mutual neutralization in collision with positive ions, associative and nonassociative detachment in collision with atoms, and electron detachment in collision with electrons [reaction 2-5, Table 1]. All reactions including in the model are summarized in the Table 1.

TABLE 1. List of the reaction taken into account in the simulation model

#	Reactions	Process	Reference
(1)	Electron dissociative attachment	$e + H_2(v) \to H + H^-$	[30]
(2)	Associative detachment with H	$H^- + H \to e + H_2$	[24]
(3)	Nonassociative detachment with H	$H^- + H \to e + H + H$	[24]
(4)	Electron detachment	$e + H^- \to H + 2e$	[24]
(5)	Mutual neutralization	$H^+ + H^- \to H + H$	[24]
(6)	Charge exchange with H	$H^- + H \to H + H^-$	[24]

At t=0, the spatial distribution of electron and positive ions density are the same, i.e. the plasma is neutral. Typical runs are performed using 10^7 macro particles and a mesh of $100 \times 100 \times 100$ cells, with a performance of 0.1 μs by day using 20 CPU at the time step $\Delta t = 3 \times 10^{-12}$ s. All ions hitting the walls or leaving the computation box are reinjected at a random position inside the volume occupied by the initial plasma.

Electrons leaving the box are also reinjected, whereas, as mentioned above, electrons hitting the PG can produce secondary electron emission.

RESULTS AND DISCUSSION

The simulation were performed with an initial plasma density filling the left region of the simulation box as shown in Fig. 2. Between x=0 and x=12 mm we assume that the plasma properties are the same as in the expansion chamber, therefore we set the same density in this region $n_0 = 10^{17} m^{-3}$, corresponding to pre-sheath. The electron and ion temperatures are respectively 3 eV and 1 eV [1, 3]. Figure 2 shows the distribution of the H^+ ions at $0.3 \mu s$. Positive ions are pushed by the external potential towards the expansion chamber, forming a structure called the meniscus visible in the figure.

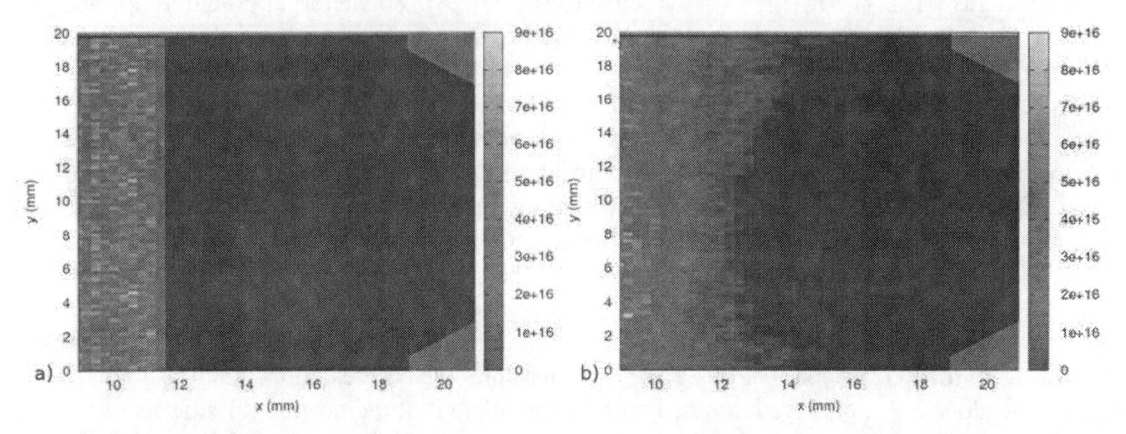

FIGURE 2. Spatial distribution of positive ion density in the plane $z = 10$ mm at the beginning of the simulation a) and at the time 0.3 μs b).

The spatial distribution of electrostatic potential at the beginning of the simulation and after $0.3 \mu s$ are shown in Fig. 3. At t=0, the external electric field deeply penetrates into the expansion chamber. However, once that plasma screening occurs, the iso-potential lines are pushed towards the PG. The neutrality of the system is maintained in the initial plasma region ($0 \leq x \leq 12$mm). The potential is constant in most of the volume, meaning that electric field vanishes.

At the beginning of the calculations negative ions are not present in the volume. They are created via 3 different processes as described above. The Fig. 4 shows the time evolution of the NI extracted current produced by different atomic process together with the co-extracted electron current. All currents are calculated at the exit plane (the right side of the simulation domain), 4mm far from the PG. One can see first that electron current grows very fast, because the extraction potential is not screened. The growing of the extracted electron current continues up to t=$0.15 \mu s$. At this time, the plasma screening of the extraction potential starts being efficient, thus reducing the fraction of the extracted electrons. After the electron and the NI currents stabilize, and system reach the steady state. The achievement of a quasi-steady state is also visible from the potential distribution evolution. For t> $0.3 \mu s$, the local value of the potential presents fluctuations smaller than 1%.

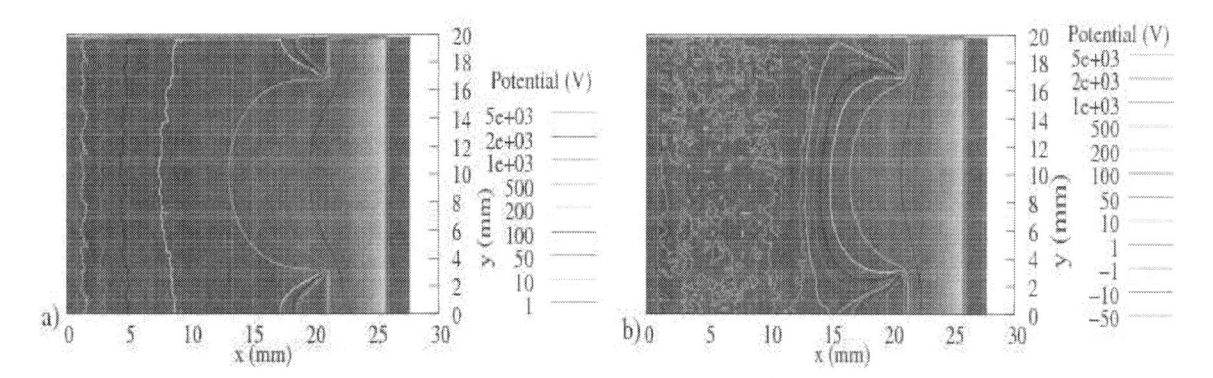

FIGURE 3. Spatial potential distribution in the plane $z = 10$ mm at the beginning of the simulation (a) and at $t = 0.3\ \mu s$ (b).

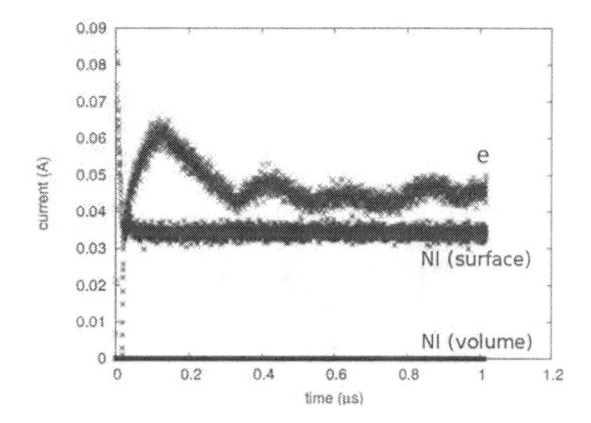

FIGURE 4. Time evolution of the extracted electron and negative ion currents.

Because of low value of the electron dissociative attachment to the $H_2(v)$ cross section and the small threshold of the reaction [table 1, reaction (1)], it is not possible to produce large amounts of the NI in the volume. Moreover, NI destruction by electron stripping, significant at electron energies of few eV, and by mutual neutralization will destroy most of the negative ion produced in the expansion chamber before being extracted (the survival length for NI is few centimeters). On the other hand, the efficient screening of the extraction field by the plasma reduces the region from the which NI can be extracted. Therefore the extracted NI current coming from volume created ions is much smaller than one extracted from the PG surface (Fig. 4). This result agrees with the strong reduction of NI current in experiments performed in the Cs-free operation regime [1].

Due to the high external potential that penetrates inside the expansion chamber and the plasma screening effect (Fig. 3) almost none positive ion reachs the PG wall, consequently the extracted NI current in the case of the interaction of positive ions with PG surface is negligible.

The dominant process in the surface creation of NI is the collision of neutral molecules with the PG wall (Fig. 4). The contributions to the extracted current of NI extracted from outer surface of the PG and from its inner surface are shown separatelly. Because of the creation of a double layer close to the PG and of the geometry of the screened electric field in this region, the contribution of NI coming from the outer surface of the PG in about 10 times smaller than the one of NI coming from the inner surface of the aperture.

The negative ion extraction from the plasma grid surface is limited by a double layer [31, 32]. Fig. 5 shows the spatial potential distribution close to the PG in the $x-y$ plane. The aperture locates between $x = 19mm$ and $x = 21mm$. It clearly visible the strong double layer behind the plasma grid at the distance $x = 18.5mm$. The double layer along the cone wall is also present but not so high, because of the compensation of the external potential in this region. NI create the negative charge field perpendicular to the PG wall, this field pushes the negative particles towards the PG and does not allow them cross the negative potential iso-surface. Reflected negative ions strike the wall and destroy, i.e. their extraction is not possible.

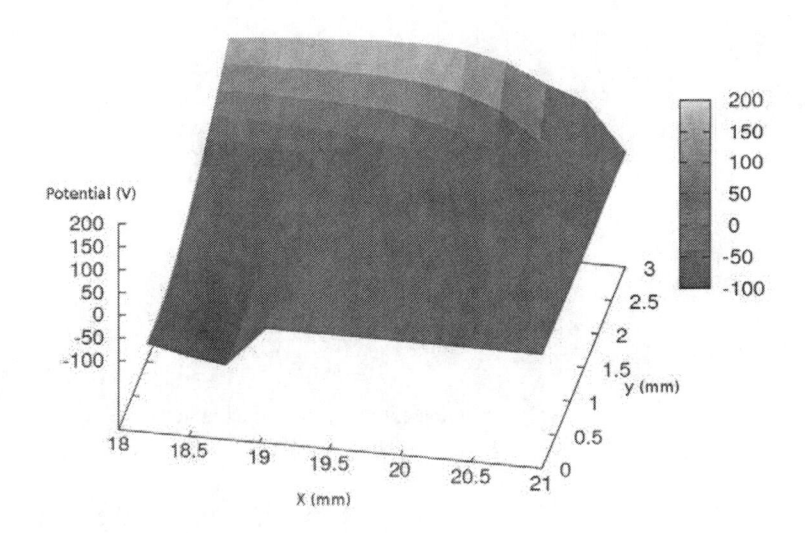

FIGURE 5. Spatial distribution of the electric potential close to the PG. The emission rate of the negative ions $2000A/m^2$

The asymptotic value of the extracted current obtained in the simulation ($I_{NI_{tot}} = 0.036A$, Fig. 4) is in good agreement with experimental results obtained in the BATMAN test bed machine at IPP Garching (0.034A) [1, 3]. The apperture geometry and the magnetic field structured implemented in this facility are close to the used in the simulations, and to the current ITER specification [2]). A well agreement is also obtained in the electron/NI current ratio (~ 1.15), close to the experimentl value.

The NI current and ratio electron/NI currents predicted by the simulations fulfill the ITER requirements [2]: $j_{NI} = 200A/m^2$, $I = 40A$, $I_e/I_{NI} = 1$. The predicted co-extracted electron current is slightly larger than the negative ion one, however we have to take into account that this current, calculated 4 mm after the PG, is much larger than actual electron current entering the accelerator. Most of electrons reaching the right side of simulation box have very large transverse velocity due to the deflecting field, and they will hit the extraction grid wall or the vessel.

CONCLUSIONS

A full 3D in velocities and space electrostatic PIC MCC code was developed and applied here to study the negative ion extraction in the ITER NBI source system. The relative contribution of volume and surface produced NI was evaluated. The extracted current from the surface of the plasma grid is dominant and it is formed mostly via the collision of energetic neutrals. Co-extracted electron current is close to NI one, in agreement with experimental data.

The limitation of NI extraction from the plasma grid surface is caused by the double layer formed close to the grid's wall. It reflects negative ions back to the plasma grid, where they are destroyed.

Results are in good agreement with experimental data. The performance predicted for the extractor fulfill the ITER NBI source requirements, in terms of extracted NI and electron currents.

ACKNOWLEDGMENTS

Useful discussion with A. Simonin and G. Maynard is truly acknowledged. This work was partially supported by EFDA (contract number WP08-HCD-04-01) and the Agence Nationale de la Recherche, France (ANR ITERNIS).

REFERENCES

1. Speth E., Falter H.D., Franzen P., Fantz U. and Bandyopadhyay M. 2006 *Nucl. Fusion* **46** S220
2. Hemsworth R. *et al* 2009 *Nucl. Fusion* **49** 045006
3. Fantz U. *et al* 2008 *Rev. Sci. Instr.* **79** 02A511
4. McNeely P. *et al* 2006 *Rev. Sci. Instr.* **77** 03A519
5. Takado N., Hanatani J., Mizuno T., Katoh K. and Hatayama A. 2009 1st International Symposium on Negative Ions, Beams and Sources, Aix-en-Provence, AIP conference proceedings **1097** 31
6. Heinemann B., Falter H., Fantz U., Franzen P. and Fröschle M. 2009 *Fusion Engineering and Design* **84** 915
7. Staebler A., Fantz U., Franzen P., Berger M. and Christ-Koch S. 2009 *Fusion Engineering and Design* **84** 265
8. Guster R. *et al* 2009, 1st International Symposium on Negative Ions, Beams and Sources, Aix-en-Provence, AIP conference proceedings **1097** 297
9. Guster R. *et al* 2010, *Plasma Phys. Control. Fusion* **52** 045017
10. Matsushita D.*et al* 2008 *Rev. Sci. Instr.* **79** 02A527
11. de Esch H.P.L., Svensoon L. and Riz D. 2009 1st International Symposium on Negative Ions, Beams and Sources, Aix-en-Provence, AIP conference proceedings **1097** 309
12. Wunderlich D., Guster R. and Fantz U. 2009 *Plasma Sources Sci. Technol.* **18** 045031
13. Taccogna F., Schneider R., Longo S. and Capitelli M. 2008 *Phys. Plasmas* **15** 103502
14. Taccogna F., Longo S., Capitelli M. and Schneider R. 2008 *IEEE Trans. Plasma Sci.* **36** 1589
15. Hatayama A. *et al* 2008 *Rev. Sci. Instr.* **79** 02B901

16. Lishev St. *et al* 2009 *Appl. Phys.* **106** 113301
17. S. Mochalskyy, A. F. Lifschitz, T. Minea 2010 *Nucl. Fusion* **50** 105011
18. Kaasschieter E.F. 1988 *J. of Comput. and Appl. Math.* **24** 265-275
19. Jomaa Z. and Macaskill C. 2005 *J. Comput. Phys.* **202** 488
20. Bacal M. 2006 *Nucl. Fusion* **46** S250-S259
21. Bandyopadhyay M. *et al* 2004 *Appl. Phys.* **96** 4107
22. Hanada M. *et al* 1990 *Rev. Sci. Instr.* **61** 499
23. Svarnas P., Annaratone B.M., Béchu S., Pelletier J. and Bacal M. 2009 *Plasma Sources Sci. Technol.* **18** 045010
24. Janev R. K. *et al* 1987 *Elementary Processes in Hydrogen-Helium Plasma*
25. Hiskes J. R. *et al* 1976 *Appl. Phys.* **47** 3888
26. Seidl M. *et al* 1996 *Appl. Phys.* **79** 2896
27. Belchenko Yu. *et al* 1974 *Nucl. Fusion* **14** 113
28. Belchenko Yu. *et al* 1993 *Rev. Sci. Instr.* **64** 1385
29. Pelletier J. *et al* 1979 *Appl. Phys.* **50** 4517
30. Celiberto R. *et al* 2001 *Atomic Data and Nuclear Data Tables* **77**, 161âĂŞ213
31. Block L. P. 1978 *Astr. and Space Sci.* **55** 59-83
32. Meige, A. 2006 *Ph.D Thesis* "Numerical modeling of low-pressure plasmas: applications to electric double layers."

Modeling of Neutrals and H⁻ Transport in a Large Negative Ion Source

N. Kameyama, D. Matsushita, S. Koga, R. Terasaki, and A. Hatayama

*Graduate School of Science and Technology, Keio University,
3-14-1 Hiyoshi, Kouhoku-ku, Yokohama 223-8522, Japan.*

Abstract. A systematic study of the extraction probability of surface-produced H⁻ ions from the ion source has been done with H⁻ and neutral Monte-Carlo transport modeling. Without the effects of the flow velocity of background H⁺ ions and neutrals, the extraction probability (E.P.) depends strongly on the strength of the magnetic filter field B_{MF} for the low gas pressure condition. On the other hand, the E.P. does not depend so much on BMF for the high gas pressure condition. Neutral transport modeling shows that H_2 neutral possibly has a relatively high velocity (~1km/s). The effect on the E.P., however, is not significant even for the high gas pressure regime.

Keywords: Negative ions, Monte-Carlo method, neutral flow velocity, extraction
PACS: 29.25.Ni, 29.27.Ac, 52.65.Pp

INTRODUCTION

Neutral beam injection based on a negative ion beam is very promising for heating and current drive in future fusion reactors. To generate intense beams of H⁻ ions and optimize the H⁻ ion sources, understanding of the transport properties of neutrals and H⁻ ions is indispensable.

In the previous study[1,2], it was observed that a relatively high beam intensity was extracted from a high-electron temperature region in the JAEA 10A H⁻ ion source under the Cs-seeded condition, i.e. the extracted H⁻ beam non-uniformity was observed. To clarify the origin of the non-uniformity of the H⁻ beam uniformity, a Monte-Carlo simulation model has been developed including the following important processes; i) H atom production (dissociation process of H_2 molecules), ii) transport of the dissociated H atoms, iii) reflection of H atoms at the wall and Plasma Grid (PG), iv) surface H⁻ production and v) H⁻ transport and extraction process. The numerical results of extracted H⁻ ion beam "*profile*" reasonably agree with the experimental results. The local enhancement of the H atom production in high-electron-temperature region and the resultant non-uniformity of H atom flux to the PG have been shown to be a possible origin of the H⁻ ion beam non-uniformity observed in the JAEA 10A H⁻ ion source. The "*absolute value*" of the H⁻ beam intensities, however, tend to be smaller than those in the experimental results[3]. In particular, the estimated H⁻ beam intensity without the magnetic filter is significantly smaller than those in the experiments.

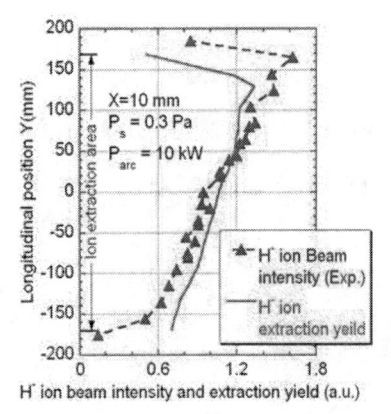

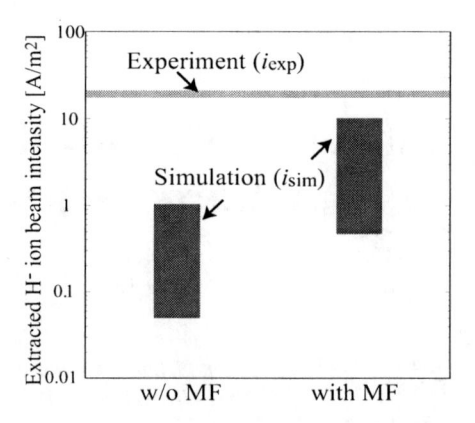

FIGURE 1. Spatial profile of H⁻ extraction yield[1,2]

FIGURE 2. Comparison of the extracted H⁻ ion beam intensity between the simulation and the experiment[3]

Therefore, further understanding of the H⁻ extraction process is required for this discrepancy. Originally, surface-produced H⁻ ions have the large velocity component not towards the extraction hole, but towards the inside of the source. It is necessary for the surface-produced H⁻ ions to change the direction of their velocities (the velocity reversal) as shown in Fig. 3. There are several possibilities for the velocity reversal of H⁻ ions towards the extraction hole and the resultant enhancement of the extraction probability: a) magnetic filter field (Larmor motion of H⁻ ions), b) collisional momentum exchange with background H^+ ions and neutrals (H atoms, H_2 molecules), c) electric-potential structure close to the PG and extraction hole.

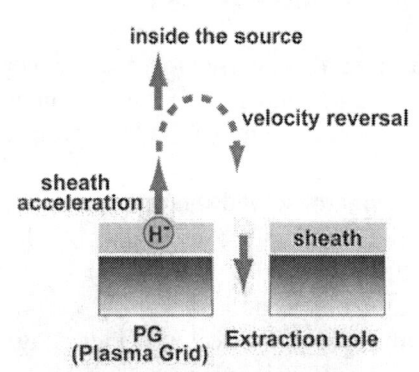

FIGURE 3. Geometry of the velocity reversal

In the present paper, we focus our attention on the mechanism: a) and b) above. The role of the a) and b) on the velocity reversal of the surface-produced H⁻ ions and the resultant H⁻ extraction are systematically studied with the 3D Monte-Carlo(MC) code of H⁻ transport code. The strength of magnetic field, the gas pressure have been varied in wide range in a model-geometry near the plasma grid (PG) of the JAEA 10 A source. The dependence of H⁻ ion extraction probability p_{ext} on these parameters is calculated and discussed for given plasma/neutral parameters. Especially, the effect of the directional flow of the background neutrals (H_2, H) towards the PG has been taken into account for the collisional momentum transfer between surface-produced H⁻ ions and these background neutrals. An estimate of the flow velocity has been done with 3D MC neutral transport code. The resultant extraction probability p_{ext} with the directional flow effect will be compared with the previous results in Ref. 2 without the effect of the flow velocity of the background neutrals.

SIMULATION MODEL

Simulation code for H⁻ ion transport

The H⁻ ion transport process has been calculated using the H⁻ ion transport code. In this code, test H⁻ ion trajectories are traced from birth points on the PG to their extraction, or destruction. H⁻ ion trajectories are traced by numerically solving the 3D equation of motion for an H⁻ ion,

$$m\frac{dv}{dt} = q(E + v \times B) + \text{(collision term)}, \quad (1)$$

where m, v, q, E and B are the H⁻ ion mass, velocity, charge, electric field and magnetic field, respectively. A leap-frog integrator[4] has been used to solve the equation of motion. Destruction processes and Coulomb collision with H⁺ ions have been considered. Collisions with the neutrals (H₂ molecules and H atoms) and destruction processes have been considered by using path length estimator algorithm[5]. The effect of Coulomb collision has been treated with the binary collision model[6]. Scattering angles by collision have been chosen by using Monte-Carlo method. The velocity of particles involving in a collision with an H⁻ ion has been chosen from a Maxwell velocity distribution with a given background temperature. Reactions taken into account in the simulation code are shown in Table 1.

TABLE 1. Reactions taken into account in the H⁻ transport code

Reactions		References
Mutual neutralization	H⁻ + H⁺ → H + H	[7]
Associative dissociation	H⁻ + H → H + H + e or H₂ + e	[7]
Electron detachment	H⁻ + e → H + e + e	[7]
Charge Exchange	H⁻ + H → H + H⁻	[8]
Elastic collision with H₂	H⁻ + H₂ → H⁻ + H₂	[9]

For good statistical accuracy, 10^5 test trajectories are followed. The birth points of H⁻ ions are determined at random on the PG. The initial energy of H⁻ ions is assumed to be 1.5 eV. They have initial velocity following cosine distribution and are accelerated by sheath potential (3 eV), i.e. initial total energy is 4.5eV. The initial total energy of H⁻ ions could depend on the assumption for the sheath potential drop. Therefore, a parametric study of the initial energy will be needed in the future to understand its effect on the extracted H⁻ current.

Model geometry, initial and boundary conditions

Fig. 4 shows the model geometry used in the following numerical calculations. The model geometry is based on the JAEA 10A ion source. The JAEA 10A negative ion source is a Cs-seeded volume type with a magnetic filter. The dimensions of plasma chamber are 240 mm wide, 480 mm high and 203 mm deep, and X, Y, Z axes are set as shown in Fig. 4(a). In the ion extraction area on the PG, which is 140 mm in width and 340 mm in height, there are extraction holes of each 9 mm in diameter in 14×27 lattice pattern.

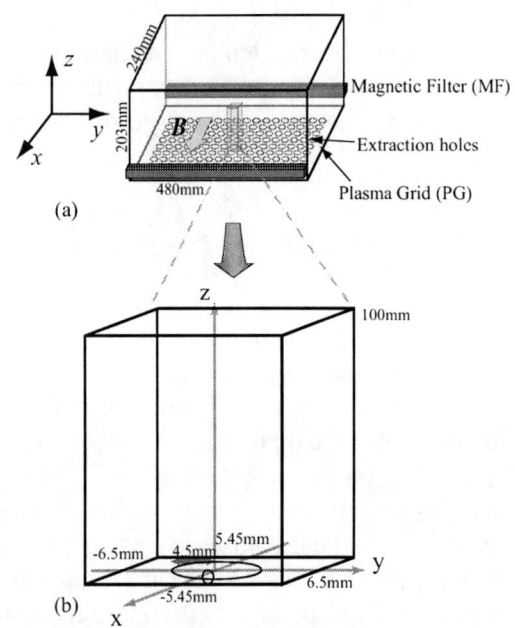

FIGURE 4. (a) Schematic diagram of JAEA 10A negative ion source.
(b) model geometry of the numerical simulation.

In the numerical calculation, only the one extraction hole is modeled as shown in Fig. 4(b) by imposing a periodic boundary condition on the side surface. The test particles across a side-boundary are re-injected from the other side-surface. The velocity is specified so as to satisfy the periodic boundary condition. For the upper boundary ($Z=100$ mm), the absorbing boundary condition is adopted. If particles reach the upper boundary, they are removed from the calculation domain. This boundary condition is reasonable, because the destruction mean free pass of H⁻ ions for the typical plasma parameters used in the following calculation is about 100 mm. A series of preliminary calculations have been done with different heights, and confirm the resultant H⁻ ion extraction probabilities does not so much depend on the height if we take the height larger than 100 mm. On the bottom boundary, i.e. at the PG, a reflection boundary condition is imposed by assuming the sheath potential drop, i.e. particles which reach the bottom boundary with the kinetic energy less than 3eV are reflected with the same speed. And particles which reach the PG with the kinetic energy more than 3eV are absorbed by the wall.

Model of magnetic filter and main numerical parameters

Spatial profile of the field intensity for the magnetic filter is given by the Gaussian profile,

$$B_x(x,y,z) = B_0 \exp\left[-\frac{z^2}{2\sigma^2}\right]. \tag{2}$$

The spatial extent of the magnetic field B_0 is systematically varied in the range $B_0=0$ to 120 G. The H_2 gas pressure in the source has been also changed in the wide range (0.3 to 1.0 Pa), while the atom/molecule density ratio (n_{H0}/n_{H2}) is fixed to be 0.2, where n_{H0} and n_{H2} are densities of H atoms and H_2 molecules, respectively. The velocities of the background H^+ ions, and H_2 molecules are sampled from a Maxwell distribution with the temperature T_{H+}, T_H and T_{H2}, respectively. These temperatures of background H^+ ions, H atoms, H_2 molecules are set to be 0.5 eV, 0.5 eV and 0.1 eV, respectively. The background electron temperature and density is fixed to be 1 eV and 5×10^{17} m^{-3}, for simplicity.

AN ESTIMATE FOR THE FLOW VELOCITY OF NEUTRALS

In order to take into account the flow velocity of the neutrals, it is necessary to estimate its order of the magnitude. We have done the neutral transport simulation with the 3D Monte-Carlo code for the JAEA 10A source to estimate the H_2-flow velocity. The directional flow could be driven by the gas pressure difference between inside and outside of the source. The model geometry used in the simulation is almost the same as in Fig. 4(a), but we have taken into account the effect of the extraction hole as shown in Fig. 5.

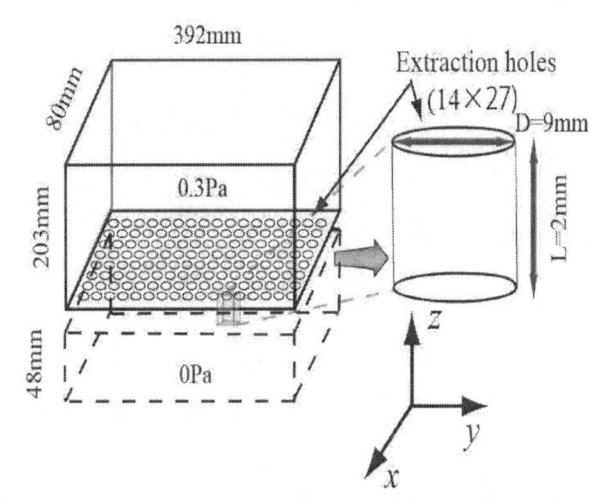

FIGURE 5. Schematic diagram of the JAEA 10A source in the neutral simulation

The collision processes in gas phase including in the analysis is summarized in Table 2.

TABLE 2. Reactions taken into account in the neutral transport code

Reactions		References
Dissociation	$e+H_2 \rightarrow H+H+e$	[7]
Ionization	$e+H_2 \rightarrow +H+e$	[7]
EV	$H_2+e \rightarrow H_2(v)+e$	[7]
Elastic collision with e	$e+H_2 \rightarrow e+H_2$	[10]

As for the collision with the wall, the cosine-law has been assumed for the angular distribution of the reflected molecules. For simplicity, the particle and energy reflection coefficients have been assumed to be 1.0. The transmission probability of the molecules through a simple cylindrical pipe with the aspect ratio $A=L/D$ (L: the axial length, D: the diameter) has been calculated for the purpose of the code check. The results are shown in Fig. 6 with the theoretical value by Santeler[11] by taking the aspect ratio A as a parameter.

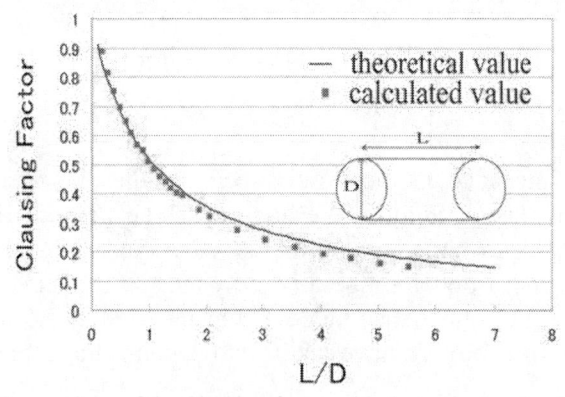

FIGURE 6. Comparison of the Clausing factor with theoretical and calculated value

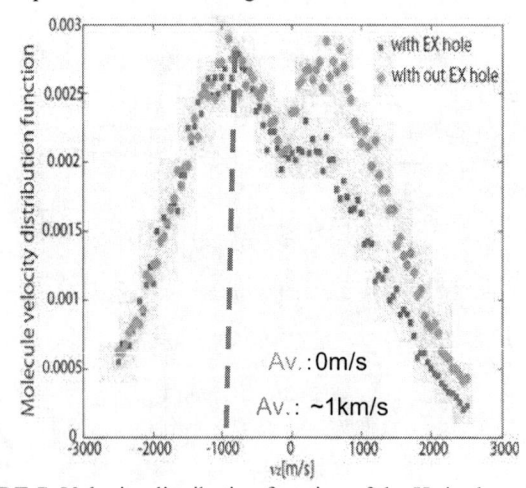

FIGURE 7. Velocity distribution function of the H_2 in the z direction

Fig. 7 shows the calculated results of the velocity distribution function of H_2 molecule in the model geometry of the JAEA 10A source shown in Fig. 5. The H_2 velocity distribution possibly has an average velocity ~1km/s in the z direction, which is the parallel direction to the extraction axis and the negative sign means that the flow velocity direction is towards the outside. Being based on the estimate, the effect of the flow velocity of the neutrals is taken into account in the following manner in the H⁻ transport simulation. Each velocity of H_2 molecules and H atoms involved in the collision event is sampled from the "shifted" Maxwellian distribution with the flow velocity $\bar{u}$ in the range of $0\,\text{km/s} \leq \bar{u} \leq 10\,\text{km/s}$. The temperature of H_2 molecules and H atoms, and other simulation conditions are the same as in the case without the effect of the flow velocity explained above.

RESULTS AND DISCUSSION

Dependent of the H⁻ ion extraction probability on the gas pressure and the filter strength without the effect of the flow velocity.

To understand the roles of the magnetic filter field and collisions with neutrals, the extraction probabilities p_{ext} have been calculated by systematically changing the magnetic field (peak filter strength B_0=0 to 120 G) and gas pressure (P=0.3 to 1.0 Pa) in the previous study in Ref. 2. The extraction probability is defined by the following equations $p_{ext} = N_{ext} / N_{tot}$, where N_{ext} is the number of extracted H⁻ ions and N_{tot} is the number of total launched H⁻ ions.

The results are briefly summarized in Fig. 8: (1) under the low gas pressure condition with P=0.3 Pa, the extraction probability is strongly dependent on the filter strength, and the extraction probabilities are relatively small when the filter strength is small, (2) with increasing the gas pressure, the dependence of p_{ext} on the filter strength becomes relatively week.

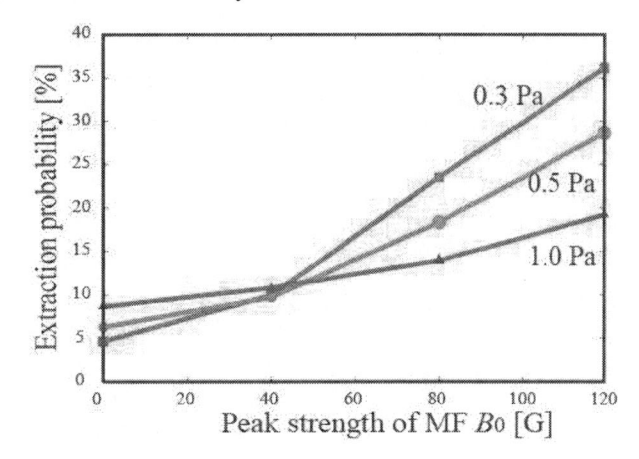

FIGURE 8. The dependence of the extraction probability of the H- ion on filter magnetic field

These tendencies are mainly explained by the pressure dependence of mean free pass (MFP) of each collision process in the model. Typical travel distance of extracted H⁻ ions from their birth points to PG due to the Larmor motion at the magnetic field B_0=40 and 120 G becomes 0.2 m and 0.06m, respectively with the H⁻ ion energy of 3 eV. Under the low gas pressure condition, MFPs are considerably larger than the travel length. In such a case, collisions with neutrals are not effective to disturb the Larmor motion of H⁻ ions. Under the high gas pressure condition, however, MFPs significantly decrease. In particular, MFP of charge exchange becomes shorter than the travel distance of extracted H⁻ ions when gas pressure is larger than 0.7 Pa. Therefore, p_{ext} is not dependent on the filter strength so much under the high gas pressure condition. This understanding well agrees with Matsumoto et al[12].

Dependence of the H⁻ ion extraction probability on the pas pressure with the effect of the flow velocity.

To understanding the roles of the flow velocity towards the extraction hole, p_{ext} has been calculated by systematically changing the flow velocity ($\bar{u} = 0$ to 10 km/s). The results are summarized in Fig. 9. For the realistic flow velocity ($\bar{u} =\sim 1$km/s), there are no significantly effects of the neutral flow velocity on the E.P., even for the high gas pressure. However, for the high flow velocity, the E.P. depends strongly on the flow velocity. Therefore, the effect of the high flow velocity (e.g. Franck-Condon H atoms) might play a role, especially for the high gas pressure case.

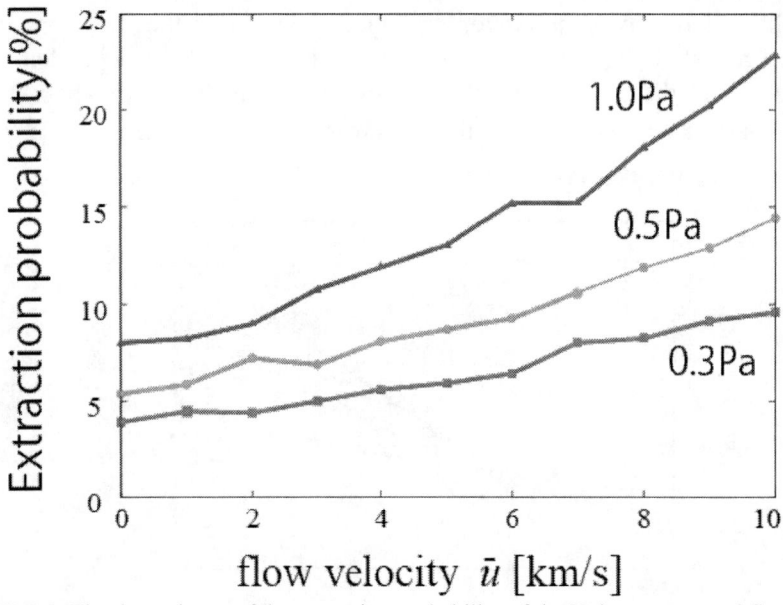

FIGURE 9. The dependence of the extraction probability of the H⁻ ions on neutral flow velocity

CONCLUSION

A systematic study of surface-produced H⁻ ion velocity reversal and extraction has been done with neutral and H⁻ Monte-Carlo transport modeling.

Without the effects of the flow velocity of H_2 and H : a) Larmor motion in the magnetic filter field has a critical effect on H⁻ ions extraction for the low gas pressure regime (0.3-0.5Pa), b) collisional momentum exchange becomes important for the velocity reversal for the high gas pressure regime (1Pa).

The neutral Monte Carlo transport modeling shows that the H_2 molecule velocity distribution possibly has an average velocity ~1km/s in the extraction direction. No significant effects of the flow velocity are shown on the extraction probability for the realistic range of flow velocity ($\bar{u}$=~ 1km/s).For higher flow velocity (e.g. Franck-Condon H atoms), the effect might play a role, especially for the high gas pressure case.

More self-consistent modeling including the effect of the potential structure close to the PG will be needed and planned for further understandings.

REFERENCES

1. N. Takado, H. Tobari, T. Inoue, J. Hanatani, A. Hatayama, M. Hanada, M. Kashiwagi, and K. Sakamoto, *J. Appl. Phys.*, **103**, 053302(2008).
2. A. Hatayama, *Rev. Sci. Instrum.*, **79**, 02B901(2008).
3. D. Matsushita, N. Takado, A. Hatayama, T. Inoue, *Rev.Sci.Instrum.* **79**, 02A527 (2008).
4. C. K. Birdsall and A. B. Langdon, *Plasma Physics via Computer Simulation* (McGraw-Hill, New York, 1985).
5. M. H. Hughes and D. E. Post, *J. Comput. Phys.* **28**, 43 (1978).
6. T. Takizuka and H. Abe, *J. Comput. Phys.* **25**, 205 (1987).
7. R. K. Janev, W. D. Langer, K. Evans, Jr., and D. E. Post, Jr., *Elementary Processes in Hydrogen-Helium Plasma, Cross Sections and Reaction Rate Coefficients* (Springer, Berlin, 1987).
8. R. K. Janev, D. Reiter, and U. Samm, *Collision Processes in Low-Temperature Hydrogen Plasmas* (Forschungszentrm-Juelich Report No. Juel-4105, 2003).
9. R. K. Janev, *Atomic and Molecular Processes in Fusion Edge Plasmas* (Plenum, New York, 1995).
10. M.Brunger, *et al, Interactions of photons and electrons with molecules* (Berlin: Springer-Verlag) (2003).
11. D. J. Santeler, *J. Vac. Sci. Technol.* A **4** 338(1986).
12. Y. Matsumoto, M. Nishiura, K. Matsuoka, M. Sasao, M. Wada, and H. Yamaoka, *Thin Solid Films* **506–507**, 522 (2006).

Fluid Model Analysis of the Distribution of the Negative Ion Density before the Extraction from a Tandem Type of a Plasma Source

Stiliyan St. Lishev[a], Antonia P. Shivarova[a] and Khristo Ts. Tarnev[b]

[a]*Faculty of Physics, Sofia University, BG–1164 Sofia, Bulgaria*
[b]*Department of Applied Physics, Technical University–Sofia, BG–1000 Sofia, Bulgaria*

Abstract. Unified description of the expansion (through a magnetic filter) and extraction regions, including also the driver, of volume-production based sources of negative hydrogen ions is presented in the study. The model is one-dimensional, developed within the fluid plasma theory. It covers description of both the second chamber of the tandem sources and a single inductively-driven (with a planar coil) discharge of a matrix source. Four parameters have been varied in the analysis of the results: type, position and magnitude of the filter field and rf power applied for the discharge maintenance. The obtained results for the spatial distribution of the plasma parameters as well as for the electronegativity, for the negative ion beam current and its ratio to the electron current at the position of the extraction are discussed regarding optimum conditions for the source operation. The conclusion is that a magnetic field extended over a large region till the extraction favors locality in the discharge behavior and, respectively, the local production of the negative ions. This leads to high density of the extracted negative ion beam current at reasonable values of its ratio to the electron current density.

Keywords: Plasma Sources of Negative Hydrogen Ions, Tandem Sources.
PACS: 52.50.Dg, 52.65.Kj

INTRODUCTION

The current development towards the use of the negative hydrogen/deuterium ion beam sources for plasma heating in big fusion machines stimulates the theoretical work on modeling of the rf-driven tandem-type two-chamber plasma source planed for ITER as well as a work directed towards new decisions for the source configuration.

The complicated design of the two-chamber rf driven source [1] and the requirements for its operation at low gas pressure ($p = 0.3$ Pa) are in the basis of the difficulties to achieving an unified description of its operation, including its three regions: the driver, the expansion plasma region (with the magnetic filter for electron cooling) and the extraction region. Combined description [2, 3] of the expansion and extraction regions of the two-chamber source is the current achievement in its modeling. Separate treatment of the driver [4], of the magnetic filter region (e.g., [5]) and of the extraction region [6-8] completes the knowledge for the discharge in the source and for the source efficiency regarding the properties of the extracted negative ion beams.

On the other hand, recent studies on low-pressure hydrogen discharges [4, 9] have outlined a concept for a new design of the source as a matrix of small-radius inductively-driven discharges. At the current stage the development of the concept is based on inductive discharges with a cylindrical coil. However, completing the matrix by inductive discharges with planar coils would be the more proper decision regarding its construction.

The model presented in this study, being one-dimensional (1D) one, covers description of both (i) the tandem source (its expansion – through a magnetic filter – and extraction regions) and (ii) a single discharge of the matrix performed as a tandem-type source (with a driver of the type of an inductive discharge with a planar coil, an expansion through a magnetic filter and an extraction). The model is within the fluid-plasma theory, with account for the volume production of the negative hydrogen ions. COMSOL Multyphysics is employed as a numerical environment. The discharge is at a low gas pressure as required for the negative ion sources used for fusion plasma heating. The conditions of low gas pressure and, respectively, the free-fall regime of discharge maintenance are specified by accounting for the inertia term in the momentum equations of the positive ions via introducing an effective electric field. The results, discussed regarding optimum conditions for the source operation, are for the spatial distribution of the plasma parameters as well as for the electronegativity, for the negative ion beam current and its ratio to the electron current at the position of the extraction. The type, the position and the magnitude of the filter field as well as the rf power applied for the discharge maintenance are the parameters varied in the analysis of the results. It is shown that a magnetic filter that ensures electron magnetization over a large plasma region extended till the extraction stimulates locality in the discharge behavior. This provides high negative ion yield that leads to high current density of the extracted negative ions at reasonable values of its ratio to the electron current density.

FORMULATION OF THE PROBLEM AND SET OF EQUATIONS

A plasma source with volume-produced negative hydrogen ions is considered within a 1D model. The description of the source includes its driver, the expansion plasma region and the extraction of the negatively-charged particles. Having a localized rf power deposition, the source is with a remote plasma maintenance. The plasma expansion is through a magnetic filter. The description of the extraction of the negatively-charged particles is at the position of the first electrode (the plasma electrode) of the extraction device.

The rf power deposition is with a half of a super-Gaussian profile $P_{ext} = P_0 \times \exp[-(1/2)(z/\sigma_P)^{2m}]$ where P_0 is its maximum at $z = 0$ and σ_P scales its width. The two configurations of the magnetic filter considered

$$B(z) = B_0 \exp\left[-\frac{1}{2}\left(\frac{z - z_{MF}}{\sigma_B}\right)^2\right] \qquad (1a)$$

and

$$B(z) = \begin{cases} B_0 \exp\left[-\frac{1}{2}\left(\frac{z-z_{MF}}{\sigma_B}\right)^2\right], z < z_{MF} \\ B_0, z \geq z_{MF} \end{cases} \quad (1b)$$

correspond, respectively, to a Gaussian profile (1a) and a step-function profile (1b) of the filter field, the latter with a Gaussian leading tail. In (1), B_0 is the maximum of the field and σ_B scales its axial variation. The magnetic field is oriented perpendicularly to the z-direction, the direction along the source axis.

The model is within the fluid-plasma theory. The set of equations

$$\frac{\partial n_\alpha}{\partial t} + \frac{\partial \Gamma_\alpha}{\partial z} = \frac{\delta n_\alpha}{\delta t} \quad (2)$$

$$\frac{\partial N_1}{\partial t} + \frac{\partial \Gamma_1}{\partial z} = \frac{\delta N_1}{\delta t} \quad (3)$$

$$\frac{\partial N_2(v)}{\partial t} + \frac{\partial \Gamma_2(v)}{\partial z} = \frac{\delta N_2(v)}{\delta t} \quad (4)$$

$$\frac{3}{2}\frac{\partial}{\partial t}(n_e T_e) + \frac{\partial J_e}{\partial z} = P_{ext} - P_{coll} + e\Gamma_e E_{dc} \quad (5)$$

$$\frac{d^2\Phi}{dz^2} = -\frac{e}{\varepsilon_0}\left(\sum_{j=1}^{3} n_j - n_e - n_n\right) \quad (6)$$

includes the continuity equations (2)-(4), respectively, of the charged particles (electrons $\alpha = e$, the three type of positive ions $\alpha = j$ with $j = 1$-3, respectively, for the H^+-, H_2^+ and H_3^+-ions, and negative ions $\alpha = n$), of the hydrogen atoms and of the vibrationally exited molecules ($v = 1$-14), the electron energy balance equation (5), the Poisson equation (6) and the equation of state $p = \kappa T_g(N_1+N_2)$. In (2)-(6), n_α, N_1 and N_2 are the densities of the different types of particles, Γ_α, Γ_1 and $\Gamma_2(v)$ are their axial fluxes, T_e and J_e are the electron temperature (in energy units) and the axial electron energy flux, Φ is the dc electric potential ($E_{dc} = -d\Phi/dz$) and T_g is the gas temperature; e and ε_0 are, respectively, the elementary charge and the vacuum permittivity, κ is the Boltzmann constant and T_g is the gas temperature.

The magnetic field is weak and, thus, only the electrons are magnetized. This is specified in the expressions for their flux $\Gamma_e = b_{\perp e}n_e(\partial\Phi/\partial z)-D_{\perp e}(\partial n_e/\partial z)-D_{\perp e}^T(n_e/T_e)(\partial T_e/\partial z)$ and for their energy flux $J_e = -\chi_{\perp e}(\partial T_e/\partial z)+(5/2)T_e\Gamma_e$, where $b_{\perp e}$, $D_{\perp e}$ and $\chi_{\perp e}$ are, respectively, the mobility-, diffusion- and thermal-conductivity coefficients across a magnetic field; $D_{\perp e}^T = D_{\perp e}$ is the thermal diffusion coefficient.

The flux $\Gamma_n = b_n n_n(\partial\Phi/\partial z)-D_n(\partial n_n/\partial z)$ of the negative ions, with b_n and D_n being the mobility and diffusion coefficients, is not influenced by the magnetic field.

The low gas pressure of the discharge imposes conditions of a free-fall regime of discharge maintenance. The latter is specified by accounting for the inertia term in the momentum equations of the positive ions by introducing an effective electric field [10] with an equation for it

$$\frac{\partial E_{\text{eff}}}{\partial t} = -\frac{1}{b_j}\left[\upsilon_j \frac{\partial \upsilon_j}{\partial z} - \frac{e}{m_j}(E_{\text{dc}} - E_{\text{eff}})\right] \qquad (7)$$

involved in the initial set of equations; υ_j is the axial ion velocity and m_j is the ion mass. In a way, the flux of the positive ions $\Gamma_j = b_j n_j E_{\text{eff}} - D_j(\partial n_j/\partial z)$ is reduced to a drift-diffusion flux; here $b_j = e/(\mu_j v_j)$ and $D_j = T_j/(\mu_j v_j)$ are the mobility and diffusion coefficients of the positive ions and $(\mu_j v_j)$ is a notation [5] combining the elastic collisions of the ions with atoms and molecules and the inelastic collisions for their production.

The processes involved in the particle production and losses on the right-hand sides of (2)-(4) are ionization of atoms and molecules, dissociation of molecules and molecular ions, dissociative recombination of H_3^+-ions, transformation of H_2^+-ions into H_3^+-ions in collisions with molecules, H^--production by electron attachment to vibrationally exited molecules $H_2(v = 4-9)$ considered separately to each of the vibrationally exited states, destruction of the negative ions in collisions with electrons and atoms and in recombination with the positive ions, wall recombination of atoms, vibrational-translational and vibrational-vibrational collisions and e-V and E-V processes. The electron energy losses in collisions P_{coll} in (5) are in elastic collisions with atoms and molecules, for atom excitation and ionization, for dissociation and ionization of molecules and their excitation.

The boundary conditions at $z = 0$ are for a metal wall whereas $E_{\text{dc}}(z = z_{\text{PG}}) = 0$ is the condition at the position $z = z_{\text{PG}}$ of the plasma grid of the extraction region. At $z = 0$, $\Gamma_{e,n} = (1/4)\upsilon_{e,n}^{\text{th}} n_{e,n}$, $\Gamma_j = (1/2)n_j \upsilon_j^{\text{th}} + b_j n_j E_{\text{eff}}$, $\Gamma_{H,H_2(v)} = (1/4)\gamma_{H,H_2(v)} N_{1,2(v)} \upsilon_{H,H_2(v)}^{\text{th}}$ and $J_e = (5/2)\Gamma_e T_e$ are the conditions for the corresponding fluxes. Here the superscript "th" refers to the thermal velocities and the coefficients $\gamma_{H,H_2(v)}$ describe the wall processes (wall recombination of the hydrogen atoms and wall deactivation of the vibrationally excited molecules). At $z = z_{\text{PG}}$, the positive ions are reflected and the fluxes of electrons and negative ions are thermal fluxes.

RESULTS AND DISCUSSIONS

The results discussed here are (i) for the axial variation of the plasma parameters (electron temperature T_e and density n_e, plasma potential Φ, density n_n and flux Γ_n of the negative ions, electronegativity n_n/n_e and local balance of the negative ions) and (ii) for the electronegativity, the current density of the negative ions and its ratio to the electron current density at the position of the extraction $z = z_{\text{PG}} = 20$ cm. Results for the two types of configurations of the magnetic field of the filter (Fig. 1), as given by (1a) and (1b), are compared. Both the position z_{MF} of the filter ($z_{\text{MF}} = (12-20)$ cm) and the magnitude B_0 of its field ($B_0 = (25-100)$ G) as well as the total rf power applied for the discharge maintenance ($Q = (0.1-4.3)$ kW) have been varied. The gas pressure is fixed at $p = 0.3$ Pa and the assumed value of the gas and ion temperatures is $T_{g,j,n} = 300$ K. The parameters of the rf power input are m = 2 and $\sigma_P = 4.7$ cm; σ_B in (1) is with the value of $\sigma_B = 1.6$ cm. Thus, when z_{MF} is not too close to $z_{\text{PG}} = 20$ cm in the case of a Gaussian profile (1a) of the filter field, the magnetic field decreases towards z_{PG} and the plasma expansion is through a spatially-localized varying magnetic field. In the case of a step-function profile (1b) of the magnetic field, a large part of the source is a region of electron magnetization.

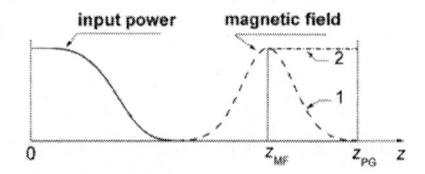

FIGURE 1. Illustration of the configuration studied with schematical presentation of the rf power deposition and of the magnetic filter field considered in two configurations: a Gaussian profile in the longitudinal direction (1) and a step profile with a Gaussian leading tail (2); z_{PG} is the position of the plasma grid of the extraction region.

Axial Variation of the Plasma Parameters

Figure 2 shows the results for the axial variation of the electron temperature T_e and density n_e and for the density n_n of the negative ions for different positions z_{MF} of the magnetic filter. The upper figures present the results obtained for a Gaussian profile of the filter field, further on called case (1), whereas the lower figures are for a step function profile, called further on case (2).

In both cases starting from a z-position in the very beginning of the region of the filter field, the electron temperature strongly drops in the filter (Figs. 2(a) and 2(d)). In case (1) T_e reaches a plateau behind the filter (Fig. 2(a)). In case (2) the decrease of T_e continues till z_{PG}, due to the magnetic field extension till the position of the plasma grid. As a result, at the z_{PG}-position T_e is lower in the latter case. In both cases T_e is higher in the driver region when the position of the filter is closer to the driver. When the z_{MF}-positions are the same, the value of T_e at $z = 0$ is the same in the two cases. The drop of T_e in the filter region is accompanied by a formation of a maximum of n_e in the filter (Figs. 2(b) and 2(e)). In the two cases the position of the maximum is before the z_{MF}-position. In case (1) the electron density drops strongly towards the position z_{PG} of the extraction. In case (2) the decrease of n_e is weaker and it appears only when the position of z_{MF} is comparatively away from the position of the extraction. In general, in case (2), the changes of n_e at $z = z_{PG}$ with varying z_{MF} are nonmonotonic. The highest value of n_e is for $z_{MF} \approx 17$ cm. The axial behavior of the density of the negative ions is quite different in the two cases. In case (1) n_n, having a maximum before the filter, decreases in the filter and after that reaches a slight maximum behind it. In case (2), the higher maximum of n_n is in the filter region preceded by a slight maximum before the filter. The highest n_n is for a z_{MF}-position which is comparatively away from the extraction. In this case the maximum n_n-value is an order of magnitude higher than the highest n_n reached in case (1).

In general, the axial variation of the density of the H^+-ions, the ions with the highest concentration among the positive ions, follows the axial variation of n_e.

The axial profiles of the electronegativity n_n/n_e (Fig. 3) are also quite different in the two cases. In case (1) n_n/n_e has only one maximum which is behind the filter. In case (2) the maxima of n_n/n_e are two: one before the filter and another one – slightly weaker – behind it. A position of $z_{MF} \approx 15$ cm of the filter in case (1) ensures $n_n/n_e > 1$ at the position z_{PG} of the extraction. The highest values of n_n/n_e in case (2) are for a position z_{MF} comparatively away from z_{PG}.

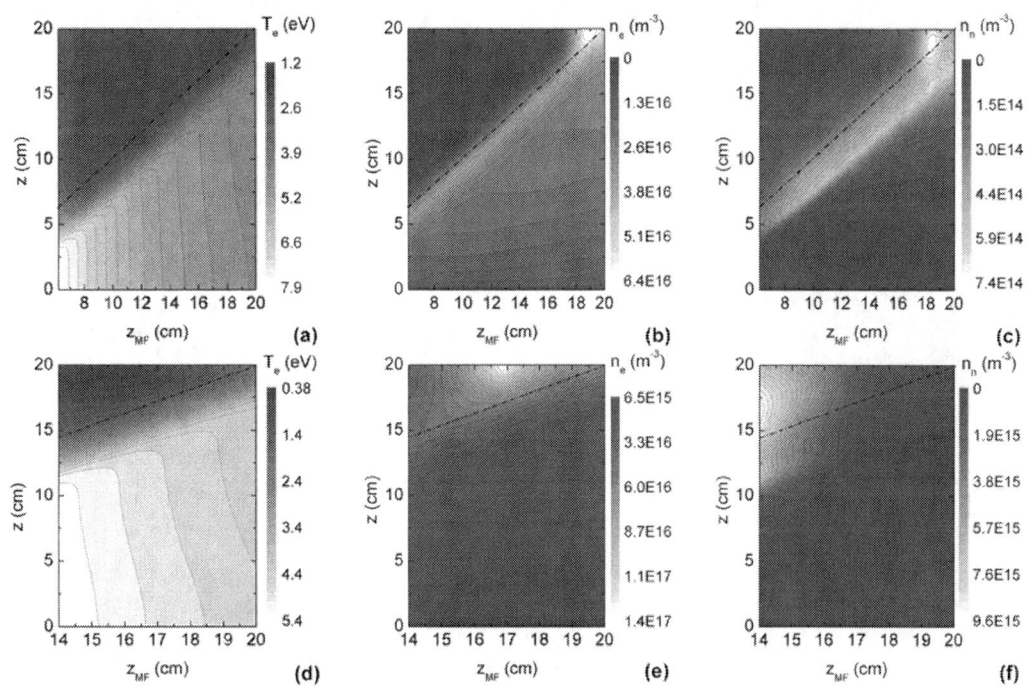

FIGURE 2. 3D plots presenting the axial (z) variation of the electron temperature T_e and density n_e and of the density n_n of the negative ions for different positions z_{MF} of the magnetic filter in the case of a Gaussian profile of the magnetic field (upper figures (a), (b) and (c)) and of a step-function profile (lower figures (d), (e) and (f)) for $B_0 = 40$ G and $Q = 2$ kW. The black dash-dotted line marks the position $z = z_{MF}$ of the magnetic filter.

In both cases, before the filter the axial variation of the plasma potential Φ is the same (a strong increase towards the filter). However, Φ behaves differently in the filter region (Figs. 3(b) and 3(e)). In case (1), Φ reaches a plateau in the filter region followed by a slight increase towards the position of the extraction. In case (2), Φ slightly decreases towards z_{PG}. Thus, in the first case the negative ions in the filter region are accelerated towards the extraction whereas in the second case the electric field tries to trap them in the filter region.

In the two cases (Figs. 3(c) and 3(f)) the n_n-production in the driver exceeds the local losses there. After the driver, in case (1) the local balance $\delta n_n/\delta t$ of the negative ions has a complicated axial behavior: predominating local losses before the filter followed by predominating negative ion production in the filter region and predominating local losses at the position of the extraction. In case (2), the local losses also predominates before the filter but, since the filter field is extended towards the position of the extraction, the region of predominating local production of n_n is extended towards the z_{PG}-position.

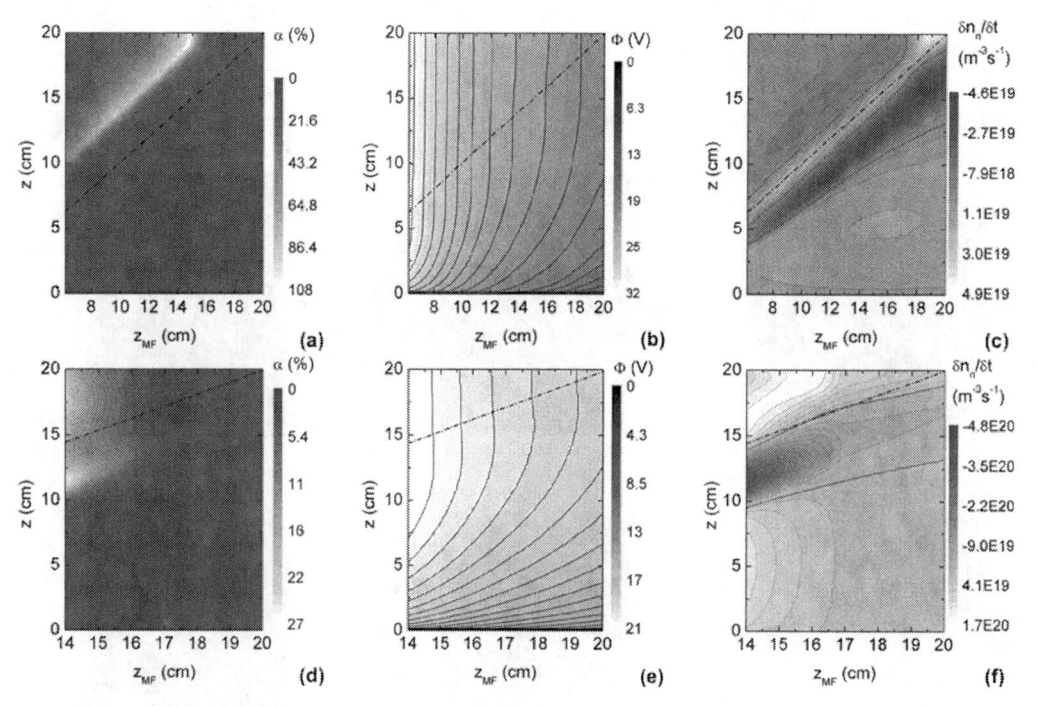

FIGURE 3. 3D plots presenting the axial (z) variation of the electronegativity $\alpha = n_n/n_e$, of the plasma potential Φ and of the local balance $\delta n_n/\delta t$ of the negative ions for different z_{MF}-position in the cases of a Gaussian profile (upper figures (a), (b) and (c)) and a step-function profile (lower figures (d), (e) and (f)) of the filter field. In (b) and (e) the equipotential curves are also given. The dotted curves in (c) and (f) trace out the positions of zero values of $\delta n_n/\delta t$.

As the comparison of the results in Figs. 2(c) and 2(f) with those in Figs. 3(c) and 3(f) shows, the axial profiles of the negative ion density do not follows the axial variation of the local balance of the ions. Thus, the fluxes are strongly involved in the formation of the axial variation of n_n in the entire source. This is shown in Fig. 4 where the axial variation of the convective and diffusion fluxes of the negative ions as well as their total flux are given. In case (1) the strong convective flux of the negative ions before the filter, due to the strong dc electric field there, carries the ions produced in the driver away from it, towards the filter region. This leads to the formation of the maxima of n_n and of the local losses before the filter. Due to this maximum of n_n, a large diffusion flux carries the ions towards the position of the extraction. The slight increase of Φ towards z_{PG} also contributes. In case (2), due to the large extension of the magnetic filed, the balance of the negative ions is more or less local. The behavior before the magnetic field is the same as in case (1). However, due to the strong local production of the ions in the filter, a diffusion flux carries away the ions towards the position of the extraction, regardless of the retarding electric field in this case. As a result, n_n is the highest one in the case of a largely-extended filter field with a step-function profile. In the opposite, the electronegativity has its highest value in the case of a Gaussian profile of the magnetic filter when it is positioned away from z_{PG}. However, this is due to the strong drop of n_e in this case.

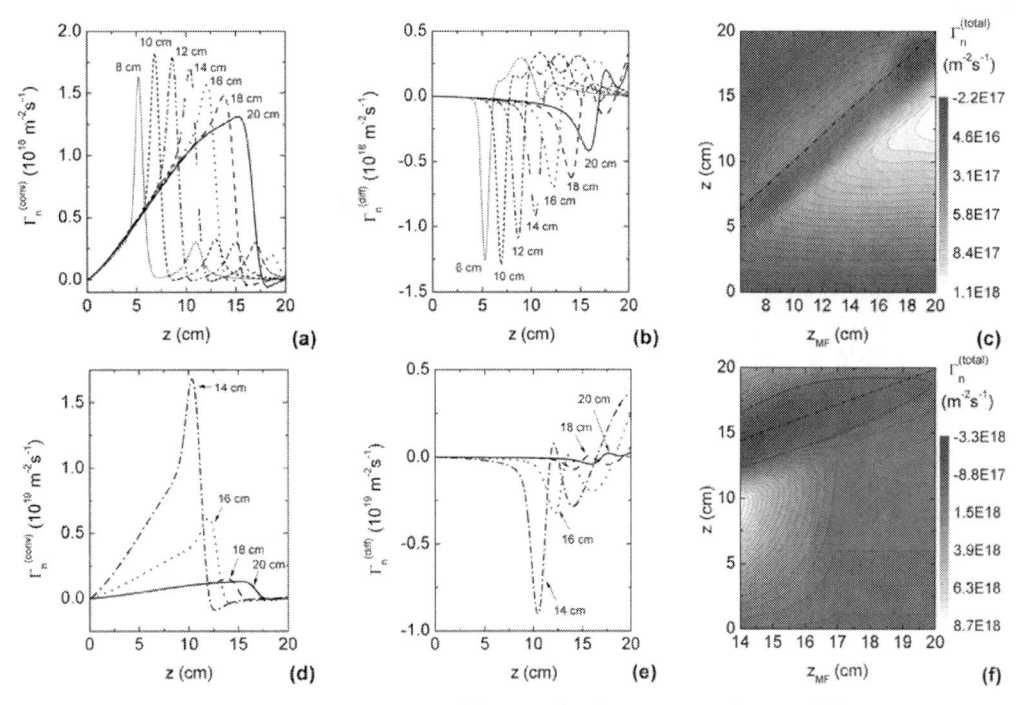

FIGURE 4. Axial variation of the fluxes of the negative ions – convective and diffusion fluxes as well as the total flux – for different z_{MF}-position, as given on the figures, in the cases of a Gaussian profile (upper figures (a), (b) and (c)) and of a step-function profile (lower figures (d), (e) and (f)).

Parameters at the Position of the Extraction

The results discussed here are for the negative ion current density j_n and for its ratio j_n/j_e to the electron current density at the position z_{PG} of the extraction since these are the parameters determining the efficiency of the source. Results for the electronegativity n_n/n_e are also given. The position z_{MF} of the magnetic filter, the magnitude of its field and the rf applied power are the parameters varied, for the two configurations (cases (1) and (2)) of the filter field.

In case (1), a filter field with $B_0 = 100$ G located comparatively close to the position of the extraction ($z_{MF} = 18$ cm) ensures the highest values of j_n and j_n/j_e (Figs. 5(b) and (c)). The highest value of n_n/n_e is at a lower magnetic field ($B_0 = 40$ G) located comparatively away from the extraction ($z_{MF} \approx 15$ cm). Thus, in this case high n_n/n_e does not mean high j_n/j_e, because of the acceleration action of the dc field close to the extraction.

In case (2) the same values of the filter field and of its position – $B_0 = 25$ G and $z_{MF} \approx 9$ cm – ensure the highest values of n_n/n_e and j_n at the position of the extraction (Figs. 5(d) and 5(e)). However, a magnetic filter with a stronger field ($B_0 = 100$ G) located close to the extraction ($z_{MF} \approx 18.5$ cm) ensures the highest value of j_n/j_e (Fig. 5(f)). In general, in this case a comparatively weak magnetic field ($B_0 = 25$ G) occupying a large volume before the extraction ($z_{MF} \approx 9$ cm) ensures high n_n/n_e and j_n at completely reasonable value of j_n/j_e ($j_n/j_e \approx 10$).

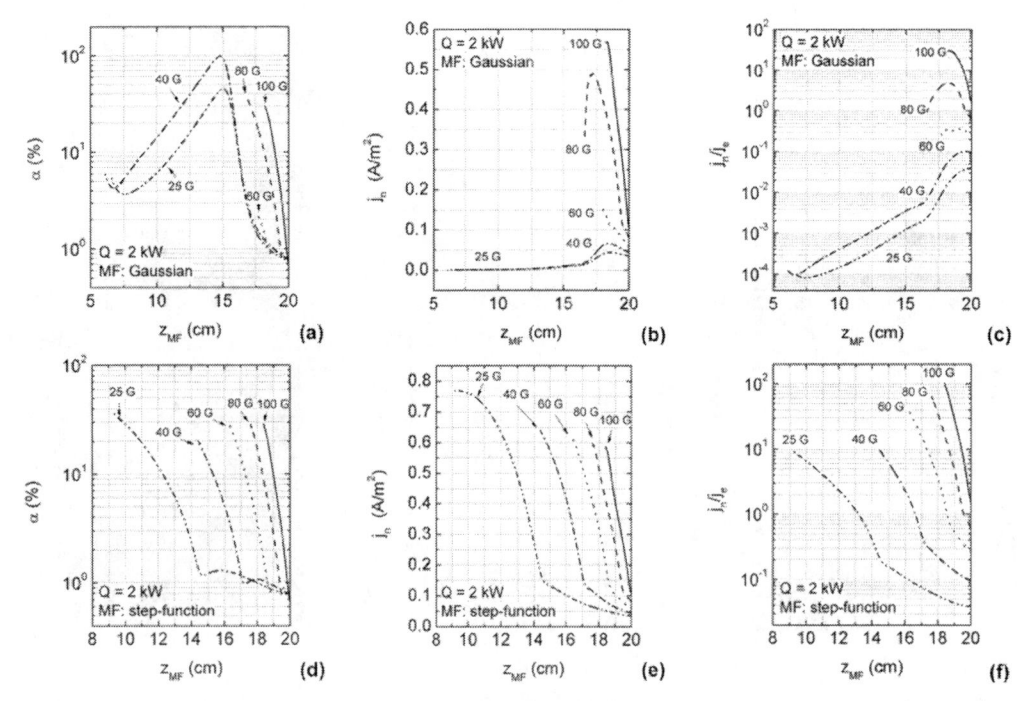

FIGURE 5. Variation – at the position z_{PG} of the extraction – of the electronegativity $\alpha = n_n/n_e$, of the current density of the negative ions and its ratio to the electron current density in the cases of a Gaussian profile (upper figures (a), (b) and (c)) and of a step-function profile (lower figures (d), (e) and (f)); $Q = 2$ kW and different values of B_0.

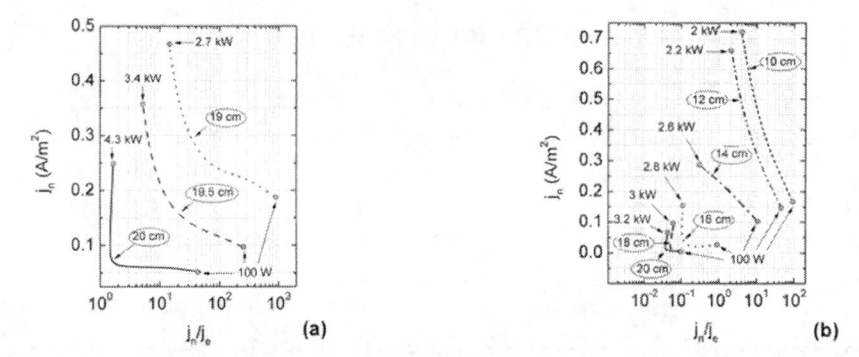

FIGURE 6. j_n vrs. j_n/j_e for different positions z_{MF} (as given on the figures) of the filter and varying applied power (in the limits given on the figures) in the cases of a Gaussian profile of the filter field ((a), for $B_0 = 100$ G) and of a step-function profile ((b), for $B_0 = 25$ G).

Figure 6 presents the variation of j_n and j_n/j_e with varying applied power, for given values of z_{MF}. Values of $B_0 = 100$ G and $B_0 = 25$ G are chosen, respectively, in cases (1) and (2) since, according to Fig. 5, these are the values ensuring the highest j_n under the conditions of high or reasonable j_n/j_e. In the two cases ((1) and (2)) when $z_{MF} = z_{PG} = 20$ cm, the leading tail of the field profile is that affecting the discharge behavior resulting in the same type of a $(j_n–(j_n/j_e))$-dependence: with increasing Q, j_n/j_e first decreases under the condition of an almost constant j_n and after that a second branch of the $(j_n–(j_n/j_e))$-dependence appears with increasing j_n for almost constant

j_n/j_e. The difference in the values of j_n and j_n/j_e in Figs. 6(a) and 6(b) is due to the difference in the B_0-values. With z_{MF} shifted away from z_{PG}, the two branches of the (j_n–(j_n/j_e))-dependence transforms into a smooth variation. In the two cases, the optimum values – high j_n at a reasonable j_n/j_e – are at high power applied for the discharge maintenance. The configuration of a step-function profile of the filter field ensures higher current density of the negative ions.

CONCLUSIONS

Both the axial variation of the plasma parameters of a low pressure hydrogen discharge and the extraction parameters of a tandem-type of negative ion source with localized rf power deposition are obtained within an 1D model covering the driver and the plasma expansion till the extraction region (its plasma grid). The model is applicable to both the second chamber of the tandem rf-driven sources and to a single inductively-driven (with a planar coil) discharge of a matrix source. A configuration of the magnetic filter providing extension of a constant field over a large volume, i.e. ensuring a large volume of electron magnetization, is introduces and compared – in terms of results for the plasma behavior and for the extraction parameters – with a localized filter field with a Gaussian profile. Such a step-function profile of the magnetic filter ensures lower electron temperature and higher electron density in the filter region that favors the local balance of the negative ions and provides high local production of negative ions and, respectively, high density of the ions close to the extraction. This results into high current density of the ions at the position of the extraction under the conditions of reasonable values of the ratio of the negative ion current density to the electron current density.

ACKNOWLEDGMENTS

Dr. Ts. Paunska is acknowledged for her contribution to the work. The work is within project DO02-267 supported by the National Science Fund in Bulgaria and it is part of the working programme of the Bulgarian Association EURATOM/INRNE (task 2.1.1).

REFERENCES

1. E. Speth et al., *Nuclear Fusion* **46**, S220-S238 (2006).
2. F. Taccogna, R. Schneider, S. Longo and M. Capitelli, *Phys. Plasmas* **15**, 103502 (2008).
3. F. Taccogna, P. Minelli, S. Longo, M. Capitelli and R. Schneider, *Phys. Plasmas* **17**, 063502 (2010).
4. Ts. Paunska, A. Shiarova and Kh. Tarnev, *J. Appl. Phys.* **107**, 083301 (2010).
5. St. Kolev, St. Lishev, A. Shivarova, Kh. Tarnev and R. Wilhelm, *Plasma Phys. Control. Fusion* **49**, 1349-1369 (2007).
6. R. Gutser, D. Wünderlich, U. Fantz, P. Franzen, B. Heinemann, R. Nocentini and the NNBI Team, *AIP Conf. Proc.* **1097**, 297-306 (2009).
7. H. P. L. deEsch, L. Svenson and D. Riz, *AIP Conf. Proc.* **1097**, 309-318 (2009).
8. P. Agostinelli et al, *AIP Conf. Proc.* **1097**, 325-334 (2009).
9. Ts. Paunska, H. Schlüter, A. Shivarova and Kh. Tarnev, *Phys. Plasmas* **13**, 023504 (2006).
10. A. Salabas, G. Gousset and L. L. Alves, Plasma Sources Sci. Technol. 11, 448-465 (2002).

Numerical Analysis of the Extraction of Volume Produced Negative Hydrogen Ions in the Extraction Region of Negative Ion Source

S. Wada[a], S. Kuppel[a], T. Fukuyama[a], K. Miyamoto[b], A. Hatayama[a] and M. Bacal[c].

[a]*Graduate School of Science and Technology, Keio University,*
3-14-1 Hiyoshi, Kouhoku-ku, Yokohama 223-8522, Japan.
[b]*Naruto University of Education,*
748 Nakashima, Takashima, Naruto-cho, Naruto-shi, Tokushima.
[c]*Laboratoirie de Physique des Plasmas,*
Ecole Polytechnique, 91128 Palaiseau,UPMC,
Universit PARIS-SUD 11, UMR 7648 du CNRS, France.

Abstract. The optimization of H⁻ extraction from the source is one of the most important issues for developing negative ion sources. The following effects on the total extracted H⁻ current I_{H^-} are systematically studied with the 2D3V PIC (Particle in Cell) modeling in a H⁻ volume production source: 1) width of the extraction slit and 2) extraction grid voltage. The PIC modeling results make clear the dependence of I_{H^-} on these important parameters. The results also give useful physics understanding about the processes leading to the extraction of H⁻ ions from the source and its optimization.

Keywords: Negative ions, particle-in-cell (PIC), weak magnetic field, extraction.
PACS: 52.50.Gj, 52.65.Rr

INTRODUCTION

Negative hydrogen ion sources are needed in two important fields: nuclear fusion and high energy accelerators. The optimization of H⁻ extraction from the source is one of the most important issues for developing negative ion sources. As in Ref.1-4, the present numerical analysis is applied to a laboratory ion source[5,6] for which numerous experimental results have been reported. This volume production based device has one extraction aperture, as the accelerator ion sources (e.g. the successful source from DESY[7]). This laboratory ion source operates steady state with much lower power than the accelerator ion source, therefore delivering lower electron and negative ion current density. The fusion oriented negative ion sources have a large number of apertures and are thought to be surface production ion sources[8]. Thus in these sources the flow of negative ions should be from the cesium seeded plasma electrode (PE) surface into the volume.

Understanding processes in volume dominated sources supports the understanding of sources working with the surface generation mechanism. In the previous study, low density sources with a single extraction aperture were focused. It has been shown experimentally in Ref. 5 and references therein that H⁻ ion extraction could be significantly improved with the combination of a weak magnetic field transverse to the extraction direction and few volts of bias voltage applied on the PE. The effects of the weak magnetic field[1-3] and plasma grid (PE bias)[4] have been studied using the particle in cell modeling. Comparison with the Camembert III experiment[5-6] has been also done. These PIC simulation works already provided some understanding about the processes leading to the extraction of H⁻ ions from the source.

The purpose of the present study is to extend the 2D3V PIC modeling in ref. 3. The following effects are systematically studied with the extended version of the 2D3V PIC code[4]: 1) The width of the extraction slit, and 2) The extraction grid voltage.

SIMULATION MODEL

In the present study, the motion of charged particles (H⁺ ions, H⁻ ions, and electrons) is solved in their self-consistent electric field using the PIC method. The trajectories are first calculated with the equation of motion for each particle

$$m\frac{dv}{dt} = q(E + v \times B), \qquad (1)$$

where m, v, q, E, and B are particle mass, velocity, electric charge, local electric field, and local magnetic field, respectively. The equation of motion is numerically solved using the Boris–Buneman version of the leap-frog method[9]. The charge density is obtained at each mesh point from the particle location by using a linear interpolation. Poisson equation is then solved at each mesh point by the Stone's strongly implicit procedure[10] to obtain the electric potential,

$$\nabla^2 \phi = -q(n_{H^+} - n_e - n_{H^-})/\varepsilon_0, \qquad (2)$$

where n_j and ε_0 are the particle density of the jth species and the permittivity of vacuum, respectively.

Figure 1,2 shows the model geometry used in the present PIC modeling. In the experiment in Refs. 5 and 6, the extracted species are accelerated by three successive electrodes: the PE, the extraction electrode (EE), and the acceleration electrode. Based on the experimental device mentioned above, the extraction region of the source is modeled with a simple 2D slab geometry. In the present model, only the PE and the EE are considered. The x-axis is taken parallel to the direction of the H⁻ extraction beam through the aperture, while the y-axis is parallel to the PE surface.

The space coordinates (x, y) and the time t are normalized as follows: $\tilde{x} = x/\lambda_{De}$, $\tilde{y} = y/\lambda_{De}$, and $\tilde{t} = \omega_{pe}t$, where λ_{De} and ω_{pe} are, respectively, the electron Debye length and the electron plasma frequency, calculated from the initial electron density n_e (taken as 10^{16} m⁻³, which is in the range of electron density observed in Refs. 5 and 6), and temperature T_e, respectively. Given the numerical values used, the Debye length is equal to 7.43×10^{-5} m. The grid-size and the timestep are used

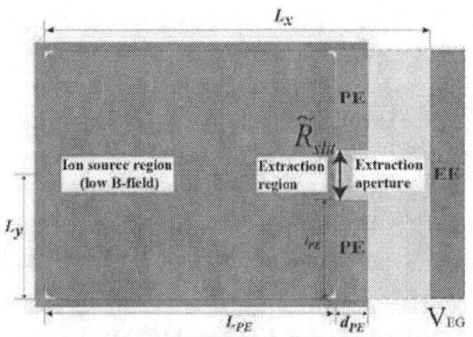

FIGURE 1. Geometry of the modeled system.

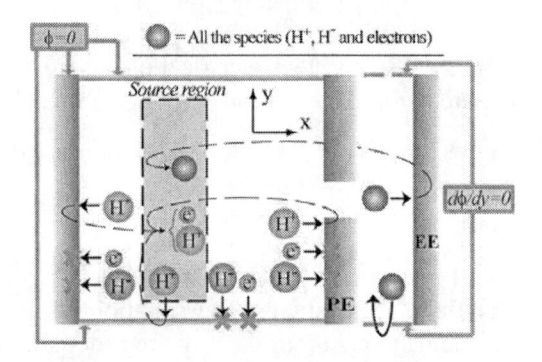

FIGURE 2. Model geometry and boundary conditions.

$d\tilde{x}=1.0$, $d\tilde{y}=0.5$, and $d\tilde{t}=0.1$, respectively. The following normalized dimensions are used in the simulation: (1) system length: $\tilde{L}_x = 85$, half-width $\tilde{L}_y = 25$, and $\tilde{L}_{PE} = 70$, (2) PE width $\tilde{l}_{PE} = 20$ and thickness $\tilde{d}_{PE} = 5$. The electrostatic potential ϕ, magnetic flux density $\boldsymbol{B}$, and current density j are also normalized as follows: $\tilde{\Phi} = e\phi/kT_e$, $\tilde{\boldsymbol{B}} = \boldsymbol{B}/(m_e\omega_{pe}/e)$, and $\tilde{\boldsymbol{J}} = \boldsymbol{j}/(n_e e v_{th})$, where e, k, and v_{th} are the unit charge, the Boltzmann constant, and the electron thermal velocity, respectively.

The transverse magnetic field applied in the extraction area is simply given by the following Gaussian shape in the present study,

$$\tilde{\boldsymbol{B}}(x,y) = \tilde{B}_0 \exp\left\{-\left(\frac{\tilde{x}-\tilde{L}_{PG}}{\tilde{l}_B}\right)^2\right\}\boldsymbol{e}_y, \qquad (3)$$

where $\boldsymbol{e}_y$ is the unit vector in the y-direction and the decay length $\tilde{l}_B$ of the magnetic field is set to be 10. The value of $\tilde{B}_0$ is specified in such a way that the resulting Larmor radii of the species satisfy the following condition:

$$\rho_e^* = \frac{\rho_e}{l_B} = \frac{\tilde{\rho}_e}{\tilde{l}_B} \ll 1, \quad \rho_i^* = \frac{\rho_i}{l_B} = \frac{\tilde{\rho}_i}{\tilde{l}_B} > 1, \qquad (4)$$

where $\tilde{\rho}_e$, $\tilde{\rho}_i$ and $\tilde{l}_B$ are the normalized Larmor radius of electrons and ions ($\tilde{\rho} = \rho/\lambda_{De}$) and the characteristic decay length of the magnetic field amplitude ($\tilde{l}_B = l_B/\lambda_{De}$), respectively. Here, we define a *magnetic trapping parameter* $\rho_j^* \equiv \rho_j/l_B$ in Eq. (4), which is one of the most important key parameters to determine the basic plasma characteristics of extraction region in the Camembert III experiment with the weak transverse magnetic field, as it was discussed in Ref. 4.

In the present simulation, the basic equations and physical quantities are normalized, and the system size ($L_x = 85\lambda_D, 2L_y = 50\lambda_D$) is very much smaller compared with the experiments in Ref. 5. In our previous 2D PIC simulations in Refs. 1 and 2, the basic equations and physical quantities were not normalized, and realistic dimensions and magnetic field were used, based on Camembert III experimental set-up. With the typical plasma parameters listed in Table 1, the magnetic trapping parameter ρ_j^* becomes $\rho_j^* = 0.06 \ll 1$ for the electrons and $\rho_{H^+}^* = \rho_{H^-}^* = 2.6 > 1$ for the H$^+$ and H$^-$ ions, i.e., in the experiments, and therefore also in the simulations in Refs. 1 and 2. This condition for ρ_j^*, i.e. Eq. (4) guarantees that electrons are magnetically trapped along the field lines, while positive and negative ions are barely affected by the magnetic field. As mentioned before, these effects determine the essential characteristics of such plasmas with the weak magnetic field close to the PE. As was discussed in Ref. 4, if the condition for ρ_j^* is the same, we have not observed any large differences in the result (for example, the potential profile, electron density and so on) between the present simulations and the experiment. These comparisons between the previous and present simulations suggest that PIC simulations with small system size are still useful to understand at least the basic physics.

Instead of a full Monte Carlo collision model, a simple diffusion model is employed for the electron diffusion across the magnetic field as in Ref. 4. The magnetic field is mainly transverse in front of the extraction aperture, hence we consider only diffusion in the *x*-direction. It is simulated as a random-walk process with step lengths $\Delta \tilde{x}$

$$\Delta \tilde{x} = \sqrt{2\tilde{D}_\perp \Delta \tilde{t}}\, \xi_x, \qquad (5)$$

where $\tilde{D}_\perp$, $\Delta \tilde{t}$, and ξ_x are the normalized perpendicular diffusion coefficient, time step, and a normal random number, respectively. The perpendicular diffusion coefficient[11] is calculated for each electron at each time step

$$D_\perp = kT_e/m_e v_e^{tot}/\left[1 + (\omega_{ce}/v_e^{tot})^2\right], \qquad (6)$$

where T_e, m_e, k, ω_{ce}, and v_e^{tot} are the electron temperature, the electron mass, the Boltzmann constant, the electron cyclotron frequency, and the total collision frequency of electrons, respectively. This total collision frequency includes three kinds of collision: electron-neutral, electron-ion and electron-electron, the latter being relevant because of the nonuniformity of the magnetic field. The diffusion coefficient defined in Eq. (6) is also normalized by $\tilde{D}_\perp = D_\perp/(\lambda_{De}^2 \omega_{pe})$.

The boundary conditions for the potential used in the present study are summarized in Fig. 2. Only half of the system $(0 \leq \tilde{y} \leq \tilde{L}_y)$ is used as calculation domain so as to reduce the time cost of the simulation. This is based on the assumption of symmetric property of the system in the y-direction, and the boundary condition $d\tilde{\Phi}/dy = 0$ is used on the middle boundary $(\tilde{y} = \tilde{L}_y)$. On the surface of the walls inside the source, $\tilde{\Phi}$ is fixed to be zero, while on the surface of the EE it is fixed to 50. In the previous study of the effect of the PE bias on the H⁻ extraction in Ref. 4, the PE bias voltage $\tilde{\Phi}_{PE}$ is taken as a parameter, while $\tilde{\Phi}_{PE}$ has been fixed in the present study as follows: $\tilde{\Phi}_{PE} = 0$.

At $\tilde{t} = 0$, a large number of particles is loaded in the source region $(0.12\tilde{L}_x \leq \tilde{x} \leq 0.3\tilde{L}_x; 0.2\tilde{L}_y \leq \tilde{y} \leq \tilde{L}_y)$. Species included are H⁺ ions, electrons and H⁻ ions, with the following initial ratio: $n_{H^+} : n_e : n_{H^-} = 10:9:1$. The initial velocity of each particle is chosen from the Maxwellian distribution with the given temperature of each species kT_e=1.0 eV, $kT_{H^+} = kT_{H^-} = 0.25$ eV. The boundary conditions for the particles in the present study are also shown in Fig. 2. All the velocities of particles reloaded in the source region are chosen in the same manner as the initial particle loading. H⁺ ions hitting the inside walls or the PE are assumed to be recycled as neutrals and ionized in the source region. Hence, a pair of H⁺ and electron is reloaded in the source region. On the contrary, electrons and H⁻ ions striking any inner wall or the PE are removed from the system. All the particles extracted (i.e., hitting the EE) are reloaded in the source region. When any of the species crosses out the boundary between the PE and the EE or the upper-half boundary, it is immediately reloaded from this point, symmetrically with respect to this boundary (the velocity in the y-direction is inverted). Finally, H⁻ ions are continuously injected in the source region with a velocity chosen from the Maxwellian based on their temperature (Table 1). This is assumed to be a very simple modeling of the H⁻ ion volume production process, and the rate of injection has been chosen empirically so as to satisfy a steady state in all the cases considered (H⁻ ion production rate is not changed between two cases).

TABLE 1. Reference plasma parameters.

Electron temperature T_e	1.0eV
H⁺ ion temperature $T_{H^+} = T_{H^-}$	0.25eV
Initial electron density n_e	10^{16}m⁻³
Initial densities ratio $n_{H^+} : n_e : n_{H^-}$	10:9:1
Electron Debye length	7.43×10^{-5}m
Electron plasma frequency	6.64×10^{9}s⁻¹

TABLE 2. Reference system size parameters and magnetic field.

Normalized system size	$\tilde{L}x = 85$ $\tilde{L}y = 50$
Normalized magnetic field $\tilde{B}_0$	1.2
Normalized extraction-slit width $\tilde{R}_{slit}$	10
Normalized EE voltage $\tilde{V}_{EE}$	50

SIMULATION RESULT AND DISCUSSION

Dependence of the H⁻ extracted current on the width of the extraction slit.

The width of the extraction slit $\tilde{R}_{slit}$ in Fig. 1 has been changed from $\tilde{R}_{slit} = 10$ to 30. Other parameters except for the width $\tilde{R}_{slit}$ are fixed as those in the reference case summarized in Table 1 and Table 2. The results on the total extracted current $\tilde{I}_{H^-}$ are shown in Fig. 3. Initially, $\tilde{I}_{H^-}$ increases almost linearly with $\tilde{R}_{slit}$ as expected. For larger values of $\tilde{R}_{slit}$ ($\tilde{R}_{slit} \geq 20$), however, $\tilde{I}_{H^-}$ starts to be saturated and is almost constant.

In order to understand these tendencies, the potential and the H⁻ density profiles are shown in Fig. 4, respectively, on the left and right hand side. Fig. 4, (a)-(a)', (b)-(b)', and (c)-(c)' correspond to the cases with the slit width of $\tilde{R}_{slit}$ =10, 15 and 25, respectively.

As seen from Fig. 4(a) for the reference case, a positive potential peak has been formed at $\tilde{x} \approx 40$ almost along the field line. The H⁻ ion density at $\tilde{x} = 40 \sim 50$ also becomes large as shown in Fig. 4(a)'. These features can be explained by the effect of the "weak" magnetic field[1-3], ($\rho_e^* \ll 1 < \rho_{H^+}^*, \rho_{H^-}^*$). More specifically, the electrons near the PE are easily magnetized ($\rho_e^* \ll 1$) and lost along the field line, while positive ions (H⁺) are not magnetized ($1 < \rho_{H^+}^*, \rho_{H^-}^*$) because of their larger Larmor radii. This difference in dynamics between electrons and H⁺ ions perturbs the plasma neutrality in this region and leads to the formation of positive electric potential. This modification of the plasma neutrality and the resultant electric potential structure lead to an increase in the H⁻ ion density near the PE and their extraction.

In Fig. 4(c) and (c)' for the case with the large $\tilde{R}_{slit}$ ($\tilde{R}_{slit}$ =25), however, the positive potential peak and H⁻ ion density enhancement disappear. This is the reason why the resultant extracted H⁻ current is not so much dependent on $\tilde{R}_{slit}$ for the larger values of $\tilde{R}_{slit}$ ($\tilde{R}_{slit} \geq 20$).

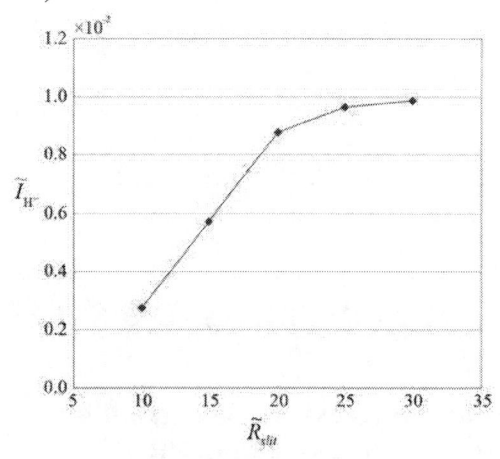

FIGURE 3. Influence of the width of the extraction slit upon the extracted H⁻ ion current.

The potential profile   The H⁻ density profile

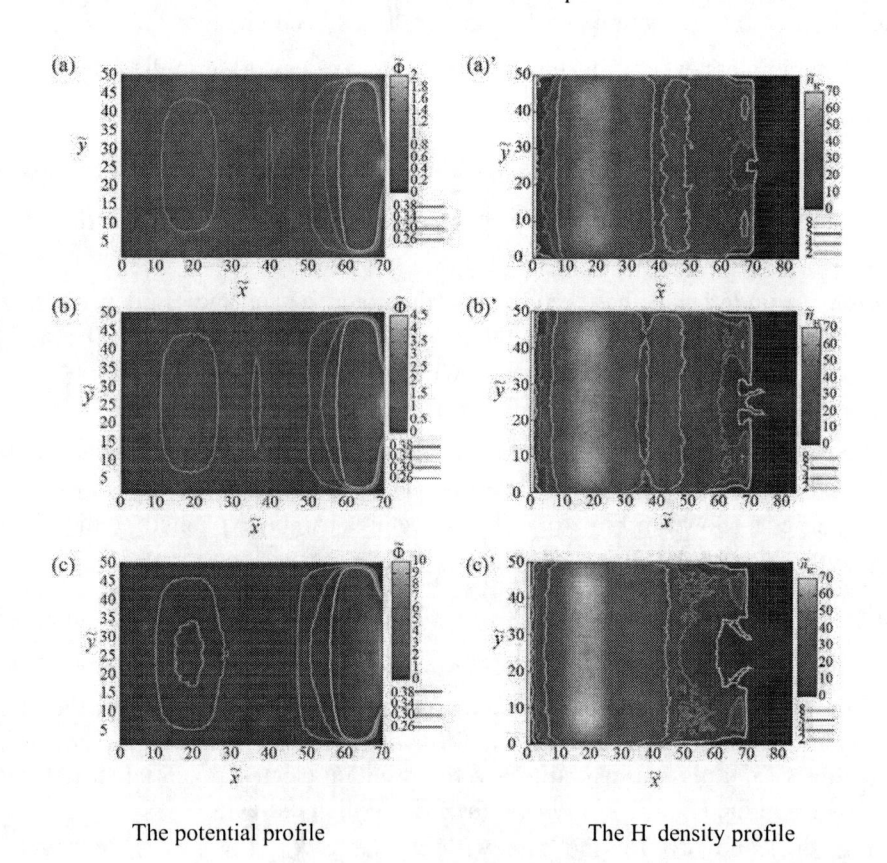

FIGURE 4. The potential profile and H⁻ density profile:
(a), (a)': the extraction slit $\tilde{R}_{slit}$ =10, (b), (b)': the extraction slit $\tilde{R}_{slit}$ =15,
(c), (c)': the extraction slit $\tilde{R}_{slit}$ =25

Dependence of the H⁻ extraction current on the extraction-grid voltage.

The extraction voltage $\tilde{V}_{EE}$ has been varied from $\tilde{V}_{EE}=20$ to 50, while other calculation parameters are fixed as those in Table 1 and Table 2. The numerical results of the extracted current $\tilde{I}_{H^-}$ do not depend so much on $\tilde{V}_{EE}$ and are almost constant as shown in Fig. 5(a). To understand the tendency, the potential profiles are shown in Fig. 6 (a)-(c), respectively, for the cases with $\tilde{V}_{EE}=20$, 30 and 40. As seen from the comparisons of these figures, the potential profiles in the x direction, especially for the small $\tilde{x}$ region with $\tilde{x} \leq 50$ are almost unchanged for the different values of voltage $\tilde{V}_{EE}$ at least within the range $20 \leq \tilde{V}_{EE} \leq 50$. The penetration of the EE voltage in the x direction is effectively shielded. This is the main reason why $\tilde{I}_{H^-}$ is almost constant in Fig. 5(a). In the future, the effect $\tilde{V}_{EE}$ on the $\tilde{I}_{H^-}$ also for the surface produced H⁻ ions should be studied.

For the purpose of the comparison, the total extracted current $\tilde{J}_{CL}$ calculated from the Child-Langmuir law, $\tilde{J}_{CL} \equiv \frac{4\sqrt{2}}{9}\left(\frac{m_e T_i}{m_i T_e}\right)^{\frac{1}{2}} \frac{\tilde{V}_{EE}^{\frac{3}{2}}}{\tilde{d}^2}$ is also shown in Fig. 5(b), where $\tilde{d}$ is the distance between the PE and EE, i.e., the right-hand-side boundary in Fig. 1. $\tilde{J}_{CL}$ is calculated for two values of $\tilde{d}$, because the PE has a finite width in the actual geometry. As shown in Fig. 5(b), the first one $\tilde{J}_{CL,1}$ is calculated with the width $\tilde{d}_1$, where $\tilde{d}_1$ is the distance between the PE "outer" surface and the right-hand-side boundary, i.e., the EE. The second one $\tilde{J}_{CL,2}$ is calculated with the width $\tilde{d}_2$, where $\tilde{d}_2$ is the distance between the PE "inner" surface inside the source and the right-hand-side boundary in Fig. 5(b).

As seen from Fig. 5(a), $\tilde{J}_{H^-}$ is smaller than both $\tilde{J}_{CL,1}$ and $\tilde{J}_{CL,2}$. The Child-Langmuir current above is obtained with the assumption that the extracted particles can be supplied infinitely at the source side. In other words, the extracted particle density becomes infinite ($n \to \infty$) for the constant current density $J(=nev=const.)$ with $v \to 0$ at the source side boundary. The numerical results suggest that the H⁻ density inside the source is still not sufficient to satisfy the above assumption of Child-Langmuir law.

It should be noted that the extracted current J_{H^+} without the magnetic field almost obeys the Child-Langmuir law in the case for the positive ion H⁺ extraction in Ref. 3 under almost the same conditions and the same geometry as in the present study except for the EE voltage polarity ($\tilde{V}_{EE} < 0$: for H⁺ extraction).

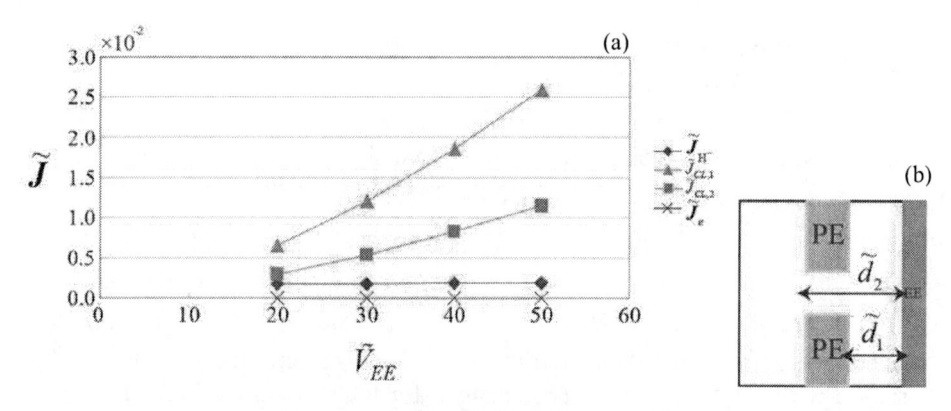

FIGURE 5. (a) Influence of the EE voltage upon the extracted H⁻ ion current and (b) the distance $\tilde{d}_1$ and $\tilde{d}_1$ to calculate Child Langmuir current $\tilde{J}_{CL,1}$ and $\tilde{J}_{CL,2}$

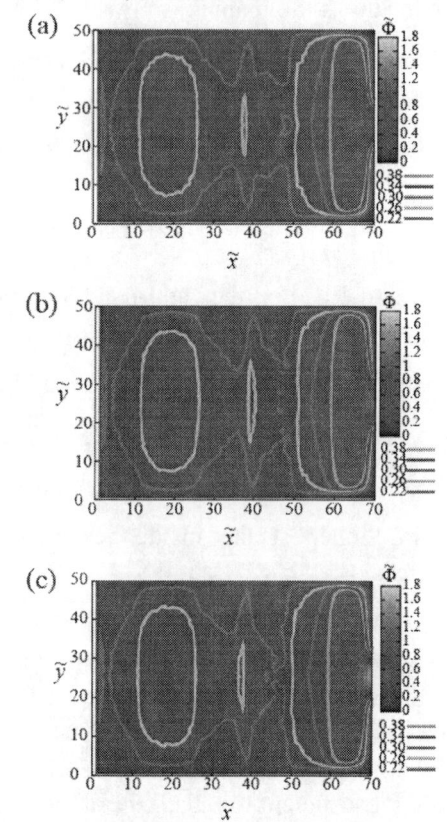

FIGURE 6. The potential profile: (a) the EE voltage = 20, (b) the EE voltage = 30, (c) the EE voltage = 40

CONCLUSION

In this study, two effects have been studied systematically with 2D3V PIC modeling (1) the width of the extraction slit and (2) the extraction grid voltage on the H⁻ extraction of the volume produced H⁻ ions.

As for the first effect, the extracted current $\tilde{I}_{H^-}$ increases almost linearly with the slit width $\tilde{R}_{slit}$ for smaller values of $\tilde{R}_{slit}$ as expected. For larger values of $\tilde{R}_{slit}$ ($\tilde{R}_{slit} \geq 20$), however, the extracted current $\tilde{I}_{H^-}$ is almost independent of the slit width. For smaller values of the slit width ($\tilde{R}_{slit} \leq 20$), the effect of the weak magnetic field is efficient. For larger values of $\tilde{R}_{slit}$ the positive potential peak and the resultant H⁻ density enhancement due to the weak magnetic field disappear in front of the extraction slit. As a result, the dependence of the $\tilde{I}_{H^-}$ on $\tilde{R}_{slit}$ becomes weak and it becomes almost constant for larger values of $\tilde{R}_{slit}$.

As for the second effect, the EE voltage does not have a significant effect on the H⁻ extracted current $\tilde{I}_{H^-}$, at least for the range in which $\tilde{V}_{EE}$ ($20 \leq \tilde{V}_{EE} \leq 50$) was varied in the present study. The potential structure near the PE does not depend on the EE voltage. The penetration of the EE potential deeply into the source region along the extraction axis is possibly suppressed due to the trapping effect of electrons by the magnetic field. As a result, the penetration depth of the EE voltage along the extraction axis is almost independent of the EE voltage. This could be a reason why the EE voltage does not affect the H⁻ extraction so much.

These results might give a useful basic understanding of the processes leading to the extraction of H⁻ ions from the volume-production based ion sources with one extraction aperture under the presence of the weak magnetic field and also some insights into the effect of above two parameters on H⁻ extraction.

REFERENCES

1. T. Sakurabayashi, A. Hatayama and M. Bacal, *J. Appl. Phys.* **95** (2004).
2. T. Sakurabayashi, A. Hatayama and M. Bacal, *Rev. Sci. Instrum.* **75**, 1770 (2004).
3. A. Hatayama, *Rev, Sci. Instrum.* **79**, 02B109 (2008).
4. S. Kuppel, D. Matsushita, A. Hatayama, and M. Bacal, *J. Appl. Phys.* to be published.
5. P. Svarnas, B. M. Annaratone, S. Béchu, J. Pelletier, and M. Bacal, *Plasma Sources Sci. Technol.* **18**, 045010 (2009).
6. P. Svarnas, J. Breton, M. Bacal, and R. Faulkner, *IEEE Trans. Plasma Sci.* **33**, 1156 (2007).
7. J. Peters, *Rev. Sci. Instrum.* **79**, 02A515 (2008).
8. S. Christ-Koch, U. Fantz, M. Berger and NNBI Team, *Plasma Sources Sci. Technol.,* **18**, 025003 (2009).
9. K. Birdsall and A. B. Langdon, *Plasma Physics via Computer Simulation* (McGraw-Hill, New York, 1985).
10. H. L. Stone, (SIAM Soc. Ind. Appl. Math.) J. *Numer. Anal.* **5**, 530 (1968).
11. F. F. Chen, Introduction to Plasma Physics (Plenum, New York, 1974).

Electric Potential Near The Extraction Region In Negative Ion Sources With Surface Produced Negative Ions

A. Fukano[a] and A. Hatayama[b]

[a] Monozukuri Department, Tokyo Metropolitan College of Industrial Technology, 1-10-40 Higashi-Ohi, Shinagawa-ku, Tokyo 140-0011, Japan
[b] Graduate School of Science and Technology, Keio University, 3-14-1 Hiyoshi, Kouhoku-ku, Yokohama 223-8522, Japan

Abstract. The potential distribution near the extraction region in negative ion sources for the plasma with the surface produced negative ions is studied analytically. The potential is derived analytically by using a plasma-sheath equation, where negative ions produced on the Plasma Grid (PG) surface are considered in addition to positive ions and electrons. A negative potential peak is formed in the sheath region near the PG surface for the case of strong surface production of negative ions or for low energy negative ions. Negative ions are reflected by the negative potential peak near the PG and returned to the PG surface. This reflection mechanism by the negative potential peak possibly becomes a factor in negative ion extraction. It is also indicated that the potential difference between the plasma region and the wall decreases by the surface produced negative ions. This also has the possibility to contribute to the negative ion extraction

Keywords: Negative ion sources, Plasma-sheath equation, Surface produced negative ions, Extraction region, Negative potential peak
PACS: 29.25.Ni, 52.40.Kh, 02.60.Nm,

INTRODUCTION

Neutral beam injection (NBI) based on a negative ion beam is a very promising method of plasma heating in future fusion reactors. In the study of negative ion sources, it is important to extract a large amount of hydrogen negative ions. Extraction of negative ions depends on the potential distribution near the extraction region. In surface production negative ion sources, negative ions produced on a plasma grid (PG) are accelerated by the potential difference between plasma and the PG. If a spatial dependence of a sheath potential formed near the PG surface is similar to that in plasma without negative ions, it accelerates surface produced negative ions toward the interior of the ion source, that is, the driver region.

One of the extraction mechanisms of surface produced negative ions is their Larmor motion around the filter magnetic field [1]. Negative ions produced on the PG surface change their direction by Larmor rotation and return to the PG surface, where they are extracted from the extraction holes. However, a recent experiment has suggested that the extracted current density of negative ions does not depend on the magnetic filter strength [2]. Thus, the physical mechanism of the extraction of surface produced negative ions from the PG has not been cleared. Since the plasma has a large amount of negative ions produced on the PG surface, there is a possibility that the potential distribution is different from the general distribution near the wall surface. Emmert *et al.* have investigated the sheath potential analytically considering both the plasma region and the sheath region self-consistently by using a plasma-sheath equation [3]. Sato *et al.* extended the method of Emmert *et al.* to the case of magnetized plasma [4]. However, these investigations are limited to plasmas that consist of positive ions and electrons.

In this paper, we will investigate the potential profile near the extraction region in the negative ion sources by using the plasma-sheath equation. In the analysis, surface produced negative ions are considered in addition to positive ions and electrons.

THE PLASMA-SHEATH EQUATION

Analytical Model And Basic Equations

In surface production negative ion sources, negative hydrogen ions H⁻ produced on the PG surface are launched to the interior of the ion source as shown in Fig. 1. The background plasma consists of electrons and positive hydrogen ions H⁺. The geometry of the analytical model of the potential is shown in Fig. 2. In the model, the PG is considered as the wall. Although the PG exists only on one side of the negative ion source practically, walls on both sides are considered for assumption that particles are lost on the wall. The electric potential $\phi(x)$ is assumed to be symmetric about $x=0$ and decreases toward the walls and equals zero at $x=0$.

Constant energies E and E_- of a positive ion and a negative ion in the x-direction are

$$E = \frac{1}{2} M v_x^2 + q\phi(x), \quad (1)$$

$$E_- = \frac{1}{2} M_- v_{x-}^2 - q\phi(x), \quad (2)$$

where M and M_- are the ion masses, v_x and v_{x-} are the velocities in the x-direction, q and $-q$ are the charges of the ion and the negative ion, respectively. The subscript _ denotes values belonging to the negative ion throughout this paper. From the Boltzmann equation and Eqs. (1) and (2), the kinetic equations for the ion and the negative ion are described by

$$\sigma v_x(x,E) \frac{\partial f(x,E,\sigma)}{\partial x} = S(x,E), \quad (3)$$

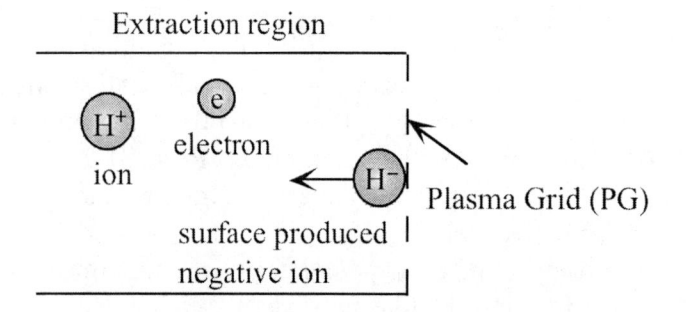

FIGURE 1. Geometry of the extraction region of the ion source. Negative hydrogen ion is launched from PG.

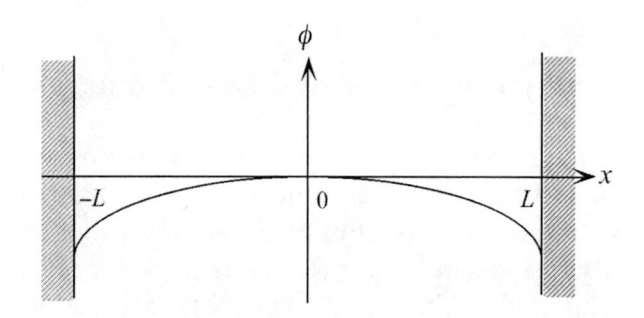

FIGURE 2. Geometry of the potential in the analysis model

$$\sigma v_{x-}(x,E_-)\frac{\partial f_-(x,E_-,\sigma)}{\partial x} = S_-(x,E_-), \quad (4)$$

where $\sigma=\pm 1$ is the direction of the particle motion, $f(x,E,\sigma)$ and $f_-(x,E_-,\sigma)$ are the distribution functions, and $S(x,E)$ and $S_-(x,E_-)$ are the source functions. From Eqs. (1) and (2), the velocities are given by

$$v_x = \left[(2/M)\{E - q\phi(x)\}\right]^{1/2}, \quad (5)$$

$$v_{x-} = \left[(2/M)\{E_- + q\phi(x)\}\right]^{1/2}, \quad (6)$$

respectively.

We assume a symmetry about $x=0$, that is, $\phi(x)=\phi(-x)$, $S(-x,E)=S(x,E)$, and $S_-(-x,E_-)=S_-(x,E_-)$. Furthermore, we assume that particles are not reflected at the wall. Then, the boundary conditions of the distribution functions are $f(-L,E,+1) = f(L,E,-1) = 0$ and $f_-(-L,E_-,+1) = f_-(L,E_-,-1) = 0$.

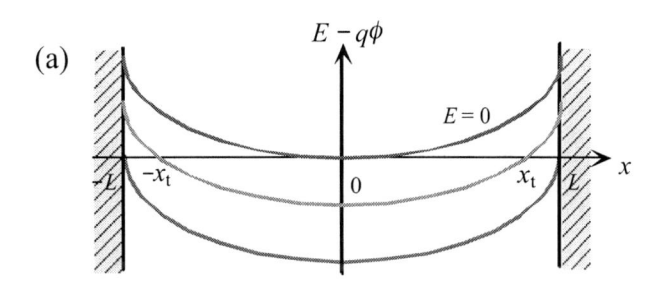

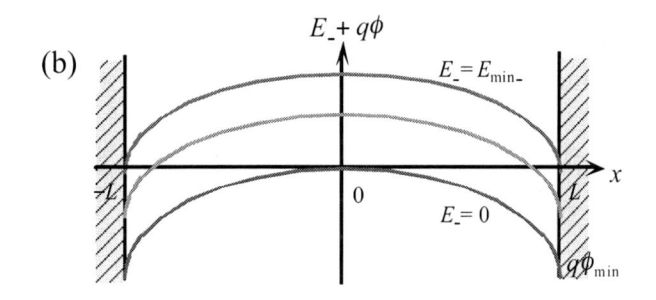

FIGURE 3. Energy spaces of (a) the positive ion ($E_{min}= q\phi(x)$) and (b) the negative ion ($E_{min\_}= -q\phi_{min}$)

Derivation Of The Plasma Sheath Equation

The energy space of the particle is divided to some regions as shown in Fig. 3, which is based on the conditions that v_x and $v_{x\_}$ must be real numbers, that is, $E-q\phi(x) \geq 0$ and $E\_+q\phi(x) \geq 0$ for the positive ion and the negative ion, respectively. Particle motion depends on its energy. All positive ions generated at the position x' are accelerated to the wall by the monotonically decreasing potential, where x' is the point of ion generation. An ion in the energy region of $E>0$ can reach and pass through the center of the plasma, whereas an ion in the energy region of $E_{min}<E<0$ cannot reach the center of the plasma and reflected at the turning point $x=x_t(E)$ for $\sigma= -1$ and $x= -x_t(E)$ for $\sigma= 1$, where $E_{min}=q\phi(x)$. An Ion in the energy region of $E<E_{min}$ cannot exist. On the other hand, negative ions generated at the position $x'=\pm L$ are accelerated to the center of the plasma. A negative ion in the energy region of $E\_ >E_{min\_}$ can reach the center of the plasma, where $E_{min\_}= -q\phi_{min}$ and ϕ_{min} is the minimum potential. A negative ion in the energy region of $-q\phi(x) < E\_< E_{min\_}$ cannot reach the wall, where a surface produced negative ion is not generated. A negative ion in the energy region of $E\_< -q\phi(x)$ cannot exist.

The distribution functions $f(x,E,\sigma)$ and $f\_(x,E\_,\sigma)$ for $\sigma=\pm 1$ are obtained by integrating Eqs. (3) and (4) for particle trajectory with the boundary conditions. The positive ion density n_i and the negative ion density $n_{i\_}$ are obtained by taking the sum of $f(x,E,+1)$ and $f(x,E,-1)$ and the sum of $f\_(x,E\_,+1)$ and $f\_(x,E\_,-1)$ for each energy region and integrating them over E and $E\_$, respectively, as

$$n_i(x) = \sum_\sigma \int dE \frac{f(x,E,\sigma)}{\upsilon_x(x,E)}$$
$$= 2\int_0^\infty dE \int_0^L \frac{1}{\upsilon_x(x,E)} \frac{S(x',E)}{\upsilon_x(x',E)} dx' + 2\int_{E_{min}}^0 dE \int_{x_t}^L \frac{1}{\upsilon_x(x,E)} \frac{S(x',E)}{\upsilon_x(x',E)} dx', \quad (7)$$

$$n_{i-}(x) = \sum_\sigma \int dE_- \frac{f(x,E_-,\sigma)}{\upsilon_{x-}(x,E_-)}$$
$$= 2\int_{E_{min-}}^\infty dE_- \int_0^L \frac{1}{\upsilon_{x-}(x,E_-)} \frac{S_-(x',E_-)}{\upsilon_{x-}(x',E_-)} dx', \quad (8)$$

By changing the order of integration, Eqs. (7) and (8) can be written as

$$n_i(x) = 2\int_0^L dx' \int_0^\infty \frac{1}{\upsilon_x(x,E)} \frac{S(x',E)}{\upsilon_x(x',E)} dE + 2\int_0^L dx' \int_{q\phi(x')}^0 \frac{1}{\upsilon_x(x,E)} \frac{S(x',E)}{\upsilon_x(x',E)} dE$$
$$= 2\int_0^L dx' \int_{q\phi(x')}^\infty \frac{1}{\upsilon_x(x,E)} \frac{S(x',E)}{\upsilon_x(x',E)} dE$$
$$= 2\int_0^L dx' \int_{E_0}^\infty \frac{1}{\upsilon_x(x,E)} \frac{S(x',E)}{\upsilon_x(x',E)} dE, \quad (9)$$

$$n_{i-}(x) = 2\int_0^L dx' \int_{E_{min-}}^\infty \frac{1}{\upsilon_{x-}(x,E_-)} \frac{S_-(x',E_-)}{\upsilon_{x-}(x',E_-)} dE_-, \quad (10)$$

where $E_0 = q\phi(x')$ for $q\phi(x') > q\phi(x)$ and $E_0 = q\phi(x)$ for $q\phi(x') < q\phi(x)$ because $\upsilon_x(x, q\phi(x')) = [(2/M)\{q\phi(x') - q\phi(x)\}]^{1/2}$ must be real number.

As the source functions, we use the same expression as Emmert et al. [3]

$$S(x,E) = \frac{S_0 h(x)}{2kT_i} \upsilon_x(x,E) \exp\left\{-\frac{E - q\phi(x)}{kT_i}\right\}, \quad (11)$$

$$S_-(x,E_-) = \frac{S_{0-} h_-(x)}{2kT_{i-}} \upsilon_{x-}(x,E_-) \exp\left\{-\frac{E_- + q\phi(x)}{kT_{i-}}\right\}, \quad (12)$$

where T_i and T_{i-} are the temperatures, $h(x)$ and $h_-(x)$ are source strengths, and S_0 and S_{0-} are the average source strengths of the positive ion and the negative ion, respectively. By integrating the sum of $S(x,E)$ and $S(-x,E)$ over $E=q\phi(x)$ to ∞ and the sum of $S_-(x,E_-)$ and $S_-(-x,E_-)$ over $E=-q\phi(x)$ to ∞, respectively, we obtain $S(x)=S_0 h(x)$ and $S_-(x)=S_{0-} h_-(x)$, where $S(-x,E)=S(x,E)$ and $S_-(-x,E_-)=S_-(x,E_-)$ from the assumption. The averages about x of $h(x)$ and $h_-(x)$ are normalized to 1.

By substituting Eqs. (11) and (12) to Eqs. (9) and (10), respectively and integrating them for E, we obtain the densities as

$$n_i(x) = \left(\frac{M\pi}{2kT_i}\right)^{1/2} S_0 \int_0^L dx' I(x,x') h(x') , \qquad (13)$$

$$n_{i-}(x) = \left(\frac{M_-\pi}{2kT_{i-}}\right)^{1/2} S_{0-} I_-(x,L) h_-(L) , \qquad (14)$$

where since the negative ion is produced only on the wall surface, $h_-(x')=h_-(L)$ and

$$I(x,x') = \begin{cases} \exp\left\{\dfrac{q\phi(x')-q\phi(x)}{kT_i}\right\} \mathrm{erfc}\left[\left\{\dfrac{q\phi(x')-q\phi(x)}{kT_i}\right\}^{1/2}\right], & q\phi(x') > q\phi(x) \\ \exp\left\{\dfrac{q\phi(x')-q\phi(x)}{kT_i}\right\}, & q\phi(x') < q\phi(x) \end{cases} , \qquad (15)$$

$$I_-(x,x') = \exp\left\{\frac{-q\phi(L)+q\phi(x)}{kT_{i-}}\right\} \mathrm{erfc}\left[\left\{\frac{-q\phi_{\min}+q\phi(x)}{kT_{i-}}\right\}^{1/2}\right] . \qquad (16)$$

For the electron density n_e, we use a Maxwell–Boltzmann distribution for simplicity

$$n_e(x) = n_0 \exp\{e\phi(x)/kT_e\}, \qquad (17)$$

where n_0 is the density at $x=0$, $-e$ is the electron charge, k is the Boltzmann's constant, and T_e is the electron temperature. Substituting Eqs. (13), (14) and (17) into Poisson's equation, we obtain the plasma-sheath equation as

$$\frac{d^2\phi(x)}{dx^2} = \frac{n_0 e}{\varepsilon_0} \exp\left(\frac{e\phi(x)}{kT_e}\right) - \frac{S_0 q}{\varepsilon_0}\left(\frac{M\pi}{2kT_i}\right)^{1/2} \int_0^L dx' I(x,x') h(x')$$

$$+ \frac{S_{0-} q}{\varepsilon_0}\left(\frac{M_-\pi}{2kT_i}\right)^{1/2} I_-(x,L) h_-(L) . \qquad (18)$$

The average source strengths S_0 and S_{0-} are determined by the equilibrium of the fluxes of the plasma particles at the wall. We consider that $j_{ew}+j_{iw}-j_{iw-}=0$, where j_{ew} is the electron current density, j_{iw} is the positive ion current density, and j_{iw-} is the negative ion current density at the wall, respectively. These current densities are given by $j_{ew}=-en_0\{kT_e/(2\pi m_e)\}^{1/2}\exp\{e\phi_w/(kT_e)\}$, $j_{iw}=q\int_0^L S(x)dx=qS_0 L$ from $\nabla\cdot j_i=qS(x)$, and $j_{iw-}=-q\int_0^L S_-(x)dx=-qS_{0-}L$ from $\nabla\cdot j_{i-}=-qS_-(x)$, where ϕ_w is the wall potential. Furthermore, we define the ratio of the production rates of negative and positive ions to be $\beta=S_{0-}/S_0$. The average source strengths are obtained as

$$S_0 = \frac{en_0}{qL(1+\beta)}\left(\frac{kT_e}{2\pi m_e}\right)^{1/2}\exp\left(\frac{e\phi_w}{kT_e}\right), \tag{19}$$

$$S_{0-} = \frac{en_0\beta}{qL(1+\beta)}\left(\frac{kT_e}{2\pi m_e}\right)^{1/2}\exp\left(\frac{e\phi_w}{kT_e}\right). \tag{20}$$

By substituting Eqs. (19) and (20) into Eq. (18), we obtain

$$\frac{d^2\phi(x)}{dx^2} = \frac{n_0 e}{\varepsilon_0}\exp\left(\frac{e\phi(x)}{kT_e}\right) - \frac{en_0}{2\varepsilon_0 L(1+\beta)}\left(\frac{MT_e}{m_e T_i}\right)^{1/2}\exp\left(\frac{e\phi_w}{kT_e}\right)\int_0^L dx' I(x,x')h(x')$$

$$+ \frac{en_0\beta}{2\varepsilon_0 L(1+\beta)}\left(\frac{M_- T_e}{m_e T_{i-}}\right)^{1/2}\exp\left(\frac{e\phi_w}{kT_e}\right)I_-(x,L)h_-(L). \tag{21}$$

NUMERICAL SOLUTION OF THE PLASMA–SHEATH EQUATION

Equation (21) is an integrodifferential equation and solved numerically. We introduce the normalized variables

$$\varphi = \frac{q}{kT_e}(\phi_w - \phi), \quad \alpha^2 = \frac{\lambda_D^2}{L^2}\exp\left(-\frac{e\phi_w}{kT_e}\right), \quad s = \frac{x}{L}, \quad \tau = \frac{T_e}{T_i}, \quad \tau_- = \frac{T_e}{T_{i-}}, \quad Z = \frac{q}{e}, \tag{22}$$

where $\lambda_D = (\varepsilon_0 kT_e/n_0 e^2)$ is the Debye length and $Z=1$ for the hydrogen plasma. Eq. (21) becomes

$$\alpha^2 \frac{d^2\varphi(x)}{ds^2} = -\exp(-\varphi) + \frac{1}{2(1+\beta)}\left(\frac{M}{m_e}\tau\right)^{1/2}\int_0^1 ds' I(s,s')h(s')$$

$$- \frac{\beta}{2(1+\beta)}\left(\frac{M_-}{m_e}\tau_-\right)^{1/2} I_-(s,1)h_-(1), \tag{23}$$

where,

$$I(s,s') = \begin{cases} \exp[Z\tau(\varphi(s)-\varphi(s'))]\mathrm{erfc}\left[\{Z\tau(\varphi(s)-\varphi(s'))\}^{1/2}\right], & \varphi(s) > \varphi(s') \\ \exp[Z\tau(\varphi(s)-\varphi(s'))], & \varphi(s) < \varphi(s') \end{cases}, \tag{24}$$

$$I_-(s,s') = \exp[Z\tau_-(\varphi(1)-\varphi(s))]\mathrm{erfc}\left[\{Z\tau_-(\varphi_{\min}-\varphi(s))\}^{1/2}\right]. \tag{25}$$

The normalized plasma-sheath equation (22) is solved numerically by transforming it into a set of finite difference equations [3-6]. The boundary conditions are $d\varphi/ds|_{s=0}=0$ and $\varphi(s=1)=0$. The wall potential is determined by $\varphi(s=0)$. The positive ion source is uniform and the negative ion is produced only at the wall.

The numerically calculated potential profile for various values of λ_D/L is shown in Fig. 4, where $\tau=2$, $\tau_-=10$, and $\beta=S_{0-}/S_0=0.4$. As the normalized potential, $\Phi = -\varphi = (q/kT_e)(\phi-\phi_w)$ is shown. It is found that the width of the sheath is about ten times the Debye length. The potential distribution for various values of β is shown in Fig. 5, where $\tau=2$, $\tau_-=10$, and $\lambda_D/L=5\times10^{-2}$. The result for a case without the negative ion, which corresponds to $\beta=0$, is also shown. The sheath potential depends on the value of β and has a negative peak near the wall for $\beta \geq 0.6$. The potential difference between the plasma and the wall also depends on the value of β and it decreases as the value of β increases. The profile of the potential for various values of the temperature

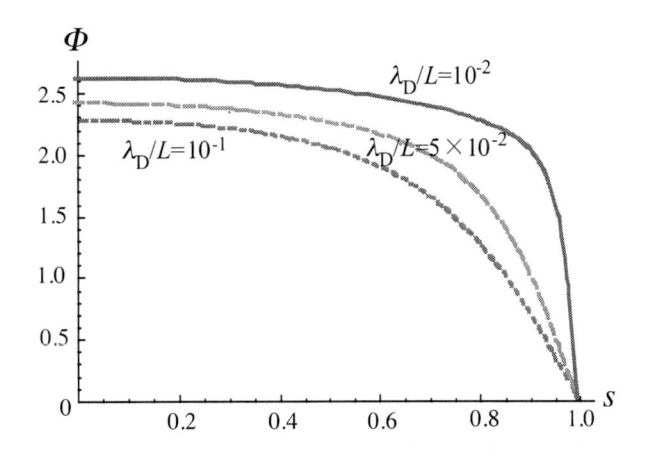

FIGURE 4. Profile of the normalized potential $\Phi=(q/kT_e)(\phi-\phi_w)$ for various values of λ_D/L with $\tau=T_e/T_i=2$, $\tau_-=T_e/T_{i-}=10$, $\beta=0.4$.

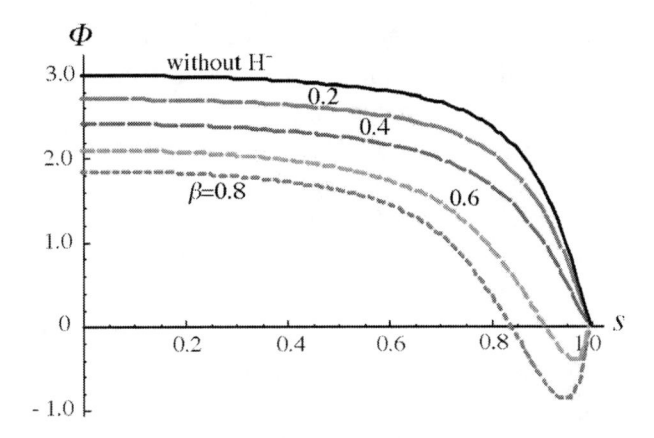

FIGURE 5. Profile of the normalized potential $\Phi=(q/kT_e)(\phi-\phi_w)$ for various values of $\beta=S_{0-}/S_0$ with $\tau=T_e/T_i=2$, $\tau_-=T_e/T_{i-}=10$, $\lambda_D/L=5\times10^{-2}$.

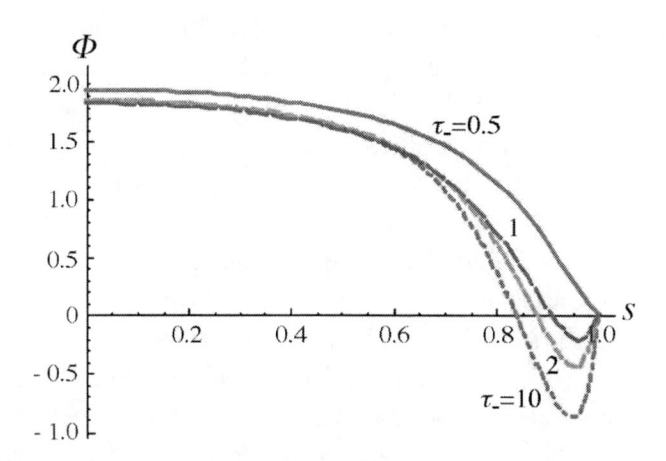

FIGURE 6. Profile of the normalized potential $\Phi=(q/kT_e)(\phi-\phi_w)$ for various values of $\tau_-=T_e/T_{i-}$ with $\tau=T_e/T_i=2$, $\beta=0.8$, $\lambda_D/L=5\times10^{-2}$.

ratio τ_- is shown in Fig. 6, where $\tau=2$, $\beta=0.8$, and $\lambda_D/L=5\times10^{-2}$. The potential depends on τ_- and has a negative peak near the wall for $\tau_- \geq 1$. This may be because low energy negative ions stay near the PG surface long time. As a result, negative ions are reflected by the negative potential peak near the PG and returned to the PG surface. This reflection mechanism by the negative potential peak possibly becomes a factor for the negative ion extraction.

CONCLUSIONS

The potential distribution near the extraction region for the plasma with the surface produced negative ions is studied analytically, where negative ions produced on the PG surface are considered in addition to positive ions and electrons. The negative ions produced on the PG surface that have the energy larger than $-q\phi_{min}$ can reach the center of the plasma, where $-q$ is the charge of the negative ion and ϕ_{min} is the minimum electric potential. The plasma–sheath equation is derived theoretically and solved numerically. The width of the sheath is about ten times as long as the Debye length. It is shown that the negative potential peak is formed near the PG surface for the case of strong surface production of the negative ions. This negative potential peak is also formed for the case of low energy negative ions. This will come from that low energy negative ions stay near the PG surface long time. Negative ions may be reflected by the negative potential peak near the PG and returned to the PG surface. This reflection mechanism by the negative potential peak possibly becomes a factor of the negative ion extraction. It is also indicated that the potential difference between the plasma and the wall becomes smaller for the plasma with the surface produced negative ions than for the plasma without negative ions. This potential distribution also has the possibility to contribute to the negative ion extraction.

REFERENCES

1. D. Riz and J. Pamela, *Rev. Sci. Instrum.* **69**, 914 (1998).
2. M. Hanada and T. Seki, *Rev. Sci. Instrum*, **77**, 03A515 (2006).
3. G. A. Emmert, R. M. Wieland, A. T. Mense and J. N. Davidson, *Phys. Fluids* **23**, 803 (1980).
4. K. Sato, F. Miyawaki, and W. Fukui, *Phys. Fluids* B **1**, 726 (1989).
5. J. H. Whealton, E. F. Jaegar, and J. C. Whitson, *J. Comput. Phys.* **27**, 32 (1978).
6. J. C. Whitson, J. Smith, and J. H. Whealton, *J. Comput. Phys.* **28**, 408 (1978).

A Sheath Model for Negative Ion Sources Including the Formation of a Virtual Cathode

R. McAdams[a], D.B. King[a] A.J.T. Holmes[b], and E. Surrey[a]

[a]*EURATOM/CCFE Association, Culham Science Centre, Abingdon, Oxfordshire OX14 3DB, UK*
[b]*Marcham Scientific, Hungerford, Berkshire RG17 0LH, UK*

Abstract. A one dimensional model of the sheath between the plasma and the wall in a negative ion source has been developed. The plasma consists of positive ions, electrons and negative ions. The model takes into account the emission of negative ions from the wall into the sheath and thus represents the conditions in a caesiated ion source with surface production of negative ions. At high current densities of the emitted negative ions, the sheath is unable to support the transport of all the negative ions to the plasma and a virtual cathode is formed. This model takes this into account and allows the calculation of the transported negative ions across the sheath with the virtual cathode. The model has been extended to allow the linkage between plasma conditions at the sheath edge and the plasma to be made. Comparisons are made between the results of the model and experimental measurements.

Keywords: negative ion source, plasma sheath, virtual cathode
PACS: 41.75.Cn, 52.50.Dg, 52.40.Kh

INTRODUCTION

Neutral beam heating and current drive will be necessary on future large fusion devices to meet the requirements of steady state plasma burn and plasma stability. These neutral beams will need to be of energies in excess of 1MeV and hence require negative ion precursors [1]. Volume production of negative ions cannot produce the required current densities and enhancement due to the addition of caesium in the source is required [2,3,4,5]. This enhancement is usually attributed to surface production of negative ions by atomic and ionic bombardment. These surface produced ions must make their way across the sheath and into the plasma before they can be extracted as a beam.

It has been shown through a variety of models that under the right circumstances this transport can be inhibited by the surface produced negative ions modifying the potential in the sheath to form a virtual cathode. The virtual cathode forms when the negative ion production at the wall is high enough that the associated space charge cannot be supported by the sheath and the space charge limit is exceeded. This is due to the production of negative ions at the wall being dominated by atoms rather than positive ions. This has been seen in PIC simulations by Wünderlich et al. [6], Hatayama [7] and Taccogna et al. [8]. Amemiya et al [9] produced an analytical model of electron injection into a plasma containing negative ions. This was simply adapted to the case of negative ions emitted from the source wall by McAdams and Bacal [10] and allowed the calculation of the maximum current density before the formation of a virtual cathode. This adaptation gave good agreement with a PIC code simulation [6].

This model has been extended to take into account the formation of a virtual cathode [11] and also the finite energy of the negative ions created at the surface. A limitation of the model was that it used the densities of the plasma particles at the sheath edge, rather than those in the bulk plasma where any measurements would take place. The model has been developed further to include the effects of the pre-sheath region to link the particle densities in the plasma to those at the sheath edge. This allows a comparison between the model results and experimental data, providing a benchmark for the method.

THE MODEL

The problem is subdivided according to Fig. 1 into a virtual cathode region, a standard sheath region, a pre-sheath region and the plasma. The cathode, from where the negative ions are emitted represents the plasma grid. The model is detailed fully in [11] but the basic parts of the process are presented here. The plasma is not modeled itself in this system, but provides values for the different particle densities. Figure 1 also shows the different parameters involved in the model; j_b is the current density emitted by the surface, V_c is the cathode potential with respect to the potential at the sheath edge, V_k is the depth of the virtual cathode, V_m is the potential between the virtual cathode and the sheath edge, j_{bmax} is the current density that is transported across the virtual cathode, U_0 is the potential between the sheath edge and the plasma and the densities $n_{i0,p}$, $n_{e0,p}$ and $n_{n0,p}$ represent the ion, electron and negative ion densities at the sheath edge (0 subscript) and in the plasma (p subscript). At low production levels of negative ions at the wall there is no virtual cathode and the emitted current can be supported by the sheath and so $V_c = V_m$ ($V_k=0$) and $j_{bmax} = j_b$. In an ion source, where the plasma grid is biased, the potential V_c is the difference between the bias potential and the plasma potential with respect to the reference of the bias voltage. The Poisson equations for the situations where the virtual cathode does not exist and when it does exist can be set up. Analysis of those equations [9,11] yields results for the transmitted current across the sheath. This analysis also allows the potentials themselves to be calculated but these are not reported here. Magnetic field effects due to the filter field are not considered in this 1d model

The maximum transmitted current across the sheath in the presence of a virtual cathode is given by

$$j_{b\max} = \frac{2n_{i0}V_0\left(\left(1+\frac{V_m}{V_0}\right)^{1/2}-1\right) + n_{e0}T_e\left(\exp\left(\frac{-V_m}{T_e}\right)-1\right) + n_{n0}T_n\left(\exp\left(\frac{-V_m}{T_n}\right)-1\right)}{\frac{2}{e}\left(\frac{M_b}{2e}\right)^{1/2}\left[(V_m+U_b)^{1/2}-U_b^{1/2}\right]} \quad (1)$$

where $T_{i,e,n}$ are the electron and negative ion temperatures, V_0 is the initial energy of the positive ions crossing into the sheath ($V_0 = U_0 + T_i/2$) and U_b is the initial energy of the emitted negative ions. This equation arises due to the condition that the electric field at the potential minimum in the virtual cathode is zero. If in this equation we set $V_m = V_c$ i.e. no virtual cathode this represents the point at which the electric field at the cathode is zero i.e. the space charge limit and j_{bmax} is the maximum transported current density before the virtual cathode is formed.

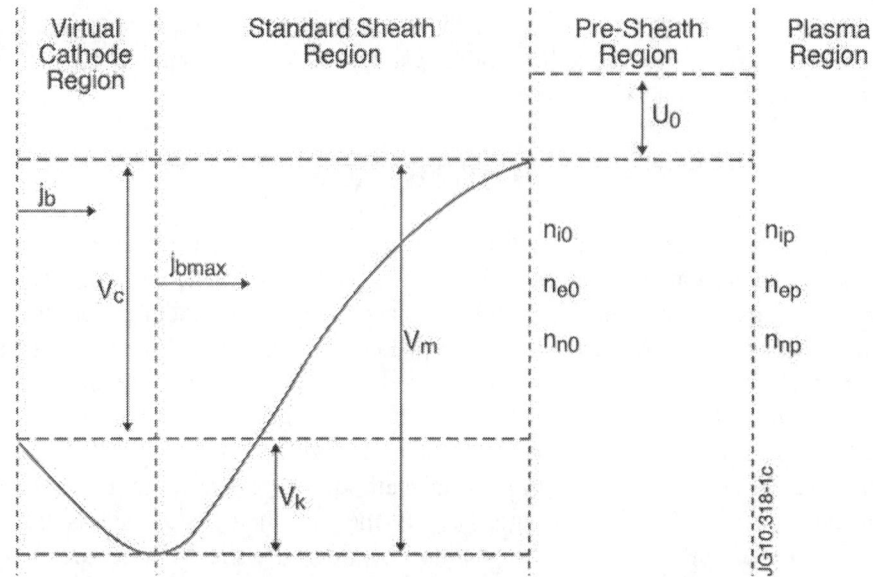

FIGURE 1. The virtual cathode, standard sheath, pre-sheath and plasma regions.

The densities and the maximum transported current density are related by the quasi-neutrality condition at the sheath edge

$$n_{e0} = n_{i0} - n_{n0} - \frac{j_{bmax}}{e}\left(\frac{M_b}{2e}\right)^{1/2}(V_m + U_b)^{-1/2} \quad (2)$$

and the initial energy of the positive ions is derived from the requirement of the derivative of the space charge density with respect to potential and given by,

$$V_0 = \frac{n_{i0}}{2\left(\frac{n_{e0}}{T_e} + \frac{n_{n0}}{T_n} - \frac{j_{bmax}}{e}\left(\frac{M_b}{e}\right)^{1/2}(2V_m + 2U_b)^{-3/2}\right)} \quad (3)$$

When a virtual cathode exists the negative ions emitted from the cathode are retarded by the virtual cathode potential V_k. It is assumed that the emitted current density, j_b, and the current density at the minimum of the virtual cathode, j_{bmax}, are related by a simple exponential barrier i.e.

$$V_k = T_b \ln\left(\frac{j_b}{j_{b\max}}\right) \qquad (4)$$

where T_b is the emitted negative ion temperature, $U_b = T_b/2$.

The equations above involve the densities at the sheath edge but these can be related to those in the bulk plasma by equations such as

$$n_{i0} = n_{ip} \exp\left(-\frac{U_0}{T_e}\right) \qquad (5)$$

This linkage between the sheath edge and the bulk plasma was not present in the original model [11].

This set of equations can be solved iteratively. To reach a solution for the system the values of n_{ip}, V_c, the temperature of each species, j_b and the ratio between volume produced negative ion density and positive ion density, D are all required. The emitted current density, j_b can be calculated from the atomic hydrogen flux and the yield of negative ions for a particular atomic hydrogen temperature and the emitted beam temperature, T_b, can be calculated from the kinetics of the conversion process [11]. The average energy $E-E_{thr}$ for a Maxwellian for atoms with a temperature T_H is calculated where E_{thr} is the threshold energy for negative ion production. This threshold energy is equal to $W-E_A$ where W is the work function, taken to be 1.5eV, and E_A is the electron affinity of a hydrogen atom (0.75eV). An energy reflection coefficient of 0.7 is also taken into account for any other energy loss in scattering from the surface. The value of D used throughout is 0.02, taken from [12]. The first part of the process is to guess a value of V_m which is then used to find the electron density at the sheath edge using the plasma neutrality condition. This is different to the process used in [11] where the value of n_{e0} was specified and n_{i0} was calculated. This method is preferable as the electrons are more mobile and will respond to the potential more than the other species. On the first iteration the value of n_{i0} is set to that of n_{ip} but on subsequent iterations a calculated value is used, similarly on the first step the value of j_b is used for j_{bmax} but on subsequent steps the calculated value is used. U_b is the initial energy of the emitted negative ions and expressed in terms of a temperature as $T_b/2$. Following this the initial energy of the positive ions is calculated according to Eq.3. Following this, the value of j_{bmax} and subsequently a value for V_k from Eq.1 and Eq.4 are calculated. The potential difference between the plasma and the sheath edge is found from

$$U_0 = V_0 - \frac{T_i}{2} \qquad (6)$$

and so the positive ion density at the sheath edge is found from Eq.5. This calculation of n_{i0} was not included in [11]. The other densities do not need to be recalculated separately from the sheath edge to plasma as they are already calculated from n_{i0} and obey quasi-neutrality, however when scaling from the sheath edge values to the plasma values for comparison with experiment they are corrected according to the same exponential as the positive ions to preserve quasi-neutrality.

All that remains of the convergence procedure is the calculation of the n_{n0} from the value of n_{i0} through Eq. 7

$$n_{n0} = D n_{i0} \qquad (7)$$

and the recalculation of V_m through Eq. 8.

$$V_m = V_c + V_k \qquad (8)$$

This value of V_m is then mixed with the previous value and the process repeated until convergence is reached. The total negative ion density in the plasma can then be found by combining the volume negative ion density n_{np} with the beam negative ion density at the sheath edge.

$$n_{b0} = \frac{j_{b\,max}}{e\sqrt{\frac{2e(V_m + U_b)}{M_b}}} \qquad (9)$$

The density from Eq. 9 is then scaled to the bulk plasma. Poisson's equation can then solved for each region for the converged parameters to give the potential through the sheath (not reported here). The entire process is coded as an IDL program that runs on a standard UNIX workstation and solves in less than a second.

APPLICATION OF THE UPDATED MODEL

Effects of the Inclusion of the Pre-Sheath

The original sheath model used the densities at the sheath edge. These are not known whereas those in the bulk plasma can be measured. The scaling method outlined above allowed the bulk plasma parameters to be used with the sheath model thus allowing comparison with experiment. The densities at the sheath edge are lower than those in the plasma by the factor $\exp(-U_0/T_e)$. The capability of the sheath to support a given flux of negative ions is mostly determined by the positive ion flux from the plasma although leakage into the sheath of plasma electrons and negative ions will reduce this. Thus using a given density as the bulk plasma density should result in a reduction in the transported flux as the positive ion density at the sheath edge is reduced. This is illustrated in Figure 2 where the virtual cathode depth is calculated for the same positive ion density at the sheath edge (omitting one step in the iteration cycle) and in the bulk plasma. The equations above are solved using a positive ion density of $n_i = 3.5 \times 10^{17} \mathrm{m}^{-3}$, $T_e = 2\mathrm{eV}$, $T_i = 0.8\mathrm{eV}$, and an emitted current density of $500 \mathrm{Am}^{-2}$ and an emitted energy of $0.7\mathrm{eV}$. No negative ions are assumed to exist in the plasma for the sake of simplicity. In the case where the conditions are applied at the sheath edge, as the cathode potential is made increasing positive with respect to the plasma, there is no virtual cathode formed until this potential difference is less than ~ 7V. When the conditions are applied in the bulk plasma the virtual cathode is found to persist out to very high cathode voltages as expected due to the lower positive ion density.

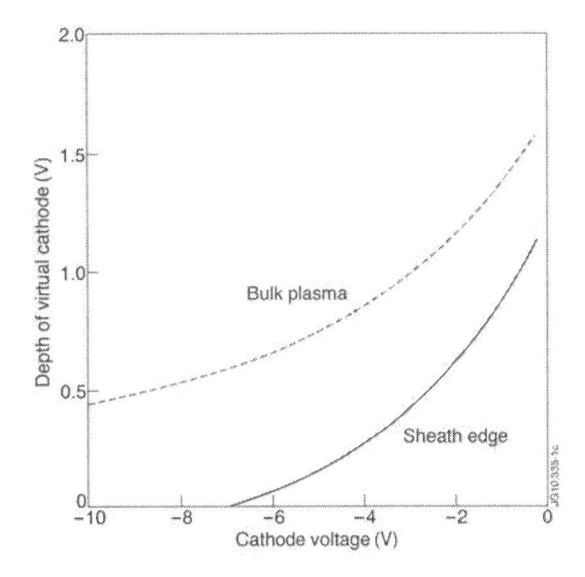

FIGURE 2. Virtual cathode depth as the cathode potential is varied for 500Am^{-2} emission at the surface for the case where the plasma conditions are imposed at the sheath edge (solid line) and in the bulk plasma (dashed line)

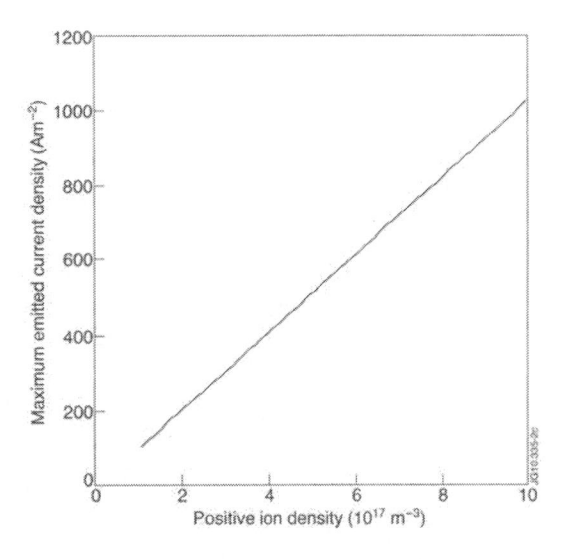

FIGURE 3. Relationship between the positive ion density in the plasma and the maximum current density before a virtual cathode can form for the plasma conditions given in the text.

To demonstrate the effect of the positive ion density on the virtual cathode directly the code was used for a range of n_{ip} and a range of j_b for a V_c of –10V, generally if the virtual cathode has not been removed for this level of bias then it will persist indefinitely. The maximum level of emitted current density j_b that could be transported without a virtual cathode being present was calculated and the results are shown in

Fig. 3 for the plasma conditions used above. As the positive ion density increases then the positive ion flux to the wall increases. This allows more negative ions to flow from the wall since each positive ion compensates for the presence of a negative. This is not the complete picture since electrons and negative ions from the plasma add to the overall negative space charge in the sheath.

Comparison with Experiment

In order to test the model against experiment, as complete a data set as possible for the IPP BATMAN source has been assembled from the literature [13,14,15]. These data cover source pressures of 0.3, 0.4 and 0.5 Pa and RF powers of 38-78kW and consists of electron, positive and negative ion densities, electron temperatures, atomic hydrogen density. The input data is shown in Figure 4.

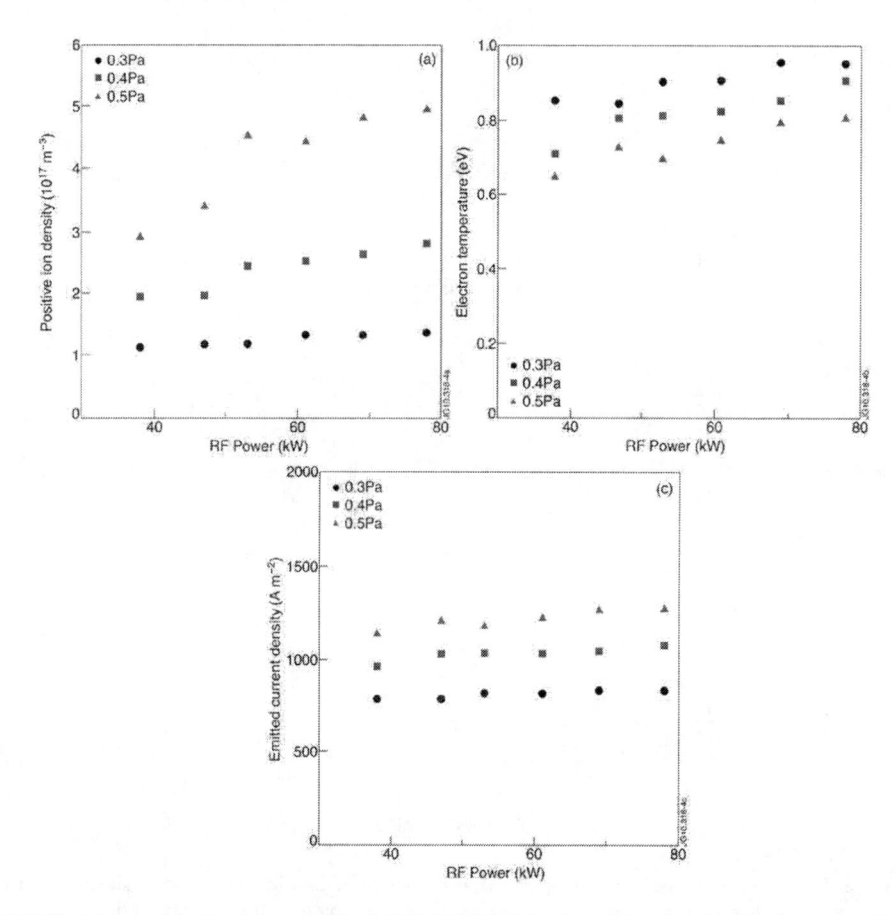

FIGURE 4. Input data for the model from IPP BATMAN: (a) positive ion density (measured), (b) electron temperature (measured) and emitted negative ion current density (calculated from data).

It was assumed that the atomic hydrogen temperature was equal to the electron temperature and that the positive and negative ions both had a fixed temperature of 0.8eV. The yield of negative ions for a given atomic temperature was taken from the work by Seidl et al [16]. For an atomic temperature of 0.8eV, which is used in previous simulations [6] and is typical of this ion source at the reported conditions, the yield is 0.12. From this data the emitted current density of negative ions from the surface could be calculated [11]; a value of U_b of 0.7eV was used. The method of comparison was to take as input to the model, the measured positive ion density, measured electron temperature and calculated emission current density. Then the electron and negative ion densities were calculated for comparison with the experimentally measured values. Also calculated were the transported current density of negative ions across the sheath and the depth of the virtual cathode; these parameters are not presently measurable. From the data it was not clear what bias voltage was used and a value of $V_c = -1V$ was assumed based on normal operating point of just below plasma potential. The comparison between the model and the experimentally measured electron and negative ion densities are shown in Figure 5. In all cases the calculated plasma neutrality was better than 0.05%. The model is in excellent agreement with the measured electron density. Agreement with the measured negative ion density is very good at the lower pressures but overestimates this density at the highest pressure. The model does not take into account any volume production of negative ions other than to fix their density as a fraction of the positive ion density. Furthermore no collision processes have been accounted for in this version; these may be important at the higher pressures. Given the uncertainty in some values such as bias voltage, and the ion and atomic temperatures the agreement with the model is very good.

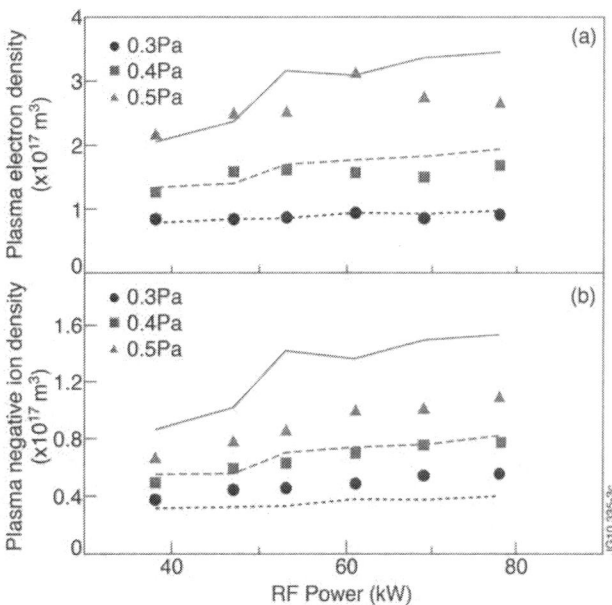

FIGURE 5. Comparison of model with measured parameters from the IPP BATMAN source: (a) electron density and (b) negative ion density. Measured values are the symbols, calculated are the lines: solid is 0.3Pa, dashed is 0.4Pa and dotted is 0.5Pa

In Figure 6 the calculated transported current density across the sheath and the depth of the virtual cathode are plotted. The virtual cathode depth is highest for the lowest operating pressure. This is due to the lower positive ion density since the emitted current density is lowest for this pressure.

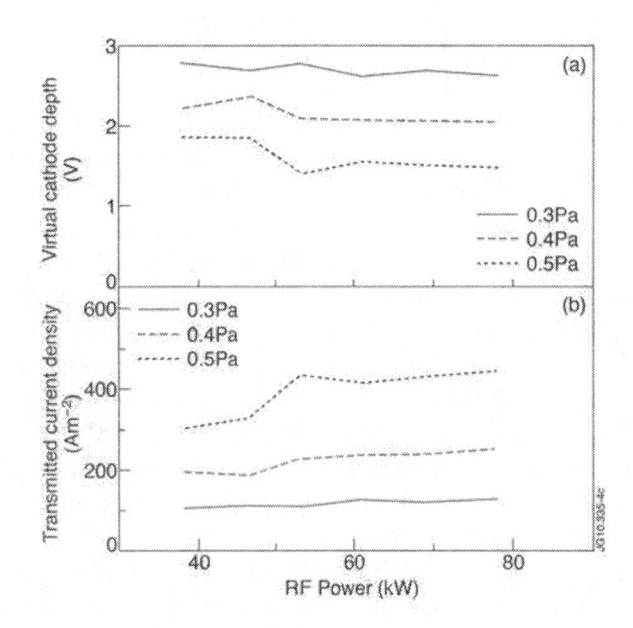

FIGURE 6. Calculated virtual cathode depth and the transported negative ion current density.

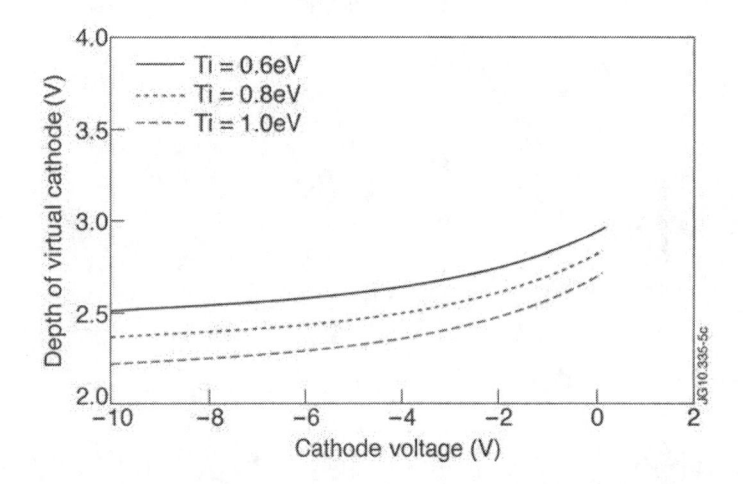

FIGURE 7. The dependence of virtual cathode depth against the cathode potential for different values of positive ion temperature for 0.3Pa and 69kW of RF power.

Positive Ion Temperature Effects

The positive ion and negative ion temperatures are not known. Thus a sensitivity study was carried out into the effect of varying the positive ion temperature since the positive ion flux is important in determining the virtual cathode depth and the transported flux of negative ions. This was done by varying the cathode voltage for different values of T_i for the case of 0.3Pa source pressure and 69kW of RF power. The results of this can be seen in Fig. 7. In these cases a variation in virtual cathode depth of around 0.5V can be seen with T_i. The positive ion flux that is transported across the sheath depends on the initial positive ion energy as given by Eq. 6; the lower the positive ion temperature the lower the flux in the sheath. At higher positive ion temperatures a higher negative ion flux can be transported across the sheath and thus the virtual cathode depth is lower.

CONCLUSIONS

A model of virtual cathode formation in the presence of surface negative ions has been further developed and then tested against experimental results. The model is able to predict experimental measurement very well given the uncertainties in some of the assumptions which have to be made. It has been shown that given a high enough surface emission of negative ions a virtual cathode will persist at very large cathode voltages. Further to this, the positive ion density has an essentially linear effect on potential negative ion flux that can be sustained from a surface without losses to a virtual cathode. The effects of positive ion temperature on virtual cathode formation have also been shown. The study has emphasized the need for comprehensive diagnostics of the ion source plasma.

ACKNOWLEDGMENTS

This work was funded by the United Kingdom Engineering and Physical Sciences Research Council under grant EP/G003955 and the European Communities under the contract of Association between EURATOM and CCFE. The views and opinions expressed herein do not necessarily reflect those of the European Commission.

REFERENCES

1. *ITER Design Description Document* 5.3, *N53 DDD 29 01-07-03 R0.1* 2003 IAEA, Vienna
2. McAdams R, King R F, Proudfoot G and Holmes A J T, *Production and Neutralisation of Negative Ions and Beams, Sixth International Symposium 1992*, eds J Alessi and A Herschovitch, AIP Conference Proceedings **287**, p353-367 (1992).
3. Fantz U et al., *Plasma Phys. Control. Fusion* **49**, B563–B580 ((2007).
4. Leung K N, Hauck C A, Kunkel W B and Walther S R, *Rev. Sci. Instr.* **60(4)** 531–538 (1989).
5. Okumura Y, Hanada M, Inoue T, Kojima H, Matsuda Y, Ohara Y, Oohara Y, Seki M, Suzuki Y, and Watanabe K, *Proceedings of the Symposium on Fusion Technology*, London 1990, 1026-1029.
6. Wünderlich D, Gutser R, and Fantz U, *Plasma Sources Sci. Technol.* **18** 045031 (2009).
7. Hatayama A, *Rev. Sci. Instr.* **79** 02B901 (2008).
8. Taccogna A, Minelli P, Longo S, Capetelli M, and Schneider R, *Phys. Plasmas* **17**, 063502 (2010).
9. Amemiya H, Annaratone B M, and Allen J E, *J. Plasma Physics* **60(1)** 81-93 (1998).
10. McAdams R and Bacal M, *Plasma Sources Sci. Technol.* **19** 042001 (2010).
11. McAdams R, Holmes A J T, King D B and Surrey E, to be *submitted to Plasma Sources Sci. Technol.*
12. Fantz U, Falter H D, Franzen P, Speth E, Hemsworth R, Boilson D and Krylov A, *Rev. Sci. Instr.* **77**, 03A516 (2006).
13. Fantz U et al, *Nucl. Fusion* **46** S297 (2006).
14. McNeely P, Dudin S V, Christ-Koch S, Fantz U et al., *Plasma Sources Sci. Technol.* **18** 014011, (2009).
15. Berger M, Fantz U, Christ-Koch S et al., *Plasma Sources Sci. Technol.* **18** 025004 (2009).
16. Seidl M, Cui H L, Isenberg J D, Kwon H J, Lee B S, and Melnychuk S T, *J. Appl. Phys.* **79**(6), 2896-2901 (1996).

About the Extraction of Surface Produced Ions in Negative Ion Sources

Francesco Taccogna[a], Pierpaolo Minelli[a], Savino Longo[a,b] and Mario Capitelli[a,b]

[a]*Istituto di Metodologie Inorganiche e di Plasmi IMIP-CNR, via Amendola 122/D, 70126 Bari, Italy.*
[b]*Dipartimento di Chimica, Università degli Studi di Bari Aldo Moro, via Orabona 4, 70126 Bari, Italy.*

Abstract. The enhancement of extracted negative ion current in cesiated sources is usually explained by the surface production of negative ions. In this contribution, the self-consistent production and transport of H⁻ in the extraction region of a radio-frequency driven negative ion source is modelled by means of a parallel two-dimensional Particle-in-Cell/Monte Carlo Collision simulation. It is shown that the number of surface-produced negative ions extracted is regulated by a potential well developed in front of the plasma grid such that the extracted current does not proportionally increase with the flux of negative ions emitted at the surface.

Keywords: Particle-in-Cell model; Monte Carlo Collision methodology.
PACS: 52.27.Cm; 52.40.Kh; 52.65.Rr

INTRODUCTION

High-energy neutral hydrogen beam injection (NBI) is needed for heating and current drive in fusion experiments like ITER. The NBI system consists of a low-pressure plasma source, from where ions are produced, extracted, accelerated and neutralized. Since the neutralization efficiency of positive ions is very low at beam energies required for ITER (1 MeV), the NBI system is based on negative hydrogen ions [1]. Here, the neutralization efficiency is still 60%. It is generally accepted [2] that two different contributions produce H⁻ ions: (i) in the volume a two-step process which involves dissociative attachment of slow electrons to vibrationally excited levels of hydrogen molecules $H_2(v)$; (ii) on the surface by neutral or positive ion electron capture. For this purpose, Caesium seeding is used which deposits Caesium on the surfaces and lowers the work function down to 1.8 eV. Nevertheless, the use of Cs remains problematic leading to many drawbacks: cleaning and maintenance, escape into the acceleration region with beam halos, *etc.*

For the international fusion experiment ITER [3], the current density of H⁻ required has been estimated to be $j_{H^-}=300$ Am⁻² with a tolerable number of co-extracted electrons ($j_e/j_{H^-}<1$), a low background pressure of $P=0.3$ Pa (in order to reduce the stripping losses of H⁻) and acceptable maintenance and lifetime. The radio frequency-driven inductively coupled plasma (RF-ICP) negative ion source developed at the Max-Planck-Institute für Plasmaphysik [4] has been recently chosen [3] by the ITER board as the new reference source for the ITER NBI system, due to the fact that it is potentially able to fulfil all ITER requirements.

The RF-ICP source consists of three parts: the driver, the expansion and the extraction regions. The driver is mounted on the back of the source body and consists of a 24.5 cm diameter alumina cylinder (length 14 cm) in which the plasma is generated by a water-cooled RF coil connected to a 1MHz oscillator with a maximum output power of 150 kW. The plasma expands and cools down into a rectangular Copper coated body (59x32x23 cm^3). The expansion and extraction regions are separated by a magnetic filter field parallel to the plasma grid (PG) with a maximum of 7 mT at a distance of 2.2 cm to PG (blue line in Fig. 1). This field is necessary in order to keep the hot electrons generated by the RF away from the extraction region. The extraction system has three grids: the plasma (PG), the extraction (EG) and the grounded (GG) grids. The extraction area on the PG is 74 cm^2 and consists of 132 holes with 0.8 cm diameter. The PG is made of Molybdenum and temperature controlled to 150 °C in order to establish a thin and homogeneous layer of Cs needed for optimal conversion. Finally, the EG equips magnets which are there to deflect the co-extracted electrons (red line in Fig. 1).

Recent experiments [5,6] have shown that an order of magnitude higher ion current density (up to 330 Am^{-2}) can be extracted when Cs is introduced into the RF-ICP source than without Cs with a simultaneous reduction of co-extracted electron (factor 10). The enhancement is generally attributed to the surface production of negative ions on the low work function cesiated surfaces close to the extraction aperture.

In this contribution, we assess the extraction of the surface-produced negative ions by using a parallel 2D-3V Particle-in-Cell / Monte Carlo Collision (PIC-MCC) plasma model [7] simulating the production and transport of H$^-$ in the self-consistent field generated in the sheath of the extraction region.

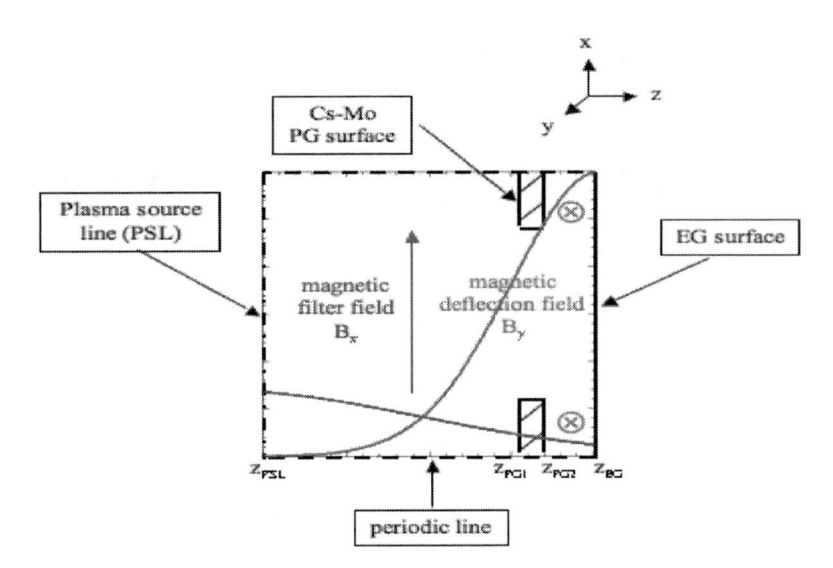

FIGURE 1. Scheme of the simulation domain used for the extraction region model. The direction and behaviour of magnetic filter and deflection fields are represented in blue and red colours respectively.

NUMERICAL MODEL

The geometry of the two-dimensional simulation domain of the extraction and part of the acceleration regions is shown in Fig. 1. The size is such that the dynamics of negative ions originating from the surface is well resolved due to the short survival length of negative ions (cm range) [8]. One single aperture is surrounded by the flat PG surface, while uniformity is assumed in the transverse not simulated coordinate y; periodic boundary conditions in x direction are imposed simulating the multi-aperture grid. Therefore, the simulation domain is a rectangle in the x-z plane with dimensions L_x=1.2 cm and L_z=2.05 cm. These dimensions are sufficient to analyze the effect of PG bias and the EG field penetration into the source since the region influenced by the bias is supposed to be several tens of Debye lengths λ_D. The plasma grid containing the orifice at the centre is located between z_{PG1} and z_{PG2}, while the orifice diameter is D_H=0.8 cm (a single aperture of the large area grid LAG extraction system). Open boundary conditions are imposed along the bulk plasma ($\partial\phi/\partial z|_{zPSL}=0$), while fixed potential is imposed on the PG (ϕ_{PG}=15 V) and EG surfaces ((ϕ_{EG}=9 kV). The two fixed, non-homogeneous magnetic fields (see Fig. 1) are directed along x and y with Gaussian profiles.

The following boundary conditions for the particles are implemented. On the plasma source layer (PSL), electrons and different ion species (H^+, H_2^+, H_3^+, H^- are considered) are created to simulate the flow from the bulk plasma to the right of the simulation domain. In particular, positive ion flux and ion molar fraction [$j_{H+}+j_{H2+}+j_{H3+}$=900 Am^{-2}, $\chi(H^+)$=0.4, $\chi(H_2^+)$=0.4, $\chi(H_3^+)$=0.2] are taken from experimental and computational results [9,10]. If particles return back to the PSL from right to left, a refluxing method is used (particles are re-injected with bulk parameters). This condition enforces equal fluxes of positive and negative charges crossing through the PSL that guarantees the neutrality in the bulk plasma. Particles crossing a periodic boundary are mapped back to the other periodic boundary. Particles, which crossed the same periodic boundary more than 10 times are removed completely from the simulation domain. When an electron hits the PG wall it is removed, while positive ions (H^+, H_2^+ or H_3^+) hitting the PG wall can create negative ions according to the energy dependent yield $Y_+(H^-)$ for a Cs/Mo surface from Ref. [11]. Moreover, on the basis of the neutral flux impinging onto the wall, a certain fraction of H^- ions are additionally launched from the wall with a constant source rate:

$$\Gamma_{H-} = \Gamma_0 Y_0(H^-) \qquad (1)$$

where $\Gamma_0 = \frac{1}{4}v_{th,H}n_H$ is the atom flux impinging the wall and $Y_0(H^-)$ is the H^- yield by neutral conversion taken from Ref. [11] (see Fig. 2):

$$Y_0(H^-) = 0.42\exp\left(-\frac{1.05}{k_B T_H}\right). \qquad (2)$$

This mechanism is dominant compared with ionic conversion (less than 1% of the total surface produced H⁻ are originated by ionic conversion), due to the fact that the ratio of neutral to ion flux at the wall is about 10 for a neutral temperature of 1 eV [12]. Due to the uncertainty in the atomic density and temperature, the negative ion flux produced on the surface by neutral conversion has been considered as a free parameter in the model and a study using five different values has been performed.

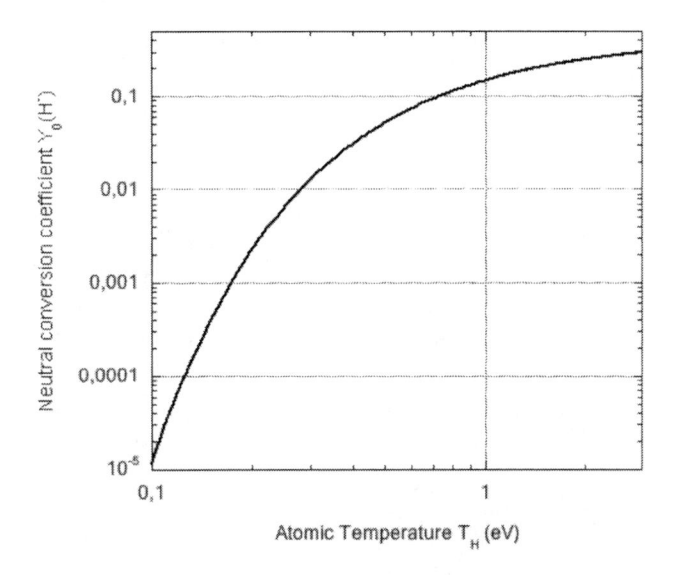

FIGURE 2. Neutral conversion coefficient Y_0(H⁻) as a function of the atomic temperature T_H [eq. (2)] in the temperature range of atoms relevant in the extraction region of the RF hybrid negative ion source developed at IPP Garching. From Ref. [11].

Among the different collision processes incorporated in the model, particular emphasis is given to those involving H⁻ production, electron dissociative attachment to vibrationally excited molecules, and H⁻ destruction: electron detachment in collisions with electrons, mutual neutralization in collisions with positive ions and associative detachment in collisions with atoms. Particularly important are also non-destructive collision reactions as elastic collisions with H⁺ and charge exchange collisions with H, because they change the transport and reduce the kinetic energy of the extracted H⁻.

Further details of the model are presented in Ref. [13].

RESULTS

The Debye length and the inverse plasma frequency for the plasma parameters used are λ_D=2.3x10⁻⁵ m and $(\omega_{pe})^{-1}$=5.6x10⁻¹¹ s, respectively. Thus, the size of the grid cells and the calculation interval for the PIC code were chosen to be 4x10⁻⁵ m and 2x10⁻¹¹ s. Approximately 80000 time steps are needed (1.6x10⁻⁶ s simulated time) to reach a quasi steady state. In this state the total number of macro-particles present in the simulation domain is about 3×10^8. Such calculation needs 2 days on a cluster of 8 Quad Core Intel Xenon X5570 (2.93 GHz, 96 GB RAM) processors.

Simulations exploring different surface production rates have been performed using seven different values of H⁻ current density originated at the surface $j_{H-,prod}$=10, 50, 100, 150, 200, 250 and 300 Am⁻². In all cases the PG bias develops at a positive voltage ϕ_{PG}=15 V with respect to the source.

The two-dimensional map of the electric potential for the case $j_{H-,prod}$=200 Am⁻² is reported in Fig. 3. The most interesting features are: a) the penetration of the EG field into the source region forming the meniscus; b) an electric potential drop between the centre of the orifice and the PG surface and c) potential wells attached to the PG surface (white regions).

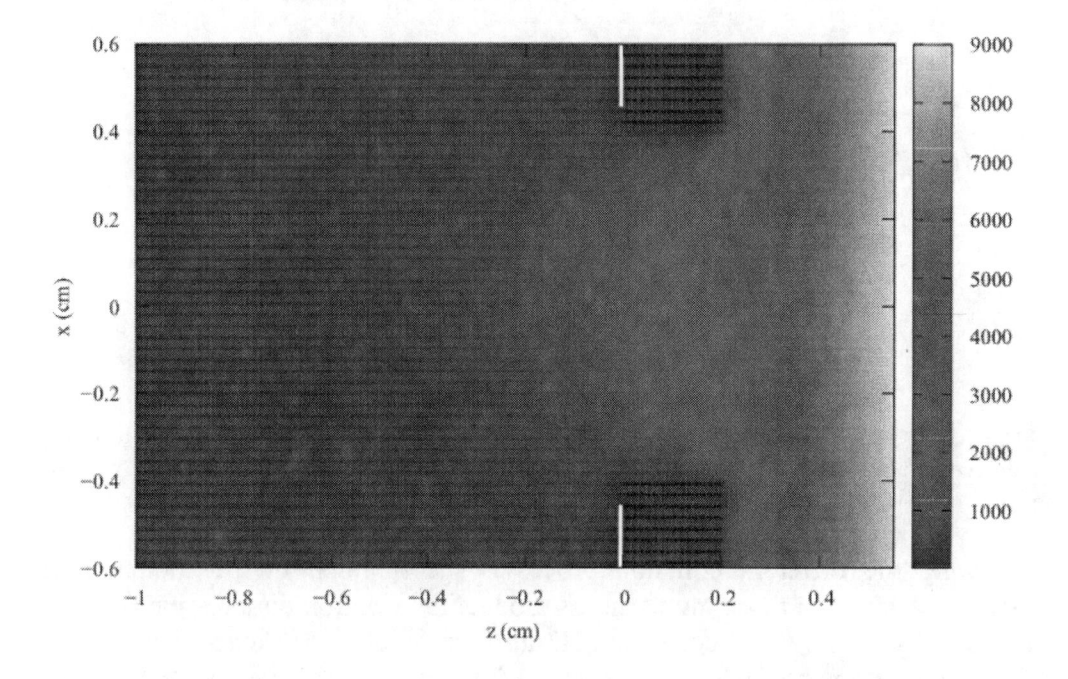

FIGURE 3. Two dimensional map of the electric potential (V) using $j_{H-,prod}$=200 Am⁻².

The negative ion current extracted at $z=z_{PG1}=0$ (see Table 1 and Fig. 4) $j_{H-,extr}$ does not increase proportionally with the surface production rate $j_{H-,prod}$ because the extraction probability of surface-produced negative ions drops down from 1 for the smallest surface production rate to 0.34 for the highest surface production rate considered. This strong reduction of the extraction probability is due to the fact that the potential wells developed in front of the PG wall reflect more and more H⁻ back to the wall. These double sheath structures just in front of the PG surface have been also observed by different numerical models [14-16]. The potential depth is larger (0.8 V) for the larger surface production rate, while it is totally absent for the smaller cases. The reason of this barrier is due to the presence of an ambipolar electric field, which tries reducing the misbalance of the charged fluxes. In fact, the total flux of net positive charges directed towards the wall ($\Gamma_+ - \Gamma_-$) is no more sufficient to neutralize the flux of negative ions emitted from the wall:

$$\Gamma_+ - \Gamma_- = Y_+(H^-)\Gamma_+ + Y_0(H^-)\Gamma_0. \qquad (3)$$

In order to re-establish the charge compensation, a potential well is formed which has a double effect: it accelerates the positive ions from the bulk to the wall and it reflects the surface produced negative ions back towards the wall.

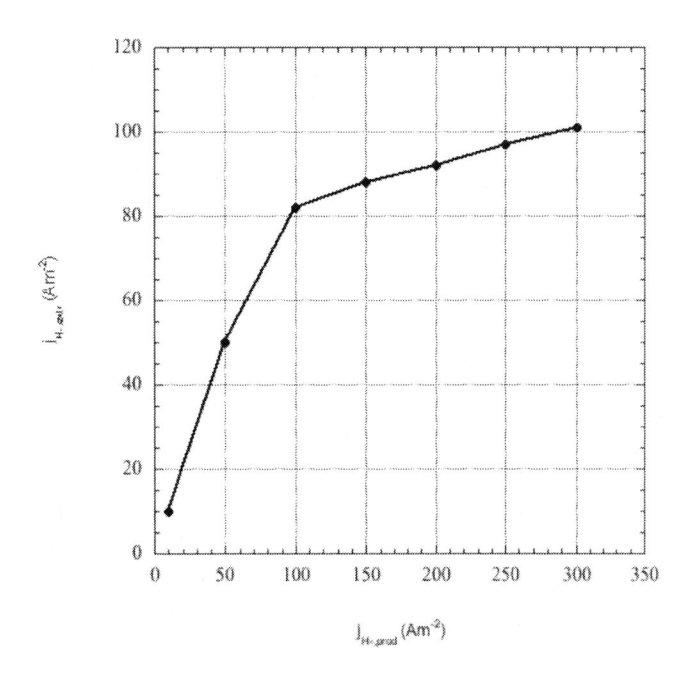

FIGURE 4. Extracted negative ion current density $j_{H^-,extr}$ as a function of surface production rates by neutral conversion $j_{H^-,prod}$.

TABLE 1. Percentage of surface-produced negative ions extracted and back-reflected to the wall and negative ion and electron current density extracted at z_{PG1} using different surface production rates by neutral conversion.

	Extraction probability of surface produced H⁻	Probability of back-reflected surface produced H⁻	Extracted negative ion current density $j_{H^-,extr}$ (Am⁻²)	Co-extracted electron current density j_e (Am⁻²)
$j_{H^-,prod}$ = 10 Am⁻²	1	0	10	335
$j_{H^-,prod}$ = 50 Am⁻²	1	0	50	272
$j_{H^-,prod}$ =100 Am⁻²	0.82	0.16	82	268
$j_{H^-,prod}$ =150 Am⁻²	0.58	0.4	88	256
$j_{H^-,prod}$ =200 Am⁻²	0.46	0.51	92	248
$j_{H^-,prod}$ =250 Am⁻²	0.39	0.58	97	240
$j_{H^-,prod}$ =300 Am⁻²	0.34	0.64	101	236

This also explains why the majority of surface produced H⁻ extracted is produced close to the orifice, while less than 20% of the H⁻ produced at a location farther than 1 mm from the orifice succeed crossing the extraction aperture (see Fig. 5).

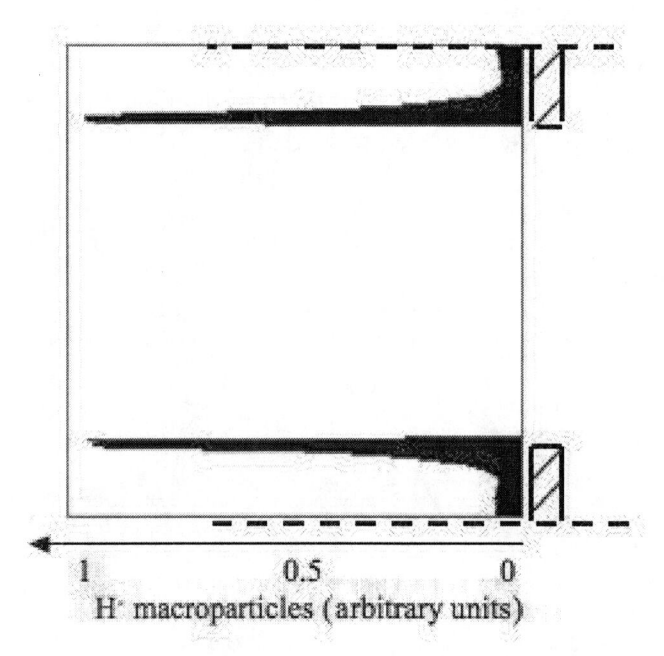

FIGURE 5. Surface birthplaces of extracted H⁻ using $j_{H^-,prod}$=200 Am^{-2}.

The electric potential map and the birthplaces of extracted H⁻ explain the shape of the negative ion beam (Fig. 6.a) that is very important for the optics. It is characterized by three peaks resulting from the overlapping of two beams of negative ions accelerated by the penetrated EG field from two point-like sources of H⁻ located at the two extremities of the extraction orifice.

The 2D spatial distribution of electron density is shown in Fig. 6.b. The electron density distribution shows the ***ExB*** drift deflecting the electron beam along the *x*-direction creating an inhomogeneity already into the source region.

Finally, it has to note that the value of extracted current is in the largest case about 70% of the experimental value [5]. The contribution from the third coordinate *y* not simulated and/or some other production mechanism or enhancement must take place, suggesting that possible near-wall effects [17] induced by the presence of Cs are still missing in the model.

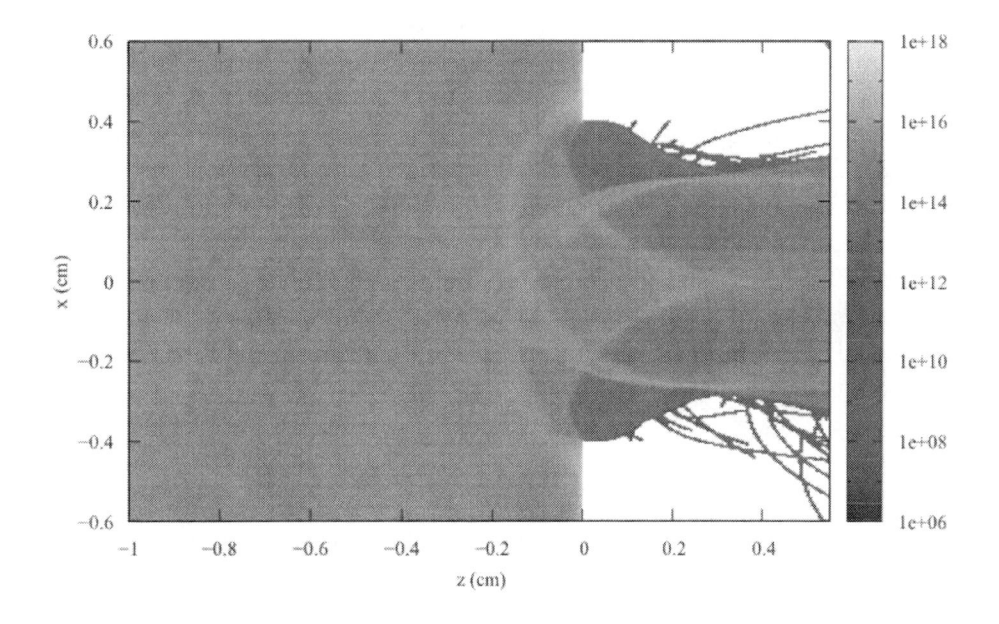

(a)

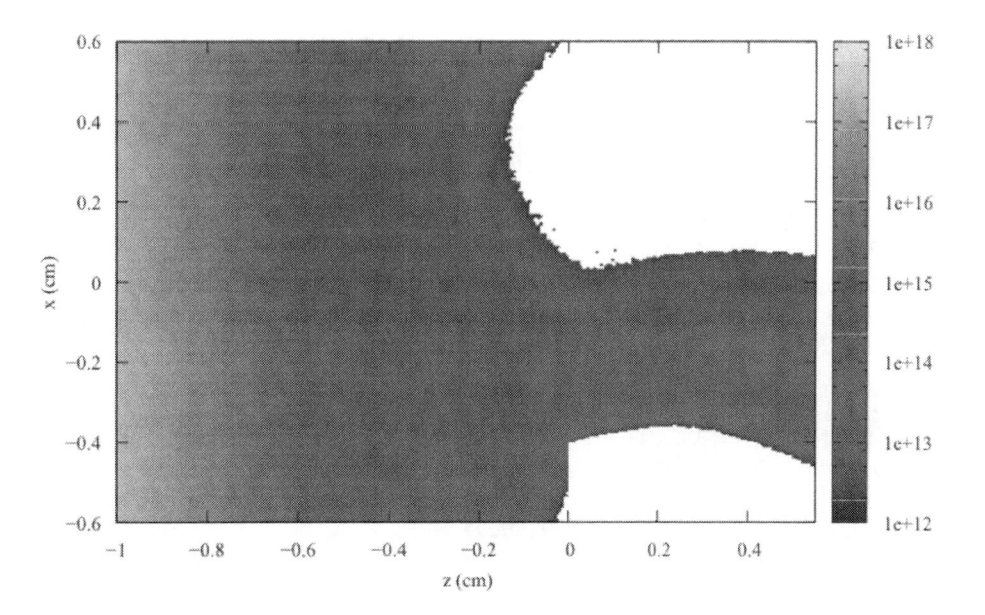

(b)

FIGURE 6. Two dimensional map of (a) negative ion and (b) electron density (m^{-3}) using $j_{H-,prod}$=200 Am^{-2}.

CONCLUSIONS

The negative ion formation and transport in the extraction region of a radio-frequency driven inductively coupled discharge have been studied using a self-consistent 2D(x,z)-3V parallel PIC-MCC simulation code. The number of extracted negative ions emitted from the surface does not proportionally increase with the surface production rate because the extraction probability decreases with a larger neutral conversion yield. This is due to the presence of a potential well formed in front of the wall established to compensate the charged fluxes (space charge compensated sheath regime). Moreover, the computational results reaches in the best case only 70% of the experimental result obtained in a Caesium seeded source showing that additional effects (third dimension or different Caesium related processes) need to be considered.

ACKNOWLEDGMENTS

This work is supported by F4E under Grant-F4E-2009-GRT-032-PMS.

REFERENCES

1. K. H. Berkner, R. V. Pyle, J. W. Stearns, *Nucl. Fus.* **15**, 249 (1975).
2. M. Bacal, *Nucl. Fus.* **46**, S250 (2006).
3. R. S. Hemsworth, A. Tanga, V. Antoni, *Rev. Sci. Instrum.* **79**, 02C109 (2008).
4. E. Speth *et al.*, *Nucl. Fus.* **46**, S220 (2006).
5. U. Fantz, *et al.*, *Plasma Phys. Control. Fus.* **49**, B563 (2007).
6. M. Berger, U. Fantz, S. Christ-Koch, and NNBI Team, *Plasma Sources Sci. Technol.* **18**, 025004 (2009).
7. C. K. Birdsall, A. B. Langdon, *Plasma Physics via Computer Simulation*, (New York: McGraw-Hill, 1985).
8. D. Riz, J. Pamela, *Rev. Sci. Instrum.* **69(2)**, 914 (1998).
9. J. P. Boeuf, G. J. M. Hagelaar, P. Sarrailh, G. Fubiani, N. Kohen, *Plasma Sources Sci. Technol.* **20**, 015002 (2011).
10. P. McNeely, S. V. Dudin, S. Christ-Koch, U. Fantz, and NNBI Team, *Plasma Sources Sci. Technol.* **18**, 014011 (2009).
11. M. Seidl, H. L. Cui, J. D. Isenberg, H. J. Know, B. S. Lee, S. T. Melnychuk, *J. Appl. Phys.* **79(6)**, 2896 (1996).
12. R. McAdams, E. Surrey, AIP Conf. Proceed. 1097, edited by E. Surrey and A. Simonin, American Institute of Physics, Melville, NY, 89 (2009).
13. F. Taccogna, P. Minelli, S. Longo, M. Capitelli, R. Schneider, *Phys. Plasmas* **17** 063502 (2010).
14. A. Hatayama, *Rev. Sci. Instrum.* **79**, 02B901 (2008).
15. D. Wünderlich, R. Gutser, U. Fantz, *Plasma Sources Sci. Technol.* **18**, 045031 (2009).
16. R. McAdams, M. Bacal, *Plasma Sources Sci. Technol.* **19**, 042001 (2010).
17. D. Pagano, C. Gorse, M. Capitelli, *IEEE Trans. Plasma Sci.* **35(5)**, 1247 (2007).

Fluid and kinetic models of negative ion sheaths

M. Cavenago

INFN-LNL, v.le dell'Universita' 2, 35020 Legnaro (PD), Italy

Abstract. Due to the presence of a large transverse magnetic field (B_x and B_y where z is the extraction axis), the extraction of electrons from a negative ion source is likely to happen with a large angle with respect to z axis. The negative ion and electron sheaths are here studied both with kinetic and with fluid models. First, Vlasov-Poisson models are reduced to one dimensional integrodifferential equations, discussing also trapped orbits. The integrodifferential equations for electron transport are analytically solved for a variety of extraction potentials (in 1D). Collision frequency dependency from electron flow speed and temperature is discussed. Then both ion and electron space charge and fluid motion are solved, using electron densities expression consistent with kinetic model. Results for the sheath charge profile and extraction field as a function of B_x are shown.

Keywords: negative ion, source, extraction, Vlasov equation
PACS: 52.55.-s, 52.20.-j, 52.59.Sa

INTRODUCTION

The modeling of the Negative Ion Sources (NIS) used, for example, for multiturn synchrotron injection[1] or in Neutral Beam Injectors (NBI) for fusion application[2], must consider magnetic field which are transverse to the extraction axis z; otherwise coextracted electron current would be much larger than ion current, making ion source inefficient (or destroying it). On the other side, even in the extraction sheath, collisions must be accounted for. Otherwise, electrons will be trapped forever with a moderate magnetic field (since Larmor radius is typically much smaller than source size). This paper discusses some basic properties and issues in modeling the extraction sheath of such sources, with kinetic models and semifluid models. Even if particle may move also in x and y, uniformity of particle densities and of potential is assumed in these coordinates for simplicity, which is the definition of one dimensional model (1D).

It should be noted that plasma NIS feature a two stage plasma: in first stage electron temperature T_e is high enough ($T \geq 4$ eV) to produce a small degree of ionization of the incoming gas; say H_2 to fix ideas. Since these electrons may easily ionize H^- (binding energy 0.75419 eV [3]), a filter magnetic field is applied between stages. Electron collision frequency v_e^m decreases when energy E_e increases; efficient transmission requires $v_e^m > \Omega_e = e|B|/m_e$ with Ω_e the electron cyclotron frequency.

Thus 2nd stage plasma, that is the extraction plasma, is colder (say $T_e \cong 1$ eV) and less dense: in the NBI sources, typically $N_{H^-} \cong 3 \times 10^{17}$ m^{-3} and $N_e < 10^{17}$ m^{-3}. On the contrary gas density ranges from 0.3 Pa for NBI application to 10 Pa and more for industrial application; so $N_g \cong 10^{20}$ m^{-3} or more. Temperature of ions is likely similar to T_e, so we assume equal temperature $T_0 = 1$ eV for all charge species; also we scale density as $n_i = N_i/N_0$ with the constant reference $N_0 = N_{H^-} + N_e$.

For comparison, in singly charged positive ion sources, the study of beam extraction was greatly simplified by the absence of magnetic field and by the presence of only two charge species, say e$^-$ and H$^+$. Still the vast majority of models is 1D; so both fluid and kinetic models have success[4, 5]. An example of kinetic model without collision, the integrodifferential complete plasma equation [6, 7, 8, 9], demonstrated that a region with a large charge unbalance (named sheath) appears at plasma borders, with thickness about ten λ_D, where $\lambda_D = (\varepsilon_0 T_0/e^2 N_0)^{1/2}$ is the Debye length.

After beam extraction, electrons are deflected away by the extraction magnetic field: in the extraction plasma we have both the filter and the extraction magnetic field, typically from 20 to 100 G; let the (local) magnetic field direction be along x so that $\mathbf{A} = \hat{y} A_y(z)$ is the vector potential. Electric field is small inside plasma and rapidly reaches values comparable to the applied field in the sheath, so that the example $\phi = (T_0/e)\{\exp(z) - 1\}$ will be used later.

To discuss transport of any particle a we find convenient to use scaled variables: velocities in units of $c_a = \sqrt{T_0/m_a}$, momentum p_z in unit of $m_a c_a$, density in units of N_0, currents in units of $q_a N_0 c_a$ with q_a the particle charge, potential $v = q_a \phi/T_0$ and vector potential $a = q_a A_y/\sqrt{m_a T_0}$. Let $\partial_z a = a_{,z}$ be the partial z-derivative of a. For electrons, $c_e = \sqrt{T_0/m_e}$ and $v = -e\phi/T_0 \equiv u$. Without collisions, the scaled Hamiltonian H/T_0 is

$$h(z, p_z; P_y) = \tfrac{1}{2} p_z^2 + \tfrac{1}{2}(P_y - a(z))^2 + v(z) \qquad (1)$$

where P_y is a canonical momentum component and is conserved; the motion along x is separable and thus is ignored here; lower letter are used for mechanical momentum as $p_y = P_y - a(z)$.

Let $P_y = P_y^D(z)$ correspond to trajectories with z constant, that is $dz/ds = 0$ with $s = c_a t$ the parameter conjugated to the Hamiltonian. Since $dz/ds = h_{,p_z} = p_z$ and $-h_{,z} = dp_z/ds = 0$ from Hamilton equation, we have

$$P_y^D(z) = a(z) + \{v_{,z}(z)/a_{,z}(z)\} \qquad (2)$$

which is constant. Then $dy/ds = P_y^D - a(z) = \{v_{,z}(z)/a_{,z}(z)\} = v_D$ is also constant and is named drift velocity.

Let $p = P_y - P_y^D(z)$ a trajectory parameter measured when $p_z = 0$ (called motion reversal point) is obtained at some z; p is a measure of deviation from drift orbit, as shown in fig 1. At any other point z' on that trajectory we have $p_z = \pm\sqrt{2V_{eff}}$, where the effective potential shown in fig 2 is

$$V_{eff}(z, z', p) = -\bar{v} + \tfrac{1}{2}(\bar{a})^2 + v_D(z)\bar{a} + p\bar{a} \qquad (3)$$

with the shorthand for potential differences $\bar{a} = a(z) - a(z')$ and $\bar{v} = v(z) - v(z')$

A kinetic model starting from Vlasov equation with a thermalized scatterer collision term and eq. 1 Hamiltonian was reduced to a 1D integral equation[10] and fully solved for H$^-$ ions. Self-consistent solutions for u and a fluid transport model were also found[11].

As alternatives, Monte Carlo simulations are much widely used to investigate plasma behavior, and resolution down to λ_D is becoming possible with parallelized computing and variance reduction techniques[12]; some regularized sampling techniques seems

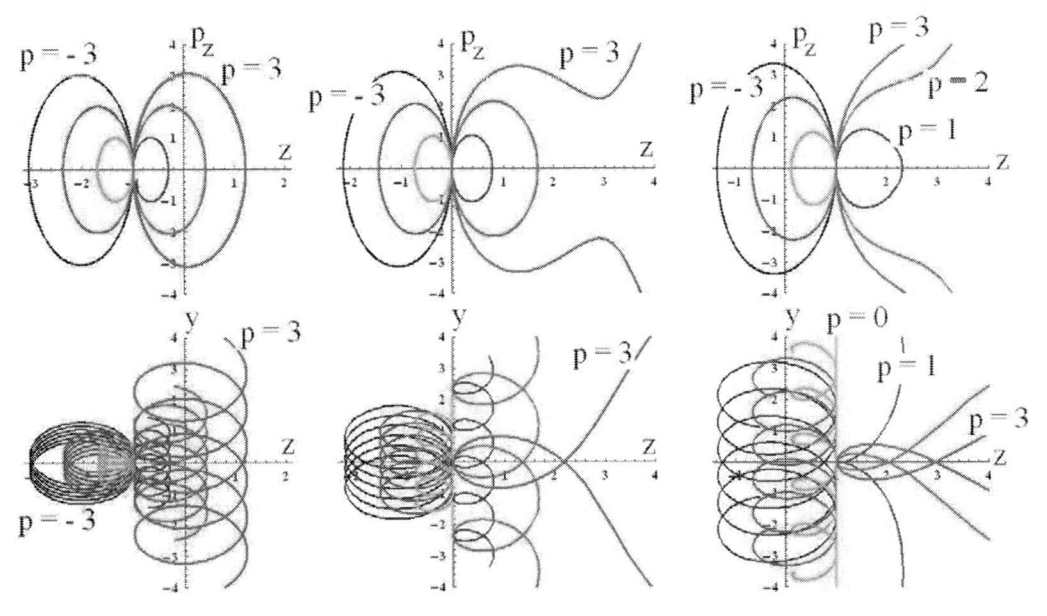

FIGURE 1. The (z, p_z) phase space orbits for several reversal points and p values; here $a(z) = 3z$ and $v(z) = 1 - \exp z$; lower plots shows the (z, y) orbit.

also promising. From 2D simulations, it is evident that electron starting angle is large, but is somewhat dominated by subsequent large deflection and acceleration in the extraction gap. Moreover, depending on the geometry of extraction hole, the emission angle can contribute to a reduction of the observed electron current. As for negative ions, starting angle is smaller, but seems to affect the following ion acceleration. A self-consistent solution of the collisional presheath and sheath transition is thus an important ingredient of any beam simulation code.

In the next section the Vlasov model is reviewed and improved, with reduction to 1D integrodifferential equations. A following section discuss possible approximated solutions with simple techniques based on Fourier Transform. A sheath model is also discussed in the last section.

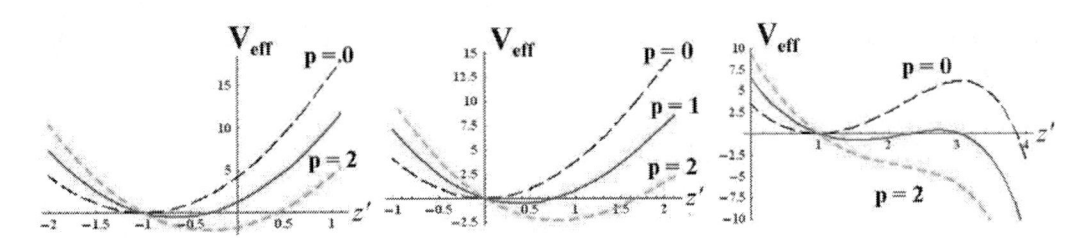

FIGURE 2. The effective potential for several reversal points and p values; other condition as in Fig 1

KINETIC MODEL

In extraction plasmas, scattering of electrons and H$^-$ from gas is typically the major collision contribution, so that a fixed scatterer model is a reasonable approximation with λ_i the mean free path (for elastic and pseudoelastic collisions) and λ_a is the attenuation length that is the mean path before destruction of a H$^-$ ion. Phase space density $f(z,p_z;P_y)$ of each species (integrated on x) satisfies a Vlasov equation $p_z f_{,z} - h_{,z} f_{,p_z} = \mathscr{S} + \mathscr{C}$ with the source $\mathscr{S}$ and collisional term $\mathscr{C} = d_t f / v_a$ approximated by

$$\mathscr{C} = |p_z|\left\{-\frac{f}{\lambda_t} + \frac{g(p_z)}{c_\pi \lambda_i} g(P_y - a) \int' |p_z| f\right\} \quad , \quad \frac{1}{\lambda_t} = \frac{1}{\lambda_i} + \frac{1}{\lambda_a} \qquad (4)$$

with $\int' = \int dp_z dP_y$ and $g(x) = \exp(-x^2/2)/\sqrt{2\pi}$ and $c_\pi = \sqrt{2/\pi}$. We neglect λ_a^{-1} and $\mathscr{S}$ in the following, so that $\lambda_t = \lambda_i = \lambda$.

In a 3D model, it will be useful to expand the velocity part of f in spherical harmonics, and to single out special direction in velocity space, as done in so-called 'deterministic neutron transport' code and models. These directions may be reduced to forward and backward in a 1D model, so that we define

$$j^+ = \int^+ p_z f, \quad j^- = \int^- |p_z| f, \quad n^- = \int^- f, \quad n^+ = \int^+ f \qquad (5)$$

with the shorthands $\int^+ = \int dP_y \int_0^\infty dp_z$ and $\int^- = \int' - \int^+$; we call j^+ the forward current, even if due to scaled variable used, it is a forward particle flow. It is convenient to define also

$$j_a(z) = \frac{j^+(z) + j^-(z)}{2} \quad ; \quad j_h = \frac{j^+(z) - j^-(z)}{2} = \tfrac{1}{2} j \qquad (6)$$

so that $2j_a = \int' |p_z| f$ is the total current (analogous to a neutron fluence) impinging on the scatterer as modeled by eq. 4, while $j = 2j_h$ is the net particle current (or flow). Note that j_h and j are uniform $\partial_z j = 0$, having neglected λ_a^{-1}.

The phase space density f at a point (z, p_z) may be related to points z' previously visited by the same unperturbed orbit (by integrating Vlasov equation on its characteristic lines), determined by the condition $h(z, p_z, P_y) = h(z', \eta_z, P_y)$ or shortly $h = h'$, where η_z is the value of p_z at point z'; moreover $\eta_y = P_y - a(z')$. It is convenient to trace back the z' integration until we reach a point where $\eta_z = 0$, called a motion reversal point, or we exit from plasma interval $[z_\alpha, z_\beta]$. Closed orbits have two reversal points called z^- and $z^+ \geq z^-$; in the case $p_z \geq 0$, we get

$$f(z, p_z, P_y) = f(z^-, 0, P_y) e^{-\frac{|z - z^-|}{\lambda}} + \int_{z^-}^z dz' e^{-\frac{|z-z'|}{\lambda}} \frac{2j_a(z')}{\lambda} G^+ \quad , \quad G^+ = \frac{g(\eta_z) g(\eta_y)}{c_\pi} \qquad (7)$$

for $z \geq z^+$. Formula for $p_z < 0$ requires obvoius changes as z^+ in place of z^-. The similar formula for the orbits which come from the boundaries was given in eq. 6 or Ref. [10]. This stated, we see that f is given in terms of $j_a(z')$, so the problem may be projected from phase space to the real z-axis, as explicitly done later (with great potential benefits for numerical solver memory). We note also the number of independent field variable is the number of separated component in the collision term (here one component, one field j_a).

As a technical point, G^+ the scattered particle distribution for $\eta_z > 0$ may be formally rewritten as

$$G^+(z,z',p_z,P_y) = \int_0^\infty d\eta_z |\eta_z| \delta(h(z,p_z,P_y) - h') g(\eta_z) g(P_y - a(z')) \qquad (8)$$

where the Dirac delta selects η_z consistent with the (z,p_z) orbit. Formula for G^+ differs only for the integration ranges. On the other hand, since we choose a Maxwellian based G and the Hamiltonian is constant on the orbit we may write

$$G^- = G^+ = G(z,z',p_z,P_y) = \frac{e^{-(\eta_z^2+\eta_y^2)/2}}{2\pi c_\pi} = \frac{\exp\{v(z') - h(z,p_z,P_y)\}}{2\pi c_\pi} \Theta(h - v(z')) \qquad (9)$$

When z is a reversal point $G = \exp(-\bar{v} - \frac{1}{2}[p + (a_{,z}/v_{,z})]^2)/(2\pi c_\pi)$.

Formula 7 contains a recirculation term proportional to $f(z^-,0,P)$ which can be neglected if $z_m \equiv (z^+ - z^-)/\lambda \gg 1$; the estimate $z^+ - z^- \cong 2p/a_{,z}$ shows that this holds for large p or weak magnetic fields. Otherwise the recirculation effect may be solved for with the following procedure. Applying formula 7 to $z = z^+$ we express $f(z^+,0,P_y)$ as a term proportional $f(z^-,0,P)$ plus j_a contribution; with the corresponding formula for backward part of the orbit, we express $f(z^-,0,P_y)$ as j_a contribution plus term proportional to $f(z^+,0,P)$; eliminating this, we get

$$f(z^-,0,P_y) = \int_{z^-}^{z^+} dz' S \frac{2 j_a(z')}{\lambda} G \quad , \quad S = \frac{\cosh(z' - z_c/\lambda)}{\sinh z_m/2} - \frac{\sinh(z' - z_c/\lambda)}{\cosh z_m/2} \qquad (10)$$

with $z_c = (z^- + z^+)/2$. By simple trigonometry $S = S_n + S_r$, where $S_n = 2e^{-|z_{dd}|}$ with the shorthand $z_{dd} = (z' - z^-)/\lambda$ is the non-recirculating part and $S_r = 2(\coth z_m - 1)\Theta(z_m - z_{dd})\cosh z_{dd}$ is the recirculation effect. Further detailed evaluation of the recirculation term is in progress and will be omitted for brevity.

Applying $\int^+$ to Vlasov equation we get

$$j_{a,z} = j^+_{,z} = -\frac{1}{2}(j/\lambda) - \int dP_y h_{,z} f(z,0;P_y) \qquad (11)$$

Note that $-h_{,z} = p a_{,z}$. By substituting f, using eq. 8 for G, noting that $\partial_z \Theta(h' - h) = -h_{,z} \delta(h' - h)$ and defining a correlation function

$$M(z,z') = M(\bar{a},\bar{v}) = \frac{2}{c_\pi} \int d\eta_y \int_0^\infty d\eta_z \eta_z \Theta(\eta_z^2 + 2\bar{a}\eta_y - 2\bar{v} - (\bar{a})^2) g(\eta_z) g(\eta_y) \qquad (12)$$

eq. 11 becomes a closed equation for j_a

$$j_h/\lambda = -j_{a,z} + j_B(z) + \int_{z_\alpha}^{z_\beta} dz' e^{-|z'-z|/\lambda} j_a(z') M_{,z}(z,z') \qquad (13)$$

where j_B that is due to particles injected at boundaries with speed not sufficient to climb $v(z)$ is known and is usually negligible. Correlation function M is displayed elsewhere[13]. Density n^+ of the forwardly directed particles is

$$n^+(z) = n_\alpha(z) + \int_{z_\alpha}^z \frac{dz'}{\lambda} e^{-|z'-z|/\lambda} F_R N(\bar{a}, \bar{v}) j_a(z') \tag{14}$$

where n_α is the known boundary term and $F_R \cong 1/(1 - \exp[-4(2/\pi)^{1/2}/(a_{,z}\lambda)])$ is a rough approximation of the recirculation effect; similarly for n^-. Here $N(0, \bar{v}) = c_2 e^{-\bar{v}} \text{erfc}\, \Re[(-\bar{v})^{1/2}]$ with $c_2 = \sqrt{\pi/2}$ and

$$N(\bar{a}, \bar{v}) = \int_0^\infty dp_z e^{-\bar{v} - (p_z^2/2)} \text{erfc} \frac{\bar{a}^2 - p_z^2 - 2\bar{v}}{|\bar{a}|\sqrt{8}} \tag{15}$$

INTEGRAL EQUATION SOLUTION

For a constant magnetic field, we get $\bar{a} = (z - z')/L$ where L is the Larmor radius $\sqrt{m_e T_0}/e|B_x|$. In the source, $\bar{a} \gg \bar{v}$ typically holds for electrons, so that M may be approximated as

$$M(z, z') \cong \exp\{-\tfrac{1}{2}[v(z) - v(z')]\} \text{erfc} \frac{|z' - z|}{\sqrt{8}L} \tag{16}$$

Defining $j_b(z) = e^{v(z)/2} j_a(z)$ and $m_b = e^{\bar{v}/2} M_{,z}$ we have

$$\frac{\partial j_b}{\partial z} + \frac{j_h e^{v(z)/2}}{\lambda} = \int_{z_\alpha}^{z_\beta} \frac{dz'}{\lambda} e^{-\frac{|z'-z|}{\lambda}} m_b(z, z') j_b(z') \tag{17}$$

$$m_b = -\tfrac{1}{2} v_{,z} \text{erfc} \frac{|z''|}{L\sqrt{8}} + e^{-(z''/L)^2/8} \frac{\text{sign}(z'')}{\sqrt{2\pi}L} \tag{18}$$

with $z'' = z' - z$. Most remarkably, when $v_{,z}$ is a rational function of z, thanks to the m_b form, the integrodifferential eq. 17 can be converted to an ordinary differential equation for $j_F(k) = \mathscr{F} j_b(z)$ with $\mathscr{F}$ the Fourier transform. For example, with a classical barrier $v = -p_2 z^2$ with $p_2 > 0$, we get

$$\frac{j_h}{\lambda} \frac{e^{-k^2/2p_2}}{p_2^{1/2}} = (ik + W_2^F) j_F + p_2 i \partial_k (1 - W_1^F) j_F \tag{19}$$

which is exactly solvable; here

$$W_1^F(k) = \int_{-\infty}^\infty \frac{dz''}{\lambda} e^{-\frac{|z''|}{\lambda}} \text{erfc} \frac{|z''|}{L\sqrt{8}}, \quad W_2^F(k) = \int_{-\infty}^\infty \frac{dz''}{\lambda} e^{-\frac{|z''|}{\lambda}} e^{-(z''/L)^2/8} \frac{\text{sign}(z'')}{\sqrt{2\pi}L} \tag{20}$$

Defining $j_c = i j_F$, the coefficients of eq. 19 are $A = j_h e^{-k^2/2p_2}/\lambda p_2^{1/2}$ and $p = p_2(1 - W_1^F)$ and $q = k - i W_2^F - p_2 W_{1,k}^F$. Its general solution j_c and integrating factor Q are

$$j_c = \frac{1}{Q} \left[c_1 + \int dk \frac{AQ}{p} \right], \quad Q = |1 - W_1^F| \exp \int \frac{dk}{p_2} \frac{k - i W_2^F}{1 - W_1^F} \tag{21}$$

Since the closed form of eq. 20 are known, its numerical quadrature is straightforward. Asymptotic expansion for $k \to \pm\infty$ shows most of solution behaviour: note that

$$W_1^F = \left[\frac{2}{\lambda^2} + \frac{c_\pi}{\lambda L}\right]\frac{1}{k^2} + O(k^{-4}) \quad , \quad iW_2^F = \frac{c_\pi}{\lambda L k}\left[1 + \frac{1}{k^2}\left(\frac{1}{4L^2} - \frac{1}{\lambda^2}\right) + O(k^{-4})\right]$$

so that

$$Q = |1 - W_1||k|^{p_3} e^{k^2/2p_2}\{1 - (\lambda^4 p_2 k^2)^{-1} + O(k^{-4})\}$$

with $p_3 = 2/p_2\lambda^2$. Finally dropping some terms

$$j_c = e^{-k^2/2p_2}\left[\frac{c_1}{|k|^{p_3}} + \frac{k}{p_3+1}\frac{j_h}{\lambda p_2^{3/2}} + O(k^{-2})\right] \tag{22}$$

Matching the asymptotic solution and the numerical solution of eq. 19 produces reasonably accurate fits of $j_c(k)$; with $\mathscr{F}$ inversion, we get solutions like

$$j_a(z) \cong e^{e_1 p_2 z^2}(d_1 - e_2 j_h z + \ldots) \quad , \quad e_1 = 2p_2 L^2/q_2 \quad , \quad e_2 = 4c_\pi L/\lambda^2 q_2^{3/2} \tag{23}$$

with $q_2 = 1 + 4L^2 p_2$ and d_1 an integration constant. Final solution grows slowly for $z \to \infty$ (anyway, less rapidly than e^{-v}); solution is fairly accurate also for large $p_2 z^2$ values, as verified numerically in fig 3.A.

In the example $v(z) = v_1 z$ we also have the translational symmetry since E_z is uniform, and Fourier transformed equation is

$$j_h \sqrt{2\pi}\, \delta(k - \tfrac{i}{2}v_1) = D_F(k) j_F(k) \tag{24}$$

with $D_F = \lambda\{ik + W_2^F + \tfrac{1}{2}v_1(1 - W_1^F)\}$. Its general solution and inversion of $\mathscr{F}$ give $j_b(z) = \sum_{n=0}^{3} c_n \exp(-ik_n z)$ where $n = 0$ is the inhomogeneous term, that is $k_0 = \tfrac{i}{2}v_1$ and $c_0 = j_h/D_F(\tfrac{i}{2}v_1)$. The other k_n are the roots of $D_F(k) = 0$; we are able to prove that no solution is nonzero and real and we found 3 solutions on the imaginary axis. In detail, $k_2, k_3 \cong \pm i/L$, so they correspond to very sharply decaying modes, which require

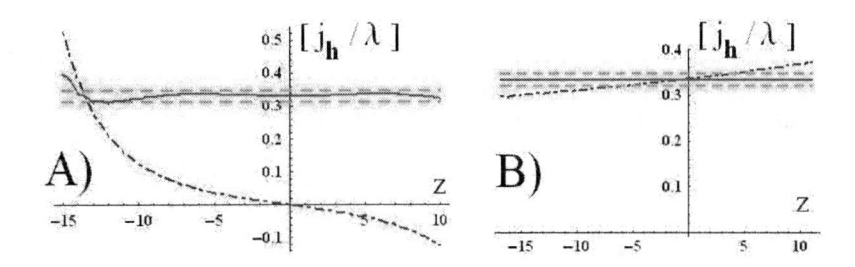

FIGURE 3. A) Dot-dashed line: $j_a/100$ for $d_1 = 0$ and $j_h = 1$ and parameters $L = 1$, $\lambda = 3$ and $p_2 = 0.04$ (from eq. 23 keeping up to z^5 term); two dashed parallel lines: 96 % and 104 % of LHS of eq. 13; solid line: RHS of eq. 13 using this j_a and eq. 16 ; B) As before, but dot-dashed line: $j_a/100$ for $c_1 = 10$ and $j_h = 1$ from eq. 26 and parameters $L = 1$, $\lambda = 3$ and $v_1 = -0.12$.

huge input currents to be maintained, so we drop them here. The solution k_1 happens to be of v_1 order, and can be computed from Taylor expansions $W_1 = W_{10} + O(k^2)$ and $W_2 = W_{21}k + O(k^3)$; we get

$$k_1 = \tfrac{i}{2}v_1(1 - W_{10})/(1 - iW_{21})$$
$$W_{21} = -ix\sqrt{32/\pi} + 8ix^2 e^{2x^2}\operatorname{erfc}(\sqrt{2}x)$$
$$W_{10} = 2 - 2x^2 e^{2x^2}\operatorname{erfc}(\sqrt{2}x) \qquad (25)$$

with $x = L/\lambda$. Final solution is

$$j_a(z) = \frac{2j_h/\lambda}{v_1(iW_{21} - W_{10})} + c_1 e^{s_1 z} \qquad (26)$$

with c_1 an integration constant and $s_1 = -\tfrac{1}{2}v_1 - ik_1 = v_1 R_1$ with the ratio $R_1 = \tfrac{1}{2}(W_{10} - iW_{21})/(1 - iW_{21})$. As apparent from fig 3.B, eq. 26 solution is exact. The ratio R_1 is clearly the modification to Maxwell density distribution in this transport problem. Moreover note that when $c_1 = 0$ we have an uniform plasma subjected to a constant electric field, so that results can be compared to Fick law (usual diffusion theory) $j = -q_e N_e \mu E_z$ with μ the mobility coefficient (positive, non scaled, in m^2/(V s) units). Now $j = 2j_h$ and, thanks to eq. 14, $n = n^+ + n^- = R_{n/j} j_a$ (in scaled variables), where $R_{n/j} \cong \sqrt{2\pi}$ is a fixed number. We get $\mu = e\lambda[(W_{10} - iW_{21})/R_{n/j}]/(m_e c_e)$ where the square bracket factor is due to magnetic field.

A SHEATH-PRESHEATH MODEL

The Poisson equation becomes

$$\lambda_D^2 u_{,zz} = n_{H^+} - n_e - n_n \quad , \quad \lambda_D = (\varepsilon_0 T_0/e^2 N_0)^{1/2} \qquad (27)$$

where $n = N/N_0$ are scaled densities with n_n for H$^-$ ions and n_{H^+} for protons H$^+$ (or the sum of all positive ions). Even if a diffusion model is appropriate for the confined protons[11], we approximate it with $n_{H^+} = n_0 e^u$ as usual, where $n_0 \cong 1$, and we set up a fluid model for electrons, still approximately retaining the effect of backward and forward currents.

Let $\mathbf{v}^F$ be the fluid velocities for electrons (or for H$^-$ ion) in scaled units, that is, $\mathbf{v}^F = \langle \mathbf{v} \rangle / c_a$ and drop the F superscript as usual. Fluid eq. for electrons are

$$v_z v_{z,z} + (n_{e,z}/n_e) = -u_{,z} - (v_y/L) - (v_e^m/c_e)v_z$$
$$v_z v_{y,z} = +(v_z/L) - (v_e^m/c_e)v_y \qquad (28)$$

where L is the particle Larmor radius and the collision frequency v^m may depend on electron speed $v = |\mathbf{v}|$. Equation for H$^-$ is obtained by changing subscript 'e' into 'n'.

Collision cross sections have complicated details, but they can be roughly modeled or fitted as sum of terms like $\sigma = \sigma_n v_R^{-n}$ where v_R is the relative velocity and here n is an index [14]. For Coulomb collision of e with H$^+$ $n = n^c \cong 3.9$; for collisions with H$_2$

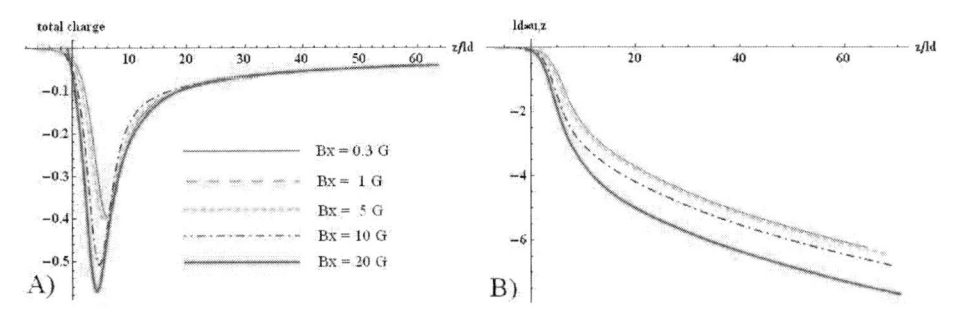

FIGURE 4. A) the (scaled) total charge vs z/λ_D for several B_x and B) The (scaled) electric field $\lambda_D u_{,z}$

molecules, $n = 1$ is a fair fit of experimental data from 0.5 to 10 eV[15]. The collision frequency is $<v_R \sigma> N_s$ where N_s is the scatterer density, so that

$$v_e^m \equiv k_g g(v) + k_c h(v) = k_g + \frac{k_c n_{H^+}}{((c_{H^+}/c_e)^2 + 1 + v^2)^{\alpha_c}} \quad (29)$$

where the gas term is $k_g = N_g \sigma_1$ and in the Coulomb collision term $k_c = N_0 \sigma_{n^c}$ and $\alpha_c = (n_c - 1)/2 = 1.45$. The denominator of eq. 29 is a power of the average square relative velocity, summing the effects of H^+ thermal motion, electron thermal motion and electron fluid motion. From collision data, $\sigma_1 = 1.01 \times 10^{-13}$ m³/s and $\sigma_{n^c} = 1.25 \times 10^{-10}$ m³/s. We take $N_0 = 3.3 \times 10^{17}$ m⁻³ and $N_g = 7.7 \times 10^{19}$ m⁻³ for a typical NIS. Equation 29 holds for H^- with obvious changes and from collision data, $\sigma_1 = 1.36 \times 10^{-15}$ m³/s and $\sigma_{n^c} = 4.29 \times 10^{-12}$ m³/s with $\alpha_c = 1.3$.

Since no absorption of electrons (or ionization) is considered (as in the previous section), we do not need the particle balance equation, discussed elsewhere[11], and simply take j_z as a parameter j. Note that using $n_e = j_z/v_z$ to close eqs. (27-28) is incorrect, since backward and forward directed particles sums in n_e, but subtracts in j_z.

To close eqs. (27-28), we need a robust relation between n_e, j and v_z^F. As a first estimate, consider a distribution $f(v_z)$ of v_z with variance 1 and mean $v_f = v_z^F$, that is

$$f(v_z) = n_e \exp[-(v_z - v_f)^2/2]/\sqrt{2\pi} \quad (30)$$

computing n^+ and j^+ by integration on $v_z > 0$, we get

$$\frac{n^+}{j^+} \cong \tfrac{1}{2}\left\{-v_f + \sqrt{v_f^2 + 4}\right\} \equiv \frac{1}{v^+(v_f)} \quad , \quad \frac{n^-}{j^-} \cong \tfrac{1}{2}\left\{v_f + \sqrt{v_f^2 + 4}\right\} \equiv \frac{1}{v^-(v_f)} \quad (31)$$

which we take as the definition of $v^{\pm}$. Rearranging, and remembering that $j^+ = j^- + j$ we get

$$n_e = \frac{j^-}{v^-} + \frac{j^+}{v^+} = \tfrac{1}{2}\left\{\sqrt{v_f^2 + 4} - v_f\right\}j + j^- \sqrt{v_f^2 + 4} \quad (32)$$

The first term is a nicely regular contribution of the net current (with no singularity at $v_f \to 0$); the second term also contains the contribution of trapped or reflected orbits, which may be the greater term in the plasma and, as better shown by eq. 14, has a nonlocal dependence on j_a (relative difference of j_a and j^- is small in magnetized plasma). Therefore we replace eq. 32 estimate by

$$n_e = \tfrac{1}{2}\left\{\sqrt{v_z^2+4}-v_z\right\}j+n_a \quad , \quad n_a = \int dz' R(z,z')\, j_a(z') \qquad (33)$$

where $R \cong NF_R$. From the kinetic model (and its solution for $v_1 \to 0$ and $L \ll \lambda$) for electrons, we approximate

$$R(z,z') = \frac{R_{n/j}}{4\Lambda} e^{-|z'-z|/\Lambda} e^{-\bar{v}/2} \quad , \quad \Lambda = L\sqrt{\frac{\pi}{2}} \qquad (34)$$

with $R_{n/j} \cong \sqrt{2\pi}$. To relate j and j_a, we note that j_a decreases when the field $u_{,z}$ increases in eq. 26, anyway rigorously valid only for constant fields; extrapolating that result to present conditions, we set here $j_a = j\lambda/(2L^2|v_{,z}|+k_e)$ with k_e a constant to match initial conditions.

Starting conditions at $z = z_{st}$ must represent a point in the quasineutral plasma. Some conditions are obvious: for example $n_0 = 1$ and $u(z_{st}) = 0$, so that $n_{H+}=1$; and k_e is adjusted so that $n_e(z_{st}) = 1$; we also set $v_z(z_{st}) = j$. Other conditions are chosen to avoid oscillations: $v_y(z_{st}) = c_e v_z/Lv_e^m$ and $u_{,z} = -v_z[(v_e^m/c_e)+(c_e/v_e^m L^2)]$; in other words, RHS of eq. 28 be zero at start.

Numerical simulations with eqs. 27-34 can cover a variety of plasma conditions and are is progress. An implementation with commercial and reliable ODE (ordinary differential equation) solvers was documented elsewhere[13], with the approximate conversion of the integral eq. 33 in ODE. In fig 4 example, we set $R_j = 10$ to better see the electron space charge effect; note that an additional amount of extraction field, proportional to the applied magnetic field, is necessary to drive the same current. Moreover, the angle between electron beam and extraction field is verified to be large as an effect of magnetic field, especially at extraction.

REFERENCES

1. M. P. Stockli et al., *Rev. Sci. Instrum.* **81**, 02A729 (2010)
2. P. Sonato et al., *Fusion Eng. Des.* **84**, 269 (2009).
3. J. C. Rienstra-Kiracofe et al., *Chem. Rev.* **102**, 231-282 (2002).
4. M. S. Benilov, *Plasma Sources Sci. Technol.* **18**, 014005 (2009), and references herewithin.
5. R. N. Franklin, *J. Phys. D* **36**, R309 (2003).
6. L. Tonks, I. Langmuir, *Phys. Rev.* **34**, 826, (1929).
7. S. A. Self, *Phys. Fluids* **6**, 1762 (1963).
8. A. T. Forrester, *Large Ion Beams*, John Wiley, NY, 1996.
9. K-U. Riemann, *J. Phys. D* **24**, 493 (1991).
10. M. Cavenago, *Rev. Sci. Instrum.* **79**, 02B709 (2008).
11. M. Cavenago, *Rev. Sci. Instrum.* **81**, 02B501 (2010).
12. D. Wunderlich et al., *Plasma Sources Sci. Technol.* **18**, 045031 (2009).
13. M. Cavenago, "Transport equation solution" in Mathematica Italia User Group Meeting 2010, Milano, (Adalta, http://www.adalta.it, Arezzo, CDROM, ISBN 978-88-96810-00-2); includes some preliminary code (and animated Mathematica 7 (TM) notebook).
14. M. A. Lieberman and A. J. Lichtenberg, *Principles of Plasma Discharges and Material Processing*, John Wiley, New York, 1994.
15. http://www-amdis.iaea.org/ALADDIN/

Monitoring Surface Condition of Plasma Grid of a Negative Hydrogen Ion Source

M. Wada[a], T. Kasuya[a], T. Kenmotsu[b], S. Tokushige[a]

[a] Graduate School of Engineering, Doshisha University, Kyotanabe, 610-0321 Kyoto, Japan
[b] Faculty of Life and Medical Sciences, Doshisha University, Kyotanabe, 610-0321 Kyoto, Japan

Abstract. Surface condition of a plasma grid in a negative hydrogen ion source is controlled so as to maximize the beam current under a discharge operation with introducing Cs into the ion source. Photoelectric current induced by laser beams incident on the plasma grid can produce a signal to monitor the surface condition, but the signal detection can be easily hindered by plasma noise. Reduction in size of a detection electrode embedded in the plasma grid can improve signal-to-noise ratio of the photoelectric current from the electrode. To evaluate the feasibility of monitoring surface condition of a plasma gird by utilizing photoelectric effect, a small experimental setup capable of determining quantum yields of a surface in a cesiated plasma environment is being assembled. Some preliminary test results of the apparatus utilizing oxide cathodes are reported.

Keywords: Negative Ion Source, Work Function, Plasma Wall Interaction
PACS: 52.27, 52.50, 52.40

INTRODUCTION

Negative ion sources for neutral beam heating system of large scale fusion experiment devices are operated with Cs introduced into H_2 plasmas [1]. In these sources, temperature of the plasma grid is controlled to enhance hydrogen negative ion (H^-) current [2]. As the measured work function of the plasma grid had changed in accordance with the temperature of the plasma grid, the observed enhancement of H^- current by the temperature control of the plasma grid in a Cs seeded ion source was attributed to H^- formation at the low work function surface of the grid [3].

The importance of temperature control of the plasma grid of an ampere-class H^- source had been pointed out by Okumura et al. [4]. They had reported that both H^- current and electron temperature near the plasma grid showed a hysteresis behavior when they changed the temperature of the plasma grid. Once the temperature of the plasma grid in their ion source had been elevated up to more than about 490 K, the H^- current had kept a larger value even after the temperature was reduced to about 450 K. As this change in H^- current can be correlated to the difference in surface conditions including work function, search for a proper method to monitor the Cs recycling in the ion source has been considered important from the view point of finding the best condition for H^- source operation. Several experimental investigations have been already started to quantify Cs flux in the ion source [5] and adsorption of atoms and molecules on the plasma grids [6]. Establishing a method to properly monitor surface conditions in these fundamental experiments, and those in real size ion source experiments should deepen the understanding of the process governing the H⁻ current enhancement due to Cs injection into H_2 discharges.

A small experimental setup is being assembled to measure photoelectric work function of plasma electrode material immersed in a hydrogen plasma environment. The surface conditions of metal sample surfaces simulating a plasma grid of a H⁻ source are investigated by measuring photoelectric current induced by an external light source. A semiconductor laser based system was used to measure work function of an oxide cathode [7], and the result had shown that the work function of oxide cathodes changed by running a discharge [8]. The present status of the experimental research to characterize the surface conditions of a cesiated metal in H_2 discharge is described.

EXPERIMENTAL APPROACH

Experimental Device for Surface Condition Evaluation

The experimental setup shown in Fig. 1 has been prepared to measure photoelectric work function of a sample surface in vacuum and in hydrogen plasma environment. A small stainless steel chamber of 25 cm³ volume served as a vacuum container for metal samples which are used to simulate the plasma grids of H⁻ ion sources. The surface of a sample can be heated with infrared radiation, and the temperature is measured with an infrared thermometer. Instead of a SAES$^{(R)}$ getter, Cs is evaporated onto the surface with an ordinary Cs oven used for compact negative ion sources [9].

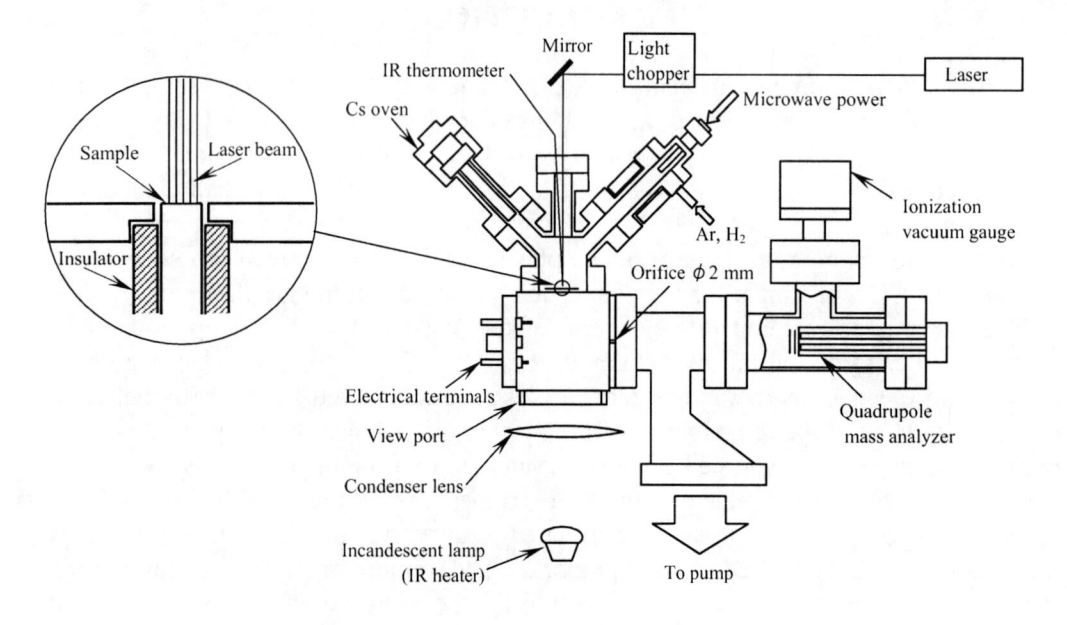

FIGURE 1. A schematic diagram of the experimental setup to measure photoelectric current of a sample in plasma.

An 18 mm diameter 40 mm long compact microwave plasma source is attached in front of the sample to expose the surface to a discharge plasma. Both microwave power and discharge gases are supplied from one end of the ion source, and a steady state discharge is maintained in a linear magnetic field created by permanent magnets. The plasma diffuses out of the magnetized region to irradiate the surface of sample. A 230 ℓ/s turbo-molecular pump evacuates the chamber through a 2 mm diameter orifice to realize pressure difference between the chamber and the downstream. The ultimate pressure downstream of the orifice is below 3×10^{-6} Pa. A quadrupole mass analyzer (QMA) monitors the gas emission from the system through the orifice.

Laser lights at 405, 525, and 633 nm wavelengths pass through a vacuum window facing against the sample. Maximum powers for these lasers are 30 mW, 50 mW and 30 mW for 405 nm, 525 nm and 633 nm wavelengths, respectively. The diameters of the lasers at the sample surface are from 1 to 3 mm, and the sample surface exposed to both laser beam and plasma is a circle of 3 mm diameter. The sample is electrically connected to the input terminal of a lock-in amplifier and biased at the same potential with the surrounding electrode simulating the plasma grid surface. With respect to the plasma potential, the sample surface is biased several volts negative and the emitted photoelectrons induced by lasers are accelerated toward the plasma across the sheath. Photoelectric current is detected phase sensitively by amplitude modulation of a laser light with a rotating wheel light chopper. A light power meter measures the incident laser power. Photoelectric yields are obtained by simply calculating the number ratios of emitted photoelectrons to incident photons.

Preliminary Measurement with an Oxide Cathode

The present Cs oven does not realize precise control of Cs evaporation, and the system is easily overloaded with Cs. As it is illustrated in Fig. 1, photoelectron emission sample is electrically isolated from the surrounding electrode with a small insulator. The excess Cs covers the surface of the insulator to make the photoelectric signal measurement impossible. Thus, the system performance had been checked by utilizing a commercially available alkaline earth oxide cathode, which did not require usage of Cs oven.

The oxide cathode was taken out from a glass container purged with Ar and was mounted in the experimental device. During this process the cathode was transported in the atmosphere and contaminated by O_2 and water in air. When the cathode was heated to bake out, peaks at mass numbers 2, 12, 13, 14, 15, 16, 17, 18, 28, 40 and 44 were observed in the mass spectrum. These peaks correspond to H_2, C, CH, CH_2, CH_3, O, OH, H_2O, CO, Ar, and CO_2. Quantity of water molecule emission was reduced as the system was baked more than several hours. Mass peaks corresponding to Ca and Sr temporarily increased and then decreased rapidly after raising the heating power to the cathode. No sizable emission of Ba was observed despite Ba was supposed to be the main constituent of the alkaline earth oxide cathode.

The photoelectric current from the oxide cathode was measured with a semiconductor laser at 405 nm wavelength in Ar plasma, as the noise excited by a H_2 plasma was too large to detect photoelectric current. The measured quantum efficiency was in the order of 10^{-7} at the largest which corresponded to the signal level of the order of 10^{-9} A for 30 mW laser power. Because of this small quantum yield, signals from other wavelength lasers near the threshold energy were not sufficient to determine the quantum yields. The quantum yield was dependent upon the cathode temperature, and showed a hysteresis behavior as shown in Fig. 2. The photoelectric current was smaller when the temperature was increased and larger when the temperature was decreased in the cathode temperature range from 370 to 450 K. At temperature higher than 600 K, the photoelectric signal was noisy and sudden changes in quantum yield were occasionally observed.

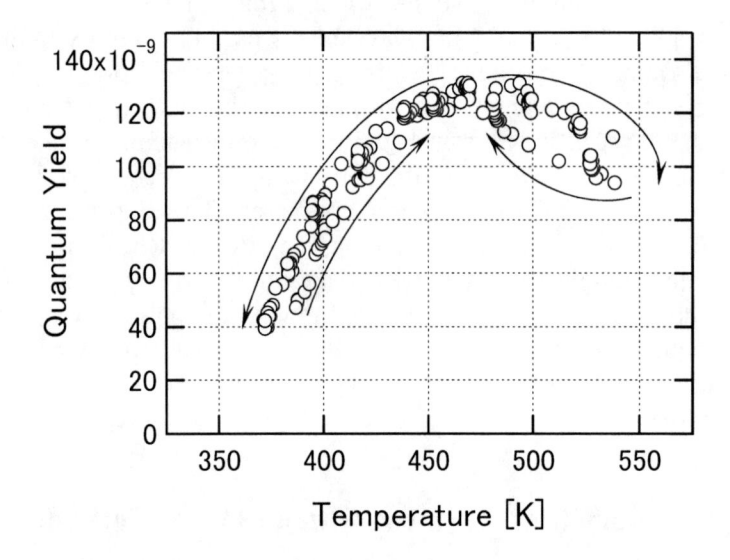

FIGURE 2. Photoelectric current from an oxide cathode in an Ar discharge by irradiation of 405 nm laser.

Below 450 K, the photoelectric yield showed a hysteresis with the magnitude larger during decrease in temperature than the magnitude during increase in temperature. A similar characteristic for electron emission from oxide cathode was observed, as shown Fig. 3. The electron emission from the oxide cathode was higher when the temperature of the cathode was elevated from a cold condition. The work function can be determined from the slope of the Richadson plot, which did not change largely in the temperature range from 600 to 700 K. The data in Fig. 3 had to be taken by turning off the Ar plasma, as the electron emission signal from the cathode became undetectable due to discharge noise in microwave plasma. Thus, Ar flow was also terminated and the QMA was turned on to analyze mass spectra of the gas in the device.

During the time the cathode was heated, gas emission of CO and that of CO_2 were found predominant. Emissions of these gases suggest that the surface of the oxide cathode had adsorbed excessive O_2 and CO before the experiment. The hysteresis characteristics had become less pronounced as the oxide cathode had been operated at higher temperature. The hysteresis characteristics appeared again when the cathode was kept in vacuum under an unheated condition for more than several days. The adsorption of excessive O_2 and CO on the surface of oxide cathode is probably the reason for observing decrease in both photoelectric yield and thermionic electron emission during the initial heating up process.

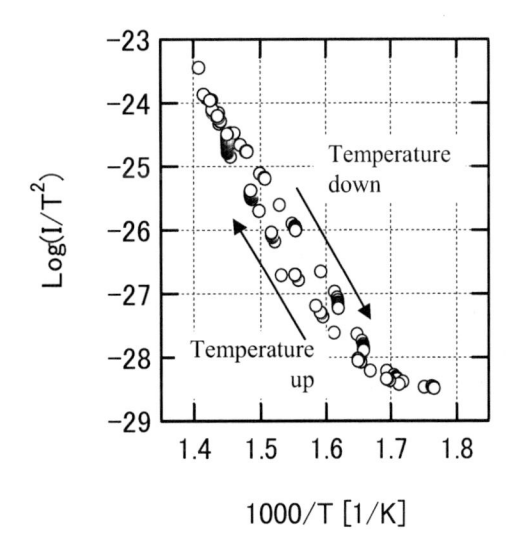

FIGURE 3. A Richardson plot for an oxide cathode showing a hysteresis behavior.

IMPROVMENTS OF MEASUREMENT SYSTEM

Photoelectric work function measurement under a plasma environment requires thorough elimination of plasma noise as quantum efficiency is quite low. This is the reason why most of the photoelectric measurements for the plasma grid surface of a large ion source have been conducted by turning off the plasma [10]. However, the Cs coverage and also the amount of adsorbed hydrogen onto the surface may be different under hydrogen plasma exposure. Particularly, excessive adsorption of hydrogen onto the plasma grid can elevate the work function and reduce the H⁻ production depending upon the coverage of Cs [11]. Or, it may not change the work function but changes the H⁻ production only by decrease in particle reflection coefficient. Thus, in-situ measurement of surface under exposure to a hydrogen plasma is indispensable to clarify the actual surface condition of the plasma grid during ion source operation.

The measurement of species and adsorbed amount of gas molecules on the sample surface is also difficult. In the present experimental configuration, gas emission from the surface cannot be monitored directly, because the sample surface does not face the QMA ionizer. Also, the present infrared radiation heating is not enough for some samples having high reflectance against infrared light. Thus, two more separate experimental systems are being prepared. Instead of measuring photoelectric work function and desorption from the surface at the same time, a target structure is moved in a vacuum system to do these measurement separately. In this design, the metal sample is mounted on a rotating motion feed through and can be directly heated by electric current. Surface Cs deposition and successive hydrogen plasma radiation will be realized by facing the target to the plasma source and a Cs oven, while it will be rotated to face the window for photoelectric measurement and to face QMA ionizer to measure a thermal desorption spectrum. These modifications, together with the system improvement so as to measure the H⁻ current through an extraction hole during hydrogen plasma exposure are being made. Also, another system composed of a 100 mm diameter 120 mm long magnetic multicusp ion source equipped with copper gasket vacuum seals is assembled to examine the surface condition of a plasma electrode in a denser plasma environment. This ion source can be attached to the present laser photoelectron measurement system.

ACKNOWLEDGMENTS

This work was supported in part by a Grant in Aid of MEXT (Ministry of Education, Culture, Sports, Science and Technology) of Japan for the Research Center of Applied Electromagnetic Energy of Doshisha University, as a part of MEXT's program to support Japanese private universities for establishing research centers of advanced sciences. The work was also supported in part by Research Center for Interfacial Phenomena at Doshisha University and LHD joint research program at the National Institute for Fusion Science.

REFERENCES

1. M. Hanada, M. Kamada, N. Akino, N. Ebisawa, A. Honda, M. Kawai, M. Kazawa, K. Kikuchi, M. Komata, K. Mogaki, K. Noto, K. Ohshima, T. Takenouchi, Y. Tanai, K. Usui, H. Yamazaki, Y. Ikeda, and L. R. Grisham, *Rev. Sci. Instrum.* **79**, 02A519 (2008).
2. D. Boilson, A.R. Ellingboe, R. Faulkner, R.S. Hemsworth, H.P.L. de Esch, A. Krylov, P. Massmann, L. Svensson, *Fusion Engg. Dsgn.* **74** (2005) 295-298.
3. T. Morishita, M. Kashiwagi, M. Hanada, Y. Okumura, K. Watanabe, A. Hatayama, M. Ogasawara, *Jpn. J. Appl. Phys.* **40**, (2001), 4709-4714.
4. Y. Okumura, M. Hanada, T. Inoue, H. Kojima, Y. Matsuda, Y. Ohara, M. Seki, and K. Watanabe, "Cs mixing in the multi-ampere volume H⁻ ion source", Production and Neutralization of Negative Ions and Beams, AIP Conference Proceedings, No. 210, (1990) 169-183.
5. M. Froschle, R. Riedl, H. Falter, R. Gutser, U. Fantz, the IPP NNBI team, *Fusion Engg. Dsgn.* **84** (2009) 788-792.
6. U. Fantz, R. Gutser and C. Wimmer, *Rev. Sci. Instrum.* **81**, 02B102 (2010).
7. M. Wada, S. Gotoh, S. Kurumada, *J. Plasma Nucl. Fusion Res.* SERIES, 8, (2009) 1366-1369.
8. H. Kozakura, S. Gotoh, T. Kasuya, M. Wada, "Photoelectric work function measurement of an electrode in a low pressure Hg discharge lamp", Proceedings of 11th International Symposium on the Science and Technology of Light Sources, (2007) 379-380.
9. A. Taniike, M. Sasao, A. Fujisawa, H. Iguchi, Y. Hamada, J. Fujita, M. Wada, Y. Mori, *IEEE Trans. Plasma Sci.* **22**, (1994) 430-434.
10. K. Shinto Y. Okumura, T. Ando, M. Wada, H. Tsuda, T. Inoue, K. Miyamoto, A. Nagase, *Jpn. J. Appl. Phys.* Part1, **35**, (1996) 1894-1900.
11. C. A. Papageorgopoulos and J. M. Chen, *Surf. Sci.* **39** (1973) 283-312.

Effect of Ion Escape Velocity and Conversion Surface Material on H⁻ Production

O. Tarvainen[a,b], T. Kalvas[a], J. Komppula[a], H. Koivisto[a], E. Geros[b], J. Stelzer[b], G. Rouleau[b], K.F. Johnson[b] and J. Carmichael[c]

[a]*University of Jyväskylä, Department of Physics, P.O. Box 35 (YFL) 40500 Jyväskylä, Finland*
[b]*Los Alamos National Laboratory, P.O. Box 1663, Los Alamos, New Mexico, 87545, USA*
[c]*Oak Ridge National Laboratory, P.O. Box 2008, Oak Ridge, Tennessee, 37831, USA*

Abstract. According to generally accepted models surface production of negative ions depends on ion escape velocity and work function of the surface. We have conducted an experimental study addressing the role of the ion escape velocity on H⁻ production. A converter-type ion source at Los Alamos Neutron Science Center was employed for the experiment. The ion escape velocity was affected by varying the bias voltage of the converter electrode. It was observed that due to enhanced stripping of H⁻ no direct gain of extracted beam current can be achieved by increasing the converter voltage. The conversion efficiency of H⁻ was observed to vary with converter voltage and follow the existing theories in qualitative manner. We present calculations predicting relative H⁻ yields from different cesiated surfaces with comparison to experimental observations from different types of H⁻ ion sources. Utilizing materials exhibiting negative electron affinity and exposed to UV-light is considered for Cesium-free H⁻/D⁻ production.

Keywords: Negative ion source, cesium equilibrium, work function, negative electron affinity.
PACS: 29.25.Ni

INTRODUCTION

Modern H⁻ ion sources are based on two ion formation channels, the surface [1] and the volume [2] production. This article concentrates on the physics of surface production of H⁻ by so-called resonant tunneling ionization, explained briefly as follows [3]: The electron affinity of free hydrogen atom is 0.75 eV. As the atom moves close to a metal surface the electron affinity level shifts and broadens. The basic mechanism of the affinity level shift is the induction of an image charge in the metal. If the electron affinity level shifts below the Fermi energy of the (often cesiated) metal surface, electrons have a finite probability to tunnel through the potential barrier forming an H⁻ ion. These ions are further emitted via two mechanisms: electron double capture followed by direct reflection of an incoming ion and sputtering of adsorbed H⁻ by energetic ions [4]. Mounting evidence suggests that the majority of the H⁻ surface emission is attributed with the sputtering process [4-6]. According to so-called *probability model* [7,3] the negative ion formation probability, p, on metal surface is proportional to the escape velocity of the ejected ion $v_\perp$ (normal to the surface) and the surface work function ϕ as shown in equation 1. So-called *amplitude model* predicts the ionization probability being proportional to $v_\perp$ and ϕ as shown in equation 2.

$$P_{probability} \propto v_\perp (\phi - S - \Delta S)^{-1}, \qquad (1)$$

$$P_{amplitude} \propto e^{c_1(S-\phi)/v_\perp}, \qquad (2)$$

where S is the electron affinity of the negative ion, ΔS the shift of the affinity level near the surface and c_1 a constant, derived from experimental data (see e.g. reference [8] for details). The probability model has been found more accurate for high ion velocities within the plasma sheath (energy > 20 eV) while the amplitude model can be applied for lower energies [3].

Both models indicate that H⁻ production on metal surfaces can be enhanced by increasing the escape velocity of the ions and/or decreasing the work function of the surface. The velocity dependence of the H⁻ formation probability stems from an interaction between the negative ion and induced image charge [3] i.e. the probability for image charge stripping of H⁻ decreases with increasing escape velocity. Reducing the work function of the surface enhances electron tunneling from the conduction band of the metal as it lowers the potential barrier between the Fermi level and electron affinity level of the hydrogen atom. In practice the work function of the conversion surface is typically reduced below 2 eV by deposition of a fractional monolayer of cesium. Being a relatively heavy element, cesium plays a dual role as bombardment by cesium ions enhances the sputtering of the H⁻ from the surface [9].

The presented ion formation models omit the role of electron density on the conduction band on electron tunneling probability. Experimental evidence suggests that materials with high Fermi energy, E_F, exhibit higher rate of electron tunneling [10]. In a simplest approximation the Fermi energy scales with density of electrons on conduction band n_{ce} (in metals) as $E_F \propto n_{ce}^{2/3}$. Higher rate of electron tunneling for materials with high E_F can be associated with increased flux of electrons impinging the surface potential barrier.

EXPERIMENTS ON H⁻ YIELD VS ION ESCAPE VELOCITY

The formation of H⁻ ions in the LANSCE negative ion source occurs on the surface of a negatively biased converter electrode, exposed to a flux of positive ions (H⁺, H_2^+, H_3^+, and Cs⁺) incident from a cusp-confined, filament-driven discharge. Deposition of cesium, injected into the plasma chamber with resistively heated oven, decreases the work function of the converter surface to a level at which the formation of H⁻ by resonant tunneling charge exchange is favored. H⁻ ions formed on the Molybdenum converter are then sputtered by incoming ions and "self-extracted" through the plasma due to the negative bias voltage (typically 225 – 275 V) and focused to the extraction by concave curvature of the converter. For a full description of the ion source see e.g. [11,12].

The claim that the H⁻ ion escape velocity is affected by the converter voltage deserves discussion. The velocity distribution of H⁻ can be separated into two components – particle reflection and ion induced desorption components. It has been shown by measuring the energy distribution of the extracted ion beam [13] that in the LANSCE ion source ion induced desorption dominates (see also [4-5]). Assuming constant plasma density, considering the plasma sheath to be parallel with the converter surface and taking into account the fluxes of different plasma species (setting the boundary conditions), it can be shown that the electric field at the converter surface increases with increasing converter voltage despite of the fact that the plasma sheath becomes thicker [14]. Under these assumptions it can be argued that the perpendicular escape velocity of the H⁻ ions is increased for two reasons – by the higher electric field and by the higher energy of incident positive ions. This implies that in order to maximize the output of H⁻ it would be beneficial to increase the (negative) converter voltage although the velocity distribution of the ions cannot be controlled precisely.

On the other hand it can be argued that larger converter bias reduces the surface coverage of adsorbed hydrogen atoms, which would presumably lower the H⁻ yield. Assessing the surface coverage is virtually impossible. However, it has been recently shown that operating the LANSCE ion source at elevated plasma chamber wall temperature has a beneficial effect on the H⁻ production. The observation has been associated with increased cesium vapor pressure in the plasma volume and, hence, enhanced sputtering of H⁻ from the converter surface by Cs^+ ions [6]. The method being effective indicates strongly that reduced surface coverage of hydrogen does not limit the H⁻ output. Moreover, the pulse shape of extracted H⁻ ion beam has not been observed to droop towards the trailing edge of the discharge pulse suggesting that hydrogen adsorption level of the converter is adequate to support the regular pulse pattern of the LANSCE ion source (865 μs pulses at 120 Hz) allowing ample time for surface coverage recovery between discharge pulses. Running the ion source at elevated temperature, however, requires converter voltage higher than normal operation in order to maintain optimal cesium coverage on the converter, i.e. equilibrium between deposition and sputtering rates. Thus, it can be argued that the observed increase of the H⁻ output associated with elevated wall temperature could as well be a consequence of increased converter voltage. These aspects motivated us to study systematically the effect of converter voltage (ion escape velocity) on H⁻ production. Due to complex nature of the H⁻ sputtering, affecting in part the initial velocity of the ions, a direct comparison between theory and experiment is virtually impossible implying that our experiment should be treated as a qualitative study.

The effect of the converter voltage on the extracted H⁻ beam current was studied at the LANSCE ion source test stand [11]. At first the discharge power and cesium oven temperature were set to values typical for operations with 120 Hz / 865 μs pulses (pulse pattern used throughout the experiments). The cesium coverage of the converter surface was optimized by adjusting the converter voltage at fixed discharge power and cesium vapor pressure - a procedure often referred as "peaking" the extracted H⁻ current. This process yields a new equilibrium coverage at the existing Cs vapor pressure determined by the temperature of the source body (plasma chamber walls) and constant supply of cesium (oven temperature). The H⁻ beam current and pulse shape were subsequently recorded. A similar procedure was carried out with different discharge powers ranging from 5.7 kW to 8.3 kW and cesium oven temperatures ranging from 223 °C to 247 °C corresponding to (negative) converter voltages from 260 V up to 440 V i.e. well beyond conventional values. The discharge power was adjusted by varying the filament heating power i.e. electron emission from the thermionic cathodes, which causes the discharge current to increase at constant discharge voltage (185 V). The feed rate of neutral hydrogen (H$_2$) was kept constant at 2.2 sccm corresponding to 2 mTorr (± 10 %) plasma chamber pressure in the LANSCE ion source test stand. Experimental results obtained with various discharge powers are illustrated in Fig. 1. Figure 1a shows the dependence of the converter voltage, $V_{c,peak}$, corresponding to the highest beam current on cesium oven temperature, T_{Cs}. Figure 1b shows the extracted H⁻ beam current corresponding to different (optimized) converter voltages.

Figure 1a is presented in order to clarify the physical processes determining the dynamical equilibrium of cesium on the converter surface. Two trends can be observed: $V_{c,peak}$ increases with increasing T_{Cs} and $V_{c,peak}$ decreases with increasing plasma density (discharge power). As the oven temperature is increased, the deposition rate of cesium between the discharge pulses increases. In order to maintain optimum coverage of the converter the voltage has to be increased to enhance sputtering of cesium from the surface. On the contrary, as the plasma density is increased the sputtering rate during the pulses overpowers the deposition rate and, as a consequence, converter voltage has to be reduced i.e. $V_{c,peak}$ decreases. Figure 1b shows that the extracted H⁻ beam current is virtually independent of the converter voltage and depends only on plasma density. The shape of the H⁻ current pulse and emittance of the extracted H⁻ beam have been found to be independent of the converter voltage [12]. We conclude that *no direct gain of extracted H⁻ beam current can be achieved by operating the converter electrode at elevated (negative) voltages*. On the other hand our result indicates that the *beneficial effect of elevated source temperature i.e. increased cesium vapor pressure* [6] *is indeed due to enhanced sputtering of H⁻ from the converter surface*.

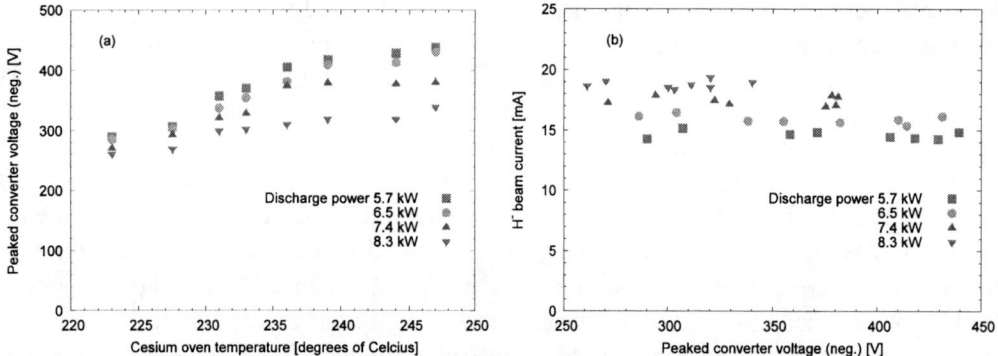

FIGURE 1. (a) Optimized converter voltage (negative) as a function of cesium oven temperature and (b) extracted H⁻ beam current as a function of optimized converter voltage.

So far, we have not accounted for the fact that, in order to become extracted, H⁻ ions created on the conversion surface have to propagate a distance of 12.5 cm through the plasma, and are subject to stripping losses. Dominating loss processes are collisions with neutral particles (hydrogen molecules) i.e. $H^- + H_2 \rightarrow H^0$ and collisions with electrons i.e. $e + H^- \rightarrow H + 2e$. Contribution of other loss mechanisms such as collisions with neutral cesium atoms can be considered insignificant as discussed in [15]. Based on experimental data obtained with the LANSCE ion source [16] we assume neutral H_2 pressure of 2 mTorr, plasma ionization degree of 1-1.7 % (electron density of $6.5 - 10 \cdot 10^{11}$ cm^{-3} increasing linearly with discharge power) and electron temperature of 5 eV. Furthermore, by ignoring collisions in the plasma sheath and assuming that H$^-$ emission from the surface by sputtering process dominates (supported by references [4-6]) we argue that the energy of the propagating ion beam can be approximated to correspond to the converter voltage. We also neglect the ion losses due to space charge effects near the extraction aperture and the effect of the cusp-magnetic field diverting H$^-$ from the ballistic orbit defined by the concave shape of the converter and the contribution of the ions produced by particle reflection. With these assumptions we can utilize cross section data from reference [17] to estimate the H$^-$ losses and plot in Fig. 2 the "corrected" conversion efficiencies of H$^-$, defined as *H$^-$ beam current from the converter / total current drawn by the converter*[1]

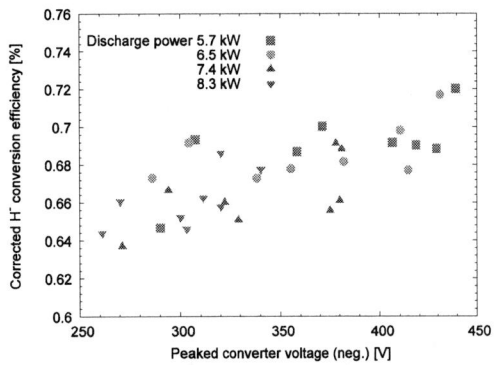

FIGURE 2. Corrected H$^-$ conversion efficiencies for various discharge powers and converter voltages.

The data presented in Fig. 2 indicates that the conversion efficiency of H$^-$ increases with converter voltage. The trend is consistent with the model suggesting that surface ionization probability should increase with ion (escape) velocity. However, it is likely that the effect is caused by enhanced desorption rate of H$^-$ with increasing incident ion energy rather than the variation of the electric field on the converter surface. As shown in Fig. 1b, increasing the converter voltage results in zero gain of extracted H$^-$ beam current due to reduced mean free path of ion-neutral collisions at higher final energy [17]. No significant difference was observed in conversion efficiencies obtained at varying plasma densities (discharge powers).

PREDICTED H$^-$ YIELD OF DIFFERENT SURFACES

Production of negative ions can be improved significantly by depositing a fractional monolayer of alkali metal, usually cesium, on the ionization surface. This is due to formation of a dipole electric field between the bulk metal and alkali layer reducing the work function and, thus, lowering the potential barrier which the electron has to tunnel through. The minimum work function ϕ_{min} of metal surface, obtained under optimized cesium coverage, can be estimated by [8]

---

[1] Or more precisely, it is the ratio $\gamma_{H^-}/(1+\gamma_e + \gamma_{H^-})$ that remains constant, where γ_{H^-} and γ_e are the sputtering coefficients for H$^-$ and converter electrons respectively due to the impinging positive ions.

$$\phi_{min} \cong 2.707 - 0.24\phi_0, \qquad (3)$$

in which ϕ_0 is the work function of clean (uncesiated) surface. The (semiempirical) equation is in good agreement with experimental data [8]. It is somewhat counterintuitive that materials with high intrinsic work function turn out to be the best candidates for H⁻ conversion surfaces under cesium deposition due to low work function. The explanation for this rather peculiar behavior is the strength of the electric field formed between the cesium layer and the metal surface.

Table 1 lists different surfaces, their intrinsic work functions in vacuum, under exposure to hydrogen gas (data for all materials were not found) and calculated minimum work functions. The table also shows a relative comparison of predicted H⁻ yields of different surfaces (of interest for H⁻ ion sources). Stainless steel is used as a reference point. The shift of hydrogen electron affinity level has been considered only as a small perturbation when applying the probability model. In the case of amplitude model the comparison has been calculated by averaging over a distribution of escape velocities corresponding to 1-5 eV (typical for surface ionization on low-bias electrode). The presented range of values for the amplitude model corresponds to uncertainty of the numerical constant c_1 (see equation 2) [8]. The data presented in the table suggests that as far as only bare metal surfaces are considered, nickel and platinum are the best candidates for cesiated H⁻ conversion surfaces. For experimental results see e.g. reference [18]. However, the situation changes when the surface is exposed to hydrogen atmosphere as described in reference [19]. Taking into account the effect of hydrogen causes molybdenum and nickel to become comparable in terms of estimated H⁻ yield. It must be stressed that the model used for the estimate fails to account for the role of surface impurities, often observed in working ion sources.

TABLE 1. Work functions of selected metals (in vacuum and hydrogen atmosphere), minimum work functions under optimized cesium coverage and relative comparison of predicted H⁻ yields.

Surface	Work function [eV]	Minimum work function [eV]	H⁻ yield probability model	H⁻ yield amplitude model
Stainless steel	4.3 [20]	1.68	1	1
W	4.55 [20]	1.62	1.07	1.07 - 1.11
Mo	4.6 [20]	1.60	1.09	1.10 - 1.16
Re	4.95 [20]	1.52	1.21	1.22 - 1.38
Ni	5.15 [20]	1.47	1.29	1.30 - 1.53
Pt	5.65 [20]	1.35	1.55	1.53 - 1.99
H₂ + W	5.05 [19]	1.50	1.24	1.25 – 1.44
H₂ + Mo	5.3 - 5.4 [19]	1.41 - 1.44	1.35 - 1.41	1.35 - 1.74
H₂ + Ni	5.3 [19]	1.44	1.35	1.35 - 1.64

In order to compare the predictions with experimental results we must make a distinction between two types of surface conversion H⁻ ion sources: sources equipped with a biased (hundreds of volts) converter electrode, e.g. the LANSCE H⁻ ion source and sources equipped with unbiased (or low-bias) collar structure adjacent to the extraction aperture, e.g. the SNS H⁻ ion source. In the case of the LANSCE ion source the best H⁻ yield under identical cesium coverage has been obtained with molybdenum, rhenium and tungsten while nickel and platinum produced only modest beam currents [13], revealing a discrepancy between experiment and prediction. However, taking into account sputtering coefficients of different metals under heavy ion bombardment shows that the best materials for H⁻ production (Mo, Rh and W) are the ones most resistant to sputtering [21] while e.g. platinum erodes easily. Xenon (atomic weight of 131) was used instead of cesium (132) for the sputter yield comparison because of ample data available for different target materials. Strong correlation between sputtering coefficient and H⁻ production leads into a conclusion that predicting the H⁻ yield of different converter materials with simple surface ionization models is extremely difficult if not impossible for LANSCE type ion source. Evaporation of filament material onto the conversion surface adds yet another complication. However, the sputtering rate of the converter surface is expected to exceed the deposition rate of the filament material.

Experiments aiming at enhancing the produced H⁻ beam current by proper selection of the conversion surface material have been conducted on the ion source test stand and front-end at SNS with a collar-type surface conversion ion source [22, 23]. The results of the tests can be summarized as follows: compared to stainless steel both molybdenum and nickel collars yield higher H⁻ beam current while significant difference between molybdenum and nickel has not been observed at the front end [23] (although the results from the test stand favor nickel [22]). The results agree well with the prediction of amplitude model, applicable for low energy ions. Although there is a significant difference in vacuum work functions of molybdenum and nickel, exposure to hydrogen makes the two materials comparable in terms of H⁻ production.

To conclude the discussion on the role of conversion surface material we estimate the stability of H⁻ production with varying cesium coverage for different metals. The work function of cesiated metal surface as a function of cesium coverage θ (measured in monolayers i.e. multiples of 4.8×10^{14} atoms/cm² [24]) can be estimated with [8]

$$\phi(\theta) = \phi_0 + \frac{6 \times (\phi_{min} - \phi_0)}{(3 - \theta_{min})\theta_{min}} \theta - \frac{3 \times (\phi_{min} - \phi_0)(\theta_{min} + 1)}{(3 - \theta_{min})\theta_{min}^2} \theta^2 + \frac{2 \times (\phi_{min} - \phi_0)}{(3 - \theta_{min})\theta_{min}^2} \theta^3, \quad (4)$$

where ϕ_{min} is the minimum work function from equation 3. Beyond a coverage by a full monolayer, the work functions remain constant so that $\phi(\theta>1) = \phi(\theta=1)$. For the majority of metals the value of θ_{min} is 0.5 – 0.7 [25,8]. In order to define a limit for the allowable deviation of the work function from its minimum value we impose somewhat arbitrary limit of 5 % deviation from maximum of H⁻ production (beam current). We restrict the discussion to collar-type ion sources allowing us to apply amplitude model and neglect the effects of material sputtering. It can be shown that the stability requirement of 5 % corresponds to deviation of 0.025 eV from the minimum work function. Calculated maximum variation of cesium coverage corresponding to this limit for different materials is presented in Fig. 3. Since the optimum cesium concentration is not unambiguous, we present the curves for three different concentrations, namely 0.5, 0.6 and 0.7 monolayers.

The allowed variation of the cesium coverage is on the order of 0.05 – 0.1 monolayers. The curves (shaded areas) are not symmetric with respect to zero, i.e. optimal coverage, which implies that maximizing the ion source output by controlling the cesium coverage is more likely to result to slight overcesiation rather than undercesiation of the conversion surface. The maximum allowed variation of the coverage increases slightly with increasing optimum coverage. Material dependent behavior can stem from this trend since the optimum cesium coverage depends on the conversion surface material typically within 0.5 – 0.7 monolayers. Controlling the cesium coverage with a precision of < 0.1 monolayers is extremely difficult as the deposition rate of cesium on the surface depends on the surface temperature, fractional cesium coverage and cesium vapor pressure in non-linear manner [6].

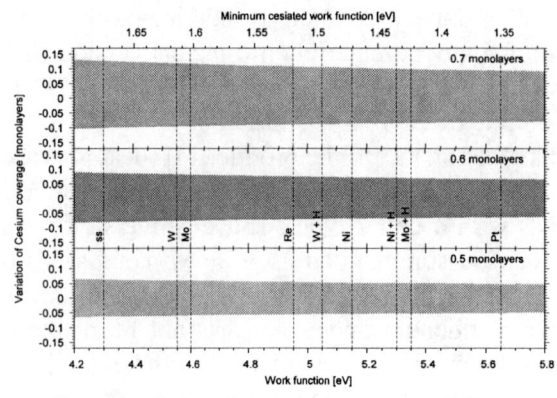

FIGURE 3. Maximum allowed variation of cesium coverage on different surfaces and cesium coverages. H⁻ ionization probability remains within 5% of the maximum within the shaded area.

UV-LIGHT ASSISTED PRODUCTION OF H⁻/D⁻ ON NEGATIVE ELECTRON AFFINITY SURFACES

Enhancing negative ion production with cesium has some drawbacks such as optimizing the cesium coverage due to varying plasma conditions, accumulation of excess cesium on cold surfaces, e.g. in the extraction and low-energy beam transport causing high voltage sparks and creating a safety hazard due to chemical properties of cesium. Nevertheless, cesiation still remains necessary for obtaining required H⁻/D⁻ beam currents for high power accelerators and neutral beam injectors. It seems unlikely that significant improvements of ion source performance, beyond the state-of-the-art, can be achieved simply by optimizing the cesium equilibrium. Material compounds possessing work functions less than 1 eV exist [19]. However, they cannot be applied for negative ion production in plasma ion sources for various reasons such as suffering from chemical and mechanical erosion. So-called *negative electron affinity (NEA) materials*, are interesting candidates for surface production of H⁻/D⁻. Hydrogen-terminated boron nitride [26] and diamond [27] are especially attractive materials despite of being insulators. We use diamond as an example of applying NEA materials for H⁻/D⁻ production since allotropes of BN (cubic and hexagonal) have slightly different band structure, complicating the discussion.

The band gap of diamond is 5.5 eV. The top of the valence band of a hydrogen-terminated diamond film is 4.4 eV below the vacuum level i.e. the bottom of the conduction band is 1.1 eV above the vacuum level. This feature is called Negative Electron Affinity. Electrons can be emitted from the conduction band of NEA-materials, which could be utilized for cesium-free production of H⁻/D⁻. The conduction band can be populated by exposing the diamond surface to ultraviolet light at wavelengths < 226 nm i.e. > 5.5 eV (210 nm, 6 eV for BN), promoting electrons from the valence band to the conduction band and turning diamond into a conductor. The excited electrons lose energy in electron-phonon scattering as they diffuse to the surface. Surface termination with positive electron affinity gas e.g. hydrogen causes favorable band bending on the surface and assures that the vacuum level is below the conduction band minimum. Thus, the electrons can be emitted forming negative ions on the surface. The process is illustrated in Fig. 4.

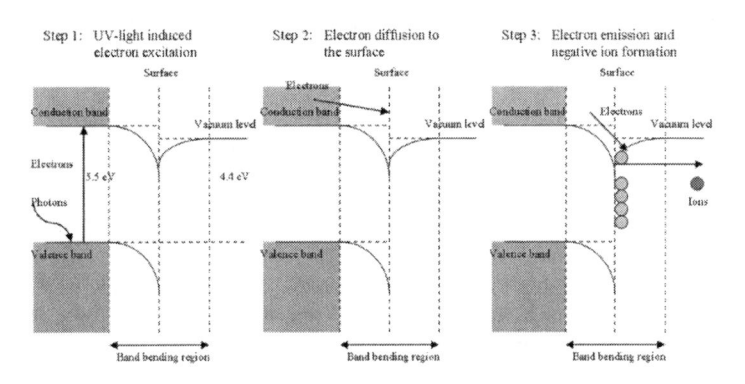

FIGURE 4. Schematic of the UV-light induced negative ion formation on NEA-surface (diamond).

In H⁻/D⁻ ion sources hydrogen termination is assured due to the presence of hydrogen gas. The molecular emission spectrum of hydrogen plasmas lies in VUV region peaking around 160 nm. The Lyman alpha transition of hydrogen emits photons at the wavelength of 121 nm, both being above the excitation limit of NEA-materials. The quantum efficiency for electron emission from diamond has been measured to be 10-15 % [28] for 121 nm. i.e. 1 W of UV translates into approximately 20 mA of electrons. This sets the lower limit for the quantum efficiency of electron excitation. The given number corresponds to electron yield of approximately 20 mA / W. Thus, an external UV-lamp is not necessarily required. Diamond is resistant to mechanical and hydrogen induced chemical sputtering below 100 °C [29]. Due to high thermal conductivity, maintaining the surface below this temperature is not problematic. Last but not least, since the Fermi level of diamond is below the vacuum level, electron tunneling back to the material is not favored reducing the stripping of H⁻/D⁻.

Producing negative ions based upon semiconductor band structure has been proposed by Akimune [30]. Wurz et al. have demonstrated that polycrystalline diamond surfaces can be used for converting energetic H_2^+ ions into H⁻ with 5.5 % efficiency and associate the observation with negative electron affinity of the surface [31]. The approach discussed in this paper is different since we propose to immerse the surface of NEA-material (not exclusively diamond) to hydrogen atmosphere, and expose it to UV-light emitted by the plasma, enhancing the electron density on the conduction band. Populating the conduction band of diamond with UV-light has been applied for fast electric switches made of diamond and controlled by a UV-laser [32]. We claim that *negative electron affinity surfaces exposed to UV-light, emitted by the ion source plasma, could presumably be applied for cesium free surface production of*

H/D⁻ ions at considerable quantities. A feasibility study of the method has been initiated at University of Jyväskylä with the first goal of characterizing the UV-light power spectrum of typical ion source (hydrogen) plasmas.

ACKNOWLEDGMENTS

This work has been supported by the Academy of Finland under the Finnish Centre of Excellence Programme 2006-2011 (Nuclear and Accelerator Based Physics Programme at JYFL) and by the US Department of Energy under contract DE-AC52-06NA25396. OT acknowledges financial support from Emil Aaltonen foundation.

REFERENCES

1. Y.I. Belchenko, G.I. Dimov and V.G. Dudnikov, Nucl. Fusion 14, 113, (1974).
2. M. Bacal, E. Nicolopoulou and H.J. Doucet, Proc. Symp. on the Production and Neutralization of Negative Hydrogen Ions and Beam (BNL, Upton, NY, USA), (1977), p. 26.
3. M. Rasser, J.N.M. Wunnik and J. Los, Surf. Sci.,118, (1982), p 697.
4. L Schiesko, M Carrère, J-M Layet and G Cartry, Plasma Sources Sci. Technol. 19, (2010), 045016.
5. C.F.A. van Os, P.W. van Amersfoort and J. Los, J. Appl. Phys., 64, 8, (1988), p. 3863.
6. O. Tarvainen, Nucl. Instrum. Meth. Phys. Res., 601, 3, (2009), p. 270.
7. J.K. Norskov and B.I. Lundqvist, Phys. Rev. B19, (1979), p. 5661.
8. G.D. Alton, Surf. Sci. 175, (1986), p. 226.
9. C.F.A. van Os and P.W. van Amersfoort, Appl. Phys. Lett. 50, (1987), p. 662.
10. C.F.A. van Os, R.M.A. Heeren and P.W. van Amersfoort, Appl. Phys. Lett. 51, 19, (1987), p. 1495.
11. J. Sherman et al., Proc. 10th Int. Symp. Production and Neutralization of Negative Ions and Beams, AIP Conf. Proc., 763 (2005) 254.
12. G. Rouleau et al. these proceedings.
13. K.N. Leung and K.W. Ehlers, J. Appl. Phys, 52, 6, (1981), p. 3905.
14. E. Chacon-Golcher and J. Sherman, Los Alamos National Laboratory, unpublished report.
15. O. Tarvainen and S. Kurennoy, Proc. ECRIS08 (JACoW.org, 2008) MOPO-03, p. 64.
16. O. Tarvainen, R. Keller and G. Rouleau, Proc. PAC07 (JACoW.org, 2007) TUPAS064, p. 1802.
17. H. Tawara et al., Nagoya University Report No. IPPJ-AM-46, (1986), unpublished.
18. H. Kashiwagi et al., Rev. Sci. Instrum., 73, 2, (2002), p. 964.
19. B.S. Rump and B.L. Gehman, J. Appl. Phys., 36, 8, (1965), p. 2347.
20. V.S. Fomenko, "*Handbook of Thermionic Properties*", Plenum Press Data Division, New York (1966), Ed. G.V. Samsonov.
21. Unpublished, available online at http://eaps4.iap.tuwien.ac.at/www/surface/script/sputteryield.html
22. R.F. Welton et al. Proceedings of the 1st International Conference on Negative Ions, Beams and Sources, Aix-en-Provence, France, AIP (2009), p. 181.
23. M.P. Stockli, B. Han, S.N. Murray, T.R. Pennisi, M. Santana and R.F. Welton, Rev. Sci, Instrum, 81, 02A729 (2010).
24. J.B. Taylor and I. Langmuir, Phys. Rev. 51 (1937) 753.
25. P.W. van Amersfoort, Thesis, FOM-Institute, Amsterdam, the Netherlands, (1987).
26. M.J. Powers et al., Appl. Phys. Lett. 67, 3912 (1995).
27. D. Takeuchi et al., Phys. Stat. Sol. (a) 202, No. 11, 2098–2103, (2005).
28. J. Himpsel, J. A. Knapp, J. A. VanVechten, and D. E. Eastman, Phys. Rev. B,20, (1979), p.624.
29. E. Salonen, "*Molecular dynamics studies of the chemical sputtering of carbon-based materials by hydrogen bombardment*", University of Helsinki, (2002), ISBN 951-45-8959-9.
30. H. Akimune, J. Appl. Phys., 54, 18, (1983).
31. P. Wurz, R. Schletti and M.R. Aellig, Surf. Sci. 373, 56, (1997).
32. "*CVD Diamond for Electronic Devices and Sensors*", J. Wiley & Sons Ltd., (2009), Ed. R.S. Sussmann, p. 285.

Towards Understanding the Cesium Cycle of the Persistent H⁻ Beams at SNS

Martin P. Stockli, B. X. Han, S. N. Murray, T. R. Pennisi,
M. Santana, R. F. Welton

Spallation Neutron Source, Oak Ridge National Laboratory, Oak Ridge, TN 37831, U.S.A

Abstract. This paper describes the accomplishments of the SNS H⁻ ion source, which delivers routinely ~50 mA at a 5.4% duty factor with ~99% availability, enabling 1 MW beams for neutron production with ~90 % availability. It discusses the need for increasing reliability and beam current. But mostly it focuses on its unexpected feature: H⁻ beams that are apparently persistent for up to 5 weeks without adding Cs after an initial dose of less than ~5 mg. Thermal emission and sputtering are qualitatively evaluated, and appear consistent with a negligible Cs sputter rate after the initial dose disappears from the Cs plasma. It concludes with a list of future experiments that can shed more light on this apparently unique Cs cycle.

Keywords: Cesium, H⁻ ions, ion source, RF ion source, multicusp ion source
PACS: 07.77.Ka, 29.25.Ni, 52.80.Pi

INTRODUCTION

In fall of 2009 the Spallation Neutron Source (SNS) reached its final project goal of neutron production being powered by a 1-MW proton beam. The desire for very short neutron pulses of very high intensity at 60 Hz required the incorporation of a ring to accumulate up to ~800 sequential beamlets into a single proton beam with peak currents of up to ~40 A and proton energies near 1 GeV. The beamlet accumulation in the ring requires the injection of negative hydrogen ions, which are double-stripped inside a dipole magnet, after which they join the trajectories of the already-stored proton beam [1]. Over the last year ~1 MW operation has become routine with availabilities near 90% for scheduled accelerator physics and neutron studies.

For this performance, the ion source and low-energy beam transport system (LEBT) have to deliver ~38 mA H⁻ linac beam current with availabilities near ~99%, which was surpassed with 40 mA and 42 mA during 2- and 4-week production runs in fall of 2009. These beam currents normally persist for up to 5 weeks without a need to add cesium after an initial injection as small as ~3 mg [2].

The 1-MW beams start with the LBNL designed and built H⁻ source [3], which was improved at SNS [2, 4]. The cesium-enhanced, multicusp source has a 2.5 turn antenna inside the plasma chamber to generate the 2-MHz plasma, as seen in Fig. 1.

Between the antenna and the 7 mm source outlet aperture is a ~200 G dipole field. Negative ions drifting into the source outlet are extracted by the -65 kV potential with respect to the normally grounded extractor featuring a 10 mm aperture. A 1.6 kG dipole field integrated into the outlet flange drives the co-extracted electrons sidewise,

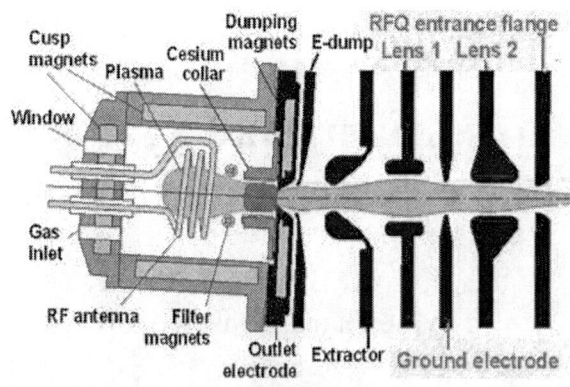

FIGURE 1. Schematic of the LBNL ion source and LEBT.

where most of them are intercepted by the electron dump - kept near -59 kV with a +8 kV supply located on the -65 kV platform. Leaving the extractor, the diverging, 65-keV beam is focused by two electrostatic lenses into the radio-frequency quadrupole accelerator (RFQ). The 12-cm long LEBT features no adequate beam current diagnostics. Leaving the RFQ with 2.5 MeV, the beam is refocused with 2 quadrupole magnets before being measured with the torroidal beam current monitor BCM02. The BCM02 measurements practically match the measured linac beam currents due to negligible losses. Using the estimated ~10% RFQ losses and ~15% LEBT losses suggest the 38 mA linac beam currents start with ~50 mH of H⁻ being extracted from the source.

With routinely ~50 mA at a 5.4% duty factor, the SNS source produces ~230 C of H⁻ ions every day. This exceeds the daily productions of the DESY RF source by a factor of 56 [5], of the BNL magnetron source by a factor of 8 [6], and the FNAL magnetron source by a factor of 41 [7]. During each degradation-free, ~monthly service cycle, the SNS source produces over 7 kC of H⁻ ions, which exceeds the yearly production of the DESY RF source by a factor of ~5. Clearly the SNS source represents a new class of high-repetition-rate, long-pulsed, high-current H⁻ sources. The unprecedented requirement for high peak current AND high duty-factor have required [2, 4] and continue to require [8, 9] the identification and mitigation of many unique issues.

Plans call for upgrading the SNS to 2-3 MW by increasing the linac energy from 1 to 1.3 GeV, and the linac beam current from 38 to 59 mA. In 2008, at the end of a 3-week source cycle, 56 mA average-, and the 59 mA peak current shown in Fig. 2, have been demonstrated for 0.7 ms-long, 60 Hz H⁻ pulses by increasing the 2 MHz power to ~60 kW [2].

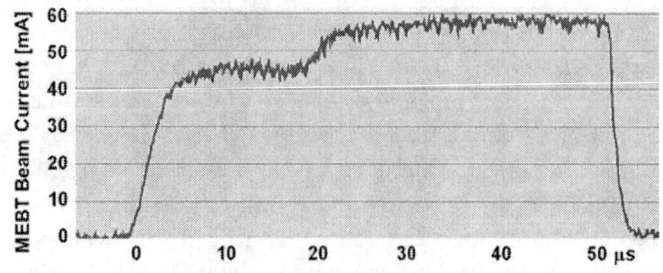

FIGURE 2. A 50 μs slice with 59 mA H⁻ beam current measured with BCM02.

To produce such beam currents routinely, we have to identify the parameters that frequently interfere with the production of high beam currents and learn how to control them with the required accuracy and repeatability. Presently we focus on better control of the cesium enhancement, because it involves procedures that can be repeated many times without requiring a source change. Having the Cs enhancement under control will help to more rapidly and more precisely assess the effect of other parameters and configurations.

In addition, improved control of the Cs should allow for further reducing the amount of Cs which is released. When releasing Cs, the load current of the first lens normally increases, apparently an indicator of Cs covering some of its surface and enhancing the local corona emission. Sometimes when ramping up to operations, especially shortly after a cesiation, and very rarely after prolonged operations, the load current of the first lens starts to grow until the lens starts to arc uncontrollably; this requires reconditioning to restore operations. In addition, sometimes when ramping up to operations a large reverse-load current develops on the e-dump supply, which appears to be streams of electrons leaving the Cs-covered edge of the electron dump and impacting on the extractor. Only gradual conditioning can restore normal operation. Such uncontrolled arcing has been reduced from several-hour durations in 2007, when releasing 20-30 mg Cs, to less than an hour in 2010 when releasing <5 mg, a trend that needs to be pursued.

PRODUCING THE NEGATIVE HYDROGEN IONS

The small 0.75 eV electron binding energy in H⁻ make its formation very unlikely due to the drastic mismatch with the roughly 15 eV ionization energy of typical atoms and molecules. In the plasma volume, the most likely process is the slow-electron induced break-up of vibrationally highly-excited H_2^v molecules (v=4-9) [10]. Hot electrons (E_e>15 eV) excite molecules in the hot plasma near the antenna. The dipole field between the antenna and the source outlet reflects these hot, H⁻-destroying electrons, but allows slow electrons and the neutral molecules to drift towards the source outlet. Colliding with slow electrons, the highly excited molecules dissociate, occasionally with the electron being captured by one of the atoms forming H⁻. Unless they recombine with the slow protons in the plasma, the H⁻ ions can reach the meniscus from which they are extracted.

Metal surfaces are better suited for H⁻ production, because their 4-5 eV work functions represent a lesser mismatch. The mismatch can be further reduced by coating the surface with alkali atoms, which reduce the work function. Alkali atoms

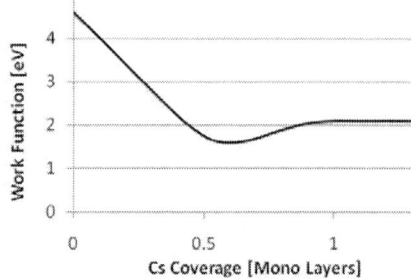

FIGURE 3. Expected work function for Mo as a function of adsorbed Cs atoms.

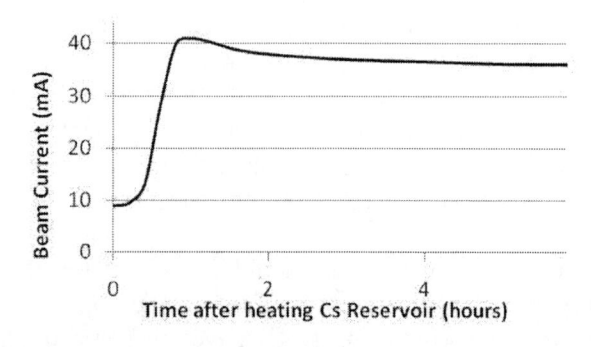

FIGURE 4. The extracted beam during a very slow cesiation of a cold Mo converter.

adsorbed on a metal surface feature a minimum in the work-function for a fractional monolayer, which can be significantly smaller than value for equilibrium value for one or more monolayers, as shown in Fig. 3 [11].

To confirm this minimum on our Cs-covered polycrystalline Mo converter surface, and to assess its effect on the H⁻ production, the heated converter was first extensively plasma conditioned. Then the converter was cooled to ~55° C, and the ampoule in the external Cs reservoir [12] was cracked and heated to 110° C to very gradually build up a Cs layer on the converter surface. As shown in Fig. 4, the extracted beam current grew and reached a maximum before it started to decay by ~10%, which is consistent with a minimum work function for an optimal coverage and a gradual increase of the work function for denser Cs layers.

We aim to develop a method for producing, verifying, and maintaining the optimal monolayer, hopefully within the time and other constraints given by our production schedules. Thermal inertia and observed variations in the cesiation process make it unlikely to be able to stop the flow of Cs when reaching the maximum. It appears more promising to first pass through the maximum, then stop and subsequently emit the slight excess of Cs to return back to the now-known maximum, and then maintain this verified, optimal layer.

THE Cs_2CrO_4-BASED CS DELIVERY SYSTEM

To reduce the risk of Cs-induced arcing, LBNL integrated slots for 8 Cs cartridges, shown in Fig. 5a, into the Cs collar, shown in Fig. 5b. The conical Mo converter is thermally attached to the Cs collar [4]. Its small outlet is ~0.5 mm from the source outlet. The temperature of the collar and the Mo converter is controlled with cold or heated compressed air. Shutting off the air allows for higher temperatures, which are then controlled with the duty factor of the plasma.

The cartridges contain a powdery mixture of Cs_2CrO_4 and St101, a Zr-Al getter [13]. Experience has shown that it is important to degas the cartridges at high temperature (>>100° C) to avoid highly-variable, unpredictable delays, apparently caused by the getter having to first absorb the surface contaminants [4]. Heating the cartridges to ~550° C activates the getter, which then reduces the cesium into its elemental form.

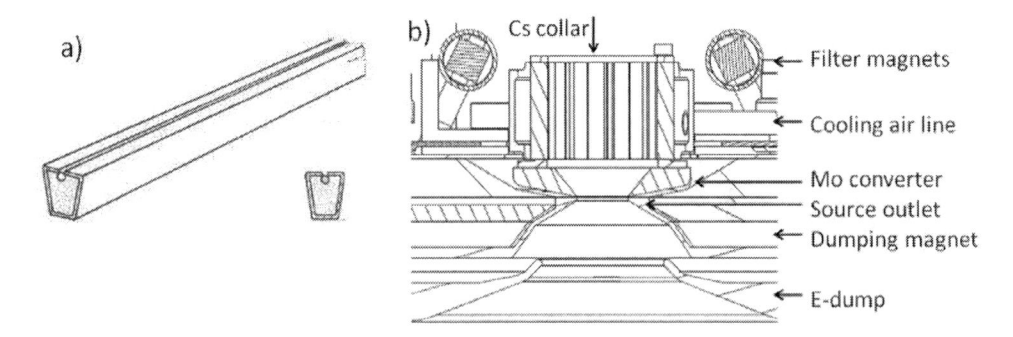

FIGURE 5. a) Cs cartridges; b) cross section through the Cs collar and the source outlet.

The tight slots in the Cs collar impede the Cs escaping along its wire-covered escape slot, while it escapes unimpeded through the open ends seen in Fig. 5a. In 2008 tiny compression springs were added to facilitate the escape of the Cs towards the converter while impeding the escape into the plasma chamber [4].

If the Cs is released into an insufficiently-conditioned source, the Cs-enhanced H⁻ beam decays rapidly, although the decay rate gradually improves with later cesiation [2]. This appears to be caused by the Cs being sputtered away with the surface adsorbates on the converter surface. If the source is conditioned for 2.5 hours with a ~50 kW, 7.2% duty factor plasma before releasing the Cs, the enhanced H⁻ beam persists up to 5 weeks without apparent decay within the ~10% uncertainty due to typical operational changes.

After a 10-minute cesiation the collar temperature is lowered to 160-180 C, where it is normally kept for the rest of the source service cycle. If the temperature is lowered during the first day or so, the beam current decreases. However, after several weeks of operation, the collar temperature can be lowered without significantly changing the beam current.

THERMAL EMISSION OF CESIUM

The thermal emission of adsorbed atoms on surfaces is characterized by their dwell time. The dwell time is characterized by the probability that the surface bond breaks, given by the exponential of the ratio of the bond energy E_{Cs} and $k \cdot T$, where k is the Boltzmann constant and T is the absolute temperature in °K.

$$\tau = \tau_0 \cdot \exp(E_{Cs}(\theta)/(k \cdot T)) \qquad (1)$$

For Cs on a clean W(110) surface τ_0 was found to be $6 \cdot 10^{-13}$ s [14]. The exponential nature of the law makes the dwell time strongly dependent on the surface temperature, as shown in Fig. 6 for a binding energy of 1.7 eV. For 200° C, the dwell time would be ~250 days, predicting a stable mono layer. But if the temperature is raised to 250° C, the dwell time drops to a few days, predicting a rapidly decaying Cs layer and consequently a decaying H⁻ beam current. This is why the temperature was kept below 200 °C for production operation.

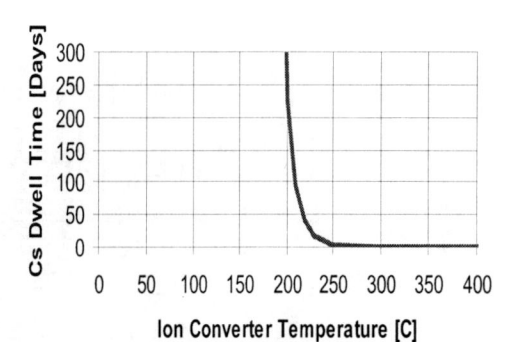

FIGURE 6. Dwell time of Cs adsorbed with 1.7 eV on a metal surface.

However, the binding energy of Cs depends on the Cs coverage θ, which has profound implications. Figure 7 shows two approximations compiled from many measurements. The dashed line shows the approximation by Kaminsky [15], which is valid from 0.06 to 0.6 monolayer of Cs on clean polycrystalline W. The solid line shows the approximation by Hansen [16], which is valid for the entire range of up to 1 monolayer of Cs on a 110 surface of a single crystal of W.

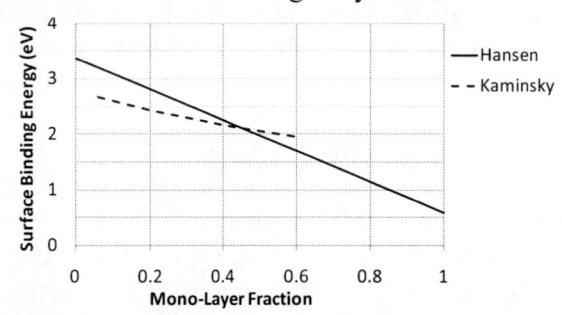

FIGURE 7. Surface binding energy of Cs as a function of surface coverage θ of W.

The two approximations suggest that the bond energies depend on the type of surface. However, in both cases the bond energy decreases with increasing coverage, likely due to the increasing average mismatch between the ideal adsorbate positions and the actual positions. This work uses the Hansen approximation due to a lack of data for Cs on poly-crystalline Mo. The absolute results will be inaccurate, but the trends are likely correct because they are governed by the fact that the bond energy appears to decrease with increasing coverage. The dwell times in Fig. 8 show that very dense layers (θ ~1) yield dwell times below 1 μs for temperatures >200 °C, whereas diluted layers (θ≤0.4) at temperatures <100 °C may outlast the life of our solar system.

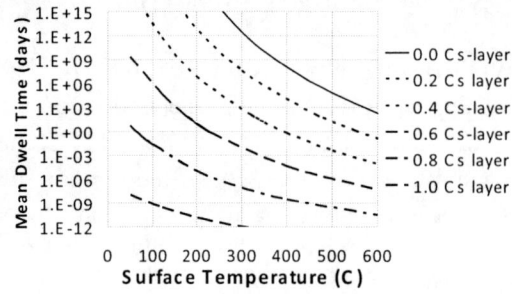

FIGURE 8. Logarithmic plot of dwell times of Cs on a W surface vs temperature.

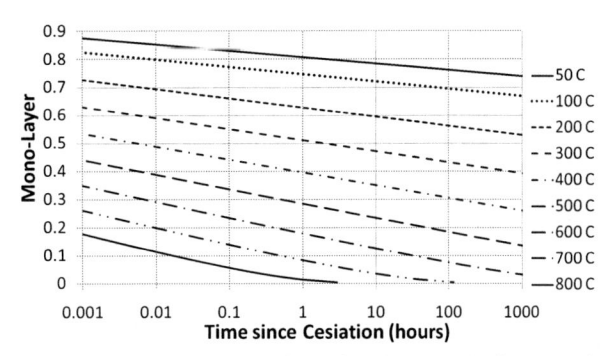

FIGURE 9. The Cs surface coverage vs time ($\theta=0.995$ at $t=0$) for several surface temperatures.

The thermal emission reduces the surface coverage θ, which increases the average bond energy, which stabilizes the system. The loss $d\theta/dt$ is equal to the amount of Cs θ divided by its dwell time τ. Integrating the loss yields the coverage versus time:

$$\theta(t) = \theta_0 - \int (d\theta/dt) \cdot dt = \theta_0 - \int (\theta(t)/\tau(T(t), E_{Cs}(\theta))) \cdot dt \qquad (2)$$

For practical purpose we assume that a cesiation yields a flash of Cs, which increases the coverage to 0.995 for $t=0$. From then on for a given surface temperature T the times for losing 0.01 monolayer are added to obtain $t(\theta)$, which is plotted in Fig. 9 for temperatures between 50 and 800 °C. The figure shows that, especially for higher temperatures, most of the Cs is emitted in the first few seconds, after which the emission gradually declines. For example, with 200 °C, the Cs drops to 0.7 monolayer within 27 s, and to 0.63 monolayer after 1 hour. From then on it loses another ~.032 monolayer over each of the next 9-hour period, the following 4-day period, and the following 37-day period.

At constant surface temperatures, Cs layers remain fairly stable after the first few days, but the initial rapid Cs loss interferes with tuning a stable, optimized beam shortly after cesiation. However, lowering the surface temperature can stabilize the system instantaneously. For example, Fig. 10a shows that lowering the surface temperature to 50 °C 10 min. after cesiation ($\theta=0.995$ at $t=0$) lowers the thermal emission to a point of almost perfect stability. This suggests using time-limited heat treatments to achieve the desired Cs monolayer before lowering the temperature to "freeze" the optimal coverage. For example, 0.6 monolayer could be generated with 250 °C for 10 min., 230 °C for 1 h, or 190 °C for 1 day, as shown in Fig. 10b.

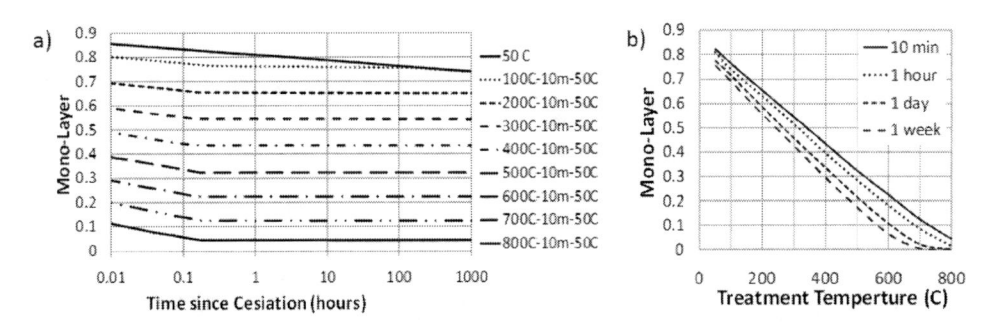

FIGURE 10. a) Cs coverage vs time when lowering the temperature to 50°C after 10 min.; b) monolayer yield versus temperature of time-limited heat treatments.

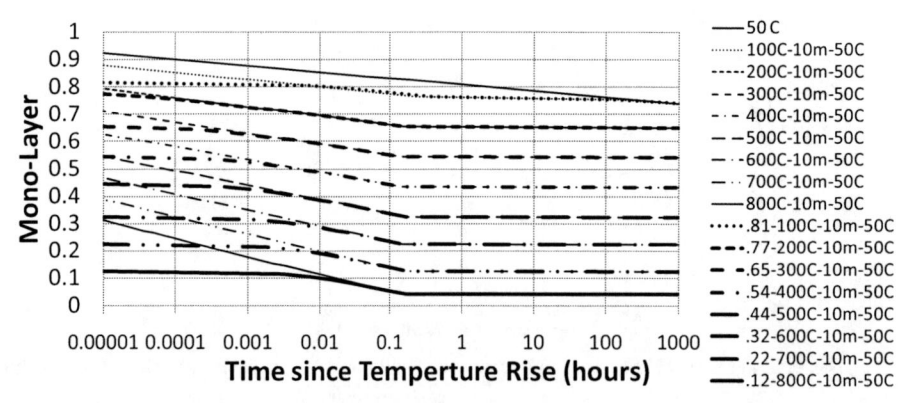

FIGURE 11. Cs coverage vs time for consecutive 10 min heat treatments with increasing temperature with and without re-cesiations.

How about sequential heat treatments without intermediate cesiations? The thin curves in Fig. 11 are identical to Fig. 10a, except that the plot starts 36 ms after each cesiation ($\theta_0=0.995$ at t=0). If one skips all but the initial cesiation, subsequently θ_0 is given by the monolayer fraction achieved with the previous heat treatment. The thick curves in Fig. 11 start at t=0 with the monolayer fraction, which was obtained one hour after starting the 10-min heat treatment at the previous lower temperature. The curves show that the Cs coverage and emission are initially significantly reduced, but within 1 s to 1 min, the coverage and emission (=slope) match the values calculated for starting with $\theta_0=0.995$. This means that as long as the heat treatments are 10 min and longer, and the temperature is increased by at least 50°C, the resulting Cs coverage depends only on the last heat treatment, showing no memory of previous heat treatments. This illustrates the rigorous control one has over the fraction of the cesium layer by adjusting the temperature of the surface.

Experiments are needed to calibrate these trend calculations. After cesiating with a cold collar in the first experiment seen in Fig 12a, the collar temperature was raised for ~12 min. by ~30° C every 1 or 2 hours, which increased the beam current every time due to decreasing the work function. In the second experiment seen in Fig 12b, the temperature of the cesiated collar was raised for 12-min. in 100 °C steps about every hour, which lowered the beam due to an increasing work function. However, after the collar cooled down, the beam current partially recovered, especially when applying high temperatures in excess of 200 °C in the latter test.

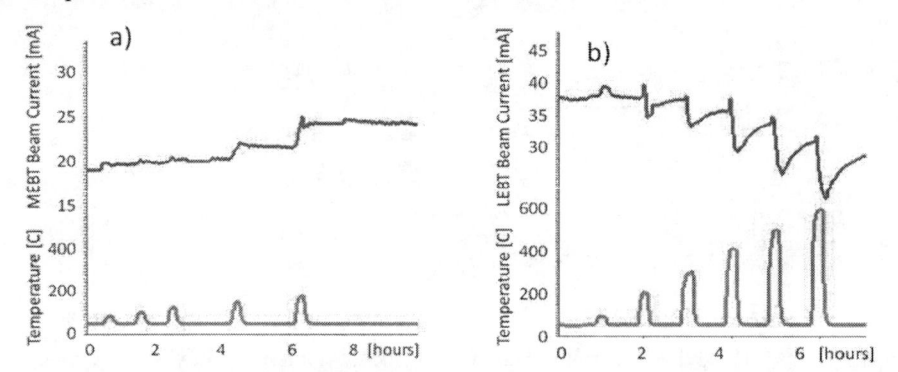

FIGURE 12. sequential 12-min. heat treatments a) first increase and b) then decrease the beam current.

Apparently, the heat treatments emit Cs into the plasma, from which it is partly readsorbed on the cold surface, unless the heat treatments are long enough for the Cs to get lost in other locations. Accordingly, we have started to optically study the relative Cs population in the plasma, and initial measurements suggest that it decays roughly exponentially with a time constant of many hours, which may preclude the use of heat treatments during our expedited 6-hour source change procedures.

PLASMA-INDUCED SPUTTERING

Ions being accelerated by the plasma potential and sputtering atoms or molecules adsorbed on surfaces play an important role. Much work has been done to measure sputter thresholds and yields for many ions, adsorbates, and surfaces. Since the plasma potential is normally rather small, most ion-adsorbate collisions are near threshold. For ions with energy E_i and mass m_i, and adsorbates with mass m_a being adsorbed with bond energy E_a, the threshold energy E_{th} and sputter yield Y can be approximated [17] with

$$E_{th} \approx 8 \cdot E_a \cdot (m_i/m_a)^{2/5} \quad \text{for} \quad m_i > 0.3 \cdot m_a \quad (3a)$$
$$E_{th} \approx E_a/(\gamma \cdot (1-\gamma)) \quad \text{for} \quad m_i \leq 0.3 \cdot m_a \quad \text{with } \gamma = 4 \cdot m_i \cdot m_a/(m_i+m_a)^2 \quad (3b)$$
$$Y \approx 0.006 \cdot m_a \cdot \gamma^{5/3} \cdot E_i^{1/4} \cdot (1-E_{th}/E_i)^{7/2} \quad \text{for } m_i/m_a < 1 \quad (3c)$$

The ion energy plays the dominant role, which can be characterized by measuring the plasma potential. In addition there is quite a strong mass-ratio effect, as shown in Fig. 13 where the relative threshold (normalized to the surface bond energy) is plotted against (the square root of) the adsorbate mass for different atoms and molecules that may be found in hydrogen plasma. To break a surface bond, the ion energy has to exceed the bond energy by at least a factor of 4 for the optimal mass ratio of $m_i \approx m_a/5$. However, much higher energies are required for highly asymmetric systems. Figure 13 shows that hydrogen ions are very unlikely to sputter ^{133}Cs and vice-versa. However, hydrogen ions, especially H_2^+, effectively sputter H_2O, CO, etc., the typical residual gases found in vacuum systems and on their surfaces. In turn, those N^+, H_2O^+, CO^+, etc. sputter adsorbed Cs atoms very effectively.

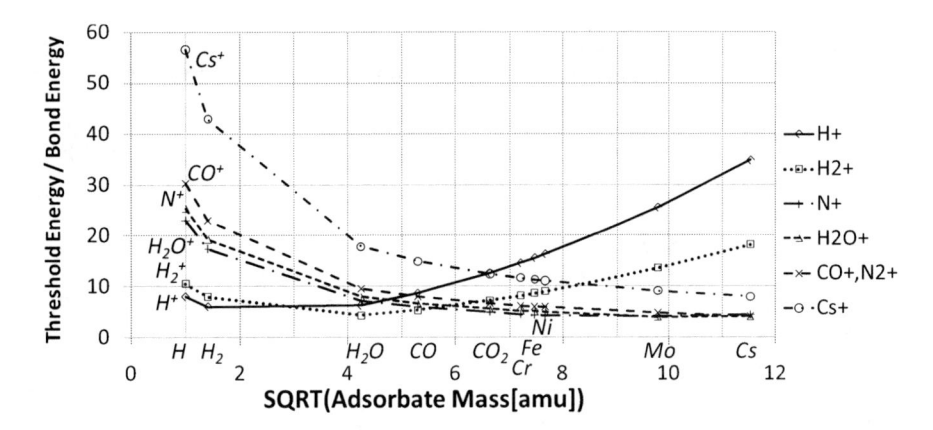

FIGURE 13. Ion threshold energy normalized to the bond energy versus adsorbate mass for different ions found in hydrogen plasma sources.

This could explain qualitatively why air and water leaks are so detrimental to H⁻ production as well as the importance of plasma conditioning. Cs may initially sputter Cs, which can be readsorbed from the plasma. By the time the Cs density in the plasma becomes too small, the plasma density of impurities may also be too small to noticeably deplete the Cs layer. However, quantitative assessments are warranted. While the mass-ratio effects appear to be consistent with the performance of our H⁻ source, the facts may be significantly more complex and involve chemical reactions with the neutral impurities, which may have higher densities than the ions.

SUMMARY AND OUTLOOK

The SNS H⁻ source routinely achieves ~50 mA with a 5.4% duty factor and a ~99% availability. Surprisingly these beam currents are produced 24/7 for up to 5 weeks without noticing a degradation and without deliberately adding Cs after the initial dose of ~5 mg. This suggests a negligible Cs sputter rate, although a small augmentation from the ~170° C Cs collar cannot be excluded at this time. To better characterize the operational conditions, and to better understand the Cs cycle, it is planned to measure the decay time of the Cs in the plasma as well as the plasma potential in our source. In addition, it is planned to measure beam decay rates as a function of impurity gases leaked into the hydrogen flow to establish impurity limits for H⁻ beams with negligible decay rates.

ACKNOWLEDGMENTS

The proofreading by P. Kite was invaluable. Work was performed at Oak Ridge National Laboratory, which is managed by UT-Battelle, LLC, under contract DE-AC05-00OR22725 for the U.S. Department of Energy.

REFERENCES

1. S. Henderson, "Spallation Neutron Source Operation at 1 MW and beyond" in *Proceedings of LINAC10, Tsukuba, Japan*, 2010, MO103.
2. M.P. Stockli, B. Han, S.N. Murray, T.R. Pennisi, M. Santana, R.F. Welton, "Ramping up the Spallation Neutron Source beam power with the H⁻ source using 0 mg Cs/day" in *Rev. Sci. Instrum.* **81**, 02A729 (2010).
3. R. Keller, R. Thomae, M. Stockli, R. Welton, "Design, Operational Experiences and Beam Results Obtained with the SNS H⁻ Ion Source and LEBT at Berkeley Lab" in *Production and Neutralization of Negative Ions and Beams*, edited by M. P. Stockli, AIP Conference Proceedings CP639, American Institute of Physics, Melville, NY, 2002, pp.47-60.
4. M.P. Stockli, B.X. Han, S.N. Murray, D. Newland, T.R. Pennisi, M. Santana, R.F. Welton, "Ramping Up the SNS Beam Power with the LBNL Baseline H⁻ Source" in *Negative ions, Beams and Sources*, edited by E. Surrey and A. Simonin, AIP Conference Proceedings CP1097, American Institute of Physics, Melville, NY, 2009, pp. 223-235.
5. D.P. Moehs, J. Peters, and J. Sherman, "Negative Hydrogen Ion Sources for Accelerators" in *IEEE Trans. Plasma Sci.* **33** (2005) 1786-1798.
6. J. Alessi, private communication (2011).
7. D. Bollinger, private communication (2011).

8. B. Han, T. Hardek, Y. Kang, S.N. Murray Jr., T.R. Pennisi, C. Piller, M. Santana, R.F. Welton, M.P. Stockli, "Performance of the H⁻ Ion Source Supporting 1 MW Beam Operation at SNS", these proceedings, 2011.
9. R.F. Welton, N.J. Desai, B.X. Han, E.A. Kenik, S.N. Murray, T.R. Pennisi, K.G. Potter, B.R. Lang, M. Santana, M.P. Stockli, "Ion Source Developments at the SNS", these proceedings, 2011.
10. R. Celiberto and A. Laricchiuta, "Electron-Impact Cross Sections for Processes Involving Vibrationally Excited Diatomic Hydrogen Molecules" in *Production and Neutralization of Negative Ions and Beams*, edited by M. P. Stockli, AIP Conference Proceedings CP639, American Institute of Physics, Melville, NY, 2002, pp. 3-12.
11. W. G. Graham, "Properties of Alkali Metals Adsorbed onto Metal Surfaces" in *Proceedings on the 2nd International Symposium on the Production and Neutralization of Negative Ions and Beams*, Brookhaven National Laboratory, New York (1980) pp. 126-133.
12. R.F. Welton, M.P. Stockli, S.N. Murray, J. Carr, J.R. Carmichael, "Initial Tests of an Elemental Cs-System with the SNS Ion Source" in *Proceedings of LINAC 2006*, edited by Ch. Horak, (2006) pp. 364-366.
13. CS/NF/3.6/11 from SAES Getters S. p. A., Via Gallarate 215, 20151 Milano, Italy.
14. T.J. Lee and R.E. Stickney, *Surf. Sci.* **32** (1972) p. 100.
15. M. Kaminsky, "Atomic and Ionic Impact Phenomena on Metal Surfaces", Springer, New York, (1965).
16. L.K. Hansen "Thermionic Converters and Low Temperature Plasma" Inform. Center / US-DOE-tr-1 (1978).
17. J. Bohdansky and J. Roth, "An analytical formula and important parameters for low-energy ion sputtering", *J. Appl. Phys.* **51** (1980) 2861-2865.

Simulation Studies of Hydrogen Ion reflection from Tungsten for the Surface Production of Negative Hydrogen Ions

Takahiro Kenmotsu and Motoi Wada

Doshisha University, Kyotanabe, Kyoto 610-0394 Japan

Abstract. The production efficiency of negative ions at tungsten surface by particle reflection has been investigated. Angular distributions and energy spectra of reflected hydrogen ions from tungsten surface are calculated with a Monte Carlo simulation code ACAT. The results obtained with ACAT have indicated that angular distributions of reflected hydrogen ions show narrow distributions for low-energy incidence such as 50 eV, and energy spectra of reflected ions show sharp peaks around 90% of incident energy. These narrow angular distributions and sharp peaks are favorable for the efficient extraction of negative ions from an ion source equipped with tungsten surface as negative ionization converter. The retained hydrogen atoms in tungsten lead to the reduction in extraction efficiency due to boarded angular distributions.

Keywords: negative ion, reflection, angular distribution, energy spectrum
PACS: 52.27.Cm, 07.77.Ka, 78.40.-q, 02.70.Uu

INTRODUCTION

Negative ions are formed through electron capture of positive ions at a low work function surface. Based on this mechanism negative hydrogen ions (H⁻) can be produced at the tungsten surface immersed in a cesiated hydrogen discharge. The production efficiency for H⁻ by hydrogen ion reflection decreases with the decreasing incident ion energy below 500 eV/proton. Thus, the produced H⁻ should possess a relatively broad energy distribution due to high incident ion energy [1]. Angular distributions of reflected ions are also important, because energy and angular distributions directly affect beam emittance of the produced negative ions. Particle reflection coefficients also strongly affect the efficiency to produced negative ions.

A Monte Carlo simulation code ACAT [2,3] has been used to calculate the energy spectra and angular distributions of the reflected hydrogen ions from tungsten surface to evaluate the suitability of the surface H⁻ production for beam formation. The effect due to hydrogen retention in the tungsten target had been also investigated.

CALCULATION MODEL OF ACAT

A Monte Carlo simulation code ACAT has been used to calculate particle reflection of hydrogen from tungsten. Since the ACAT code has been described in detail elsewhere, the main features of the code are briefly outlined here. The code numerically calculates trajectories of atoms colliding in an amorphous target based on binary collisions approximation with the target atom positions assigned by a Monte Carlo method. The target atoms are thus, randomly distributed in each unit cubic cell having the lattice constant calculated from $R_0 = N^{-1/3}$, where N is the atomic density of the target material. To see the effect due to hydrogen retention in tungsten target, tungsten and retained hydrogen atoms are assigned in the cells in accordance with their atomic fractions assuming a uniform distribution over the target material. The lattice constant of the target material is determined from the density composed of target tungsten and retained hydrogen atoms.

RESULTS

Figure 1 shows the schematic illustrations of a surface conversion type negative hydrogen ion source. The hydrogen plasma consists of different positive hydrogen ions such as H^+, H_2^+ and H_3^+. Some of these positive hydrogen ions move toward the converter surface and irradiate the surface after being accelerated across the sheath between the converter and the plasma. Negative hydrogen ions are formed though electron capture of incident positive ions in the surface of the converter, and the negative hydrogen ions formed at the surface of the converter are extracted back by the sheath electric field. Angular distributions and energy spectra of reflected ions are determined by this surface collision processes.

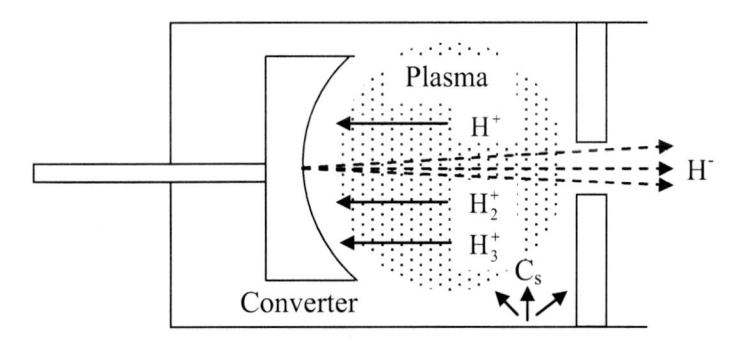

FIGURE 1. Schematic of a surface conversion H⁻ source.

Figure 2 (a) and (b) show the calculated angular distributions of reflected ions from tungsten due to 100 eV hydrogen and cesium ions at 0 degree from normal to the surface. The ACAT results indicate that the angular distribution of reflected hydrogen ions tend to be an over-cosine distribution in the low-energy incidence. Meanwhile the angular distribution of reflected cesium ions shows an under-cosine distribution. The mass of hydrogen is much smaller than that of tungsten. Thus, most incident hydrogen ions are backscattered easily due to the collision with tungsten atoms. This is why the angular distribution of reflected hydrogen ions becomes over-cosine. Cosidering the mass ratio of cesium to tungsten, the backscattering process is not dominant but incident cesium ions form collision cascade to leave small portion of back scattered conponent with an under-cosine distribution.

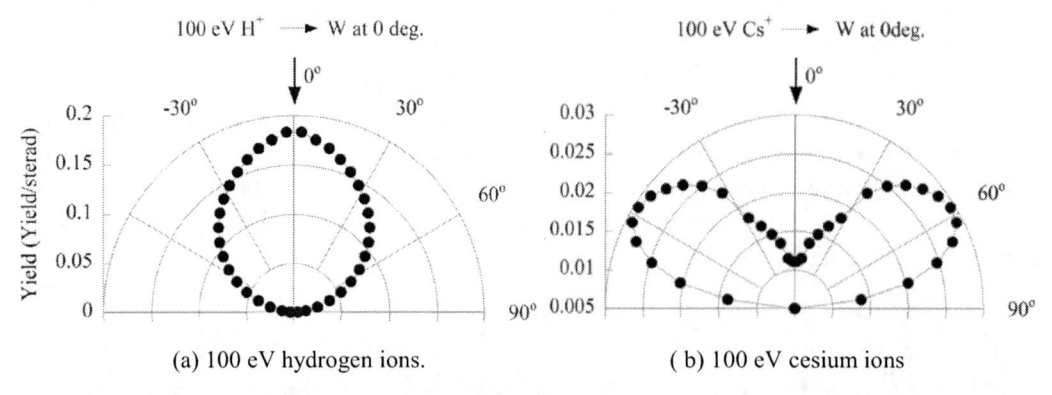

FIGURE 2. Angular distributions of reflected (a) H$^+$ and (b) Cs$^+$ ions from tungsten at normal incidence.

Shown in Fig.3 (a) and (b) are the calculated angular distributions of reflected hydrogen ions from tungsten bombarded by 100 eV hydrogen ions at 50 and 80 degrees. The ACAT results have indicated that the angular distributions of reflected hydrogen ions are strongly influenced by the change in incident angle. The distribution for the incident angle of 80 degree shows the narrow distribution compared with that at 50 degree incidence. Narrower distributions are suitable for the efficient extraction of negative ions. When particles are injected with right angle or 80° from normal to the surface narrow angular distributions as indicated in Fig.3 (b) are observed, while the intermediate angles such as 50 degree result in a broader distribution. An angular distribution for an intermediate angle consists of a distribution due to fewer collisions, and that produced from collision cascade process, which is shown schematically in Fig.4.

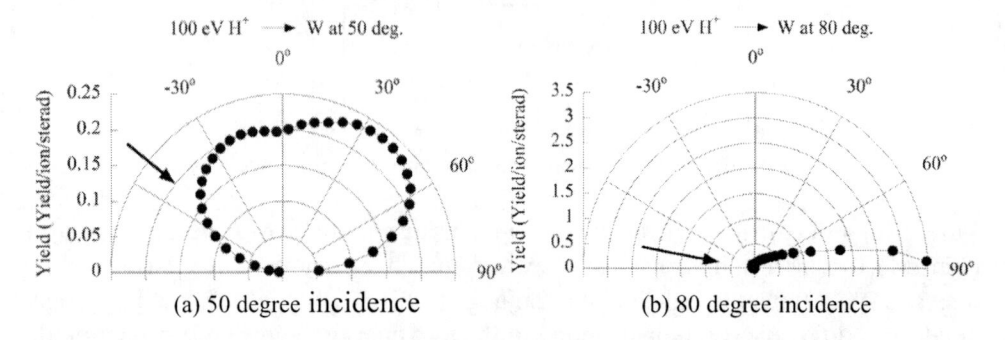

FIGURE 3. Angular distribution of reflected hydrogen ions calculated with ACAT. (a) 50 degree incidence. (b) 80 degree incidence.

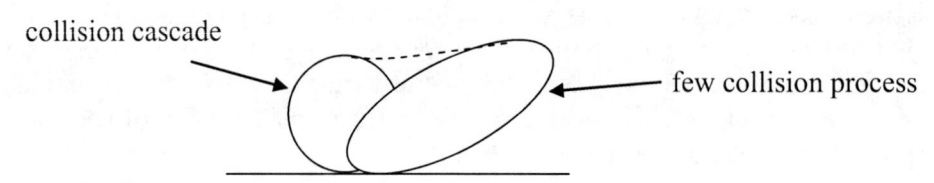

FIGURE 4. Schematic of the angular distribution for medium degree incidence.

Figure 5 (a) and (b) show that the calculated energy spectra of reflected hydrogen ions from tungsten due to 100 eV hydrogen ions at 0 and 80 degree incident angles. These spectra show sharp peaks around 90 eV. The sharp peaks on energy spectra are favorable for the efficient extraction of negative ions. The hydrogen ions colliding with tungsten atoms result in these sharp peaks on energy spectra of reflected hydrogen ions.

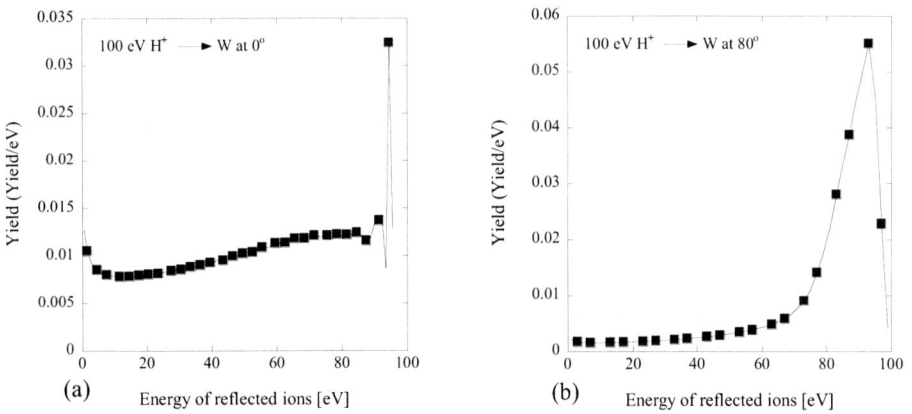

FIGURE 5. Energy spectra of reflected hydrogen ions from tungsten due to 100 eV H^+ ions. (a) 0 degree incidence and (b) 80 degree incidence.

DISCUSSION

Narrower angular distributions or over-cosine distributions of reflected ions are favored for the production of better quality negative ion beam. The under-cosine distributions are expected in low-energy bombardment for heavy ion incidences. However, the ACAT results have indicated that the distribution of reflected hydrogen ions from tungsten surface produced by 50 eV hydrogen ions injected from the surface normal becomes over cosine as shown in Fig. 6 due to the large mass ratio of hydrogen to tungsten. Considering the sheath potential in front of the converter, angular distributions of reflected ions becomes even more peaked toward the forward direction. Shown in Fig. 7 is the calculated angular distribution of reflected hydrogen ions from tungsten due to 100 eV hydrogen ion bombardments with the sheath potential of 200 eV. The sheath potential leads to the enhancement for the extraction of negative ions by making an angular distribution narrower.

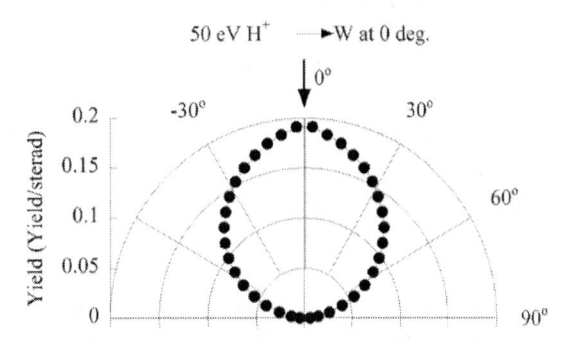

FIGURE 6. Calculated angular distribution of reflected hydrogen ions from tungsten due to 50 eV H^+ ions at normal incidence.

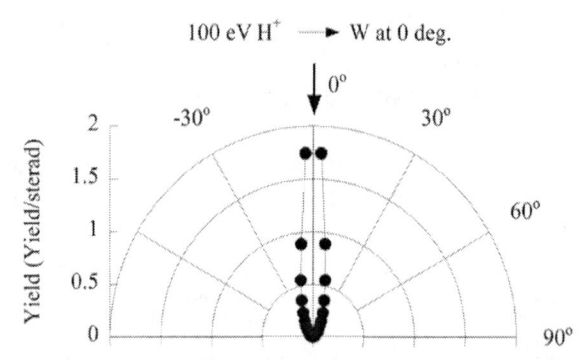

FIGURE 7. Calculated angular distribution of reflected hydrogen ions from tungsten due to 100 eV H$^+$ ions at normal incidence with the seath potential energy of 200 eV.

Considering the operation of the surface conversion source for a long time, hydrogen atoms are retained in the converter. Figure 8 indicates the angular distributions of reflected hydrogen ions from pure tungsten and tungsten containing 10% hydrogen at the topmost layer of the tungsten target due to 100 eV hydrogen ions at 0 degree incidence. The ACAT results indicate that the angular distribution from the tungsten containing hydrogen looses the component directed toward the surface normal compared with that for pure tungsten. The reflection coefficient is also reduced due to the retention of hydrogen atoms in the tungsten target. Thus, retention of hydrogen in tungsten leads to the loss of the production efficiency of negative ions from the converter surface.

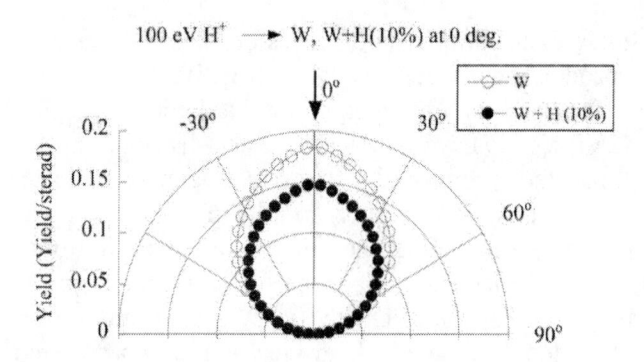

FIGURE 8. Angular distributions of reflected hydrogen ions from pure tungsten and tungsten containing 10% hydrogen at the topmost layer of the tungsten target due to 100 eV H$^+$ ions incident on surface at 0 degree.

CONCLUSION

In order to quantify the production efficiency of negative ions at tungsten surface, the ACAT code has been used to calculate angular distributions and energy spectra of reflected hydrogen ions from tungsten surface. The ACAT results have indicated that narrower angular distributions of reflected hydrogen ions are expected when incident hydrogen ions are injected at 0 degree or near the right angle from the surface normal. These characteristics of surface reflection are suitable for the production of low emittance negative ion beams. Setting the incident angle at some intermediate value broadens the angular distribution of surface produced H$^-$. The sheath potential in front of the converter leads to an apparently narrow angular distribution that effectively transports negative ions to the extraction hole. The retention of hydrogen atoms in tungsten results in the reduction of the production efficiency of negative ions due to a broader angular distribution, a lower mean energy of reflected particles and a reduced particle reflection coefficient.

REFERENCES

1. P. J. Schneider, "Negative Ion Production by Backscattering from Alkali-Metal Surface Bombarded by Ions of Hydrogen and Deuterium", Ph.D. Thesis, University of California, (1980).
2. W. Takeuchi, Y. Yamamura, *Radiat. Eff.* **71** (1983) 53.
3. Y. Yamaura and Y. Mizuno, IPPJ-AM-40, Inst. Plasma Physics, Nagoya Univ., (1985).

A discharge with a magnetic X-point as a negative hydrogen ion source

Tsanko Tsankov and Uwe Czarnetzki

Institute for Plasma and Atomic Physics, Ruhr-University Bochum, Bochum 44780, Germany

Abstract. The study presents first results from investigations of a novel low-pressure plasma source, intended for a negative hydrogen ion production. The source utilizes a dc magnetic field, shaped to form a cusp with a magnetic null-point (X-point). Beside the common role of filtering out the high energy electrons, this magnetic field configuration ensures in the present case also an interesting mechanism of coupling the RF power to the plasma. Investigations performed using radio frequency modulation spectroscopy (RFMOS) reveal that the main power coupling to the electrons is confined in the region on one side of the X-point. The modulation of the light intensity indicates also the presence of a strong dc drift close to the plane of the X-point. Several hypothesises for its explanation are raised: an azimuthal diamagnetic drift due to strong axial gradients of the electron energy, the excitation of a standing helicon wave, which couples to the radial magnetic field in the plane of the X-point, or a Trivelpiece-Gould wave which is resonantly absorbed near the plane of the X-point.

Keywords: negative hydrogen ion source, magnetic X-point, power coupling in magnetized plasma, radio frequency modulation spectroscopy
PACS: 52.25.Xz, 52.50.Qt, 52.70.Kz, 52.80.Pi

INTRODUCTION

Negative hydrogen ions are rarely considered to play a leading role in the industrial plasma processing due to the ease of destruction and inherently their low densities encountered in plasmas under normal conditions. Negative hydrogen ion sources are however of paramount importance for the field of fusion research as an essential part of the neutral beam injection systems [1, 2]. The necessity for improving the efficiency drives intensive investigations of these sources. Their optimization follows the strategy of improving the conditions for the widely accepted production channels of the negative hydrogen ions. These could be separated in two groups – volume and surface production [3]. Volume production is a two-step process – excitation of a vibrational level of the ground electronic level of the hydrogen molecule by an electron impact $\left(e+H_2\left(X^1\Sigma_g^+, v=0\right) \to e+H_2\left(X^1\Sigma_g^+, v\right)\right)$ followed by dissociative attachment $(e+H_2\left(X^1\Sigma_g^+, v\right) \to H+H^-)$. Surface production relies on conversion into negative ions of the hydrogen atoms and atomic ions impinging on a low work function surface $(H+e_{wall} \to H^-, H^+ + 2e_{wall} \to H^-)$. For improvement of the surface conversion the walls are covered usually with cesium. In comparison with volume production, the use of the surface mechanism ensures parameters of the extracted negative ion currents close and in some aspects even exceeding the ITER requirements [4]. However, application of cesium seeding is inherently connected with technical difficulties, which stimulates an ongoing search of new ways to improve the volume production of H^-.

Recent investigations suggest the importance of the walls and the wall material for the balance of the excited hydrogen molecules [5, 6]. Therefore the source, investigated here, has a design, that aims at taking advantage of this surface production mechanism of excited molecules. Since the production of the negative ions still relies on the dissociative attachment process, which is enhanced by the presence of low energy electrons and hindered by high-energy electrons, the source incorporates also a magnetic filter field, here shaped as a cusp. The magnitude and the direction of this magnetic field in the region of the RF coil allow for launching of helicon waves by the flat RF coil, as demonstrated in [7, 8]. However, earlier investigations with a B-dot probe in similar magnetic field configurations [8, 9] reveal that at certain conditions the helicon waves cannot propagate beyond the plane of the X-point. The results discussed here confirm this, which suggests that it could be possible to combine the efficiency of the helicon wave heating with the localization of the power deposition region, necessary for the negative hydrogen ion production.

This study presents results from radio frequency modulation spectroscopy (RFMOS) measurements [10] of the modulation of the Balmer H_α line in a low-pressure discharge in hydrogen. Unlike previous investigations in argon discharges with such magnetic field configuration [8, 9], here we stress on the case when the plane of the X-point is too close to the RF coil to allow a helicon wave to develop and propagate, which leads to a peculiar coupling of the high-frequency field to the plasma. Also, unlike the case of a standard helicon wave where the wave is concentrated mostly on the discharge axis, here the light modulation is strongest close to the walls, which indicates that the electric field amplitude is largest in this region. An attempt to explain this has been made, based on a concept for a standing helicon wave or for a helicon wave coupled to a surface wave. The results are also compared with results obtained when the source is operated in an inductively coupled (ICP) mode which is used as a baseline case.

EXPERIMENTAL SET-UP AND DIAGNOSTIC METHOD

The experimental set-up consists of a quartz dome having an inner radius of 6.5 cm and a depth of 12 cm, mounted on a stainless steel chamber with a radius of 11 cm and a length of 14 cm (Fig. 1). The quartz dome walls are 0.5 cm thick on the side and 1 cm at the bottom. The far end of the steel chamber is a quartz window, allowing direct observations from the front of the source. The system is evacuated to a base pressure of a few 10^{-5} Pa by a turbomolecular pump backed up by a membrane pump. The pressure is monitored by a Pfeiffer Vacuum Compact full-range gauge PKR 251 and by a Pfeiffer Vacuum ceramic capacitance gauge CCR 272. The chamber is filled with hydrogen through a MKS flow controller to a pressure of 1 Pa. During the measurements the flow has been kept constant at 5.5 sccm. Two water-cooled coils with an inner radius of 15 cm produce the static magnetic field. The coils are positioned 7 cm apart (side to side) and positioned roughly at the two ends of the quartz dome (Fig. 1). They are connected in series and powered by the same power supply. The coils have a quadratic cross-section with a side of 4.4 cm, which is filled with 6 by 6 copper coil windings. This position has been chosen to ensure a mostly axial magnetic field at the plane of the RF coil and at the same time providing the largest possible unobscured view on the quartz

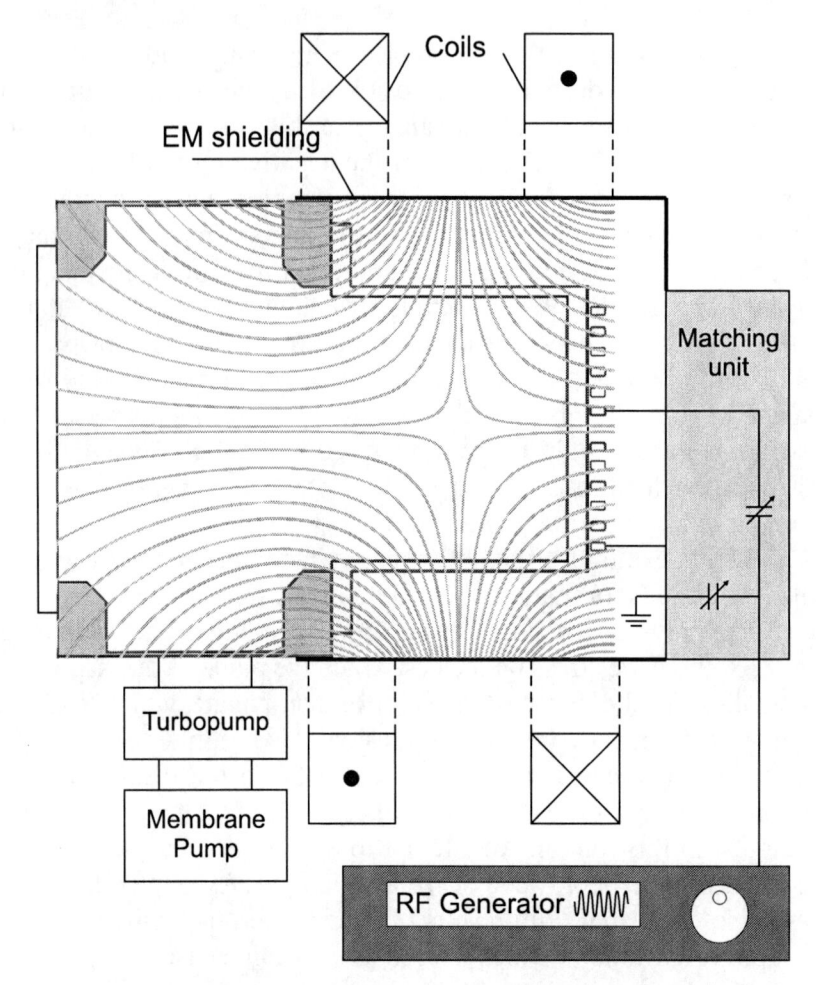

FIGURE 1. Schematic representation of the experimental set-up. Superimposed are the magnetic field lines.

dome side. The field lines of the resulting magnetic field are calculated using COMSOL and given also in Fig. 1.

The plasma is produced by coupling high-frequency power at 13.56 MHz to a flat spiral antenna. The antenna consists of two water cooled windings with the power fed to the center and the side ends are grounded. A capacitive matching unit matches the output resistance of the RF generator (Dressler Cesar 1320 200V, maximal output power 1200 W) to the coil impedance. A cylinder made of copper sleet 2 mm in thickness serves as a RF shield. An opening in this cylinder covered with a movable metal mesh provides the observation port at the side of the quartz.

Phase resolved measurements of the emission modulation have been performed using a Roper Scientific ICCD camera operating at fast gating mode. In this mode the illumination of the CCD chip of the camera is gated at 2 ns synchronous with the output of the RF generator, thus allowing a total of 38 frames per RF cycle. Each frame is obtained by taking $\approx 10^5$ images. To further improve the sensitivity and reduce the noise, a hardware binning of 2×2 pixels is used. After the measurements each frame is additionally binned by the software to further improve the signal to noise ratio. A band-pass optical filter for the wavelength of the H$_\alpha$ line is positioned in front of the camera, thus rendering the camera insensitive to emission at other wavelengths.

The diagnostics method of RFMOS that has been used is relatively new [10]. The method allows – by recording of the time modulation of a given line during the RF period – to infer information about the oscillation and drift velocities in the investigated plasma. The method assumes that the excitation $\Gamma(t)$ of an upper atomic state is modulated in the RF period due to the oscillation of the electron energy. The excitation is assumed to be direct and the depopulation of the state is only by spontaneous emission:

$$\dot{n}_{ex} = \Gamma(t) n_g - n_{ex}/\tau \tag{1}$$

where τ is the lifetime of the excited state and n_g is the ground state atom density. The emission intensity is then assumed to be proportional to the excited atom density n_{ex} (with a proportionality constant T). Equation (1) can be solved by using a Fourier series expansion which gives for the n-th harmonic of the registered intensity I_n the following expression:

$$I_n = T\Gamma_n n_g \tau \frac{\cos(n\omega t + \theta_n - \phi_n)}{\sqrt{1+(n\omega\tau)^2}}, \quad \phi_n = \arctan(n\omega\tau). \tag{2}$$

Here Γ_n is the amplitude of $\Gamma(t)$ at frequency $n\omega$ and θ_n is its phase. The excitation rate at frequency $n\omega$ is connected with the modulation $\langle f_n \rangle_\Omega$ of the electron velocity distribution function (EVDF) at the same frequency via:

$$\Gamma_n = n_e \int_{v_{exc}}^{\infty} \sigma_{exc}(v) v^3 \langle f_n \rangle_\Omega \, dv, \tag{3}$$

where $\langle \ldots \rangle_\Omega$ denotes averaging over the solid angle. The components $\langle f_n \rangle_\Omega$ of the EVDF can be obtained by assuming a time independent isotropic distribution $f(v)$ displaced by a small oscillating velocity $\vec{u}(t) = \vec{u}_d + \vec{u}_{osc} \sin(\omega t + \delta)$, consisting of a dc drift velocity $\vec{u}_d$ and an oscillation velocity $\vec{u}_{osc}$. The EVDF could then be expanded in power series of $\vec{u}(t)$:

$$f(|\vec{v}-\vec{u}|) = f(v) + \vec{u} \cdot \nabla_v f(v) + \frac{1}{2}(\vec{u} \cdot \nabla_v)^2 f(v) + O(\vec{u}^3). \tag{4}$$

The second term in (4) vanishes after integration over the solid angle. If a Maxwellian EVDF with a temperature T_e is assumed, the corresponding harmonics $\langle f_n \rangle_\Omega$ obtained from (4) have the following form:

$$\langle f_0 \rangle_\Omega = f(v) \tag{5}$$

$$\langle f_1 \rangle_\Omega = 2\left(\frac{2}{3}\frac{\varepsilon}{\kappa T_e} - 1\right) f(v) \frac{\vec{u}_d \cdot \vec{u}_{osc}}{v_{th}^2} \tag{6}$$

$$\langle f_2 \rangle_\Omega = \frac{1}{2}\left(\frac{2}{3}\frac{\varepsilon}{\kappa T_e} - 1\right) f(v) \frac{\vec{u}_{osc}^2}{v_{th}^2} \tag{7}$$

Here $\varepsilon = mv^2/2$ is the electron energy and $v_{\text{th}} = \sqrt{2\kappa T_e/m}$ is the thermal velocity. It is now obvious that the emission from levels with higher excitation energy will show stronger modulation. Equation (2) indicates that the level should additionally be short-lived to avoid damping of this modulation [10]. For this reason the H_α line is particularly well suited for the RFMOS measurements due to the characteristics of the upper emitting level in addition to being the strongest line in the plasma spectrum.

The recorded time variation of the emission intensity $I(t)$ is normalized to the time-averaged emission $\langle I \rangle$ to obtain the pure modulation $\eta(t)$:

$$\eta(t) = \frac{I(t)}{\langle I \rangle} - 1. \tag{8}$$

Using such normalization avoids any complications associated with the sensitivity of the camera and the strong background light as well as eliminates the effects connected with the inhomogeneous electron density distribution. The modulation is then extracted by performing a discrete Fourier transform on $\eta(t)$ to get the amplitudes $\eta_{1\omega}$ of the first and $\eta_{2\omega}$ of the second harmonic. Using equations (3), (5)–(7) these could be expressed as follows:

$$\eta_{1\omega} = \frac{\Gamma_1}{\Gamma_0} \frac{1}{\sqrt{1+(\omega\tau)^2}} = \frac{2}{\sqrt{1+(\omega\tau)^2}} \left(\frac{2}{3}\frac{\varepsilon_{exc}}{\kappa T_e} - 1\right) \frac{\vec{u}_d \cdot \vec{u}_{osc}}{v_{\text{th}}^2}, \tag{9}$$

$$\eta_{2\omega} = \frac{\Gamma_2}{\Gamma_0} \frac{1}{\sqrt{1+(2\omega\tau)^2}} = \frac{1}{2\sqrt{1+(2\omega\tau)^2}} \left(\frac{2}{3}\frac{\varepsilon_{exc}}{\kappa T_e} - 1\right) \frac{\vec{u}_{osc}^2}{v_{\text{th}}^2} \tag{10}$$

This shows that the emission modulation at the second harmonic $\eta_{2\omega}$ is connected with the oscillation velocity amplitude which could be related to the induced electric field strength and the modulation at the first harmonic $\eta_{1\omega}$ represents the drift velocity in the direction of the oscillation velocity [10].

RESULTS AND DISCUSSION

The results presented in this section are obtained in a discharge in hydrogen, operating at a pressure of 1 Pa and absorbed power of 1 kW. The gas flow was kept constant at 5.5 sccm. The magnetic field configuration was produced by flowing a current of 100 A through both coils, which resulted in a magnetic field strength in the source chamber of the order of 5 mT ($\omega_{ce} \approx 10\omega_{RF}$, ω_{ce} and ω_{RF} being the electron cyclotron and the RF frequency, respectively). Results from an ICP discharge (i.e. with no external magnetic field applied) operated in the same setup and at the same plasma conditions (in terms of power, pressure and gas flow) have also been obtained. When the intensity was registered from the side, the two coils for the dc magnetic field have obscured partially the view. Thus, these regions have been covered by gray overlays in the the figures shown below. Nevertheless, these areas are included in the figures to allow a better orientation within the setup geometry. The position of the X-point plane is at $z \approx 5.5$ cm and is marked by

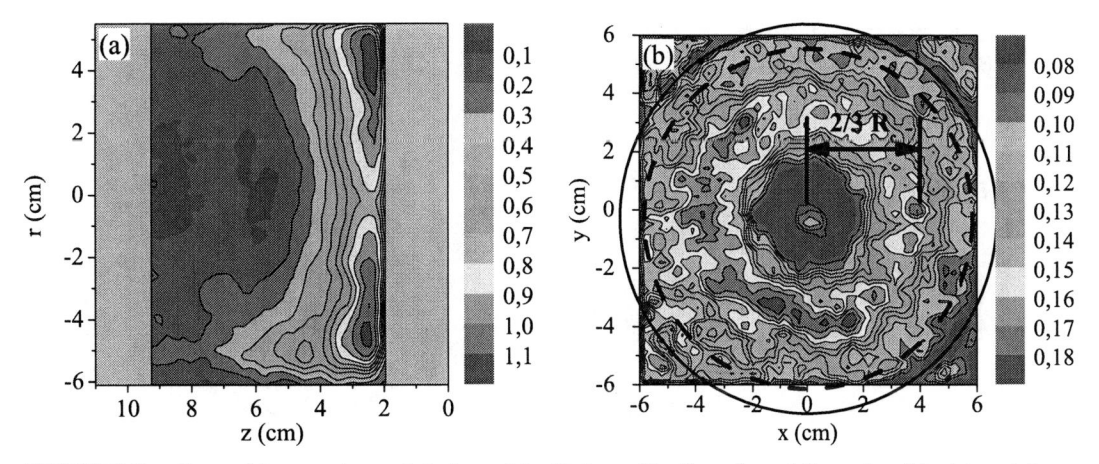

FIGURE 2. Second harmonic modulation of the Balmer H_α-line viewed from the side (a) and from the front (b) of the source, operated in ICP mode. The greyed out overlays in (a) indicate the positions of the coils and the RF coil is to the right. The circles in (b) indicate the position of the wall (full curve) and the extent of the RF coil (dashed curve).

a line where appropriate. The RF coil creating the discharge is to the right at $z = -1$ cm. Additionally, in figures representing a view from the front of the source, a circle has been drawn to show the position of the side walls and a dashed curve indicates the extent of the RF coil.

Figure 2 presents the modulation on the second harmonic in the ICP case. As equation (10) shows, this modulation represents the amplitude of the electron oscillation velocity u_{osc} which in turn represents the induced electric field amplitude. The side view of the modulation reveals the profile typical for an ICP discharge. The modulation peaks at the antenna and decays strongly with increasing distance due to the approximately exponentially decreasing field amplitude. The enhanced amplitude at the radial edges are at least partly an artefact of the line integration in observation. No attempt of removing this artefact by Abel inversion was made here. A further but probably smaller contribution might result from a reduced density near the walls. The view from the front (Fig. 2(b)) shows almost no modulation in the center of the coil which is to be expected, as the induced azimuthal field should be zero on the axis. The maxima in the modulation is observed at roughly 2/3 of the coil radius R_c. An analytical solution for the distribution of the electric field reveals that this is the position where the induced electric field is maximal [10]. All of these give confidence in the method of RFMOS.

Figures 3 and 4 present the second and first harmonic modulation as seen sideways (Figs. 3(a) and 4(a)) and head on (Figs. 3(b) and 4(b)), respectively. The amplitude of the modulation at the second harmonic (Fig. 3) reveals that the modulation, i.e. the electron oscillation and thus the RF field, are concentrated primarily on one side of the X-point. The plasma on this side looks also brighter as compared to the plasma on the other side of the cusp, indicating that this is the region where the RF power is primarily absorbed. Considering the necessity of spatial separation in the source [11], this result is favorable for the application of the discharge as a negative hydrogen ion source. Some modulation, which follows roughly the magnetic field lines, can also be seen on the other side of the X-point plane. This can be understood in terms of a helicon wave since the discharge

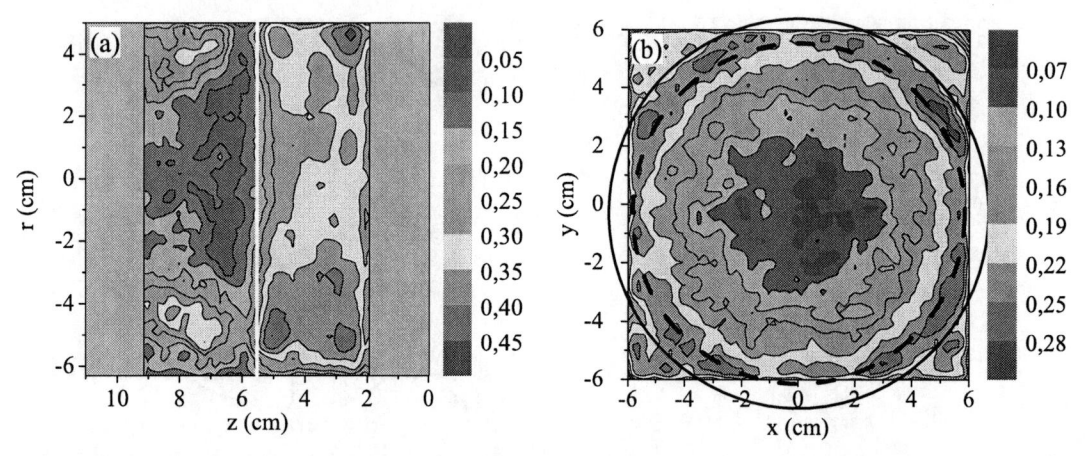

FIGURE 3. Second harmonic modulation of the Balmer H$_\alpha$-line viewed from the side (a) and from the front (b) of the source. The greyed out overlays in (a) indicate the positions of the coils and the RF coil is to the right. The circles in (b) indicate the position of the wall (full curve) and the extent of the RF coil (dashed curve).

conditions allow for the excitation of such a wave, as it will be discussed below. It has been shown by measurements and simulations [8], that under certain conditions the helicon wave can penetrate through the X-point plane which could explain the observed oscillations. This, however, could be easily avoided by reducing the magnetic field strength. The head-on view of the second harmonic modulation (Fig. 3(b)) reveals that it is concentrated close to the walls (but outside the coil radius) which is a striking difference from what was observed in the ICP case (Fig. 2(b)). This unexpected result could be attributed to the presence of a pseudosurface wave as discussed later on.

The emission modulation at the first harmonic (Fig. 4) has an even more puzzling distribution – a strong oscillation concentrated close to (but not exactly at) the plane of the X-point. The amplitude of this oscillation is bigger than the amplitude of the second harmonic modulation. According to equation (9), the modulation at the first harmonic indicates dc drifts parallel to the electron oscillation direction. The presence of an external magnetic field offers a rich variety of drifts. In the present case the most suitable candidate is a diamagnetic drift. Since the plasma emission is azimuthally homogeneous, i.e. no azimuthal gradients, required for an axial drift, the only possible direction for this drift is the azimuthal direction. The strong gradients in the plasma emission in axial direction (most pronounced in the vicinity of the X-point plane) speak for the presence of strong gradients of the temperature and the role of this magnetic field configuration as a filter for the energetic electrons. In this case the observed modulation would also indicate the presence of a strong electron oscillation in the direction of the drift. If the observed modulation is really due to such drift, this would then indicate also the presence of strong azimuthal electric field.

Somewhat puzzling however is the asymmetry with respect to the X-point plane and the concentration of this modulation close to the walls. A possible explanation could be the fact that the modulation on the second harmonic (Fig. 3), and thus the electric field, are also concentrated close to the walls. The asymmetry could be connected with either the lack of electron oscillation in the direction of the drift, or to the much weaker

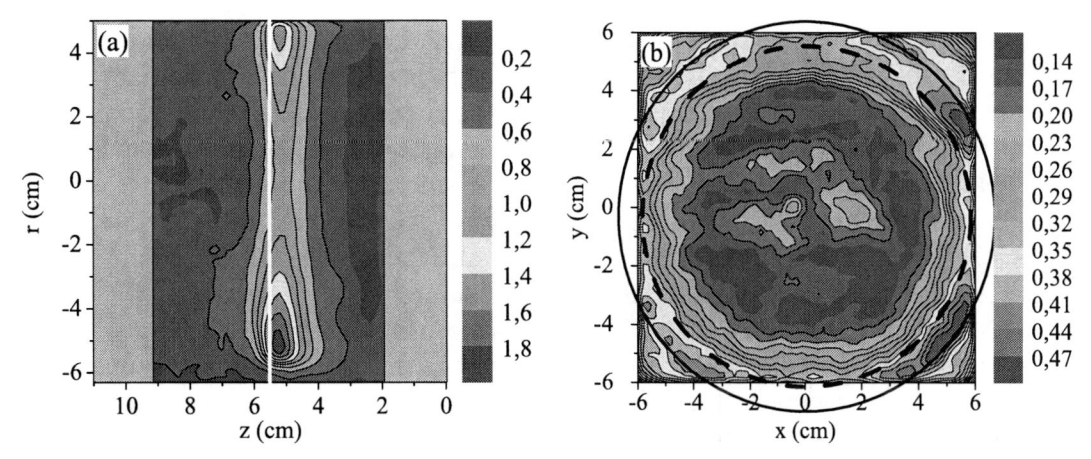

FIGURE 4. First harmonic modulation of the Balmer H$_\alpha$-line viewed from the side (a) and from the front (b) of the source. The greyed out overlays in (a) indicate the positions of the coils and the RF coil is to the right. The circles in (b) indicate the position of the wall (full curve) and the extent of the RF coil (dashed curve).

gradients of the temperature and pressure on the other side of the X-point plane.

To explain these results, one could also reason along the following lines. To ensure quasineutrality a dc field perpendicular to the walls builds up, causing also a drift in this direction. For this drift to be registered by the RFMOS technique, the electrons have to oscillate in the same directions. In our case this requires a radial RF field to be present. The presence of such field is not totally unjustified and different reasons could be given for its presence.

One of them is the formation of a standing helicon wave. As it is shown experimentally [8], at certain conditions the wave is reflected from the X-point plane and a standing-wave-like pattern is formed. The formation of such a standing helicon wave has been also observed in other cases [12]. Since the azimuthally symmetric $m = 0$ helicon mode (as pointed out, no azimuthal inhomogeneities are observed) is circularly polarized, during part of the RF period the electric field has a component in the r-direction. This radial electric field could easily couple to the radial magnetic field in the plane of the X-point (Fig. 1), forcing the electrons to oscillate along the magnetic field lines. This creates a kind of capacitive-like discharge but with an induced electric field. This modulation should be concentrated at the sheath edges, i.e. on both ends of the magnetic field lines. These lines extend from the X-point plane to the axis of the discharge, ending close to the center of the RF coil. Looking at Fig. 4(b) one does see some modulation in the middle of the RF coil, i.e. in the region where the magnetic field lines end and where the other end of this "capacitive-inductive" discharge should be. As the plasma density is expected to be highest in the region close to the coil, the oscillation amplitude and, thus the emission modulation, should be weaker as compared to the plane of the X-point.

Another possibility would be the presence of a surface wave, which in magnetized plasmas are pseudosurface. It is suggested [13] that in magnetized plasmas the helicon mode could sometimes couple to the Trivelpiece-Gould (TG) pseudosurface modes. These two waves – the helicon wave and the TG wave – are the different roots of the same equation and in a way they are dual to each other. Under normal conditions the TG mode has all six field components. At resonance the radial component of the electric field increases strongly [14, 15] which could lead to the suggested radial oscillations of the electrons. The calculated dc magnetic field distribution shows that practically at the point where the modulation maximum is observed, the condition $\omega_{ce} \cos(\alpha) = \omega_{RF}$ (α being the angle between the dc magnetic field and the discharge axis) is fulfilled, which is the necessary resonance condition. This hypothesis is also tempting as it could easily explain the observed concentration of the electron oscillations close to the walls (Figs. 3(b) and 4(b)). To give preference to one or the other of these proposed hypothesises it is obvious that further studies will be necessary. However, even at this point one could see the enormous amount of the information and the insight in the processes that the RFMOS technique has to offer.

CONCLUSIONS

The study presents results from phase-resolved optical emission spectroscopy measurements of a novel type of discharge. The discharge is in hydrogen and with an external magnetic field in the form of a cusp. The modulation of the Balmer H_α-line emission shows that the primary heating of the electrons is concentrated on one side of the cusp and close to the walls. In the vicinity of the X-point plane a strong modulation at the first harmonic is observed. The reasons for this modulation are still unclear but it is suggested that this effects could be either a diamagnetic drift due to strong axial gradients of the electron temperature or due to the presence of an oscillating radial electric field. The latter could be justified by either assuming a standing helicon wave which effectively couples energy to the radial magnetic field structure at the plane of the X-point or by the presence of a magnetized surface wave, which is resonantly absorpbed. To better understand the effect and fully utilize its benefits for the source in terms of efficiency of the negative hydrogen ion production, further studies will be necessary.

ACKNOWLEDGMENTS

Support by the Alexander von Humboldt Foundation is gratefully acknowledged. The authors are indebted to Th. Zierow, St. Wietholt, F. Kremer, B. Becker for the expert technical assistance.

REFERENCES

1. R. S. Hemsworth, A. Tanga, and V. Antoni, *Rev. Sci. Instrum.* **79**, 02C109 (2008).
2. U. Fantz, P. Franzen, W. Kraus, M. Berger, S. Christ-Koch, M. Fröschle, R. Gutser, B. Heinemann, C. Martens, P. McNeely, R. Riedl, E. Speth, and D. Wünderlich, *Plasma Phys. Control. Fusion* **49**, B563–B580 (2007).
3. M. Bacal, *Nucl. Fusion* **46**, S250–S259 (2006).
4. U. Fantz, P. Franzen, W. Kraus, M. Berger, S. Christ-Koch, H. Falter, M. Fröschle, R. Gutser, B. Heinemann, C. Martens, P. McNeely, R. Riedl, E. Speth, A. Stäbler, and D. Wünderlich, *Nucl. Fusion* **49**, 125007 (2009).

5. M. Bacal, A. Ivanov Jr., M. Glass-Maujean, Y. Matsumoto, M. Nishiura, M. Sasao, and M. Wada, *Rev. Sci. Instrum.* **75**, 1699–1703 (2004).
6. J. Amorim, J. Loureiro, and D. C. Schram, *Chem. Phys. Lett.* **346**, 443–448 (2001).
7. Y. Celik, D. L. Crintea, D. Luggenhölscher, and U. Czarnetzki, *Plasma Phys. Control. Fusion* **51**, 124040 (2009).
8. S. Takechi, S. Shinohara, and A. Fukuyama, *Jpn. J. Appl. Phys.* **38**, 3716–3722 (1999).
9. S. Takechi, S. Shinohara, and Y. Kawai, *Surf. Coat. Technol.* **112**, 15–19 (1999).
10. D. L. Crintea, D. Luggenhölscher, V. A. Kadetov, C. Isenberg, and U. Czarnetzki, *J. Phys. D: Appl. Phys.* **41**, 082003 (2008).
11. S. Kolev, S. Lishev, A. Shivarova, K. Tarnev, and R. Wilhelm, *Plasma Phys. Control. Fusion* **49**, 1349 (2007).
12. D. L. Crintea, C. Isenberg, D. Luggenhölscher, and U. Czarnetzki, *IEEE Trans. Plasma Sci.* **36**, 1406–1407 (2008).
13. F. F. Chen, and R. W. Boswell, *IEEE Trans. Plasma Sci.* **25**, 12454–1257 (1997).
14. A. Shivarova, and K. Tarnev, *Plasma Sources Sci. Technol.* **10**, 260–266 (2001).
15. H. Schlüter, A. Shivarova, and K. Tarnev, *Plasma Sources Sci. Technol.* **10**, 267–275 (2001).

Cusp loss width in multicusp negative ion source: A rigorous mathematical treatment

Ajeet Kumar and V.K.Senecha

Ion Source Laboratory, Proton Linac and Superconducting Cavity Division, Raja Ramanna Centre For Advanced Technology, P.O.-RRCAT, Indore(M.P.)-452013, India

Abstract. Cusp leak width (CLW) is an important parameter used in designing of H- ion source as it helps in determining the total power requirement of the source by considering particle loss at the multicusp regions. This parameter has been derived by many workers based on certain assumptions and approximations but it does not take into account the curved nature of magnetic lines of force in the cusp region[1-3]. This statement is vindicated by the fact that above method derives same expression for CLW irrespective of different cusp geometries. Similarly, the final expression of CLW depends on ion acoustic velocity, magnetic field at the cusp and half-length of magnetic lines of force [2,3]. The last parameter is the only geometrical parameter in the expression. However, it does not define which particular half-length of magnetic lines of force has been considered, thus, leading to insensitivity to geometrical aspects of the multicusp: planar,cylindrical etc.

In the present analytical study, we report on a rigorous mathematical treatment considering geometrical aspects of the cusp leak width taking into account the appropriate geometrical factors for the cylindrical and planar line cusp. Our results show that apart from the reported term by others, there is another term that is dependent on the geometrical aspects of the multicusp and become quite dominant in the low pressure region (≤ 1 mTorr) contributing nearly 90% to CLW and for region typically applicable for negative ion sources (≈ 10 mTorr) it contributes to nearly 48%, with the assumption that particles at the cusp are lost with ion acoustic velocity (Cs) along the field line at the cusp.

Keywords: Multicusp magnetic field, H^- Ion Source,Magnetic Cusp, Cusp leak width
PACS: 52.75.-d; 29.25.Ni

INTRODUCTION

The multicusp type ion sources have found wide applicability due to its capability to confine large volume of high density uniform plasma. Although there exist different cusp magnetic geometries for plasma confinement but the line cusp is the most commonly employed as it give better confinement efficiency of plasma. In the H negative ion source, magnetic multipole confinement of plasma coupled with the filter field has become indispensable. Such multicusp kind negative ion source has demonstrated its capability for high current low emittance and stable H- ion beams which is required for the use in accelerators.

The two important parameters associated with such kind of the magnetic field design are the knowledge of field free region and line CLW. The detail magnetic field analysis for such multicusp has been carried out in [6] and concluded that the magnetic field intensity, varies as power law rather than decaying exponentially as the best approximation for cylindrical geometry and accordingly the expression of field free region has been determined. The cusp loss has been investigated through analytical calculation by authors [1, 2, 3, 4, 5] etc. The method rely on the steady state analysis of plasma volume, bounded by magnetic field lines on two side and plasma wall and magnetic field symmetry line on other two side. Merlino et al. [3] has calculated cusp loss width for even point cusp, using similar method with slight modification. They have derived the loss width of the cusp field in terms of the Diffusion coefficient, ion acoustic velocity and the half length of the magnetic line of force between two consecutive magnets. They have considered the gross feature of the magnetic field in the region of their analysis, and since the magnetic field varies considerably in that region, an effective diffusion coefficient has been considered perpendicular to the magnetic field lines. But no method has been suggested to compute the effective diffusion coefficient and value rather rely on the experimental evidence. Similar method has been used by [7] to estimate the width of electron energy loss at cusp in negative ion source, which has half-length of magnetic line of force as one of its parameter. The half-length of magnetic line of force is not a well-defined parameter, as it is not clear which magnetic line of force is being considered. Further their analysis shows that the plasma density decay as $\exp(-|x|/\Delta s)$, perpendicular to the magnetic field line at cusp, where Δs is the leak width. But this functional form is not possible as the symmetry requires that the gradient of plasma density must be continuous at the cusp (Here at $x = 0$ the gradient of plasma density is not continuous). Consequently, ion flux which is proportional to the gradient of plasma density will have non-unique value at the discontinuity, which is physically impossible. Therefore mathematical rigor was lacking in these derivations.

We have addressed these issues in this article and present a derivation of the line CLW and plasma density profile in the narrow region of cusp. Diffusion of plasma across the magnetic field lines has been considered similar to the authors of [1, 2, 3, 4], but the diffusion equation has been written in the curvilinear coordinates, such that two coordinate axes are parallel and perpendicular to magnetic field lines. This has been made possible by using magnetic vector potential and decomposing it in its various harmonics. With few approximation, the diffusion equation become variable separable, and can be solved to get analytical expression of plasma density and its leak width at the cusp. The expression of the CLW are compared with those obtained by [1, 2, 3, 4] and important differences and similarities are highlighted. The possibility of ionization has been incorporated using simple model by considering it proportional to plasma density.

MAGNETIC FIELD PROFILE OF MULTIPOLE CONFIGURATION

In a typical magnetic multipole configuration a set of alternating rows of North Pole and South Pole, permanent magnets are placed around the surface of the plasma chamber. The alternating rows of permanent magnets generate a line cusp magnetic configuration, in which magnetic field strength is higher near the surface of the plasma chamber and decreases towards the centre of the plasma chamber. The magnetic field generated by the multipole configuration are periodic; hence it can be expressed in terms of Fourier series, i.e. by decomposing the magnetic field in terms of its various harmonics. Further since the magnetic field in the plane perpendicular to the chamber axis is independent of the z coordinate, for sufficiently long cylinder(*length* >> *diameter*), in this plane the

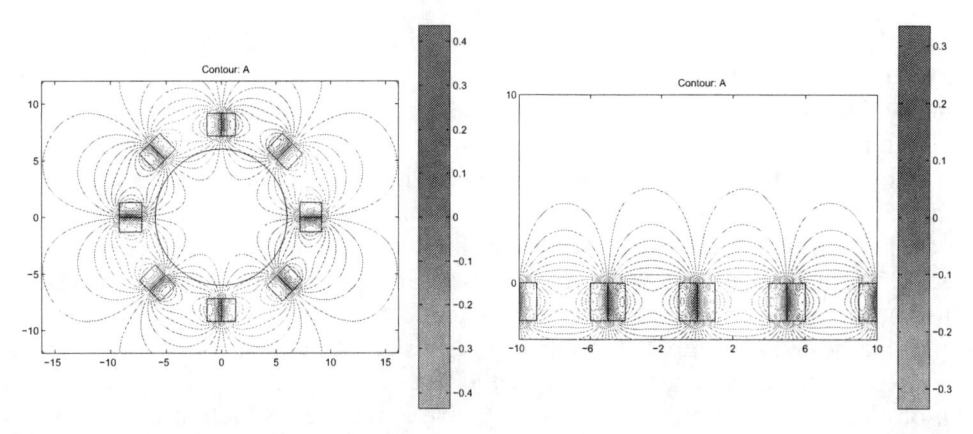

FIGURE 1. Magnetic Field lines due to Permanent Magnets. The Plot shows the constant contour line of Vector potential, in the mid plane perpendicular to the chamber axis. **Left**: Permanent magnets arranged on the cylindrical chamber parallel to the chamber axis. **Right**: Permanent magnets arranged on a plane surface with alternate polarity.

vector potential has only z component i.e. $\vec{A} = (0,0,A_z)$ and in the current free region $\nabla^2 \vec{A} = 0$ i.e $\nabla^2 A_z = 0$ and with the knowledge of A_z, $\vec{B}$ can be computed as $\vec{B} = \vec{\nabla} \times \vec{A}$.

It has been done in [6] for typical cases of permanent magnets arranged around the cylindrical chamber (referred as cylindrical configuration Fig. 1-Left and rectangular chamber (referred as planar configuration Fig. 1-Right. If N is the number of pair of permanent magnets (PM) arranged on the cylindrical chamber of radius R and l is the distance between two neighbouring PM in planar arrangement, then the vector potential for the two cases can be written respectively as

$$A_z = \sum_{n=1}^{\infty} a_n \left(\frac{r}{R}\right)^{Nn} \sin Nn\theta \; ; \; A_z = \sum_{n=1}^{\infty} a_n \exp\left(-\frac{n\pi y}{l}\right) \sin \frac{n\pi x}{l} \quad (1)$$

Where r is measured from the axis of the cylindrical plasma chamber and in planar case x and y are measured from the centre of the pole face (i.e from cusp). Hence in cylindrical case the cusp is located at $(r = R, \theta = 0)$ and in planar case cusp is located at origin $(x = 0, y = 0)$.

It has been shown that $|B|$ inside the cylindrical plasma chamber varies as $(r/R)^{N-1}$ and in the planar case, it varies as $\exp\left(-\frac{n\pi y}{l}\right)$ as the higher harmonics would vanish much rapidly inside the chamber and also because the coefficient $a_n \propto 1/n^2$ in most of the practical cases.

MATHEMATICAL MODELLING OF PLASMA LEAK

The magnetic multicusp geometry are usually employed in H- ion source which operates at neutral gas pressure (1 to 10 mTorr), and with only 10% ionization. According to the classification given in [8] it falls in intermediate to high pressure regime depending on the actual value of the neutral gas pressure, ion and electron temperature and dimension of the plasma chamber. For neutral gas pressure of 1mTorr, with electron and ion temperature of 0.5eV and 5 eV respectively and plasma chamber radius of 5 cm, the system would be in intermediate pressure regime as mean free path of ions $(\lambda_i) \approx 3cm$, whereas, if neutral gas pressure of 10 mTorr $\lambda_i \approx 0.3cm$, hence the system would be in high pressure regime. In low pressure regime the ion transport is collision-less where as in intermediate and high pressure regime the ion transport is diffusive. Particularly in the high pressure regime the transport is well described by the ambipolar diffusion equation and to a good approximation the ionization can be assumed to be proportional to plasma density. But the presence of hot electrons in the plasma, either created by Arc, RF or Microwave generated, complicate the situation as ionization is strongly influenced by the hot electrons. In the present analysis we will study the plasma density profile at the cusp using diffusion equation, and assume that the ionization is proportional to the plasma density in the region.

In presence of magnetic field, the diffusion parallel to field line and perpendicular to the field line are different due to difference in the mobility of the ions in the two directions. The diffusion coefficient perpendicular to B is much smaller than the diffusion coefficient parallel to B and is related as $D_\perp = \frac{D_\parallel}{1+w_c \tau_m}$, where w_c is cyclotron frequency and τ_m is collision time of ions. The exact dependence of diffusion coefficient at the cusp is governed by various other factors apart from magnetic field like degree of ionization, gas pressure, short circuit condition etc, which has been appropriately elaborated in [1, 4]. In order to incorporate the difference in $D_\perp$ and $D_\parallel$, it is natural to write the ion flux and diffusion in a curvilinear coordinate system whose axis are aligned parallel to $\vec{B}$ and perpendicular to $\vec{B}$. As shown in the Appendix, that in our case, the contours of constant A_z is the magnetic field lines, hence a natural choice of curvilinear coordinate would be A_z and its orthogonal trajectories. Together they can be written for cylindrical and planar case respectively as:

$$v = \sum_{n=1}^{\infty} a_n \left(\frac{r}{R}\right)^{Nn} \sin Nn\theta \; ; \; u = \sum_{n=1}^{\infty} a_n \left(\frac{r}{R}\right)^{Nn} \theta \qquad (2)$$

$$v = \sum_{n=1}^{\infty} a_n \exp\left(-\frac{n\pi y}{l}\right) \sin \frac{n\pi x}{l} \; ; \; u = \sum_{n=1}^{\infty} a_n \exp\left(-\frac{n\pi y}{l}\right) \cos \frac{n\pi x}{l} \qquad (3)$$

In the cylindrical case, this is a valid transformation for $|\theta| \leq \pi/2N$ and $(r,\theta) \neq (0,0)$ and in planar case it is valid transformation for $|x| \leq l/2$. Because of symmetry (always even number of magnets are present and are periodically arranged) when n is even $a_n = 0$ in both the cases [6]. The typical plot of (u,v) in (r,θ), for cylindrical geometry has been shown in the Fig. 2 and the plot of (u,v) in (x,y) for planar geometry has been shown in Fig. 3. Note here since $v = A_z$, the above transformation is conformal mapping to complex vector potential($= v + iu$).

Here $\hat{u}$ is the unit vector which is parallel to the contours of constant v and hence $\hat{u}$ is parallel to the field line and similarly $\hat{v}$ is parallel to the contours of constant u and hence $\hat{v}$ is perpendicular to the field line. Thus the ion flux at the cusp, in terms of $D_\parallel$ and $D_\perp$ is:

$$\vec{\Gamma} = -D_\parallel \frac{\partial n}{h \partial u} \hat{u} - D_\perp \frac{\partial n}{h \partial v} \hat{v} \qquad (4)$$

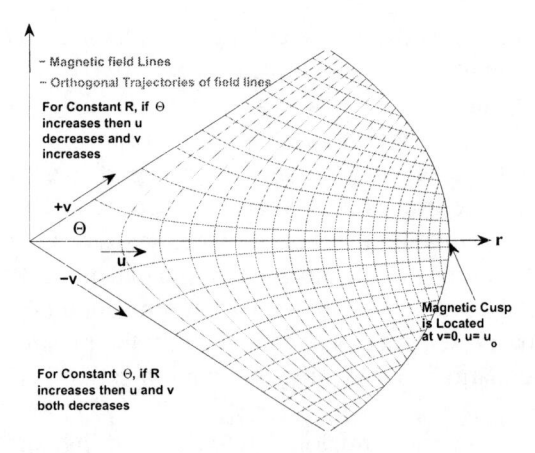

FIGURE 2. Representation of u and v in polar coordinates for cylindrical case

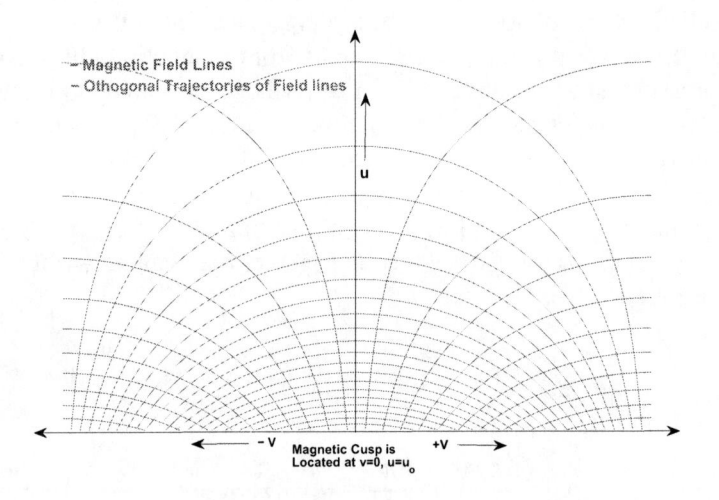

FIGURE 3. Representation of u and v in Cartesian coordinates for planar case

Here n is plasma density and h is metric coefficient for curvilinear coordinates such that $\vec{ds} = hdu\,\hat{u} + hdv\,\hat{v}$ (Appendix). It is generally accepted that the loss of plasma at the cusp along the field line takes place with ion acoustic speed. Hence the ion flux can be written as

$$\vec{\Gamma} = nC_s\hat{u} - D_\perp \left(\frac{\partial n}{\partial s}\right)_u \hat{v} = nC_s\hat{u} - D_\perp \frac{\partial n}{h\partial v}\hat{v} \qquad (5)$$

Since for constant u, $ds = hdv$. In references [1, 2, 3, 4], while deriving the leak width, authors have considered steady state solution of the diffusion equation and absence of ionization in the cusp region. But as demonstrated through various examples in [8], a simple model of including ionization would be to assume it to be proportional to plasma density. And for the pressure range and degree of ionization we are considering, this is quite reasonable assumption. More on its validity and other issues can again be found in [8]. With this assumption the divergence of ion flux becomes

$$\vec{\nabla} \cdot \vec{\Gamma} = K_{iz} n \tag{6}$$

Where K_{iz} is the ionization constant, which can be calculated if cross-sections of various ionization reaction are known. In our further calculation we will always assume this simple model for ionization. The solution derived by this can always be extrapolated for case of no ionization by putting $K_{iz} = 0$ in the solution. The plasma diffusion equation in the cases when flux is given by Eq. 5 and that when given by Eq. 6 would respectively be written as:

$$-\frac{\partial}{\partial u}\left(D_{\|}\frac{\partial n}{\partial u}\right) - \frac{\partial}{\partial v}\left(D_{\perp}\frac{\partial n}{\partial v}\right) = h^2 K_{iz} n \tag{7}$$

$$\frac{\partial}{\partial u}(hnC_s) - \frac{\partial}{\partial v}\left(D_{\perp}\frac{\partial n}{\partial v}\right) = h^2 K_{iz} n \tag{8}$$

The solution of above diffusion equations, are quite similar as has been shown subsequently. The validity of the above diffusion equation is primarily governed by the fact that the mean free path should be less than the region of our interest, which would hold true in the high gas pressure region.

PLASMA DENSITY PROFILE AND CUSP LEAK WIDTH

We will solve the above mentioned equations, first by assuming $h = |B|^{-1}$ (Appendix) to be constant in the cusp region, which will allow us to solve the diffusion equation in quite general manner. Next we will relax this assumption, which will make the partial differential equation quite complicated, and then solve it assuming u and v to be given only by first harmonic, in both cylindrical and planar case.

1. Assuming h to be constant: With this assumption of h as constant, $D_{\perp}$ also turns out to be a constant since w_c will be a constant. Hence the diffusion Eq. 7 modifies to the form:

$$D_{\|}\frac{\partial^2 n}{\partial u^2} + D_{\perp}\frac{\partial^2 n}{\partial v^2} = -h^2 K_{iz} n \tag{9}$$

The structure of equations allows us to look for the solution of plasma density in form of variable separation as $n = n_1(u) n_2(v)$ which reduces the above equation to

$$\frac{d^2 n_1(u)}{n_1(u) h^2 du^2} + \frac{D_{\perp}}{D_{\|}} \frac{d^2 n_2(v)}{n_2(v) h^2 dv^2} = -\frac{K_{iz}}{D_{\|}} \tag{10}$$

Hence let

$$\frac{d^2 n_1(u)}{n_1(u) h^2 du^2} = k; \frac{D_{\perp}}{D_{\|}} \frac{d^2 n_2(v)}{n_2(v) h^2 dv^2} + \frac{K_{iz}}{D_{\|}} = -k \tag{11}$$

It is important to note here that k is a constant and its value can be determined, if density profile and density gradients are known to us at any one point. For convenience, we assume it to be at $u = u_o, v = v_o$, then $k = \frac{d^2n_1(u_o)}{n_1(u)h^2du^2} = \left(\frac{\partial^2 n}{nh^2\partial u^2}\right)_{v,u=u_o}$. Here we are not using boundary condition to find the possible value of k. Instead we assume k to be a given constant and we are more interested in finding the expression of plasma density and leak width. From physical consideration it is clear that as u increases then n decreases for $v = 0$ and from symmetry of cusp geometry $\left(\frac{\partial n}{\partial v}\right)_u = 0$ at $v = 0$, we get the solution as $n_1 = n_{1o}\exp\left(-\sqrt{k}hu\right)$ and $n_2 = n_{2o}\cos\left(\sqrt{\frac{D_{||}k+K_{iz}}{D_\perp D_{||}}}hv\right)$. Further if the plasma density at $u = u_o$ and $v = 0$ is n_o then the solution can be expressed as:

$$n = n_o\exp\left(-\sqrt{k}h(u-u_o)\right)\cos\left(\sqrt{\frac{(D_{||}k+K_{iz})}{D_\perp D_{||}}}hv\right) \quad (12)$$

It is to be noted that $k < 0$ can be ruled out on physical consideration as this would make $n_2 = n_{2o}\cosh\left(\sqrt{\frac{D_{||}k+K_{iz}}{D_\perp D_{||}}}hv\right)$ which would imply n increases as v increases or decreases at $v = 0$ which is not acceptable, as plasma density is maximum at $v = 0$. Now to compute the plasma density profile in terms of r and θ or x and y, we substitute the expression of u and v in terms of r and θ or x and y. The contributions of higher harmonics of u and v, may be ignored as they add very little to the overall magnetic field, as the primary contribution comes from the first harmonic of u and v. Further it also simplifies the analytical expressions of plasma density at cusp. In order to compute the leak width, we need to know the variation in plasma density along the surface of plasma chamber at the cusp, for this we express the plasma density at $r = R$, hence dependent on θ or at $y = 0$ and hence dependent on x alone. With this the expression of plasma density for cylindrical and planar case, respectively are:

$$n = n_o\exp\left(-\sqrt{k}ha\left(\cos(N\theta)-1\right)\right)\cos\left(\sqrt{\frac{(D_{||}k+K_{iz})}{D_\perp D_{||}}}ha\sin(N\theta)\right) \quad (13)$$

$$n = n_o\exp\left(-\sqrt{k}ha\left(\cos\left(\frac{-\pi x}{l}\right)-1\right)\right)\cos\left(\sqrt{\frac{(D_{||}k+K_{iz})}{D_\perp D_{||}}}ha\sin\left(\frac{-\pi x}{l}\right)\right) \quad (14)$$

Now to find the leak width, defined as FWHM, we need to find the value of x or θ such that $n/n_o = 1/2$. This results in a transcendental equation, which can only be solved numerically. But as the cusp is quite narrow, we are interested in variation in density for small x or θ, hence we can expand the expression of n/n_o in terms of x or θ and retaining the term up to second order, we get the expression of leak width in cylindrical and planar case respectively as:

$$\Delta s_c = \frac{2}{\sqrt{\frac{D_{||}}{D_\perp}k+\frac{K_{iz}}{D_\perp}}-\frac{N}{R}\sqrt{k}} \ ; \ \Delta s_p = \frac{2}{\sqrt{\frac{D_{||}}{D_\perp}k+\frac{K_{iz}}{D_\perp}}-\frac{\pi}{l}\sqrt{k}} \quad (15)$$

Here $k = L^{-2} = \frac{\partial^2 n(r,\theta)}{n\partial r^2}\Big|_{r=R,\theta=0}$ for cylindrical case and $k = L^{-2} = \frac{\partial^2 n(x,y)}{n\partial y^2}\Big|_{x=0,y=0}$ for planar case. Thus it is proportional to the second derivative of plasma density at cusp along the magnetic field line. Note that N/R for cylindrical case is equivalent to π/l for planar case, as N/R is nothing but π divided by distance between two consecutive magnets in cylindrical case. Also note that there is a negative term in the denominator under square root. It appears that by substantially decreasing the distance between the two consecutive magnets one can make the term under the square root, over all negative. But by decreasing the distance between the two consecutive PM the approximation that h is constant will be no more valid. In subsequent analysis it is shown that, by relaxing this approximation, that h is constant, we get one more positive term, which is always greater than the absolute value of negative term present here, hence together they will always be positive, hence square root will never be imaginary.

Now if we assume plasma to be lost at cusp with ion acoustic velocity along the magnetic field direction then, in order to solve for density profile we need to start with the Eq.8. Following the same procedure with assumption that h to be constant and using separation of variable for $n = n_1(u)n_2(v)$ we get

$$\frac{dn_1(u)}{n_1 h du} = \frac{D_\perp}{h^2 C_S} \frac{d^2 n_2(v)}{n_2 dv^2} + \frac{K_{iz}}{C_s} = -\tilde{k} \equiv Constant \tag{16}$$

Using similar physical argument as done before we get the solution for plasma density given by the expression:

$$n = n_o \exp\left(-\tilde{k}h(u-u_o)\right) \cos\left(\sqrt{\frac{(C_s\tilde{k}+K_{iz})}{D_\perp}} hv\right) \tag{17}$$

Where n_o is the plasma density at $u = u_o$ and $v = 0$. And again to get the CLW we first substitute the expression of u and v in terms of x and y or r and θ and then considering $x = 0$(or $r = R$), we expand the expression at $y = 0$(or $\theta = 0$) up to second order in y(or θ). Finally we get the value of y(or θ) for $n/n_o = 1/2$, and we get cusp width as $\Delta s = 2y$(or $\Delta s = 2R\theta$) as:

$$\Delta s_c = \frac{2}{\sqrt{\frac{C_s}{D_\perp L} + \frac{K_{iz}}{D_\perp} - \frac{N}{RL}}} \quad ; \quad \Delta s_p = \frac{2}{\sqrt{\frac{C_s}{D_\perp L} + \frac{K_{iz}}{D_\perp} - \frac{\pi}{lL}}} \tag{18}$$

But here $L^{-1} = \tilde{k} = \frac{dn}{ndr}\Big|_{cusp}$ and not as half length of the magnetic field line, as specified by the authors [1, 2, 3, 4]. L is dependent on the density gradient at the cusp point, and hence a local parameter. And if it is assumed that the relation between the density gradient and ion acoustic velocity at the plasma edge is given by $D_\parallel \frac{\partial n}{n\partial r} = C_s$ $\left(or\ D_\parallel \frac{\partial n}{n\partial y} = C_s\right)$ at the cusp then $L = \frac{D_\parallel}{C_s}$ and $L^{-2} = \frac{\partial^2 n}{n\partial r^2} = \frac{\partial^2 n}{n\partial y^2} = \left(\frac{C_s}{D_\parallel}\right)^2$, with assumption that both C_s and $D_\parallel$ are constant. If this assumption is made, it turns out that the expression for leak width is same, whether we assume, plasma to be lost at cusp with ion acoustic velocity or diffuses along field line having coefficient $D_\parallel$. And we get the expression of leak width for cylindrical and planar case as:

$$\Delta s_c = \frac{2}{\sqrt{\frac{C_s^2}{D_\perp D_{||}} + \frac{K_{iz}}{D_\perp} - \frac{NC_s}{RD_{||}}}} \quad ; \quad \Delta s_p = \frac{2}{\sqrt{\frac{C_s^2}{D_\perp D_{||}} + \frac{K_{iz}}{D_\perp} - \frac{\pi C_s}{lD_{||}}}} \quad (19)$$

2. Assuming only first harmonic of u and v to be non-zero: The assumption that h is constant i.e it is independent of u and v, is strictly not true. We can relax these assumption, but then we need the partial derivative of $D_\perp$ and h (hence magnetic field) with respect to to u and v. So to keep the things simpler and to get analytical results we restrict the analysis to the first harmonic of u and v. This is not bad approximation as the major contribution to u and v (hence B) comes from first harmonic [6]. Since the expression of u and v for planar and cylindrical case are different hence it will be required to deal with them separately. We will assume that the plasma is lost at the cusp, along the field line with ion acoustic velocity. The expression of $D_\perp = D_{||}/(1+(w_c\tau)^2)$ but for our case $(w_c\tau)^2 >> 1$ hence we assume $D_\perp = D_{||}/(w_c\tau)^2 = \Omega/B^2$, where Ω is constant independent of u and v, and B is magnetic field. With this the diffusion equation reads:

$$C_s n \frac{\partial h}{\partial u} + C_s h \frac{\partial n}{\partial u} - \Omega \frac{\partial h^2}{\partial v} \frac{\partial n}{\partial v} - \Omega h^2 \frac{\partial^2 n}{\partial v^2} = h^2 K_{iz} n \quad (20)$$

The multicusp geometry create magnetic field in such a manner that the magnetic field show rapid variation along y (or r) but not along the x(or θ). Hence at the cusp, the variation in B due to change in v is very small compared to that of change in u, which will also be evident from the analytical result of CLW up to the second order, which will be independent of $\frac{\partial h^2}{\partial v}$. In planar and cylindrical case the expression of h in terms of u and v are

$$h_p = \frac{l}{\pi}\left(u^2+v^2\right)^{-\frac{1}{2}} \quad ; \quad h_c = \frac{R}{Na^{1/N}}\left(u^2+v^2\right)^{-\frac{1}{2}\left(1-\frac{1}{N}\right)} \quad (21)$$

We have restricted u and v up to first harmonic for above derivation. We then calculate the derivative of h with respect to u and h^2 with respect to v. Since at cusp $v \ll u$, so we ignore v^2 term in u^2+v^2, where ever it occurs. With these consideration the above equations for planar and cylindrical case respectively will become

$$\frac{\pi C_s}{l\Omega} u \frac{\partial n}{n\partial u} - \frac{\pi C_s}{l\Omega} = \frac{\partial^2 n}{n\partial v^2} - 2\frac{v}{u^2}\frac{\partial n}{n\partial v} + \frac{K_{iz}}{\Omega} \quad (22)$$

$$\frac{C_s u^{-\frac{1}{N}+1}}{c_o \Omega} \frac{\partial n}{n\partial u} - \frac{c_1 C_s u^{-\frac{1}{N}}}{c_o^2 \Omega} = \frac{\partial^2 n}{n\partial v^2} - \frac{2c_2}{c_o^2} \frac{v}{u^2}\frac{\partial n}{n\partial v} + \frac{K_{iz}}{\Omega} \quad (23)$$

Where $c_o = \frac{R}{Na^{1/N}}, c_1 = \frac{(N-1)R}{N^2 a^{1/N}}$ and $c_2 = \frac{(N-1)R^2}{N^3 a^{2/N}}$. Note that the second term on RHS has a factor $\frac{v}{u^2}$ and since near cusp, $v \approx 0$ hence $v \ll u$, this factor will be more sensitive to change in v rather than u, hence we can replace u with u_o, which is the value of u at the cusp. With this approximation we can do variable separation as $n = n_1(u)n_2(v)$, which will make LHS as function of u and RHS as function of v. Hence equating above two equation with a constant $-k$, we get in cylindrical case,

$$\frac{dn_1(u)}{du} = \left(\frac{c_1}{c_o} - \frac{k\Omega c_o}{C_s}u^{1/N}\right)\frac{n_1(u)}{u} \tag{24}$$

$$\frac{d^2n_2(v)}{dv^2} - 2\frac{c_2}{c_o^2}\frac{v}{u_o^2}\frac{dn_2(v)}{dv} + \left(\frac{K_{iz}}{\Omega}+k\right)n_2(v) = 0 \tag{25}$$

The solution of above differential equation can be found compatible with the physical situation and plasma density $n = n_1(u)n_2(v)$ can be written as (Appendix) :

$$n = n_o\left(\frac{u}{u_o}\right)^A \exp\left(-B1\left(u^{\frac{1}{N}}-u_o^{\frac{1}{N}}\right) - B2v^4\right)\cos(Cv) \tag{26}$$

Where $A = c_1/c_o$, $B1 = \left(\frac{K_{iz}}{\Omega}+k\right)\frac{\Omega c_o}{C_s}$ and $B2 = \frac{c_2(K_{iz}+\Omega k)}{c_o^2 6u_o^2\Omega}$ and $C = \sqrt{\frac{K_{iz}}{\Omega}+k}$. Now in order to get the analytical expression of leak width, we substitute the expression of u and v in terms of r and θ, then put $r = R$ and Taylor expand the resulting expression at $\theta = 0$ up to second order in θ. And solving for θ such that $n/n_o = 1/2$, we get

$$\Delta s_c = \frac{2}{\sqrt{\frac{C_s}{D_\perp}k + \frac{K_{iz}}{D_\perp} + \frac{Nk}{R}\left(\frac{(N-1)D_\|}{ND_\perp}-1\right)}} \tag{27}$$

Where $k = \left.\frac{\partial n}{n\partial r}\right|_{at\ cusp}$. If we assume that $D_\|\frac{\partial n}{n\partial r} = C_s$ then $k = C_s/D_\|$ and $A = \frac{N-1}{N}$, $B1 = \frac{RC_s}{D_\|} + N - 1$ and $B2 = \frac{(N-1)R^2}{6N^3}\left(\frac{C_s^2}{D_\|D_\perp} + \frac{(N-1)C_s}{RD_\perp}\right)$ and $C = \frac{R}{N}\sqrt{\frac{C_s^2}{D_\|D_\perp} + \frac{K_{iz}}{D_\perp} + \frac{(N-1)C_s}{RD_\perp}}$
hence with this simplification the expression of leak width become:

$$\Delta s_c = \frac{2}{\sqrt{\frac{C_s^2}{D_\perp D_\|} + \frac{K_{iz}}{D_\perp} + \frac{NC_s}{R}\left(\frac{N-1}{ND_\perp}-\frac{1}{D_\|}\right)}} \tag{28}$$

Beginning with the Eq. 22, the expression equivalent to Eq.27 and Eq. 28 (Cylindrical case), can be derived for planar case, giving CLW as below:

$$\Delta s_p = \frac{2}{\sqrt{\frac{C_s}{D_\perp}k + \frac{K_{iz}}{D_\perp} + \frac{\pi k}{l}\left(\frac{D_\|}{D_\perp}-1\right)}} \ ; \ \Delta s_p = \frac{2}{\sqrt{\frac{C_s^2}{D_\perp D_\|} + \frac{K_{iz}}{D_\perp} + \frac{\pi C_s}{l}\left(\frac{1}{D_\perp}-\frac{1}{D_\|}\right)}} \tag{29}$$

The first term in the square root $\frac{C_s}{D_\perp}k$ (Eq. 27) or $\frac{C_s^2}{D_\|D_\perp}$ (Eq. 28) is dependent on the diffusion coefficients and ion acoustic velocity (or plasma gradient at cusp). These factor depends on the plasma property like collision frequency, electron and ion temperature and $|B|$ at the cusp. It contains no information of cusp geometry. The third term is dependent on the distance between two consecutive permanent magnets. Since the geometry of the cusp (the shape and curvature of the field lines) is mainly decided by the distance between the consecutive permanent magnet, the geometrical information on the cusp is contained in the third term. The second term again is independent of cusp geometry.

Thus only third term captures the geometry of the cusp. To get an idea of relative contribution of the various terms in the leak width, we compute the value of these terms for two typical case of H_2 gas pressure of $1mTorr \& 10mTorr$. We assume in both the pressure, the value of T_e and T_i of $5eV$ and $0.5eV$ respectively, and cusp magnetic field of $0.2Tesla$, with distance between the two consecutive magnet is equal to $1.5cm$. The % contribution of the term in the leak width has been tabulated in Table1.

TABLE 1. Comparing the contribution of various terms in leak width

	$\frac{C_s^2}{D_\perp D_\parallel}$	$\frac{K_{iz}}{D_\perp}$	$\frac{NC_s}{R}\left(\frac{N-1}{ND_\perp} - \frac{1}{D_\parallel}\right)$
1	8.54%	1.32%	90.17%
10	45.23%	7.1%	47.8%

Hence the contribution of third term is most dominant in low gas pressure regime. In high pressure regime contribution of first and third term is almost equal. The term associated with the ionization contribute only moderately to the overall leak width. Note that ionization in the cusp region reduces the leak width. This is somewhat counterintuitive, but can be understood as the ionization increases the number of ion particles at the cusp, but due to restricted movement of particle perpendicular to the field line, more number of particles are lost parallel to the field line than perpendicular to it, hence the field gradients are more steep perpendicular to the field line and half of the maximum density is reached at relatively smaller distance. Further note that, because of the factor $K_{iz}/D_\perp$, if $D_\perp$ increases i.e. flow of the ions perpendicular to the field line increases, this factor do not contribute much in such case the contribution due to ionization can be ignored. To convince the analytical expression derived for Δs using Taylor expansion and retaining the term up to second order, is reasonably close to $n/n_o = 1/2$, we directly solve the transcendental equation for $n/n_o = 1/2$ for different value of B and compare the result with those obtained from leak width formula (Eq. 28). It has been tabulated in Table 2. To find the roots of transcendental equation, bisection method has been used.

Here $\Delta t_{1/2}$ is half leak width computed numerically by finding roots using transcendental equation and $\Delta s_{1/2}$ is half leak width computed using Eq. 28.

TABLE 2. Comparing Half Leak Width (HWHM) computed through transcendental equation and second order Taylor expansion at two different gas pressures.

	1mTorr			10 mTorr		
$\|B\|$ Gauss	$\Delta t_{1/2}$ cm	$\Delta s_{1/2}$ cm	Ratio $\frac{\Delta t_{1/2}}{\Delta s_{1/2}}$	$\Delta t_{1/2}$ cm	$\Delta s_{1/2}$ cm	Ratio $\frac{\Delta t_{1/2}}{\Delta s_{1/2}}$
100	0.2234	0.2122	1.0526	0.2131	0.2041	1.0440
500	0.0455	0.0434	1.0474	0.0955	0.0912	1.0465
1000	0.0228	0.0217	1.0473	0.0511	0.0488	1.0470
2000	0.0114	0.0109	1.0472	0.0260	0.0249	1.0471
3000	0.0076	0.0072	1.0472	0.0174	0.0166	1.0472
5000	0.0046	0.0043	1.0472	0.0105	0.0101	1.0472

As can be seen the ratio between the numerically calculated and analytically calculated never exceed more than 1.053. Hence the analytical formulae are quite accurate.

The formula of leak width for cylindrical and planar case is similar. It can be expressed in general form as:

$$\frac{1}{\Delta s_{1/2}^2} = \frac{1}{\Delta d^2} + \frac{1}{\Delta k_{iz}^2} + \frac{1}{\Delta g^2} \quad (30)$$

where Δd is diffusion term, Δk_{iz} is ionization term and Δg is geometric term. For example, for Eq. 29, $\Delta d^2 = \frac{D_\perp D_\parallel}{C_s^2}$, $\Delta k_{iz}^2 = \frac{D_\perp}{K_{iz}}$ and $\Delta g^2 = \frac{l D_\perp D_\parallel}{\pi C_s (D_\parallel - D_\perp)}$

DISCUSSION ON CUSP LEAK WIDTH

The cusp loss width for multipole geometry by the authors [1, 2] has been derived as

$$\Delta s = 2\sqrt{\frac{D_\perp L}{C_s}} \quad (31)$$

Here $D_\perp$ is the diffusion coefficient perpendicular to the magnetic field lines, C_s ion acoustic velocity and the L is the half length of magnetic field line(or called as scale length of cusp). While both of them have based their analytic calculation method given in [3] but it has not been made clear that which magnetic field line is being considered in their derivation as their final expression clearly depends on the half length of the magnetic field line between the two consecutive poles of magnets. In ref [2] author has argued for the magnetic surface that is boundary between plasma and magnetic field for the computation of the half length of the magnetic field line, and that magnetic surface is decided by equating magnetic pressure and plasma pressure. This argument can't be true in the sense that there is no such magnetic surface which can be considered as strict boundary between plasma and magnetic field, as plasma collisionally diffuses across the magnetic field lines. Further the ion flux perpendicular to the field line has been assumed to be constant over the field line since $\Gamma_\perp = -D_\perp \frac{\partial n}{\partial x}$ hence it implies the diffusion coefficient has been assumed to be constant over the length of the magnetic field line, which is not true as the field strength varies considerably along the field line and more so for the field line originating near the pole face (i.e at the cusp) as they penetrate deep into the plasma chamber. Moreover author [2] has shown that the best fit for $|B|$ gives the value of $|B|$ as 48 Gauss, in contrast the $|B|$ at the cusp is as large as 2kG, as it has been argued that the average value of $|B|$ over the magnetic field line is much smaller than the $|B|$ at the cusp, but without giving any appropriate method for the computation of the average magnetic field for evaluation of diffusion coefficient, that can be used in the Eq. 31 for the computation of CLW.

The CLW of $2\sqrt{r_{ce}r_{ci}}, 2\sqrt{r_{ce}r_{ci}}, \sqrt{r_{ce}r_{ci}}, 4r_{ci}$ etc. has been reported, but all of them are convenient fit to the experimental results and not analytically derived result. To compare the results of the leak width formula obtained and other analytic result of leak width formula, we have used the experimental results of Horiike et al. [12] as a reference. The table 1 of [12] has listed the value of leak width vs magnetic field at the cusp. The gas pressure used in [12] is 3 mTorr, with H^+ temperature at around 0.3eV, hence the discharge is at intermediate pressure, the formula derived by Matthieussent and Pelletier is worth comparing. According to them leak width is

$$\Delta s = \frac{2d}{\pi} \sqrt{\frac{r_{ce} r_{ci}}{\lambda_{me} \lambda_{mi}}} \qquad (32)$$

where λ_{me} and λ_{mi} are the electron and ion mean free path. Also Koch et al in [4] has characterized the plasma diffusion transverse to magnetic field through parameter $\gamma \beta B^2$, where β is the ratio of electron and ion production rate, and has been assumed to be constant for the plasma. The field line starting from the cusp and tangent to the region defined by $\gamma \beta B^2 = 1$ has been taken as the field line which contains the plasma (Fig 4 in [4]). Hence, field line position at the cusp determines the leak width. To keep the things simple, if we apply the same method to the first harmonic of the multipole magnetic field, then the leak width evaluates to

$$\Delta s_c = \frac{R}{N} \left(\frac{B_o}{B_s} \right)^{\frac{N}{N-1}} ; \quad \Delta s_p = \frac{l}{\pi} \left(\frac{B_o}{B_s} \right) \qquad (33)$$

respectively for cylindrical and planar case. R and l has the usual meaning and B_o is the field determined by $\gamma \beta B_o^2 = 1$. The leak width computed using the above mentioned formula and the result of experiment quoted in [12] has been plotted in Fig. 4. It is clear

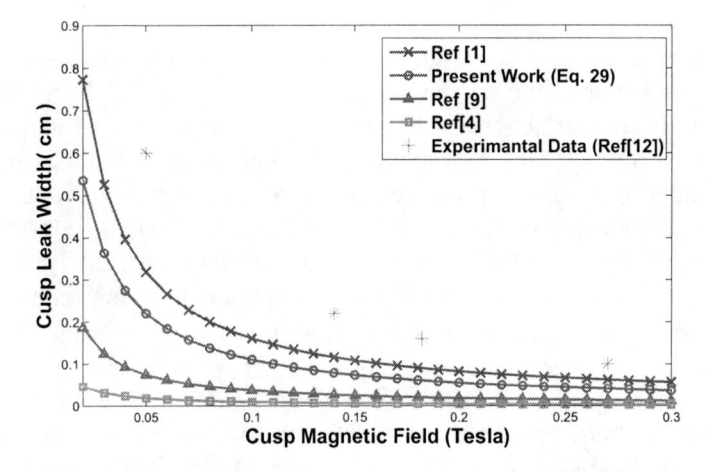

FIGURE 4. Comparing CLW obtained using different analytic formulae derived by various authors, including current result, with that of experimental result

from this figure that none of the formula, derived on the basis of the classical diffusion theory, or geometrical consideration, predicts the correct leak width. The ambiguities present in the formula given by Eq. 31 has been removed by the Eqs. 28 & 29, derived in this paper, but even then the experimental leak width estimated by this formula is off by a factor of around 2.2. The Eq. 33 which has been derived on the basis of the geometrical consideration, predicts much smaller leak width, even when β has been taken equal to 1which is in general greater than 1.

Few possibilities can be considered, like the gradient of the plasma density at the cusp is much smaller than what is being assumed here, or it might be that the classical diffusion across the magnetic field at cusp is not sufficient to account for the leak width experimentally obtained and the phenomena like turbulence, trapping of hot electron, instability etc. might be playing some role in the cusp field region.

APPENDIX

General characteristic of the magnetic field

Magnetic field has been computed using vector potential ($\vec{A}$). Since $\vec{A} = (0,0,A_z)$ which simplifies the expression for $\vec{B}$ as

$$\vec{B} = \frac{1}{r}\frac{\partial A_z}{\partial \theta}\hat{r} - \frac{\partial A_z}{\partial r}\hat{\theta} \tag{34}$$

Using the functional form of A_z, the B has been evaluated for cylindrical and planar case respectively as:

$$B_r = \frac{N}{R}\sum_{n=1}^{\infty} na_n \left(\frac{r}{R}\right)^{Nn-1} \theta \ ; \ B_\theta = -\frac{N}{R}\sum_{n=1}^{\infty} na_n \left(\frac{r}{R}\right)^{Nn-1} \sin Nn\theta \tag{35}$$

$$B_x = \frac{-\pi}{l}\sum_{n=1}^{\infty} na_n \exp\left(-\frac{n\pi y}{l}\right)\sin\frac{n\pi x}{l} \ ; \ B_y = \frac{-\pi}{l}\sum_{n=1}^{\infty} na_n \exp\left(-\frac{n\pi y}{l}\right)\cos\frac{n\pi x}{l} \tag{36}$$

When $\vec{A} = (0,0,A_z)$ then the contours of A_z is same as magnetic field lines as:

$$d\vec{s} \times \vec{B} = 0 \Rightarrow B_\theta dr - B_r r d\theta = 0$$
$$\Rightarrow \frac{\partial A_z}{\partial r}dr + \frac{\partial A_z}{\partial \theta}d\theta = 0_z = 0 \tag{37}$$

The metric parameter h can be evaluated for the coordinate transformation $(r,\theta) \rightarrow (u,v)$ related through Eq. 3, as follows

$$\begin{aligned} dv &= \frac{\partial v}{\partial r}dr + \frac{\partial v}{r\partial \theta}rd\theta = \eta dr + \xi rd\theta \\ du &= \frac{\partial u}{\partial r}dr + \frac{\partial u}{r\partial \theta}rd\theta = \xi dr - \eta rd\theta \end{aligned} \tag{38}$$

But $B_r = \frac{\partial v}{r\partial \theta} = \frac{\partial u}{\partial r} = \xi$ and $B_\theta = -\frac{\partial v}{\partial r} = \frac{\partial u}{r\partial \theta} = -\eta$ hence

$$dv^2 + du^2 = (\xi^2 + \eta^2)(dr^2 + r^2 d\theta^2) = (\xi^2 + \eta^2) ds^2 \tag{39}$$

thus

$$h = |B|^{-1} \tag{40}$$

Same result follows for the planar case. Some other useful properties of two dimensional multipolar magnetic field has been given in [13]

Solution of differential equation

In solving for plasma density profile, we will frequently encounter the differential equation of following form

$$\frac{d^2y}{dx^2} - \alpha x \frac{dy}{dx} + \beta y = 0 \qquad (41)$$

where α and β are constant. When $\alpha = 0$ the general solution of the above equation can be written as

$$y = A\cos\left(\sqrt{\beta}x\right) + B\sin\left(\sqrt{\beta}x\right) \qquad (42)$$

and when $\alpha \neq 0$ the general solution can be written as

$$y = A \; {}_1F_1\left[\frac{-\beta}{2\alpha}, \frac{1}{2}, \frac{\alpha}{2}x^2\right] + B\, H_{\frac{\beta}{\alpha}}\left[\sqrt{\frac{\alpha}{2}}x\right] \qquad (43)$$

where A and B are constant and ${}_1F_1$ is Hyper geometric function and H is Hermite function. The physical requirement of our problem is such that, $\left.\frac{dy}{dx}\right|_{x=0} = 0$, hence for our case, $B = 0$ in both the Eq. 42, 43. Since in our case we are interested in variation of y for small x, near $x = 0$, it is better to approximate function $Hypergeometric_1F_1$, which is quite difficult to comprehend in terms of well-known function. One such approximation of $Hypergeometric_1F_1\left[\frac{-\beta}{2\alpha}, \frac{1}{2}, \frac{\alpha}{2}x^2\right]$ is $\exp\left(-\frac{\alpha\beta}{12}x^4\right)\cos\left(\sqrt{\beta}x\right)$. This approximation matches, up to 4th power in x, hence for small x this is excellent approximation, which has been used in all the analytical expressions of plasma density in this paper. Further note that when $\alpha = 0$ this expression conveniently reduces to $\cos\left(\sqrt{\beta}x\right)$, which is the correct solution. The term $\exp\left(-\frac{\alpha\beta}{12}x^4\right)$, can be viewed as correction, to the well-known solution of $\cos\left(\sqrt{\beta}x\right)$, if there is small perturbation to the system, which gives rise to the term $-\alpha x\frac{dy}{dx}$ in the differential equation, where the strength of the perturbation being determined by the coefficient α.

REFERENCES

1. A. Fukano, T. Mizuno, A. Hatayama, and M.Ogasawara, *Rev. Sci. Instrum.* 77, 03A524(2006).
2. T. Morishita, M. Ogasawara and A. Hatayama, *Rev. Sci. Instrum.* 69, 968(1998).
3. R. A. Bosch and R. L. Merlino, *Phys. Fluid* 29,1998(1986).
4. C. Koch and G. Matthieussent, *Phys. Fluid* 26,545(1983).
5. R. Jones, *Phys. Fluid* 27(11),2780(1984).
6. Ajeet Kumar and V. K. Senecha, *AIP Conf. Proc.* vol. 1097,137 (2009).
7. A. Fukano, A. Hatayama, and M.Ogasawara, *Japanese Journal of Applied Physics* 46,4A 1668(2007).
8. Michael A. Liberman and Allan J. Lichtenberg, *Principal of Plasma Discharge and Material Processing*,Publisher: John Wiley and sons, 1994.
9. C. Gauthereau, G. Matthieussent, J.L. Rauch and J. Godiot, *Phys. Letters* 101A,8,394(1984).
10. C. Gauthereau and G. Matthieussent, *Physics Letters* 102A 5,6 (1984)
11. K.N.Leung, Noah Herskowitz and K.R.Mackenizie, *Phys. Fluid* 19,1045(1976).
12. Hiroshi Horiike, Masato Akiba, Yoshihiro Ohara, Yoshikazu Okumura and Shigeru Tanaka, *Phys. Fluids* 30(10), 3268(1987).
13. D E Lobb, *Nuclear Instrument and methods* 64, 251(1968).

Low-Pressure Small-Radius Hydrogen Discharge as a Volume-Production Based Source of Negative Ions

Tsvetelina V. Paunska[a], Antonia P. Shivarova[a] and Khristo Ts. Tarnev[b]

[a]*Faculty of Physics, Sofia University, BG-1164 Sofia, Bulgaria*
[b]*Department of Applied Physics, Technical University–Sofia, BG-1000 Sofia, Bulgaria*

Abstract. The two-dimensional model of low-pressure discharges presented in the study provides description of hydrogen discharge maintenance in a free-fall regime with account for the volume produced negative ions in the discharge. The factors determining the spatial distribution of the plasma parameters, including that of the negative ions, in discharges with a localized rf power deposition and, respectively, with regions with remote plasma maintenance, are discussed. In agreement with previous one-dimensional models of hydrogen discharges maintained in a free-fall regime, the results show that small radius discharges sustain high density of volume-produced negative ions in the discharge center. This accumulation of the negative ions in the central part of the discharge where the rf power deposition is localized results from substantial influx of negative ions produced all over the discharge volume, i.e. from a nonlocal behavior of the discharge strongly predominating over the local balance of the ions associated with the elementary processes (collisions).

Keywords: 2D Discharge Modeling, Hydrogen Discharges, Sources of Negative Hydrogen Ions
PACS: 52.27.Cm, 52.50.Dg, 52.80.Pi

INTRODUCTION

The study is in the course of current work [1-7] on a negative hydrogen-ion rf plasma source performed as a matrix of small-radius discharges [2]. A discharge with a simple configuration – a rf discharge in a straight-tube discharge vessel – is considered as a single element of the matrix. The studies carried out up to now are mainly on inductively driven discharges with a cylindrical coil.

In general, the discharge behavior is governed by local processes (collisions) and fluxes resulting from transport processes. The efficiency of the one-chamber dc (filament) plasma source [8-10] regarding maintenance of negative hydrogen ions is due to predominating local balance [11] assisted by the magnetic filter that favors the volume production of the ions. On the other hand, one-dimensional (1D) models [1,2] of small-radius rf-driven hydrogen discharges sustained at low gas pressures, in a free-fall regime, have shown high efficiency regarding volume-produced negative ions due to predominating nonlocality in the discharge behavior. The result is a strong accumulation of the negative ions – produced all over the discharge cross section – in the on-axis region of the discharge. It is due to their flux in the radial dc electric field and to the small radius (2-3 cm) of the discharge ensuring possibility for the ions to reach the discharge axis without being destroyed in collisions. A two-dimensional (2D) model [3] of diffusion-controlled hydrogen discharges with localized (in the axial direction) rf power deposition, like in inductive discharges with a cylindrical coil, has displayed, as it should be expected, more considerable role of the transport processes due to the presence in the discharge of regions with remote plasma maintenance. The conclusion is that the driver (the region of the rf power deposition) of small-radius discharges sustains high density of negative ions due to the fluxes of the ions in the dc electric field. High density of negative ions in the driver and in its vicinity is a result also from experiments [5].

This study is an extension of the previous 1D models [1,2] of hydrogen-discharge maintenance in a free-fall regime and of the 2D model of diffusion-controlled hydrogen discharge [3] towards 2D description of hydrogen discharges (with volume-produced negative ions) sustained in a free-fall regime. As it is known, the description of the free-fall regime is an arduous task, in particular in the 2D modeling, because of necessity of accounting for the nonlinear inertia term in the momentum equations of the positive ions. Acting as a retarding force it limits the increase of the ion velocity in the wall sheaths and, respectively, reduces the charged-particle losses ensuring reasonable values of the electron temperature and, respectively, adequate description of the discharge. In the model presented here the inertia-term action is described within an approximation recently suggested [12]. The presented results are for the spatial distribution of plasma parameters. The analysis stresses on the factors determining the maintenance of the negative ions. The conclusion is for a high density of the negative ions in the discharge center resulting from a substantial ion influx which has orders of magnitude higher contribution than the local production of the ions there. Thus, contrary to the concept for the tandem source which is based on locality, a discharge with overlapped regions of the vibrational excitation of the hydrogen molecules and of the volume production of the ions by electron attachment to vibrationally excited molecules is an efficient source of negative ions when its maintenance is governed by nonlocality.

DESCRIPTION OF THE MODEL

The study presents fluid-plasma model description of low-pressure hydrogen discharges maintained in straight-tube gas-discharge vessels by applying axially-varying rf power deposition $P_{ext}(z) = P_0 \exp[-(1/2)(z/\sigma)^{2m}]$, homogeneous in the radial direction; $P_0 = P_{ext}(z=0)$ and σ is the parameter of the super-Gaussian profile

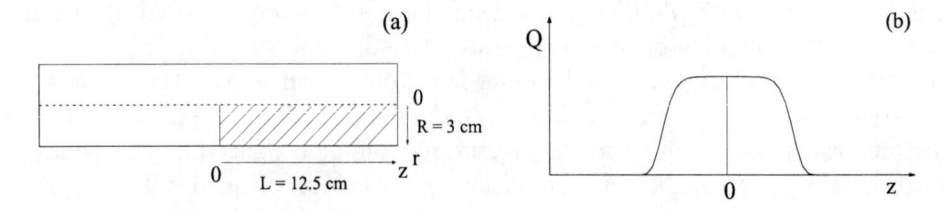

FIGURE 1. Schematical presentation of the gas-discharge vessel (a) and of the axial variation of the input power. A quater of the vessel is the modeling domain.

of P_{ext} (Fig. 1). Electrons, the three types of positive ions (H^+, H_2^+ and H_3^+), negative ions (H^-), hydrogen atoms (H), ground-state molecules (H_2) and vibrationally excited molecules $H_2(v=1-14)$ are the species in the model.

The set of equations

$$\frac{\partial n_\alpha}{\partial t} + \vec{\nabla}.\vec{\Gamma}_\alpha = \frac{\delta n_\alpha}{\delta t} \qquad (1)$$

$$\frac{\partial N_1}{\partial t} + \vec{\nabla}.\vec{\Gamma}_1 = \frac{\delta N_1}{\delta t} \qquad (2)$$

$$\frac{\partial N_2(v)}{\partial t} + \vec{\nabla}.\vec{\Gamma}_v = \frac{\delta N_2(v)}{\delta t} \qquad (3)$$

$$\frac{3}{2}\frac{\partial}{\partial t}(n_e T_e) + \vec{\nabla}.\vec{J}_e = P_{ext} - P_{coll} - e\vec{\Gamma}_e.\vec{E}_{dc} \qquad (4)$$

$$\Delta \Phi = -\frac{e}{\varepsilon_0}(\sum_{j=1}^{3} n_j - n_e - n_n) \qquad (5)$$

$$p = \kappa T_g (N_1 + N_2) \qquad (6)$$

consists of the continuity equations (1)-(3), respectively, of all types (α) of charged particles (5 equations), of the hydrogen atoms and of the vibrationally excited molecules (14 equations), the electron-energy balance equation (4), the Poisson equation (5) and the equation of state (6). In (1)–(6), n_α and $\vec{\Gamma}_\alpha$ (with $\alpha = e$ for electrons, $\alpha = j$ for the positive ions (with $j = 1-3$, respectively for H^+, H_2^+ and H_3^+) and $\alpha = n$ for H^-) are the densities and the fluxes of the charged particles, N_1, $\vec{\Gamma}_1$ and $N_2(v)$, $\vec{\Gamma}_v$ refer to hydrogen atoms and vibrationally excited molecules, T_e is the electron temperature (in energy units), $\vec{J}_e$ is the electron energy flux, E_{dc} and Φ are the dc field in the discharge and its potential, e, ε_0 and κ are the electron charge, the vacuum permittivity and the Bolzmann constant, p is the gas pressure and T_g is the gas temperature.

The fluxes of the electrons ($\vec{\Gamma}_e = -b_e n_e \vec{E}_{dc} - D_e \vec{\nabla} n_e - D_e^T n_e (\vec{\nabla} T_e / T_e)$) and of the negative ions ($\vec{\Gamma}_n = -b_n n_n \vec{E}_{dc} - D_n \vec{\nabla} n_n$) are drift-diffusion fluxes with thermal diffusion ($D_e^T \equiv D_e$) accounted for the electrons; b_α and D_α are the corresponding mobility and diffusion coefficients. The description of the free-fall regime of the discharge maintenance is specified by solving the stationary form of the momentum equations of the positive ions (with account for the nonlinear inertia term)

$$m_j n_j (\vec{v}_j.\vec{\nabla})\vec{v}_j = -e\vec{\nabla}\Phi - T_j \vec{\nabla} n_j - (\mu_j \nu_j)\vec{v}_j \qquad (7)$$

according to the procedure outlined in [12]; here m_j and T_j are, respectively, the ion mass and temperature and $(\mu_j v_j)$ is a notation for elastic collisions with both atoms and molecules. The simplifications involved [12] concerning the inertia term in (7) are accounting only for the velocity component that is perpendicular to the corresponding wall (since – due to the strong drop of the dc potential close to the walls – the inertia term acts mainly in the wall sheath where the dc field is almost perpendicular to the wall) and using the collisionless-case energy-concentration law of the ions. In a way, the ion fluxes are reduced again to drift-diffusion fluxes $\Gamma_{j(r,z)} = -b^{eff}_{j(r,z)} n_j \partial_{(r,z)} \Phi - D^{eff}_{j(r,z)} \partial_{(r,z)} n_j$, however, with effective mobility $(b^{eff}_{j(r,z)})$ and diffusion $(D^{eff}_{j(r,z)})$ coefficients defined via effective collision frequencies $v^{eff}_{j(r,z)} = \partial_{(r,z)} v_{j(r,z)} + (\mu_j v_j)/m_j$ (where $\partial_{(r,z)} v_{j(r,z)} = -\sqrt{e/2m_j}[(\partial_{(r,z)}\Phi)/\sqrt{\Phi_{max} - \Phi}]$ with $\Phi_{max} = \Phi(r=0, z=0)$, i.e. at the discharge center). In a way, the two retarding forces in the momentum equations of the positive ions – the collisions and the inertia term – are combined in effective collision frequencies, specified for each direction.

The electron energy flux $\vec{J}_e = -\chi_e \nabla T_e + (5/2) T_e \vec{\Gamma}_e$ includes both the conductive and convective fluxes; $\chi_e = (5/2) n_e D_e$ is the thermal conductivity coefficient. The electron energy losses in elastic and inelastic collisions are summarized in P_{coll}.

The processes of particle production and losses, summarized in the right-hand sides of (1)–(3), are ionization of molecules and atoms, dissociation of molecules and molecular ions, H_3^+ production in H_2^+-collisions with molecules, dissociative recombination of H_3^+-ions, H^--production by electron attachment to vibrationally excited molecules $H_2(v = 4-9)$ considered separately to each of the vibrationally excited states, destruction of H^- in collisions with electrons and atoms and in recombination with positive ions, wall recombination of atoms, vibrational-translational and vibrational-vibrational collisions, e-V and E-V processes.

The boundary conditions are for symmetry at $r = 0$ and $z = 0$ (Fig. 1), for the fluxes at the walls (thermal flux of the electrons, drift-diffusion fluxes of the positive ions, a zero flux of the negative ions, wall recombination losses of H, wall deactivation of $H_2(v)$ and convective electron energy flux) and for a zero dc potential (metal walls).

RESULTS AND DISCUSSIONS

The results from the model presented here are for the spatial distribution of the plasma parameters and for the factors determining its formation. The discussions is based on results for three values of the gas pressure ($p = 6, 8, 10$ mTorr) and three values of the total applied power ($Q = 315, 1200, 1600$ W). The size of the discharge vessel is: $R = 3$ cm for the radius and $L = 12.5$ cm for half of its length (Fig. 1). The value $R = 3$ cm of the discharge radius is according to the conclusions from the 1D model [2] that a small radius of the discharge is a requirement for reaching high density of the negative ions. The gas temperature is $T_g = 300$ K and the parameters of the rf power input are m = 4 and $\sigma = 5$ cm.

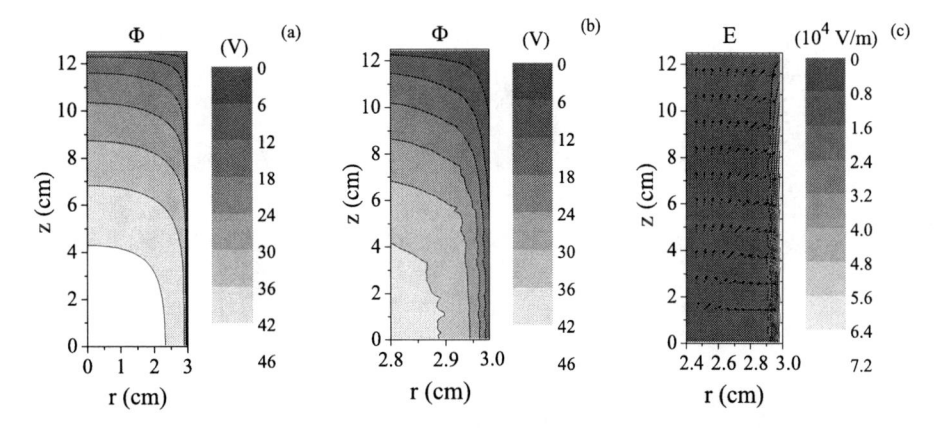

FIGURE 2. Spatial distribution of the potential ((a) and (b)) of the dc electric field and its intensity (c); $p = 6$ mTorr and $Q = 1600$ W. In (a) a quater of the discharge vessel is presented whereas (b) and (c) show the region close to the side wall of the discharge.

Figure 2 which shows the spatial distribution of the dc electric field intensity and of the dc potential displays the main feature of the free-fall regime of the discharges at low-gas pressure: well pronounced wall sheath with strong drop of the dc potential and, respectively, sharp increase of the dc electric field leading to a strong increase of the velocities of the positive ions close to the walls. The latter motivates the requirement to have the inertia term in the momentum equations of the positive ions since it is the factor that limits their velocities. Figure 2 justifies also the approximations made in the inertia term: collisionless sheath, according to its extension compared to the mean free path, and field orientation towards the wall.

Spatial distribution of the plasma parameters

In addition to the results for the dc field and its potential (Fig. 2), the spatial distributions of the electron density (Figs. 3(a), 4 and 5) and temperature (Fig. 3(b)) also display behavior specifying the general features of both free-fall-regime discharge maintenance and discharge maintenance by localized power deposition, respectively, discharges with remote plasma regions. The strong drop of the dc potential in the wall sheaths in the free-fall regime is accompanied by flattening of the radial profiles of n_e (Figs. 4(a) and 5(a)) and widening of the wall sheath (Fig. 5(a)) with the gas pressure decrease. Although the electron temperature is high (Fig. 3(b)), as should be expected because of the low gas pressure, the obtained values are reasonable high owing to accounting for the inertia term in the momentum equation of the positive ions. Limiting the increase of the ion velocity in the wall sheath, it limits the particle losses and, respectively, the increase of T_e. Behavior of the spatial distribution of n_e and T_e related to plasma maintenance by a localized power deposition shows evidence in the axial decrease (Fig. 3(b)) of T_e (which is comparatively slight due to strong thermal conductivity) and in the strong axial drop of n_e (Figs. 3(a), 4(b) and 5(b)) outside the region of the rf power deposition. As should be expected, n_e increases with both Q (Fig. 4) and p (Fig. 5), the latter, when the input power is the same, as it is here.

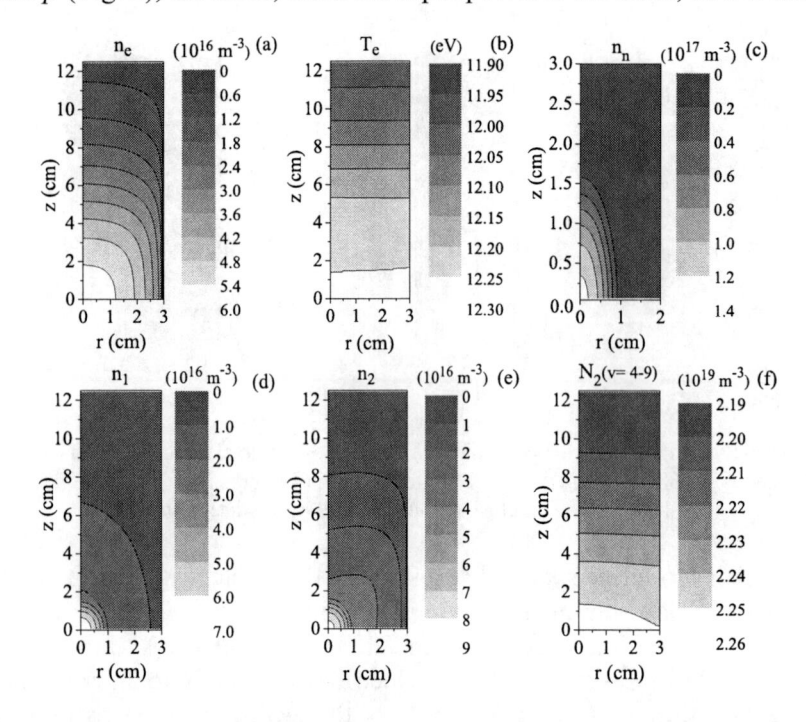

FIGURE 3. Spatial distribution of the electron density (a) and temperature (b), of the densities of the negative ions (c) and of the H⁺ (d) and H_2^+ (e) positive ions as well as of the density of the vibrationally excited molecules (f). A quater of the discharge vessel is presented everywhere except for n_n in (c) which presents the central part of the vessel. The same gas discharge conditions as in Fig. 2.

In agreement with the results from the previous models [1-3], the negative ions are accumulated in the discharge center (Figs. 3(c), 6 and 7). Their density increases with Q, respectively, with n_e, and also with p, in the studied gas pressure range. However, the main conclusion is that the driver, the region of the rf power deposition, where T_e is high, sustains high density of negative ions. The variation – over the entire discharge – of the density of the vibrationally excited molecules is very slight (Fig.

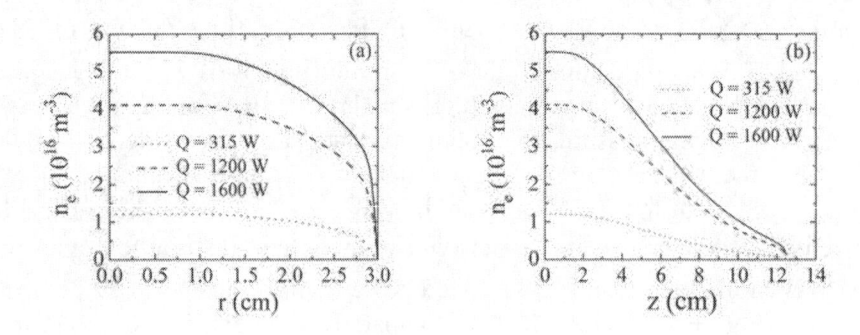

FIGURE 4. Radial (at $z = 0$) and axial at ($r = 0$) variations of the electron density at $p = 6$ mTorr for different values of the applied rf power.

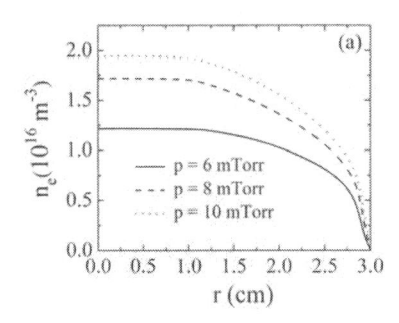

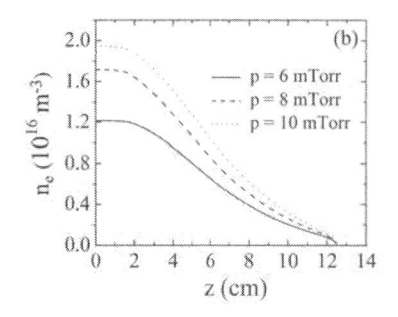

FIGURE 5. Radial (at $z = 0$) and axial (at $r = 0$) variations of the electron density at $Q = 315$ W for different values of the gas pressure p.

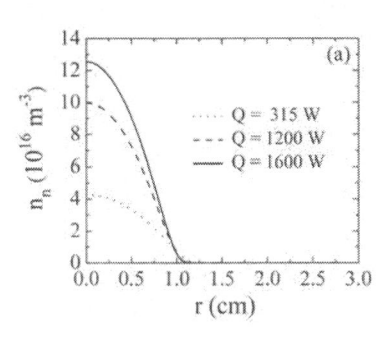

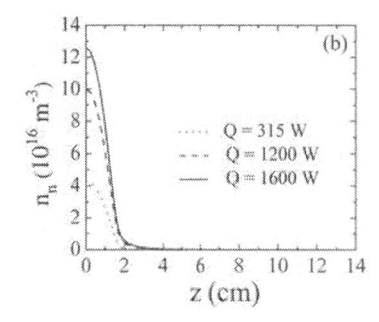

FIGURE 6. The same as in Fig. 4 but for the density of the negative ions.

3(f)), with a decrease outside the region of the rf power deposition. Thus, owing to the high T_e over the total volume of the discharge the vibtational excitation is efficient also in the entire discharge.

At the highest value of the applied power ($Q = 1600$ W), the H_2^+ and H^+ ions are with the highest concentrations (Figs. 3(d) and 3(e)). With the decrease of Q, the density of the H^+ ions decreases, due to the n_e-decrease, and at $Q = 315$ W the density of the H_3^+ ions exceeds that of the H^+ ions. The H_2^+ ions are again with the highest density. As the comparison of Figs. 3(d) and 3(e) with Fig. 3(c) show, the central region of the discharge, where the negative ions are accumulated, shows up

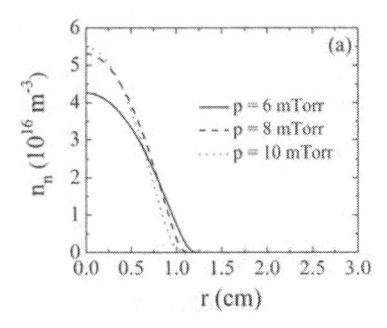

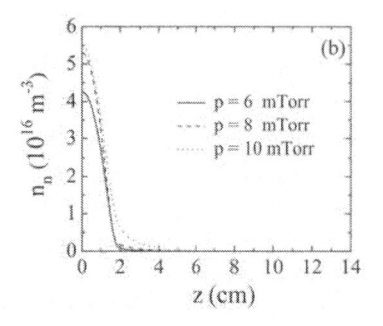

FIGURE 7. The same as in Fig. 5 but for the density of the negative ions.

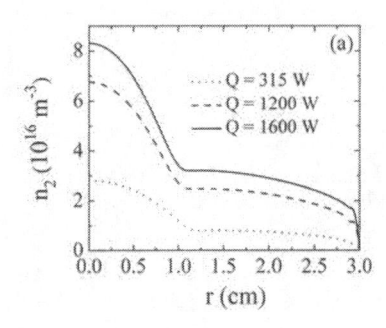

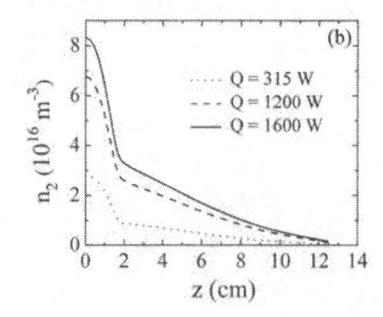

FIGURE 8. The same as in Fig. 4 but for the density of the H_2^+- ions.

with high values of the densities of the positive ions. Thus, as it has been commented before [1-3], the positive ions take the role to keep the plasma quasineutrality when negative ions are present in the discharge. In more details this could be seen in Fig. 8 (where the radial and axial profiles of n_2 are given) and its comparison with Figs. 4 and 6: in the central region of the discharge the shape of the radial and axial variations of n_2 follows that of n_n whereas outside the region with negative ion accumulation the radial and axial variations of n_2 follow those of n_e.

Concerning the negative ion balance, the conclusion from the spatial distribution of the plasma parameters is that whereas the vibrational excitation of the molecules is in the entire discharge volume, the negative ions are concentrated in the central region of the discharge.

Factors determining the negative ion balance

With vibrational excitation of the molecules in the entire discharge and negative ions concentrated in its central region, the question is whether the negative ions are produced locally in the discharge center (due, e.g., to the high electron density there) or they are produced in the entire discharge and then are transported to the discharge center via their flux in the dc field, as it has been shown before [1-3].

The results for the spatial distribution of the local production and of the local losses of the negative ions and for their flux (Fig. 9) confirm the latter. The local losses in the discharge center (Fig. 9(b)) exceed with two orders of magnitude their local production. Thus, as has been discussed also before, the sharp maximum of n_n in the discharge center and the complete accumulation of H$^-$ there is due to the transport process in their balance, i.e. to their flux (Fig. 9(c)) towards the central part of the discharge, not to local production there. This flux is in the dc electric filed (Fig. 2(c)) that accelerates the ions produced all over the discharge towards its center. The radial (Fig. 10(a)) and axial (Fig. 10(b)) variations of the local production of H$^-$ follow those of n_e (Fig. 4): smooth radial and axial variations, not giving indication for the sharp drop of n_n outside the central region of the discharge (Fig. 6). As should be expected, the shape of the radial and axial variations of the local losses of H$^-$ (Fig. 11) follows the spatial changes of n_n (Fig. 6).

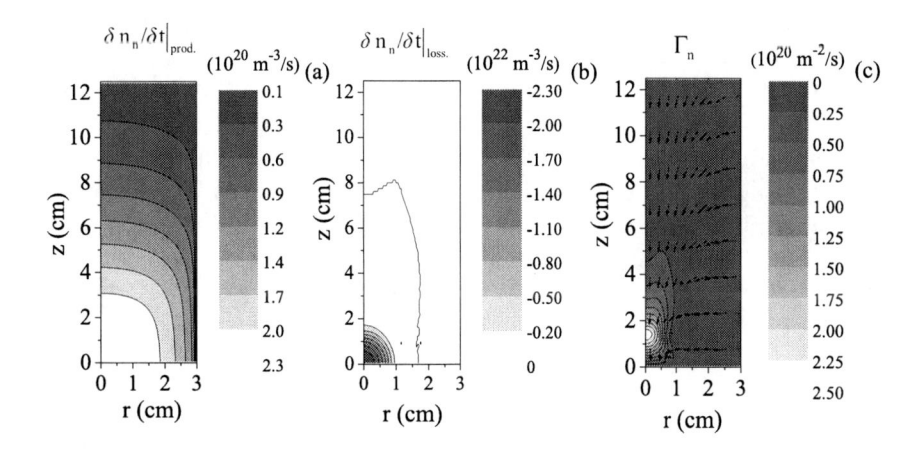

FIGURE 9. Spatial distribution of the local production of the negative ions (a) and their local losses (b) and of the flux of the negative ions (c); $p = 6$ mTorr and $Q = 1600$ W.

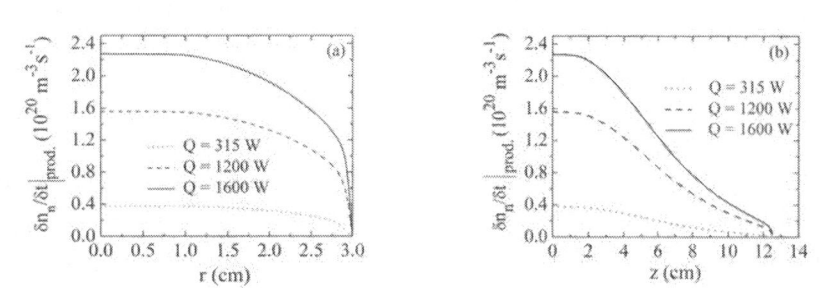

FIGURE 10. The same as in Fig. 4 but for the local production of the negative ions.

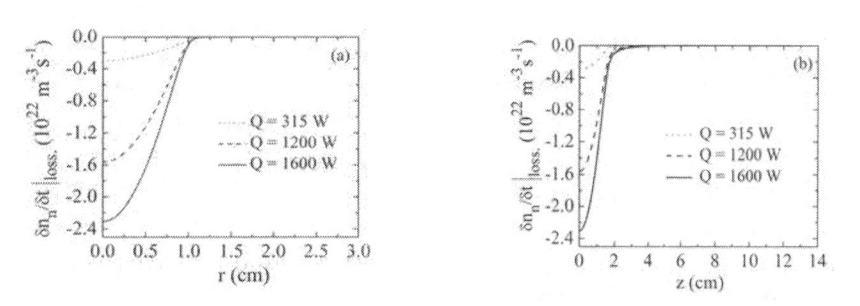

FIGURE 11. The same as in Fig. 4 but for the local losses of the negative ions.

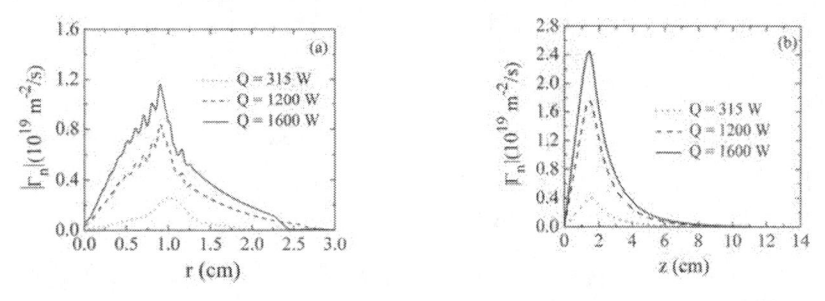

FIGURE 12. Radial variation (at $z = 0$) of the radial component of the negative ion flux (a) and axial variation (at $r = 0$) of its axial component at $p = 6$ mTorr and different values of the applied power.

The flux of the negative ions (Fig. 9(c)), being a flux in the dc electric field in the discharge, is directed towards the discharge center. The maxima in the magnitude of the total flux and its components (Figs. 9(c) and 12) result from a balance of a convective flux towards the discharge center and an outward diffusion flux connected with the accumulation of the ions in the central region of the discharge. The flux of the negative ions is higher at higher applied power and the density of the negative ions (Fig. 6) is higher in this case. Concerning the changes with p, the higher flux at higher gas pressure results in a higher negative ions density.

CONCLUSION

The study deals with the arduous task of 2D description within the fluid-plasma theory of free-fall-regime sustained discharges, moreover, hydrogen discharges with account for the negative ions in the discharge, which tremendously enlarges the number of the differential equations in the initial set of equations. The model provides description of inductive discharges with a cylindrical coil. The main conclusion from the results is that the driver, the region of the rf power deposition, sustains high density of the negative ions due to their accumulation in the discharge center owing to predominating nonlocality in their balance: Negative ions produced all over the discharge volume are accelerated towards the discharge center in the dc electric field in the discharge. However, small size of the discharge vessel (small radius of the discharge) is the requirement since only in this case the ions can reach the center without being destroyed in collisions.

ACKNOWLEDGMENTS

Dr. St. Lishev is acknowledged for his contribution to the work. The work is within project DO02-267 supported by the National Science Fund in Bulgaria and it is part of the work of the Bulgarian Association EURATOM/INRNE (task 2.1.1).

REFERENCES

1. Ts. Paunska, H. Schlüter, A. Shivarova and Kh. Tarnev, *Phys. Plasmas* **13**, 082110 (2006).
2. Ts. Paunska, A. Shivarova and Kh. Tarnev, *J. Appl. Phys.* **107**, 083301 (2010).
3. Ts. Paunska, A. Shivarova, Kh. Tarnev and Ts. Tsankov, *Phys. Plasmas* **18**, 023503 (2010).
4. Ts. Paunska, A. Shivarova, Kh. Tarnev and Ts. Tsankov, 20th Eur. Conf. Atomic and Molecular Physics (ESCAMPIG, Novi Sad, Serbia, 2010), topic number: № 6.
5. St. Lishev, A. Shivarova and Ts. Tsankov, *J. Phys.: Conf. Series* **223**, 012001 (2010).
6. Ts. Paunska, A. Shivarova, Kh. Tarnev and Ts. Tsankov, 36th EPS Conf. Plasma Phys. (Sofia, 2009), ECA vol. 33E, P-2.120 (2009).
7. St. Kolev, Ts. Paunska, A. Shivarova, Kh. Tarnev and Ts. Tsankov, 36th EPS Conf. Plasma Phys. (Sofia, 2009), ECA vol. 33E, O-5.064 (2009).
8. M. Bacal, A. M. Brunetau, and M. Nachman, *J. Appl. Phys.* **55**, 15-24 (1984).
9. M. Bacal, *Nucl. Fusion* **46**, S250-S259 (2006).
10. A. J. T. Holmes, *Plasma Phys. Controll. Fusion* **34**, 653-676 (1992).
11. O. Fukumasa, *J. Phys. D: Appl. Phys.* **22**, 1668-1679 (1989).
12. St. Lishev, A. Shivarova and Kh. Tarnev, *J. Plasma Phys.* (2011), in press.

Vibrational States of Hydrogen Molecules in One Third LHD Negative Ion Source

M. Nishiura[a], Y. Matsumoto[b], K. Tsumori[a], M. Wada[c], T. Inoue[d]

[a]National Institute for Fusion Science, 322-6 Oroshi-cho, Toki 509-5292, Japan
[b]Tokushima Bunri University, Yamashiro-cho, Tokushima, 770-8514 Japan
[c]Doshisha University, Kyotanabe, 610-0321, Japan.
[d]Japan Atomic Energy Agency, 801-1, Mukoyama, Naka 311-0193, Japan

Abstract. The effect of the cesium on hydrogen negative ions is discussed using the vacuum ultraviolet emissions (VUV) from vibrational states of hydrogen molecules. The VUV spectrum from 90 to 165 nm is related to the H⁻ production from the dissociative attachment of electrons to hydrogen molecules. The VUV spectra in hydrogen plasmas are measured in the extraction region of a negative ion source for neutral beam injector. Under the pure hydrogen discharge, the cesium vapor is introduced into the ion source to enhance the hydrogen negative ion density. When the same arc power of ~99kW is applied in both with and without cesium admixture cases, the ratio of the observed VUV spectrum without to with cesium is not changed clearly. Therefore the production process of H⁻ related to the wall/electrode surface would be enhanced rather than the volume production process.

Keywords: negative ions, vibrational states, VUV spectrum, ion source, hydrogen molecules
PACS: 33.20.-t, 33.70.-w, 52.50.Dg, 52.20.-j, 52-27Cm

INTRODUCTION

Neutral beam injector (NBI) is one of most essential devices for fusion plasma heating and sustainment. NBI consists of usually positive/negative ion source and neutralizer. Negative ion based NBIs have been developed and been utilized in the Large Helical Device (LHD) [1], JT-60U [2], and ITER [3]. The production of hydrogen negative ions is considered to be a competitive production between so called "volume" and "surface" production in a negative ion source. For the comprehension and optimization of ion sources, the diagnostic of hydrogen atoms and molecules is one of effective approaches, because those atoms and molecules give an origin to negative hydrogen ions. In particular, negative hydrogen ions as volume production are produced by the dissociative electron attachment through vibrationally excited hydrogen molecules.

In a small negative ion source, the emission spectrum originated from Lyman band ($B^1\Sigma_u^+ \rightarrow X^1\Sigma_g^+$) and Werner band ($C^1\Pi_u \rightarrow X^1\Sigma_g^+$) transitions of hydrogen molecules were measured by a vacuum ultraviolet (VUV) spectroscopy. The vibrational temperature of hydrogen molecules is found to be 300 K [4]. The Argon admixture in the hydrogen negative ion source does not enhance the Lyman and Werner bands. The H$^-$ density in low gas pressure region (less than 0.1 Pa) is enhanced because the electron density is increased due to Ar additive in hydrogen plasmas, and it leads to the enhancement of dissociative electron attachment to H$_2$ molecules [5, 6].

A large size negative ion source for NBI usually uses cesium vapor admixture in it to enhance negative ion beam currents. The behavior of vibrationally excited states of hydrogen molecules is not clearly understood under the cesium admixture operation. We prepared the VUV spectrometer for the one third scale negative ion source for NBI at NIFS. Since the input power and the volume are comparable to the real size negative ion source for LHD-NBI system, it is expected that the progress will be made in understanding of the negative ion production for real scale ion sources.

NEGATIVE ION PRODUCTION IN AN ION SOURCE THROUGH VOLUME PROCESSES

In gas discharge plasma, various collisional processes with electrons, atoms, and molecules can be considered [7, 8]. Among these processes, the dissociative attachment of electrons was considered to be a possible reaction for the production of negative ions of hydrogen isotopes, as follows:

$$e + AB \rightarrow A^- + B . \quad (1)$$

However the cross section for this reaction was thought to be very small. In 1959, Schulz et al. [9] measured the total negative ion production in H$_2$ as a function of electron energy. A few years later, Rapp et al. [10] obtained the cross sections for negative ion production in H$_2$, HD, and D$_2$. Schulz and Asundi [11] in 1965 also measured the cross sections for the dissociative attachment of electrons, as well as their isotope effect, in a lower electron energy region than that studied by Rapp.

According to the energy diagram [12] of H$_2^-$ and H$_2$, as shown in Fig. 1, if the ground state of H$_2$ is excited to one of the resonant states H$_2^-$ ($^2\Sigma_u^+$ and $^2\Sigma_g^+$), it can dissociate to H + H$^-$. This reaction is called dissociative attachment (DA). If the initial molecule is vibrationally excited, the reaction has a larger cross section as has been shown by Wadehra and Bardsley [13-15],

$$H_2(X^1\Sigma_g^+, v") + e(slow) \rightarrow H^- + H . \quad (2)$$

If the H$_2$ molecule is excited to a higher electronically excited level of c$^3\Pi_u$, the dissociative attachment leads to the formation of a negative ion and an excited atom [16]:

$$H_2 (c^3\Pi_u) + e(slow) \rightarrow H^- + H(2p) . \quad (3)$$

The cross section can also be relatively high, as was shown theoretically in Ref. [16].

Vibrationally excited hydrogen molecules can be produced by collisions with low energy electrons (e-V),

$$H_2 + e \text{ (slow)} \rightarrow H_2(X^1\Sigma_g^+, v'') + e, \qquad (4)$$

and followed by the dissociative attachment of Eq. (2).

The cross sections for dissociative attachment to excited H_2 and D_2 molecules were calculated theoretically by Bardsley and Wadehra for D_2 [15] to explain anomalously high H^- densities in ion sources. They showed that the rate for dissociative attachment of electrons is enhanced several orders of magnitude by the vibrational and rotational excitation of the $X^1\Sigma_g^+$ ground state of H_2 and D_2. Therefore, the vibrationally excited molecules of hydrogen are necessary to produce H^- ions.

Vibrationally excited molecules, $H_2(X^1\Sigma_g^+, v'')$, can be produced by the collisions with energetic electrons via excited singlet states (E-V); [17, 18, 19]

$$H_2 + e \text{ (fast)} \rightarrow H_2(B^1\Sigma_u^+, C^1\Pi_u) \rightarrow H_2(X^1\Sigma_g^+, v'') + h\nu. \qquad (5)$$

The vibrationally excited molecules lead to the production of H^- ions through the process in Eq. (2). Hiskes has calculated the cross sections for the vibrationally excited molecules, $H_2(X^1\Sigma_g^+, v'')$, via the B and C states initiated from the ground state[17, 18]. The population of $H_2(B^1\Sigma_u^+)$ is also increased by the following cascade transitions:

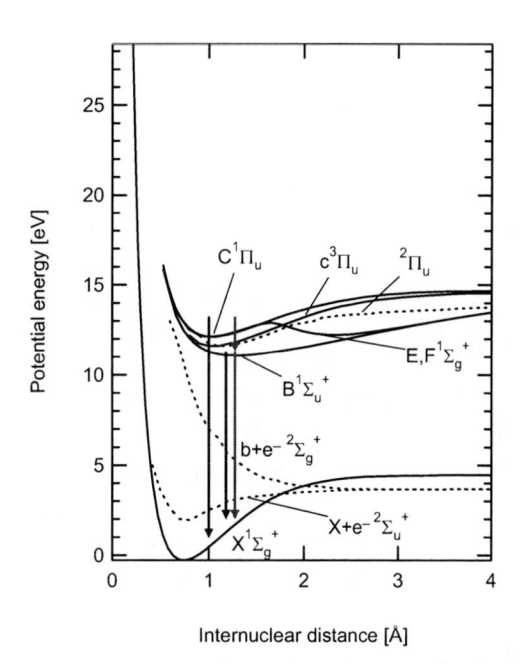

FIGURE 1. Potential energy curves for H_2 ($B^1\Sigma_u^+$, $C^1\Pi_u$, $c^3\Pi_u$ and $X^1\Sigma_g^+$) and H_2^- ($X + e^{-2}\Sigma_u^+$, and $b + e^{-2}\Sigma_g^+$) and H_2^- taken from Sharp in Ref. [12]. The potential curve for $H_2^-(^2\Pi_u)$ taken from Bottcher and Buckley in Ref. [16].

$$H_2(X^1\Sigma_g^+, \upsilon=0) + e \rightarrow H_2(E, F\ ^1\Sigma_g^+), \tag{6}$$

$$H_2(E, F\ ^1\Sigma_g^+) \rightarrow H_2(B^1\Sigma_u^+). \tag{7}$$

Ajello et al. [19] pointed out that the cross section for cascading to the *B* state cannot be ignored, compared with that for cascading to the *C* state, and that this cascade process accounts for 10 – 20 % of the total *B* state. It is thus essential to examine the contribution of the cascade transition to the hydrogen ground state in a negative ion source by VUV spectrum.

NEGATIVE ION PRODUCTION BY SURFACE EFFECT

Cesium effect on H⁻

When the cesium vapor is injected into a volume production type negative ion source, the H⁻ beam current is enhanced. It is essential to add the cesium vapor in a negative ion source for NBI. This phenomenon is considered to be due to the H⁻ production on the metal surface with a low work function. The following processes may occur,

$$H^+ + 2e_{surface} \rightarrow H^-\ [20\text{-}22] \tag{8}$$

$$H^0 + e_{surface} \rightarrow H^-\ [22\text{-}25] \tag{9}$$

Shinto et al. [26] have investigated the relation between the work functions of the cesium adsorbed molybdenum surface and the extracted H⁻ beam currents using the photoelectron emission method assisted with a laser light. Employing the probability for the surface production of H⁻ on a wall of an ion source, P_{cs}, Fukumasa et al. have solved the numerical balance equations in steady state hydrogen plasma [27]. Their result shows a good agreement in the dependence of plasma parameters on hydrogen gas pressure.

In the extraction region in front of the beam extraction hole, the n_-/n_e ratio is investigated as a function of the electron temperature in a comparison between the pure hydrogen and the cesium added cases. Here n_- and n_e are the negative ion density and the electron density, respectively. The electron temperature is controlled by the discharge current. It has already been proved that the n_-/n_e ratio [28, 29] and H⁻ beam current [30] are well correlated with the work function of a plasma facing electrode, which is called a plasma electrode, in the cesium admixture case. The results in references [28, 29] seem that the volume production component is not influenced by the surface modification. However the results were only the collateral evidence, because the population of vibrationally excited hydrogen molecules is unknown in cesium admixture case.

Surface effect on H⁻

In addition to the volume processes and the cesium effect described above, we cannot ignore the effect of chamber wall condition(filament and cesium depositions) because of the recombinative desorption (RD) [31-33]. The process would also influence the population distribution of the vibrational states. Bacal *et al.* [33] reported that hydrogen molecules with an initial vibrational level v less than or equal to 2 produced by RD are pumped up to the higher level v'' of $H_2(X^1\Sigma_g^+, v < v'')$ via the *B* and *C* states.

$$H + H + wall \rightarrow H_2(X^1\Sigma_g^+, v = 1, 2) \qquad (10)$$

followed by the *E-V* excitation of the $X^1\Sigma_g^+$ state with the low v:

$$H_2(X^1\Sigma_g^+, v = 1, 2) + e\ (fast) \rightarrow H_2(B^1\Sigma_u^+, C^1\Pi_u) \rightarrow$$
$$H_2(X^1\Sigma_g^+, v'' \geq 1, 2) + h\nu \qquad (11)$$

Thus the RD process changes the original v distribution. The increase of the $X^1\Sigma_g^+$ states with higher vibration levels enhances the production of H⁻ by DA, whereas it is essential to take into account the destruction processes of H⁻ by the electron, ion, and neutral collisions. Moreover, in the case of adding the cesium vapor to an H⁻ ion source, the surface condition is significantly important for the enhancement of the H⁻ density. Although it has been found that the work function of the plasma electrode is strongly correlated to the n_-/n_e ratio and the H⁻ beam current described in the previous section, the DA and the RD processes in cesium admixture hydrogen plasma have not been clearly understood yet. The measurement of the VUV spectra becomes a strong tool for estimating the population distribution of vibrationally excited molecules and for understanding the feature of the H⁻ production mechanism in ion sources.

Based on the intensities of the $B \rightarrow X$ band and $C \rightarrow X$ band spectra in the negative ion source, Graham [34] concludes that there exist vibrationally excited H_2 with a level $v'' > 5$ rich enough to produce H⁻ in the ion source by integrating the intensities of the *B* and the *C* band spectra by the de-excitation from the *B* and the *C* states to the vibrationally excited H_2 molecules with the level $v'' > 5$. Fukumasa *et al.* [35] confirmed by means of VUV spectroscopy that the fast electrons lead to hydrogen molecule excitation. In the following sections, we introduce the VUV spectra in cases of pure and cesium admixture plasmas in the one-third negative ion source as examples to explain the H⁻ production mechanism under a variety of situations.

EXPERIMENTAL SETUP FOR VUV SPECTROSCOPY IN ONE THIRD LHD NEGATIVE ION SOURCE

The one-third negative ion source is shown in Fig. 2. The cross section of the extraction region is the rectangular shape of 742 mm x 460 mm. The VUV spectrometer is installed into the ion source to diagnose the extraction region. The monochromator (Acton Research VM-502, nominal 0.2 m focal length) is pumped down differentially to keep the vacuum pressure of less than 10^{-3} Pa. A photomultiplier (Hamamatsu 721-01) with a fluorescent material (sodium salicylate) pasted on a vacuum seal window was used as a detector of the VUV spectrometer in the former experiments. For this experiment, the detector is replaced into a CCD camera (Princeton Instruments PI-SX100), which can reduce the acquisition time from ~ ten minutes to ~ millisecond in the wavelength range of 90-160 nm. The VUV emission is integrated over the sightline, and most of the detected intensity would be dominated by the longitudinal extraction region. The H$^-$ ion source is operated at the arc power of less than about 100 kW. The arc current is started up and kept about 6 sec during a discharge. The H$^-$ beam is extracted about 0.5 A at 70 kW with hydrogen plasmas and about 10 A at 100 kW with cesium admixture plasmas. The H$^-$ beam current from the cesium admixture plasmas is roughly enhanced by one order of magnitude higher than that from the pure hydrogen plasmas.

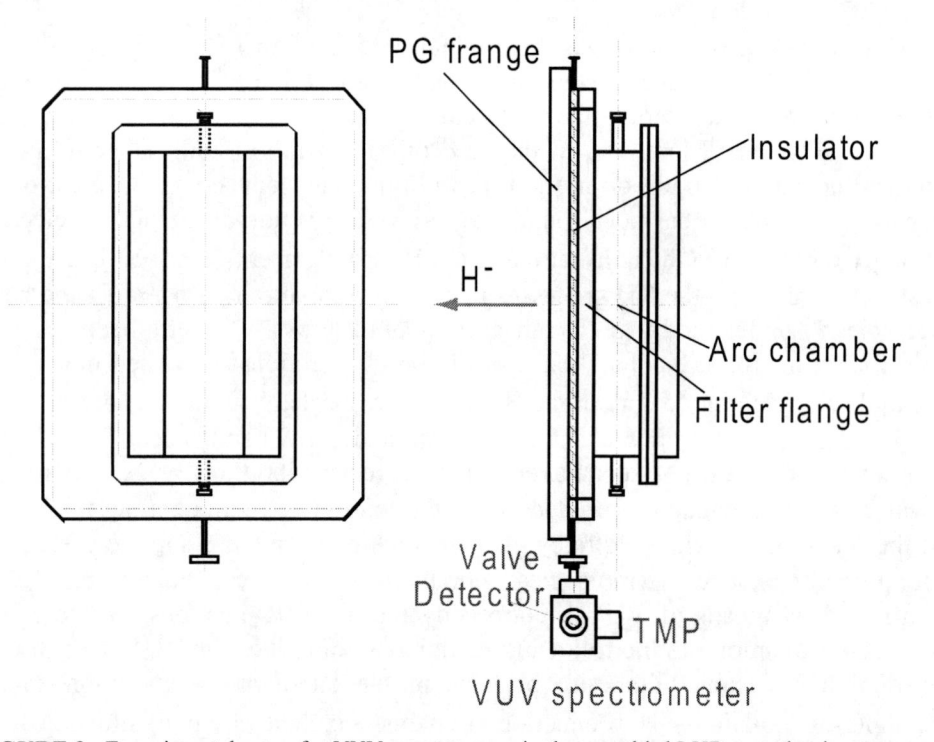

FIGURE 2. Experimental setup for VUV spectroscopy in the one-third LHD negative ion source. The left figure shows the schematic viewing from the up-stream of the extraction electrode. The right one shows the side view. The VUV spectrometer measures the emission in the extraction region.

EXPERIMENTAL RESULT OF VUV SPECTROSCOPY IN ONE THIRD LHD NEGATIVE ION SOURCE

The VUV spectra are shown in Fig. 3 in a pure hydrogen discharge of one third LHD negative ion source. The same spectrometer with the CCD detector is used for two experiments (present experiment and the former experiment in the small negative ion source [6]). The observed atomic lines at 121.6, 102.7, 97.2, and 94.9 nm correspond to Lyman-α, -β, -γ, and -δ, respectively. The sensitivity does not change in two systems. The ratio of the intensities of atomic Lyman-β, -γ, and -δ lines to molecular bands in one-third negative ion source is relatively higher than that measured in the small negative ion source [6]. The difference would be caused by the ionization degree and the electron temperature. The typical two lines originated from *C-X* (123 nm) and *B-X* (161.5 nm) transitions and the atomic Lyman-γ line in Fig. 4 are chosen from the Fig. 3 to know the characteristics. These lines are increased linearly with increasing the arc power. The negative ion source tends to the saturation of H$^-$ beam current extracted from pure hydrogen plasmas at high arc power more than about 70 kW, although the intensity of molecular hydrogen lines related to the volume production of H$^-$ increases linearly. The destruction of H$^-$ by high electron temperature or other causes should be considered at high arc power region for the explanation.

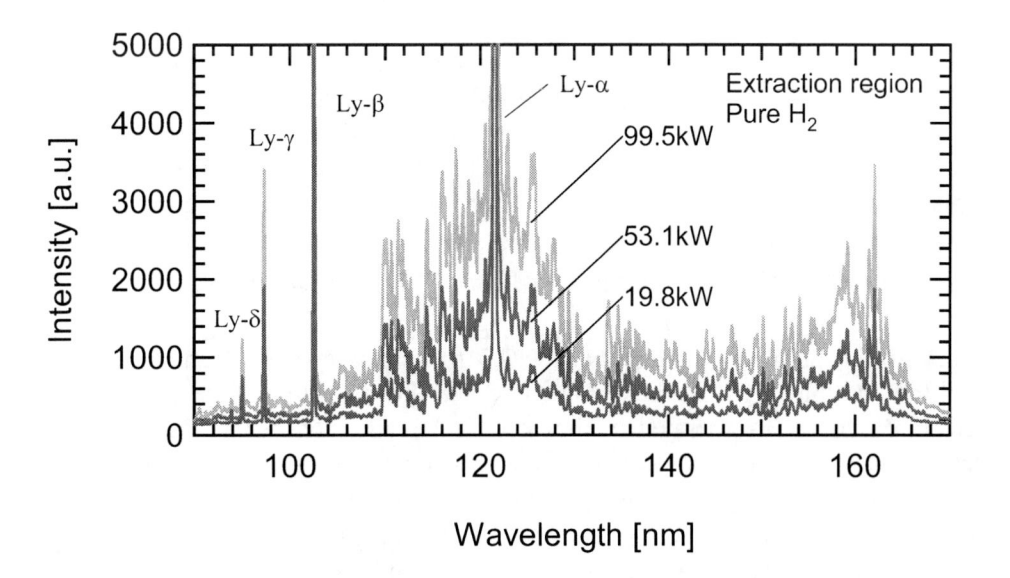

FIGURE 3. VUV spectra in hydrogen plasmas in the extraction region of the one-third LHD negative ion source. Three spectra are measured at the arc power of 19.8, 53.1, and 99.5 kW. The Lyman series from α to δ are indicated.

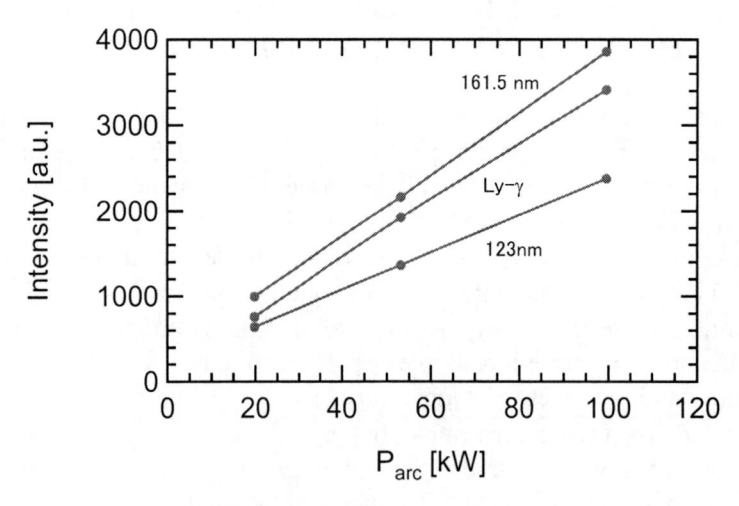

FIGURE 4. The intensity variation at the wavelengths of 123 nm, 161.5 nm, and Ly-γ. These lines are chosen from Fig. 3.

VUV spectra are measured under the cesium admixture hydrogen plasmas. Fig. 5 shows the ratio of VUV spectrum at low arc power to that at high arc power in pure hydrogen in a black curve (19.8kW/99.1kW). The ratio of Lyman-γ is decreased to ~20 %, compared with the arc power of ~ 100 kW. Other molecular bands and atomic lines in VUV spectrum are also decreased to 20~50% in this wavelength region. This would be caused by the reduction of electron density in the extraction region. When the arc power ratio is set nearly to the same value at 53.1kW/99.5kW and 51.1kW/99.1kW, the ratio of VUV spectrum is compared between without (black curve) and with cesium (gray curve) cases. Both curves do not change obviously, and this result suggests that the cesium admixture has no clear effect on the population of vibrationally excited hydrogen molecules. That is, there is no clear correlation with volume production of H$^-$.

Fig. 6 shows the H$^-$ beam current as a function of input arc power in the cases of with and without cesium admixture hydrogen plasmas. The H$^-$ beam current is measured by a thermocouple array at the 4.3 meter downstream from the grounded grid of the one-third LHD negative ion source. In this experiment, the one-third LHD negative ion source is operated at the feeding H$_2$ pressure of 0.2 Pa and the bias voltage of ~1 V. Here the bias voltage is applied between the plasma electrode and the ion source body (=reference potential). In cesium admixture case, the H$^-$ beam current increases by ten times or more for the one-third LHD negative ion source, although we showed the no clear effect on the vibrationally excited hydrogen molecules in Fig. 5. Therefore the cesium admixture would enhance the surface production of H$^-$ in the extraction region rather than the volume one. This result is also supported by the result for the small negative ion source that the work function reduction is correlated with the increase of the n_-/n_e ratio [29]. However we still need further experiment and discussion, because no clear correlation in the absolute n_- is observed in the small ion source.

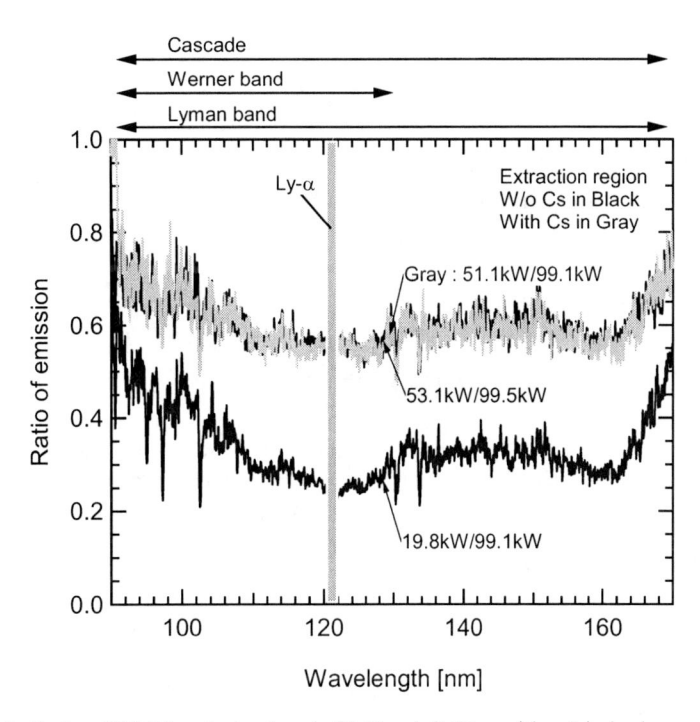

FIGURE 5. Ratio of VUV emission bands (*B-X* and *C-X* transitions) in hydrogen plasmas (without cesium in black) and cesium admixture hydrogen plasmas (with cesium in gray) in the extraction region of the one-third LHD negative ion source. The spectra are normalized by those in the cases of the arc power of 99.1 or 99.5 kW with and without cesium. The cesium does not affect the spectrum in gray curve(51.1kW/99.1kW), compared with that in black(53.1kW/99.5kW). The wavelength of Lyman-α at 121.6nm is hatched.

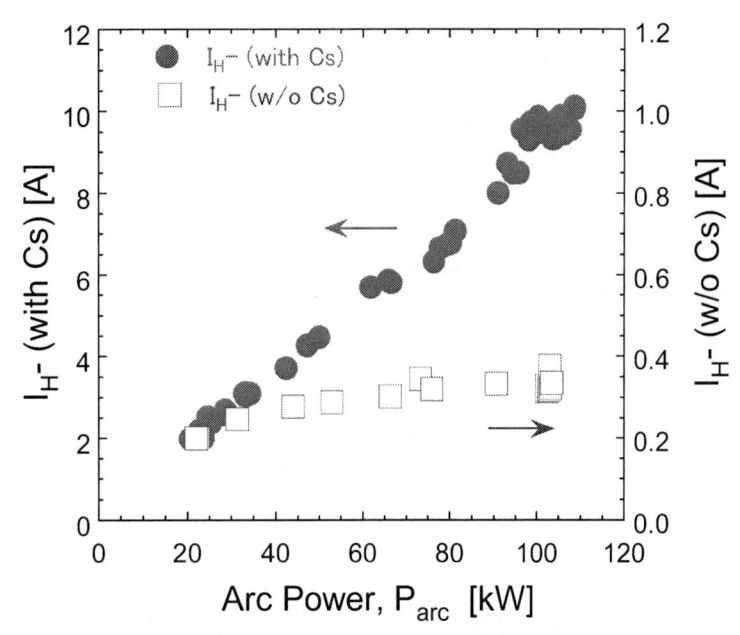

FIGURE 6. Comparison of H⁻ beam currents extracted from plasmas with (solid circles) and without (open squares) Cs seeding. The H⁻ beam current is shown as a function of input arc power.

ACKNOWLEDGMENTS

The authors would like to thank Dr. M. Bacal at Ecole Polytechnique and Dr. M. Glass-Maujean for their beneficial discussions. We also thank Professor M. Sasao for her support. This work is supported by the National Institute for Fusion Science under NIFS10KOAR011.

REFERENCES

1. O. Kaneko, Y. Oka, M. Osakabe, Y. Takeiri, K. Tsumori, R. Akiyama, E. Asano, T. Kawamoto, A. Ando and T. Kuroda; "Negative-Ion-Based Neutral Beam Injector for the Large helical Device," in *Proc. of the 16th International Conference on Fusion Energy*, IAEA Vienna, (1997) Vol.3 pp.539 - 545.
2. M. Kuriyama et al. *Fusion Engineering and Design*, **26**, 445-453 (1995).
3. R. Hemsworth, H. Decamps, J. Graceffa, B. Schunke, M. Tanaka, M. Dremel, A. Tanga, H.P.L. De Esch, F. Geli, J. Milnes, T. Inoue, D. Marcuzzi, P. Sonato and P. Zaccaria, Nuclear Fusion, 49, 045006 (2009).
4. M. Bacal, A. A. Ivanov, M. Glass-Maujean, Y. Matsumoto, M. Nishiura, M. Sasao, M. Wada, *Rev. Sci. Instrum.* **75**, 1699-1703 (2004).
5. M. Bacal, M. Nishiura, M. Sasao, M. Hamabe, M. Wada, H. Yamaoka, *Rev. Sci. Instrum.* **73**, 903-905 (2002).
6. Masaki Nishiura, *J. Plasma and Fusion Res.* **80**, 757-762 (2004).
7. B. M. Smirnov, 'Negative Ions', McGraw-Hill Inc., 1982.
8. S. H. Massey, 'Negative Ions 3rd Ed.', Cambridge Univ. Press., 1976.
9. G. J. Schulz, Phys. Rev. **113**, 816(1959).
10. D. Rapp, T. E. Sharp, and D. D. Briglia, *Phys. Rev. Lett.* **14**, 533-535 (1965).
11. G. J. Schulz, and R. K. Asundi, *Phys. Rev. Lett.* **15**, 946-949 (1965).
12. T. Sharp, *Atomic Data* **2** 119 (1971).
13. J. M. Wadehra and J.N. Bardsey, *Phys. Rev. Lett.* **41**, 1795-1798 (1978).
14. J. M. Wadehra, *Appl. Phys. Lett.* **35**, 917-919 (1979).
15. J.N. Bardsley and J.M. Wadehra, *Phys. Rev.* **A20**, 1398-1405 (1979).
16. C. Bottcher and B. D. Buckley, *J. Phys.* **B12**, L497 (1979).
17. J. R. Hiskes *J. Appl. Phys.* **51**, 4592-4594 (1980).
18. J. R. Hiskes *J. Appl. Phys.* **70**, 3409-3417 (1991).
19. J. M. Ajello, S. K. Srivastava, Yuk L. Yung, *Phys. Rev.* **A25**, 2485-2498 (1982).
20. P. J. Schineider, K. H. Berkner, W. G. Graham, R.V. Pyle, and J. W. Stearns, *Phys. Rev.* **B23**, 941-948 (1981).
21. J. N. M. Van Wunnick, J. J. C. Geerlings, E. H. A. Granneman, and J. Los, *Surf. Sci.* **131**, 17-33 (1983).
22. K. N. Leung, C. F. A. Van Os, and W. B. Kunkel, *Appl. Phys. Lett.* **58**, 1476-1469 (1991).
23. W. G. Graham, *Phys. Lett.*, **73**A, 186-188 (1979).
24. S. T. Melynychuk, M. Seidl, W. Carr, J. Isenberg, and J. Lopes, *J. Vac. Sci. Technol.*, **A7**, 2127-2131 (1989).
25. T. Okuyama and Y. Mori, Rev. Sci. Instrum. **63**, 2711-2713 (1992).
26. K. Shinto, Y. Okumura, T. Ando, M. Wada, H. Tsuda, T. Inoue, K. Miyamoto and A. Nagase, *Jpn. J. Appl. Phys.* **35**, 1894-1900 (1996).
27. O. Fukumasa and E. Niitani, *Jpn. J. Appl. Phys.*, Part2, **35**, L1528-1531, (1996).
28. M. Nishiura, M. Sasao, M. Wada, IEEE transaction 318(1999).

29. M. Nishiura, M. Sasao, M. Wada, in Proceedings of the 9th international symposium on the production and neutralization of negative ions and beams, (29~31 May 2002, Gif-sur-Yvette, France), AIP Conference Proceedings **639**, pp.21-27.
30. T. Morishita, M. Kashiwagi, M. Hanada, Y. Okumura, K. Watanabe, A. Hatayama, M. Ogasawara, *Jpn. J. Appl. Phys.* **40**, 4709(2001).
31. R. I. Hall, I. Cadez, M. Landau, F. Pichou, and C. Schermann, *Phys. Rev. Lett.* **60**, 337(1988).
32. T. Inoue, Y. Matsuda, Y. Ohara, Y. Okumura, M. Bacal, P. Berlemont, *Plasma Sources Sci. Technol.* **1**, 75(1992).
33. M. Bacal, M. Glass-Maujean, A. A. Ivanov Jr, M. Nishiura, M. Sasao, M. Wada, in Proceedings of the 9th International Symposium on the Production and Neutralization of Negative Ions and Beams, Gif-sur-Yvette, France, 2002. AIP Conference Proceedings **639**, pp. 13-20.
34. W. G. Graham, *J. Phys.* D: *Appl. Phys.* **17**, 2225-2231 (1984).
35. O. Fukumasa, N. Mizuki, E. Niitani, *Rev. Sci. Instrum.* **69**, 995-997 (1998).

Factors Affecting VUV Emission Spectrum near Lyman-α from a Hydrogen Plasma Source

K. Ogino[a], T. Kasuya[a], Y. Kimura[b], M. Nishiura[c], S. Shimamoto[a] and M. Wada[a]

[a]*Graduate School of Engineering, Doshisha University, Kyotanabe, Kyoto 610-0321, Japan*
[b]*Faculty of Science and Engineering, Doshisha University, Kyotanabe, Kyoto 610-0321, Japan*
[c]*National Institute for Fusion Science, Toki, Gifu 509-5292, Japan*

Abstract. Vacuum ultra violet (VUV) emission spectra from plasmas near walls of different metallic materials were measured to estimate the effect upon the local production rate of vibrational excited hydrogen molecules due to plasma wall interaction. Among Cu, Mo, Ni, Ta and Ti, the intensity of band spectrum around Lyman-α had become the largest when Cu wall was used while it was the smallest for Ti. The role of particle reflection from the plasma electrode surface upon the H⁻ production by a pure electron volume process is discussed.

Keywords: Vacuum Ultraviolet Spectrometry, Vibrational Excitation, Negative Ion
PACS: 52.27

INTRODUCTION

The so-called "electron volume process" to produce negative hydrogen ion (H⁻) in hydrogen plasma had been proposed [1] and confirmed [2] by Bacal and their coworkers. It is a two step process starting from a ground state hydrogen molecule excited to vibrational states, H_{2v}^*, by a collision with fast electron, e_f, in the plasma. The produced vibrationally excited molecule collides with a slow electron, e_s, to be a H⁻ through a dissociative electron attachment. As the produced H⁻ can be easily destroyed by e_f, the region where H_{2v}^* are produced is geometrically separated from the region where H⁻ are produced and extracted. An ion source plasma is usually separated into these two regions by inducing an external magnetic field [3], which is often called a "magnetic filter field." The region to produce H_{2v}^* is called driver region, and the region to extract H⁻ is called the extraction region. The overall efficiency of the volume production type ion source can be separated into the efficiencies of the two electron collision processes, and the efficiency to transport H_{2v}^* from the driver region to the extraction region.

Apart from the transport efficiency of H_{2v}^*, electron "volume" process is affected by plasma-wall, or "plasma-surface" interaction, because it determines species composition of plasma particles and other plasma parameters including electron temperature and density. Like the photodetachment method [4] which had become a standard method to evaluate H⁻ density in the plasma, vacuum ultraviolet (VUV) spectroscopy is commonly employed to estimate the rate of formation of H_{2v}^* in the plasma volume [5]. Thus, in this report, the VUV spectra near the ion source wall of the different materials are compared.

EXPERIMENT

Experimental Apparatus

A schematic diagram of the experimental apparatus is shown in Fig. 1. Hydrogen plasmas are excited in a 150 mm diameter 180 mm long cylindrical plasma source made of stainless steel. Attached to the cylindrical side wall of the source are 16 columns of Sm-Co magnets to form multi-cusp magnetic field to enhance plasma confinement. The volume of the discharge chamber is 2.8×10^3 cm^3, while lobes of confinement magnetic fields extend about 3 cm from the wall. A steady state discharge is maintained between a pair of 0.3 mm diameter tungsten filament serving as the cathode and the chamber wall serving as the anode. The tip of the tungsten cathode is located 10 mm from the center of the discharge chamber where the line of sight of the VUV spectrometer and the axis on which the probe tip travels cross each other.

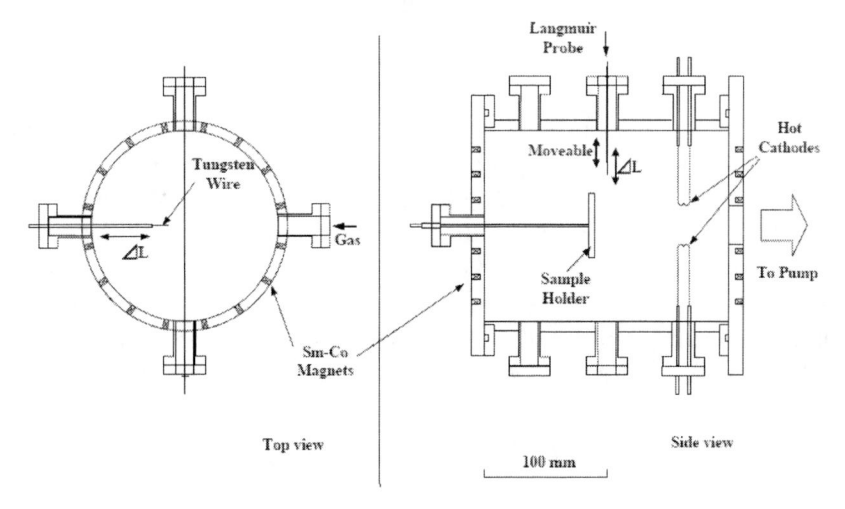

FIGURE 1. Main part of the H$^-$ ion source.

Samples of Mo, Ni, Ti, Ta, and Cu can be attached onto a stainless steel made sample holder. The distance from the sample surface to the center of the plasma, where the tip of the Langmuir probe is located to measure local plasma parameters, is 5 mm. The Langmuir probe has a coax structure with an outside insulation alumina tube, an electrical noise shielding copper tube, an inner insulation alumina tube, and a tungsten line serving as the probe tip. The diameter of the tungsten probe tip is 0.3 mm while the length of the probe tip is 3 mm. The probe can be moved in all directions because it is installed on 3 direction manipulation stage. The maximum travel distance is 20 mm in the radial direction of the ion source, while it is 5 mm on each side in other two directions.

A VUV spectrometer observes the center of the ion source plasma through a 15 mm diameter conduit connected to a pumping port evacuated by a 160 l/s turbo molecular pump. The spectrometer is independently pumped down by another 160 l/s turbo pump, while the ion source is evacuated by a 350 l/s turbo pump. The experimental configuration for VUV spectrum measurement is schematically shown in Fig. 2.

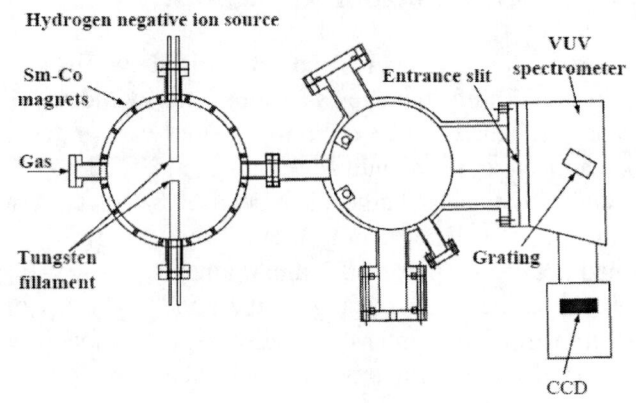

FIGURE 2. Vacuum system arrangement for VUV spectrometry study.

Experimental Procedure

The effect due to difference of wall material to the electron temperature and electron density was investigated by a Langmuir probe. While keeping the H_2 pressure from 0.2 to 0.7 Pa, H_2 plasma had been maintained with 1 A current and 40 V discharge voltage. The probe bias voltage was changed from -20 to +15 V to determine electron temperature and density from the *I-V* characteristics. The influence upon the rate of H_{2v}^* production due to the difference of wall material was investigated by changing H_2 pressure. Spectra of VUV emission were recorded at each gas pressure from 75 nm to 160 nm, setting Lyman-α at the center of the wavelength spectrum.

EXPERIMENTAL RESULTS

VUV Emission Spectrum

In Fig. 3 is shown a typical VUV emission spectrum obtained in the present series of experiments. As shown in the figure, both Lyman bands and Werner bands are observed. The band spectra are integrated and then normalized to the intensity of Lyman-α emission line so as to obtain the relative concentration ratio of H_{2v}^* by the following procedure. The stray light due to Planck radiation of tungsten filament was subtracted from the signal by drawing a smoothed baseline connecting minima in the measured spectrum. After subtracting the continuum component, the band signals are integrated from 109 to 160 nm. The relative Lyman-α emission rate was determined from the intensity peak at 121.7 nm. As discussed by Graham [5] this signal normalization procedure should underestimate the ratio as there is some contribution from band spectrum at 121.7 nm depending upon the spectrometer resolution. Apart from this error, this procedure should give us an estimation of the local ratio in production rate of hydrogen atoms to that of vibrationally excited molecules.

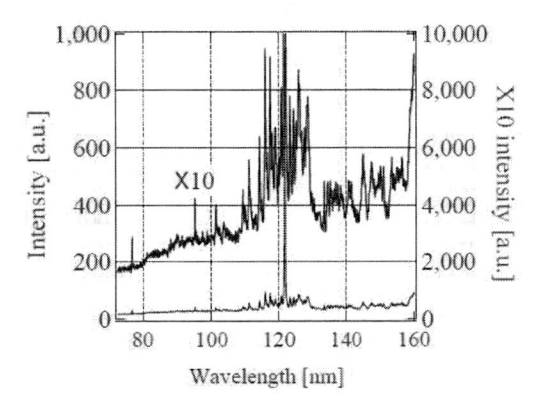

FIGURE 3. Typical VUV emission spectrum from a H_2 discharge.

Figure 4 shows the obtained intensity ratio for all the wall materials tested. As shown in the figure, the largest difference in the ratio was observed between copper and titanium samples. Thus, more precise measurements have been done for these two materials.

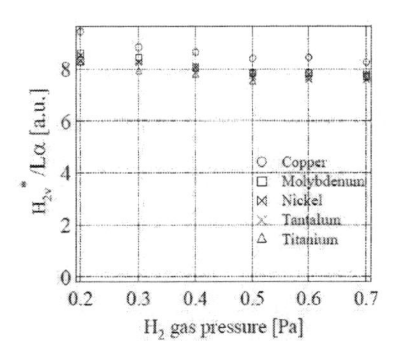

FIGURE 4. Band intensity/$L\alpha$ ratio plotted for different materials.

Effect due to Pressure

The experimental results of plasma parameter measurements as changing the H_2 pressure are summarized in Fig. 5. In Fig. 5 (a) is shown the electron temperature for copper and titanium measured by a Langmuir probe. The electron temperature was almost constant at 1 eV, and may be slightly lower for the titanium wall. In Fig. 5 (b) is shown the electron density plotted as a function of the gas pressure. The density varied similarly for two different materials. The corresponding intensity ratio of band spectrum to Lyman-α intensities is plotted as a function of pressure in Fig. 6. The result shows a gradual increase against H_2 pressure for both materials. The measured ratio was higher for copper wall against titanium wall in the low pressure region. This result may signify the difference due to wall materials.

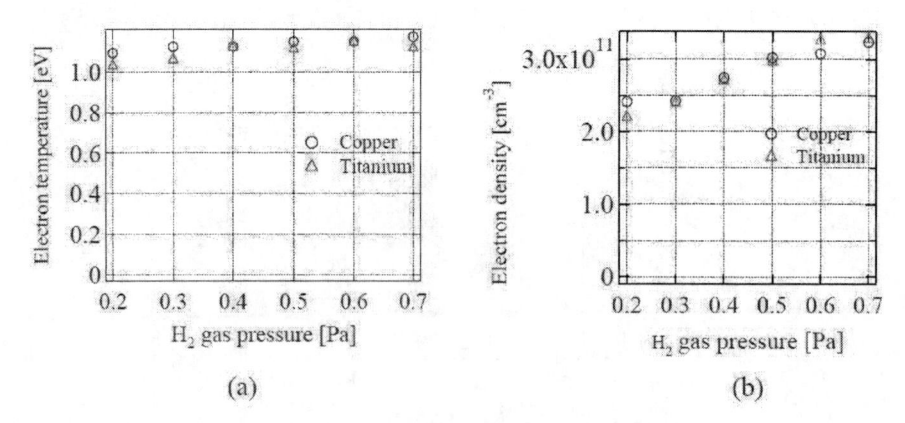

FIGURE 5. (a), Difference in electron temperature between copper and titanium walls. (b), difference in electron density between two electrode materials.

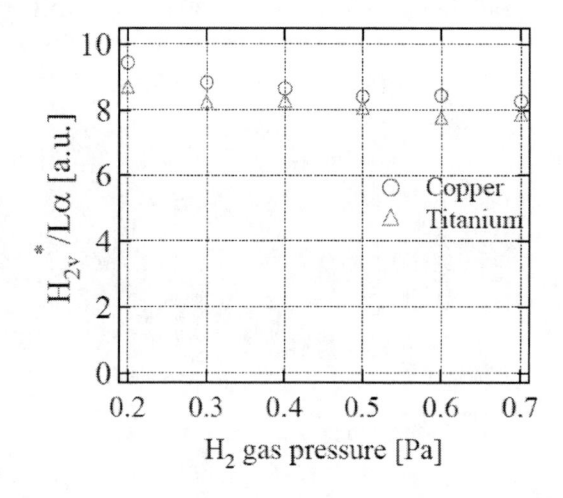

FIGURE 6. Band spectrum to Lyman-α intensity ratio against H_2 pressure.

DISCUSSION

The results show that the difference in wall material can affect the H_{2v}^* production rate near the wall surface with almost the identical plasma parameters. A possible reason for observing these results may be attributable to the difference in neutral species composition near the wall arising from reflection and desorption of hydrogen from the wall surface. Titanium and tantalum adsorb hydrogen once, and release them in the atomic form. This will result in increase of Lyman-α emission. Meanwhile, copper wall can reflect hydrogen molecules back to the plasma, which are vibrationally excited in the plasma volume.

CONCLUSIONS

An experimental system to observe VUV emission spectrum from H_{2v}^* molecules in H_2 discharge has been designed and assembled. The system is now being utilized to see if the wall material put into a H_2 plasma enhances or reduces the local production of H_{2v}^* by some process. More efforts will be made to further investigate which factor is influential on the enhancement of H^- volume production by choosing proper material for a plasma grid.

ACKNOWLEDGMENTS

This work has been supported in part by LHD cooperative program sponsored by National Institute of Fusion Science, and in part by Doshisha University's High-Tech Research Center Project "Interfacial Science".

REFERENCES

1. M. Bacal, H.J. Doucet, *IEEE Trans. Plasma Sci.* PS-1, pp. 91-99, (1973).
2. M. Bacal, G.W. Hamilton, *Phys. Rev. Lett.* 42, pp. 1538-1540, (1979).
3. R.L. York, R.R. Stevens Jr., K.N. Leung, K.W. Ehlers, *Rev. Sci. Instrum.* 55, pp. 681-686, (1984).
4. M. Bacal, *Rev. Sci. Instrum.* pp. 3981-4006, (2000).
5. W.G. Graham, *J. Phys. D.* 17, pp. 2225-2231, (1984).

Laser-photodetachment and Faraday-cup Measurements in the Expansion Region of a Tandem-type Plasma Source

Stiliyan St. Lishev and Antonia P. Shivarova

Faculty of Physics, Sofia University, BG–1164 Sofia, Bulgaria

Abstract. The study combines experiments on probe diagnostics with laser-photodetachment-technique and Faraday-cup measurements directed towards determination of the position of the extraction device and its influence on the discharge structure. The measurements have been carried out in the second chamber of an inductively-driven tandem plasma source performed as small scale arrangements, with a magnetic filter located just after the transition between the two chambers of the source. Results for the axial profiles of the plasma parameters display the correlation of the ratio n_-/n_e of the densities of the negative hydrogen ions and of the electrons and of the concentration of the negative ions with the electron density and temperature: The maxima of the (n_-/n_e)-ratio and of the density of the negative ions obtained are located at the position of maximum of the electron density behind the filter, in the region of the low electron temperature. Results from probe diagnostics and laser photodetachment measurements at a given axial position for different positions of the Faraday cup show the changes in the spatial distribution of the electron density and temperature and the reduction of the (n_-/n_e)-ratio and of the density of the negative ions caused by the extraction device.

Keywords: Plasma Sources of Negative Hydrogen Ions, Tandem Sources, Probe Diagnostics, Laser Photodetachment Technique, Extraction of Negative Ions.
PACS: 25.50.Dg, 52.70.-m, 52.50.Pi

INTRODUCTION

The small scale experimental arrangements with their flexibility with regards to applicability of diagnostic techniques provide possibility for studying both different regions of the plasma sources and the mechanism governing their operation. This is the case with the small-scale two-chamber inductively-driven plasma source [1] completed in analogy to the BATMAN source at IPP, the rf tandem plasma source of negative hydrogen ions [2] build regarding the neutral beam injection plasma heating for ITER.

The recent active experimental work on plasma diagnostics at BATMAN [3-8] has provided plenty of results for different plasma parameters – electron temperature and density, concentration of the negative ions, degree of dissociation, gas temperature – by applying different diagnostic techniques: probe diagnostics, optical emission spectroscopy combined with a collisional-radiative model, laser photodetachment diagnostics and cavity ring-down spectroscopy. These measurements have been carried out partially in the driver and mainly in the vicinity of the extraction region of the source which, indeed, is of main importance with regards to determining the source efficiency, for the given design of the source. On the other hand, the rf plasma source of negative hydrogen ions consisting of two chambers – a smaller-size chamber where the driver (the region of the rf power deposition) is located and a bigger-size chamber for plasma expansion from the driver – is with complicated configuration and it is governed, as it has been theoretically shown [9, 10] by nonlocality, i.e. by charge-particle and electron-energy fluxes. This means that the local values of the plasma parameters in the source are not locally determined. They result from the entire discharge behavior. Thus, optimization of the source design regarding, e.g., the position of the extraction, requires knowledge on the spatial distribution of the plasma parameters in the entire source. This could be easily achieved in small-scale experimental arrangements where the entire discharge is accessible for diagnostics.

This study is an extension of previous work on plasma diagnostics in small-scale experimental arrangements of an inductively driven two-chamber plasma source sustaining hydrogen discharges [1, 11-17]. Probe and spectroscopy diagnostics and laser photodetachment technique have been employed in the experiments. Probe diagnostics [1] carried out in the second chamber of the source, without a magnetic filter, has shown the pattern of the plasma expansion from the driver: axial drop of the electron temperature and structuring of the axial profile of the electron density composed by three regions: a decrease starting from the transition between the two chambers, a plateau region and a strong drop further on. Probe diagnostics performed with a magnetic filter located in the second chamber, just behind the transition between the two chambers [11], shows the effect of the filter to reduce both the electron temperature and density. Simultaneously acting plasma expansion through the magnetic filter and plasma expansion from the driver in the bigger volume of the second chamber of the source have been shown in details through the axial profiles of the electron temperature and density measured also by probe diagnostics, at different positions of the magnetic filter [15]. The results display not only the electron temperature drop but also a formation of a maximum of the electron density behind the filter. Laser photodetachment technique measurements of the ratio n_-/n_e of the negative-ion and electron densities and of the negative ion density n_- at a given axial position in the second chamber and different positions of the magnetic filter show that the highest concentrations of the negative ions are at the positions of the lowest electron temperature and of the highest electron density measured [15]. The guess regarding the optimization of the source was that the position of the extraction should "hit" the maximum of the electron density behind the filter which, anyhow, is in the region of low electron temperatures. Furthermore, in order to have this maximum as high as possible, the magnetic filter should be close to the driver. In general, the conclusion from the probe diagnostics and the photodetachment technique measurements is that the registered negative ions exist locally at the position of the measurements and its vicinity. However, the conditions for their effective production are nonlocally formed. Furthermore, measurements of the axial discharge structure in the driver show that the driver is more effective – compared to the second chamber – source of negative ions [16]. These results have been completed with optical spectroscopy diagnostics [13, 17] giving results for the gas temperature and the degree of dissociacion, the latter measured by using an argon gas as an actinometer.

This study completes the experiments from Ref. [15, 16] with results for the axial structure of the discharge mainly in the second chamber of the source and, partially, in its driver. Probe diagnostics and laser photodetachment technique measurements provide the axial profiles of the (n_-/n_e)-ratio and of the concentration of the negative ions in their relation to the axial variation of the electron density and temperature. The magnetic filter in the second chamber of the source is located close to the transition to its first chamber, as it has been outlined in Ref. [15] as the most proper position. The results confirm the supposition from Ref. [15] that the maximum of the (n_-/n_e)-ratio and of the concentration of the negative ions coincides with the maximum of the electron density behind the filter, the latter located in the region of low electron temperatures. Measurements with a Faraday-cup type of device for extraction of the negative ions have been also carried out together with probe and laser-photodetachment diagnostics. Being with the traditional design of an extraction device, the Faraday cup consists of three electrodes: a plasma electrode, a second electrode with magnets incorporated in it for deviation of the electrons and a Faraday cup for collecting the negative ions. Although the axial variation of the ratio of the currents to its third and second electrodes is also given, the stress here is on the results for the influence of the extraction device on the discharge structure. The latter is shown by measurements with probe- and laser-photodetachment diagnostics performed at a given axial position for different positions of the Faraday cup. The obtained reduction in the (n_-/n_e)-ratio and in the density of the negative ions close to the extraction device correlates with the increase of the electron temperature and the decrease of the electron density there due to the additional losses in the discharge introduced by the Faraday cup.

FORMULATION OF THE PROBLEM, EXPERIMENTAL SET-UP AND METHODS

Results for the axial variation of the (n_-/n_e)-ratio and of the negative ion density in the expanding plasma volume of a rf driven tandem plasma source as well as on the extraction of the negative ions are aimed at in the study. Thus, it includes measurements employing the laser photodetachment technique and measurements with a Faraday-cup type of extraction device combined with extended involvement of probe diagnostics. The probe diagnostics have been used for determination of the negative ion density from the measured – with the laser photodetachment method – values of the (n_-/n_e)-ratio as well as for determination of the influence of the extraction device on the discharge structure, the latter also in a combination with measurements of the (n_-/n_e)-ratio. Thus, the experiment includes:

(i) determination of the axial variations of the (n_-/n_e)-ratio and of the negative ion density,

(ii) determination of the axial variation of the currents to the electrodes of the Faraday cup and,

(iii) determination of the influence of the extraction device on the plasma parameters (the electron density and temperature as well as the ratio of the negative-ion and electron densities and the density of the negative ions).

The experimental set-up (Fig. 1) is a two-chamber inductively-driven plasma source, performed as small-scale arrangements (the size of its two chambers is given in the figure). The main components of the set-up are described in more details in [14]. The driver is an inductive discharge with a cylindrical coil (9-turn coil). The construction of the magnetic filter (MF) in the second chamber of the discharge vessel is as given in [11]. The filter is positioned in the second chamber, close to the transition ($z = 0$) between the two chambers. The magnetic field is perpendicular to the plasma flow coming out from the first chamber. The results in the next section are for a filter field with a maximum $B_0 = 100$ G at $z_{MF} = 4$ cm. The discharge is in hydrogen, in a flowing gas. It is maintained at 27 MHz, by applying a rf power $P = 700$ W. The results given further on are at two values of the gas-pressure: $p = 6$ mTorr and $p = 8$ mTorr. The measurements are mainly in the second chamber of the source and, partially, in the first chamber (in the end part of the driver, close to the transition between the two chambers).

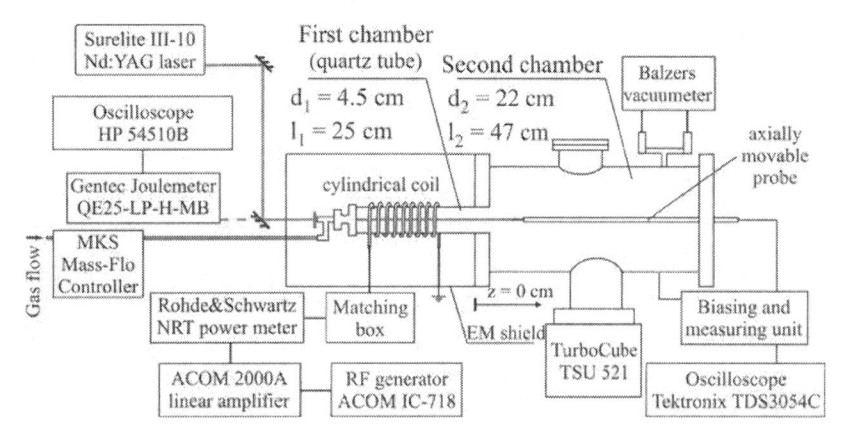

FIGURE 1. Experimental set-up with the arrangements for the laser photodetachment technique measurements.

Figure 1 shows also the experimental arrangements for the measurements of the (n_-/n_e)-ratio by employing the laser photodetachment method. Since results for the axial variation of the (n_-/n_e)-ratio are aimed at (item (i) given above), the laser beam path is axially aligned, centered at the discharge axis. The second harmonics (wavelength of 532 nm) of a Surelite III-10 Nd:YAG laser is used. The laser beam diameter is fixed at 1 cm by a diaphragm and the laser pulse energy is measured by a Gentic joulmeter. The detection of the electrons detached by the laser photons from the negative ions ($H^- + h\nu \to H + e$) is via a probe. The probe (as it is given in Fig. 1) is axially movable, centered also at the discharge axis. The radius of the probe is 0.25 mm and its length is 10 mm. The probe bias is 30 V above the plasma potential, well located in the electron-saturation-current part of the probe characteristics. A measured shape of the photodetachment signal – a pulse in the electron saturation current to the probe due to the extra electrons produced at the destruction of the negative ions – is shown in Fig. 2(a). As it is known [18-20], when the laser pulse energy density W is in the saturation region of the photodetachment fraction (well fulfilled for $W = 50$ mJ cm^{-2} in this experiment) the ratio (n_-/n_e) of the negative ion density and the electron density is directly related ($n_-/n_e = I_{ph}/I_{e(dc)}$) to the ratio of the plateau amplitude I_{ph} of the pulse in the electron current to its stationary value $I_{e(dc)}$. For determination of the axial variation of the negative ion density from the (n_-/n_e)-ratio, a second – also axially-movable – probe has been used, for measuring the axial profiles of the electron density and temperature. The probe is passively compensated, for avoiding the influence of the rf fluctuations on the probe characteristics. Its radius and length are, respectively, 0.3 mm and 6.0 mm.

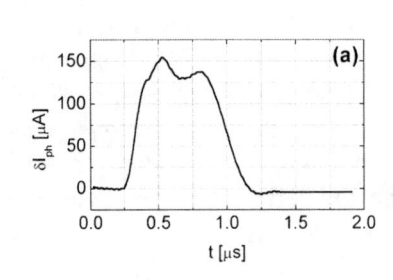

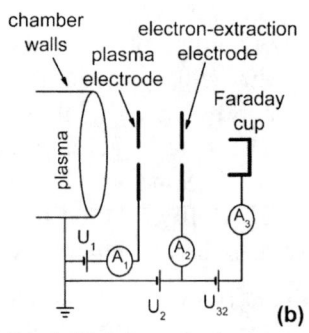

FIGURE 2. (a) Measured shape of the photodetachment signal. (b) Schematical presentation of the electrodes of the Faraday cup.

The Faraday cup-type of an extraction device, also involved in the experiment, is with the construction schematically shown in Fig. 2(b). It consists of a plasma electrode (PE), an electron-extracting electrode (EE), with magnets installed in it, and a negative-ion-collecting electrode (a Faraday cup (FC)). Each electrode is biased separately, using different voltage supplies. The materials of the electrodes are molybdenum for PE and cupper for EE and FC and the front cover of the device is made of stainless steel. Teflon is used for the isolation between the electrodes. The aperture of the device is with a diameter of 5 mm. The spatial variation of the magnetic field associated with EE is between 260 G and 320 G. The device, being axially movable, is positioned at the axis of the discharge vessel replacing the axially movable probe shown in Fig. 1. For determining the values of the voltages U'_{PE}, U'_{EE} and U'_{FC} applied to the electrodes, the V-A characteristics of the three electrodes (Fig. 3) have been measured at each given axial position. As it should be expected, the V-A characteristics of PE (Fig. 3(a)) measured at $U_{EE} = 0$ V and $U_{FC} = 0$ V is with the shape of a probe characteristics showing the value of the plasma potential ($U_{PE} = U'_{PE}$) by the "kink" between the transition and electron-saturation regions. The saturation in the other two V-A characteristics (Fig. 3(b) and 3(c)) marks the values U'_{EE} and U'_{FC}, respectively, of EE and FC. The V-A characteristics of EE (Fig. 3(b)) is measured for $U_{PE} = U'_{PE}$ and $U_{FC} = 0$ V whereas that of FC (Fig. 3(c)) is for $U_{PE} = U'_{PE}$ and $U_{EE} = U'_{EE}$. Thus, the results for the axial variation of the ratio of the currents to the second and the third electrodes given in the next section are for the values U'_{PE}, U'_{EE} and U'_{FC} of the voltages of these electrodes, specified for each axial position.

Laser-photodetachment-technique measurements and probe diagnostics have been employed in the determination of the influence of the extraction device on the discharge structure (item (iii) in the beginning of this section). The measurements are on the cylindrical axis of the source, at a given axial position: $z' = 12$ cm. The Faraday cup have been moved axially, away from this position (z'). In the laser photodetachment measurements, the laser beam path is again axially aligned. However, a radially movable probe positioned at $z' = 12$ cm has been used for measuring the electron saturation current $I_{e(dc)}$. The radius of the probe is 0.2 mm and its length is 10 mm. Like in the measurements to item (i), the laser pulse energy density is in the saturation region of the photodetachment fraction and, thus, the (n_-/n_e)-ratio is determined directly from the relation $n_-/n_e = I_{ph}/I_{e(dc)}$ where I_{ph} is the plateau amplitude of the photodetechment signal.

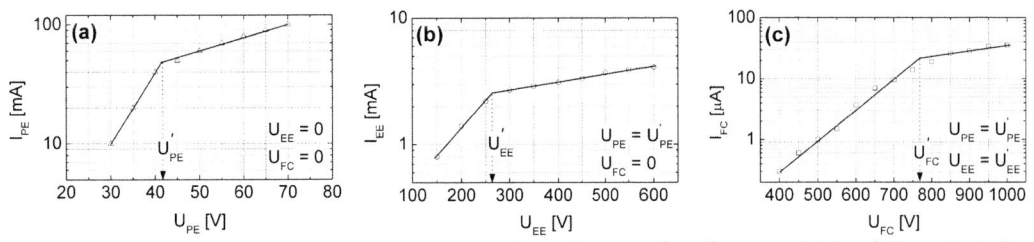

FIGURE 3. Measured V-A characteristics of the electrodes: (a) of PE for $U_{EE} = 0$ V and $U_{FC} = 0$ V, (b) of EE for $U_{PE} = U'_{PE}$ and $U_{FC} = 0$ V and (c) of FC for $U_{PE} = U'_{PE}$ and $U_{EE} = U'_{EE}$.

The changes in the plasma parameters – electron density and temperature – with varying the axial position of the Faraday cup are obtained from the characteristics of another radial probe, positioned also at $z' = 12$ cm. Being passively compensated (against the influence of the rf field), it is with a radius of 0.25 mm and a length of 6.0 mm. The obtained results for the (n_-/n_e)-ratio (from the laser photodetachment measurements) and for the plasma density (from the probe diagnostics) are used also for determination of the changes in the negative ion density caused by the Faraday cup.

RESULTS AND DISCUSSIONS

As it has been already mentioned, the results presented in this section are for two values of the gas pressure: $p = 6$ and 8 mTorr. The rf power applied for the discharge maintenance is $P_w = 700$ W.

The experiments are with a magnetic filter, in the second chamber of the source, with a maximum value $B_0 = 100$ G. However, in order to have the role of the magnetic filter checked, laser photodetachment technique measurements have been carried out also without the filter. The results for the axial variation of the (n_-/n_e)-ratio without a magnetic filter, obtained at $p = 8$ mTorr, are shown in Fig. 4. In the second chamber of the source ($z > 0$) the (n_-/n_e)-ratio stays almost constant, at a low value. This is associated with the axial variation of the electron density and temperature – a decrease of both of them – determined by the plasma expansion from the first chamber of the source to its second chamber. However, the first chamber of the source located at $z < 0$ in Fig. 4 (and in the figures further on) shows up with high values of the (n_-/n_e)-ratio. This is in accordance with the experimental results in Ref. [16] showing that the driver is an efficient source of negative ions.

The comparison of the obtained axial profiles of the (n_-/n_e)-ratio in Figs. 4 and 5, in the latter with a magnetic filter in the second chamber of the source, shows the effect of the filter. In the second chamber, the (n_-/n_e)-ratio is not anymore axially constant.

It has strong axial variation with a maximum located at $z = 10$ cm. This variation is closely related to the axial changes of the electron density (Fig. 6(a)). In agreement with former results [11, 15], the filter field causes a strong drop of the electron density followed by a maximum behind the filter. This maximum of n_e is in the region of the low electron temperature behind the filter (Fig. 6(b) which shows – by the strong drop of T_e in the filter region – the operation of the filter as an electron "cooler"). The maximum of the electron density is at $z = 10$ cm and this is exactly the position of the maxima of both the $(n\_/n_e)$-ratio (Fig. 5) and of the density of the negative ions (Fig. 7) behind the filter. With a magnetic filter in the second chamber of the source, the values of the $(n\_/n_e)$-ratio in its first chamber ($z < 0$) is lower (Fig. 5 compared to Fig. 4). This is due to the changes which the magnetic filter causes to the entire discharge structure. At the lower gas pressure ($p = 6$ mTorr) the electron density is lower (Fig. 6(a)). This is expected since the applied power is kept the same and the charged particle losses at lower pressure are higher requiring higher electron temperature (Fig. 6(b)). Both the $(n\_/n_e)$-ratio (Fig. 5) and the negative ion density (Fig. 7) are lower at the lower pressure. Also the structuring of their axial profiles is suppressed and the maxima behind the filter are less pronounced.

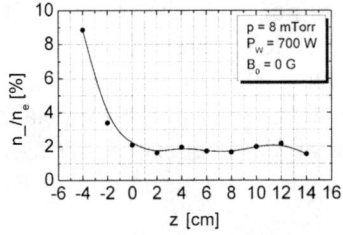

FIGURE 4. Axial variation of the $(n\_/n_e)$-ratio obtained without a magnetic filter in the second chamber of the source.

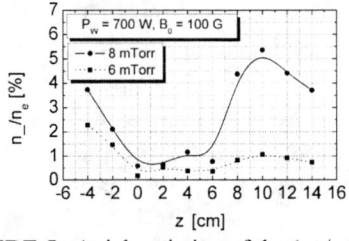

FIGURE 5. Axial variation of the $(n\_/n_e)$-ratio obtained with a magnetic filter for two value of the gas pressure ($p = 6$ and 8 mTorr).

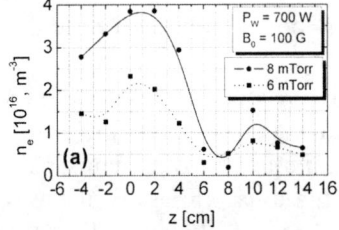

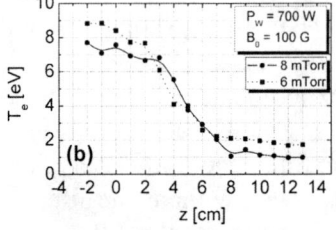

FIGURE 6. Axial variation of the electron density (a) and temperature (b) obtained with a magnetic filter for two values of the gas pressure ($p = 6$ and 8 mTorr).

Strong structuring of the axial profiles of the electron density caused by the magnetic filter, with appearance of a maximum of the electron density behind the filter in the region of the low – almost constant – electron temperature there, is a result known from former theoretical and experimental studies [15, 21, 22]. On that basis, a guess was made that the location of the extraction should "hit" the position of the maximum of the electron density behind the filter. The results for the axial variations of the $(n\_/n_e)$-ratio and for the negative ion density obtained here (Figs. 5 and 7) confirm this supposition.

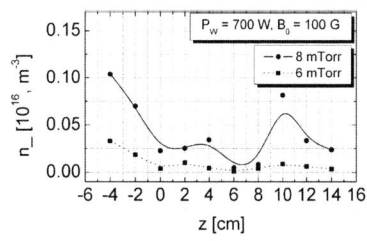

FIGURE 7. Axial variation of the density of the negative ions obtained with a magnetic filter for two values of the gas pressure (p = 6 and 8 mTorr).

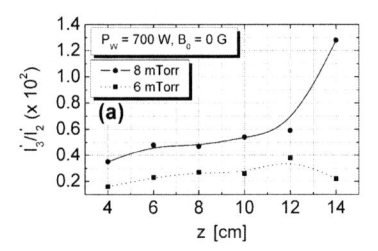

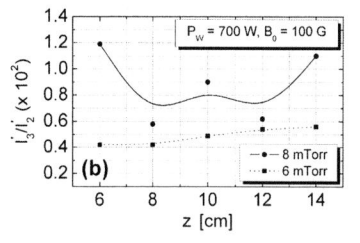

FIGURE 8. Axial variation of the ratio I'_3/I'_2 of the currents to the second and third electrodes of the extraction device obtained without (a) and with (b) magnetic filter in the second chamber of the source.

However, the results from the measurements of the currents I'_2 and I'_3 to the second and the third electrodes of the extraction device (the Faraday cup) – the axial variation of the ratio I'_3/I'_2 (Fig. 8) – do not show a behavior similar to that of the ($n\_/n_e$)-ratio and of the negative ion density (Figs. 4, 5 and 7). This can be attributed to lack of optimization of the extraction device. However, it was supposed that this could be due to the changes in the discharge structure introduced by the extraction device. This is confirmed by the results shown in Figs. 9-11.

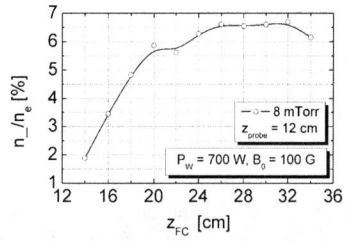

FIGURE 9. Values of the ($n\_/n_e$)-ratio measured at z' = 12 cm for different positions z_{FC} of the Faraday cup (at p = 8 mTorr and with a magnetic filter present in the second chamber of the source).

Probe and laser-photodetachment diagnostics have been carried out at a given axial position (z' = 12 cm) for different positions of the extraction device (z_{FC}, the position of the front cover of the Faraday cup). In fact, the Faraday cup has been moved away from the position z' = 12 cm of the measurements, towards the back wall of the second chamber of the source. Figure 9 shows the results for the ($n\_/n_e$)-ratio: low values of $n\_/n_e$ when the Faraday cup is close to the position z' = 12 cm of the measurements and its increase – up to more than 3 times – when the Faraday cup is away from it. In the latter case the value of the ($n\_/n_e$)-ratio is even a little bit higher than that measured at the same position (z = 12 cm) without having the Faraday cup in the second chamber of the source (Fig. 5). The changes in the values of the (n_-/n_e)-ratio in the discharge caused by the extraction device might be associated with the changes it causes in the spatial distribution of the plasma parameters. This is confirmed by Fig. 10 which shows the changes in the electron temperature and density at $z' = 12$ cm when the extraction device is moved away. Close to the Faraday cup the electron temperature is higher (Fig. 10(a)) and the electron density is lower (Fig. 10(b)). This results from the additional wall losses in the discharge introduced by the extraction device. As it has been discussed before [15, 16] the detected negative ions exist locally at the position of the measurements and its vicinity. Higher electron temperature and lower electron density when the Faraday cup is close to the position $z' = 12$ cm of the measurements lead to lower negative ion yield as Fig. 11, which presents the variation of the density of the negative ions, also shows. The higher electron density under the conditions of low electron temperature when the Faraday cup is away from the position of the measurements results into higher density of the negative ions (Fig. 11).

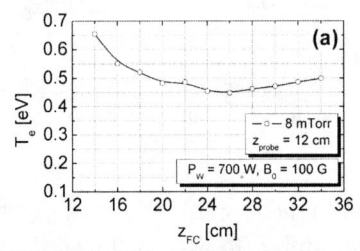

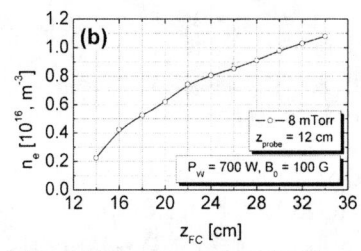

FIGURE 10. Values of the electron temperature (a) and density (b) measured at $z' = 12$ cm for different positions z_{FC} of the Faraday cup (at $p = 8$ mTorr and with a magnetic filter present in the second chamber of the source).

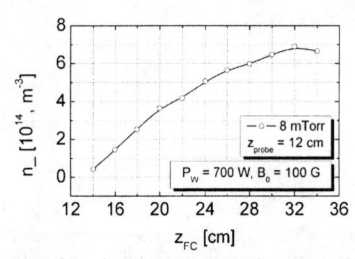

FIGURE 11. Values of the negative ion density measured at $z' = 12$ cm for different positions z_{FC} of the Faraday cup (at $p = 8$ mTorr and with a magnetic filter in the second chamber of the source).

CONCLUSION

The study presents experimental results on the axial variation of the ratio n_-/n_e of the negative-ion and electron densities and of the negative ion density – mainly in the second chamber and, partially, in the first chamber of a tandem-type of a plasma source – and their correlation with the axial profiles of the electron temperature and density. The obtained maximum of the (n_-/n_e)-ratio and of the negative ion density behind the magnetic filter is exactly at the position of the maximum of the electron density, located in the region of the low electron temperatures, which leads to the conclusion that this is the proper position of the extraction device. However, in order to get use of this the construction of the extraction device should be optimized regarding reduction of its influence on the discharge structure.

ACKNOWLEDGMENTS

The work is within the programme of the Bulgarian Association EURATOM/INRNE (Task 2.1.1)

REFERENCES

1. Zh. Kisss'ovski, St. Kolev, A. Shivarova and Ts. Tsankov, *IEEE Trans. Plasma Sci.* **35**, 1149-1155 (2007).
2. E. Speth et al., *Nuclear Fusion* **46**, S220-S238 (2006).
3. U. Fantz, H. D. Falter, P. Franzen, E. Speth, R. Hemsworth, D. Borilson and A. Krylov, *Rev. Sci. Instrum.* **77**, 03A516 (2006).
4. U. Fantz and D. Wünderlich, *New J. Phys.* **8**, 301 (2006).
5. D. Wünderlich, S. Dietrich and U. Fantz, *J. Quant. Spectros. Radiat. Transfer* **110**, 62-71 (2009).
6. M. Berger, U. Fantz, S. Shrist-Koch and NNBI Team, *Plasma Sources Sci. Technol.* **18**, 025004 (2009).
7. P. McNeely, S. V. Dudin, S. Shrist-Koch, U. Fantz and the NNBI Team, *Plasma Sources Sci. Technol.* **18**, 014011 (2009).
8. S. Shrist-Koch, U. Fantz, M. Berger and NNBI Team, *Plasma Sources Sci. Technol.* **18**, 025003 (2009).
9. Ts. V. Paunska, A. P. Shivarova, Kh. Ts. Tarnev and Ts. V. Tsankov, *AIP Conf. Proc.* **1097**, 12-21 (2009).
10. Ts. V. Paunska, A. P. Shivarova, Kh. Ts. Tarnev and Ts. V. Tsankov, *AIP Conf. Proc.* **1097**, 99-108 (2009).
11. I. Djermanov, St. Kolev, St. Lishev, A. Shivarova and Ts. Tsankov, *J. Phys.: Conf. Series* **63**, 012021 (2007).
12. Zh. Kiss'ovski, St. Kolev, A. Shivarova and Ts. Tsankov, *J. Phys.: Conf. Series* **113**, 012021 (2008).
13. S. Iordanova, *J. Phys.: Conf. Series* **113**, 012005 (2007).
14. Zh. Kiss'ovski, St. Kolev, S. Müller, Ts. Paunska, A. Shivarova and Ts. Tsankov, *Plasma Phys. Control. Fusion* **51**, 015007 (2009).
15. St. St. Lishev, A. P. Shivarova and Ts. V. Tsankov, *AIP Conf. Proc.* **1097**, 127-136 (2009).
16. St. St. Lishev, A. P. Shivarova and Ts. V. Tsankov, *J. Phys.: Conf. Series* **223**, 012001 (2010).
17. S. Iordanova, I. Koleva and Ts. Paunska, *Spectroscopy Lett.* **44**, 8-16 (2011).
18. M. Bacal, G. W. Hamilton, A. M. Bruneteau, H. J. Doucet and J. Taillet, *Rev. Sci. Instrum.* **50**, 719-721 (1979).
19. M. Nishiura, M. Sasao and M. Bacal, *J. Appl. Phys.* **83**, 2944-2949 (1998).
20. M. Bacal, *Rev. Sci. Instrum.* **71**, 3981-4006 (2000).
21. St. Kolev, St. Lishev, A. Shivarova, Kh. Tarnev and R. Wilhelm, *Plasma Phys. Control. Fusion* **49**, 1349-1369 (2007).
22. St. Lishev, A. Shivarova and Kh. Tarnev, *Plasma Phys. Control. Fusion* (2011), submitted.

ION SOURCES FOR ACCELERATORS

Latest Results from the Front End Test Stand High Performance H⁻ Ion Source at RAL

D. C. Faircloth, S. R. Lawrie, A. P. Letchford, C. Gabor, M. Whitehead, T. Wood and M. Perkins

STFC, Rutherford Appleton Laboratory, Chilton, Didcot, Oxfordshire OX11 0QX, United Kingdom.

Abstract. The aim of the Front End Test Stand (FETS) project is to demonstrate that chopped low energy beams of high quality can be produced. FETS consists of a high power Penning Surface Plasma Ion Source, a 3 solenoid LEBT, a 3 MeV RFQ, a chopper and a comprehensive suite of diagnostics. This paper briefly outlines the status of the project, hardware installation and modifications. Results from experiments running the H⁻ ion source at 2 ms pulse length are detailed: the discharge current is varied between 20 A and 50 A. The discharge repetition rate is varied between 12.5 and 50 Hz. Hydrogen and Caesium vapour flow rates are varied. The effect of electrode surface temperature and beam current droop are discussed. Peak beam currents of over 60 mA for 2 ms pulse length can be achieved. Normalised r.m.s emittances of 0.3 πmm.mrads at the exit of the LEBT are presented for different source conditions.

Keywords: High performance negative hydrogen ion sources
PACS: 07.77.Ka

INTRODUCTION

The Front End Test Stand (FETS)

FETS is being developed as a generic injector for future high power proton particle accelerators. The aim of the FETS is to demonstrate the production of a 60 mA, 2 ms, 50 pps chopped H⁻ beam at 3 MeV with sufficient beam quality for future applications. FETS consists of a high power ion source, a 3 solenoid magnetic Low Energy Beam Transport (LEBT), a 324 MHz, 3 MeV, 4-vane Radio Frequency Quadrupole (RFQ), a fast electrostatic chopper and a comprehensive suite of diagnostics.

Present Status

At the time of writing the ion source, LEBT and some of the diagnostics are operational. Figure 1 shows the present status of the FETS beam line. A moveable diagnostics vessel is mounted on the end of the beam line directly after the third LEBT solenoid. The position of four beam current toroids labelled T1 to T4 are shown in fig. 1.

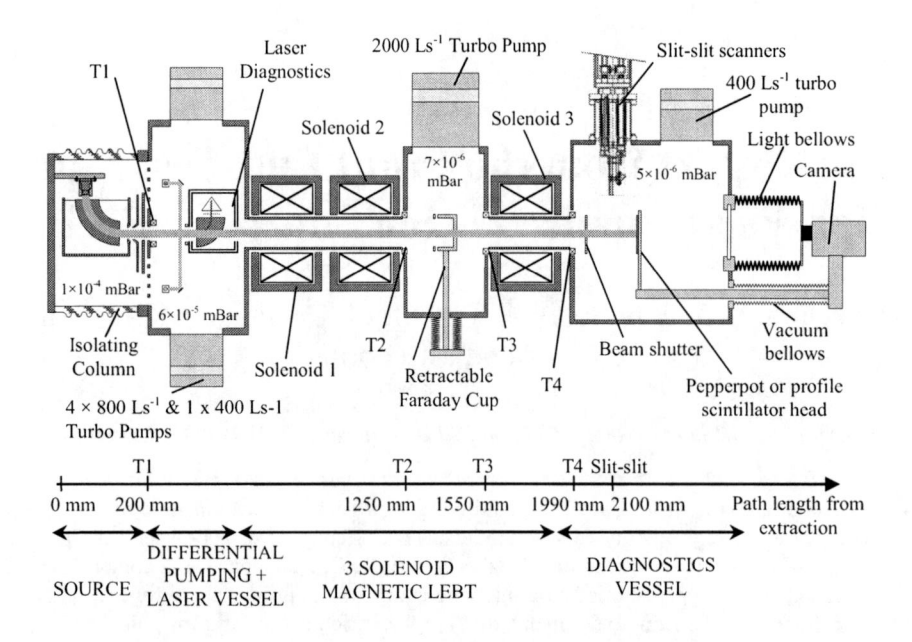

FIGURE 1. Schematic showing the present status of the FETS beam line.

Beam Line Development

RFQ

An RFQ cold model has been manufactured and tested and used to validate an electromagnetic model of the RFQ. Tuners and associated control systems have been built and tested. An integrated CAD/FEA system has been developed to design the RFQ vane modulations. Machining, bonding and vacuum sealing tests have allowed a strategy for manufacturing the RFQ to be developed. A detailed thermal model has validated the cooling scheme. High temperature stability cooling plant has been installed along with the 2 MW klystron RF power supply. The time span for manufacture of the RFQ is now dependent on the uncertain financial climate.

Chopper and MEBT

The design of the chopper is nearly complete, several prototypes of chopper components have been manufactured. Particle tracking in the MEBT has been undertaken and initial quadrupole magnet designs produced.

Laser Beam Diagnostics

A laser on loan from Frankfurt University has been used to progress the development of the laser profile measurement system. There have been significant problems with the acceptance of the electron detection system making it very sensitive to beam fluctuations. A proof of principle experiment has demonstrated photo-detachment, but more development work is required before a working profile measurement system can be achieved.

Ion Source

FETS ion source

The FETS ion source has been described in detail previously [1]. The source is of the Penning type. A schematic of the plasma source and the extraction system is shown in fig. 2.

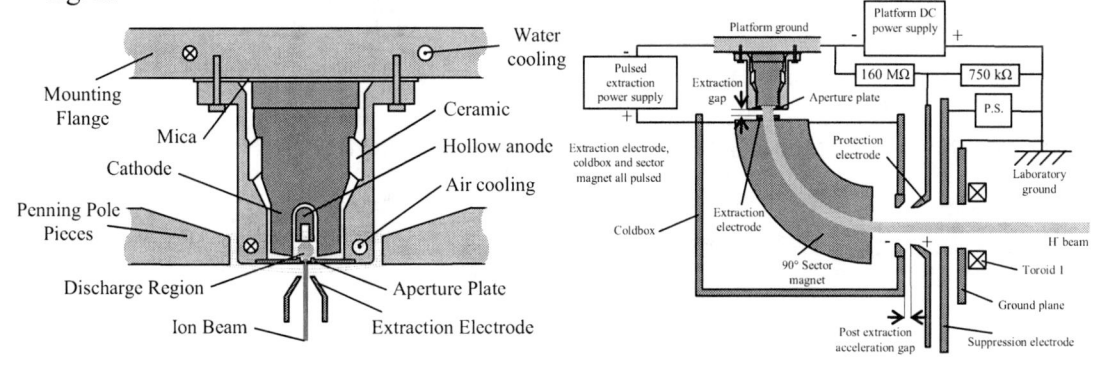

FIGURE 2. A schematic of the FETS ion source and extraction system.

Discharge Power Supply Modifications

The existing source discharge power supply is a pulsed current source capable of delivering up to 90 A, 1 ms long pulses at 50 Hz. The discharge power supply is designed and constructed in house and has successfully driven the plasma on the ISIS ion source for about 10 years. The power output stage consists of 4 IGBTs driven in linear mode to produce a current regulated output. FETS requires beam current pulse lengths of up to 2 ms. To achieve this, a discharge longer than 2 ms must be produced. This provides a settling time for the discharge and power supply so that beam can extracted from a stable plasma. By adding an extra 2 IGBTs to the output stage the power supply can generate current pulses up to 2.2 ms long at 50 Hz. The output current however is limited to about 55 A at this duty cycle.

EXPERIMENTS

Vary Discharge Current

Figure 3 shows the discharge current and voltage oscilloscope traces for different discharge currents for a 2.2 ms long discharge pulse. The pulsed discharge power supply is current regulated so that flat current pulses can be produced. For discharge currents below 30 A the plasma takes up to 300 μs to stabilise which produces a substantial amount of ringing on the front of the discharge pulse due to the lower plasma density.

The discharge voltage slowly drops along the length of the discharge. The discharge voltage increases for lower discharge currents.

For each discharge current, the ion source cooling is modified to keep the electrode temperatures approximately constant and the source is given time to stabilise.

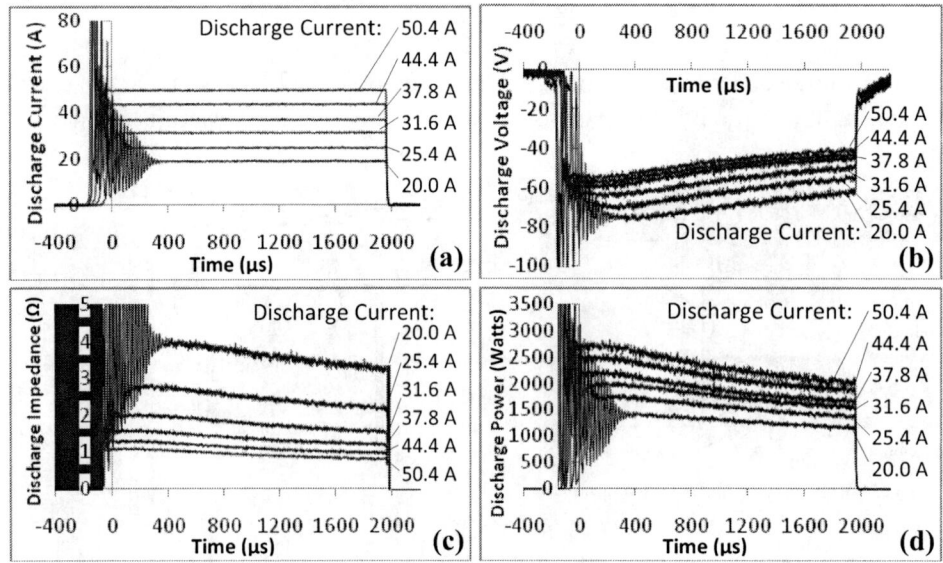

FIGURE 3. The source discharge current and voltage oscilloscope traces for different discharge currents. The discharge impedance and power are calculated from the current and voltage.

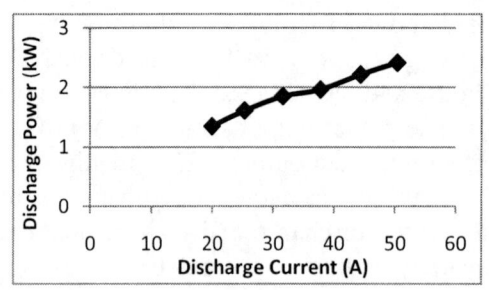

FIGURE 4. The variation of discharge power with discharge current.

The extraction voltage power supply is voltage regulated and is kept constant at 18.6 kV for all the experiments presented in this paper. The beam energy is also kept at 65 keV and the solenoid settings are left unchanged at 0.19 T, 0.17 T and 0.30 T. Figure 5 shows how the extraction current increases as the discharge current is increased.

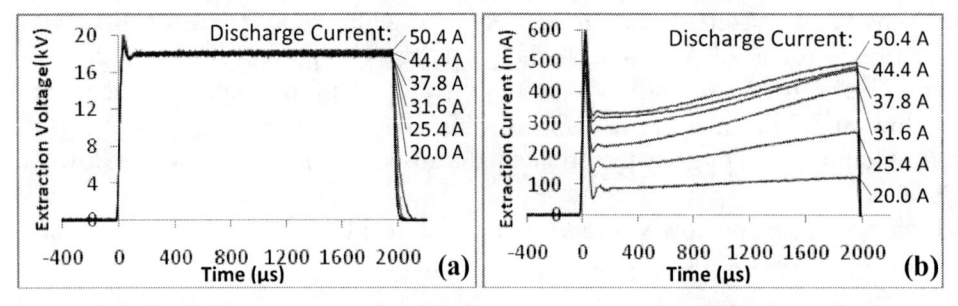

FIGURE 5: Extraction voltage pulse and current for different discharge currents.

Figure 6 shows the beam current measured at T1 for different discharge currents for a 2 ms long beam pulse. As the discharge current is reduced below 50 A the average beam current decreases but the droop decreases also, and for currents below 25 A the beam current is almost flat. The noise on the front of the H⁻ beam current is caused by noise at the start of the discharge becoming more severe and creeping into the extraction region. The noise present throughout the pulse is pickup from the high-frequency switching of the LEBT solenoid power supplies, and is not a real feature of the beam current.

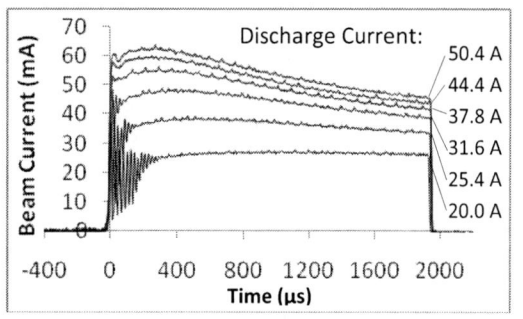

FIGURE 6. Variation in H⁻ beam current pulse shape for different discharge currents for 2 ms pulse lengths.

Figure 7 shows how the H⁻ beam current as measured by each of the toroids increases as the discharge current increases. For higher discharge currents the rate of increase of beam current becomes slower.

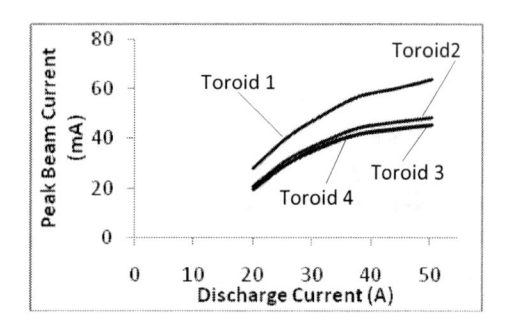

FIGURE 7. Variation in peak beam current measured at each of the 4 toroids for different discharge currents for 2 ms pulse length.

Vary Discharge Repetition Rate

The effect on the H⁻ beam current of reducing the discharge repetition rate is shown in fig. 8. As the repetition rate is reduced the beam current increases and the droop decreases. For a 50 Hz discharge the beam current starts at 50 mA but droops to 30 mA at 2 ms. The shape of the beam current changes as the repetition rate is reduced; for lower repetition rates the beam current first rises then droops. At each repetition rate, source cooling is modified to keep the electrode temperatures approximately constant and the source is given time to stabilise.

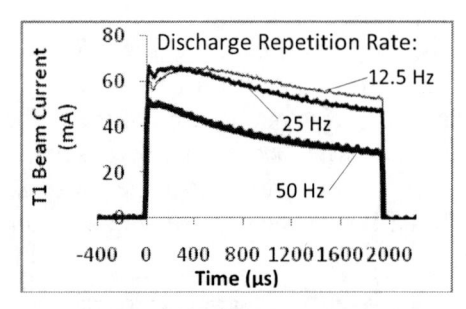

FIGURE 8. Variation in beam current in T1 for different discharge repetition rates.

Vary Hydrogen Flow Rate

Hydrogen at a static pressure of 600 mBar is pulsed into the ion source via a piezoelectric valve. The timing is kept constant: a 200 μs long pulse, triggered 800 μs before the start of the discharge and 1000 μs before the start of the extraction voltage pulse. By varying the amplitude of the voltage applied to the piezoelectric valve, the valve can be made to open wider and the amount of hydrogen delivered to each discharge pulse varied. The average hydrogen flow rate is measured using a flow meter near the hydrogen bottle.

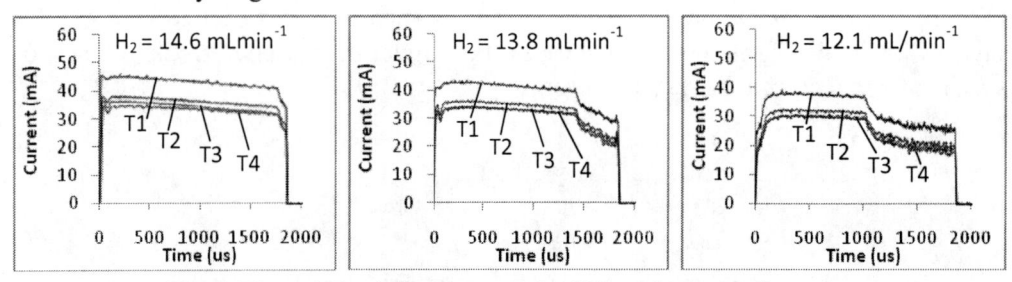

Fig. 9. The variation in beam current for different hydrogen flow rates.

Figure 9 shows how the H⁻ beam current as measured by the toroids varies as the hydrogen flow rate is decreased. As the amount of hydrogen delivered to the discharge is reduced, the peak beam current starts to drop and a notch appears in the back of the pulse indicating hydrogen starvation. This phenomenon is only observed for discharges longer than 1000 μs. Increasing the hydrogen flow rate beyond 15 mLmin⁻¹ does not further increase the beam current.

Emittance Scans

A pair of slit-slit emittance scanners in the diagnostics vessel after the LEBT are used to measure the horizontal and vertical emittances during the 2 ms beam pulse. Figure 10 shows emittance phase space plots at the beginning, middle and end of the 2 ms beam pulse for a 50 A 25 Hz discharge. The horizontal and vertical normalised r.m.s. emittances of approximately 0.3 πmm.mrad. Emittances are calculated using the SCUBEEx algorithm [2].

Figure 11 shows how the horizontal and vertical emittances, Twiss alphas and betas, vary along the 2 ms pulse for three different source conditions.

(a) Start: 210 μs   (b) Middle: 1050 μs   (c) End: 1943 μs

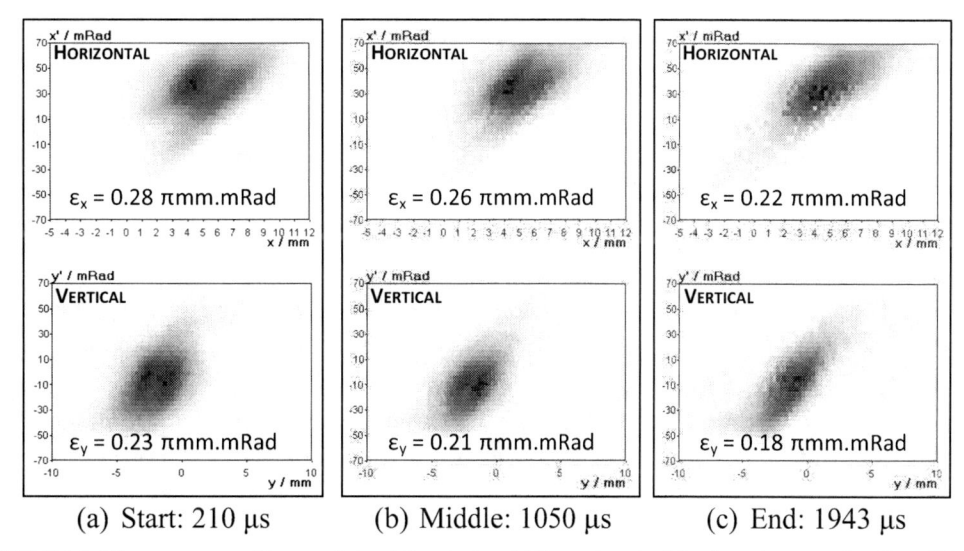

FIGURE 10. Phase space emittance plots at the start, middle and end of the 2 ms beam pulse for a 50 A 25 Hz discharge. All emittances are normalised r.m.s. The horizontal offset is caused by an alignment problem that has now been fixed.

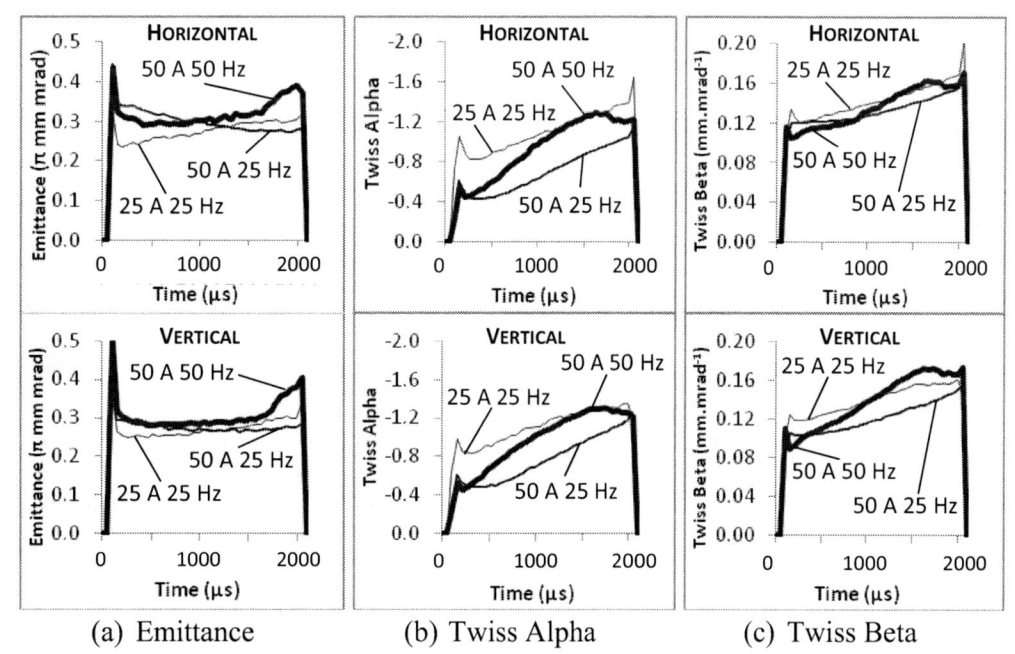

(a) Emittance   (b) Twiss Alpha   (c) Twiss Beta

FIGURE 11. Emittance and Twiss parameters versus time, for three different combinations of discharge current and discharge repetition rate.

DISCUSSION

Vary Discharge Current

The discharge current has a significant effect on both the magnitude and shape of H⁻ beam current produced as shown in fig. 6. For higher discharge currents the beam current droop becomes significant. There are two main factors that could cause this droop:

Increased plasma density

Higher discharge currents produce an increased plasma density, which leads to more Cs being desorbed from the cathode surfaces by increased bombardment of H^+ and Cs^+. This pushes the level of Cs coverage away from its optimum for H⁻ production. Also H⁻ that are produced at the cathode surface are more likely to be stripped of their extra electron as they pass through the increased plasma density on their way to the extraction region.

Greater surface temperature change during the discharge

Figure 12 shows how the cathode surface temperature rises during the discharge then falls after the discharge is over. (See fig. 4 for actual measured discharge powers.) The cathode surface temperature shown is calculated using transient 3-dimensional finite element analysis [3]. As the power in the discharge increases, the cooling is increased to maintain the same steady state temperature, but the transient surface temperature change increases. For higher discharge currents this change can be as high as 100°C for a 2 ms long discharge; this could be enough to push the level of Cs coverage away from the optimum for H⁻ production.

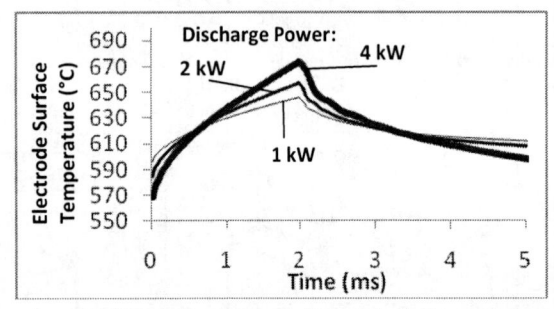

FIGURE 12. Transient 3D finite element modelling calculations of rise in cathode surface temperature during the discharge for different discharge powers.

Increasing the discharge current also leads to an increase in the rate at which the extraction current rises during the pulse (see fig. 5). As the H⁻ beam current does not increase during the pulse this implies that more electrons are extracted near the end of the pulse which suggests that the electron density in the plasma also increases near the end of the discharge. This could possibly be caused by the depletion of surface caesium and hence the extracted H⁻/e⁻ ratio goes down.

Vary Discharge Repetition Rate

The amount of droop in the beam current can also be altered by changing the discharge repetition rate as shown in fig. 8. Leaving longer between pulses reduces the beam current droop. Cs vapour is continuously fed into the source, so leaving longer between pulses gives more time for Cs to build up on the electrode surfaces before the discharge comes on again.

Vary Caesium Oven Temperature

Figure 13 shows how the Cs vapour pressure varies with temperature [4]. Increasing the Cs oven temperature from 160°C to 190°C should produce a fourfold increase in vapour pressure. However initial experiments have failed to show an improvement in the amount of droop when the Cs oven temperature is varied in this range.

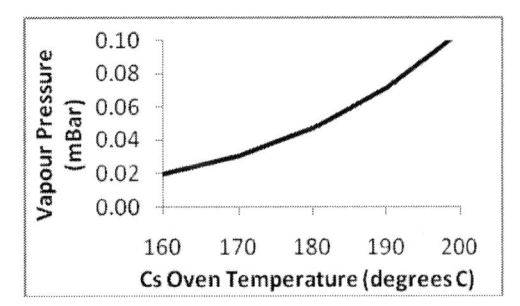

FIGURE 13. Cs vapour pressure versus temperature.

Vary Hydrogen Flow Rate

For a typical flow rate of 15 ml/minute of H_2 into the source at a static pressure of 0.6 Bar above atmosphere at a temperature of 15°C this gives a flow rate of:

$$\frac{\left(\frac{15 \times 10^{-3}}{60 \times 50}\right) \times 1.6 \times \left(\frac{273}{273+15}\right) \times 2}{22.414} = 4 \times 10^{17} \text{protons per pulse}$$

For a 50 A 2.2 ms discharge, the total number of charge carriers in the discharge is:

$$50 \times 2.2 \times 10^{-3} \times 1.602 \times 10^{-19} = 7 \times 10^{17} \text{charge carriers per pulse}$$

The charge carriers in the discharge are predominantly electrons and protons, with a small contribution from caesium and molybdenum ions and other ion species. The similarity between the number of protons delivered to the discharge and the number of charge carriers in the discharge is striking. The notch in the back of the H⁻ beam current pulse shown in fig. 9 is most likely caused by a lack of protons available for the discharge or H⁻ production mechanism in the source.

Emittance Scans

Figure 11 shows how the beam emittance varies slightly along the length of the discharge. Phase rotation during the beam pulse can be seen in fig. 10 which can be confirmed by the decreasing value of Twiss alpha in fig. 11.

Space charge neutralisation of the beam occurs in the first 100 μs or so of the beam. For long beam pulse lengths phase rotation and a slight increase in emittance can be explained by loss of compensating particles. The neutralised beam no longer traps the compensating particles and they drift out of the beam, causing the beam to slowly blow up.

Figure 14 shows the beam brightness: $\frac{I_b}{8\pi\epsilon_x\epsilon_y}$ for 3 different source conditions. When the source is running at 50 Hz with a 50 A discharge there is a significant drop in source brightness along the 2 ms beam pulse. Halving the repetition rate allows for an almost constant brightness throughout the pulse. Reducing the current to 25 A at 25 Hz yields a similar drop in brightness which implies that at 25 A the plasma density in the discharge is not high enough to sustain a high brightness beam.

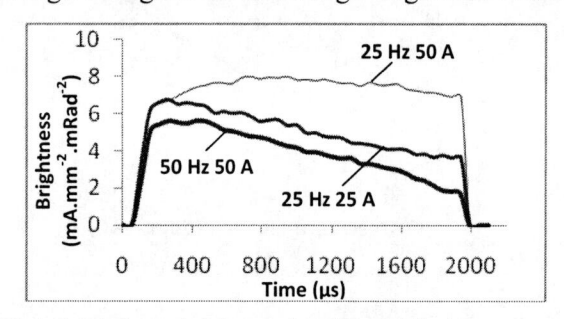

FIGURE 14. Variation in beam brightness for different discharge rep. rates and currents.

CONCLUSIONS

The physics of the operation of the source is complex and dynamic and requires further work to fully explain all the results presented in this paper. To produce a 50 Hz, 2 ms flat beam pulse it is clear that something must be done to keep the electrode surfaces in the optimum condition for H⁻ production. Increasing Cs flux does not seem to reduce the beam current droop. A possible way forwards is to increase the electrode surface area. Increasing the electrode surface area reduces the surface power density and reduces the transient temperature rise. It will also reduce the bombardment flux that causes Cs desorption from the electrodes surfaces. This has been successfully demonstrated by Smith and Sherman [5] with a x4 scaled source.

REFERENCES

1. D.C. Faircloth S. Lawrie, A. P. Letchford, C. Gabor, P. Wise, M. Whitehead, T. Wood, M. Westall, D. Findlay, M. Perkins, P. J. Savage, D. A. Lee, J. K. Pozimski, *Review of Scientific Instruments*, Volume 81, Issue 2, (2010).
2. Stockli M. P. et al., *AIP Conf. Proc.* 639, 135 (2002).

3. D.C. Faircloth, J. W. G. Thomason, M. O. Whitehead, W. Lau and S. Yang, *Review of Scientific Instruments,* Volume 75, Number 5, (2004).
4. J. B. Taylor, I. Langmuir, *Phys. Rev.* 51, 753 (1937)
5. H. V. Smith and J. Sherman, "H– and D– scaling laws for Penning Surface-Plasma Sources", *Review of Scientific Instruments*, Vol 65 (1), pp. 123-128 (1994).

Performance of the H⁻ Ion Source Supporting 1-MW Beam Operations at SNS

B.X. Han, T. Hardek, Y. Kang, S.N. Murray Jr., T.R. Pennisi, C. Piller, M. Santana, R.F. Welton, and M.P. Stockli

Spallation Neutron Source, Oak Ridge National Laboratory, Oak Ridge, TN 37831, USA

Abstract. The Spallation Neutron Source (SNS) at Oak Ridge National Laboratory reached 1-MW of beam power in September 2009, and now routinely operates near 1-MW for the production of neutrons. This paper reviews the performance, operational issues, implemented and planned mitigations of the SNS H⁻ ion source to support such high power-level beams with high availability. Some results from R&D activities are also briefly described.

Keywords: H⁻ ion source, RF ion source, Low energy beam transport
PACS: 07.77.Ka, 29.25.Ni, 52.80.Pi, 41.85.Ja

INTRODUCTION

The injector for the Spallation Neutron Source (SNS) accelerator system consists of a Cs-enhanced, radio-frequency (RF) driven, multi-cusp H⁻ ion source and an electrostatic 2-lens low-energy beam transport (LEBT) section. It was designed and built at Lawrence Berkeley National Laboratory (LBNL) and delivered to the SNS site as part of the accelerator front-end system [1]. Continuous ~300 W from a 600-W, 13.56-MHz RF supply provide a low-power plasma to facilitate the high-power breakdown conditions [2]. Pulsed high power RF from an 80-kW, 2-MHz amplifier drives the high-density hydrogen plasma to produce the high-current H⁻ beam pulses. A magnetic dipole field in the outlet region separates the co-extracted electrons from the H⁻ beam, as shown in Fig. 1.

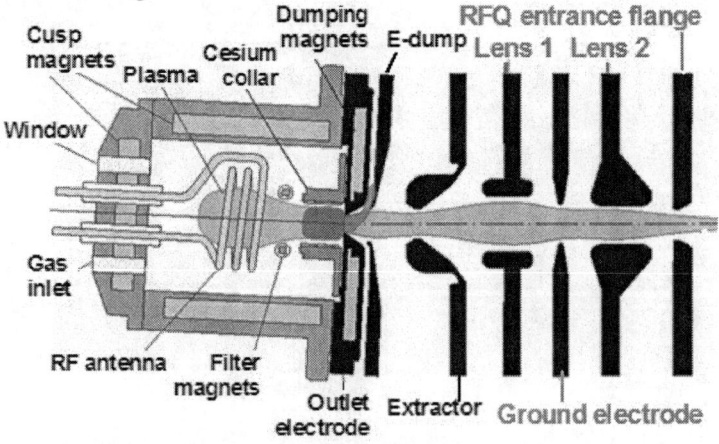

FIGURE 1. A schematic view of the SNS H⁻ injector.

The electron-dumping magnetic dipole field bends the H⁻ beam by ~3° by the time the beam enters the field-clamping extractor, which is made from 430 stainless steel. The bend can be compensated by tilting the source body in the range between 0 and 5.5°. The associated transverse motion of the beam can be compensated by moving the source body transversely with respect to the extractor and LEBT axis.

The second lens of the LEBT is split into four electrically-isolated quadrants to blank, chop, and steer the H⁻ beam. The 12-cm-long electrostatic LEBT lacks space for beam diagnostics. The radio frequency quadrupole (RFQ) accelerates the beam to 2.5 MeV and two magnetic quadrupoles refocus the beam before the beam current is measured with the beam current torroid BCM02, which practically matches the linac peak current.

SNS started neutron production in 2006 and reached the milestone of 1-MW of beam power in September 2009. The SNS beam power ramp-up plan required the beam current and duty-factor to be raised during each biannual run cycle until reaching 38 mA linac beam current and ~5 % duty factor, while increasing the ion source and LEBT availability to ~99%. Several design changes and many procedural changes had to be implemented to meet the requirements of this beam power ramp-up plan [3, 4]. This paper reviews recent beam current performance, issues, implemented and planned mitigations of the ion source required to support 1-MW beam operation with high availability. Some R&D testing results of the ion source are also given.

ION SOURCE BEAM OUTPUT PERFORMANCE

For 1-MW beams, the ion source runs with ~0.9 ms long pulses at 60 Hz. The bottom of Fig. 2 shows the typical shape of the beam pulse measured with BCM02. About 0.05 ms after the start the beam peaks and then decays a few mA to a steady value throughout the rest of the pulse. The archived maximum BCM02 current is an overestimation due to significant noise issues, while the archived averaged BCM02 current depends mostly on the chopping fraction. For this reason we normally quote the flat part of the pulse reached after 0.5 ms or near the end of the pulse, e.g. ~40 mA at the top of Fig. 2.

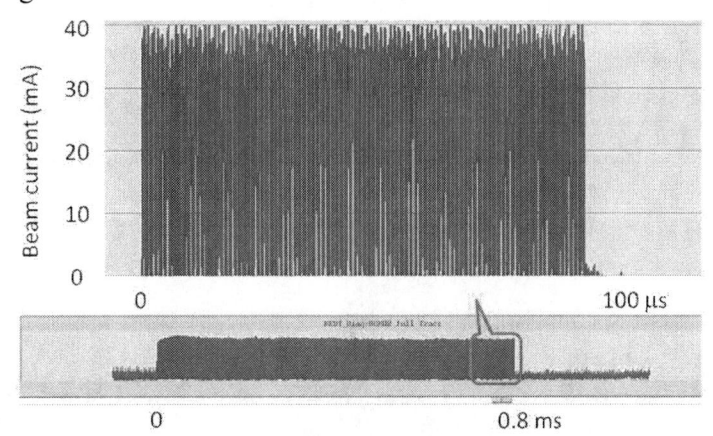

FIGURE 2. Typical beam pulse measured with BCM02. The lower part shows the full beam pulse with both ends blanked by the chopper. The upper part shows the last 90 μs of the pulse enlarged, revealing the chopping pattern.

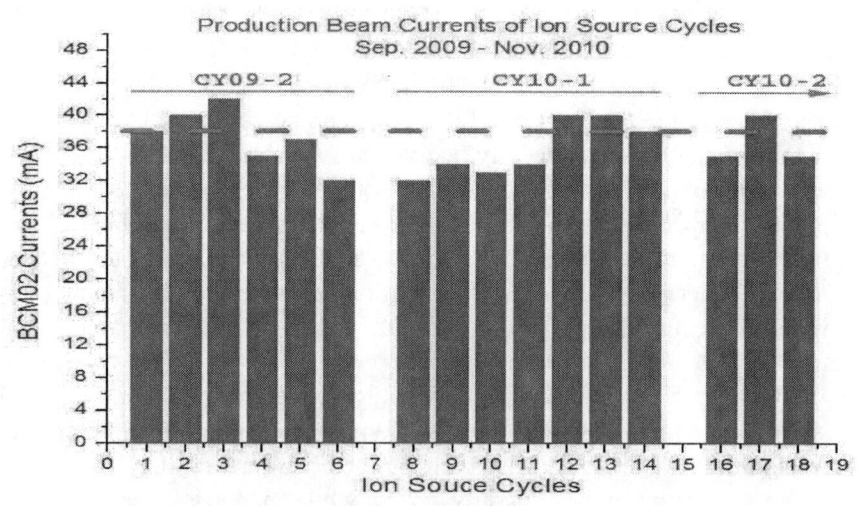

FIGURE 3. Summary of beam current performance of ion source cycles

Figure 3 shows the linac peak beam current of the 16 ion source service cycles spanning Sep. 2009 – Nov. 2010, typically established within 8 hours of starting the source replacement. It is rare that the schedule permits retuning and aligning the source on the following day, which often increases the beam current. After this the beam current normally sustains for the entire 3-5 weeks of the source service cycle without adding Cs beyond the initial release of ~5mg. Typically the matching capacitance requires lowering over the first few days to keep the reflected power near the minimum. Sometimes the RF power is slightly increased to adjust for small perceived or real decreases in beam current. Figure 4 shows the best example, the third ion source in the CY09-2 run, which delivered 42 mA beam current with perfect persistence during its 3-week source service cycle. Late in 2009 the normal source service cycle was increased from 3 to 4-5 weeks because there were and continue to be no old-age failures and no obvious performance degradations besides the plasma outages discussed in the next section.

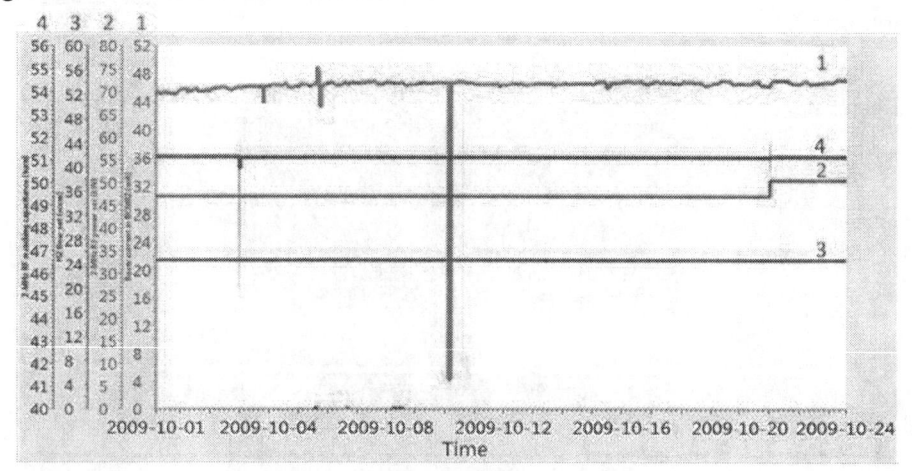

FIGURE 4. An example of the ion source performance with high beam current and perfect persistence. Axis 1: BCM02 beam current; axis 2: requested 2-MHz RF power; axis 3: requested H_2 flow; axis 4: 2-MHz RF matching capacitance.

PLASMA OUTAGES AND MITIGATIONS

Plasma outages have become an increasingly frequent problem as RF power was increased to increase the H⁻ beam current. Outages that are immediately detected can often be easily overcome by temporarily lowering the capacitance of the 2-MHz matcher. If the outages are prolonged the out-of-resonance 2-MHz amplifier cools down and requires a longer procedure to restore the high-power performance.

The rate of plasma outages appears to gradually increase with the age of each source. In addition the rate appears to increase throughout the run-cycle. When plasma outages occur frequently it takes up to 0.05 ms for the reflected 2-MHz power to level out at a low value, which needs to be compared to ~0.01 ms with a recently-installed source.

Initially-successful mitigations included increasing the 13.56-MHz RF power, increasing the H_2 gas flow, lowering the capacitance of the 2-MHz matching network and increasing the 2-MHz RF power, all of which are consistent with generating more favorable breakdown conditions. However, the H_2 gas flow had to be restricted to 28 sccm because larger H_2 flows occasionally caused the RFQ cooling regulation to go out of bounds [5]. In addition, the power of the RF amplifiers was frequently restricted because of reliability concerns and associated downtime. This often led to lowering capacitance as the only viable option. Figure 5 shows the worst example, the last source of the CY09-2 run, where the matching capacitance was frequently lowered to prevent plasma outages. The detuning lowered the antenna current, and even more drastically the H⁻ beam current, despite eventually pushing the 2-MHz power and the H_2 flow to their limits.

The root cause of these plasma outages is the changing plasma inductance, which significantly shifts the RF resonance every time the plasma power is drastically changed. This is a fundamental issue of pulsed high-power RF ion sources.

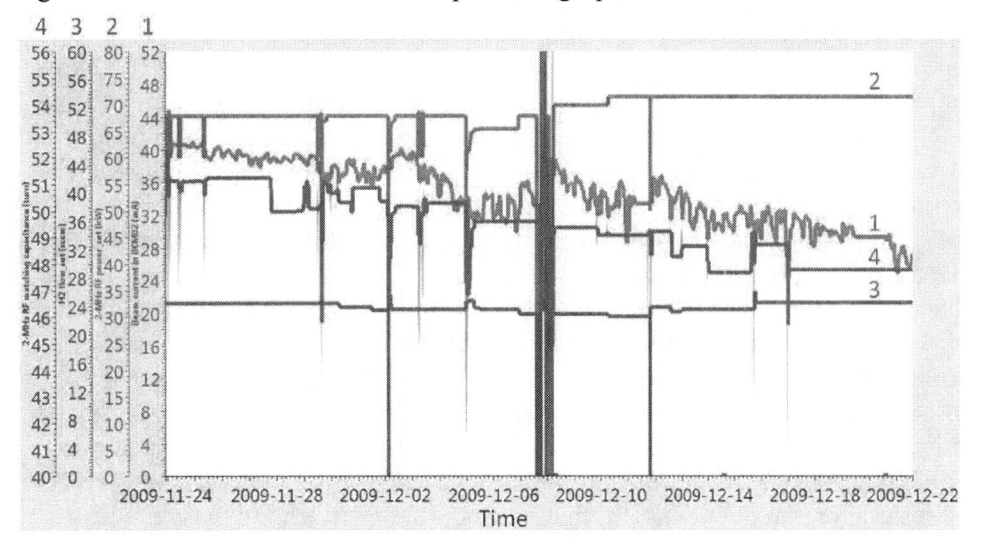

FIGURE 5. At the end of run CY09-2 the beam current degraded drastically due to detuning the match in order to prevent plasma outages; Axis 1: BCM02 beam current; axis 2: requested 2-MHz RF power; axis 3: requested H_2 flow; axis 4: 2-MHz RF matching capacitance.

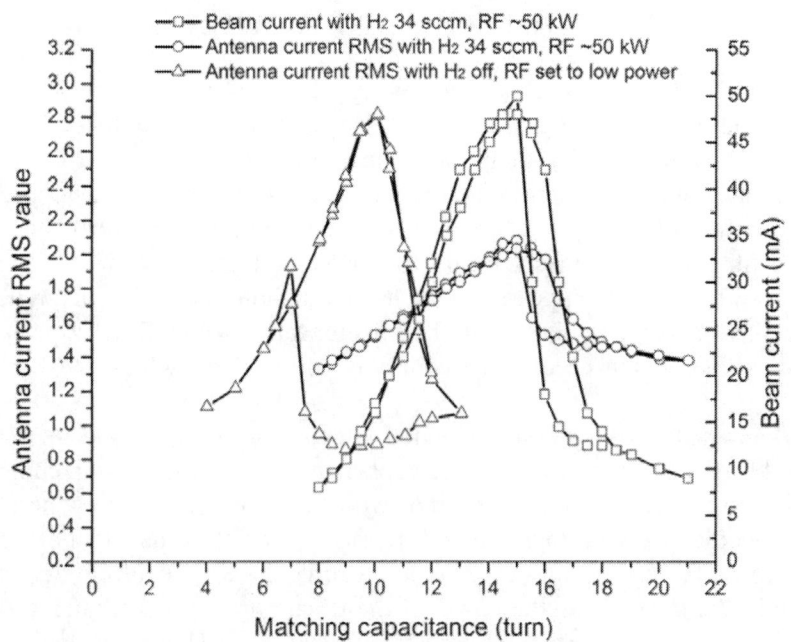

FIGURE 6. BCM02 beam current- and RF antenna current resonances without and with plasma.

To shed light on this issue, our production 2-MHz system was characterized by measuring the BCM02 H⁻ beam current and the antenna current for 0.9 ms long pulses at 60 Hz as a function of the 2-MHz matching capacitance, as shown in Fig. 6. The characterization started with a low capacitance, which was increased in small steps until the beam current peaked near 50 mA at 15-turns and then dropped to ~10 mA. Next the capacitance was lowered in small steps, which yielded lesser values for both currents until reaching resonance. This hysteresis appears to be caused by a temperature change in the 2-MHz amplifier, because the hysteresis seems to vanish when one waits for a while between each step, during which the current values settle somewhere between those two points. The antenna current is less resonant than the strongly-peaked H⁻ beam current because the plasma power is proportional to the square of the antenna current in a well-tuned system. Afterwards the hydrogen flow was stopped and the 2-MHz power was lowered to keep the antenna current in a safe range. The measured resonance peaked near 10-turns, but showed a more pronounced hysteresis, likely for the same cause.

The key is the 5-turn shift between the two resonances, which corresponds to 66 pF, or a ~4% change. The figure also illustrates that when the tune is optimized for maximum beam production, the electric fields available at the beginning of the pulse with no or low-power plasma are ~70% below their resonant values. Obviously the electric fields can become marginal for ionizing the neutral gas, which will increase the time it takes to develop full-power plasma. The electric fields can even drop below the threshold required to ionize the neutral gas, which will lead to plasma outages.

The net inductance L of our matching network can be calculated from the equation for the resonant frequency f (= 2 MHz):

$$f = (2 \cdot \pi)^{-1} \cdot (L \cdot C)^{-1/2} \qquad (1)$$

where C is the net capacitance, which is dominated by the tuning capacitor set near 1.6 nF. The measured 4% change of the resonant capacitance corresponds to a plasma inductance of ~0.15 µH, a significant fraction of the ~0.5 µH of the immersed 2.5-turn antenna. Measuring the shift of the resonance requires long pulses for the plasma to fully develop, which may be the reason that this method was not considered for measuring the inductance of the DESY source [6].

The mismatch can be reduced by increasing the inductance of the matching network so that the plasma inductance becomes a smaller fraction. However, increasing the inductance increases the high voltages generated in the matching network, which in the past apparently led to discharges in our high-power RF system [4]. Discharges sap energy from the matching network, which can also lead to plasma outages.

To increase the inductance without increasing the high voltages with respect to the source body, the original LBNL-supplied 1.6-µH inductor was replaced with two 1-µH inductors connected to both ends of the antenna, as shown in Fig. 7. This has reduced the voltage on one side of the capacitor, while it increased the near-ground voltages on the other side of the capacitor and the transformer. So far the change seems to have reduced the number of plasma outages, although the end-of-the-run will be the ultimate test.

The pattern of the rate of plasma outages suggests that the breakdown voltage of the hydrogen gas has a significant dependence on the levels of impurities, which tend to feature lower ionization energies and higher electron yields. While the impurity levels are ultimately limited by the 99.9995 purity of our hydrogen supply, the actual levels are significantly higher due to gradual outgassing during each source service cycle. In addition, a gradual decay throughout the run is observed, most likely due to gradual outgassing of the LEBT chamber, which is only briefly vented with nitrogen whenever a source is changed. There is no plan to add gas with low ionization energies and/or high electron yields, because the mass of such atoms and molecules have a low threshold for sputtering the Cs, and could therefore negatively affect the persistence of our H⁻ beams [7].

Rather than reducing the mismatch by increasing the inductance, the problem could be eliminated by compensating the plasma inductance in real time. The inductance, which changes on a µs time scale, could be compensated by varying the system's inductance, capacitance, or frequency. The current plan calls for shifting the frequency down by ~2% for the first ~0.01 ms of each pulse, which may not only eliminate the plasma outages, but could also render the 13-MHz low-power plasma obsolete.

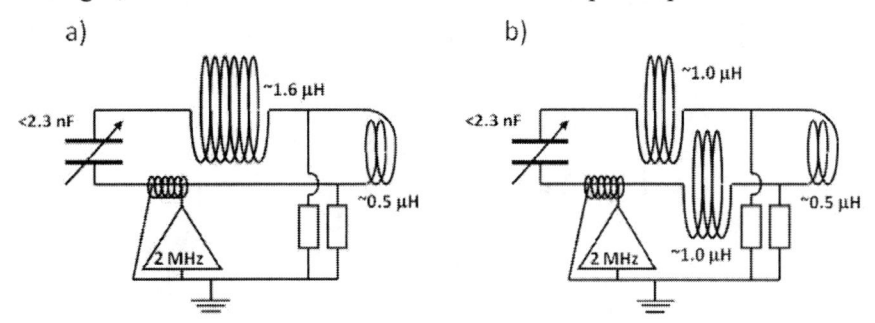

FIGURE 7. 2-MHz RF matching network a) for run CY10-1 and b) for run CY10-2.

INCREASING THE RELIABILITY OF THE 2-MHz SYSTEM

The 6% 80-kW 2-MHz amplifier was initially located on the 65 kV platform of the ion source potential. This caused many trips, likely due to high voltage discharges. As the system conditioned and as problem areas became apparent and were mitigated, the trips became less frequent.

However, increasing the beam current by raising the RF power and simultaneously increasing the duty factor over the last few years raised the heat generated by the amplifier. This increased the air temperature in the high-voltage enclosure to over 40 °C. The amplifier itself must have been considerably hotter, which likely led to a new set of trips which could rarely be identified exactly. During a 6% 80-kW 24-hour test, the tube turned black, showing that amplifier cooling was marginal at best.

The high-voltage enclosure, normally closed for operation, had to be open for troubleshooting as well as for tuning, to provide access for the bulky load on the outside. After being tuned with the amplifier providing high power into the test load, when providing high power to the source on high-voltage sometimes the amplifier would become unstable or trip. This contributed to the restriction at the end of run CY09-2, which ended near 600 kW after starting with 1 MW.

Apparently our 2-MHz system lacks the robustness needed for placement on a high-voltage platform. For this reason a 2-MHz isolation transformer was developed [8]. Before run CY10-2 a spare 2-MHz amplifier was installed on ground and its output was connected through the isolation transformer to the 2-MHz matching network on the 65-kV platform as shown in Fig. 8. When starting up the grounded amplifier the resonance needs to be tracked until the amplifier can generate the desired high output power, likely due to the more-efficient cooling. So far the grounded 2-MHz amplifier has caused no downtime, although the operational power has typically been limited to ~55 kW to gain experience. This is part of the reason that not every source was started up with ≥38 mA in run CY10-2, as seen in Fig. 3.

The original amplifier was left on the 65-kV platform and can be reemployed in case of a prolonged failure of the grounded amplifier.

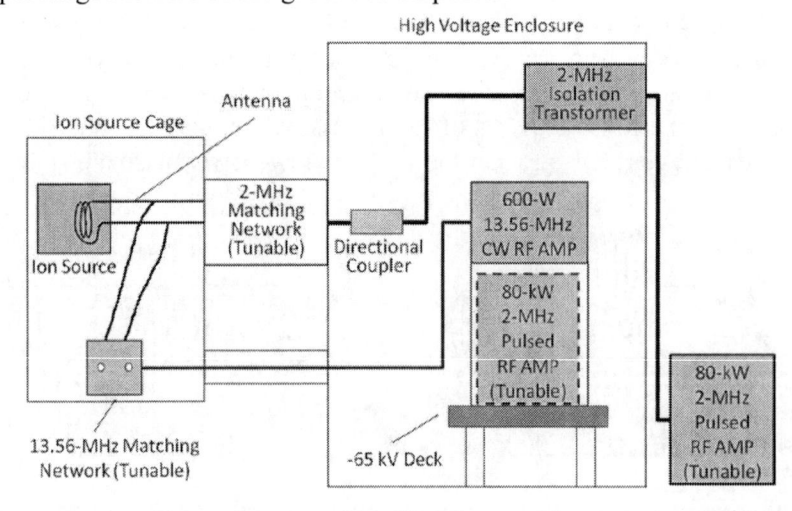

FIGURE 8. Recent installation of the ground-based 2-MHz amplifier, replacing the 2-MHz amplifier on the 65 kV deck, which now serves as the spare shown as a dashed box.

ION SOURCE TILT AND OFFSET

Using the MEBT beam stop when preparing for run CY09-2, optimizing the source position yielded significantly more BCM02 beam current when the source was not tilted (0°) compared to the source being tilted at the 3° design angle. This unexpected finding was subsequently confirmed several times with different sources over a period of time.

A more systematic study was undertaken after run CY10-1 was completed. Figure 9 shows the correlation between the optimal horizontal and vertical source position for tilt angles of 3°, 1.5°, and 0°. After setting the tilt angle and finding the optimal position, both LEBT lenses (L1 and L2) were tuned without active steering in the second lens (L2) until obtaining the maximum BCM02 beam current indicated in Fig. 9. The optimized beam current increased by more than 10% for both 1.5° decreases of the tilt angle.

In order to understand this unexpected behavior, the 3-D ion optical code SIMION [9] was used for beam dynamics simulations. The electron dumping field in the ion source outlet region was calculated using the Infolytica MagNet [10] and imported into the SIMION beam simulation with user programming. An estimated ion source output of 60 mA H$^-$ beam with the same current of electrons was simulated under two different ion source tilt angles with the position offsets and lens potentials found in the experiment.

The resulting beam traces and plots of the horizontal emittances at the RFQ entrance are shown in Fig 10. The simulation suggests in Fig. 10a that with a 0° tilt angle, the beam is significantly off-center through the entire LEBT, which bends the beam back towards the axis, reaching the axis inside the RFQ. The other simulation suggests in Fig 10b that with a 3° tilt, the beam passes through the extractor and the LEBT and enters the RFQ as a well-centered beam. Well-centered beams should have beam transmissions that equal or exceed the transmissions of less-centered beams, and

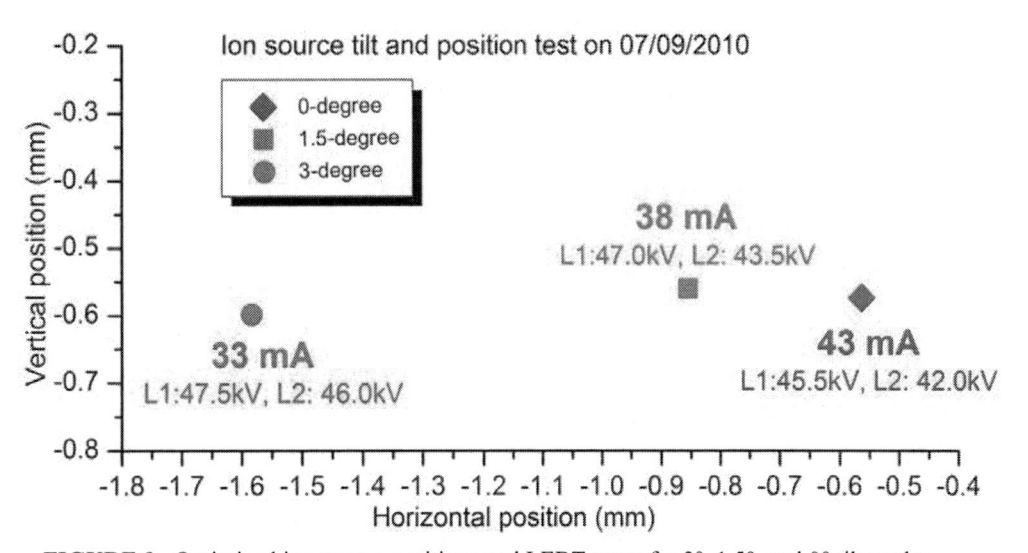

FIGURE 9. Optimized ion source positions and LEBT tunes for 3°, 1.5°, and 0° tilt angles.

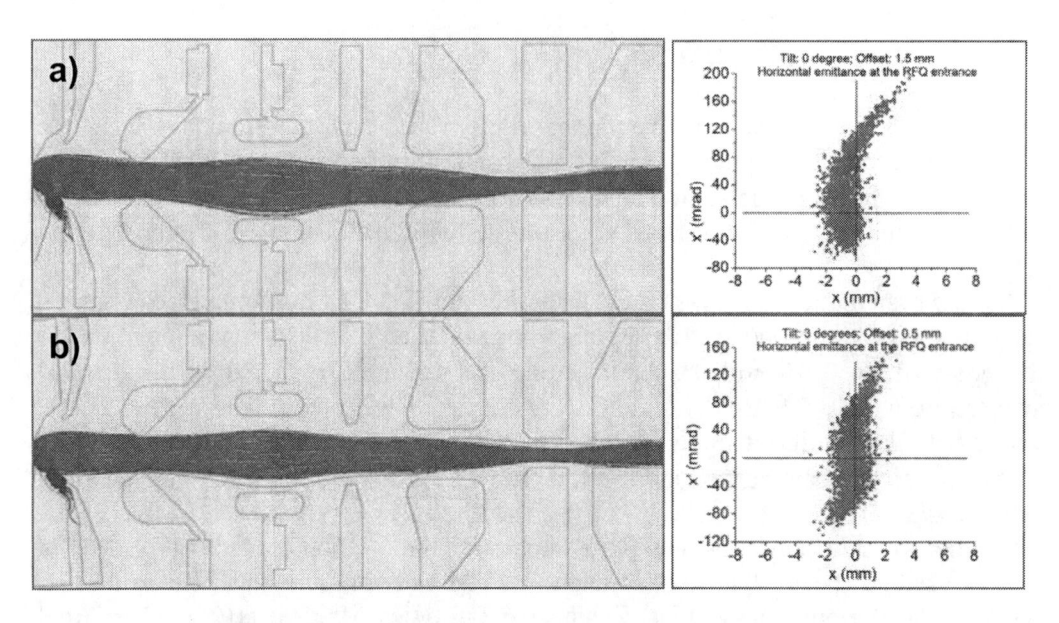

FIGURE 10. Beam simulations for a) 0° and b) 3° ion source tilt angles.

so these simulations fail to explain why 0° angles yield more BCM02 beam current. The discrepancy may be due to an inadequate calculation for the space charge of the electron beam and/or due to a beam with non-uniform distribution. Further studies, including beam simulation with more comprehensive computer codes, are underway to better understand those issues.

ION SOURCE PERFORMANCE OUTLOOK

Before the start of each biannual run, the three production ion sources are normally cycled through the Frontend, in part testing the newest upgrades, and in part demonstrating that the requirements of the next run can be met. This time allows for pushing the limits without risking excessive damage to the equipment. Pushing the RF power to 60 kW and above has allowed several demonstrations of 50 mA MEBT beam currents, and 46 mA sustained for over 2 days. Figure 11 shows the most recent such demonstration. Given those results, we feel confident that we can gradually increase the delivered beam current to support power levels beyond 1 MW.

ACKNOWLEDGMENTS

The proofreading by P. Kite was invaluable. Work was performed at Oak Ridge National Laboratory, which is managed by UT-Battelle, LLC, under contract DE-AC05-00OR22725 for the U.S. Department of Energy.

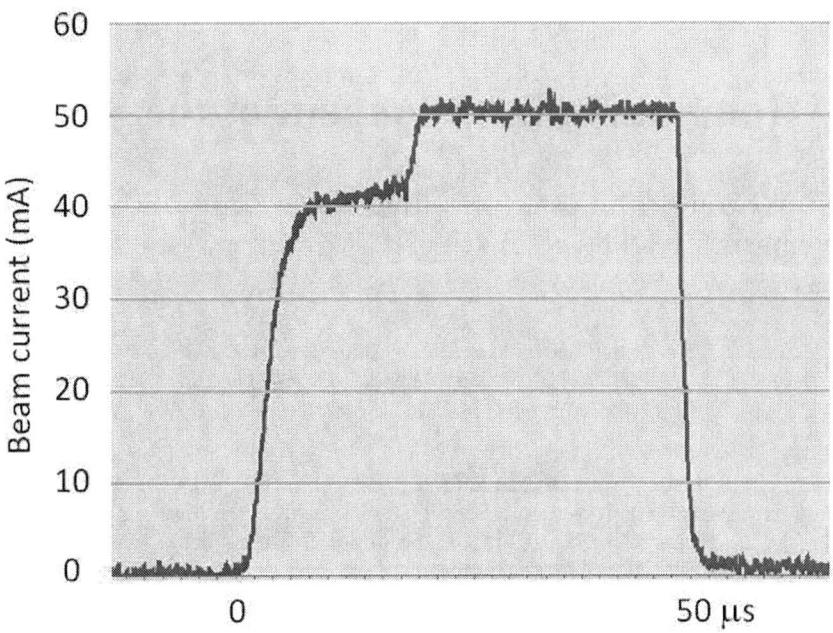

FIGURE 11. Demonstration of 50 mA beam before run CY10-2. In the MEBT BEAM STOP mode, a 50 µs slice of the beam in the middle of the 0.9 ms ion source pulse was accelerated by the RFQ and measured with BCM02. The initial ~20 µs are an artifact of the RF feed-forward control.

REFERENCES

1. R. Keller, R. Thomae, M. Stockli, R. Welton, "Design, Operational Experiences and Beam Results Obtained with the SNS H⁻ Ion Source and LEBT at Berkeley Lab" in *Production and Neutralization of Negative Ions and Beams*, edited by M. P. Stockli, AIP Conference Proceedings CP**639**, American Institute of Physics, Melville, NY, 2002, pp.47-60.
2. J. Staples, T. Schenkel, "High-Efficiency Matching Network for RF-Driven Ion Sources" in Proceedings of the 2001 PAC, edited by P. Lucas, S. Webber, 2001, pp. 2108-2110.
3. M.P. Stockli, B.X. Han, S.N. Murray, D. Newland, T.R. Pennisi, M. Santana, R.F. Welton, "Ramping Up the SNS Beam Power with the LBNL Baseline H⁻ Source" in *Negative ions, Beams and Sources*, edited by E. Surrey and A. Simonin, AIP Conference Proceedings CP1097, American Institute of Physics, Melville, NY, 2009, pp. 223-235.
4. M.P. Stockli, B. Han, S.N. Murray, T.R. Pennisi, M. Santana, R.F. Welton, "Ramping up the Spallation Neutron Source beam power with the H⁻ source using 0 mg Cs/day" in *Rev. Sci. Instrum.* **81**, 02A729 (2010).
5. S.H. Kim, A. Aleksandrov, M. Crofford, J. Galambos, P. Gibson, T. Hardek, S. Henderson, Y. Kang, K. Kasemir, C. Peters, D. Thompson, M. Stockli, D. Williams, "Stabilized operation of the Spallation Neutron Source radio frequency quadrupole" in *Phys. Rev. ST Accel. Beams* **13**, 070101 (2010).
6. J. Peters, HH. Sahling, I. Hansen, "Beam characteristics of the new DESY H⁻ source and investigations of the plasma load" in *Rev. Sci. Instrum.* **79**, 02A523 (2008).
7. M.P. Stockli, B. Han, S.N. Murray, T.R. Pennisi, M. Santana and R.F. Welton, "Towards Understanding the Cesium Cycle of the Persistent H⁻ Beams at SNS" in these proceedings.
8. Y.W. Kang, R. Fuja, R.H. Goulding, T. Hardek, S.W. Lee, M.P. McCarthy, M.C. Piller, K. Shin, M.P. Stockli, R.F. Welton, "rf improvements for Spallation Neutron Source H⁻ ion source" in *Rev. Sci. Instrum.* **81**, 02A725 (2010).
9. SIMION, http://www.simion.com.
10. MagNet, http://www.infolytica.com.

Ion Source Development at the SNS

R.F. Welton[a], N.J. Desai[b], B.X. Han[a], E.A. Kenik[a], S.N. Murray[a], T.R. Pennisi[a], K.G. Potter[a], B.R. Lang[a], M. Santana[a] and M.P. Stockli[a],

[a] *Spallation Neutron Source, Oak Ridge National Laboratory, P.O. Box 2008, Oak Ridge, TN, 37830-647, USA*
[b] *Worcester Polytechnic Institute, Worcester, MA, 01609, USA*

Abstract. The Spallation Neutron Source (SNS) now routinely operates near 1 MW of beam power on target with a highly-persistent ~38 mA peak current in the linac and an availability of ~90%. The ~1 ms-long, 60 Hz, ~50 mA H⁻ beam pulses are extracted from a Cs-enhanced, multi-cusp, RF-driven, internal-antenna ion source. An electrostatic LEBT (Low Energy Beam Transport) focuses the 65 kV beam into the RFQ accelerator. The ion source and LEBT have normally a combined availability of ~99%. Although much progress has been made over the last years to achieve this level of availability further improvements are desirable. Failures of the internal antenna and occasionally impaired electron dump insulators require several source replacements per year. An attempt to overcome the antenna issues with an AlN external antenna source early in 2009 had to be terminated due to availability issues. This report provides a comprehensive review of the design, experimental history, status, and description of recently updated components and future plans for this ion source. The mechanical design for improved electron dump vacuum feedthroughs is also presented, which is compatible with the baseline and both external antenna ion sources.

Keywords: negative ion sources, particle accelerators, ion formation
PACS: 29.25.Ni; 52.75.Di; 52.77.Dq; 52.80.Pi; 52.80.Vp

INTRODUCTION

In September 2009 the U.S. Spallation Neutron Source (SNS) achieved the milestone of delivering ~1 MW of beam power on target and now operates routinely near this level [1]. This requires 60 Hz, ~1-ms long, 38 mA linac beam pulses, which are accepted from the >50 mA beam pulses extracted from the source. Over the next several years the SNS will gradually increase the neutron production, maybe up to ~1.4 MW of beam power on target. A power upgrade to 2-3 MW is planned by increasing the beam energy from 1 to 1.3 GeV and the beam current from 38 to 59 mA. Presently, the SNS uses an RF-driven, Cs-enhanced, multicusp, H⁻ ion source and electrostatic LEBT (Low Energy Beam Transport) developed at Lawrence Berkeley National Laboratory (LBNL) and improved at Oak Ridge National Laboratory (ORNL) [2-5]. The source now routinely produces 35-42 mA with a 5.4 % duty factor (0.9 ms, 60Hz) with a combined LEBT and ion source availability of ~99% during typical operation. However, in the past the availability was occasionally significantly lower due to problems uncovered by ramping up the duty factor.

Both continuous (~300W of 13.56 MHz) and pulsed (40-70 kW of 2 MHz) RF power are simultaneously applied to a Cu internal antenna coated with porcelain enamel [5]. Since the fall of 2008 the beam current was increased by ~20% and the source duty factor was increased by 30%. At the same time the source service cycle was extended from 2, to 3 to 4 and 5 weeks, without encountering old-age performance degradation or failures of the source.

Since 2008 about one antenna failure per ~20-week run causes ~8 hours of downtime, limiting the availability to ~99.8%. Typically the antennas fail from chipped or melted porcelain, which exposes the copper tube. This drastically reduces the extracted beam current and/or impairs the 65 kV source potential. Sometimes the color of the plasma is changed to yellow or red. Despite several efforts with an optical spectrometer and a residual gas analyzer, no optical line or neutral gas specie could be found that would predict or confirm antenna failures. Accordingly, when the source operation or yield is seriously impaired, and all other components appear to work flawlessly, the source is replaced. Except in one case over the last two years, these source changes always confirmed antenna failures.

Since 2008 antennas always failed in the first half of the source service cycle after being used between less than 1day and 11 days, with an average of 4.2 days. This is consistent with infant mortality due to undetectable defects, possibly increased porosity at certain locations. Four of those antenna failures occurred during scheduled source changes and therefore did not cause downtime.

The reliability issue of the internal antenna drove the development of external antenna sources since 2003 [6]. An external antenna source with an Al_2O_3 plasma chamber was not able to withstand the stress of 6% duty factors [7]. This led to the development of the external antenna source with an AlN plasma chamber, which was implemented as production source early in 2009. However, due to its low 96.6 % availability, the external antenna source had to be taken out of service, and the internal antenna source was redeployed [4]. The next section will discuss the design, performance history and reliability problems associated with the AlN external antenna source. Unreliable source components were revised and incorporated in the design of a 2nd generation AlN external antenna source, which is almost ready for testing in preparation for a possible implementation at the SNS in summer of 2011. The design of this source and a testing plan will be discussed in section III of this report.

Efficient ion extraction requires very high fields between the ion source and the extractor. In combination with the locally high neutral density, this leads to occasional arcs. The e-dump is positively biased to +6.2 kV with respect to the outlet of the source and collects co-extracted electrons deflected through a transverse ~1600 Gauss magnetic field. With the e-dump being between the source and the extractor, arcs between the e-dump and the extractor can cause huge overvoltage spikes between the source and the e-dump, far exceeding the strength of the insulators between the e-dump and the source. Not surprisingly, e-dump failures are the second frequent cause of failures. The last section of this report will discuss a revised design of the electron dump feedthrough to make such failures less likely.

PROTOTYPE AlN EXTERNAL ANTENNA SOURCE

The prototype ORNL Aluminum Nitride AlN external antenna source has been previously described and will be summarized here [8-10]. Figure 1 shows a cross-sectional view of the source which consists of a cylindrical, high-purity, sintered, AlN ceramic plasma chamber with an inside diameter of 6.8 cm, a length of 18 cm, and a wall thickness of 0.7 cm. The outer surface of the AlN chamber is directly water-cooled by an annealed Lexan serpentine jacket. Computationally, the AlN chamber was designed to withstand isotropic plasma heat-loads of 100kW at 7% duty-factor while maintaining a thermal stress safety-margin of ~2x using finite elemental analysis. No chamber has failed so far.

The external antenna itself, also shown in Fig. 1, is wound with 4.5-5.5 turns of polyolefin/Teflon-encased Cu tubing arranged in two layers which are separated by a Teflon spacer. The antenna is located as close as possible to the outlet aperture of the source to maximize extracted beam current as well as allow clearance for 8 multicusp magnets symmetrically placed around the cooling jacket a distance of 1.9 cm from the inner wall of the plasma chamber. The water-cooled source body, as shown in Fig. 1, functions to mechanically support the plasma chamber assembly as well as the outlet aperture assembly which, is interchangeable, as a unit, with internal antenna sources. The water-cooled source back flange contains a port for a plasma ignition device and small optical window viewing the plasma. In this work a hollow-anode, DC glow-discharge (800V, 5 mA) gun has been employed which utilized a Cu or Mo cathode and feeds H_2 into the source (20-40 SCCM) [10].

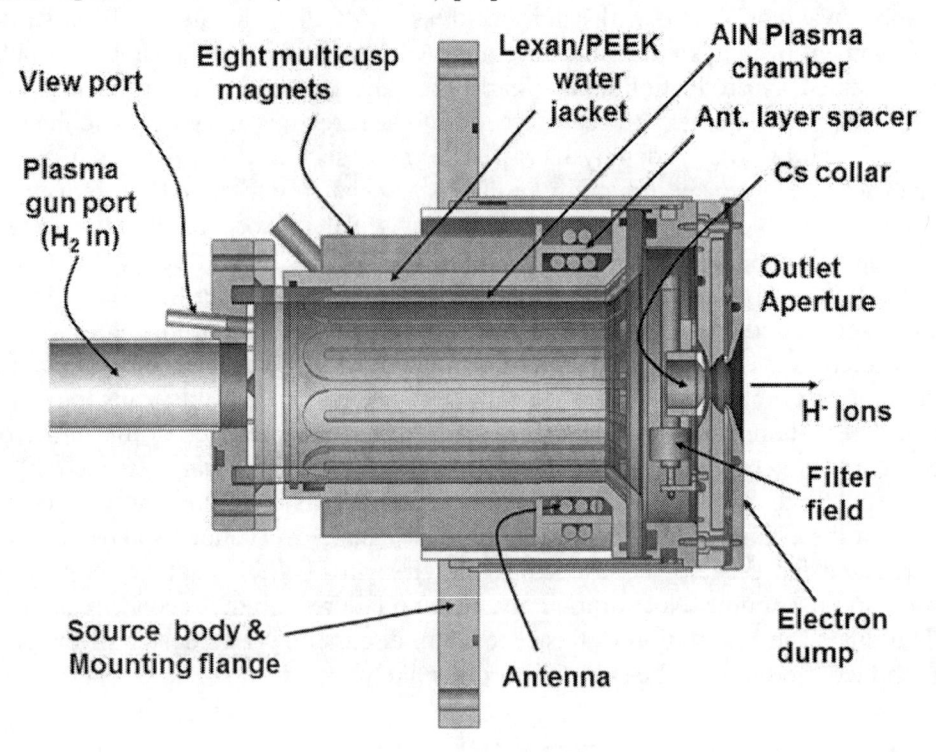

FIGURE 1. Cross sectional view of the prototype source. The AlN chamber is shown as transparent to allow viewing of the Lexan cooling jacket.

The following is a brief review of the testing history of this source [9]. The source was initially tested on the SNS test stand in early 2008 primarily using an elemental Cs system where up to ~100 mA of beam current was extracted [8]. During the maintenance period in the summer of 2008 we demonstrated this source could transport ~40 mA through the SNS RFQ for several hours using either the elemental Cs system or the baseline Cs_2CrO_4 system. In the later part of 2008, four sources were fitted with baseline Cs_2CrO_4 collars and plasma guns with Mo cathodes and rotated through ~week-long tests on the test stand. It culminated with implementation of this source for the beginning of the CY09-1 neutron production run. It generally produced the required 35 mA with an availability of ~97%, significantly below the ~99% for the baseline source. Figure 2 shows the beam current record during source testing, accelerator start-up, and neutron production run as well as scheduled and unscheduled source changes (see caption). After experiencing 5 failures leading to unscheduled source changes within a 4-week period, the baseline source was reinstated as production source.

This experience helped identify the following reliability issues: (i) twice, the DC plasma gun with Mo cathodes extinguished and could not be restarted; (ii) twice, the antenna failed due to burning of the polyolefin insulation likely due to excessive RF electric fields exceeding the break down strength of air (~3kV/mm) between turns due to insufficient inner and outer layer separation; (iii) one startup was prevented due to water leaks in the cooling jacket which was later ascribed to lose bolts. Other reliability issues which were noted on the test stand included the occasional overheating of the air-cooled multicusp magnets and water leaks in the Lexan through small spider cracks.

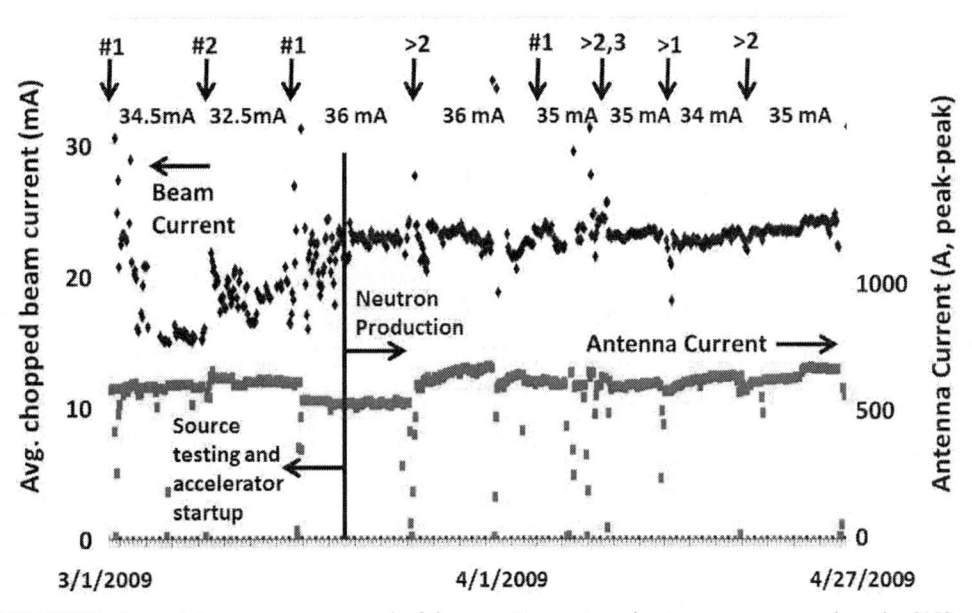

FIGURE 2. 8-week beam current record of the prototype external antenna source used on the SNS accelerator during the first 5-weeks of the CY09-1 run period. Vertical arrows indicate source changes, symbols show the identification number of the installed source and, > symbols indicate an unscheduled source change.

We also note that some external antenna sources exhibited beam degradation with time requiring increasing the RF power to maintain constant source output, requiring an up to 10% increase of the antenna current per week. This could be related to a poisoning of the cesium layer either from emission from the plasma gun or from the AlN chamber.

After this experience several changes we made to the prototype source: (i) Cu cathode was reinstated over the Mo version; (ii) antennas were rewound using a thicker 2.4 mm Teflon spacer increasing separation between the inner and outer layers of the antenna, see Fig. 1 (this change was implemented for the last two sources used for neutron production) and (iii) the Lexan cooling jacket was replaced by a PEEK version. Beginning in December of 2009 a test of this configuration was performed on the test stand which showed that >45 mA could be sustained for ~6 weeks (~1mA/kW) with similar performance degradations as noted above between cesiations.

To shed additional light on this attenuation process elemental analysis of coatings deposited on the conical Mo ionization surface between the outlet aperture and Cs-collar were analyzed using Scanning Electron Microscopy / Energy Dispersive Spectroscopy (SEM/EDS) using 15kV electrons sampling to a depth of 2-3μm. In one case, an external antenna source with a DC plasma gun featuring a Cu cathode operated at 800 V was cesiated and 45-50 mA beam were extracted for ~40 days on the test stand with 3 intermittent cesiations. After being exposed to air for four days, the surface analysis of the Mo cone yielded 63% O, 26% Al, 7% Mo, 2% Fe, 1% Cs, and 1% Cr, all in atomic %. In the other case baseline source #2 was installed on the front-end, cesiated, and produced similar beam current for 28 days, before it was removed. The cone was exposed to air for one day before the surface analysis yielded 41% O, 23% Fe, 13% Mo, 12% Si, 7% Cr, and 4% Na. In both collars surface resistance measurements showed the presence of tenacious insulating layers.

The prevalence of oxygen may have been enhanced by the prolonged air exposures. The different fractions of Mo suggest that the external antenna source produced a thicker coating on the Mo cone. Also the prevalence of Al and the lack of Cu from the external antenna source suggest that the AlN chamber is the likely source of the observed poisoning. This may limit the suitability of AlN chambers for the production of persistent high-duty factor H⁻ beams. Such sources may require frequent cesiations, requiring regular maintenance and possible failures from Cs induced discharges [11].

The coating found on the cone of the baseline source seems to be dominated by sputtering from the stainless steel plasma chamber. The Si and Na appear to be from the porcelain antenna coating. After 4-5 week runs, the antennas normally exhibit black carbon coatings, which is inconsistent with heavy sputtering. Furthermore it is unclear how to reconcile the Fe and Si coating with the persistent H⁻ production, after H⁻ drastically increases after applying the initial, small dose of Cs [11].

SECOND-GENERATION AlN EXTERNAL ANTENNA SOURCE

In order to address the reliability issues as well as to test potential performance improvements, a 2nd generation source has been designed, constructed, and is in the final stages of fabrication. Figure 3 shows a schematic, cross-sectional view of the source as well as photographs of both external antenna sources. No changes have been made to the source body/mounting flange, the outlet aperture assembly, the AlN plasma chamber and the PEEK cooling jacket used in the prototype source. Upgraded components include: (i) the plasma ignition gun, (ii) the multicusp magnet array and, (iii) the antenna. Most of these upgraded components are interchangeable with those employed earlier in the prototype source allowing direct performance comparisons. This will allow for testing with a known, fully operational source, systematically swapping in new components individually. Some aspects of these components have been previously discussed [10].

(i) The prototype source backflange and DC plasma ignition gun have been replaced with an integrated RF-driven plasma gun. This system should take advantage of the existing continuous 13 MHz RF system decoupling it from the high-power 2 MHz system as well as require lower RF powers as compared to the baseline source due to a smaller plasma volume (ϕ=1.3 cm) and more favorable operating pressures due to restricted H_2 flow, which is baffled by a ϕ~2 mm aperture into the ion source. The new backflange also contains an interchangeable W or AlN plate facing the main plasma chamber of the source. Testing shall include offline water pressure and vacuum leak tests of the ceramic plasma chamber. Low-level RF will then be used to configure the antenna to match impedance of baseline source, after which the fully assembled source can be tested. The primary challenge will be to determine the operating parameters, coupling geometry and bias (if any) needed to first ignite the RF gun plasma and then reliably ignite the main source plasma. The effect of W versus AlN back plates on source efficiency and beam persistence is another desirable test if time permits.

(ii) The air-cooled multicusp array comprised of eight 1x1x9 cm Nd-Fe-B magnets (~900G max at the plasma chamber wall) surrounding the plasma chamber of the prototype source was replaced by a water-cooled set of eight 1.3x2.5x5 cm magnets (~1600G @ wall). The Cu magnet holder was designed computationally to maintain the magnet temperature within limits in the presence of RF heating [12,13]. As shown in Fig. 3, the axial length of the magnets were shortened by ~2 cm on each end to improve clearance for larger, ferrite-backed antenna configurations and for a water cooling passage shown on the left. Lack of axial space for the antenna had been problematic during assemblies of the prototype source and having plasma confinement ~15cm from the outlet aperture was deemed expendable. The two magnet configurations will be compared for their beam production. Finding significant differences may justify an optimization of the strength of the magnets.

(iii) In later runs the prototype source utilized a two-layer, water-cooled, 4.5 turn, round Cu antenna (ϕ=4.8 mm) surrounded by ~1 mm of polyolefin and Teflon shrink tubing positioned as shown in Fig. 1. Inner and outer turns were separated by a 2.4 mm thick, T-shaped, Teflon ring separating inner and outer windings. Although we have not yet experienced a failure of this configuration, estimates of peak RF electric field in the immediate region of the antenna suggest operation close to the breakdown limit of air in the absence of plasma. In order to mitigate this risk and consequent damage to the antenna, as well as eliminate variations between antennas, the 2nd generation source employs a helical Teflon antenna guide which locates the antenna inside a 9x5.5 mm groove. This guarantees inter-layer spacing of the antenna conductors is > 7 mm and intra-layer spacing > 6.6 mm. The initial design calls for a lower inductance 3.5 and 4.5-turn antenna compared with 4.5 and 5.5 turn antennas already tested, the latter two have shown similar beam current performance [10]. To further reduce inter-layer field strength the Cu antenna has been flattened to 5.7 x 4 mm and oriented as shown in Fig. 3. Estimates suggest the new 3.5-turn antenna should have ~50% lower peak RF electric field for equal antenna current when compared to those employed in the prototype source. In order to enhance inductive RF coupling to the plasma, the design includes a provision for testing an L-configuration of ferrites surrounding the antenna (NiZn Fair-Rite material 52, 6.3 mm thick) oriented as shown in Fig. 3. This approach has been used successfully in other H$^-$ ion sources [14]. Matching should not be problematic for the new 3.5-turn antenna with inductance between antenna geometries currently supported by our

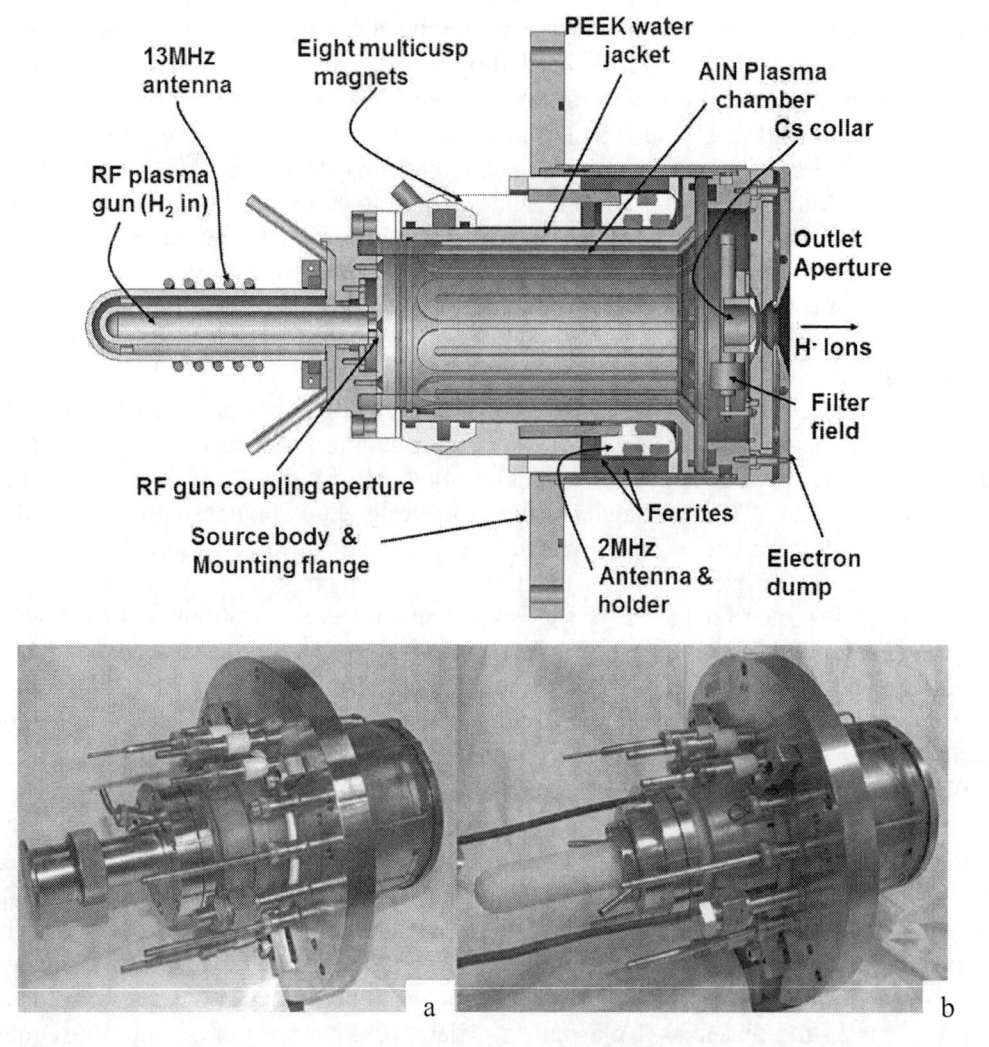

FIGURE 3. Cross sectional view of 2nd generation AlN external antenna source (above). Photographs of the prototype source (a) and the 2nd generation source (b).

network (4.5- 2.5- turn antennas). Since the revised prototype source antenna will support initial testing of the components discussed above, this change can be implemented last. If time permits, the ferrites will be optimized in a separate test.

Overall test plan: Testing will be performed with a single source to eliminate source-to-source variations. After a successful test with the new RF plasma gun, the performance will be base lined against the pervious external antenna source by reinstalling the DC plasma gun. Then the source will be retrofitted with the new plasma gun to check consistency. Then the original multicusp array will swapped with the new multicusp array and again tested for yield performance. Then the source will be retrofitted with the new antenna configuration and tested for performance variations. If significant differences are found, optimization will be considered. This will be followed by a lifetime test aimed at measuring changes in yield performance. If significant degradations are found, we will evaluate recesitions, which also require significant developments [11].

If the performance of the initial test source is satisfactory, two more sources will be readied as a replacement and as a spare. These sources need to be ready for short tests on the front-end before the first source is implemented for neutron production. If the Frontend source test can start early in July, and all tests are successful, the external antenna source will become the new baseline source for run CY11-2.

DEVELOPMENT OF THE SNS EXTRACTION SYSTEM

Several efforts are also being directed towards improving the extraction system employed in all SNS ion sources. Areas of focus include (i) improving voltage holding capability of the electron dump electrode, (ii) developing a robust, cooled extraction electrode/electron target capable of better withstanding higher electron loads and source misalignments and, (iii) developing a remote, downstream electron dumping capability separated from the source outlet region.

As discussed, the electron dump insulators can be subjected to very high voltage spikes, and therefore can fail as the one shown in Fig. 4. Analysis of the existing design shows that excessive electric fields as well as trapped gases can exist in the gap between the electron dump water tube and the source body, typically held at $\Delta V=6.2\text{kV}$, shown in Fig. 4. Figure 4 also shows the design of a modified feedthrough which increases the minimum voltage gap from 0.8 to 3.9 mm and insulator thickness from 1.3 to 2.2 mm in addition to offering a many fold improvement in vacuum conductance. Since e-dump arcs to ground can greatly exceed 6.2kV we are also considering modifications to our startup procedure, reducing the values of and/or relocating the load resistors, and/or optimizing spark gap spacing. If this system functions well on the test stand it will be rotated into SNS operation.

ACKNOWLEDGMENTS

The proofreading by P. Kite was invaluable. Work was performed at Oak Ridge National Laboratory, which is managed by UT-Battelle, LLC, under contract DE-AC05-00OR22725 for the U.S. Department of Energy.

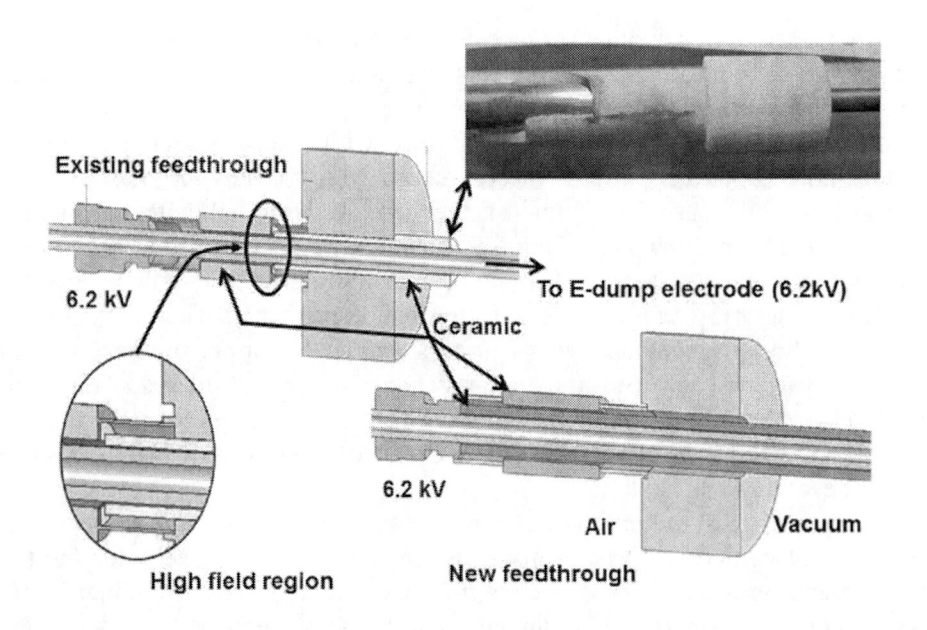

FIGURE 4. Both the new and existing electron dump feedthroughs, picture shows an actual insulator damaged during SNS operation. These insulators are also shown near the top of the photos in Fig. 3.

REFERENCES

1. E.g.: S. Henderson, Proceedings of the 25th LINAC Conference, Tsukuba, Japan, Sept 2010
2. R. Keller, et al, *Rev. Sci. Instrum.* **73** 914 (2002).
3. B.X. Han, M.P. Stockli, R.F. Welton, S.N. Murray, T.R. Pennisi and M. Santana, these proceedings.
4. M.P. Stockli, B. Han, S.N. Murray, T.R. Pennisi, M. Santana, R.F. Welton, *Rev. Sci. Instrum.* **81** 02A729 (2010).
5. R. F. Welton, M.P. Stockli, Y. Kang, M. Janney, R. Keller, R.W. Thomae, T. Schenkel and S. Shukla, *Rev. Sci. Instrum.* **73** 1008 (2002).
6. R. F. Welton, M. P. Stockli, R.T. Roseberry, Y. Kang, R. Keller, *Rev. Sci. Instrum.* **75** 1789 (2004).
7. R.F. Welton, M. P. Stockli, S. N. Murray, T. R. Pennisi, B. Han Y. Kang, R. H. Goulding, D. W. Crisp, N. P. Luiciano, J. R. Carmichael, J. Carr, Proceedings of PAC 2007 (Albuquerque), p. 3774.
8. R.F. Welton, M.P. Stockli, S.N. Murray, D. Crisp, J. Carmichael, R.H. Goulding, B. Han, O. Tarvainen, T. Pennisi, M. Santana, AIP Conf. Proc. #1097 (American Institute of Physics, New York, 2009) pp. 181-190.
9. R.F. Welton, M.P. Stockli, S.N. Murray, D. Crisp, J. Carmichael, R.H. Goulding, B. Han, T. Pennisi M. Santana, Proceedings of the Particle Accelerator Conference, Vancouver, CA 4-8 May 2009.
10. R.F. Welton, J. Carmichael, N. J. Desai, R. Fuga, R.H. Goulding, B. Han, Y. Kang, S.W. Lee, S.N. Murray, T. Pennisi, K. G. Potter, M. Santana, M.P. Stockli, *Rev. Sci. Instrum.* **81** 02A727 (2010).
11. M.P. Stockli, B. Han, S.N. Murray, T.R. Pennisi, M. Santana and R.F. Welton, these proceedings.
12. S.W. Lee, R.H. Goulding, Y.W. Kang, K. Shin, R.F. Welton, *Rev. Sci. Instrum.* **81** 02A726 (2010).
13. Y.W. Kang, R. Fuja, R.H. Goulding, T. Hardek, S.W. Lee, M.P. McCarthy, M.C. Piller, K. Shin, M.P. Stockli and R.F. Welton, *Rev. Sci. Instrum.* **81** 02A725 (2010).
14. J. Peters, AIP Conf. Proc. #1097 (American Institute of Physics, New York, 2009) pp. 171-180.

Operation Status of the J-PARC Negative Hydrogen Ion Source

H. Oguri, K. Ikegami, K. Ohkoshi, Y. Namekawa, and A. Ueno

J-PARC Center, Tokai-mura, Naka-gun, Ibaraki-ken 319-1195, Japan

Abstract. A cesium-free negative hydrogen ion source driven with a lanthanum hexaboride (LaB$_6$) filament is being operated without any serious trouble for approximately four years in J-PARC. Although the ion source is capable of producing an H$^-$ ion current of more than 30 mA, the current is routinely restricted to approximately 16 mA at present for the stable operation of the RFQ linac which has serious discharge problem from September 2008. The beam run is performed during 1 month cycle, which consisted of a 4-5 weeks beam operation and a few days down-period interval. At the recent beam run, approximately 700 h continuous operation was achieved. At every runs, the beam interruption time due to the ion source failure is a few hours, which correspond to the ion source availability of more than 99 %. The R&D work is being performed in parallel with the operation in order to enhance the further beam current. As a result, the H$^-$ ion current of 61 mA with normalized rms emittance of 0.26 πmm.mrad was obtained by adding a cesium seeding system to a J-PARC test ion source which has the almost same structure with the present J-PARC ion source.

Keywords: negative hydrogen ion source, lanthanum hexaboride, proton accelerator, J-PARC
PACS: 29.20.Ej, 29.25.Ni, 41.75.Cn

INTRODUCTION

The Japan Proton Accelerator Research Complex (J-PARC) is a multipurpose facility with a 1 MW class proton beam power. It is jointly operated by the Japan Atomic Energy Agency (JAEA) and the High Energy Accelerator Research Organization (KEK). The overview of J-PARC is shown in Fig. 1. J-PARC consists of a linac, a 3-GeV rapid cycling synchrotron (RCS), a 50-GeV main ring synchrotron (MR), and experimental facilities such as the Materials and Life Science Experimental Facility, the Hadron Beam Facility, and a neutrino production facility. In the first stage of the J-PARC, an H$^-$ ion beam with a peak current of 30 mA and a pulse length of 500 μs is accelerated up to 181 MeV by the linac and then injected into the RCS at a pulse repetition rate of 25 Hz. The beam commissioning of the accelerator started in November 2006. The first neutron and muon beams were produced in May and September 2008, respectively. The neutron beam became available to users in December 2008. The 30-GeV beam was delivered to the Hadron Beam Facility in February 2009 and to the neutrino production target in April 2009. Then the construction of all facilities planned in the first stage of the J-PARC was completed. The operation of 120 kW at 3 GeV is in progress since November of 2009. From now we plan to strengthen the accelerator beam power to 1 MW for the final stage.

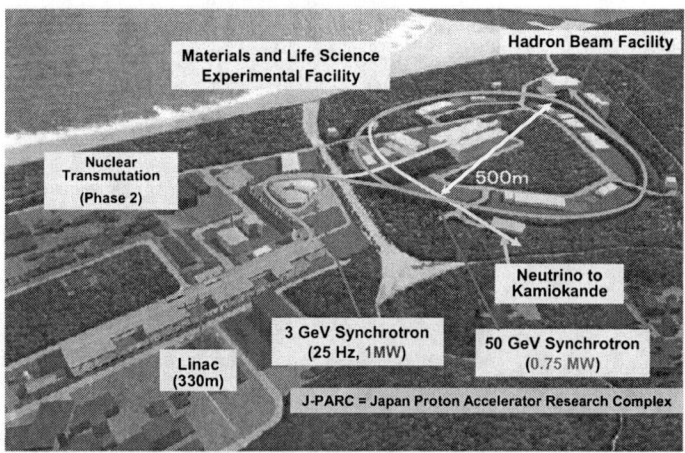

FIGURE 1. Overview of the J-PARC.

Before starting the J-PARC, several types of H⁻ ion source were developed at JAEA and KEK independently [1, 2]. On the basis of the results of various studies, a cesium-free ion source driven by a lanthanum hexaboride (LaB$_6$) filament was adopted as the ion source for the first stage of J-PARC [3]. Experiment is being performed with an ion source test bench in order to produce the H⁻ ion current of more than 60 mA, which is the requirement of the J-PARC final stage. Requirements of the J-PARC ion source for the first and the final stage are summarized in Table 1.

TABLE 1. Requirement of the J-PARC ion source for the 1st and the final stages.

Parameter	Value
H⁻ ion current	36 mA (1st stage), 60 mA (final stage)
Beam energy	50 keV
Pulse length	500 µs
Pulse repetition	25 Hz
Emittance	< 0.25 πmm.mrad (norm. rms)
Maintenance cycle	> 500 hours
Cesium	Free (1st stage), free or micro amount (final stage)

ION SOURCE

A cross-sectional view of the present J-PARC ion source is shown in Fig. 2. The ion source consists of a cylindrical plasma chamber, a beam extractor, an ejection angle correction (EAC) electromagnet, and a large vacuum chamber with 2 turbo molecular pumps (TMPs) of 1500 L/s for differential pumping. The basic structure of the ion source is not changed since the operation was started.

The source plasma is produced by an arc discharge using the LaB$_6$ filament having a cylindrical double-spiral structure (DENKA [4]). The filament was originally developed at KEK and used in the KEK-PS H⁻ ion source. In order to make high arc power and duty factor operation possible, the diameter, length, and thickness of the filament are gradually increased in the case of 3 different sources as follows; 15, 42, and 2.25 mm for the KEK-PS source; 20, 45, and 2.50 mm for the JHF source [2]; and

29.5, 49, and 3.25 mm for the J-PARC source, respectively. At the beginning of the operation, a short circuit between the adjacent turns of the filament led to its failure. The failure was caused by filament deformation due to arc power heating. In order to prevent such failures, the gap between the turns of the filament was increased from 0.3 to 0.6 mm. We have not encountered similar problems since then.

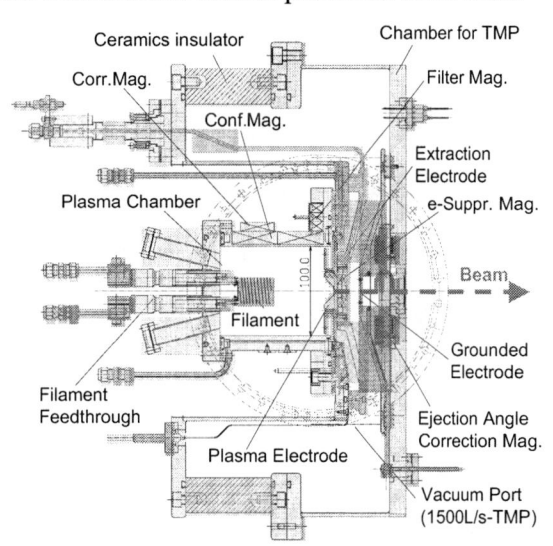

FIGURE 2. Cross-sectional view of the present J-PARC ion source.

Figure 3 shows the detailed layout of the ion source plasma chamber. The plasma chamber is made of oxygen-free copper (OFC). The inner diameter and length of the chamber are 100 mm and 133 mm, respectively. The source plasma is confined by the multicusp magnetic field produced by the 18 rows of Nd-Fe-B permanent magnets and 4 rows of magnets lining the side wall and the upper flange of the chamber, respectively. Filter magnets, which produce a magnetic filter field, are attached to 2 of the confinement magnets attached to the side wall near the downstream end. Correction magnets, which produce a field-free region around the filament without changing the magnetic filter field, are attached to 4 of the confinement magnets attached to the side wall near the upstream end [2, 5]. Figure 4 shows that the length of the correction magnets has a significant effect on the beam current; the beam current is maximum when the length of the magnet is around 50 or 55 mm.

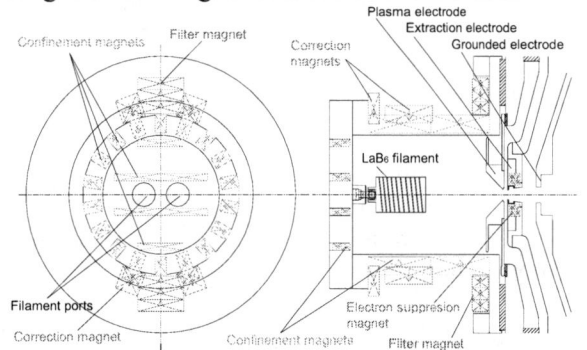

FIGURE 3. Detailed layout of ion source plasma chamber.

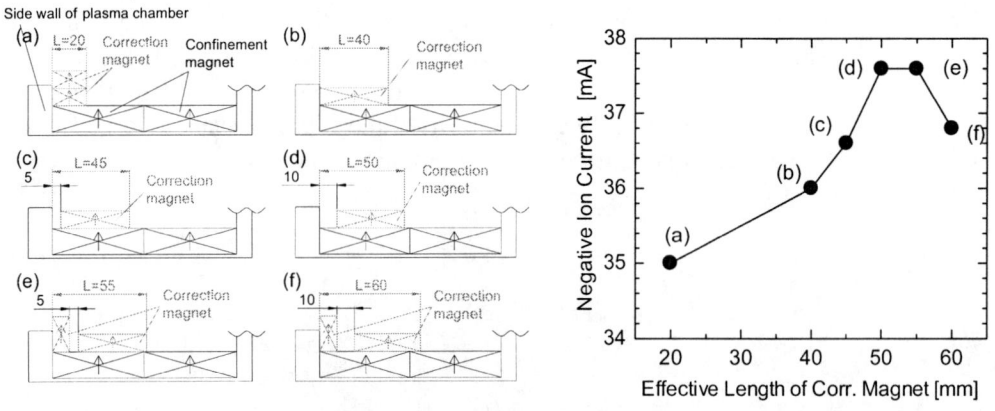

FIGURE 4. Dependence of beam current on effective length of correction magnet.

Figure 5 shows the detailed layout of the beam extractor. The beam extractor consists of 3 electrodes: a plasma electrode (PE), an extraction electrode (EXE), and a grounded electrode (GE). The shape of the electrodes has been determined by carrying out a two-dimensional beam simulation using BEAMORBT [6]. The H⁻ ion beam is extracted from a single aperture with a diameter of 9 mm bored at the center of the PE. The 50 keV beam required for the Radio Frequency Quadrupole linac (RFQ) is produced by applying 10 kV to the extraction gap between the PE and the EXE and 40 kV to the acceleration gap between the EXE and the GE, typically. The extraction and acceleration gap length are 3.0 mm and 12.0 mm, respectively. The acceleration gap length is optimized to obtain maximum beam transmission rate through the RFQ [7].

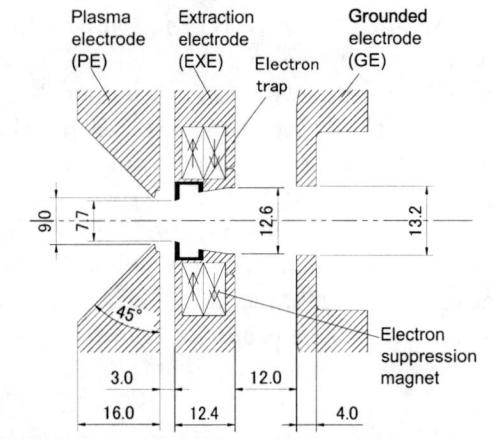

FIGURE 5. Detailed layout of ion source beam extractor.

The PE is fabricated by boring a 45 degree tapered hole with a diameter of 9 mm on a molybdenum plate with a thickness of 16 mm. The taper angle is optimized in order to produce the highest beam current [2]. The EXE is an OFC plate with a thickness of 12.4 mm. A copper pipe is brazed on the EXE and is used as a water cooling channel. Since the beam simulation result shows that the beam divergence angle at the exit of the ion source decreases with the diameter of the beam hole of EXE, we set the hole diameter to 7.7 mm, which is approximately 10 % larger than the simulated beam diameter at the EXE surface. Two pairs of Nd-Fe-B permanent magnets are mounted inside the EXE in order to produce a dipole magnetic field which deflects the electrons extracted along with the H$^-$ ion beam. These magnets are called as electron suppression magnets. The deflected electrons are dumped on the electron trap made of molybdenum, which is brazed on the EXE. The GE is fabricated by boring a hole with a diameter of 13.2 mm on a molybdenum plate with a thickness of 4 mm. A copper pipe is brazed on the base of the GE and is used as a water cooling channel. The base of the GE is made of stainless steel. The ejection angle error of the H$^-$ ion beam mainly produced by the electron suppression magnets is corrected by an EAC electromagnet, which is located just behind the GE. This electromagnet has 4 poles to deflect the beam both horizontally and vertically [2].

ION SOURCE PERFORMANCE

The present ion source is being operated without any serious trouble for approximately four years as of October 2010. Typical operation parameters of the ion source at the recent beam run are listed in Table 2.

TABLE 2. Typical operation parameters of the present J-PARC ion source. (RUN#35)

Parameter	Value
H$^-$ ion current	17 mA
Beam energy	50 keV
Beam pulse length/ repetition	50 - 500 µs / 1 - 25 Hz
Arc pulse length/ repetition	800 µs / 25 Hz
Filament voltage/ current	8.1 V/ 131 A
Arc voltage/ current	144 V/ 125 A
Arc power	18 kW
Bias voltage (PE - plasma chamber)	5.5 V
H$_2$ gas flow rate	11 SCCM
Extraction voltage/ current	7.8 kV/ 770 mA
Acceleration voltage (constant)	31.5 kV
Acceleration voltage (modulation)	12.6 kV
Frequency of H$^-$ ion current tuning	Once a day or less

At present, the H$^-$ ion current from the ion source is routinely restricted approximately 16 mA for the stable operation of the RFQ which has serious discharge problem from September 2008 [8], although the ion source is capable of producing an H$^-$ ion current of more than 30 mA [7]. The H$^-$ ion current is tuned by changing the arc power. The maximum arc voltage is limited to be approximately 150 V to prevent the filament from inducing the arcing which causes adverse effect to its lifetime. During the beam run, the H$^-$ ion current is monitored at a low energy beam transport (LEBT: beam transport line between the ion source and the RFQ) with a beam current transformer. At the first week of the beam run, the beam current decreases approximately 1 mA/day. After that the current decreases very slightly. Indeed, the ion source operator tunes the ion source parameters once a day or less not to drop the beam current of approximately 0.1 mA, it was much smaller than the variation level allowing to the accelerator operation.

The repetition rate of the H⁻ ion beam is changed frequently depending on the accelerators or experimental facilities commissioning task. To keep the condition of the ion source constant, the repetition rate of the arc discharge is kept constant at 25 Hz, although the repetition rate of the H⁻ ion beam is changed. If the beam is not required, the arc discharge is delayed with respect to the time when the extraction voltage is turned off. The arc pulse length is also kept constant at 800 μs, although the pulse length of the beam is changed as on the same reason.

The PE is biased negatively with respect to the plasma chamber. At the beginning of the operation, the bias voltage cannot be set to the optimum value because the PE is biased automatically (approximately 30 V), probably due to the surface oxidation of the PE. Therefore, the surface of the PE is conditioned by applying the maximum bias voltage of 40 V intentionally for a few hours. During this process, the PE surface probably gets cleaned due to positive ion bombardment. Then, the bias voltage is adjusted to the optimum value in order to obtain the beam current required for the beam run. The optimum voltage decreases gradually with the operation time and saturates approximately 6 V (5.5 V in this case).

The acceleration voltage applying to the acceleration gap is produced by 2 different types of power supply; a constant voltage power supply and a modulation one. The former is always turned on and applies to 31.5 kV. When the extraction voltage of 7.8 kV is applied to the extraction gap, pre-beam extraction, which the beam energy is less than 40 keV; at this energy level no beam acceleration takes place in the RFQ, is started. After the pre-beam extraction for 100 μs, the modulation voltage power supply is turned on and applies to 12.6 kV. Then the beam is given the full energy of approximately 50 keV and accelerated by the RFQ. The pre-beam extraction is necessary for the rise-time of space charge neutralization effect by beam plasma in the LEBT. In order to minimize the beam loss in the linac, especially in the high-energy section, the beam emittance should not be changed during the beam pulse.

Operation status of the ion source in the last one year is summarized in Table 3. The beam run is performed during 1 month cycle, which consisted of a 4-5 weeks beam operation and a few days down-period interval. At the recent beam run, approximately 700 hours continuous operation was achieved, which is satisfied with the requirement of the ion source lifetime for the first stage.

TABLE 3. Operation status of the present ion source in the last one year.

RUN #	Date	Ion Source Operation Time [hour]	Typical Beam Current [mA]	Availability [%]	Failure ( ): Accelerator Stop Time
27	Nov-09	642	16	99.23	Vacuum gauge (2h) Vacuum pump (3h)
28	Dec-09	568	16	99.65	Vacuum gauge (1h) Power supply (1h)
29	Jan-10	605	16	99.75	Arc voltage OV (1h) Power supply (0.5h)
30	Feb-10	492	16	99.69	Vacuum pump (1.5h)
31	Mar-10	323	16	99.94	Interlock error (0.2h)
32	Apr-10	704	16	~100	(non)
33	May-10	609	16	99.93	Power supply (0.4h)
34	Jun-10	613	16	99.89	Arc current OC (0.7h)
35	Oct-10	734	17	99.42	Arc voltage OV (4.3h)

At every runs, the beam interruption time due to the ion source failure is a few hours, which correspond to the ion source availability of more than 99 %. The ion source failure was mainly caused by ion source peripheral equipments such as the vacuum pump, the gauge and so on.

After each beam run, the ion source is overhauled, cleaned and inspected for any damage to its internal elements. The filament, the plasma chamber and the PE are exchanged every maintenance. The plasma chamber and the PE are reused by polishing their surface with alumina powder. The filament, on the other hand, is replaced by a brand-new one every maintenance to prevent the filament occurring the failure during the beam run. Figure 6 shows photographs of the filament before (a) and after (b) operation (approximately 600 hours). Most area of the used filament surface became darkly-discolored although the original color of the LaB_6 is violet, which is shown in Figure 6 (a). The result of the surface elemental analysis by using the EDS (Energy-Dispersive x-ray Spectroscopy) method shows that the discolored area is covered with boron [9]. Because the low temperature area around the connections with electric current terminals became also dark, it is recognized the discolored area is electron un-emission area.

Figure 7 shows the variation of the filament and H⁻ ion currents with operation time during the beam run #35. In the beginning of the operation, the filament current increased gradually. The increase is probably caused by the process of covering the boron on the filament surface. The temperature of the filament should be increase to compensate the reduction of the electron emission area in order to keep the H⁻ ion current constant. The filament current began to decline after approximately 20 days operation. The decline is caused by consumption of the coil due to its evaporation. In the case of the ion source driven with tungsten filament, the filament was usually broken when the filament current decreased by approximately 10 % in comparison with the initial one [10]. If the index can be applied to the LaB_6 case, the lifetime of the LaB_6 filament is expected to be more than 1,000 hours.

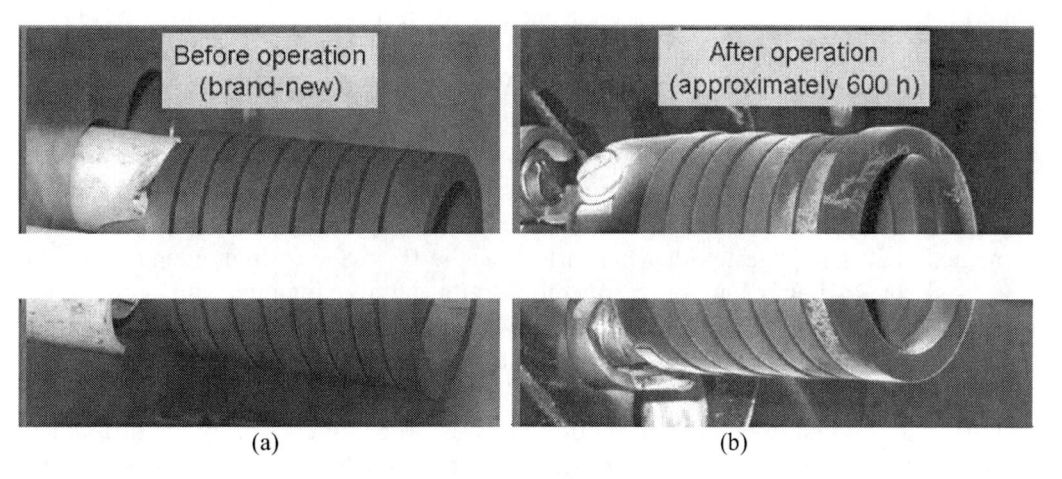

FIGURE 6. Photographs of the LaB_6 filament before (a) and after (b) operation.

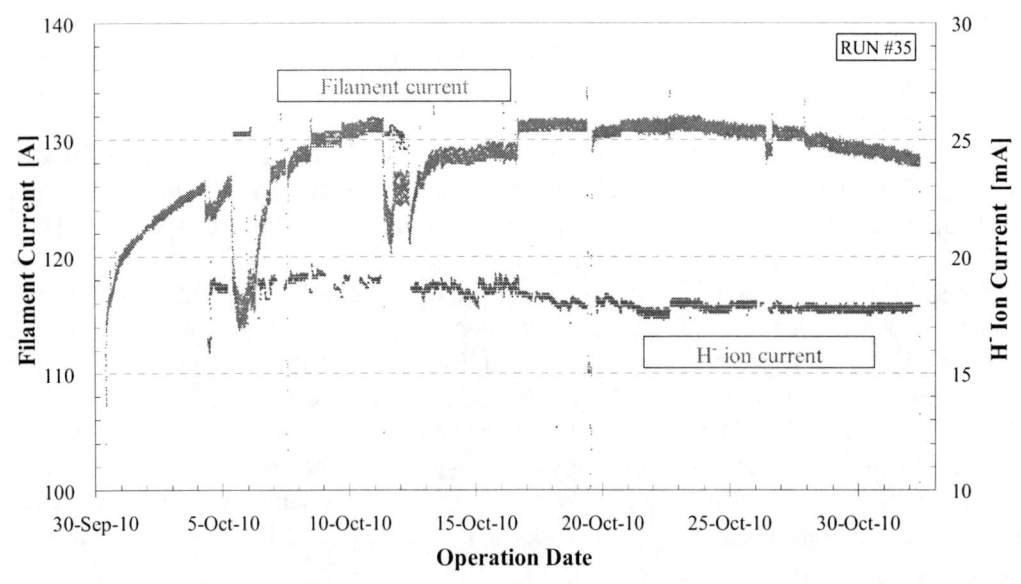

FIGURE 7. Variation of the filament and H⁻ ion currents with operation time during the beam run #35.

EXPERIMENT OF H⁻ CURRENT ENHANCEMENT

In order to satisfy the J-PARC final stage requirement, the H⁻ ion current from the ion source should be increased to more than 60 mA. Therefore, the capability of the ion source when seeded with Cs to produce an H⁻ ion current of more than 60 mA is examined.

We performed the beam test of the Cs-seeded ion source by adding a Cs seeding system to a J-PARC test ion source which has the almost same structure with the present J-PARC source [10]. The Cs seeding system consists of a Cs reservoir, a valve, and a tube. These are individually heated using mantle heaters. The injection rate of vaporized Cs onto the PE surface is controlled by the reservoir temperature.

Figure 8 shows the measured dependence of the H⁻ ion current on the arc discharge power driven with a LaB_6 or a tungsten (W) filament for Cs-seeded and Cs-free conditions. In the Cs-free condition, the maximum H⁻ ion currents were around 18 mA for the cases with the LaB_6 and the W filament. In the Cs-seeded condition, on the other hand, the H⁻ ion current increased by less than 45 % for the case with the LaB_6 filament and by approximately 4 times for the case with the W one. The surface elemental analysis suggested that the surface of the PE made of molybdenum, which was used for 24 hours ion source operation with a LaB_6 filament and without Cs seeding, was covered with boron [9]. Although we have not carried out the analysis of the PE used for ion source operation with the W filament, it should be completely covered with W because the evaporation rate of W filament is much higher than that of LaB_6 one. The large H⁻ ion current difference is probably caused by the difference between the H⁻ ion production efficiency. The adhered boron on the PE seems to prevent the cesium effect of decreasing the work function of the PE surface.

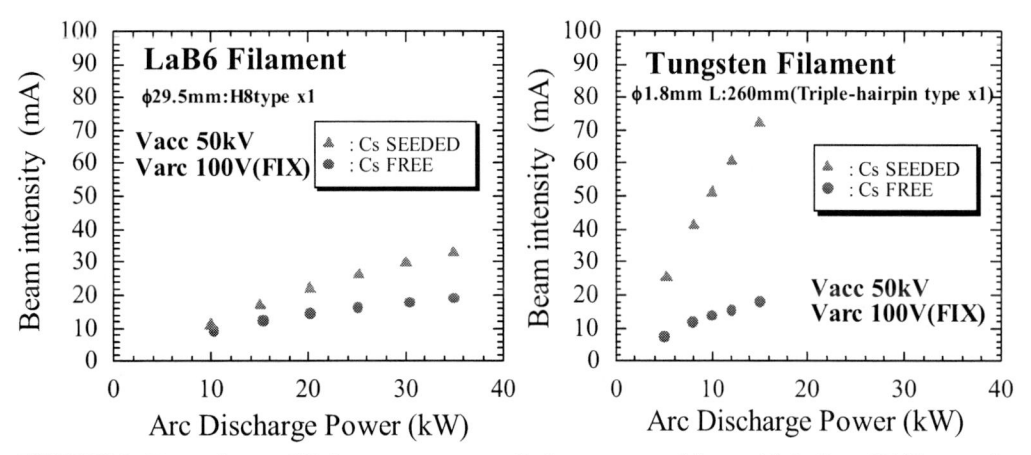

FIGURE 8. Dependence of H⁻ ion current on arc discharge power driven with LaB$_6$ or W filament for Cs-seeded and Cs-free condition.

We measured the beam emittance by using two sets of double slit scanners. Figure 9 shows the horizontal and vertical emittance diagrams for 50 keV beams with H⁻ ion currents of 61 mA (a, b) and 76 mA (c, d), respectively. The beam was focused by a solenoid magnet located immediately after the ion source. The horizontal/vertical normalized rms emittances were measured as 0.26/0.15 and 0.32/0.19 πmm.mrad for the 61 and 76 mA beams, respectively. Although the 61 mA beam almost satisfies the J-PARC requirement of less than 0.25 πmm.mrad, the horizontal emittance of the 76 mA beam exceeds the requirement by 28 %. It can probably be improved by optimizing the gap lengths and electrode shapes of the extractor.

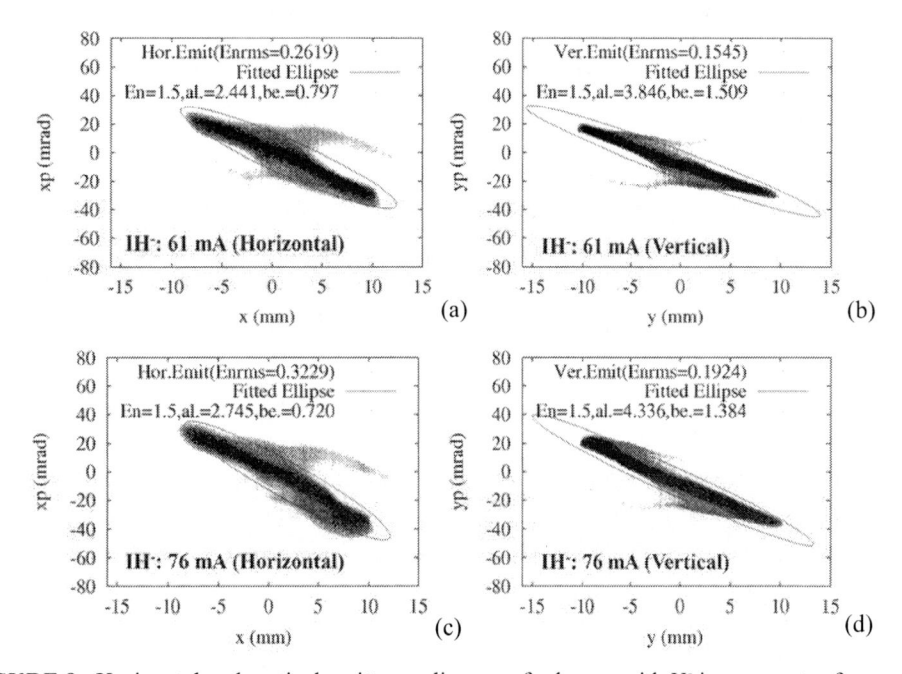

FIGURE 9. Horizontal and vertical emittance diagrams for beams with H⁻ ion currents of 61 mA (a, b) and 76 mA (c, d).

SUMMARY

The present J-PARC ion source is being operated for approximately four years without any serious trouble. At the recent beam run, approximately 700 hours continuous operation with the H⁻ ion current of 17 mA was achieved. In order to enhance the current required for the final stage of the J-PARC in the near future, we started to develop a Cs-seeded ion source. As a result, the H⁻ ion current of 61 mA with the normalized rms emittance of 0.26 πmm.mrad was obtained, which is satisfied with the requirement. We consider the cesium is indispensable for 1MW source. We decided that Cs-seeded source driven with radio frequency (RF) is the first candidate for 1MW source because the RF source is superior to filament one for reduction of the amount of cesium.

REFERENCES

1. H. Oguri, Y. Okumura, K. Hasegawa, Y. Namekawa and T. Shimooka, *Rev. Sci. Instrum.* **73** (2) 1021 (2002).
2. A. Ueno, K. Ikegami, and Y. Kondo, *Rev. Sci. Instrum.* **75** (5) 1714 (2004).
3. H. Oguri, A. Ueno, Y. Namekawa, K. Ohkoshi, Y. Kondo, and K. Ikegami, *Rev. Sci. Instrum.* **79**, 02A506 (2008).
4. Denka Corporation, Chuo-ku, Tokyo, JAPAN, www.denka.co.jp.
5. H. Oguri, A. Ueno, Y. Namekawa, and K. Ikegami, *Rev. Sci. Instrum.* **77** (3) 03A517 (2006).
6. Y. Ohara, Japan Atomic Energy Research Institute Report, JAERI-M 6757 (1976) [in Japanese].
7. H. Oguri, A. Ueno, K. Ikegami, Y. Namekawa, and K. Ohkoshi, *PHYSICAL REVIEW SPECIAL TOPICS - ACCELERATORS AND BEAMS* **12**, 010401 (2009).
8. K. Hasegawa, T. Kobayashi, Y. Kondo, T. Morishita, H. Oguri, Y. Hori, C. Kubota, H. Matsumoto, F. Naito, M. Yoshioka, *Proceedings of IPAC '10*, Kyoto, Japan, p621 (2010).
9. H. Oguri, A. Ueno, K. Ikegami, Y. Namekawa, and K. Ohkoshi, *Rev. Sci. Instrum.* **81** 02A715 (2010).
10. K. Ohkoshi, Y. Namekawa, A. Ueno, H. Oguri, and K. Ikegami, *Rev. Sci. Instrum.* **81** 02A716 (2010).

Measurement of optical emission from the hydrogen plasma of the Linac4 ion source and the SPL plasma generator

J. Lettry, S. Bertolo, A. Castel, E. Chaudet, J.-F. Ecarnot, G. Favre,
F. Fayet, J.-M. Geisser, M. Haase, A. Habert, J. Hansen, S. Joffe,
M. Kronberger, D. Lombard, A. Marmillon, J. Marques Balula, S. Mathot,
O. Midttun, P. Moyret, D. Nisbet, M. O'Neil, M. Paoluzzi, L. Prever-
Loiri, J. Sanchez Arias, C. Schmitzer, R. Scrivens D. Steyaert,
H. Vestergard, M. Wilhelmsson

CERN, 1211 Geneva 23, Switzerland.

Abstract. At CERN, a non caesiated H⁻ ion volume source derived from the DESY ion source is being commissioned. For a proposed High Power Superconducting Proton Linac (HP-SPL), a non caesiated plasma generator was designed to operate at the two orders of magnitude larger duty factor required by the SPL. The commissioning of the plasma generator test stand and the plasma generator prototype are completed and briefly described. The 2MHz RF generators (100 kW, 50 Hz repetition rate) was successfully commissioned; its frequency and power will be controlled by arbitrary function generators during the 1 ms plasma pulse. In order to characterize the plasma, RF-coupling, optical spectrometer, rest gas analyzer and Langmuir probe measurements will be used. Optical spectrometry allows direct comparison with the currently commissioned Linac4 H⁻ ion source plasma. The first measurements of the optical emission of the Linac4 ion source and of the SPL plasma generator plasmas are presented.

Keywords: Plasma Generator, Negative Ion Source, Hydrogen.
PACS: 52.25.Os, 52.27.Cm.

INTRODUCTION

The ion source of CERN's future 160 MeV H⁻ linear accelerator (Linac4) [1] is a non caesiated volume H⁻ ion source derived from the DESY ion source [2]. The High Power Superconducting Proton Linac (HP-SPL) [3] project requires 40-80 mA, 0.4-0.8 ms pulses of H⁻ ions after suppression of typically 100 µs head and tail instabilities at 50 Hz repetition rate. A non caesiated plasma generator was designed to operate at the two orders of magnitude larger duty factor required by the SPL [4]. Shielding of the permanent magnets from RF induced eddy currents heating is included [5].

The 2MHz RF generator (100 kW, 50 Hz repetition rate) and dedicated impedance matching circuit were successfully commissioned. The plasma generator is designed to operate at a 7% time duty factor and up to 6 kW average RF-power [4] and to provide plasmas comparable to the one of the Linac4 volume ion source. The 5.5 windings solenoid RF antenna is surrounded by ferrites and permanent magnets; its specific coupling to the plasma is being modeled via equivalent electrical circuits [6]. The measurement of the RF power coupling to the plasma and characterization of the RF-induced plasma are essential information required to compare the Linac4 and SPL plasmas and to optimize key features of plasma generators namely geometry, electrical and magnetic configuration. Furthermore, it provides a non invasive method that should become integral part of the on-line monitoring of the RF-coupling and ion source plasma. A laboratory was equipped with 3D magnetic field measurement, optical spectrometer, plasma light detection, residual gas analyzer quadrupole mass spectrometer and pressure gauges while a Langmuir probe is being selected. A new test stand was designed and produced to support the plasma generator prototype and its measurement systems.

PLASMA GENERATOR TEST STAND

The plasma generator laboratory is sketched in Fig. 1. The laboratory ancillary equipments are power, cooling, gas distribution vacuum and safety systems. The plasma chamber is located in the center of the room, mounted on a test stand and connected to the power supplies, gas distribution, vacuum and monitoring systems. The measurement systems are installed on the six port test stand vacuum chamber presented in Fig. 2.

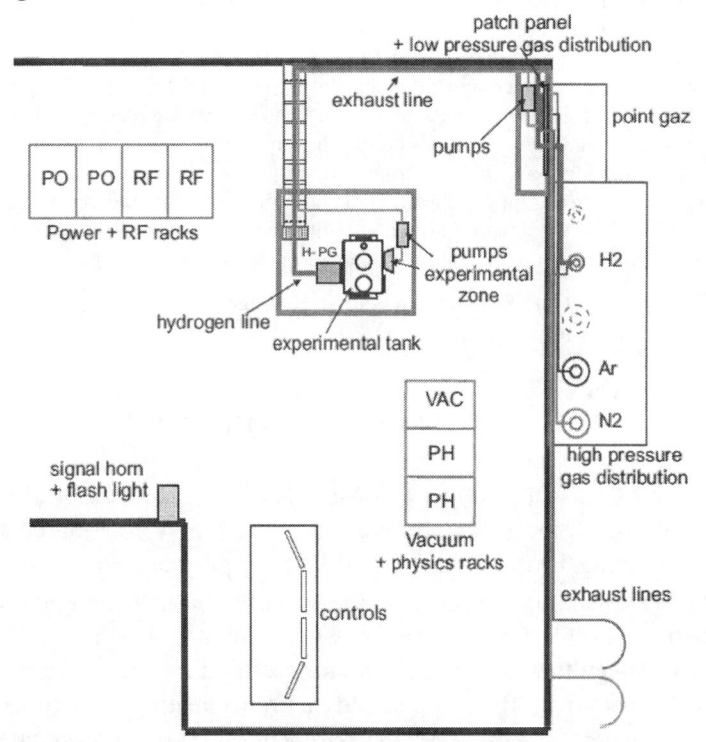

FIGURE 1. Laboratory layout, the gas distribution, pumping and safety systems are shown. The high tension (PO) RF amplifier (RF) and the vacuum controls (VAC), monitoring, delay generator and analysis racks (PH) labels are indicated.

ANCILLARY LABORATORY EQUIPMENT

Separate demineralized water cooling circuits are equipped with individual flow-regulation, flow-interlocks and return temperature monitoring. The main cooling circuit is dedicated to the cooling of the RF amplifier and an array of four units distributes the demineralized water to the plasma generator prototype's cooling circuits: spark gap ignition, plasma chamber, antenna and extraction plate. As sources are know to be susceptible to temperature variations, the piezo valve which controls the H_2 gas injection, is temperature stabilized with a Lauda ALPHA RA8 system, which should allow systematic investigation of small density changes in the injection line. The temperature controller is also used to calibrate the infra red emissivity of the plasma generator components that will be monitored via IR-imaging.

The gas distribution panel is equipped with precision manometers; a second pressure reduction stage allows reproducible gas injection pressure conditions. The flammable gas safety is provided by monitoring hydrogen concentration above the gas distribution panel and test setup. Two flammable gas detectors (type OLC-50) are installed in metallic hoods above the low pressure gas distribution panel and the experimental zone. Both are coupled to a central unit (MX-32) that generates alarms and interlocks at 10% and 20% of the hydrogen lower explosive limit (LEL) respectively. In addition an interlocked minimal flow of nitrogen flushes the exhaust of the hydrogen pumping system.

The vacuum system is composed of three pumping units: A 60 l/s turbo molecular pump (TMP) and an Explosive gas-rated fore-pump are ensuring the purity of the H_2 injection line but also offers the possibility to operate from 0 to 2.5 bars absolute H_2 injection pressure. Another 60 l/s TMP and fore-pump are dedicated to the ultra high vacuum quadrupole mass spectrometer and a unit composed of 500 l/s TMP and Ex-rated fore pump provides pumping of the experimental setup. In view of the expected mass flow range and of the low compression factor of TMPs for hydrogen, the system includes the possibility to add a 60 l/s TMP at the exhaust of the 500 l/s TMP before the fore pump. The vacuum gauges were selected to cover from UHV to the unusual high range of pressures expected close to the plasma chamber.

Power Supplies

The test stand is equipped with a 1.9-2.1 MHz RF amplifier described in reference [6] capable of producing millisecond pulses of 100 kW peak power at a 50 Hz repetition rate. The ignition pulser supply delivers a pulse of up to 1000V limited to 10A for a duration of up to 100 microseconds at a maximum rate of 50Hz. A +1000V DC supply (Fug MCP 140-1250) is backed by capacitors, the output is switched using a high voltage bipolar MOSFET. A selectable series resistor limits the output current.

The Piezo valve is driven by an in house designed amplifier capable of delivering -10 to +100V. The amplifier system has two inputs which are both amplified 20 times and then summed. Using an arbitrary waveform generator on one input and a DC level on the other any combination of DC offset and pulse shape is possible to drive the valve.

The Plasma extraction plate and Collar electrode bias supplies consist of Fug NTN 140-65 +/-65V 2A DC supplies. The output of the power supplies is backed by a

22000µF capacitor to allow a peak current of 10A for 1.2 ms during the pulse and to limit the voltage droop to 2% when operated at 30V.

Trigger Controls and Monitoring

Arbitrary function generators (AFG) (Agilent 33220 and TTi TG5011) are used to generate the piezo valve pulse shape, the RF-frequency and the RF-power control signals. The RF frequency can be switched at plasma ignition from the frequency at which plasma ignition is fastest to the one where plasma coupling is optimum.

Pulse generators (Quantum 9528) are dedicated to the triggering of the gas injection piezo valve AFG, spark gap ignition start and stop, RF warning, RF-on and RF power signal. A light amplification system is described in the next section that while passing a light threshold provides a TTL trigger signal to switch the RF frequency.

The control system is based on Simatic PLCs and is operated by step7 and Labview applications. All pressures and temperatures are recorded and stored on a slow control database. The PLC handles hardware protection conditions (e,g, cooling water temperature or flow) and delivers an enable TTL signal to the triggering system if all temperatures, pressure and flow values are within their nominal range of operation. The flammable gas 10% LEL warning is included in the chain, vetoing operation of the pulsed power supplies and RF pulse trigger.

Measurement Systems

A visible to near ultra violet (200-1100 nm) spectrometer (Ocean optics USB4000) is equipped with a 25 µm entrance slit. A database of atomic and molecular transitions is included in the analysis software. Together with an absolute calibration system allowing optical fiber transmission measurement, it enables quantitative measurements. The plasma light is detected by an Infineon SFH250V photodiode and converted to a voltage by a pre amplifier stage. This is followed by a second amplifier stage with user adjustable gain. The total rise time is approximately 3 µs. The signal is buffered and made available for display on an oscilloscope and is also fed to a comparator which detects when the light signal is above a preset threshold. This then triggers a monostable stage to produce a digital TTL level pulse which is adjustable in delay and duration that can be used as plasma detection trigger and sent to the RF amplifier.

A Gauss meter (THM1176-HF-PC) is available; its precision is 1% from 30 µT to 3T with a range up to 20 Tesla. The active probe dimension is 150x150x10 µm and the maximum measurement rate is 2 kHz. The Hall probe is controlled via Labview software jointly with a set of motorized linear slides and rotation from Zaber Technologies. The magnet measurement system is mounted on a table and provides simple and efficient mapping of the complex 3D magnetic configuration as well as monitoring of the permanent magnets homogeneity or magnetization changes before and after operation.

An infra red camera (Fluke Ti32) will be used to measure the temperature distributions of the plasma generator hardware during high power operation; the thermal images will be directly compared to the results of thermal simulation. A germanium window is used to measure temperatures of material inside the vacuum chamber. The window is equipped with a protective plate.

The quadrupole mass spectrometer (1-100 amu) is a QMG 700 equipped with a QMA 125 analyzer. Its role is to identify impurities and to investigate potential sputtering of the plasma chamber material. It is equipped with a differential pumping aperture reduction allowing a reduced gas sampling during high hydrogen pressure operation. Its position on axis of the ion source aperture enables the measurement of condensable elements.

LeCroy WaveSurfer 24MXs-A 4 channel oscilloscopes, 200 MHz bandwidth and 10 Mpts per channel are dedicated to the measurement of the injected and reflected RF power voltages and current, the measurement of the plasma emitted light and also the monitoring of the timing signals. The resolution is sufficient to precisely measure the phase within each RF cycle over pulse duration of up to 1.2 ms.

Plasma Generator Test Stand

The test stand vacuum chamber is illustrated in Fig. 2. It is a simple industrial three axis six DN 200 ports cross mounted on a steel support. The plasma chamber prototype is connected via a reduction flange horizontally and the 6mm hole of the plasma electrode is on axis facing all detection systems that will be installed on the opposite flange.

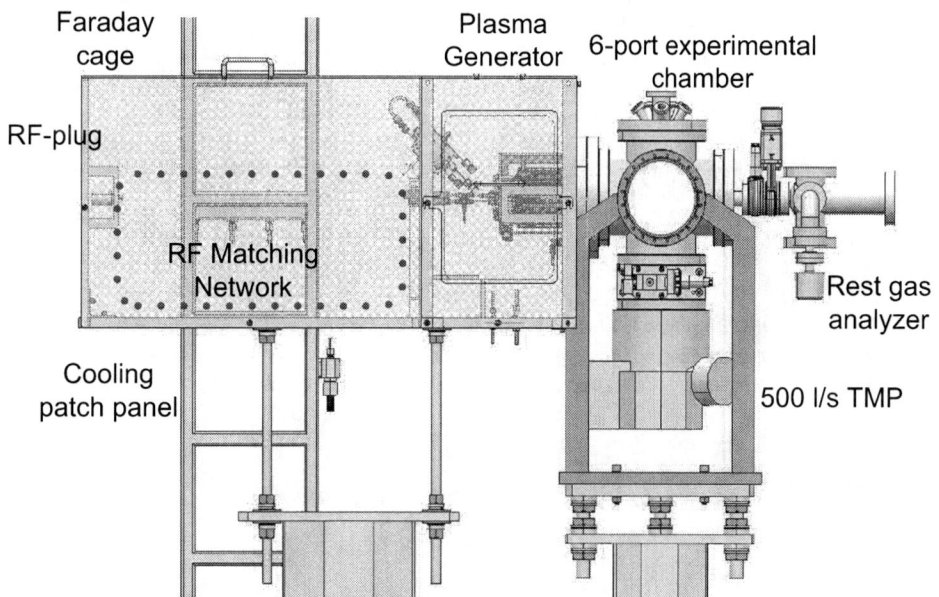

FIGURE 2. View of the plasma generator test stand, the prototype is mounted on its 6-axis pumping port. All diagnostics equipment will be coupled in the DN200 port facing the plasma generator. As an illustration, the rest gas analyzer quadrupole mass spectrometer is shown.

The Vacuum gauges are located on the top and the TMP connected to the bottom flange. The two remaining ports can host a viewing window, mechanical supports or electrical feed through. The impedance matching circuit and the plasma generator are housed within a Faraday cage. A mockup dedicated to ignition studies where the plasma generator is replaced by a quartz tube can be mounted onto the test stand. A system dedicated to the validation of the shielding of permanent magnets from eddy current heating can be installed in the Faraday cage of the test stand; its RF-heating is provided by an antenna identical to the one of the plasma generator. Ferrites and magnets are positioned in a plexiglass support reproducing the relative positions in the plasma generator. Temperatures will be recorded via IR imaging and Pt-100 resistance thermometer.

Plasma Chamber Prototype

The plasma chamber prototype is shown and illustrated in Fig. 3. The design of the copper RF-shielding of the multi-cusp permanent magnets is described in reference [5]. The ignition spark gap distance was reduced to 2.4 mm as the original design (5 mm) lead to ignition voltages close to the maximum of the power supply.

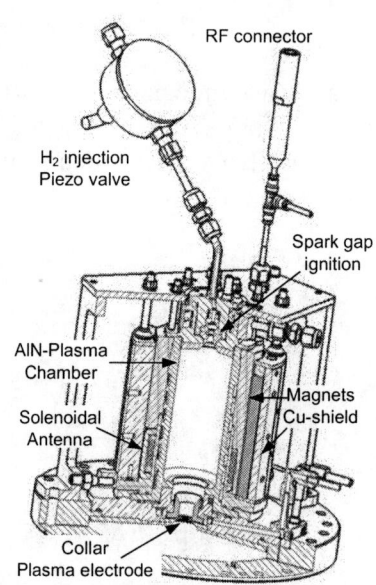

FIGURE 3. Photograph and scheme of the SPL Plasma chamber the main components are the H_2-injection and the individually cooled spark gap ignition, plasma chamber, antenna and base plate supporting the plasma and collar electrodes.

HYDROGEN PLASMA CHARACTERIZATION AND MONITORING

Invasive methods such as Langmuir gauges provide key information on the plasma. we would like to develop plasma monitoring relying on optical emission complemented by RF-transmission monitoring and implement them into the control system of CERN's H⁻ ion source. The Optical Emission Spectroscopy (OES) technique applied to the determination of H⁻ ion densities in hydrogen plasma ion sources was proposed by Fantz and Wünderlich [7]. The EOS technique relies on measurement of the H_α, H_β and H_γ lines analyzed by collisional radiation model and provides a simple formula relating the H⁻ ion density to the electron density and the

H_α, H_β, H_γ lines intensities. For these EOS measurements, the short pulse duration motivated the selection of a trigger able fast spectrometer and a photodiode amplifier.

The photodiode (SFH250) light detection is coupled to a tunable amplifier. H_α was measured with a narrow band filter. For H_β, a light detector covering the 400-500 nm range should provide pulse by pulse real time information on the intensity of the H_β lines. The commissioning measurement geometry is illustrated in Fig. 4. The Quartz light viewing port of the Linac4 source copied form the design of DESY proved to be very effectively collecting the plasma light. The differences of light collection geometries are significant and should be addressed prior to detailed analysis.

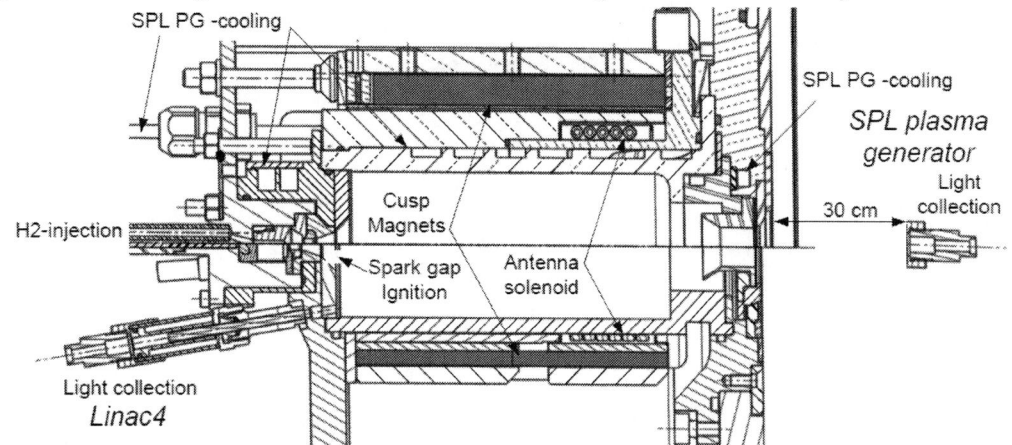

FIGURE 4. Illustration of the differences induced by the cooling and magnet shielding between the SPL Plasma Generator (top) and the linac4 ion source (bottom). The quartz viewing port of the linac4 plasma chamber and the direction of light collection of the SPL plasma generator are indicated.

All measurements presented here were performed without high voltage applied to the extraction in order to minimize the differences between the Linac4 ion source and the SPL plasma generator. In view of its reproducibility, it is anticipated that optical diagnostics should allow measurement the impact on the plasma light emission of the few amperes of electrons co-extracted from the ion source during H⁻ pulses. The first results of light measurement obtained on the Linac4 ion source and during the commissioning of the test stand are presented in the following sections.

Plasma Light Measurement Results

The potential of the method is illustrated via measurements made on the Linac4 ion source operated without ion or electron extraction and on the SPL plasma generator. A much larger effort is required to produce a plasma physics analysis of the measurement. The effect of the RF frequency switch operation mode optimized during the commissioning of the Linac4 ion source is illustrated in Fig. 5. The frequency at which best ignition conditions are met (1.946 MHz) differs from the frequency at which the best energy transfer is obtained (2.0 MHz). The most stable H⁻ beam is obtained while starting the RF at 1.946 MHz and on plasma light appearance switching to the plasma tuned frequency. This is confirmed by the light measurement which also illustrates the plasma ignition delay resulting from non optimal conditions.

Figure 6 presents the piezo valve and RF power control signals, the plasma ignition spark gap current and the clearly separated light signals issued form the spark and the plasma.

In Fig. 7, the Linac4, light intensity is shown as a function of the RF power; the RF coupling according to reference [5] was 70%. A similar behavior was observed in the plasma generator.

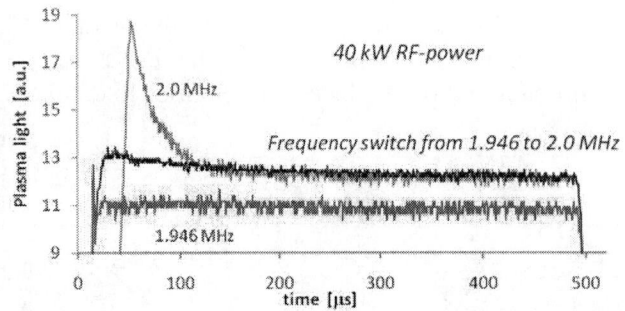

FIGURE 5. Light emission of the fastest ignition RF frequency (1.946 MHz), best energy transfer RF frequency (2.0 MHz) and switch from 1.946 to 2 MHz few microseconds after plasma ignition (note offset scale).

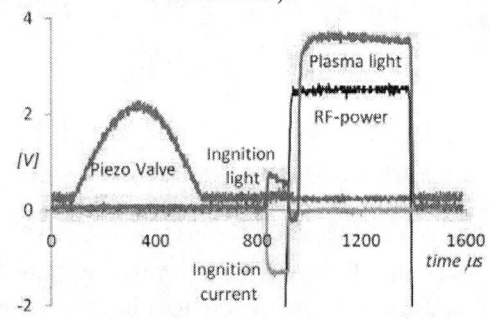

FIGURE 6. Ignition and plasma generator light emission at the SPL plasma generator, the gas valve signal, the ignition spark gap current and the 20 kW (2.06 MHz) RF power control pulse are shown.

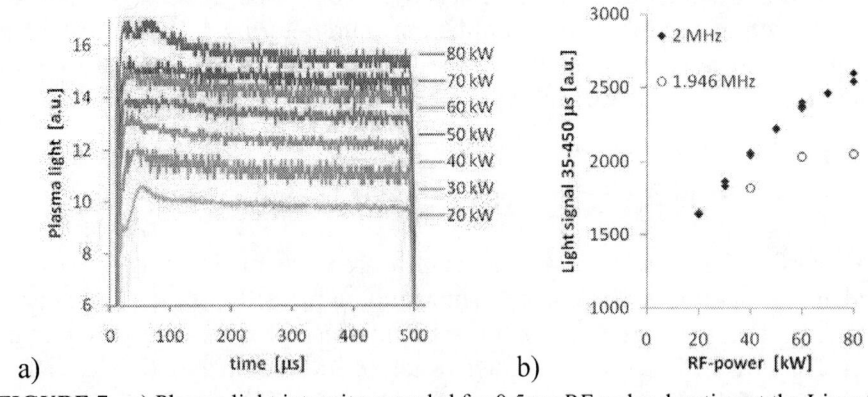

FIGURE 7. a) Plasma light intensity recorded for 0.5 ms RF-pulse duration at the Linac4 ion source. The RF frequency is switched from 1.946 to 2 MHz at plasma ignition, its power is indicated. b) Evolution of the light intensity (integral from 350-450 μs) at RF-frequencies of 1.946 and 2 MHz.

Spectrometer Measurements

The dynamic range of the spectrometer was slightly below two orders of magnitude during our measurements, the Hα line collected in the linac4 setup was saturating the spectrometer. A two orders of magnitude aperture reduction obtained by a 0.2 mm diameter diaphragm provided good measurement conditions for the Hα and Hβ lines. A notch filter suppressing the Hα line is required for the spectrometry of low intensity lines. The RF-power dependant Hα and Hβ intensities and their ratio are presented in Fig. 8 a) and b). The Hβ/Hα ratio measured on the flat section of 0.5 ms pulses at the Linac4 rose with RF power (20-80 kW) from typically 6 to 14 %.

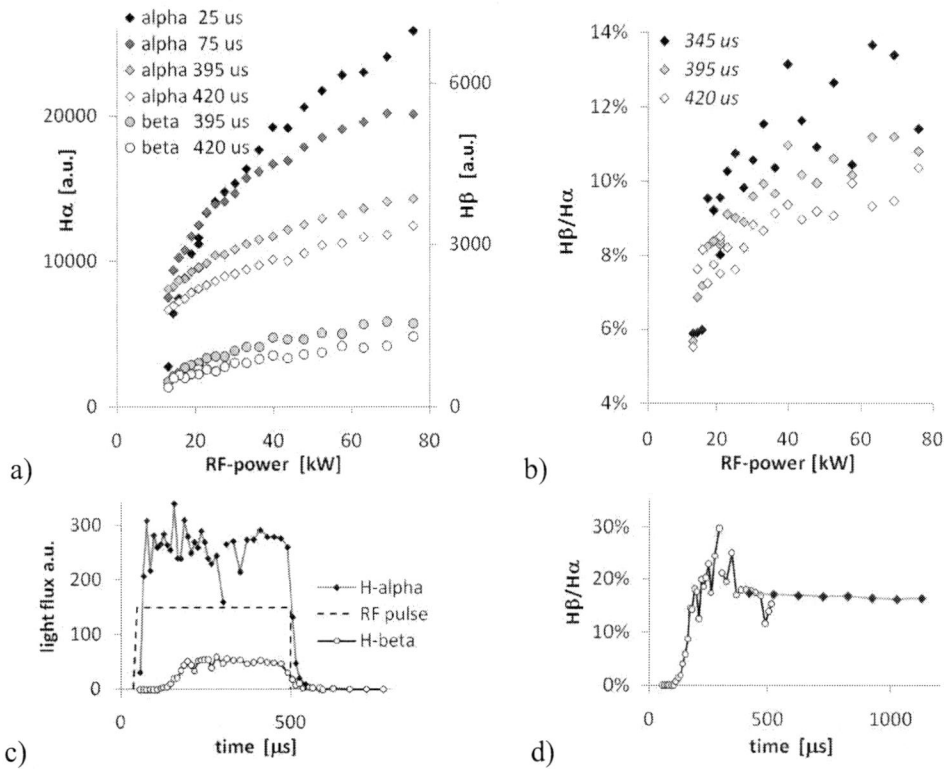

FIGURE 8. (a) Hα and Hβ lines intensities measures at the Linac4 ion source. The Hβ/Hα ratios extracted from these data are shown in (b). The average integration time [μs] is indicated. (c) Hα and Hβ line profiles measured on the SPL plasma generator heated by a 25 kW RF power pulse of 500 μs duration. (d) The Hβ/Hα ratio derived form the line profiles (white circle) and Hβ/Hα ratio measured over the last 150 μs of plasma pulses form 500 μs up to 1.2 ms duration (black diamond) are presented.

The minimal integration time of the spectrometer is 4 ms and the pulse duration 0.5 ms; therefore an integral measurement is obtained via variable delay trigger of the 4 ms acquisition window across the pulse. The rise of the plasma light is measured by approaching the end of the measurement window and the plasma light tail by moving the start of the acquisition window. The Hα and Hβ line intensities obtained with the plasma generator prototype for a 20 kW RF power heated plasma are presented in Fig. 8 c). A delayed raise of the Hβ line of typically 100 μs is observed. The Hβ/Hα ratio is presented in Fig. 8 d), the average ratio between 200 and 500 μs is 18%; it drops to 17% at the end of a 1.2 ms pulse.

CONCLUSION AND OUTLOOK

The SPL plasma generator prototype test stand was successfully commissioned. The preliminary tests of spectrometric and fast light acquisition are presented and deemed successful in terms of light monitoring reproducibility and sensitivity. Calibration of the optical acquisition chain is required. Optical measurements combined with precise measurement of the RF coupling is the base of diagnostic that will be applied to optimize the plasma generator features and to investigate the variation of parameters such as pressure and magnetic filed strength or cusp configuration. Narrow band filters will be implemented to upgrade the light detection system to wave length selective recording of the Hα and Hβ lines. On Linac4, the stabilization of the H$^-$ pulse intensity droop by tuning the RF power and frequency will be further tested and correlated to its light emission.

ACKNOWLEDGMENTS

The authors are grateful and wish to acknowledge the contributions and support from: Laszlo Abel, Alessandro Bertarelli, Oliver Bruning, Maryse Da Costa, Alain Demougeot, Paolo Chiggiato, Dan Faircloth, Ramon Folch, Philippe Frichot, Roland Garoby, Jonathan Gulley, Christophe Jarrige, Erk Jensen, Emmanuel Koutchouk, Detlef Kuchler, Robert Mabillard, Marina Malabaila, Cristiano Mastrostefano, Sophie Meunier, Catherine Montagnier, Jose Monteiro, Mauro Nonis, Julien Parra-Lopez, J. Peters, Stephen Rew, Miguel Riesgo Garcia, Ghislain Roy, Franck Schmitt, Alain Stalder, Laurent Tardi, Dominique Trolliet, Donatino Vernamonte and Fredrik John Carl Wenander.

This project has received funding from the European Community's Seventh Framework Programme (FP7/2007-2013) under the Grant Agreement no 21211M.

REFERENCES

1. D. Küchler, T. Meinschad, J. Peters, and R. Scrivens, *Rev. Sci. Instrum.* 79 (2008), 02A504.
2. J. Peters, *The HERA Volume H$^-$ Source*, PAC05 Conference Proceedings, TPPE001, p. 788 (2005).
3. M. Baylac, F. Gerigk (ed.), E. Benedico-Mora, F. Caspers, S. Chel, J.M. Deconto, R. Duperrier, et.al., CERN Report No. 2006-006, (2006).
4. J. Lettry, M. Kronberger, R. Scrivens, E. Chaudet, D. Faircloth, G. Favre, J.-M. Geisser, D. Küchler, S. Mathot, O. Midttun, M. Paoluzzi, C. Schmitzer, and D. Steyaert, *Rev. Sci. Instrum.* 81 (2010) 02A723.
5. M. Kronberger, E. Chaudet, G. Favre, J. Lettry, D. Küchler, P. Moyret, L. Prever-Loiri, C. Schmitzer, R. Scrivens, D. Steyaert, Magnetic cusp configuration of the SPL plasma generator, these proceedings.
6. M. Paoluzzi, M. Haase, J. Marques Balula, D. Nisbet, CERN LINAC4 H$^-$ Source and SPL plasma generator RF systems, RF-power coupling and impedance measurements, these proceedings.
7. U. Fantz and D. Wünderlich, *New Journal of physics* 8 (2006), 301.

Magnetic Cusp Configuration of the SPL Plasma Generator[1]

Matthias Kronberger, Elodie Chaudet, Gilles Favre, Jacques Lettry, Detlef Küchler, Pierre Moyret, Mauro Paoluzzi, Laurent Prever-Loiri, Claus Schmitzer, Richard Scrivens, and Didier Steyaert

CERN, 385 Route de Meyrin, 1211 Geneva, Switzerland

Abstract. The Superconducting Proton Linac (SPL) is a novel linear accelerator concept currently studied at CERN. As part of this study, a new Cs-free, RF-driven external antenna H⁻ plasma generator has been developed to withstand an average thermal load of 6kW. The magnetic configuration of the new plasma generator includes a dodecapole cusp field and a filter field separating the plasma heating and H⁻ production regions. Ferrites surrounding the RF antenna serve in enhancing the coupling of the RF to the plasma. Due to the space requirements of the plasma chamber cooling circuit, the cusp magnets are pushed outwards compared to Linac4 and the cusp field strength in the plasma region is reduced by 40% when N-S magnetized magnets are used. The cusp field strength and plasma confinement can be improved by replacing the N-S magnets with offset Halbach elements of which each consists of three magnetic subelements with different magnetization direction. A design challenge is the dissipation of RF power induced by eddy currents in the cusp and filter magnets which may lead to overheating and demagnetization. In view of this, a copper magnet cage has been developed that shields the cusp magnets from the radiation of the RF antenna.

Keywords: Accelerators, sLHC upgrade, SPL, Negative ion sources, H⁻ sources, plasma sources, ohmic heating
PACS: 29,25.Ni, 52,50.Dg, 52,50.Qt

INTRODUCTION

In the framework of the future LHC luminosity upgrade, an RF-driven non-cesiated H⁻ source [1] has been installed in 2009 at the 3MeV test stand of Linac4, which will replace the currently employed Linac2 in 2014. The RF power used to generate and heat the plasma is supplied by an external antenna that operates in pulsed mode at a frequency close to 2MHz. During operation at Linac4, this antenna will deliver up to 100kW of pulsed RF power in order to achieve the nominal H⁻ current of 80mA. The length of the RF pulse is 400µs, yielding a maximum duty factor of 0.08% for a repetition rate of 2Hz [2]. As shown in [3], the resulting maximum average heat load of 80W is already close to the limit of operation for some of the components of the Linac4 source. In view of the possible integration of Linac4 as the front-end of the Superconducting Proton Linac (SPL) [4], where pulse lengths of 1ms and repetition

---

[1] This project has received funding from the European Community's Seventh Framework Program (FP7/2007-2013) under the Grant Agreement no 21211

rates of 50Hz are expected, it is therefore necessary to upgrade the Linac4 type source to a high power version.

THE SPL PLASMA GENERATOR PROTOTYPE

In order to account for the massively increased thermal load expected during ion source operation in SPL, a novel Cs-free plasma generator for H⁻ production that can operate at time average RF powers of up to 6kW [5] has been designed and constructed, and is ready for testing (Fig. 1). Its main components are shown in Fig. 2. The plasma generator is currently being commissioned and first plasma production was achieved at the end of October 2010. In the following, the design and latest improvements of the plasma generator are briefly reviewed. A description of the test stand and of the RF system is given in other contributions to this volume [6,7].

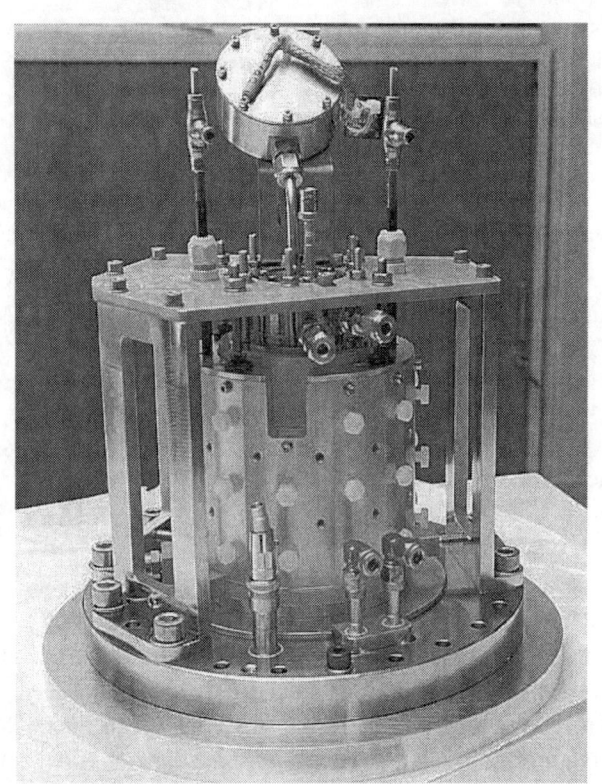

FIGURE 1. The SPL plasma generator prototype.

H$_2$ Gas Injection and Ignition and Plasma Chamber

The injected H$_2$ gas is ignited by a spark gap at the back end of the source. A cooling circuit ensures an efficient removal of the thermal load generated by the electrical discharge. From the ignition region, the plasma expands into the ceramic plasma chamber with inner diameter of 48mm and outer diameter of 64mm. In order to optimize heat transport in the region of the plasma volume, where the highest heat loads are expected, the plasma chamber is made of Aluminum Nitride (AlN) ceramic (thermal conductivity = 180W/mK) and has a helical cooling channel machined into the outer surface.

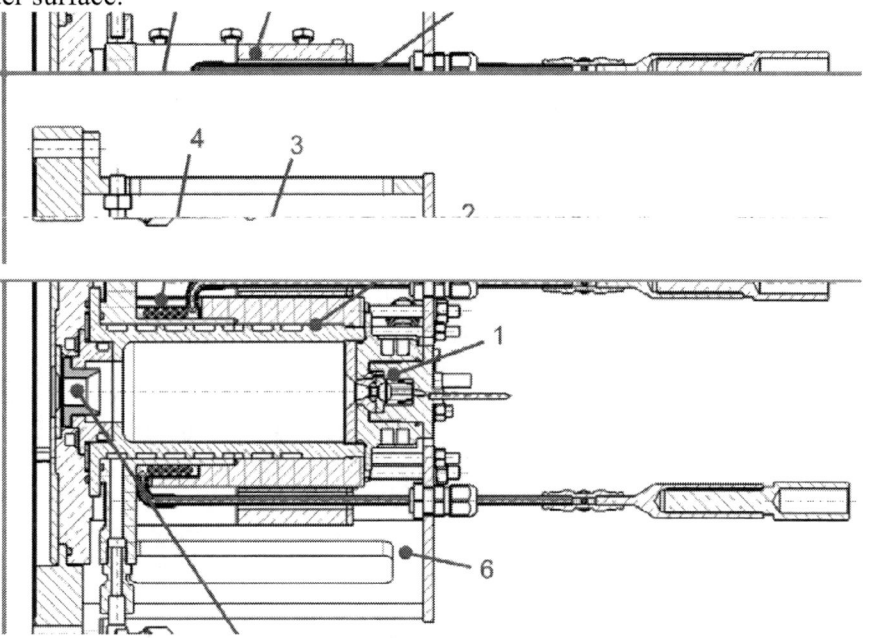

FIGURE 2. Cross-sectional view of the SPL plasma generator. 1: Gas injection and ignition region. 2: Ceramic plasma chamber and cooling sleeve. 3: Magnet cage. 4: RF antenna. 5: Extraction region. 6: Mechanical support.

Similar as in the H⁻ source developed as SNS [8], the ceramic plasma chamber is surrounded by a jacket made of Polyether ether ketone (PEEK) that confines the cooling circuit on the outside and contains connections for the cooling circuit in- and outlet. The cooling sleeve is split into a front part and a back part in order to allow for the RF antenna to be installed. Leak tightness of the cooling circuit is assured by three o-rings at the front and back end of the PEEK sleeve and at the contact surface between the two parts. The design of the cooling sleeves presented in [5] was improved in order to improve the rigidity of the sleeve and to facilitate the installation of the cooling circuit in- and outlets. On the outside, the plasma chamber and cooling sleeve are confined by the magnet cage containing the cusp magnets.

RF Antenna and Ferrites

The RF field used to heat the plasma is produced by a water-cooled, hollow-tube Cu antenna with 5 1/2 turns that operates at a frequency close to 2MHz and can run at up to 100kW RF power. This antenna may be exchanged with a stacked version with up to 10 windings in future experimental runs. In order to assure antenna uniformity, to prohibit sparks between the windings, and to improve its rigidity and mechanical stability, the antenna is embedded in a molded epoxy cylinder. Simultaneous supply with cooling water and electrical power is achieved by T-connectors at the end of the antenna tips. NiZn ferrites (type 8C11) surrounding the antenna coil enhance the coupling of the RF field to the plasma.

Extraction Region

The extraction region contains the filter magnets that produce the magnetic filter field, and a funnel and an aperture electrode that can be biased individually for optimized H⁻ generation. A cooling circuit that runs close to the inner perimeter is used to control the temperature in the collar region. For efficient heat transport, only materials with high thermal conductivity were selected, and electrodes and interleafing insulators brazed together to minimize thermal contact conductances.

MAGNETIC CONFIGURATION OF THE SPL PLASMA GENERATOR

As the Linac4 source, the SPL plasma generator employs a dodecapole magnetic cusp field for plasma confinement that, at present, is generated by standard N-S magnetized NdFeB magnets. However, the space requirements of the plasma chamber cooling circuit imply a significantly larger distance of the magnets to the plasma volume (SPL: 22mm; Linac4: 11mm), which leads to a reduction in cusp field strength of $\approx$ 40% compared to Linac4 despite a larger magnet size. For this reason, the TOSCA module of Vector Fields Opera™ was used in order to find alternative cusp magnet configurations for which the cusp field strength, gradient and orientation inside the plasma chamber were comparable with the Linac4 source. The best compromise between magnetic field strength and space requirements was found for a cusp configuration employing offset-Halbach elements, of which each consists of three single NdFeB magnets (remanence B_R = 1.33T) with different magnetization directions (Fig. 3).

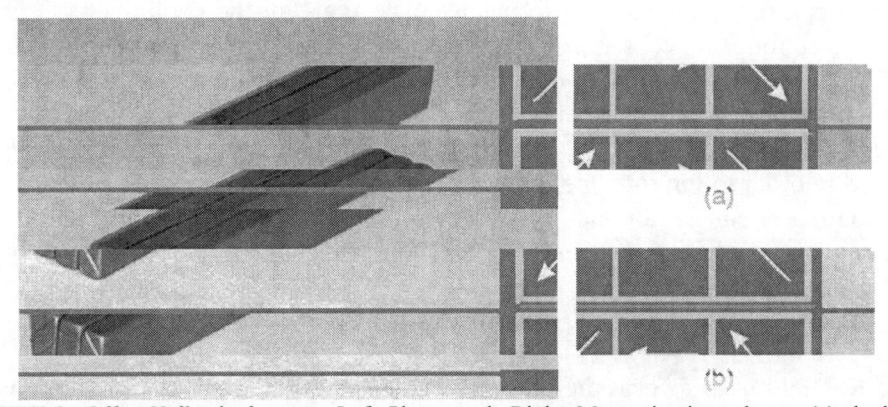

FIGURE 3. Offset Halbach elements. Left: Photograph. Right: Magnetization scheme. (a) clockwise orientation. (b) counter clockwise orientation. The arrows indicate the direction of magnetization.

By aligning the Halbach elements in clockwise (cw) and counterclockwise (ccw) orientation (Fig. 4), we found an improvement in cusp field strength of up to 50% compared to standard N-S magnetized magnets of the same size, radial distance, and remanence (Fig. 5). Variations in the cusp field strength are due to the missing ferrites at the position of the antenna tips.

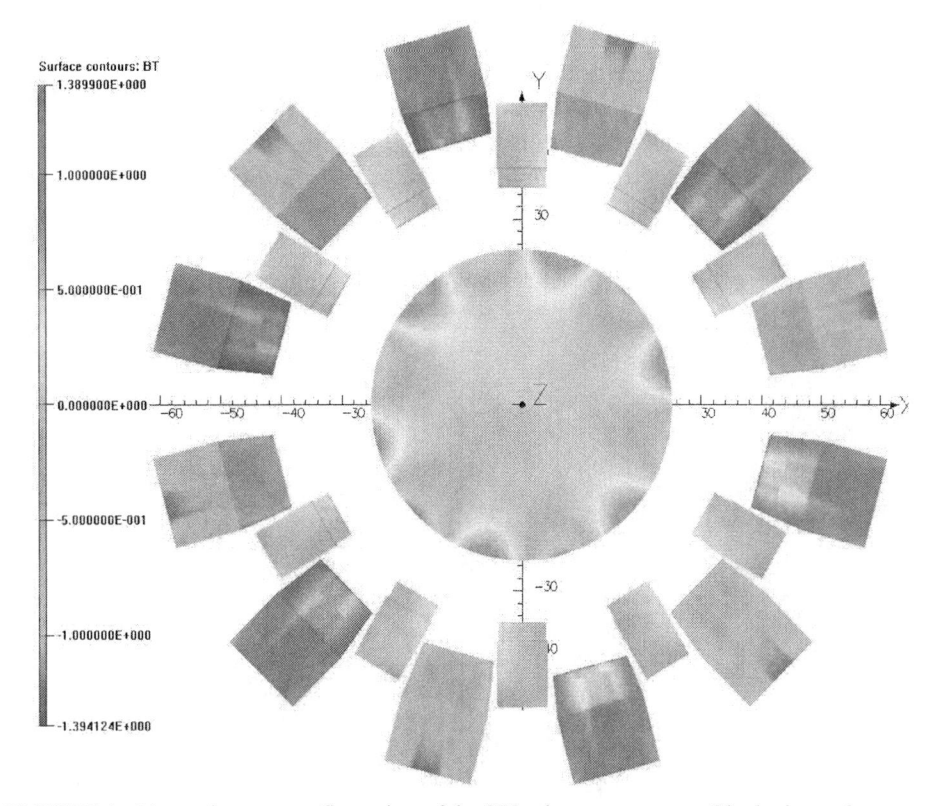

FIGURE 4. Magnetic cusp configuration of the SPL plasma generator. Clockwise and counter clockwise oriented Halbach elements are shown in darker red and brighter turquoise tones, respectively. Ferrites are located between the magnets (green). The circular inset shows the tangential magnetic field B_t inside the plasma region at the center of the antenna.

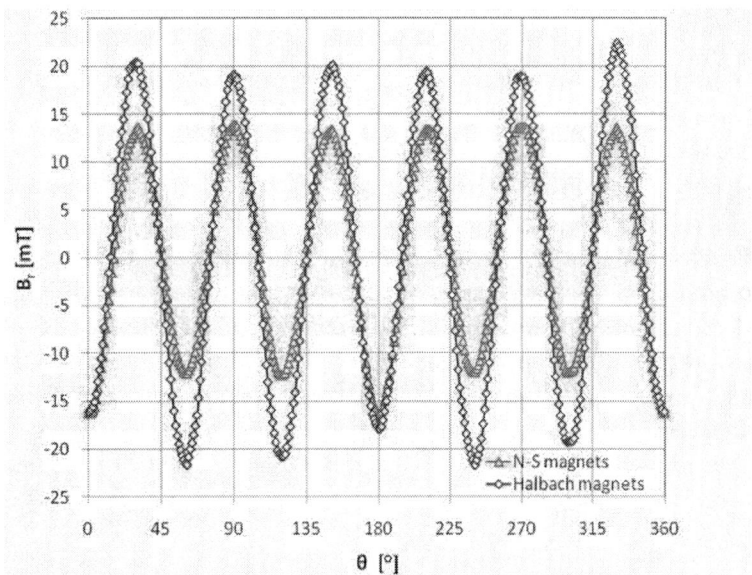

FIGURE 5. Variation of B_r along a circle at the plasma chamber wall for two dodecapole cusp configurations. Red triangles: N-S magnetized magnets. Black circles: offset Halbach elements of the same size, radial distance, and remanence.

The extraction region of the SPL plasma generator contains two filter magnets serving as a low-pass energy filter for electrons. The use of arc-shaped instead of cuboid magnets allows for the installation inside the ceramic plasma chamber in a dedicated AlN magnet holder. A further advantage of the arc shape is that it increases the central field strength and improves the radial uniformity of the filter field. In the current configuration, the B-field strength increases from 11mT in the center to up to 40mT at the wall (Fig. 6). Stronger filter fields may be generated by using magnets of larger size, which however requires a redesign of the magnet holder.

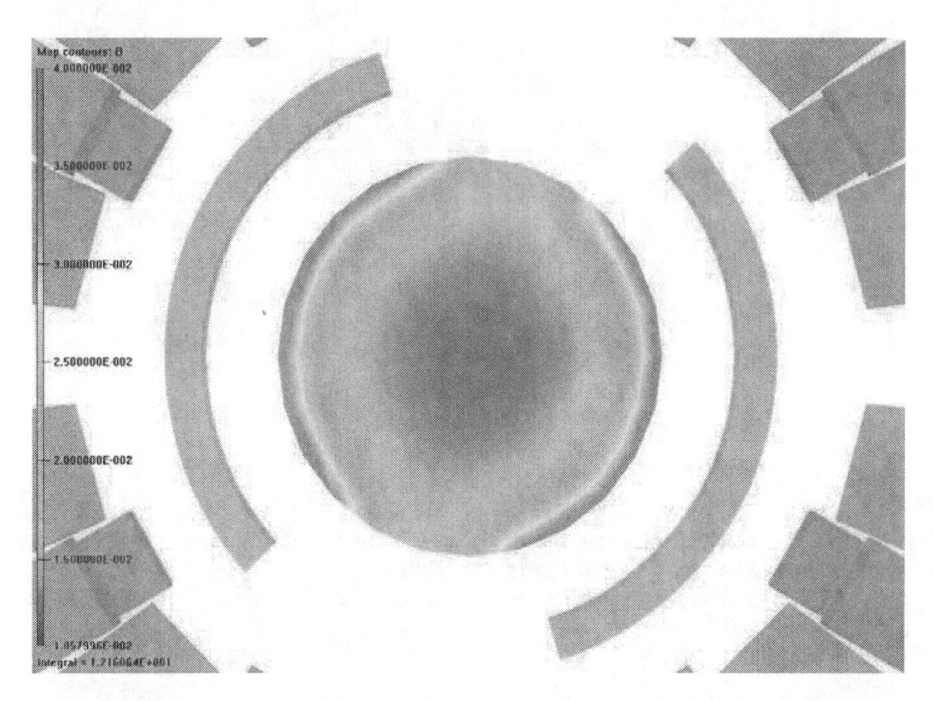

FIGURE 6. B-field produced by the filter magnets in the extraction region. The maximum and minimum values correspond to 11mT and 40mT, respectively.

RF SIMULATION OF THE SPL PLASMA GENERATOR

In the SPL plasma generator, the cusp and filter magnets are located close to the RF antenna, and ohmic heating may lead to a partial or a complete loss of B-field strength due to overheating. In view of this, the dissipation of RF power induced by eddy currents was estimated by a simplified EM model of the SPL plasma generator. The model consisted of four different components: the RF antenna, a NdFeB permanent magnet with dimensions and position corresponding to an SPL cusp magnet, the plasma, and an optional NiZn ferrite. The relative permeabilities μ_R and electrical conductivities σ of the materials are summarized in Table 1 together with their skin depths δ,

$$\delta = \left(\pi f \mu_0 \mu_R \sigma\right)^{-1/2} \qquad (1)$$

where $\mu_0 = 4\pi \times 10^{-7}$H/m is the magnetic constant and f the RF frequency, $f = 2$MHz. The properties of the ferrites and NdFeB magnets were taken from data sheets provided by the suppliers. μ_R and σ of the plasma were derived from measurements at the Linac4 source [7]. Hysteresis losses of the ferrites were not taken into account. The relative permeability μ_R of NdFeB was set to 1 to account for the saturation of the material. The thin Ni coating of the magnets was not included in the simulations as the skin depth δ exceeds the thickness of the layer ($\approx 10\mu$m) by one order of magnitude. To simplify the model, no permanent magnetic field was included in the simulations.

TABLE 1: Skin depths δ calculated for the materials in the simplified source model at $f = 2$MHz.

	NdFeB	Copper	Ferrite	Plasma
μ_r	1	1	variable	0.84
σ (Sm^{-1})	6.67x10^5	6.0x10^7	1.0x10^{-5}	22
δ (μm)	440	46	$\approx 10^6 - 10^7$	7.8x10^4

Figure 7 compares the B-field distribution around a standalone cusp magnet with the B-field distribution around a cusp magnet flanked by two ferrites. The peak antenna current I_{peak} of 350A for which the distribution was determined corresponds to the value measured at Linac4 for 100kW operation. Due to the strong B-field close to the antenna, the ferrites are saturated but still lead to a significant reduction of magnetic field strengths at the position of the magnets. In addition, they increase the B-field strength inside the antenna by more than 10%, which effectively leads to an improvement of RF coupling with the plasma.

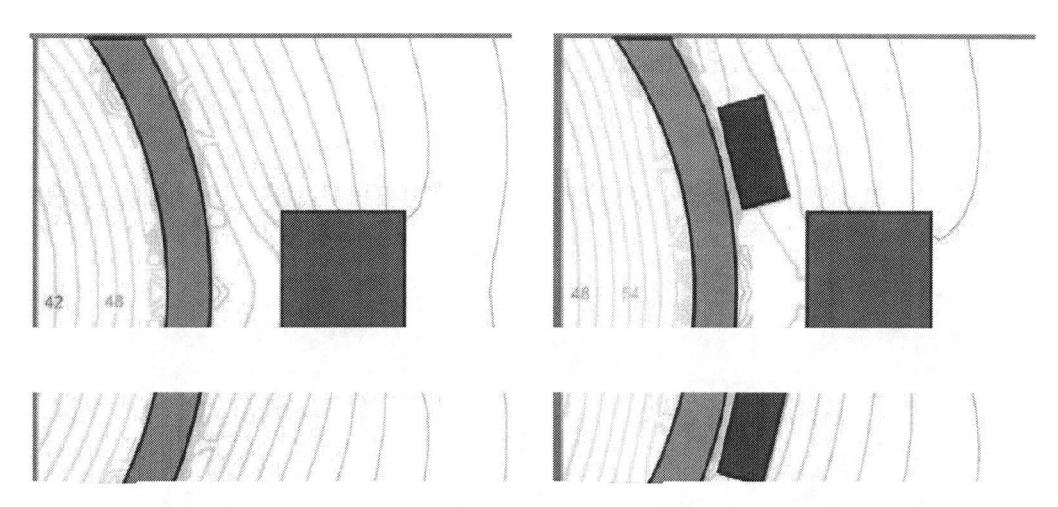

FIGURE 7. RF B field distribution around one cusp magnet for an antenna current of 350A. Left: Setup without ferrites. Right: Setup with ferrites. The numbers give the RF B-field strength in mT. Antenna, ferrites, and magnet are shown from left to right in orange, dark blue and grey, respectively.

Table 2 lists, for both simulated cases, the power losses per cusp and filter magnet. The given numbers correspond to the power dissipated on average at a duty cycle of 6%. Losses in the cusp magnets are reduced by more than 50% when ferrites are placed around the antenna. However, even then they are still in the range of several tens of watts per magnet and may result in magnet temperatures above the Curie temperature T_C during SPL operation, which is about 200°C for the presently used magnets. Similar numbers are observed for the filter magnets, with the exception that the positive effect of the ferrites is smaller as they do not influence the field distribution in the front section of the source significantly.

TABLE 2: Average dissipated power P_{diss} in NdFeB cusp and filter magnets for $f = 2$MHz, peak antenna current $I_{peak} = 350$A, and 6% duty factor.

	P_{diss} **per cusp magnet [W]**	P_{diss} **per filter magnet [W]**
Without ferrites	78	40
With ferrites	34	36

For the ferrites, ohmic losses do not pose a problem since their low conductivity effectively suppresses the generation of eddy currents. However, the saturation of the ferrites may lead to significant hysteresis losses. A rough estimation of the losses can be obtained by the Steinmetz equation,

$$P = k \cdot f^\alpha \cdot B_{max}^\beta \cdot V \qquad (2)$$

where α, β and k are material dependent parameters and B_{max} and V the maximum B-field and ferrite volume, respectively. The assumptions $\alpha=1$, $\beta=2$ and $k=\pi\varphi_m/(\mu_0\mu_R)$, where φ_m is the loss angle at $f = 2$MHz, yields losses in the range of 10W per ferrite, which would be sufficient to heat them to temperatures close to the Curie temperature ($T_C = 125$°C for 8C11). Correspondingly, the shielding effect by the ferrites may be lower or even diminishing.

The results of the Opera simulations triggered the design of a Cu magnet cage shielding the magnets from the RF field. A photograph of the magnet cage is shown in Fig. 8.

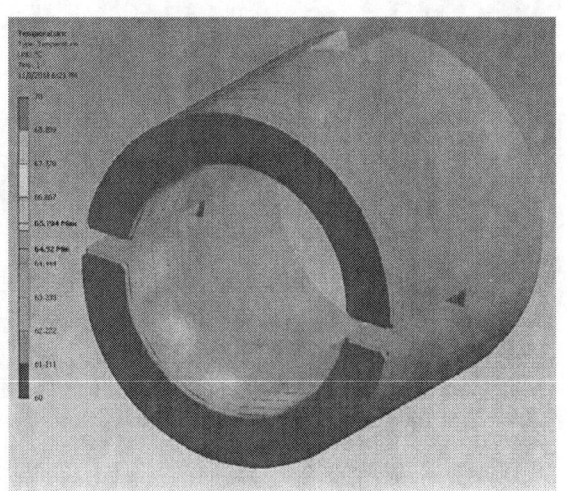

FIGURE 8. Cu magnet cage of the SPL plasma generator. Left: photograph taken before installation. Right: thermal model.

The cusp magnets are contained in rectangular holes with a cross-section of 14.2 × 10.5mm. Small Cu wedges at the back end of the holes protect the magnets from falling out during handling. Plastic screws for magnet alignment assure that the magnetic cusp field is as homogeneous as possible. Additional threads on the outside may be used to couple a cooling circuit onto the magnet cage, in case that active cooling is deemed to be necessary during operation. However, a combination of Opera RF power loss calculations and ANSYS thermal modelling for a worst case scenario (100kW RF power, 6% duty cycle, no ferrites) showed that even without active cooling, the temperature of the cage is expected to stay below 70°C, equivalent to a modest reduction in cusp magnet remanence of < 5%. Pulsed power deposition was not taken into account since the temperature rise time is » the pulse length of 1ms.

In the case of the filter magnets, the design of the AlN magnet holder prevented the production of a dedicated Cu piece. Instead, each filter magnet is contained in a box made of 0.5mm thick Cu (Fig. 9) which is sufficient to reduce RF losses in the magnets by several orders of magnitude.

FIGURE 9. View of the extraction collar, showing the AlN magnet holder and a Cu-shielded filter magnet.

EXPERIMENTAL DETERMINATION OF RF HEATING

In order to study the power dissipation by eddy currents in the magnets experimentally, a test stand has been fabricated (Fig. 10). The test stand contains a number of slots for shielded and unshielded cusp and filter magnets and allows also the insertion of ferrites to assess their influence on magnet heating and to estimate hysteresis losses. The RF field will be produced by an antenna of the same size and number of windings as the one used in the SPL plasma generator. Except for the base plate and the Faraday cage, all test stand components are made of poly-methyl methacrylate (PMMA) in order to avoid unwanted shielding effects. The power dissipation will be assessed by temperature measurements corrected for the emissivity of the materials with a thermal camera.

FIGURE 10: Layout of the test bench for power dissipation measurements. 1: cusp magnet. 2: antenna. 3: ferrite. 4: filter magnets. 5: Faraday cage.

CONCLUSIONS AND OUTLOOK

A novel Cs-free H⁻ plasma generator for SPL has been successfully designed and is now being commissioned. First plasma production was achieved at the end of October 2010. A dodecapole cusp magnetic configuration employing offset-Halbach type magnets has been designed to maintain the required magnetic field strengths in spite of the increased distances from the magnets to the ceramic plasma chamber wall. A Cu magnet cage and Cu boxes have been designed to shield the cusp and the arc-shaped filter magnets from RF-induced ohmic heating. A test stand will serve for experimental validation of the simulated results.

ACKNOWLEDGMENTS

The authors would like to thank S. Bertolo, D. Faircloth, C. Mastrostefano, O. Midttun, M. O'Neil, and J. Sanchez Arias for their help and support.

REFERENCES

1. D. Küchler et al., *Rev. Sci. Instr.* 79, 02A504 (2008).
2. L. Arnaudon et al., CERN Report No. CERN-AB-2006-084 (2006).
3. D. Faircloth et al., *Rev. Sci. Instr.* 81, 02A722 (2010).
4. M. Baylac et al., CERN Report No. CERN-AB-2006-006 (2006).
5. J. Lettry et al., *Rev. Sci. Instr.* 81, 02A723 (2010).
6. J. Lettry et al., this volume
7. M. Paoluzzi et al., this volume
8. R. F. Welton et al., *Rev. Sci. Instr.* 81, 02A727 (2010).

CERN LINAC4 H⁻ Source and SPL plasma generator RF systems, RF power coupling and impedance measurements[1]

M. M. Paoluzzi, M. Haase, J. Marques Balula, D. Nisbet

CERN Rte de Meyrin, 1200 Geneva - Switzerland

Abstract. In the LINAC4 H⁻ source and the SPL plasma generator at CERN, the plasma is heated by a 100 kW, 2 MHz RF system. Matching of the load impedance to the final amplifier is achieved with a resonant network. The system implements a servo loop for power stabilization and frequency hopping to cope with the detuning effects induced by the plasma. This paper provides a detailed description of the system, including the pulse rate increase to 50 Hz for use in the SPL plasma generator. The performances, measurements of RF power coupling, contribution of the plasma to the impedance as well as first operation are reported.

Keywords: Plasma heating, RF heating, SPL, Plasma generator, RF power coupling, Plasma impedance, H⁻ source, Plasma detuning.
PACS: 52.50.Qt

GENERAL DESCRIPTION

In the LINAC4 H⁻ source and the SPL plasma generator at CERN, the plasma is heated by a 100 kW, 2 MHz, 10 % bandwidth RF system. The block diagram is shown in fig. 1. The system works in pulsed mode delivering bursts of up to 2 ms length. Repetition rate is 2 Hz for the LINAC4 H⁻ source and 50 Hz for the SPL plasma generator. The RF power is provided by a vacuum tube final stage requiring a 50 Ω load. The high voltage power supplies and overall cooling differ in the two applications and are adapted to the different repetition rates.

In the plasma chamber, the power is inductively coupled to the hydrogen gas or plasma using a coil surrounded by ferrites and located inside a multi-cusp arrangement of permanent magnets. In this paper the EM equivalent of this assembly is called antenna. The losses in the antenna are mainly due to the coil resistivity but a contribution from the ferrites and magnets introduces some non-linearity at high power. This saturation effect affects both the losses and the antenna inductance. When

---

[1] This project has received funding from the European Community's Seventh Framework Program (FP7/2007-2013) under the Grant Agreement no 21211

driven without gas in the plasma chamber the H⁻ source antenna has the behavior shown in fig. 2.

In the LINAC4 H⁻ source the antenna is floating and biased at 45 kV while the rest of the RF system is at ground level. The galvanic isolation required to transfer the RF power is provided by a wideband RF transformer with a 1÷1 transformation ratio. No such transformer is required in the SPL plasma generator because there is no beam extraction and thus no HV bias.

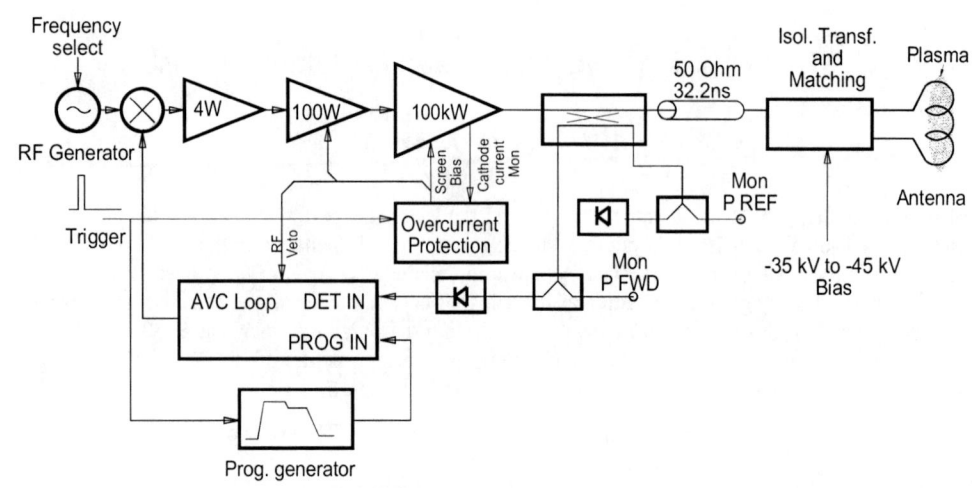

FIGURE 1. RF system block diagram.

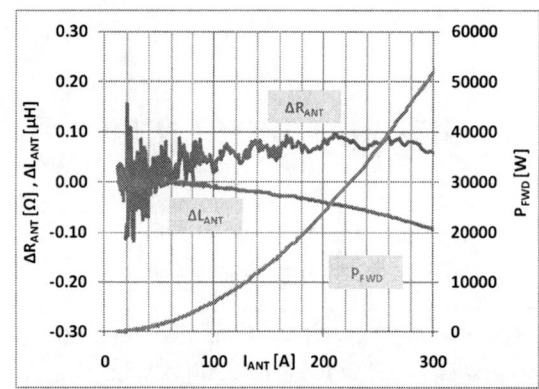

FIGURE 2. Saturation effects in the H⁻ source antenna (see table 1 for reference values)

Both the real and imaginary parts of the antenna impedance drastically change when the plasma is excited. Coupling the energy into the antenna and maximizing its transfer to the plasma, implies a detailed knowledge of the system and its dynamical behavior as well as means of measuring the plasma induced perturbations. The core of the measuring system is a high directivity (>30 dB) directional coupler present at the power stage output. It provides measurement of the forward and reflected signals that allow derivation of the main plasma RF characteristics.

Impedance matching to the 50 Ω amplifier output is achieved with a resonant network which has a different response with and without plasma. Careful setting of the tuning conditions and frequency hopping allows compensating the detuning effect induced by the plasma and optimizing the system response. Switching between two predefined frequencies is driven by a plasma light detector.

Agile power control along the pulse is provided by an automatic voltage control system (AVC). The forward signal from the directional coupler is detected using a logarithmic detector and used to stabilize the forward power/voltage according to a dedicated program. Figure 3 plots the programmed and detected signals as well as 50 Hz pulsing of the amplifier working on a resistive 50 Ω load. Scaling is 10 dB/V with 0 V≡100 V_{peak} or, in a 50 Ω system, 0 V≡100 W. Figure 4 shows the forward voltage monitor sine wave signal at 100 kW (higher harmonics are filtered out).

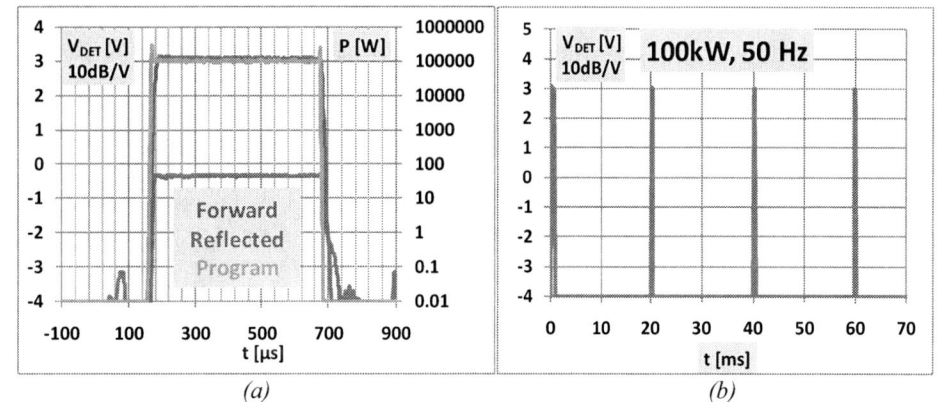

FIGURE 3. Servo loop response and 50 Hz pulsing.

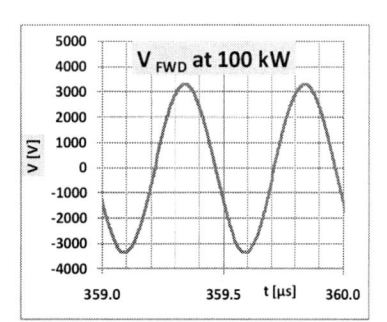

FIGURE 4. 2 MHz, 100 kW forward signal (filtered).

FIGURE 5. SPL Plasma Generator RF system.

ANTENNA, MATCHING CIRCUIT AND CONNECTION LINE

As discussed above, the antenna is basically a coil and thus characterized by its own inductance and losses (L_{ANT}, R_{ANT}). In series with this impedance we can model the plasma contribution as additional inductance and resistance (L_{PLASMA}, R_{PLASMA}). To provide adequate loading to the final amplifier the sum of the two individual contributions must be transformed into $R_M = 50\,\Omega$ (see fig.6). As the system is almost working at fixed frequency this can be achieved adding a series and a parallel capacitor (C_S, C_P) provided the circuit inductance exceeds a given minimum value L_{Tmin}. Good matching conditions can only be obtained either with or without plasma. To make the circuit easily tuneable an additional lumped coil is added (L_S, R_S). Figure 6 shows the overall circuit and the corresponding relations.

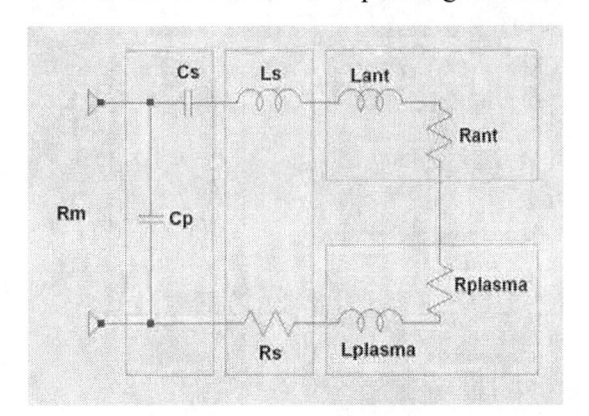

$$L_T = L_S + L_{ANT} + L_{PLASMA}$$

$$R_T = R_S + R_{ANT} + R_{PLASMA}$$

$$C_P = \frac{\sqrt{\frac{R_M}{R_T} - 1}}{\omega_0 \cdot R_M}$$

$$C_S = \frac{1}{\omega_0 \cdot (\omega_0 \cdot L_T - R_T \cdot \sqrt{\frac{R_M}{R_T} - 1})}$$

$$Q \approx \frac{\omega_0 \cdot L_T}{R_T} \qquad L_{Tmin} = \frac{R_T \cdot \sqrt{\frac{R_M}{R_T} - 1}}{\omega_0}$$

FIGURE 6. Antenna and matching circuit.

The circuit is then connected to the isolation transformer (H⁻ source only), the transmission line and the final amplifier output. The component values of the H⁻ source and SPL plasma generator matching networks are listed in table 1.

TABLE 1. Measured values

Parameter	H⁻ Source	SPL Plasma generator
Antenna dimensions (OD, ID, Length)	62.5/58.5/28.5mm	79/71/28.5mm
Antenna turns number	5.5	5.5
L_{ANT}	3.2 µH	2.53 µH
R_{ANT}	0.4 Ω	0.2 Ω
L_S	1.98 µH	2.27 µH
R_S	0.5 Ω	0.16 Ω
C_S	1.61 nF	1.505 nF
C_P	6.5 nF	6.3 nF
Q	~70 no plasma	~170 no plasma
	~30 with plasma	~150 with plasma
Cable Z_0	50 Ω	50 Ω
Cable electrical length	32.2 ns	30.0 ns

In the H⁻ source and in absence of plasma ($L_{PLASMA}=0$ and $R_{PLASMA}=0$), the circuit resonates at ~1.945 MHz where $R_M \approx 180\ \Omega$. When plasma develops, the resonance shifts towards higher frequency and the additional losses reduce R_M close to *50 Ω*. The new matching situation depends of course on the plasma characteristics which, in turn, depend upon many parameters, i.e. gas pressure, deposited power, time. Detailed measurements will be developed in the following section.

MEASUREMENT OF L_{PLASMA}, R_{PLASMA} AND DEPOSITED POWER

The directional coupler available at the final stage output provides a calibrated sample of the forward and reflected waves. This information allows calculating the load impedance at the coupler plane from which L_{PLASMA} and R_{PLASMA} can be derived. The circuit is then completely known and can be solved for voltages, currents and all other relevant parameters.

For this V_{FWD}, V_{REF} and a signal proportional to the light emitted by the plasma are acquired at high sampling rate. A program[2] then extracts the peak values of the two sine waves and the phase difference from which all the calculations will be done. The light information has an arbitrary scale.

Figure 7 shows typical measurement taken on the H⁻ source system (no extraction voltage was applied). The ignition system is not activated so that the plasma is only started by the RF heating. The left plot shows the voltages and phase fitted from measured data. A sharp transition can be seen at ~ 250 μs when the light signal rises at plasma formation. The right plot shows the corresponding change of the plasma resistance and inductance as well as power dissipated by the plasma. The plasma inductance has a negative value. This corresponds to a reduction of the antenna inductance due to the change of the electrical characteristics of the volume inside the antenna coil. The low conductivity of this volume changes when filled with plasma and conductivity strongly increases. From the right plot it is interesting noticing that a negligible amount of RF power is transferred to the gas present in the plasma chamber before plasma formation.

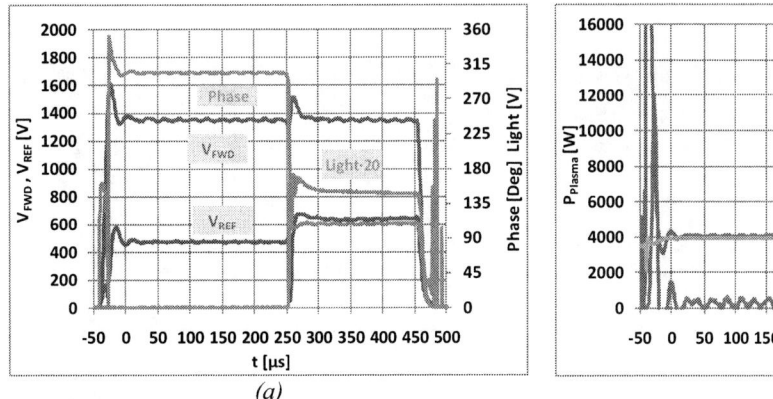

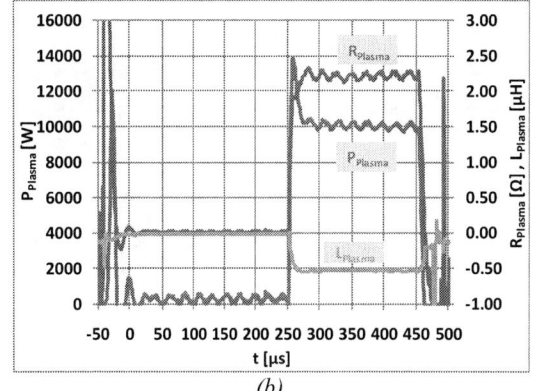

(a) *(b)*
FIGURE 7. Typical response without frequency change

---
[2] Thanks to Claus Schmitzer for writing the code.

Figure 8 and 9 show the effect of shifting the frequency at the plasma formation time. In this case the different conditions existing inside the plasma chamber let plasma formation take place much earlier than in the previous case. The main differences are a higher gas pressure and the activated ignition system. Figure 8 shows how the plasma formation generates a transient that brings the reflected voltage through a minimum. In the left plot no frequency shift is applied and the reflected voltage rises again. On the right plot the situation is stabilized by shifting the working frequency from 1946 kHz to 2000 kHz. Despite the relatively small differences of L_{PLASMA} and R_{PLASMA} in the two cases (Fig. 9) the different matching conditions are such that only with good matching the power amplifier can deliver the full power.

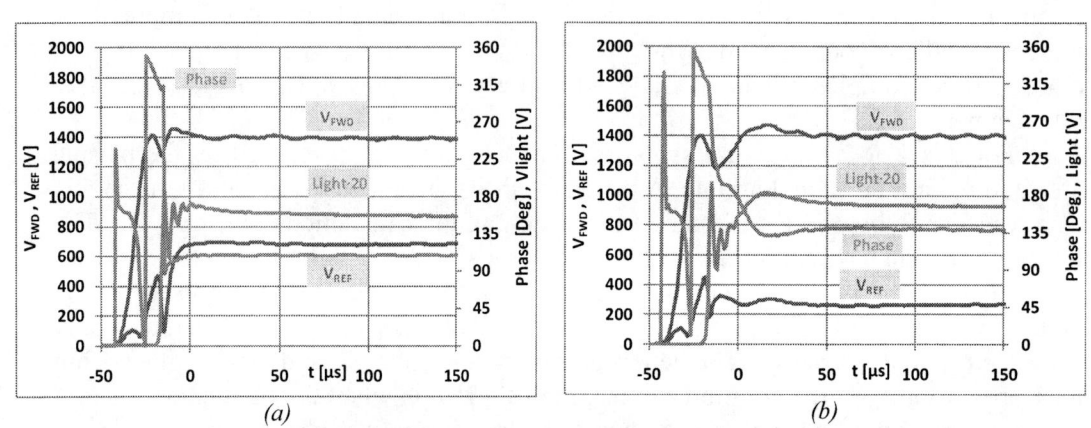

FIGURE 8. Measured V_{FWD}, V_{REF} and Light
Left: fixed frequency 1946 kHz – Right: frequency change from 1946 kHz to 2000 kHz

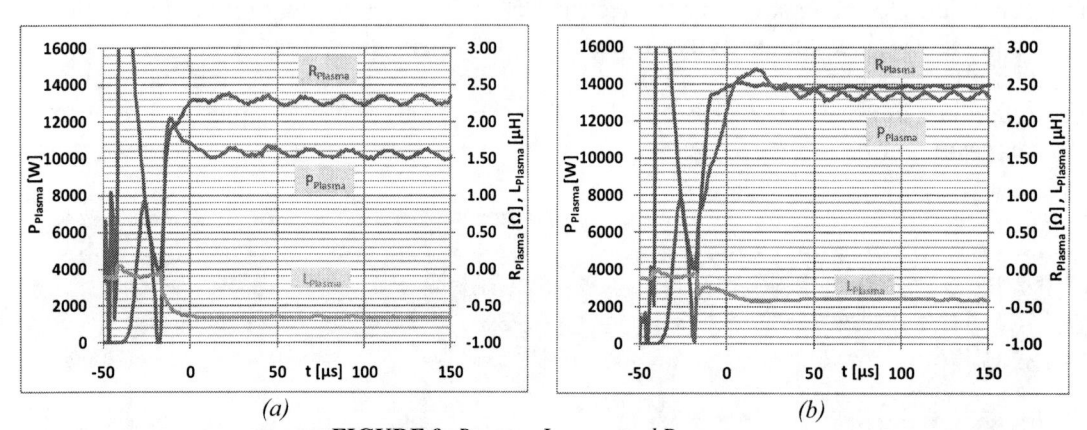

FIGURE 9. R_{PLASMA}, L_{PLASMA} and P_{PLASMA}
Left: fixed frequency 1946 kHz – Right: frequency change from 1946 kHz to 2000 kHz

Making use of the system ability to start the process with constant conditions and thus change the forward power once a stable situation has been reached, a set of measurements has been done to establish various parameters as function of RF power. Figure 10 shows a typical power modulation program (with frequency hopping). The results plotted in figure 11 show L_{PLASMA}, R_{PLASMA} and the light intensity measured at time 300 μs. As expected they exhibit good linearity when plotted vs. the antenna current.

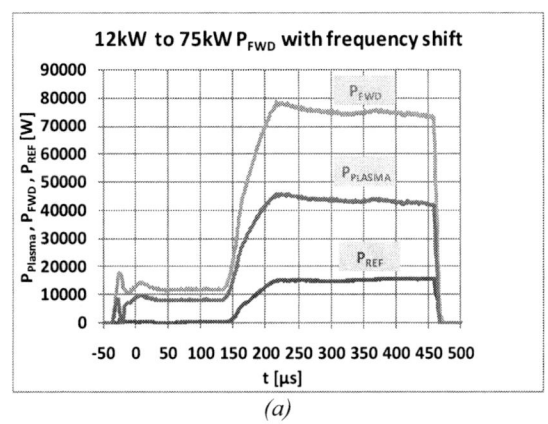

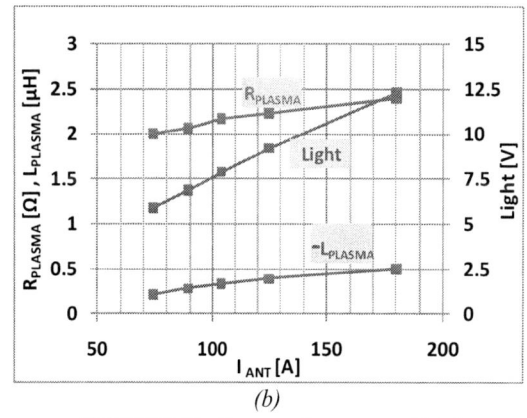

(a)
FIGURE 10. Typical power modulation program.

(b)
FIGURE 11. R_{PLASMA}, L_{PLASMA} and Light vs. Antenna current

CONCLUSIONS

RF systems for the LINAC4 H⁻ source and the SPL plasma generator at CERN have been designed, built and commissioned. They showed good reliability and ease of operation. In particular the power stabilization servo loop, the frequency hopping and the ability of working on very mismatched load proved very useful to cope with the wide load impedance changes occurring at plasma formation. A sophisticated plasma impedance measurement system has also been implemented. It was very useful for the design of the impedance transformation network. The ability of dynamically measuring the RF power coupling to the plasma as well as other plasma characteristics will provide valuable help during the source and plasma generator development period and interesting information during operation.

Operation of Negative Ion Sources at the Cooler Synchrotron COSY/ Jülich

R. Gebel, O. Felden, and R. Maier

Forschungszentrum Jülich GmbH, IKP, 52425 Jülich, Germany
Jülich Center for Hadron Physics (JCHP)

Abstract. The Institute for Nuclear Physics at the Forschungszentrum Jülich is dedicated to fundamental research in the field of hadron, particle and nuclear physics. Main activities are the development of the High Energy Storage Ring for the Facility for Antiproton and Ion Research at Darmstadt and the operation and improvement of the cooler synchrotron COSY at Jülich. The injector, a cyclotron with polarized and unpolarized H⁻ and D⁻ sources, has exceeded 7000 hours availability per year, averaged over the last decade. Work in progress is the investigation of production, extraction and transport of the low energy 4.5 keV/u ion beams. A brief overview of the activities is presented.

Keywords: polarized beams; cyclotrons; synchrotrons; polarized ion sources; negative ion sources; atomic and molecular beam sources and techniques; beam extraction.
PACS: 29.20.Hm, 29.20.20.Lq, 29.25.Lg, 29.25.Ni, 29.27.Hj, 39.10+j, 41.85.Ar.

INTRODUCTION

The Institute for Nuclear Physics (IKP) at the Forschungszentrum Jülich (FZJ) is dedicated to fundamental research in the field of hadron, particle and nuclear physics. The aim is to study the properties and behavior of hadrons in an energy range that resides between the nuclear and the high energy regime. Main activities are the development of the high energy synchrotron ring HESR, operation and improvement of the Cooler Synchrotron COSY-Jülich [1], with the injector cyclotron JULIC [2-4], as well as the design, preparation, and operation of experimental facilities at this large scale facility, and theoretical investigations accompanying the scientific research program.

The accelerator HESR, part of the GSI FAIR project [5], a synchrotron and storage ring like COSY, is dedicated to the field of high energy antiproton physics with high quality beams over the broad momentum range from 1.5 to 15 GeV/c. The mission is to explore the research areas of hadron structure and quark-gluon dynamics. An important feature of the new facility is the combination of phase space cooled beams with internal targets which opens new capabilities for high precision experiments. The tools to reach the required quality are tested at COSY. The cooler synchrotron COSY offers excellent research opportunities for hadron physics experiments and for essential preparatory studies for the machine development of HESR. The 2 MeV electron cooler project for COSY is funded and is expected to boost the luminosity in the presence of beam heating by high density internal targets. In cooperation with the Budker Institute of Nuclear Physics, BINP Novosibirsk (Russia), the manufacturing has been started. The straight COSY target section has been prepared to insert the cooler at the former position of the accelerating rf cavity [6]. Theoretical investigations of stochastic momentum cooling for the HESR clearly reveal that the strong mean energy loss induced by the interaction of the beam with an internal pellet target can not be compensated by cooling alone. At COSY stochastic cooling and internal targets are available similar to those which will be operated at the HESR. A barrier bucket cavity is routinely in operation. The COSY machine is exquisitely suited for beam dynamic experiments in view of the HESR. Important feasibility studies to compensate the large mean energy loss induced by an internal pellet target similar to that being used by the PANDA experiment at the HESR with a barrier bucket cavity (BB) have been carried out [7, 8].

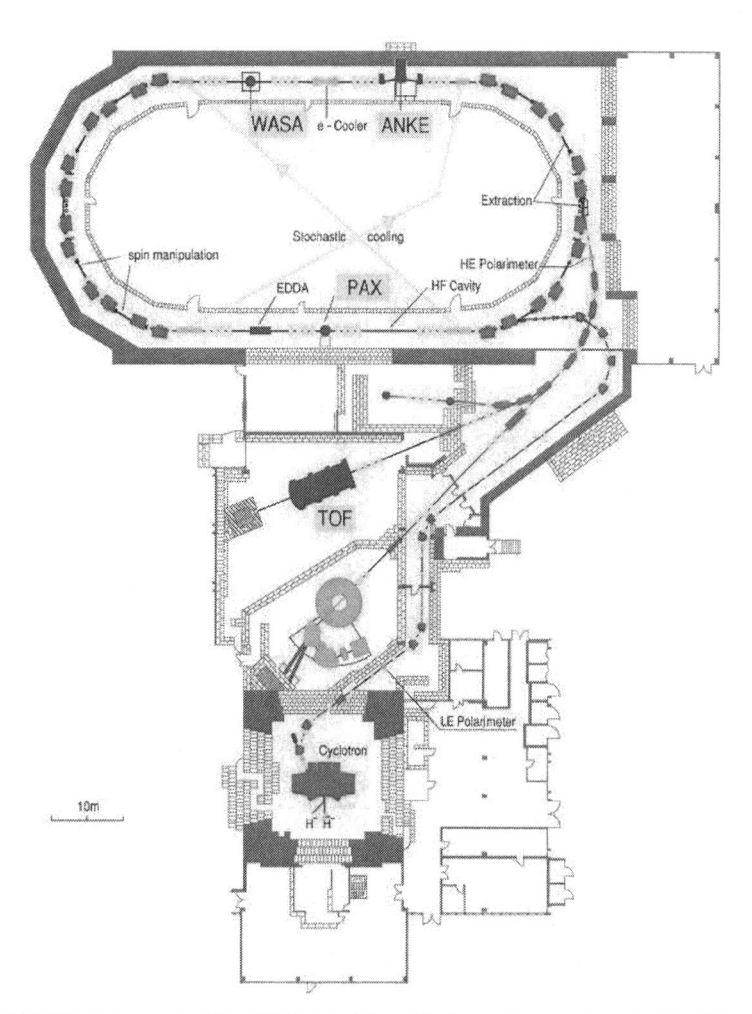

FIGURE 1. Layout of the COSY facility with the main experimental facilities.

The cooler synchrotron and storage ring COSY delivers unpolarized and polarized beams of protons and deuterons with momenta up to 3.7 GeV/c for three internal experiments — ANKE, PAX and WASA — and one experiment — TOF — at an external target position. All four detection systems are operated by large international collaborations [9]. ANKE, Apparatus for Studies of Nucleon and Kaon Ejectiles, is a large acceptance forward magnetic spectrometer at an internal target station in the COSY ring. First double polarized experiments have been performed with a polarized internal target with a storage cell. The 4 pi spectrometer for neutral and charged particles, WASA, Wide Angle Shower Apparatus, is operated also with the internal COSY beam. The barrier bucket cavity of COSY was successfully used to optimize the compensation of the main energy loss, which is introduced by the WASA frozen-pellet target.

In addition, the unique COSY capabilities are used by the SPIN@COSY-, dEDM- and PAX-collaborations to investigate spin-manipulations, to build a dedicated EDM-storage ring experiment, and to prepare experiments on polarization build-up in storage rings. An advanced grant of the European Research Council has been obtained for the polarization of anti-protons. Additionally installed quadrupoles in the mid of the straight target section provides the needed low beta values for the newly installed former HERMES polarized target as a polarized internal gas target. This is an important step to provide polarized antiprotons for FAIR.

INJECTOR AND ION SOURCE OPERATION

The operation and development of the accelerator facility COSY is based upon the availability and performance of the isochronous cyclotron JULIC, as the pre-accelerator, and its ion sources. The cyclotron has been commissioned in 1968 and exceeded 241000 hours of operation in 2010. A fraction of the run time since 2000 is shown in the upper part of Fig. 2.

The lower part shows the distribution of the available beam species, protons and deuterons, polarized or unpolarized. Since 1996 the cyclotron delivers negative light ions for charge exchange injection into the synchrotron. The injector shows availability for synchrotron operation in excess of 7000 hours a year, averaged over the last decade. Beam times for experiments at COSY have been delivered successfully for over 90 % of the scheduled periods.

Two filament driven volume sources deliver unpolarized H⁻ or D⁻ alternately. A charge exchange colliding beams source provides polarized H⁻ or D⁻ [10-15]. For the ion sources uninterrupted operation on the order of 4 weeks is standard and short swap is realized by providing two operational ion sources. Beam losses for the 4.5 keV/u ion beam from source extraction to first acceleration of the ions have been investigated. Based on the investigation, the ion optical elements and vacuum chambers have been modified and the vacuum condition in the low energy beam line has been improved significantly. This enables a change from polarized to unpolarized ions from one synchrotron injection to another on the order of one second. The switching process from one source to another is depicted in Fig. 3.

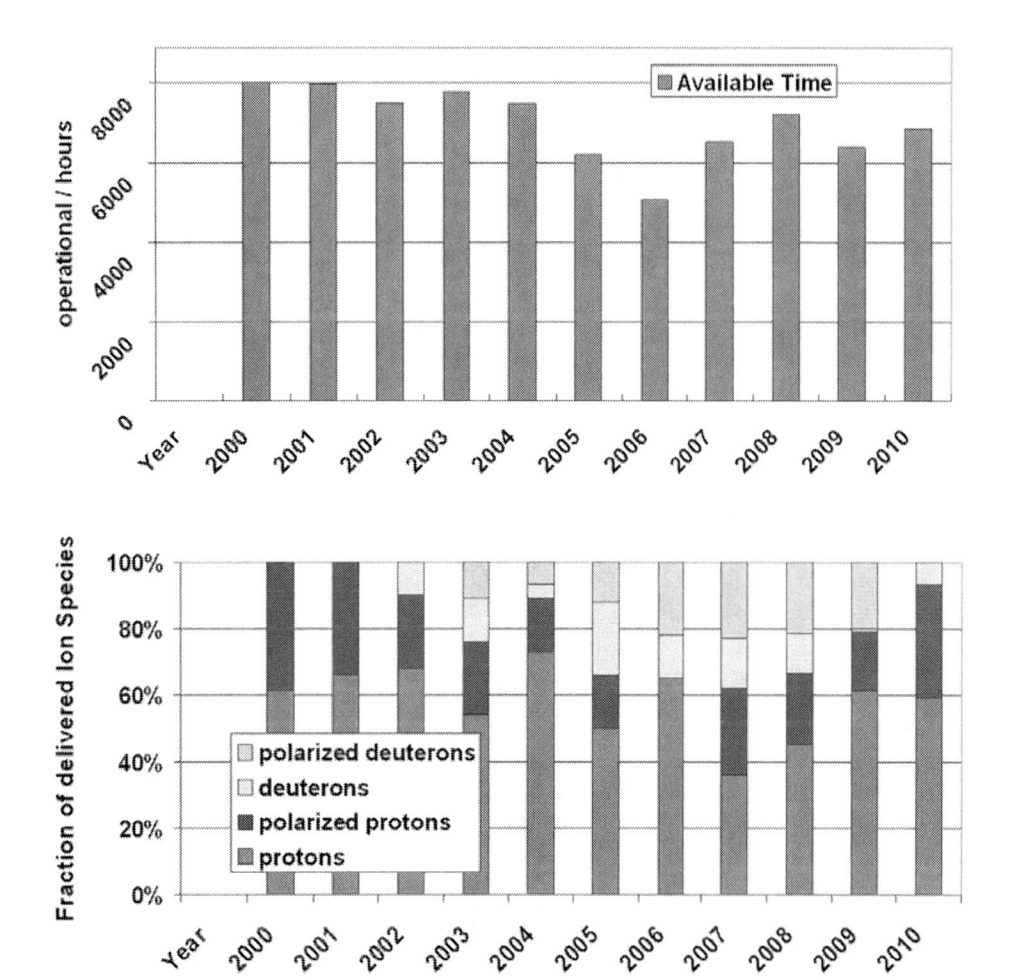

FIGURE 2. Beam usage statistics.

The power supplies for the two ion sources provide parallel operation. For each of the sources, the setup of the power supplies for the source beam line elements was stored by the computer control system. This feature was used to test the fast change of source beam line settings. To switch the settings for 36 power supplies took less then a second. The time constant was dominated by the response of the massive magnetic dipole at the junction of the three source beam lines. This feature is essential for precision experiments with polarized particles, which need a real unpolarized reference, not available from a polarized source, as well as for a very precise momentum reference from forced polarization losses by a resonance method [16].

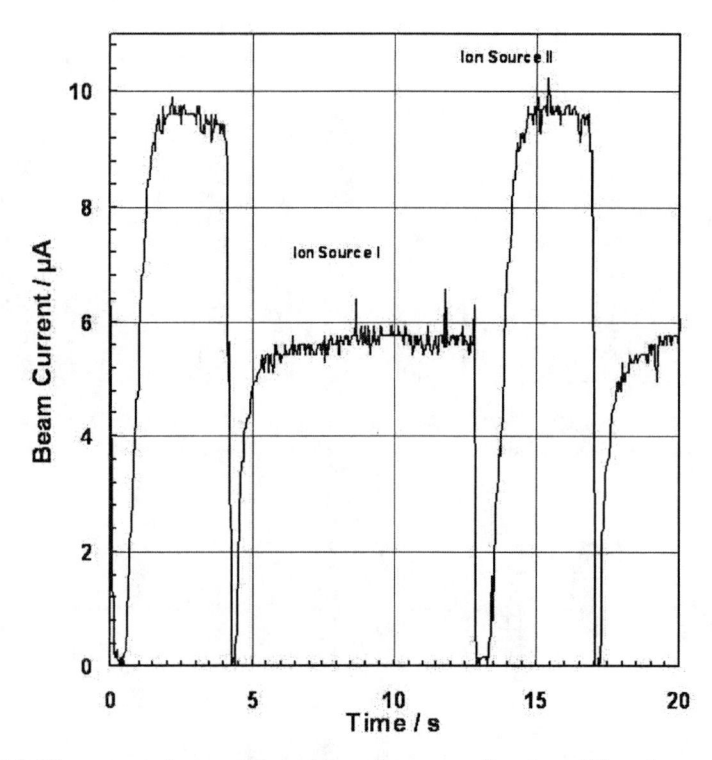

FIGURE 3. The extracted current from the cyclotron as a function of time demonstrates the fast switching between two unpolarized ion sources and source beam line settings.

Table 1 reflects the high availability of the injector from 2007 to mid 2010. Only about 3 % of the scheduled beam time could not be provided to experiments, due to failures at the injector. Excluded were short events, like spark recovery, which are fixed automatically or by the operator. The most common reasons for such events were power drops, shortage in water cooling, septum exchange and failures in the radio frequency (rf) subsystems of the cyclotron.

TABLE 1. Runtime and failures of the COSY injector including ion sources.

Year	Available / h	Unavailable / h
2007	6590	300
2008	7303	156
2009	6716	164
2010 (until August)	3621	40

IMPROVEMENT OF POLARIMETRY

The operation of the polarized ion source includes beam delivery to experiments and the preparation and diagnostics of the polarized beam states. Obviously the measurement of the polarization prior to beam delivery is a very useful option. During the past years the demand for precise polarization measurements, especially for deuterons with various polarization states was increasing. Another motivation has been an observed reduction of proton polarization to about 30 % for experiments at the synchrotron at maximum beam momentum. To exclude effects in the vicinity of the cyclotron and in the path to the synchrotron additional polarimeters have been realized and installed at the start and end of the transfer beam line. No significant polarization losses have been found there. Spin precession in the source beam line and the working point of the synchrotron with respect to the magnet's timing have been identified as the main causes for the polarization reduction.

Since 2009, data from the routine polarimeter, in the beam line between the cyclotron and the synchrotron, can be collected by controlled transmission of the ion beam from the source to the polarimeter. The beam is dumped behind the polarimeter. This monitoring feature enables the judgment of stability up to the polarimeter. Figure 4 shows an example for polarized deuterons. The data is collected over a period of 11 days and demonstrates the overall stability behind the injector.

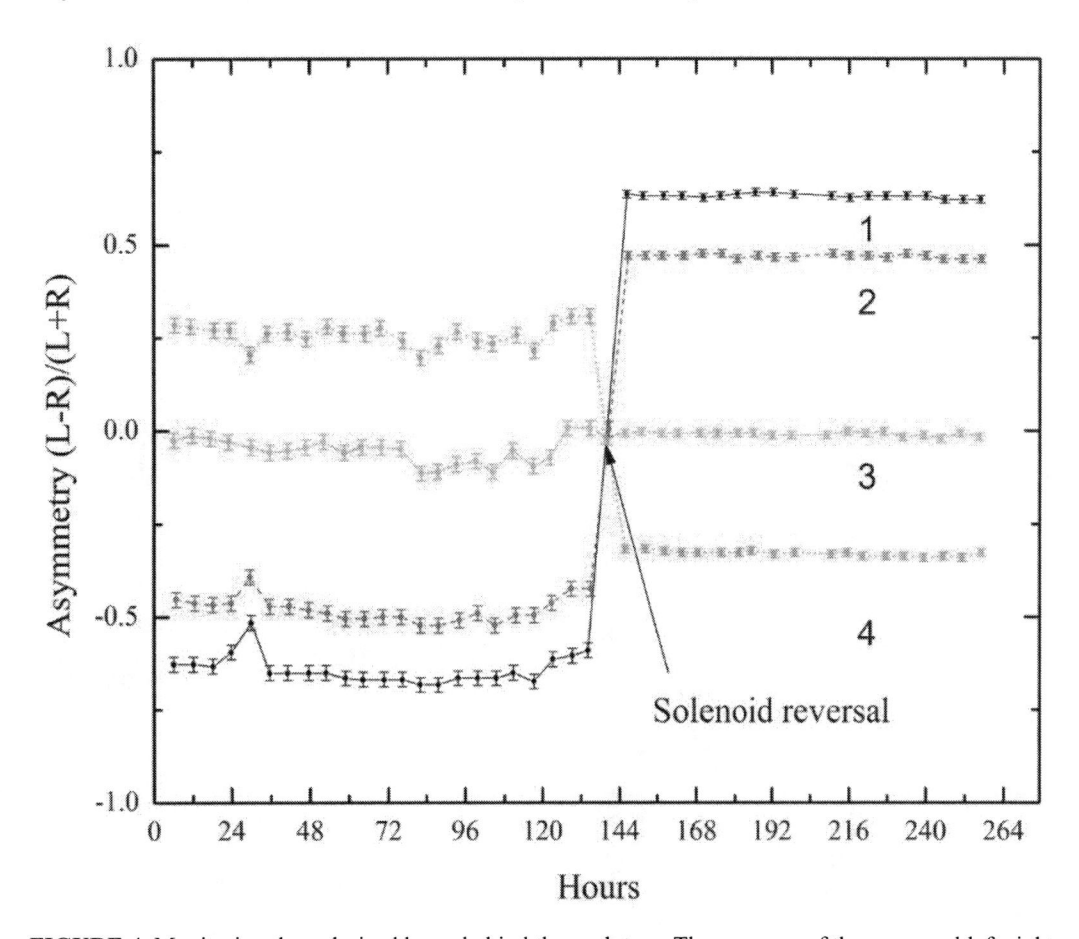

FIGURE 4. Monitoring the polarized beam behind the cyclotron. The averages of the measured left-right asymmetries from four different states are plotted.

The available expertise at our institute allowed us to realize a lamb shift polarimeter to be used for beam preparation. The beam transport from the polarized source, one floor level below, to the new polarimeter had been a challenge, because the deflectors needed to be installed in the gaps of the existing dipole magnets. The transfer line is depicted in Fig. 5. Two spherical electrostatic 135° deflectors have been designed and installed inside the existing 45° dipole magnets. The beam line is commissioned and beam has been delivered to the polarimeter with an acceptable transmission of about 50 %. The polarimeter is now accessible and operation is possible during unpolarized beam delivery.

The current status of the polarized ion source and polarimetry at the injector were described in a contribution to the workshop for Polarized Sources, Targets and Polarimetry at Ferrara, Italy [17].

IMPROVEMENT OF SOURCE OPERATION

Due to changing installations at COSY the requirements and expectations for the ion sources also have to be adjusted. The operation with electron-cooled beams, with storage-cells and low-beta-sections inside COSY needs highest possible intensity in short pulses and fast change between polarized and unpolarized ions without large losses. The possible causes for beam loss have been investigated. E.g. the extraction from the ion sources, the transmission losses by poor vacuum conditions and an improper beam alignment have been identified as main contributors to limited performance.

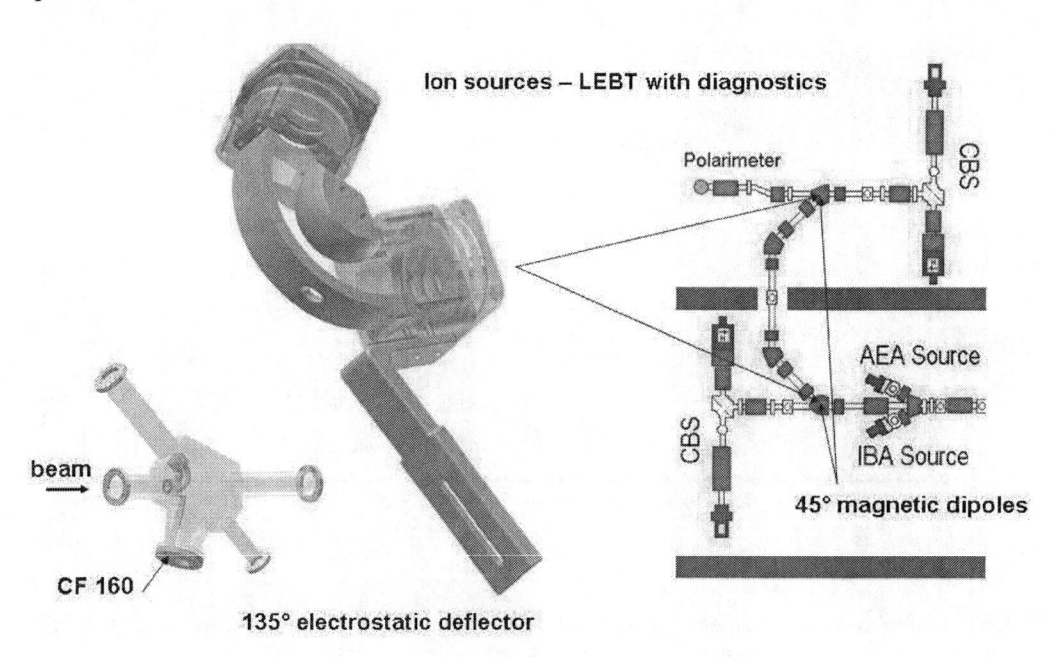

FIGURE 5. The new 135° deflectors and a fraction of the low energy transfer beam line.

Extraction from the unpolarized sources

The operation of the two filament driven cusp sources for H⁻ requires stable and long term operation, this is realized by moderate settings with respect to discharge power and matched emittance. However, an increased gas flow increases in principle the performance and beam quality, also the extracted amount of electrons is reduced.

The operation of the unpolarized ion sources reveals additional limitations caused by the very low beam energy of 4.5 keV/u. These limitations became obvious during tuning for routine operation. A satisfactory matching and performance is obtained only at low intensities up to 0.3 mA extracted ion current. At higher intensities, the extracted beam becomes unstable and is truncated by the extraction elements. To understand this behavior, field and ion beam tracking simulations were started, using the program IES Lorentz-2EM [18]. A simplified model can reproduce the observed beam features qualitatively; the model neglects electrons moving along, magnetic filter fields and attenuation or space charge compensation by the residual gas. Figure 6 shows the results of simulation runs for the original arrangement of the ion source and a modified one. However, the simulation indicates measures, how to improve the performance of the ion sources: By matching the voltage settings of the extraction, the formerly earthed electrode and the einzel lens, the extracted beam current could be increased from 0.3 to 3 mA.

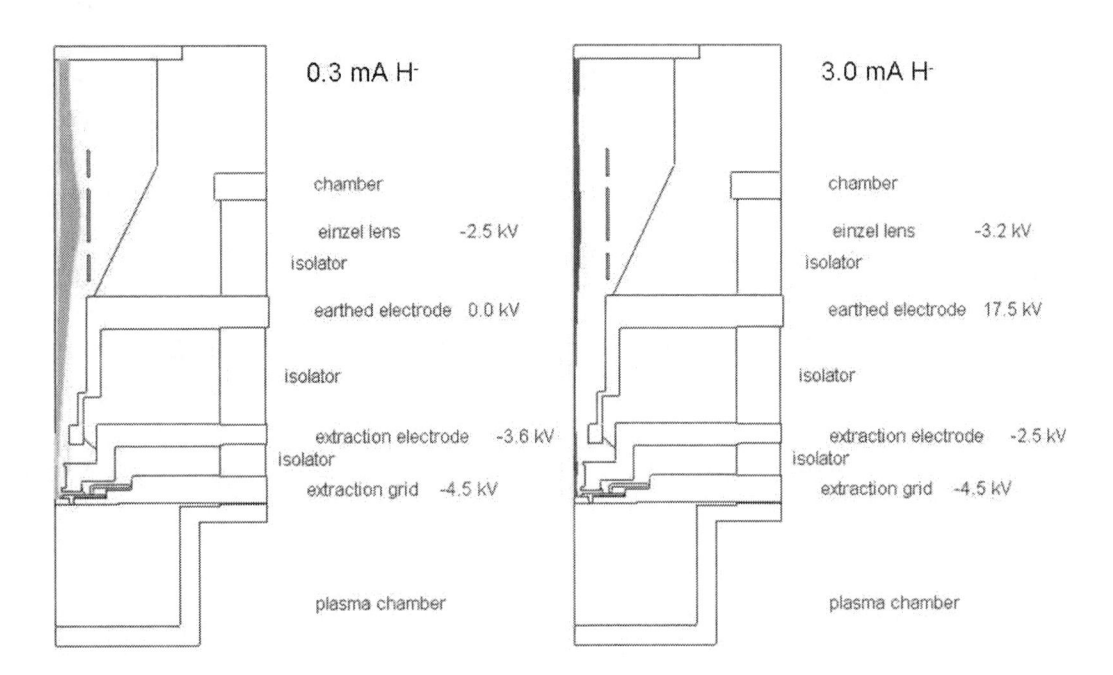

FIGURE 6. Simulated increase of the extracted beam intensity from 0.3 mA to 3.0 mA, by introduction of an additional isolator and matching the settings of the ion optical elements. These figures are results of the IES Lorentz-2EM [18] program, from a model with rotational symmetry and space charge calculation enabled.

Improved conditions for the beam transfer

For parallel operation the fast switch time for ion optical elements had been demonstrated. The unpolarized ion sources had been operated with low pumping speed and the separation of vacuum sections was realized by low conducting diaphragms. The present gas flow causes a beam loss to about 50 % of the available intensity for parallel operation of polarized and unpolarized sources. Figure 7 shows the layout of the beam line with its main elements from the ion sources to the cyclotron. The overall performance was reduced in principle by losses after extraction and during beam transport to the cyclotron. At a level of 1 mA extracted beam from an unpolarized source only 20 % H⁻ ions reached the last faraday cup before the cyclotron. The cyclotron's transmission is routinely on the order of 5 % and as a result the extracted beam intensity is about 10 μA. The transmission inside the cyclotron is reduced by stripping losses. During the last decade the vacuum situation has been improved significantly. The actual operating pressure is below $1 \cdot 10^{-7}$ mbar and the analysis of beam intensity versus pressure inside the cyclotron chamber reveals that no significant improvement is possible. Losses during acceleration are mainly caused by emittance and misalignment or coherent beam oscillations.

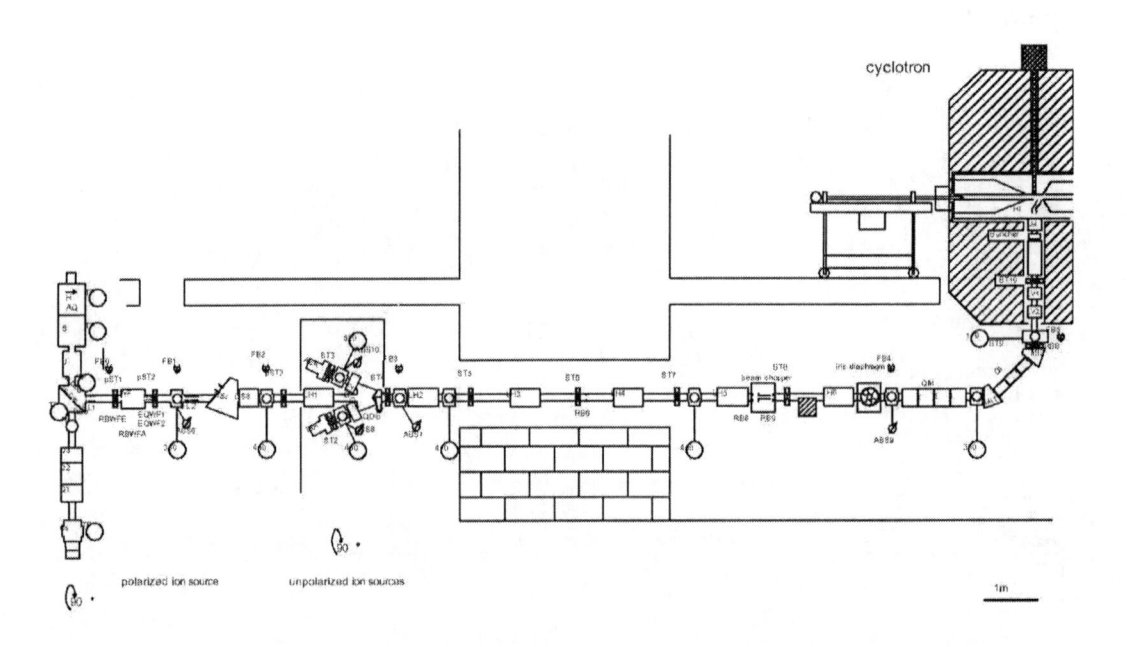

FIGURE 7. Layout of the beam line from the ion sources to the centre of the cyclotron.

New connection to the beam transfer line

The complete low energy beam transfer line (LEBT) with ion sources and the cyclotron is depicted in Fig. 7. The old set-up of the first section of the source beam line with the two unpolarized ion sources is compared to the new one in Fig. 8. The diagnostic tools used for routine operation are combined function beam stops with a viewer (KBr on Cu) and a Pearson 110A current monitor. An additional copper plate with a pepper pot structure in front of the first cup (FB3) in the straight section is used for alignment of the beams. The alignment of the beam is essential for a telescopic transport through the solenoid transfer. The beam intensity is controlled by an iris diaphragm for continuous operation and by a chopper during pulsed injector operation.

Beam diagnostics revealed a misalignment of the unpolarized beam behind the switching dipole, without a simple measure to correct it. The position of the electro magnetic switching dipole is critical. The existing chamber did not allow minimizing the deviation from the beam line axis due to the diameter of the support structure. The new four way cross piece, which was manufactured for improving the vacuum situation, allows the movement of the dipole towards the solenoid (LH2). Experimentally a new position has been found, which minimizes the deviation from the beam line axis with respect to position and angle. The behavior of the telescopic set of solenoid lenses (LH2-6) is now less sensitive and fewer corrections by steering magnets are needed. In parallel to the installation of pumping speed and the realization of increased conductance the four way cross piece has been exchanged. New vacuum pumps increased the pumping speed from 500 l/s to about 2000 l/s. The size of the chambers changed from 160 mm diameter to about 480 mm. A viewer and a current monitor have been included inside the new chambers. The viewer allows monitoring of the passing beam and to judge beam properties and focusing, before injection into the straight section. All viewers are equipped with CCD cameras, connected to a video distribution system. The beam can be judged by eye.

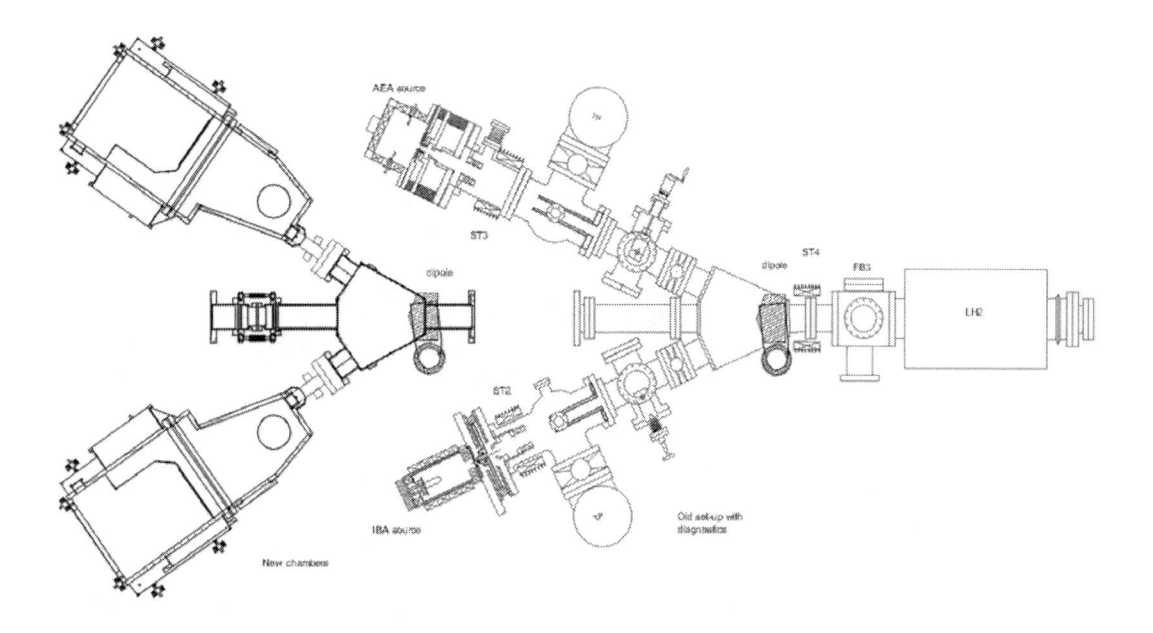

FIGURE 8. Ion source section of the low energy transfer beam line: The new chambers and four way cross with dipole (left) and the old set-up with ion sources, pumps and diagnostic elements.

CONCLUSION AND OUTLOOK

Besides improving the situation for routine operation with the new set-up for the unpolarized sources it is now possible to investigate losses caused by the very low beam energy of 4.5 keV/u. The first section of the beam transport behind the ion sources is dominated by space charge effects, stripping losses due to the relatively high pressure and an inherent misalignment of the beam. In order to investigate these effects and to improve the transmitted beam intensity new vacuum chambers and a four way cross piece have been installed. The grounded electrode of the ion sources has been isolated from the first chamber. It is possible to extract with higher field, and transport the accelerated beam for a certain distance. An additional electrostatic lens allows further matching to the beam line. These changes were initiated by early discussions about improvement of the overall transmission, and stimulated after some simulations of a simplified model of the source. The first operation revealed, as expected, an improved vacuum separation, increased intensities and improved beam alignment.

ACKNOWLEDGMENTS

The authors are grateful to L. Barion, A.S. Belov, R. Engels, G. d'Orsaneo and M. Westig for their help and advice, and thank the injector staff members R. Brings, A. Kieven and N. Rotert, as well as H. Hadamek and his staff of the IKP workshop for their effective support of our endeavors in keeping the ion sources and cyclotron operating reliably and to improving their performance.

References

1. R. Maier, Cooler Synchrotron "COSY – performance and perspectives", NIM A 390, (1997) p. 1.
2. W. Bräutigam et al; "H⁻ operation of the cyclotron JULIC as injector for the cooler synchrotron COSY-Jülich; Proceedings of Cyclotrons 1998, Caen, IOP (1999) p. 654.
3. W. Bräutigam et al.; "Extraction of D⁻ beams from the cyclotron JULIC for injection into the cooler synchrotron COSY". Proceedngs of Cyclotrons 2001, East Lansing, AIP 600, (2001) 123.
4. R. Gebel et al.; "Negative ion sources at the injector cyclotron JULIC for the cooler synchrotron COSY". Proceedings of Cyclotrons 2004, Tokyo, Particle Accelerator Society of Japan, (2005) p. 287.
5. http://www.gsi.de/fair.
6. J. Dietrich et al., "Status of the 2 MeV Electron Cooler for COSY-Jülich", Proceedings of COOL 2009, Lanzhou, China.
7. H. Stockhorst et al., "Compensation of Mean Energy Loss due to an Internal Target by Application of a Barrier Bucket and Stochastic Momentum Cooling at COSY", Proceedings of COOL 2009, Lanzhou, China.
8. R. Stassen et al., "COSY as ideal Test Facility for HESR RF and stochastic cooling hardware", Proceedings of PAC09, Vancouver, Canada.
9. http://www.fz-juelich.de/ikp/en.
10. W.Haeberli, *NIM*, 62, 1968, pp. 355.

11. P.D. Eversheim et al., "The Polarized Ion-Source for COSY", Proceedings of the SPIN 1996 Symposium, Amsterdam.
12. R. Weidmann et al., Rev. Sc. Instr., Vol. 67 p.II, 1996, pp. 1357.
13. R. Gebel et al., "Polarized negative light ions at the cooler synchrotron COSY/ Jülich", 10th Symp. on PNNIB, Kiev, Ukraine, AIP Conference Proceedings 763, New York, 2005, pp. 282.
14. O. Felden et al., Proceedings of 11th EPAC, Edinburgh, 2006, pp. 1705.
15. R. Gebel et al., "Tripling the total charge per pulse of the polarized light ion source at COSY/Jülich", 11th Symposium on PNNIB, Santa Fe, USA, AIP Conference Proceedings 925, 2007, pp. 105-113.
16. P. Goslawski et al., "High precision beam momentum determination in a synchrotron using a spin-resonance method", *PRST-AB*, 13, 022803 (2010).
17. O. Felden et al., „The COSY/ Jülich Polarized H⁻ and D⁻ Ion Source", Proceedings of PSTP 2009, Ferrara, 7.9.-11.9.2009.
18. http://www.integratedsoft.com

Ion Source Development For The Proposed FNAL 750keV Injector Upgrade

D.S. Bollinger

Fermi National Accelerator Laboratory, Box 500, Batavia, IL 60543 U.S.A.

Abstract. Currently there is a Proposed FNAL 750 keV Injector Upgrade for the replacement of the 40 year old Fermi National Laboratory (FNAL) Cockcroft-Walton accelerators with a new ion source and 200 MHz Radio Frequency Quadruple (RFQ). [1] The slit type magnetron being used now will be replaced with a round aperture magnetron similar to the one used at Brookhaven National Lab (BNL). Operational experience from BNL has shown that this type of source is more reliable with a longer lifetime due to better power efficiency [2]. The current source development effort is to produce a reliable source with >60 mA of H- beam current, 15 Hz rep-rate, 100 μs pulse width, and a duty factor of 0.15%. The source will be based on the BNL design along with development done at FNAL for the High Intensity Neutrino Source (HINS) [3].

Keywords: Ion Source, Magnetron, RFQ, HINS, Cockcroft-Walton
PACS: 29.25.Ni

INTRODUCTION

The Fermilab preinjector system consists of two Cockcroft-Walton accelerators that were installed in 1968. They have been a reliable source of H- ions since 1978 [4]. The reliability stems from having redundancy by having two Cockcroft-Walton accelerators and the help of FNAL skilled technicians that maintain them. However, with the retirement of critical personnel and aging parts that are no longer produced, the possibility of significant downtime has increased. To address these problems, there is a proposed plan to replace the existing preinjectors with a round aperture magnetron, a low energy beam transport (LEBT), RFQ and a medium energy beam transport (MEBT) similar to BNL. Another benefit to the design is the quality of beam entering the Drift Tube Linac (DTL) may be much higher. The BNL injector, which is similar in design, has an emittance out of the RFQ of $\varepsilon(n,rms) \sim 0.4$ πmm mrad [2], The FNAL Linac emiitance at the entrance to tank one of the DTL is $\varepsilon(n,rms) \sim 0.83\pi$ mm mrad vertical and $\varepsilon(n,rms) \sim 0.40$ π mm mrad horizontal.

CURRENT OPERATIONS

The current operational sources are surface plasma magnetrons that have a slit aperture. The sources are mounted so that the aperture points down with a 90 degree bend magnet that helps sweep away electrons and shape the beam for injection into the accelerating column (Fig. 1).

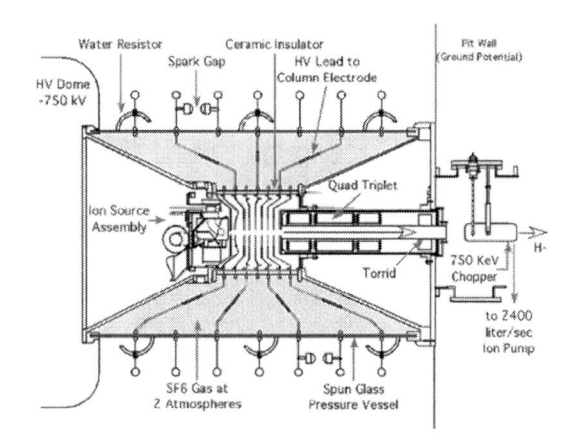

FIGURE 1. H- ion source Cockcroft-Walton assembly. (Image from FNAL Linac rookie book)

The H- ions are extracted through a slit opening in the anode cover plate by an H shaped extractor electrode with a positive potential of 12 kV to 20 kV. The extraction scheme is shown in Fig. 2.

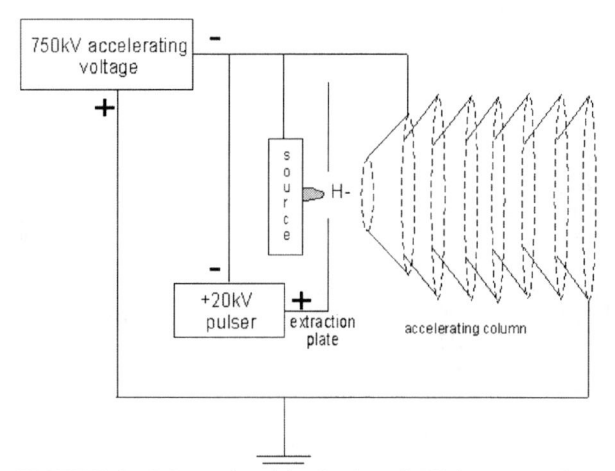

FIGURE 2. Schematic of the Cockcroft-Walton extraction.

The low extraction voltage requires the source to run with a high arc current to achieve the required H- beam current (Table 1). With the high arc current and voltage, the efficiency is on the order of 20 mA/kW. The high arc current and low efficiency contribute to a source lifetime of 3 to 4 months. Typical aging of sources is caused by cathode erosion and cesium hydride blocking the hydrogen inlet aperture. Once a source is removed from operation, it is cleaned and worn parts replaced.

TABLE 1. Current ion source operational parameters

Parameter	Value
H⁻ current	45-55 mA
Arc current	40-50 A
Arc Voltage	120-140 V
Extraction Voltage	15-18 kV
Rep Rate	15 Hz
Duty Factor	0.12%
Pulse Width	80 μs
Power efficiency	20 mA/kW

THE PLAN

The 750 keV linac injector upgrade plan [1] for replacing the existing Cockcroft-Walton accelerators with a 200 MHz RFQ and two round aperture direct magnetron ion sources as shown in Fig. 3. It was decided to have two ion sources for redundancy that will be mounted on a rail system which allows one source to slide into the operational position, connected to the first solenoid, while the other has maintenance performed on it, off to the side. This allows the LEBT to be as short as possible which empirical evidence from BNL shows is desirable.

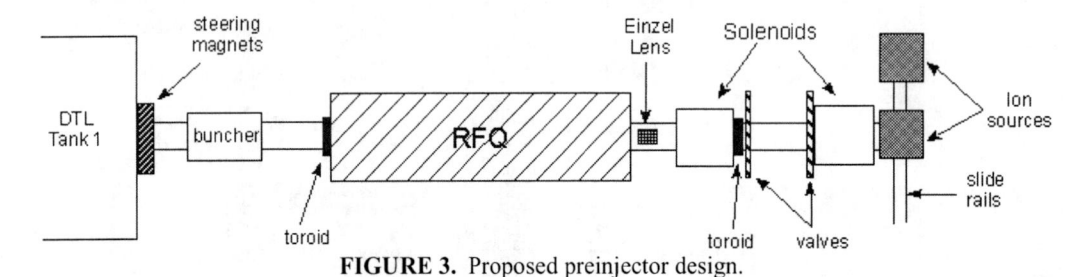

FIGURE 3. Proposed preinjector design.

Magnetron Source

The round aperture magnetron (Fig. 4) has been used quite successfully at BNL since 1989 [2]. It consists of a dimpled cathode that spherically focuses the H⁻ and electrons at the anode circular aperture. The ions are extracted through a 3.2 mm opening in both the anode cover plate and extractor cone, across a 2.5 mm gap. The cone tip is made out of molybdenum to reduce erosion due to extracted electrons that get bent due to the stray magnetic field of the source and strike the cone.

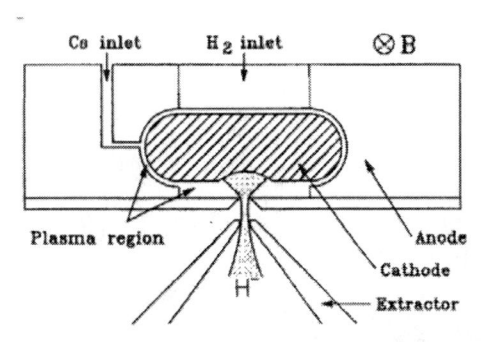

FIGURE 4. Round aperture magnetron schematic. (Image from reference [2])

The spherical focusing along with low arc current and high extraction voltage contribute to the power efficiency of the source which is 67 mA/kW, as shown in Table 2. BNL experience has shown that the circular aperture ion source is very reliable, often running 6 months before needed maintenance.

TABLE 2. BNL operational parameters

Parameter	Value
H- current	90-100 mA
Arc current	8-18 A
Arc Voltage	140-160 V
Extraction Voltage	35 kV
Rep Rate	7.5 Hz
Duty Factor	0.5%
Pulse Width	700 µs
Power efficiency	67 mA/kW

Ion source data taken to this point at FNAL has been using a modified version of the HINS ion source [3] mounted on the source test stand. The source has been modified for a grounded extraction cone, and to match the BNL source geometry, apertures and cone spacing. The testing to this point has been a proof of principle before building the operational source and was not intended to optimize the source.

For current testing the rep rate is 7.5 Hz, with an arc pulse width 300 µs, and an extraction pulse width of 200 µs, which are very close to how BNL operates. With these parameters the maximum H- beam current witnessed so far in the test stand has been 70 mA at 35 kV extraction. Fig. 5 shows a typical beam pulse measured on the toroid with 35 kV extraction and arc current of 23 A. The noise on the beam current flattop is about 20 mA and is very dependent of the arc current. Higher arc currents tend to have less noise.

For this project a new source vacuum chamber and source mounting can are currently being designed and should be installed on the test stand within a couple of months. The design is a mixture of the HINS and BNL designs with ease of maintenance in mind. Fig. 6 shows the preliminary layout of the vacuum chamber and source. The vacuum pumping will be 1200 L/S for hydrogen and will be separated from the LEBT vacuum by a 3 mm extraction aperture. This should allow for better vacuum in the beamline and reduce H- stripping losses.

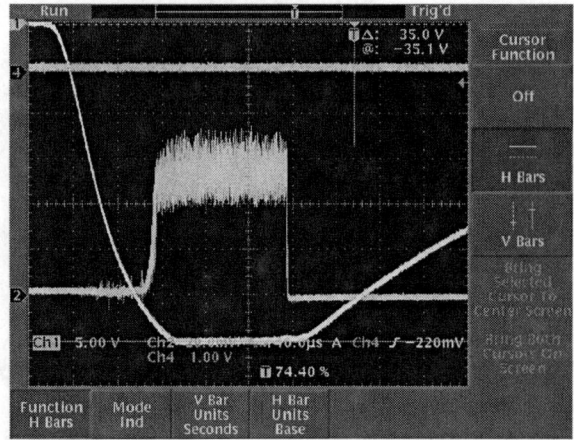

FIGURE 5. Typical scope trace of the source beam current and extractor voltage on the test stand operating at 35 kV.

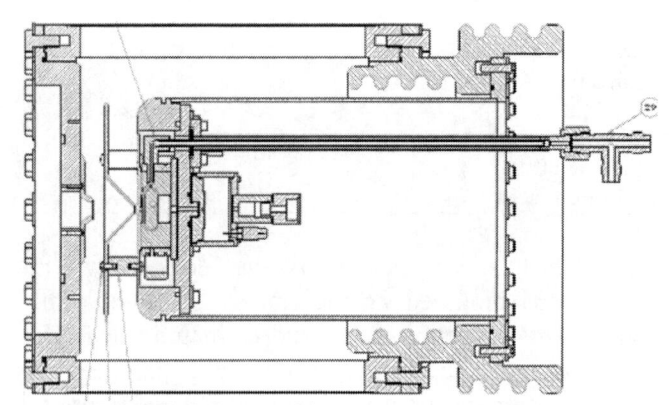

FIGURE 6. Proposed ion source vacuum chamber.

Extraction

The source extraction shown in Fig. 7 is different than the current operational system in that the extraction voltage is the acceleration voltage. This higher extraction voltage is more affective at pulling H⁻ out of the source.

The negative 35 kV extraction pulser design is a modified version of the FNAL extractor and similar to the one used at BNL. The pulser is capable of delivering 40 kV, 400 mA pulses at 15 Hz. It pulses floating relay racks that contain source electronics that are tied to the source body (anode) at -35 kV. This provides the difference of potential for the extraction/accelerating voltage since the extractor is tied to ground.

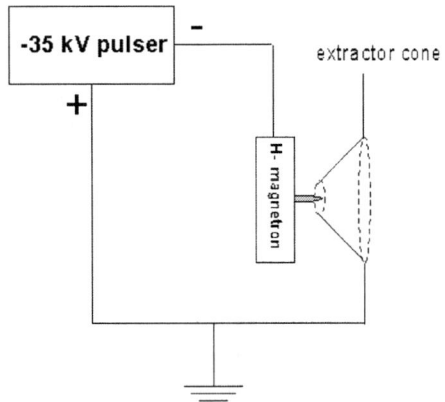

FIGURE 7. BNL extraction scheme. The extraction voltage is the accelerating voltage.

Einzel Lens Chopping

Chopping is needed for changing the width of the beam pulse for different aspects of the physics program. An Einzel lens located near the entrance to the RFQ will be used as a chopper for the leading edge of the beam [1]. As shown in Fig. 8, the lens will initially be at 37 kV which will stop the beam coming out of the source. The lens will then be set to 0 V allowing the beam to pass through to the entrance of the RFQ. After the beam passes through the lens it will be charged back up to 37 kV prior to the next beam pulse. Xenon gas will be introduced into the LEBT for space charge compensation. BNL has demonstrated that with the introduction on xenon gas into the beam pipe, the neutralization rise time is about 40 μs and the H⁻ transport increases by 30% [5].

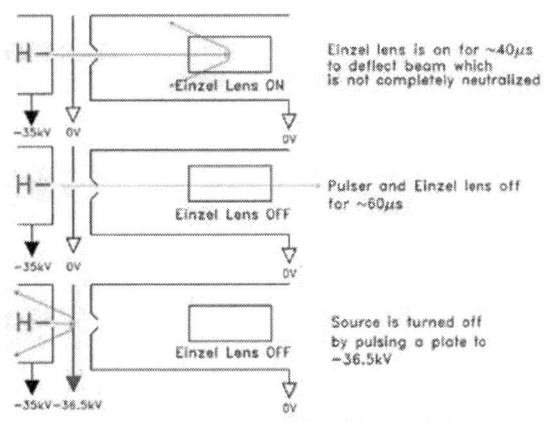

FIGURE 8. Proposed chopping using an Einzel lens. Figure is from reference [1].

There is an Einzel lens located just downstream of the ion source in the test stand that is used to focus the beam coming out of the source so that it will fit through a toroid and allow for emittance and Faraday cup measurements. This will allow measurement of the Einzel lens rise time which needs to be < 1 μs for chopping that is needed for sharp beam edges and reduced beam loss. Fig. 9 shows SIMION^{TM} simulations of the test stand with different voltages on the Einzel lens with a constant 35 kV on the extractor.

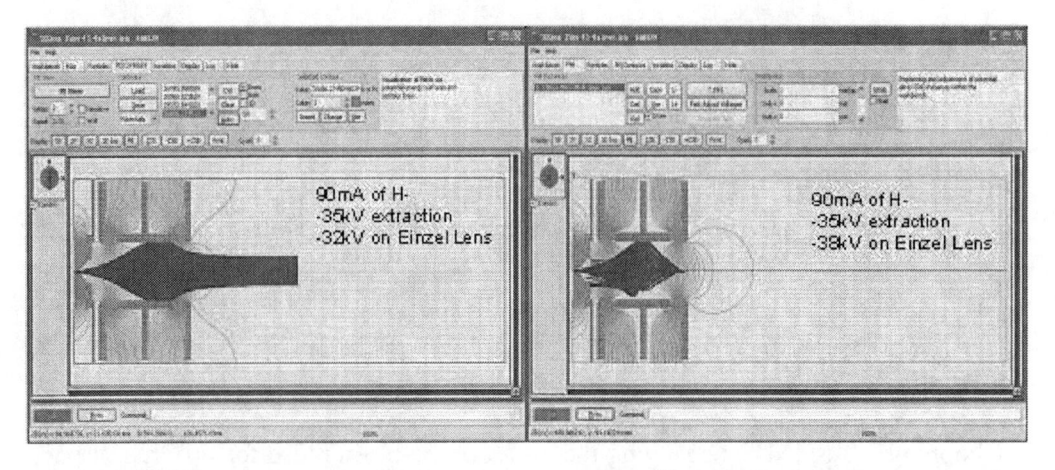

(a) SIMION^{TM} simulation results for focused beam in the test stand. (b) SIMION^{TM} simulation for stopped beam in the test stand.

FIGURE 9. SIMION^{TM} simulation results for Einzel lens test on the test stand.

For -35 keV a lens voltage of -38 kV will stop the beam coming out from making it through the lens. A high voltage pulser has been built and is now being tested that will short the Einzel lens to ground with a thyratron tube and should have a rise time of ~50 ns, which is less than the required time for beam chopping.

CONCLUSION

Before the new preinjector can be installed the magnetron ion source will need to be optimized. The source anode and extractor cone openings will need to have apertures that allow for enough H- to be extracted while maintaining the smallest emittance possible. Also, the extractor gap spacing will need to be large enough and the stray magnetic field optimized to sweep away the electrons (while not affecting the H$^-$ trajectories) to support at least 35 kV extraction with a minimum amount of arcing.

ACKNOWLEDGMENTS

I would like to thank Chuck Schmidt for all of his help and mentoring for this project. His decades of experience and willingness to help have always been welcome. I would also like to thank the FNAL Preacc Group technicians for their help with the modifications to the HINS source.

REFERENCES

1. C.Y. Tan, D.S. Bollinger, C.W. Schmidt, "750 keV LINAC INJECTOR UPGRADE PLAN" *Beams-doc*-3646-v, 2 2010.
2. J.G. Alessi, "Performance of the Magnetron H- Source on the BNL 200 MeV Linac", *AIP Conf. Proc.*, Vol. 642, pf 279-286, 2002.
3. R. Webber, " H⁻ Ion Source requirements for the HINS R&D Program", *Beams-doc*-3056-v1, 2008
4. C.W. Schmidt and C.D. Curtis, " A 50 mA Negative Hydrogen Ion Source", *IEEE Proc. Nucl. Sci. NS*-26, pg 4120-4122, 1979.
5. D. Raparia *et al*, "Results of LEBT/MEBT reconfiguration at BNL 200 MeV LINAC" *Part. Acc. Conf.*, 2009.

Impedance Measurement of an Antenna with Hydrogen Plasma Driven by 13.56MHz-rf for J-PARC H$^-$ ion source

A. Ueno, Y. Namekawa, K. Ohkoshi, K. Ikegami, and H. Oguri

J-PARC Center, Tokai-mura, Naka-gun, Ibaraki-ken 319-1195, Japan

Abstract. In order to satisfy the Japan Proton Accelerator Research Complex (J-PARC) second stage requirements of an H$^-$ ion beam current of 50 mA and a life-time of 2000 hours, the development of a 13.56MHz-rf-driven H$^-$ ion source was started by using an antenna developed at the Spallation Neutron Source (SNS). As the first step, the impedance of an antenna with hydrogen plasma, which is one of the most important parameters, was measured. For an rf-power of 0 kW, 1.25 kW or 7.45 kW, it was measured as j84.19+5.107 Ω, j80.22+11.82 Ω or j58.24+21.16 Ω, respectively. Its real-part had a rather large value and rather large rf-power dependence unexpectedly. By using the measured impedance, the circuit to produce rf-plasma in the plasma chamber on a -50kV potential with a 13.56MHz-rf-source on the ground potential was designed by the circuit simulation code LTSpice IV.

Keywords: ion source, negative hydrogen, rf-driven, J-PARC
PACS: 29.20.Ej, 29.25.Ni, 41.75.Cn

INTRODUCTION

A cesium (Cs) free H$^-$ ion source driven with a lanthanum hexaboride (LaB$_6$) filament is being operated at the Japan Proton Accelerator Research Complex (J-PARC) [1]. Although it satisfies the J-PARC first stage requirements of an H$^-$ ion beam current of 30 mA and a life-time of 500 hours, it was proven that the current was not increased by seeding Cs [2]. In order to satisfy the J-PARC second stage requirements of an H$^-$ ion beam current of 50 mA and a life-time of 2000 hours, the development of a 13.56MHz-rf-driven H$^-$ ion source was started by using an antenna developed at the Spallation Neutron Source (SNS) [3]. In this paper, the measured results of the impedance of an antenna with hydrogen plasma driven with a 13.56MHz-rf are presented. The design of the circuit to produce rf-plasma in the plasma chamber on a -50kV potential with a 13.56MHz-rf-source on the ground potential by using the measured impedance is also presented.

EXPERIMENTAL SETUP

The experimental setup to measure the impedance of an antenna with hydrogen plasma is shown in Fig. 1.

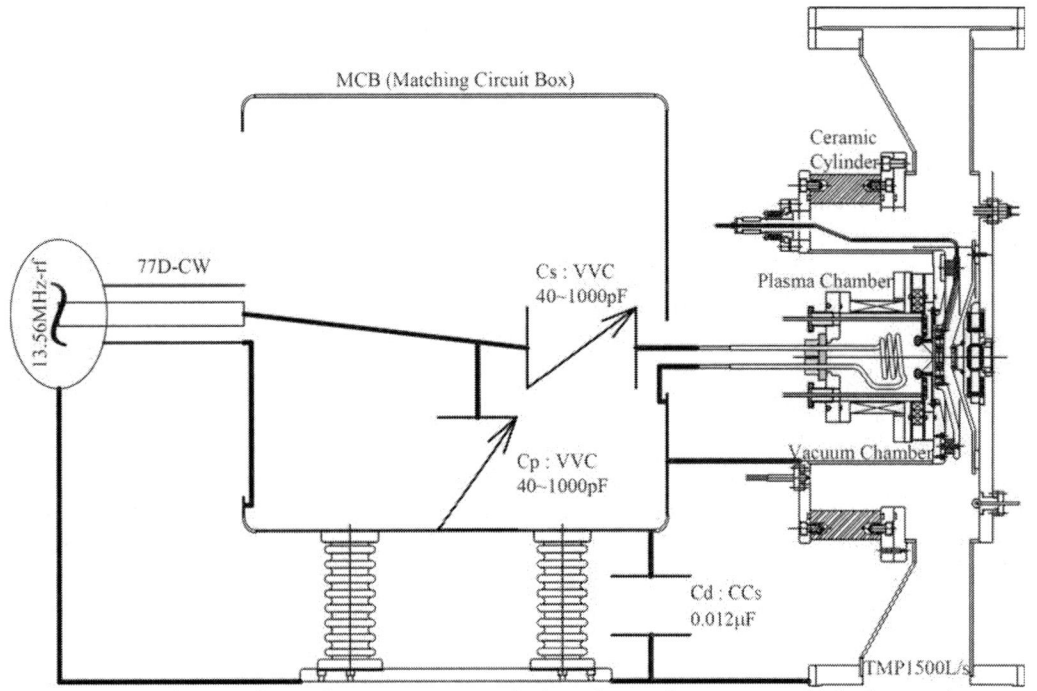

FIGURE 1. The experimental setup to measure the impedance of an antenna with hydrogen plasma.

The 13.56MHz-rf is transferred to the matching circuit box (MCB) through 77D coaxial waveguide (77D-CW). At the entrance of the MCB, the outer conductor of the 77D-CW is connected to it by using a copper plate. The inner conductor of the 77D-CW is connected to the terminals of two vacuum variable capacitors (VVCs) shown as C_p and C_s by using copper plates. The other terminal of C_p is connected to the MCB wall by a bolt. The same VVC is used for C_p and C_s, whose tuning range of capacitance and peak working voltage are from 40 to 1000 pF and 6 kV, respectively. The other terminal of C_s is connected with one terminal of an antenna, which is developed by the SNS, by using a copper plate. The averaged loop diameter of the antenna is 55 mm, which is slightly smaller than that (65.4 mm) of the original SNS antenna. It is cooled by flowing water its inside. The other terminal of the antenna is connected to the MCB wall by a copper plate. The MCB wall is connected to the plasma chamber, whose inner diameter and inner length are 100 mm and 124.5 mm, respectively, through the medium of a copper plate and a vacuum chamber wall. The vacuum chamber is pumped out by using a 1500 L/s turbo molecular pump (TMP). The measurement was done for the H_2 gas flow rate of 14 SCCM. Since the vacuum pressure of the vacuum chamber was measured as 8.9×10^{-3} Pa and the beam hole diameter of the plasma electrode is 9 mm, the vacuum pressure inside of the plasma chamber is estimated to be 0.85 Pa. In order to dump the local rf-resonance around the MCB, a set of ceramic capacitors (CCs) was used as C_d. Since three 4000 pF CCs were connected in parallel, the capacitance of C_d is 0.012 µF.

EXPERIMENTAL RESULTS

The measured waveforms of the forward (trace 1) and reflected (trace 2) 13.56MHz-rf are shown in Fig. 2, when the pulsed and CW rf-powers were 7.45 kW and 0.2 kW, respectively. The former was measured by a wave detector at the end of the pulse. The reflected rf-power at the end of the pulse, which was smaller than the reflected rf-power of the CW rf as understandable from the figure, was measured as 0 kW with the wave detector. In this measurement, C_d was essential to input rf-power to the plasma by dumping the local rf-resonance around the MCB. After a short period success of inputting 15 kW rf-power, the matched condition for the rf-power more than a few kW has not been reproduced, probably due to the withstand voltage decrease of C_p and C_s caused by the sparking in them. Although the CW rf-power of 0.2 kW was inputted to maintain thin plasma and reduce the initial impedance for the pulsed rf, the plasma seems to vanish since all of the CW rf and the initial part of the pulsed rf is reflected. The rf-power higher than 15 kW, which produced an H^- ion beam current of 70 mA in a Cs-seeded J-PARC ion source driven with a tungsten filament [2], will be inputted after C_p and C_s are replaced to new VVCs with a tuning range of capacitance from 60 to 1000 pF and a peak working voltage of 30 kV.

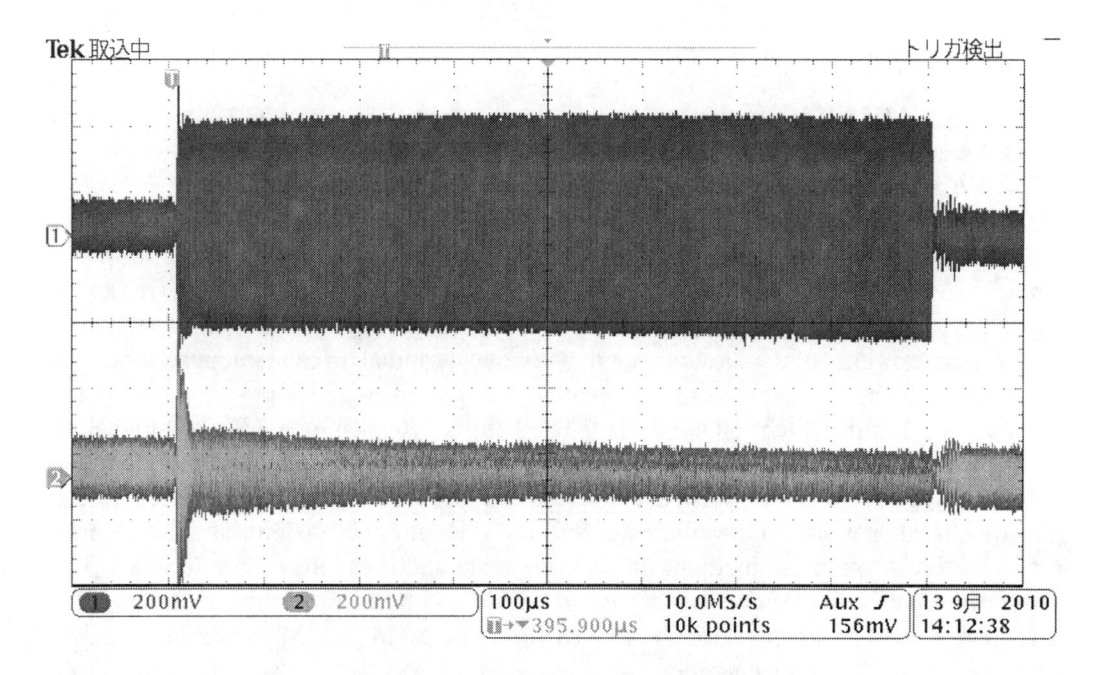

DPO3054 - 14:16:17   2010/09/13

FIGURE 2. The measured waveforms of the forward (trace 1) and reflected (trace 2) 13.56MHz-rf, when the forward pulsed and CW rf-powers were 7.45 kW and 0.2 kW, respectively.

The impedance of the antenna with hydrogen plasma can be presented as $j\omega L_{ap}+R_{ap}$. It is calculated by solving the following complex equation, since the remaining two parameters of C_p and C_s can be measured.

$$1/(j\omega C_p+1/(1/j\omega C_s+j\omega L_{ap}+R_{ap}))=50 \qquad (1)$$

L_{ap} and R_{ap} were calculated as 0.6836 µH and 21.16 Ω, respectively. By using the circuit simulator code LTSpice IV [4], the matched condition of the matching circuit was confirmed for the parameters acquired with the 7.45 kW rf-power measurement as shown in Fig. 3. It is the result of a two-terminal pair network analysis when the internal series resistance and amplitude of the rf-source were set to 50 Ω and 2 V, respectively. The lower and upper figures of Fig. 3 show the equivalent circuit and the amplitude and phase of the current through R_{ap} as functions of frequency around the operating one. The power dissipation on R_{ap} is calculated from the simulated amplitude of -30.245 dB ($=20\log_{10}(I(R_{ap}))$) as $R_{ap}*I(R_{ap})^2=21.16[\Omega]*(10^{(-30.245/20)}[A])^2=0.02$ W, which is the same as the power dissipation on the real matched resistance of 50 Ω directly connected to the rf-source of $(1[V])^2/50[\Omega]=0.02$ W. The simulation is useful to estimate the steady state voltage and current on C_p and C_s and so on.

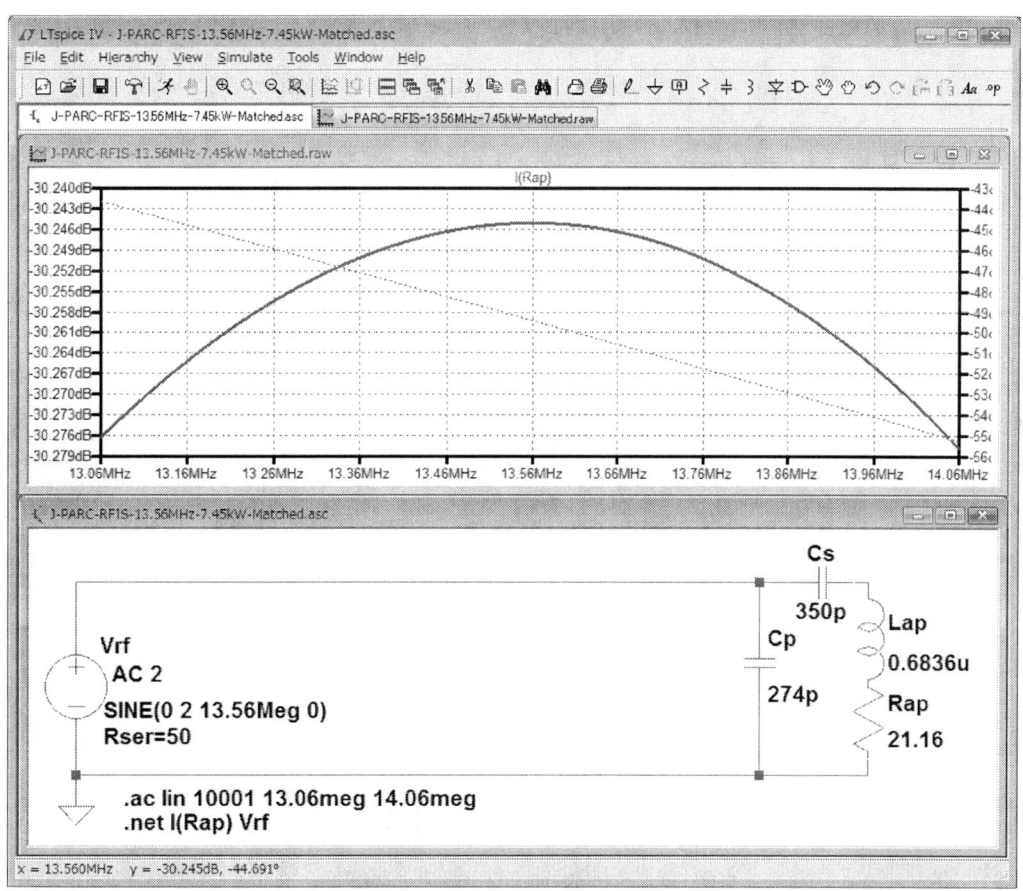

FIGURE 3. The result of a LTSpice IV two-terminal pair network analysis for the parameters of C_p, C_s, L_{ap} and R_{ap} acquired for the 7.45 kW rf-power measurement, when the voltage and series resistance of the rf-source was set to 2 V and 50 Ω, respectively.

The results of the impedance measurements including those for two different forward rf-powers of 1.25 kW and 0 kW are summarized in TABLE 1. In the case of 0 kW rf-power, C_p and C_s are measured as the setting to minimize the reflection from the matching circuit when a network analyzer was connected to 77D-CW instead of the 13.56MHz-rf-source. In this case, there should be no plasma in the plasma chamber. The values of R_{ap}, L_{ap} and $j\omega L_{ap}$ calculated with Eq. (1) are also presented in TABLE 1. Although a small R_{ap} around 0.6 Ω estimated for an antenna with hydrogen plasma driven with 2MH-rf [5] was presumed, the rather large values of R_{ap} and its rather large positive dependence on the rf-power P_{rf} were unexpectedly measured. The measured values are shown by the digits with underline in TABLE 1. The measured dependences of C_p and C_s on the rf-power P_{rf} are presented together with fitted curves in Fig. 4.

TABLE 1. Summary of the impedance measurements. Measured values are shown by the digits with underline and the values extrapolated with the fitted curves are shown with digits in parentheses.

P_{rf} [kW]	C_p [pF]	C_s [pF]	R_{ap} [Ω]	L_{ap} [µH]	$j\omega L_{ap}$ [Ω]
<u>0</u>	<u>696</u>	<u>170</u>	5.107	0.9881	j84.19
<u>1.25</u>	<u>422</u>	<u>199</u>	11.82	0.9416	j80.22
<u>7.45</u>	<u>274</u>	<u>350</u>	21.16	0.6836	j58.24
(15)	(250)	(533)	(23.48)	(0.5514)	(j46.98)
(30)	(236)	(896)	(24.82)	(0.4472)	(j38.10)

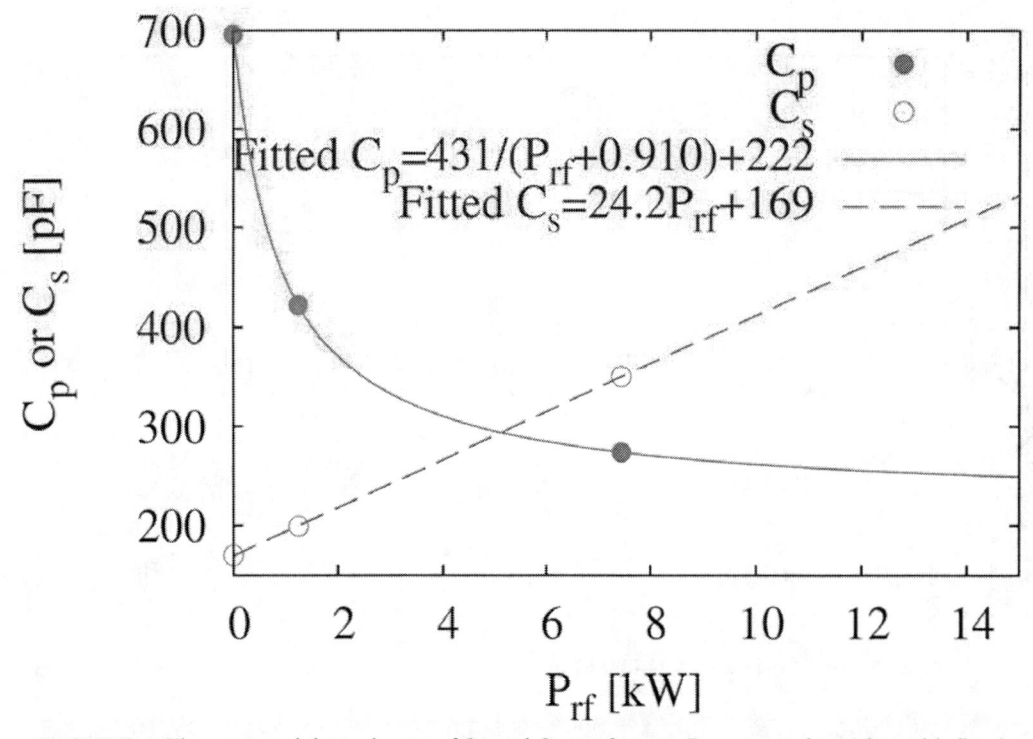

FIGURE 4. The measured dependences of Cp and Cs on rf-power P_{rf} presented together with fitted curves.

In TABLE 1, the values extrapolated with the fitted curves are shown with digits in parentheses. By using the new VVCs, the rf-power of more than 30 kW will be inputted. The maximum voltage on VVCs is simulated as 3.7 kV on C_s by using the LTSpice IV, for the condition of C_p and C_s setting matched for 0 kW rf-power and 15 kW rf-power is inputted, which is realized when there is no plasma. In the resonant circuit, the withstand voltage of the VVC (6 kV) should be twice of the possible steady state voltage as the margin for the transient phenomena. The withstand voltage of the new VVC (30 kV) is higher than the twice of the maximum voltage on C_s (8.6 kV) for the rf-power of 80 kW, which is the maximum power of the rf-source.

-50 KV ISOLATION MATCHING CIRCUIT DESIGN

By using the LTSpice IV, the matching circuit with the -50 kV isolation transformer circuit composed of L_{t1}, L_{t2}, C_t, C_r was designed for the parameters acquired with the 7.45 kW rf-power measurement as shown in Fig. 5. The lower and upper figures of Fig. 5 show the equivalent circuit and the amplitude and phase of the current through R_{ap} as functions of frequency around the operating one.

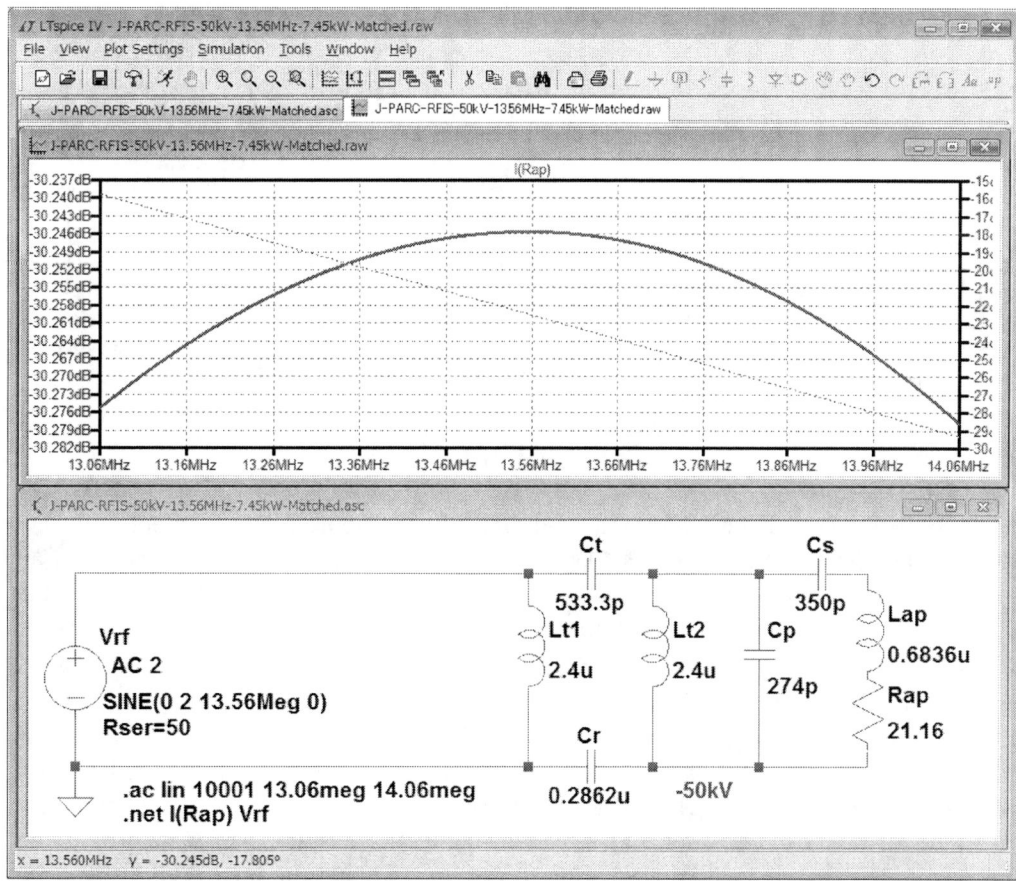

FIGURE 5. The result of a LTSpice IV two-terminal pair network analysis for the parameters of C_p, C_s, L_{ap} and R_{ap} acquired for the 7.45 kW rf-power measurement, with isolation transformer circuit composed of L_{t1}, L_{t2}, C_t, C_r.

L_{t1} and L_{t2} are the same inductors with 2.4 µH made by bending a copper hollow conductor. C_t is a set of 24 vacuum capacitors (VC) with a capacitance of 200 pF and a withstand voltage of 20 kV, in which eight of three VCs connected in series are connected in parallel. C_r is a set of 424 ceramic capacitors (CC) with a capacitance of 2700 pF and a with stand voltage of 30 kV, in which 212 of two CCs connected in series are connected in parallel. A schematic drawing of the experimental setup is shown Fig. 6.

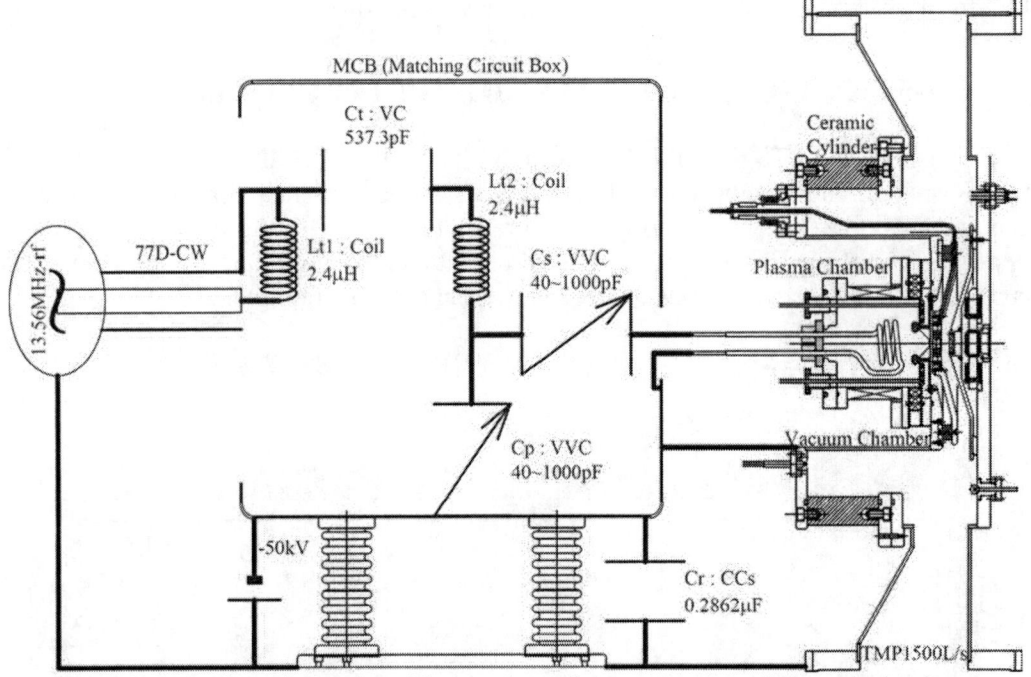

FIGURE 6. A schematic drawing of the experimental setup to produce rf-plasma in the plasma chamber on a -50kV potential with a 13.56MHz-rf-source on the ground potential.

CONCLUSIONS

As the first step of the 13.56MHz-rf-driven H^- ion source development using an antenna developed at the SNS to satisfy the J-PARC second stage requirements of an H^- ion beam current of 50 mA and a life-time of 2000 hours, the impedance of the antenna with hydrogen plasma was measured. For an rf-power of 0 kW, 1.25 kW or 7.45 kW, it was measured as j84.19+5.107 Ω, j80.22+11.82 Ω or j58.24+21.16 Ω, respectively. Since a small value of the real-part around 0.6 Ω estimated for an antenna with hydrogen plasma driven with 2MH-rf [5] was presumed, the measured values were unexpectedly large and had unexpectedly large positive dependence on the rf-power. Although the rf-power was limited to 7.45 kW due to the insufficient withstand voltage (6kV) of the VVCs, it will be increased more than 30 kW by using new VVCs with a peak working voltage of 30 kV. By using the measured impedance acquired with the 7.45 kW rf-power measurement, the matching circuit with the -50 kV isolation transformer circuit was designed with the LTSpice IV. As the next step, an H$^-$ ion beam will be extracted with the circuit in near future.

REFERENCES

1. H. Oguri, A. Ueno, K. Ikegami, Y. Namekawa, and K. Ohkoshi, *Phys. Rev. ST Accel. Beams* **12**, 010401 (2009).
2. A. Ueno, H. Oguri, K. Ikegami, Y. Namekawa, and K. Ohkoshi, *Rev. Sci. Instrum.* **81**, 02A720 (2010).
3. R. F. Welton, M. P. Stockli, Y. Kang, M. Janney, R. Keller, R. W. Thomae, T. Schenkel, and S. Shukla, *Rev. Sci. Instrum.* 73, 2, 1008 (2002).
4. http://www.linear.com/designtools/software/ltspice.jsp.
5. J. Staples, and T. Schenkel, Proceedings of the 2001 Particle Accelerator Conference, Chicago, 2108 (2001).

ION SOURCES FOR FUSION

Dependence of the Performance of the Long Pulse RF Driven Negative Ion Source on the Magnetic Filter Field

W. Kraus, U. Fantz, D. Wünderlich and the NNBI team

Max-Planck-Institut für Plasmaphysik, Boltzmannstr. 2, D-85748, Garching, Germany

Abstract. In the long pulse RF source the plasma drift caused by the magnetic filter field leads to different plasma parameters close to the top and bottom halves of the plasma. In order to investigate the effect of the plasma density on the extracted currents, the bottom grid halve was covered and the direction of the drift was converted by changing the polarity of the magnets. The asymmetry of the plasma density was measured by H_β spectroscopy. By biasing the plasma grid the asymmetry could be enlarged, but the current of co-extracted electrons could be minimized in a similar way for both polarities. The extracted currents of H⁻ ions as well as of the electrons, however, depend only weakly on the direction of the drift i. e. on the plasma densities. The results indicate that the source performance is determined to a greater degree by the plasma sheath close to the plasma grid which is not yet accessible by diagnostics.

Keywords: H⁻, Ion Source Development, Neutral Beam Heating, ITER, Radio Frequency Ion Source
PACS: 52.50.Dg, 52.50.-b, 52.59.-f, 52.70.-m, 52.80.Pi

INTRODUCTION

On the long pulse "MANITU" test facility (multi ampere negative ion test unit) the prototype of the RF source for the ITER neutral beam system is being tested. The target is to reach accelerated ion current densities of 24 mA/cm² (H⁻) and 20 mA/cm² (D⁻) respectively in long pulses of up to 3600 s. The current of the co-extracted electrons has to be smaller than the ion current.

The previous development efforts were concentrated on the improvement of the testbed and source design in order to achieve the necessary long pulse capability together with high operation reliability. The electrical insulation of the RF coil was improved by immersing it into an SF_6 filled case, which enables reliable high power operation in long pulses without RF breakdowns. The last remaining copper surface on the source back plate was coated with Molybdenum. This prevents poisoning of the Caesium layer on the plasma grid by sputtered copper and had led to more stable ion and electron currents in long pulses [1]. The last steps were to raise the extraction voltage to 10 kV and to improve the control of the Caesium evaporation rate.

After these improvements the maximum extracted H⁻ current density obtained was 35 mA/cm² in short pulses and with a weak magnetic filter field. In long pulses; however, only approx. 20 mA/cm² was achievable with moderate electron current [2].

The magnetic filter field has a great impact on the currents of negative ions and electrons as well as on the plasma density profiles close to the plasma grid.

If the horizontal magnetic field was sufficiently strong, the plasma drifted in vertical direction resulting in different plasma densities on the two halves of the plasma grid [2]. Probe measurement showed that the plasma in the side with low density consists mainly of negative ions with only a small electron fraction [3]. It was observed by Doppler shift measurements that for some filter field configurations the divergence of the beam extracted from the top and bottom halves of the extraction system was also different [2, 4]. In most cases the divergence on the side with lower plasma density was smaller. As the beam extraction is normally under perveant, the measured lower divergence would indicate that the H⁻ current density was closer to the perveance optimum and hence higher than on the other grid half.

For further source development it is important to find out, when the plasma density profile is asymmetric, do the two grid halves contribute differently to the total H⁻ current? Therefore, one half of the grid was masked and only half was used for the beam extraction. By changing the polarity of the filter field the direction of the plasma drift inverted.

The experiments were carried out in pulses of 100s – 500s length. An advantage of long pulses is that parameter scans can be performed within one pulse and in this way a minimum of change of the other parameters is achieved. This paper will concern itself mainly with the investigation of the effect of the plasma drift caused by filter field variations on the extracted ion and electron currents.

THE MAGNETIC FILTER FIELD

In the prototype source the RF power at a frequency of 1 MHz is inductively coupled into a circular volume of 25 cm diameter and 14 cm height ("driver"), out of which the plasma is flowing into the main chamber (b x l x d = 30 x 59 x 25 cm^3). . The negative ion production is based on surface conversion of atoms and positive ions on a Caesium covered plasma grid. Details of the design are described in [1]. The ITER source will have eight of these "drivers".

The magnetic filter field is generated by 8 rows of Co-Sm magnets which are mounted on both sides into the diagnostic flange at a distance of 2.5 cm from the plasma grid (s. Fig. 1). Because these rows are interrupted at several locations for the diagnostic feedthroughs the resulting field is not very homogeneous. One weak point of the design is that there is no access to the magnets from outside and so the initial field can only be weakened or strengthened without breaking the vacuum by superposing on it a field from additional external magnets. These were placed 5 cm higher outward from the magnets in the diagnostic flange. The resulting field is very inhomogeneous and has a different 3D topology inside the source and the driver for each of the field variations [2].

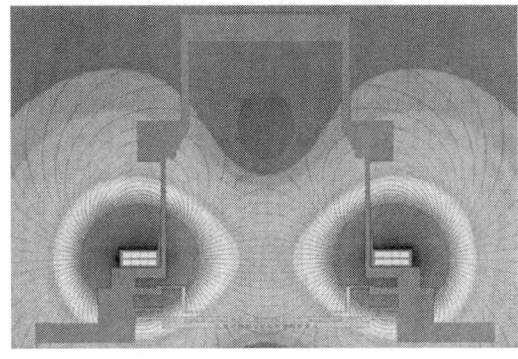

FIGURE 1. Filter field generated by CoSm magnets inside the diagnostic flange (top) and new configuration with magnets outside the source (bottom)

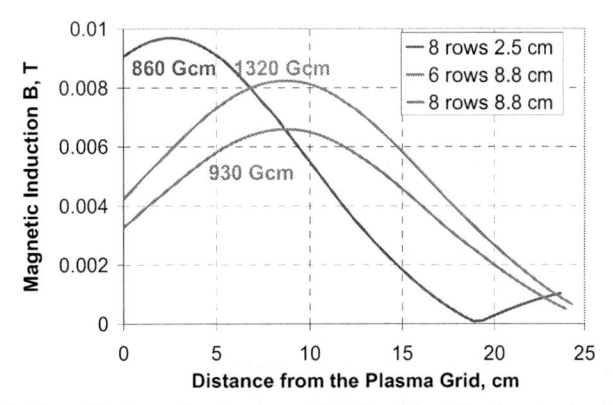

FIGURE 2. B field profile from the plasma grid to the driver for the standard field close to the plasma grid and the new field with 8 or 6 rows of magnets on each side

For the experiments it was necessary to change the polarity of the filter field without opening the source in order to avoid restarting the conditioning procedure again. For that reason the filter magnets were removed and mounted outside the source on top of the diagnostic flange at a distance of 8.8 cm from the plasma grid. The field strength can now be adjusted by the number of magnets and the polarity can be changed by turning the magnet pack. However, the maximum of the B field is now shifted further away from the plasma grid and the B field on the plasma grid is reduced to about 40 G which is less than half the previous strength. The field integrated in the driver direction is considerably higher with the same number of magnets (Fig. 2). In the experiments described below only the configuration with 6 rows of magnets was used which has almost the same integrated B-field as the standard field.

MODIFIED EXTRACTION SYSTEM AND DIAGNOSTICS

In order to extract only one half of the beam, the bottom half was masked by a water cooled Copper plate (Fig. 3). On the top half the part outside the projection of the driver was covered as in previous experiments and so the total extraction area was reduced to 65 cm^2.

The plasma homogeneity was measured by H_β spectroscopy in two lines of sight (LOS) of 1 cm diameter in 2.5 cm distance from the plasma grid on both sides at the lower and uppermost positions (Fig. 3). It should be mentioned that because the cover of the bottom grid halve reaches 1 – 2 cm into the source, the plasma is also shifted into the source and hence on the bottom half the plasma density in the LOS of H_β is expected to be lower compared to the top side even without drift.

Another reason for asymmetries of the source plasma is related to the position of the Caesium oven. Because it is mounted at the top of the source back plate the upper part of the plasma grid has a high deposition rate, which may cause differences in the plasma parameters, the H$^-$ production and the electron currents between top and bottom.

FIGURE 3. Partly covered plasma grid with the LOS of the H_β spectroscopy

RESULTS

The H_β signals in Fig. 4 show that in case of the drift upwards the plasma density was indeed much higher at the top than at the bottom position. When the polarity of the magnets was changed the measured value at the top decreased rapidly but the bottom signal did not increase in the same extend (s. Fig. 5). This may be caused by the small distance between the bottom LOS and the cover of the bottom grid halve. The dependence of the H_β intensities on the bias current was very similar for both drift directions; raising the bias current increased the H_β signal at the side, where the drift is directed to, and decreased it at the opposite side. This corresponds to an enhancement of the drift by a higher bias voltage.

The ion current and the co-extracted electron current, however, are not very much affected by the direction of the drift. There was, in particular, no correlation between the currents of co-extracted electrons and the H_β intensities as a function of the bias current. With drift upwards a slightly higher bias current was necessary to suppress the electron current and the ion current was higher at low bias voltage.

When the bias current was adjusted such that the electron current was minimal, there was no significant change in the ion and electron currents observed when the polarity of the magnets was changed.

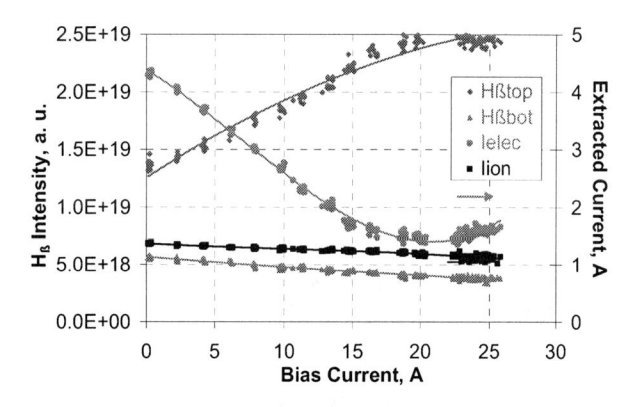

FIGURE 4. Extracted ion and electron current and top and bottom H_β signals measured at 50 kW RF power and 0.4 Pa. The plasma drift is directed upwards.

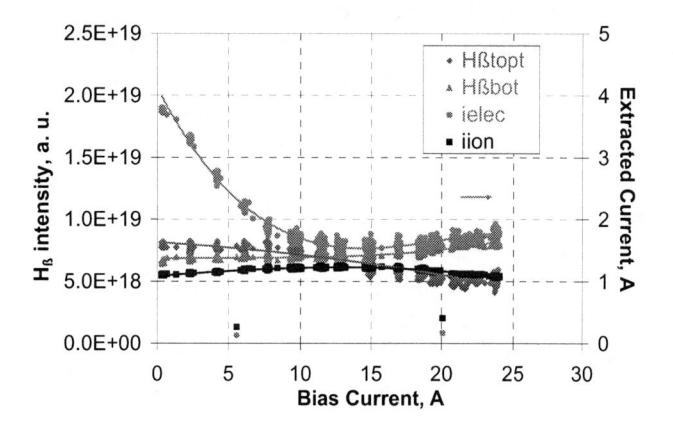

FIGURE 5. Extracted ion and electron current and top and bottom H_β signals measured at 50 kW and 0.4 Pa. The polarity of the filter magnets is reversed and the plasma drift is directed downwards.

This was also valid at different power. The difference of the extracted currents as a function of the RF power for both polarities shown in Fig.6 is not significant, when the bias current was optimized to suppress the electrons. The results are within the spread caused by different Caesium conditions on the plasma grid surface.

The efficiency of the H⁻ production in terms of H⁻ current density /RF power was approximately the same compared to the source with the standard filter and reached approx. 0.4 mA/cm^2.

For both polarities using the new filter the fraction of co-extracted electron current was significantly higher compared to what was observed with the standard filter and a much higher bias current was required to suppress the electrons. It was not possible to achieve an ion current/ electron current ratio below 1. A possible reason for this result is the much lower B field strength close to the plasma grid.

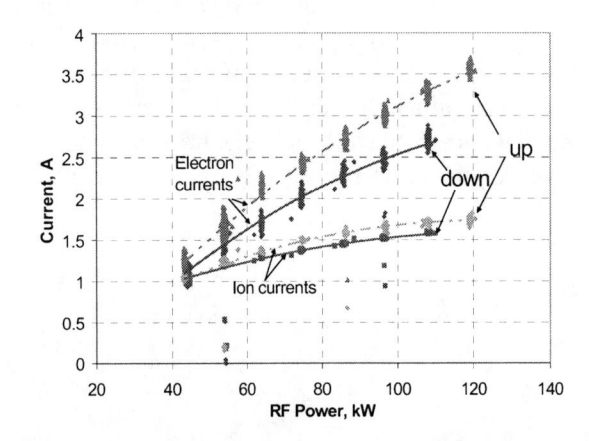

FIGURE 6. Extracted ion and electron currents and as a function of the RF power at 0.4 Pa for drift up and drift down of the source plasma.

CONCLUSIONS

The used external magnetic filter has a lower B field near the extraction area but with the same integrated B-field inside the source. The electron current could in these experiments not be reduced to the same extent as with the filter field previously employed. This indicates that the field strength close to the plasma grid is more important for the electron suppression than the integrated field.

The plasma drift changes direction when the polarity of the filter field is changed. The resulting differences of the plasma densities at top and bottom position can be enlarged by biasing the plasma grid. The dependence of the drift on the bias potential suggests that the observed density changes are caused by an ExB drift, but the B field would force the positive and negative charges in opposite directions which is not possible due to the quasi neutrality of the plasma. The theoretical studies dealing with cross-field diffusion of plasmas clearly indicate the situation may be of considerable complexity [3].

Without the PG bias the co-extracted electron current is high in both cases of the drift, i. e. for high or for low plasma density at the top and can be reduced by the biasing in the same way. Also the H- ion current depends only weakly on the direction of the drift. The assumption that one grid halve would deliver a higher H- current could not be confirmed by these experiments. On the contrary, the extracted electron and ion currents seem to depend only weakly on the bulk plasma but much more on the plasma sheat close to the plasma grid.

ACKNOWLEDGMENTS

This work was party supported by a grant from Fusion for Energy (F4E-2008-GRT-07). The opinions expressed herein are those of the authors only and do not represent the Fusion for Energy's official position.

REFERENCES

1. W. Kraus, et al., Rev. Sci. Instrum. 79 (2008) 02C108
2. W. Kraus, et al., Rev. Sci. Instrum. 81 (2010) 02B110
3. P. McNeely, et al., Rev. Sci. Instrum. 81 (2010) 02B111
4. P. Franzen et al., Beam Homogeneity Dependence on the Magnetic Filter Field at the IPP Test Facility MANITU, these proceedings

Beam Homogeneity Dependence on the Magnetic Filter Field at the IPP Test Facility MANITU

P. Franzen, U. Fantz and the NNBI Team

Max-Planck-Institut für Plasmaphysik, EURATOM Association, PO Box 1533, 85740 Garching, Germany

Abstract. The homogeneity of the extracted current density from the large RF driven negative hydrogen ion sources of the ITER neutral beam system is a critical issue for the transmission of the negative ion beam through the accelerator and the beamline components. As a first test, the beam homogeneity at the IPP long pulse test facility MANITU is measured by means of the divergence and the stripping profiles obtained with a spatially resolved Doppler-shift spectroscopy system. Since MANITU is typically operating below the optimum perveance, an increase in the divergence corresponds to a lower local extracted negative ion current density if the extraction voltage is constant. The beam H_α Doppler-shift spectroscopy is a rather simple tool, as no absolute calibration — both for the wavelength and the emission — is necessary. Even no relative calibration of the different used lines of sight is necessary for divergence and stripping profiles as these quantities can be obtained by the line broadening of the Doppler-shifted peak and the ratio of the integral of the stripping peak to the integral of the Doppler-shifted peak, respectively. The paper describes the H_α MANITU Doppler-shift spectroscopy system which is now operating routinely and the evaluation methods of the divergence and the stripping profiles. Beam homogeneity measurements are presented for different extraction areas and magnetic filter field configurations both for Hydrogen and Deuterium operation; the results are compared with homogeneity measurements of the source plasma. The stripping loss measurements are compared with model calculations.

Keywords: Neutral Beam Injection, ITER, negative ion source, RF source
PACS: 28.52.Cx, 52.27.Cm, 52.50.Dg, 52.50.Gj

INTRODUCTION

The ITER neutral beam system [1] requires a large source for negative hydrogen ions (1 m width and 2 m height). In order to minimize the losses within the acceleration system the source pressure has to be 0.3 Pa at maximum. At this low pressure, Cs seeding is needed for a sufficiently large negative ion production. The RF driven ion source is now the ITER reference source, as it is principally maintenance free. The development at IPP Garching during the last decade was successful [2,3,4]: the required extracted and accelerated current densities together with an amount of co-extracted electrons less than the extracted negative ions have been achieved with the small IPP prototype source (~1/8 of the area of the ITER source), both in Hydrogen and Deuterium, but for short pulses only. The long pulse test facility MANITU (Multi

Ampere Negative Ion Test Unit) equipped with small IPP prototype source demonstrated stable pulses up to one hour; the parameters however, are still slightly below the ITER requirements [5].

A still open issue is the homogeneity of the extracted beam. In order to avoid excessive losses in the accelerator and in the beamline ITER requires an inhomogeneity of the extracted current density across the 1x2 m^2 source of less than 10% [1]. This topic will be addressed primarily by the large ion source test facilities ELISE at IPP Garching [6,7] and SPIDER/MITICA in Padua [8] which are presently under construction. The extracted current density distribution, however, cannot be easily accessed by diagnostic tools. It depends on the negative ion density at the plasma meniscus of each extraction aperture, which is not measurable at all. Quantities that can be measured — with a certain space resolution — are plasma parameters and the negative hydrogen ion density in some distance from the apertures, i.e. the plasma grid (PG), and the divergence of the resulting beamlets. The link between the plasma parameters near the PG and the extracted current density is not straightforward and depends on the various complex processes in the boundary layer near the PG. Under certain circumstances, namely underperveant beam conditions, the divergence profile reflects the distribution of the extracted current density, and hence the profile of the negative ion density at the meniscus. Therefore, the measurement of the beamlet divergence profile by means of Dopplershift spectroscopy and the comparison with the plasma parameters near the PG offers the possibility of a better understanding of the processes in the boundary layer [9].

The paper reports on the results of measurements at MANITU where a spatially resolved Dopplershift spectroscopy system is now routinely operating, discusses the influence of various parameters (magnetic filter field, ion mass, size of extraction area) on the divergence profile and shows the correlation with plasma homogeneity measurements. Furthermore, the fraction of stripped ions in MANITU is discussed.

H$_\alpha$ DOPPLERSHIFT SPECTROSCOPY FOR NEGATIVE HYDROGEN IONS

The beam Doppler-shift spectroscopy is a rather simple tool, as no absolute calibration — both for the wavelength and the emission — is necessary. It allows gaining relevant information of the beam properties (energy spectrum, divergence) and the homogeneity of the beam properties if several lines of sight (LoS) are used. Even no relative calibration of the different LoS is necessary for divergence and energy spectra profiles. Typically, the (n=3) excitation of the hydrogen Balmer emission (H$_\alpha$) is used due to its largest emissivity. H(n=3) atoms are created due to collisions of the ion beam with the residual gas (or the neutralizer gas, if present) on their way through the beamline to the calorimeter. For a negative hydrogen ion beam, the main reactions are:

$$H_f^- + H_2 \rightarrow H_f^0(n=3) + H_2 + e, \tag{1}$$

$$H_f^0 + H_2 \rightarrow H_f^0(n=3) + H_2, \tag{2}$$

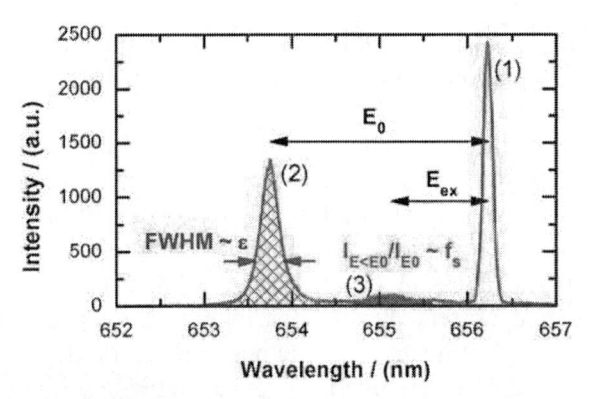

FIGURE 1. Typical H$_\alpha$ Dopplershift spectrum for a negative hydrogen ion beam. The beamlet divergence ε is proportional to the width of the full energy peak (corresponding the total beam voltage E_0), the stripping fraction f_s is proportional to the integral ratio of the full energy peak and the stripping peak (corresponding to the extraction voltage E_{ex})

$$H^0_f + H_2 \rightarrow H^0_f + 2H^0_s (n=3), \qquad (3)$$

where the indices f and s denote fast and slow ions, respectively. Fast neutral atoms (reactions 2 & 3) are always present in a negative ion beam as neutralization occurs everywhere along the beam path. Other reactions — collisions between beam particles or collision of positive ions (generated by ionization along the beam path) — can be neglected here due to the small relative velocity of the particles within the beam or the small amount of positive ions generated in MANITU due to the rather low tank pressure during the pulse (~10^{-5} mbar). The intensity of the Doppler shifted emitted light, $\xi_{n=3}(E)$, is then given by

$$\xi_{n=3}(E) \propto n_{H_2}(\sigma_{0\alpha}(E)j_{H^0}(E) + \sigma_{-1\alpha}(E)j_{H^-}(E)), \qquad (4)$$

$\sigma_{-1\alpha}$ and $\sigma_{0\alpha}$ being the cross sections of reactions 1 & 2 [10,11,12], E being the beam energy, and j_{H^0} and j_{H^-} being the current densities of neutrals and negative ions, respectively.

As the lifetime of an H(n=3) atom is rather short (~10^{-8} s), the typical distance between the places of excitation and deexciation is only a few cm for beams of up to 30 kV, which is the maximum beam energy at MANITU. Hence, the measurement is local with respect to the excitation conditions.

Figure 1 shows a typical H$_\alpha$-spectrum of a central LoS at MANITU with a summary of the kind of information that can be gathered. The spectrum shows three peaks: (1) the unshifted H$_\alpha$ peak at the nominal wavelength at 656.3 nm (reaction 3); (2) the Doppler-shifted peak, emitted by fast neutrals having the full accelerated energy; and (3) the so-called stripping peak in-between the latter two, caused by neutrals having less than the full energy. Those neutrals are generated by collisions of not fully accelerated ions with the residual gas in the accelerator and are therefore a measure of the amount of stripped ions in the acceleration system. The stripping peak location always corresponds to the extraction voltage; hence, most of the stripping at MANITU occurs mainly in the first gap of the extraction system, i.e. between PG and extraction grid (EG).

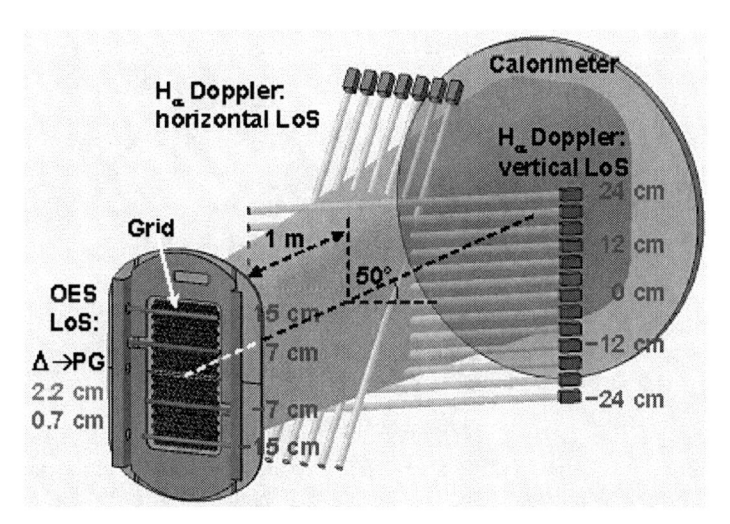

FIGURE 2. Arrangement of the Line-of-Sights for the H_α Dopplershift spectroscopy system at MANITU. Also indicated are the Line-of-Sights for the optical emission spectroscopy (OES) near the plasma grid.

From the Doppler width, the divergence can be calculated as the Doppler shift $\Delta\lambda$ depends on the viewing angle α via

$$\Delta\lambda = \lambda_0 \frac{v}{c} \cos\alpha, \quad (5)$$

v being the beam velocity and λ_0 the rest wavelength of H_α emission. The viewing angle however changes along the line-of-sight due to the divergence ε of the beam

$$\alpha_0 - \varepsilon \leq \alpha \leq \alpha_0 + \varepsilon, \quad (6)$$

α_0 being the nominal angle of the line-of-sight to the beamlet center. For small angles, the resulting Doppler shifted peak is Gaussian and the divergence — being an average divergence for the beamlets within the respective line-of-sight — can be obtained from the measured width of the Doppler peak λ_{FWHM} by

$$\varepsilon = \frac{\lambda_{FWHM}}{\Delta\lambda \tan\alpha_0}, \quad (7)$$

taking various experimental line broadening effects into account. For the MANITU system, we assume furthermore no focusing effects due to simplicity. Since MANITU is typically operating below the optimum perveance, an increase in the Doppler width corresponds to a lower negative ion current density for constant extraction voltage.

The stripping fraction f_s is defined by the fraction of neutral atoms in the beam with less than the full energy:

$$f_s = \frac{j_{H^0}(E < E_0)}{j_{H^0}(E < E_0) + j_{H^0}(E_0) + j_{H^-}(E_0)}. \quad (8)$$

It can be calculated by the ratio of the integral of the stripping peak $I_{E<E_0}$ to the integral of the fully shifted peak I_{E_0}, weighing with the neutralization fraction f_n and the emission cross section at the different energies (see Eq. 4). The broadening of the

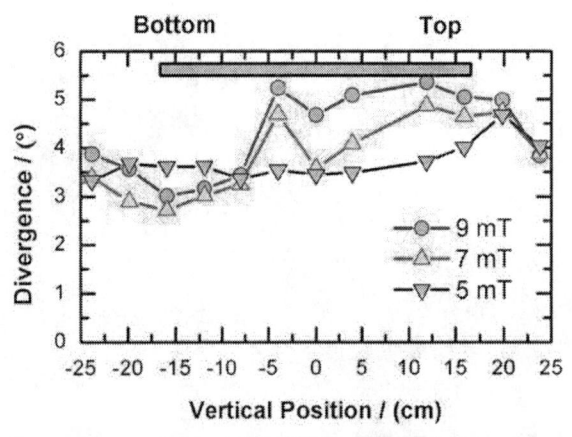

FIGURE 3. Vertical divergence profiles at MANITU for different filter field configurations for hydrogen pulses. The extraction area — indicated by the grey rectangle — is 204 cm², the plasma drift upwards. All pulses have similar normalized perveance of about 0.05.

stripping peak is both due to the divergence of the beamlets and the energy distribution of the stripped neutrals. For simplicity, we assume that all stripped neutrals have the same energy, i.e. the extraction energy E_{ex}; this is justified as the cross sections are more or less constant in the energy range of interest (6 - 8 kV for E_{ex}, 20 - 30 kV for E_0). Hence:

$$f_s = \frac{I_{E<E_0}/\sigma_{0\alpha}(E_{ex})}{I_{E<E_0}/\sigma_{0\alpha}(E_{ex}) + I_{E_0}(f_n/\sigma_{0\alpha}(E_0) + (1-f_n)/\sigma_{-1\alpha}(E_0))}. \quad (9)$$

The neutralization fraction f_n was estimated taking both the source and beamline pressure of MANITU into account; it ranges between 0.07 and 0.15.

With the divergences and stripping factors obtained, a profile of these quantities across the beam and hence a measure of the beam inhomogeneity can be obtained. Such a system is installed in the MANITU vacuum tank and is routinely operating (see the scheme in figure 2); it consists of 7 horizontal channels, viewing in vertical direction and 13 vertical channels, viewing in horizontal direction. The optic heads are separated by a typical distance of 4 cm and have a viewing angle of about 50° at a distance of about 1 m from the grounded grid towards to the beam. Hence, the Doppler-shifted peaks are blue-shifted (see Figure 1). Due to the small background pressure in the MANITU tank, the H_α emission is rather weak; hence a large integration time is needed (typically 2 s). For the evaluation of the Doppler spectra, a numerical limit was set for the height of the full energy Doppler-shifted peaks in order to discriminate peaks above the noise level. Hence, missing data for certain LoS correspond to rather weak beam intensity and the error increases with decreasing peak height (typically at the beam edges). Furthermore, there are no data from the LoS at 8 cm (vertically) due to a broken light fibre inside the beam line.

Also shown in figure 2 are the diagnostic tools for the source plasma homogeneity near the PG: different line-of-sights for optical emission spectroscopy allow the estimation of the distribution of plasma parameters across the PG at two different distances from the PG and two different vertical positions [9]. For the experimental campaigns reported here, the magnetic filter field was created by magnets inside the

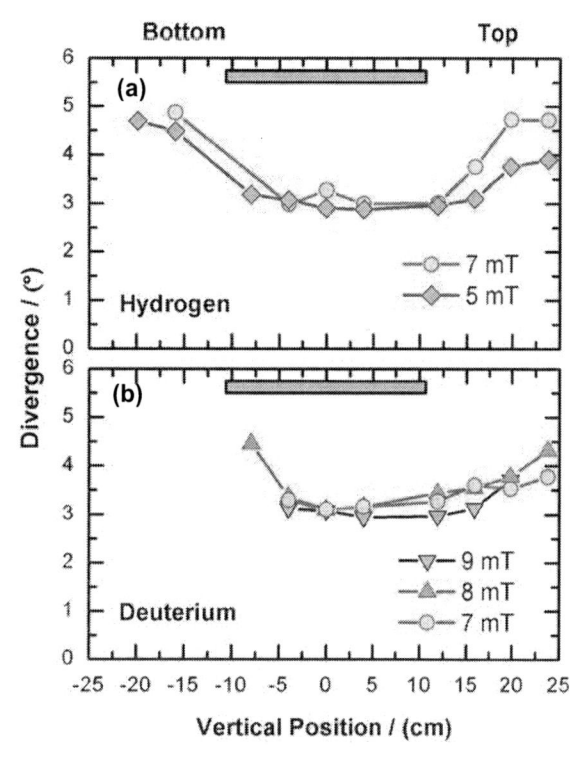

FIGURE 4: Vertical divergence profiles at MANITU for different filter field configurations for hydrogen (a) and deuterium pulses (b). The extraction area — indicated by the grey rectangle — is 132 cm², the plasma drift downwards. All pulses have similar normalized perveance of about 0.06.

source body with a horizontal magnetic filter field strength near the PG of 7 mT. The field strength could be changed by attaching magnets outside the source body, by choosing the polarity of the external magnets with respect to the filter field, the field near the PG could be increased or decreased by 1 mT or 2 mT, respectively. This method, however, has the drawback, that it also changes the 3D structure of the magnetic filter field [5]. The polarity of the standard field was reversed during the experiments changing the direction of the observed vertical plasma inhomogeneity across the PG. As the latter is most probably caused by a drift of the plasma [9], the field polarity is indicated in the following either with "Drift up" and "Drift down", according to the maximum of the plasma density.

DIVERGENCE PROFILES VS. MAGNETIC FILTER FIELD

Figure 3 shows the dependence of the vertical divergence profiles for hydrogen operation on the magnetic field configuration. These data were obtained for a large extraction area (204 cm²) and with the polarity of the magnetic filter field so that the plasma drifts to the upper part of the PG.

It can be clearly seen that for the low magnetic filter field (5 mT) the divergence profile is more or less symmetric and rather homogeneous. For the larger magnetic filter fields the divergence profile becomes more inhomogeneous: the divergence at the bottom part of the beam is lower than the divergence of the upper part of the beam.

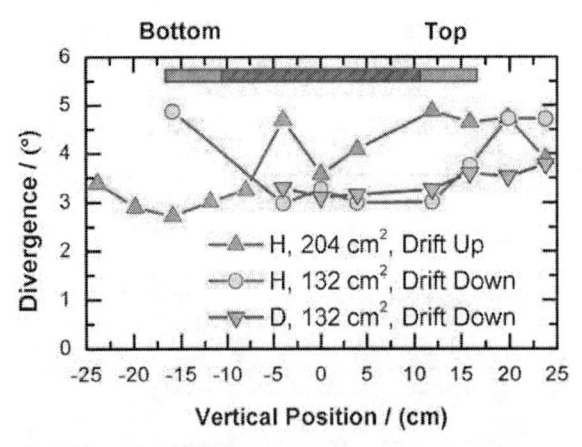

FIGURE 5. Dependence of the beamlet divergence for hydrogen and deuterium pulses for the standard magnetic filter field (7 mT) for the two different cases of plasma drift and extraction area.

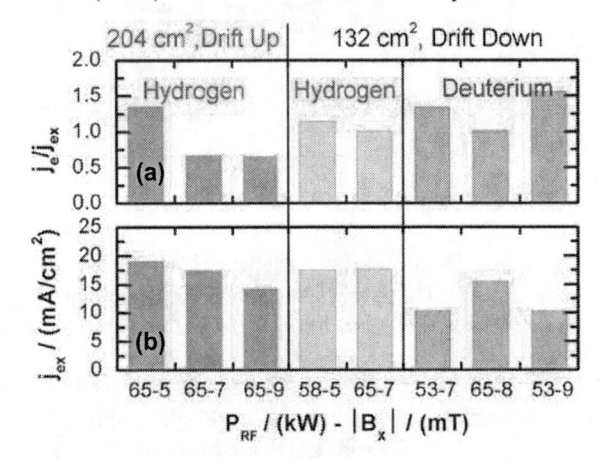

FIGURE 6. Amount of co-extracted electrons (a) and extracted current density (b) for the pulses shown in figure 3 and 4.

This indicates that the local current density in the upper part of the PG is lower than in the lower part. The 'dip' in the vertical profiles at the center of the beam for the stronger filter field cases is most probably caused by the overlap of the two partial beams emitted from the MANITU grid system due to the inclination of the two grid halves of the MANITU grid system. The detailed mechanism, however, is still not understood.

Due to these results, the magnetic field polarity was reversed for the next experimental campaign (drift down), as the Cs oven is also located at the upper part of the source. Hence an improved performance was expected. Additionally, the size of the extraction area was reduced to 132 cm^2 in order to minimize the area outside the driver projection [5].

Figure 4 shows the dependence of the vertical homogeneity on the magnetic field configuration for hydrogen and deuterium pulses with the smaller extraction area and a reversed magnetic filter field. For deuterium operation, only pulses with increased magnetic filter field strength have been performed. Operation in deuterium at 5 mT is not possible due to a non-sufficient suppression of the co-extracted electron fraction.

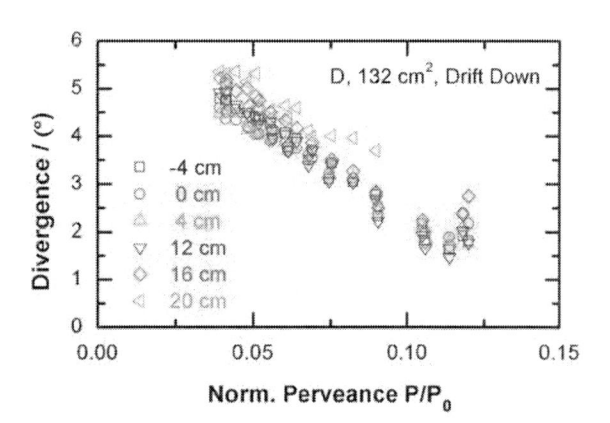

FIGURE 7. Dependence of the beamlet divergence for deuterium pulses on the normalized perveance for different LoS. Extraction area was 132 cm², drift downwards.

In contrast to the results shown in figure 3, the divergence profiles for these experimental conditions do not show a dependence on the magnetic filter field, neither for hydrogen or deuterium. The homogeneity is also much better. As expected due to the field reversion, the beam is concentrated in the upper part of the PG; hence the majority of the negative ions are extracted again from the part of the PG opposite from the plasma drift direction.

But the concentration of the beam at the "low plasma" region is much more pronounced in the case of the small extraction area and plasma drift downwards. This is indicated in figure 5, where a comparison of the three cases for the standard filter field configuration (7 mT) is shown. Even data of the LoS at -8 cm and -12 cm are missing for the "drift down, small extraction area" case, although these LoS are still within the projection of the extraction area. As discussed above, this indicates that the beam intensity within these LoS is too weak for H_α excitation above the noise level. The minimum divergence (about 3°) is more or less the same in all cases, but this is due to the fact that all pulses shown here have roughly the same normalized perveance P/P_0, P being the overall beam perveance (= $I_{ex}/E_{ex}^{1.5}$), and P_0 the perveance of a planar diode. The measured minimum divergence agrees also rather well with results of ion beam calculations [13] and with results of the BATMAN test facility [14].

The reason for the large discrepancy in the dependence of the beam inhomogeneity on the magnetic filter field for the cases shown is not clear. Unfortunately, due to technical and operation time limitations, data for corresponding cases — for example deuterium data or "drift down" data for the large extraction area — are missing. In all cases, the source was well conditioned with cesium; the respective data (amount of co-extracted electrons and extracted current density) are shown in figure 6. There are two main differences, which might affect the plasma dynamics in front of the PG and/or the negative hydrogen ion distribution at the meniscus: (1) In the case of the large extraction area (204 cm²), some grid areas on top and on bottom of the PG are outside the projection of the driver; and (2) there is an intrinsic asymmetry of the MANITU source as the Cs oven is located on top of the source. Neither of these effects, however, is present at the large sources foreseen for ELISE and ITER.

Figure 7 shows the dependence of the beamlet divergences for several LoS for deuterium pulses on the normalized perveance. It shows that due to the particularities of

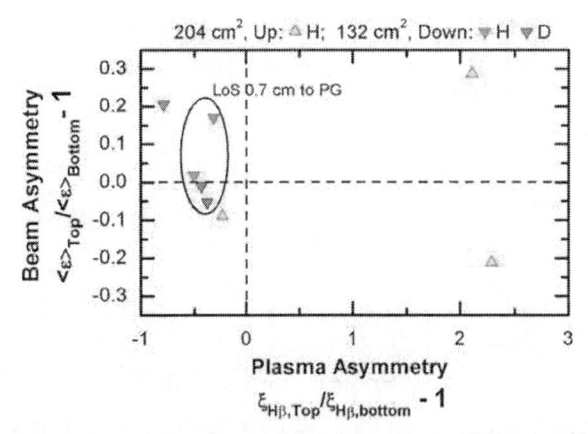

FIGURE 8. Comparison of beam and plasma asymmetry for the pulses shown in figures 3 and 4.

the MANITU grid system[1] operation is done mostly in the in the underperveant region. Hence, as it was discussed above, the divergence depends linearly on the perveance. But there is no strong dependence of the inhomogeneity of the divergence on the perveance; if at all, the inhomogeneity improves with increasing perveance. Furthermore, the beam is concentrated more and more in the center, indicated by the disappearance of the Doppler peak for the LoS at 20 cm.

BEAM VS. PLASMA HOMOGENEITY

Figure 8 shows a comparison of the beam inhomogeneity for the pulses shown in figures 3 and 4 with the plasma inhomogeneity. The respective inhomogeneities (plasma and beam) are given by the asymmetry factor that is defined by the deviation of the ratio of suitable parameters from unity. With that definition, a perfectly homogeneous beam or plasma has an asymmetry factor of zero. Positive values of the asymmetry factor indicate plasma/beam concentration at the upper part, negative values at the lower part of the source/beam line. For the plasma asymmetry, the LoS-integrated H_β emission is evaluated in regions near the PG for the upper and lower parts of the source (see figure 2). The choice of suitable parameters for the beam, however, is not straightforward. In order to reduce numerical artifacts due to missing or large erroneous data, the average divergences of the upper and lower beam regions are evaluated, even if especially for the lower beam part of the small extraction area only a few data points are available.

Figure 8 indicates no clear correlation between plasma and beam asymmetry. It can be also seen that the plasma asymmetry cannot be correlated clearly with the assumed drift direction: the "204 cm^2, Up" data point with a plasma asymmetry of -0.3 corresponds to a filter field with 5 mT at the PG. This might be caused by the change of the 3D structure of the magnetic filter field due to the applied method of changing the magnetic field. Hence, for the upcoming campaigns, the internal magnets have been removed and a magnet frame was installed. This allows flexible changes of the filter field without change of the 3D structure.

---

[1] MANITU is equipped with the "large area grid" system built for an optimum perveance at 5 kV and 200 A/cm^2 due to historical reasons; but it turned out in the last years that 8 to 9 kV are needed for maximum negative hydrogen ion extraction.

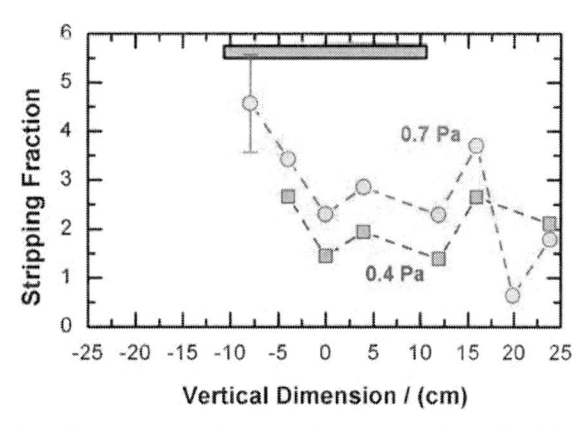

FIGURE 9. Vertical profile of stripping losses for hydrogen pulses with different source pressure. Extraction area was 132 cm^2, drift downwards.

The numerical problems in defining the beam asymmetry are highlighted for the hydrogen data with the small extraction area: although both profiles agree rather well, the calculated asymmetry factor differs by 20%. A final conclusion might only be possible after a relative calibration of the MANITU LoS so that also the Doppler peak intensities can be evaluated or eventually by measurements at the large ion sources at ELISE or SPIDER where more LoS near the PG for source OES are available. But, and this is an important result, the beam seems to be much more homogeneous than the plasma in a certain distance to the PG. This indicates that the negative hydrogen ion density distribution at the plasma meniscus is linked only weakly with the plasma parameters at a certain distance from the PG. It is in agreement with modeling results showing that the negative ions are mainly produced by — intrinsically symmetrically — impinging neutrals [15]. Asymmetries can only be caused by Cs coverage of the PG with inhomogeneous conversion yields and by the positive ion distribution needed for space charge compensation and leading to negative hydrogen ion losses due to mutual neutralization. This decoupling of beam and plasma homogeneity in RF sources is one advantageous difference to arc driven sources where due to the primary electron distribution also the neutral hydrogen flux to the plasma grid is inhomogeneous and hence a strong coupling between plasma and beam homogeneity is observed [15, 16].

STRIPPING LOSSES

No dependence of the stripping fraction from the magnetic field configuration is observed — it is also not expected. The most important dependence is on the source pressure: this is shown in figure 9 by a comparison of vertical profiles of the stripping losses for two pulses with different source pressures. The profiles are very similar, the shapes themselves, however, are still a subject for further investigation.

Figure 10 shows the dependence of the average stripping losses on the source pressure. The well known linear dependence can be seen, which is also predicted by calculations. The measured stripping losses at MANITU, however, are much lower (factor 5) than model calculations with a simple analytical model [14] predict. This model calculates the background gas density profile within accelerator and beam line by taking into account just the respective conductance of the grid apertures as well as the

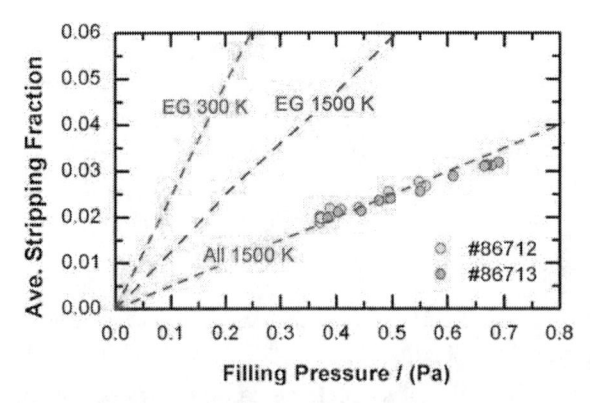

FIGURE 10. Dependence of the average stripping losses on the source pressure for two consecutive pulses at MANITU. The pressure was changed during the pulse. E_{ex} 8 kV, E_0 28.6 kV, tank pressure 1×10^{-5} mbar. Also shown are the calculated stripping fractions with a simple model for a source gas temperature of 1500 K, gas temperatures in the extraction aperture of 300 K and 1500 K, and a gas temperature of 1500 K in the whole accelerator.

source and beam line pressures. This model was benchmarked against a Monte Carlo model for the ITER NBI accelerator system [18]; it agrees with that to within 20%. In both models a gas temperature of 1000 - 1500 K in the source and in the first gap (PG-EG) is assumed, but temperature accommodation takes place quickly within the extraction aperture.

The MANITU data, however, show a much lower stripping fraction than the results of this standard calculation. One possibility for the large discrepancy might be that not all of the stripped ions can leave the accelerator, as the ion trajectories at the place where stripping occurs might not point towards the aperture. In this case, however, the stripping fraction should depend strongly on the beam optics which is not the case. On the other hand, a larger gas temperature in the accelerator can much better reproduce the data (see figure 10), but the reason for this larger temperature is not clear up to now. No accommodation, i.e. no gas collisions in the extraction apertures, is rather unlikely, but perhaps also beam heating of the gas within the extraction aperture takes place as it was observed in the neutralizer of the JET NBI system [19]. This might also to some extent explain the shape stripping loss profiles shown in figure 9: less beam power at the edges of the beam corresponds to lower gas temperature in the extraction apertures and thus to larger stripping fractions.

CONCLUSIONS

The properties of the extracted ion beam have been measured at MANITU by means of Doppler-shift spectroscopy. There is a strong dependence of the homogeneity measured divergence profile on the magnetic filter field for a large extraction area, but no dependence when the extraction area is reduced and the filter field polarity is reversed. The reasons are not clear. The beam homogeneity, however, is much better than the plasma homogeneity is some distance (~ cm's) from the plasma grid.

The measured stripping losses are much lower (factor 5) than models with standard gas temperature profiles within the accelerator show. A possible explanation might be an enhance gas temperature also within the extraction aperture due to beam heating.

ACKNOWLEDGEMENTS

The work was partly supported by a grant from Fusion for Energy (F4E-2008-GRT-07). The opinions expressed herein are those of the authors only and do not represent the Fusion for Energy's official position.

REFERENCES

1. R. Hemsworth et al., Nuclear Fusion 49 (2009) 045006.
2. E. Speth et al., Nuclear Fusion 46 (2006) 220.
3. A. Stäbler et al., Fusion Engineering and Design 84 (2009) 265.
4. P. Franzen et al., Nuclear Fusion 47 (2007) 264.
5. W. Kraus et al., these proceedings.
6. P. Franzen, AIP Conference Proceedings 1097 (2008) 451.
7. B. Heinemann et al., Proc. of the 26th SOFT Conf., Porto, 2010; to be published in Fus. Eng. Des.
8. P. Sonato, Proc. of the 26th SOFT Conf., Porto, 2010; to be published in Fus. Eng. Des.
9. U. Fantz et al., Nuclear Fusion 49 (2009) 125007.
10. J Geddes, J Hill and H B Gilbody, J. Phys. B: At. Mol. Phys. 14 (1981) 4837.
11. I D Williams, J Geddes and H B Gilbody, J. Phys. B: At. Mol. Phys. 15 (1982) 1377.
12. C.F. Barnett, ORNL-6086/V1, Oak Ridge National Laboratory, 1990
13. U. Fantz et al., Nucl. Fusion 49 (2009) 125007.
14. A. Lorenz, U. Fantz, P. Franzen, Technical Report IPP 4/285, 2006.
15. D. Wuenderlich et al., Plasma Sources Sci. Technol. 18 (2009) 045031.
16. K. Ikeda et al., Rev. Sci. Instrum. 75 (2004) 1744.
17. H. Tobari, et al., Rev. Sci. Instrum. 79 (2008) 02C111.
18. A. Krylov, R.S. Hemsworth, Fusion Engineering and Design 81 (2006) 2239.
19. E. Surrey, Nuclear Fusion 46 (2006) S360.

Characteristics of Hydrogen Negative Ion Source with FET based RF System

A.Ando[a], T.Matsuno[a], T.Funaoi[a], N.Tanaka[a], K.Tsumori[b] and Y.Takeiri[b]

[a] *Graduate School of Engineering, Tohoku University, Aoba-yama, Sendai, 980-8579, Japan*
[b] *National Institute for Fusion Science, Oroshi-cho, Toki, 509-5292, Japan*

Abstract. Characteristics of radio frequency (RF) plasma production were investigated using a FET inverter power supply as a RF generator. High density hydrogen plasma was obtained using an external coil wound a cylindrical ceramic tube (driver region) with RF frequency of lower than 0.5MHz. When an axial magnetic field around 10mT was applied to the driver region, an electron density increased drastically and attained to over 10^{19}m^{-3} in the driver region. Effect of the axial magnetic field in driver and expansion region was examined. Lower gas pressure operation below 0.5Pa was possible with higher RF frequency. H$^-$ density in the expansion region was measured by using laser photo-detachment system. It decreased as the axial magnetic field applied, which was caused by the increase of energetic electron from the driver.

Keywords: ion source, RF plasma, FET inverter, neutral beam injector.
PACS: R52.50.Dg , 52.25.Jm , 52.80.Pi

INTRODUCTION

Neutral beam injection (NBI) system is the most reliable and powerful heating methods for fusion devices. Neutral hydrogen isotopes are injected with the energy of more than 100keV for heating core plasmas and driving plasma current in the magnetically confinement devices. As neutralization efficiency of positive ions of hydrogen with the high energy is extremely low, negative ion based NBI systems have been investigated for years.

The high power negative NBI systems utilizing hydrogen/deuterium ion beam have been successfully developed in National Institute for Fusion Science (NIFS) and Japan Atomic Energy Agency (JAEA) [1-5]. In NIFS, high current negative ion sources with 45A of H$^-$ current per source and the ion energy of 180keV are installed in the Large Helical Device (LHD)-NBI system, where total injected energy more than 16MW has been successfully achieved. In JAEA, 10A D$^-$ ion sources are operated with the beam energy of more than 300keV for plasma heating and current drive.

The ion source for ITER NBI is required to be driven in the condition of the long pulse (1000 sec) acceleration of D$^-$ ion beams, whose the current density is more than 200 A/m^2 and the beam energy is 1 MeV. For a long pulse operation, it is requisite to develop RF driven ion sources without any electrode for plasma production.

Researches of negative ion sources using high power RF have been performed in various groups. A large area RF source was developed using a vacuum-immersed RF coil and a reliable high H⁻ ion current was obtained [6]. Ion sources with external RF antennas has been successfully developed for the production of positive hydrogen ions at IPP (Max-Plank-Institut für Plasmaphysik), Garching. A RF plasma was produced in a cylindrical tube of insulator with a RF coil wound in several turns externally in order to avoid the direct interaction with produced plasmas. The source was modified and intensively developed for negative ion production with Cs additive [7-10].

RF waves are usually supplied by high power RF amplifier composed of vacuum tubes and transmission line with matching impedance of 50 Ohm. Recent development of a high current semiconductor device enables us to utilize it as a fast switching device in an inverter circuit. We can utilize metal-oxide-semiconductor field-effect transistors (MOSFET) and a high power RF source with a frequency around 1MHz can be realized. Although plasma production in low pressure below 1 Pa becomes easier in the frequency of over 10 MHz range, lower frequency of around 1 MHz is attractive for high density plasma production because of larger skin depth. We have developed a FET-based high power RF amplifier with the frequency below 0.5MHz and utilized it as a RF source [11,12]. In this paper, we report characteristics of RF plasma production using the RF generator based on a FET inverter circuit.

EXPERIMENTAL SETUP

Experiments were performed using a small ion sources which consists of a driver and diffusion chamber. The schematic of the experimental setup is shown in Fig.1. A plasma was produced in the driver region, where a RF coil was wound around ceramic (Al$_2$O$_3$) cylindrical tube (inner diameter: 70mm, outer diameter: 80mm, length: 170mm) in several turns, typically ten turns in the experiments. The driver is attached to a cylindrical chamber (diffusion chamber). Axial magnetic field up to 150 G can be applied by Helmholtz magnetic coils attached at the driver. The expansion chamber has two Langmuir probe ports upstream and downstream and a pair of permanent magnets that forms a filter magnetic field. Plasma parameters were measured by

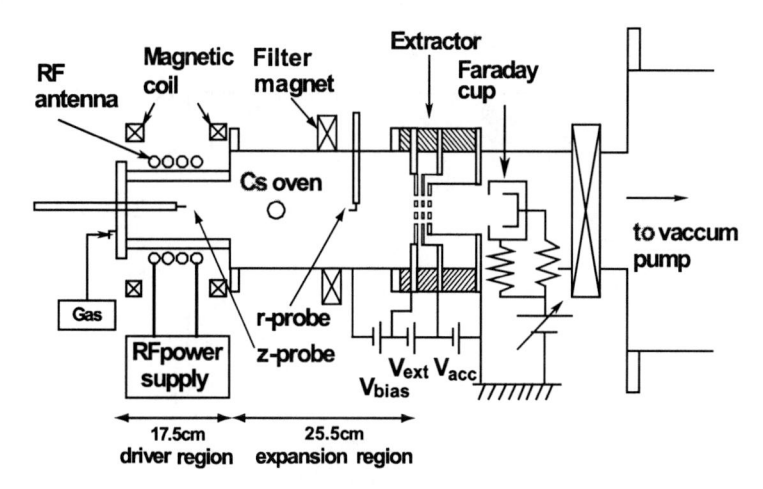

FIGURE 1. Schematic of the RF ion source.

Langmuir probes set at the expansion chamber. Ion beams can be extracted using acceleration electrodes, which consist of three grids, a plasma grid, an extraction grid and an acceleration grid. The source was installed at a test stand with a large cylindrical vacuum chamber (1m in diameter and 2m in length). Characteristics of ion sources were investigated using the test stand.

An inverter circuit of the RF generator was used for RF plasma production. We utilize the MOSFET (PDM755HA) as a switching device in the inverter circuit. An operation frequency can be adjusted by external signal generator in the range of 0.1-0.5MHz. A dc voltage up to 250V was fed to the inverter units, which compose a full bridge inverter circuit.

The output lines of the FET units were combined and connected to a primary winding of an impedance transformer, which is employed a good impedance matching between the FET inverter and the low impedance antenna and transmission lines. A parallel plate balanced line made of copper was used as the transmission line to the RF antenna. The secondary winding of the transformer was connected to a capacitor and RF antenna for plasma production. The output voltage of transmission lines becomes lower than that of the primary voltage. The secondary circuit composed a LC matching circuit. The RF frequency and capacitance of the capacitor are tuned to maximize the antenna current at LC matching condition. Total RF power of more than 15kW can deliver to the antenna with a short duration of 5ms. In order to isolate the plasma source from the RF power supply in beam experiments, isolation transformer was inserted between the RF antenna and the transmission line.

EXPERIMENTAL RESULTS

High Density Operation with Axial Magnetic Field

Applying the RF power to the antenna with the frequency of 0.34MHz, hydrogen plasma was produced in the driver region and diffused into the expansion region [12]. Electron density n_e and temperature T_e was measured at in the expansion region (8cm apart from the driver exit) and slightly increased with input RF power. Although

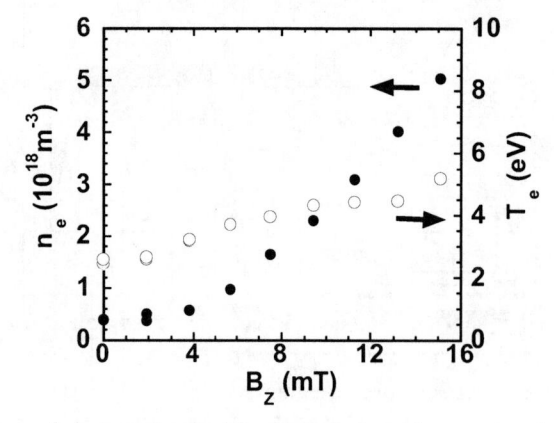

FIGURE 2. Dependence of electron density (closed circles) and temperature (open circles) on the axial magnetic field. P_{RF}=13kW. f_{RF}=0.34MHz, p=2 Pa.

plasma was produced, they reached 0.4×10^{18}m^{-3} and 1.5eV, respectively, with the RF power of 13kW. In order to increase the density of hydrogen plasma, we have applied an axial magnetic field in the driver region. With the increase of the axial magnetic field, the electron density drastically increased and attained to 5×10^{18}m^{-3} as shown in Fig.2.

Figure 3 shows spatial profiles of n_e, T_e and space potential V_p. They were measured by a Langmuir probe immersed from the top end position (Z=0) of the driver, where the driver region corresponds to Z=0 to 175mm. The n_e, T_e became maximum at the center of the driver and attained to 1×10^{19}m^{-3} and 6.5eV, respectively. One possible reason why the high density hydrogen plasma was obtained is helicon mode excitation with the help of the axial magnetic field. Simple calculation of the

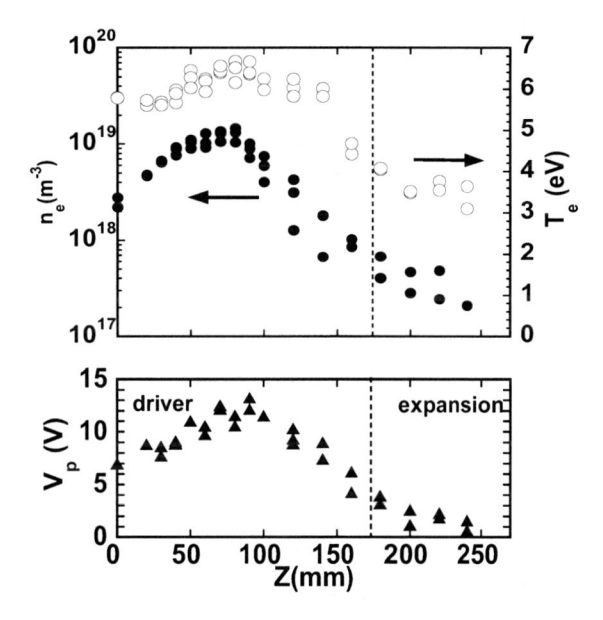

FIGURE 3. Axial profiles of electron density (closed circles), temperature (open circles) and space potential (closed triangles). P_{RF}=10kW. B_z=11.2mT, f_{RF}=0.34MHz, p=2 Pa.

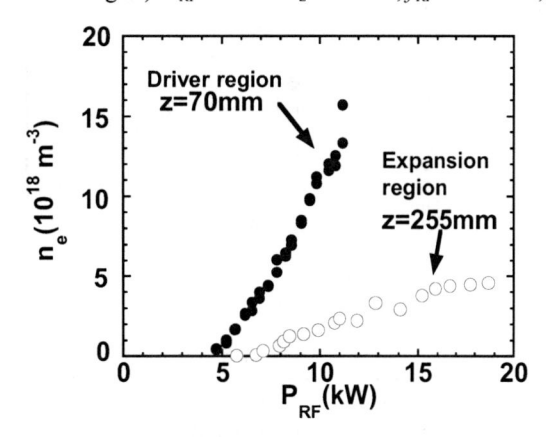

FIGURE 4. Dependences of electron density on RF power in the driver and expansion regions. P_{RF}=10kW. B_z=11.2mT, f_{RF}=0.34MHz, p=2 Pa.

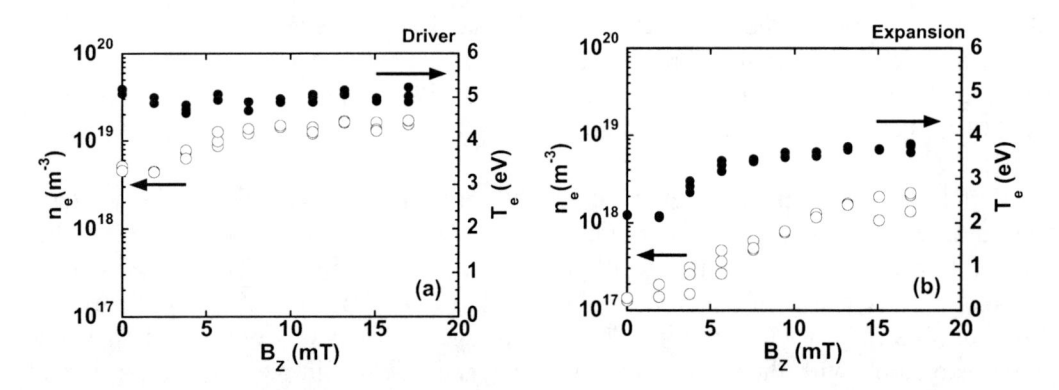

FIGURE 5. Dependences of electron density and temperature on axial magnetic field in (a) driver and (b) expansion regions. P_{RF}=13kW, f_{RF}=0.34MHz, p=2 Pa.

helicon mode dispersion relation shows the mode can be excited at low frequency around 0.5MHz in such high density plasma of more than 1×10^{19}m^{-3}. It should be pursued further to utilize the helicon mode discharge for high density operation in RF plasmas.

The density shown in Fig.3 was affected by the insertion of the probe into the driver region. When the density was measured from the other end of the expansion chamber, n_e is more than 1×10^{18}m^{-3} at Z=255mm (expansion region).

Figure 4 shows dependences of n_e on the input RF power at the driver and expansion regions. In the both regions n_e increased linearly to the input power and attained to nearly 1.5×10^{19}m^{-3} in the driver region and 0.5×10^{19}m^{-3} in the expansion region. No data was obtained in the driver with the power of more than 11kW, because plasma became unstable with the probe insertion in the driver.

The dependences of measured electron density n_e and temperature T_e on the applied axial magnetic field are shown in Fig.5. As the axial magnetic field increased, n_e increased more than twice in the driver region and more than one order of magnitude in the expansion region. In a lower magnetic field below 10mT, n_e increased both in the driver and expansion region due to the increase of plasma production rate and

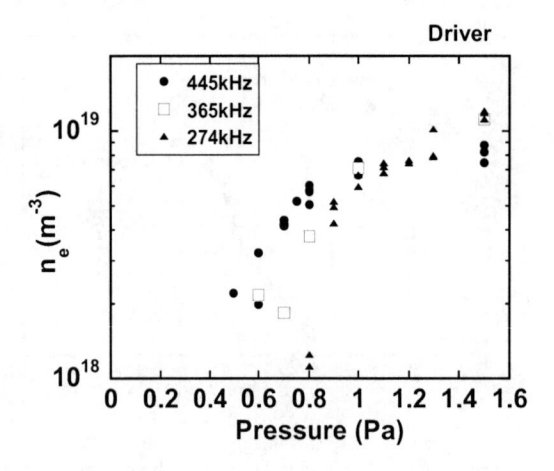

FIGURE 6. Dependences of electron density on filling gas pressure in the driver region with several RF frequencies. P_{RF}=10kW. B_z=11.2mT.

reduction of wall loss. In a higher magnetic field, n_e increased only in the expansion region, which indicates the increase of diffusion rate from the driver to the expansion region.

Figure 6 shows dependences of n_e on the filling gas pressure at different RF frequencies. Although the high density plasma was produced at more than 1 Pa, the density decreased at lower pressure below 1 Pa. When the RF frequency changed slightly (from 0.27 to 0.45MHz), plasma production occurred in lower pressure region with higher frequency. It should be related with mean free path of electrons accelerated by RF induced electric field. Additional small amount of Cs enable to reduce the operational pressure lower than 0.3Pa in the same source. Further investigation is necessary for more efficient RF plasma production in the low pressure region.

H⁻ Ion Density Measured by Laser Photo Detachment

In order to measure the negative ion density in the RF source, photo-detached electrons were measured by using YAG laser (λ=1064nm) photo-detachment system.

Schematic of the laser system and typical waveform of electron saturation current I_{es} measured by a Langmuir probe is shown in Fig.7(a) and (b), respectively. The sudden increase of I_{es} corresponds to H⁻ ion density in the plasma. The dependence of increment ratio of electron saturation current I_{es} on the gas pressure is also shown in Fig.7(c). No Cs was added and the measured position is Z=255mm in the expansion region. As the axial magnetic field applied, the ratio decreased due to the increase of

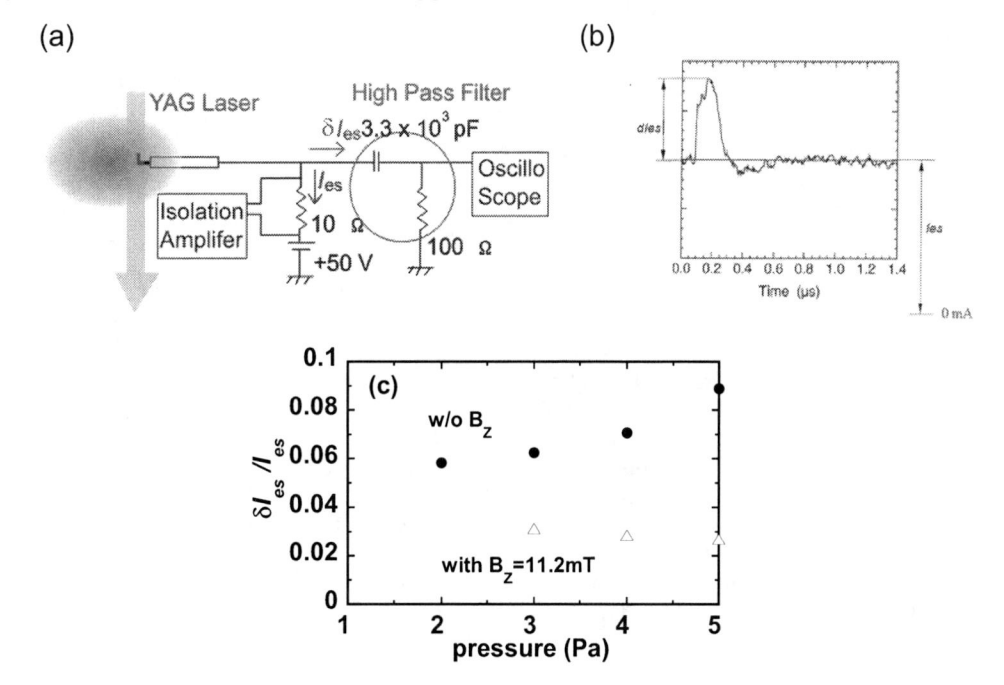

FIGURE 7. (a) Schematic of laser photo-detachment system, (b) typical waveform of electron saturation current I_{es} with laser irradiation, and (c) increment ratio of I_{es} as a function of gas pressure with (open triangles) and without (closed circles) axial magnetic field. P_{RF}=13kW, f_{RF}=0.42MHz. Z=255mm.

energetic electron from the driver. We are going to measure H⁻ density by the photo-detachment method in the extraction region near the plasma grid and to investigate the effect of Cs vapor injection.

SUMMARY

We have investigated characteristics of a small RF ion source by using RF power supply operated with a FET inverter circuit. In the non 50 Ohm matching system, a matching circuit in the RF system is simple to be operated. High density hydrogen plasma was obtained using RF frequency of lower than 0.5MHz with the help of axial magnetic field at the driver region. Effect of the axial magnetic field was examined and the density in the expansion region increased more than one order of magnitude. Axial profiles were measured in the source and electron density more than $10^{19} m^{-3}$ was obtained at the driver region. It was caused by the increase of plasma production rate and reduction of wall loss, and also increase of diffusion rate from the driver to the expansion region affected the increase in higher magnetic field. The working pressure was still high nearly 1Pa. Operation of lower pressure region below 0.5Pa was possible using higher RF frequency and the help of Cs additive.

Negative ion density was measured by using laser photo-detachment technique. H⁻ signal decreased as the axial magnetic field applied, which was caused by the increase of energetic electron from the driver.

ACKNOWLEDGMENTS

This work was supported in part by Grant-in-Aid for Scientific Researches from Japan Society for the Promotion of Science, and under the auspices of the NIFS Collaborative Research Program (NIFS07KOAB010).

REFERENCES

1. Y. Takeiri, *et al.*, *Nuclear Fusion* **47**, 1078–1085 (2007): *ibid.* **46**, S199-S210 (2006).
2. K. Tsumori, *et al.*, *Rev. Sci. Instrum.* **75**, 1847-1850 (2004): *ibid.***81**, 02B117_1-4 (2010).
3. Y. Takeiri, *et al.*, *Rev. Sci. Instrum.*.**81**, 02B114_1-7 (2010).
4. Y. Ikeda, *et al.*, *Nuclear Fusion* **46**, S211-S219 (2006).
5. T. Inoue, *et al.*, *Nuclear Fusion* **46**, S379-S385 (2006).
6. T.Takanashi, *et al.*, *Jpn. J. Appl. Phys.* **35** 2356-2362. (1996).
7. E. Speth, *et al.*, *Nuclear Fusion* **46**, S220-S238 (2006).
8. P. Franzen, *et al.*, *Nuclear Fusion* **47**, 264-270 (2007).
9. W. Kraus, *et al.*, *Rev. Sci. Instrum.* **79**, 02C108 (2008): *ibid.***81**, 02B110_1-3 (2010).
10. U. Fantz, *et al.*, *Rev. Sci. Instrum.* **77**, 03A516 (2006).
11. A.Ando, *et al.*, *AIP conf. Proceedings,* 1097, 291-296 (2009).
12. A.Ando, *et al.*, *Rev. Sci. Instrum.* **81**, 02B107_1-3 (2010)..

Improvement of Plasma Production for Large Area Multi-antenna RF Ion Source

Y. Oka[a], T. Shoji[b], and NIFS-NBI Group

[a]*National Institute for Fusion Science*
-Inter-University Research Institute Corporation NINS-,
Oroshi, Toki 509-5292, Gifu, Japan
[b]*Nagoya University, Chikusa, Nagoya 464-8603, Japan*

Abstract. A multi-antenna RF ion source for negative ions has been studied. In a previous experiment, basic characteristics of a Faraday-shielded multi-antenna RF ion source with a high RF power up to 300kW at 9MHz for 10msec pulse duration were studied. Present studies are focused on a study of the plasma production and/or an improvement of the RF power efficiency with hydrogen. Plasma parameters of T_e, N_{pl}, and V_{pl} were measured with Langmuir probes along with their dependence on the net RF power and the gas pressure. The parameters are calculated also by a simple transport model. On the basis of comparative discussions of two results and the experimental ones, a new Faraday-shielded multi-antenna system with a close coupling configuration is designed and tested for the improvement of plasma production. In a preliminary test of the new multi-antenna system, the positive ion saturation current density of over several amperes per cm^2 is achieved at ~30kW of the net RF power. Correspondingly the power efficiency improves drastically on the order of a tenfold dense plasma production compared to that of an existing antenna system.

Keywords: RF source, negative ion source, NBI.
PACS: R07.77.Ka, 52.50.Dg

INTRODUCTION

An ion source for a neutral beam injection system based on negative ions (NNBI) has been successfully realized by applying a filament-driven arc discharge plasma source in the National Institute for Fusion Science (NIFS) and the Japan Atomic Energy Agency [1, 2]. A large area radio frequency (RF-) driven negative D$^-$/H$^-$ ion source without filaments is required and under development for a long lifetime of NNBI in the International Thermonuclear Experimental Reactor. The large area RF ion source is composed of a multi-cusp chamber as a plasma expansion section and multiple RF plasma source on a small scale [3, 4, 5]. Another type of RF ion source which is a Multi-Antenna Type Ion SourcE (MATISE) has been developed [6]. The multi-antenna is installed like a multi-filament for a large area ion source. An advantage of this source is that high density plasmas for large plasma chamber can be produced by putting the highest RF input power possible without suffering a voltage breakdown across the RF elements, and that an all metal antenna is good from the point of the structurally toughness.

A brief review of our studies of MATISE RF ion sources is undertaken as follows:

- In the negative ion source (NIS) 10 pieces of multi-antenna were arranged in a circle: The chamber of the plasma source reused the 1/5th scaled NIFS-LHD NIS after completing R&D on the filament-driven NIS. The first plasma was produced successfully with the RF power up to 20-30kW for a 10ms pulse duration, and H$^+$ ions were extracted with a small extractor. Studies of RF systems for a parallel connection of the multi-antenna, number of the antenna (1-10 pieces) / RF frequency (2-14MHz), and the operational parameters of the discharge were carried out [7].

- An NIS with linear current conductors (1-4 pieces) for a multi-antenna [8, 9]: A system for high RF power up to 40kW was tested, and the uniformity control of the plasma was studied. Extracting the H$^-$ and H$^+$ ions was tested.

- An NIS with linear current conductors for the multi-antenna with a Faraday shield [6]: First test of the Faraday-shielded multi-antenna was made on studies of the plasma production and the uniformity control under operation with a high RF power. The high RF power up to 300kW was transferred to the antenna without the RF breakdown, and high density plasma with a positive ion current density of 0.148 A/cm^2 was produced at 174kW for 10msec. H$^-$ ions were extracted, too.

Present research on the MATISE RF ion source with the Faraday-shielded multi-antenna aims to study the plasma parameters, and improve the plasma production especially the RF power efficiency. Ion source and operation of the plasma source at a high RF power level are described in a previous paper [6]. In the following first section, the experimental system is described simply. In next section of experiments, the plasma parameters, the power dependences and the gas pressure dependences are measured by Langmuir probes. The density distribution along the z-axis (i.e., beam axis) and the effects of the plasma production on the error magnetic field are studied. In the following section on calculations, the plasma density and the electron temperature are evaluated by a simple particle transport model. Calculated and experimental results are comparatively discussed from the view point of the

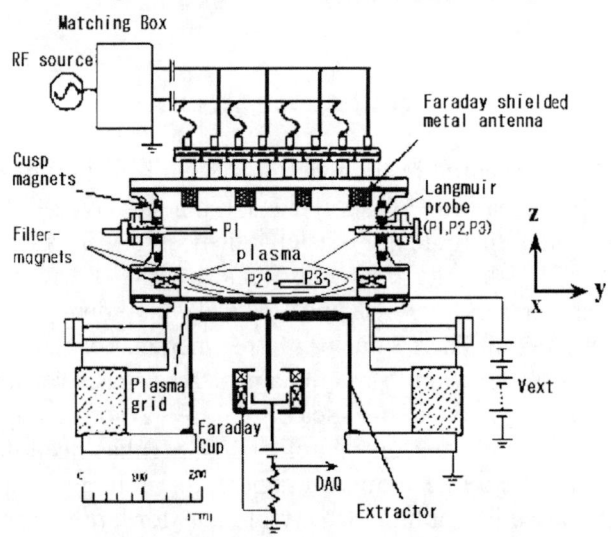

FIGURE 1. A schematic diagram of a multi-antenna type rf ion source (MATISE) and the instruments.

improvement of plasma production and power efficiency. In the next section, a new multi-antenna for the purpose of improved plasma production is tested in terms of a working hypothesis about a strong coupling of the RF field to the electrons. The achieved ion saturation current density is raised drastically in a preliminary test. The power efficiency seems to be improved by a factor of over 10 compared to the one in an existing multi-antenna system.

EXPERIMENTAL SYSTEM

The ion source, RF system, and the probes are described in the following paragraphs. The new antenna will be described in a later section. A schematic diagram of the MATISE RF ion source and the major instruments are shown in Fig.1 [6]. A multi-antenna system is installed in the 1/5th scaled LHD-NNBI ion source (of which size is 35 x 35 x ~18 cm^3) with a small single hole extractor (0.5 cm in diam.) for H$^+$/H$^-$ -beam. An existing RF antenna is composed of 4 linear current-conductors and a Faraday shield, as shown in Fig.2.

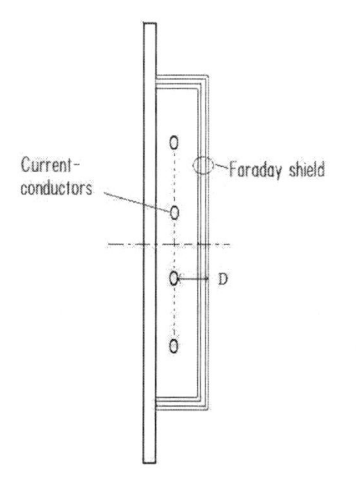

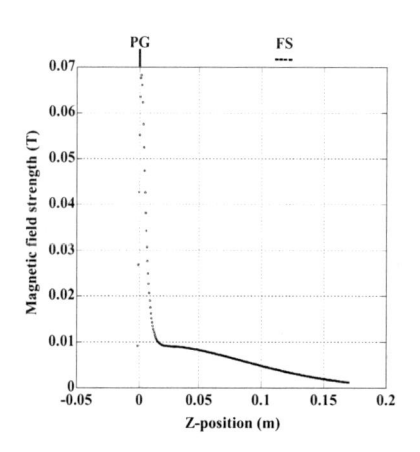

FIGURE 2. A cross-sectional sketch of the multi-antenna system.

FIGURE 3. A distribution of the magnetic field strength along the z-axis (i.e., beam axis) in a superposed magnetic filter.

Each conductor is a copper pipe of 0.6 cm in diam. and 20cm in length, and their ends pass through feed-through ports on the back plate of the plasma source. The separation of parallel antenna conductors is 7cm each. The Faraday shield (FS) for shielding the electrostatic field is composed of 9 plates covering the current conductors. The RF power up to 300kW at 9MHz for a 10msec pulse duration was transferred to the antenna via matching network circuits. The net RF power in this paper is defined by differences of forward-power minus reflected-power at the output of the RF generator. The number of the antennas adopted for normal operation was 4 ones, which were connected electrically in parallel.

Plasma parameters were measured with Langmuir probes. The origin of the coordinate of the probe is taken at the center of the plasma grid (PG), i.e., the center of the extraction hole (in Fig.1). Positive ion saturation current / the current density (J^+_{sat}) are measured by a cylindrical probe P1 and P3. The former is movable along the y-axis (i.e., upward at the bottom port) around the center of the plasma source in the z=7.5cm plane. The probe is biased at -180 to -220V with respect to the plasma source chamber and the PG. The electron temperature (T_e) is measured by a triple probe P2 having three collecting needles. It is movable along the x-axis (horizontally) at the side port in the z~1.7cm plane from the PG. The plasma potential is deduced from the floating potential and T_e. The distribution of J^+_{sat} along the z-axis (i.e., beam axis) is measured by a newly attached probe, P3, which has a long L-hand of 11cm in length for rotatable scanning. The probe is set at the side port in the z~1.7cm plane.

For a magnetic filter to control T_e, there is a superposed magnetic filter (of 1245Gcm) consisting of the external filter plus the localized virtual magnetic filter (LMF) embedded inside the PG over ~5x5cm^2 [10]. Figure 3 shows the magnetic field distribution along the z-axis from the PG to the back plate. Around the location of the multi-antenna, the error field coming from the tail of the external filter field for the most part ranges from 44 to 32G. The orientation of the current conductor to the error field lines is orthogonal in normal operation. In another test configuration it is parallel.

EXPERIMENTS

Plasma parameters

When the net RF power is changed, plasma parameters are shown in Fig.4. J^+_{sat} increases with power. Inductively coupled dense plasmas are produced efficiently beyond approximately 100kW of the net power (which corresponds to a threshold power to an inductive mode of operation at high densities). The maximum value of

FIGURE 4. Dependence of the plasma parameters (T_e, J^+_{sat}, V_{pl}) on the net rf power at 3.7mTorr.

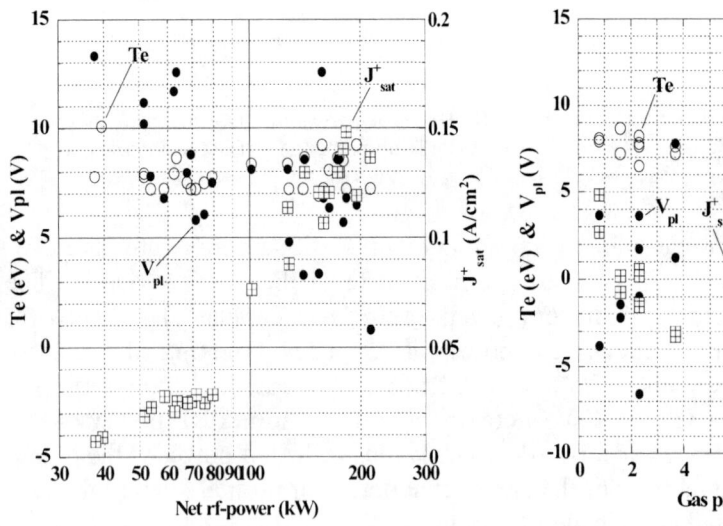

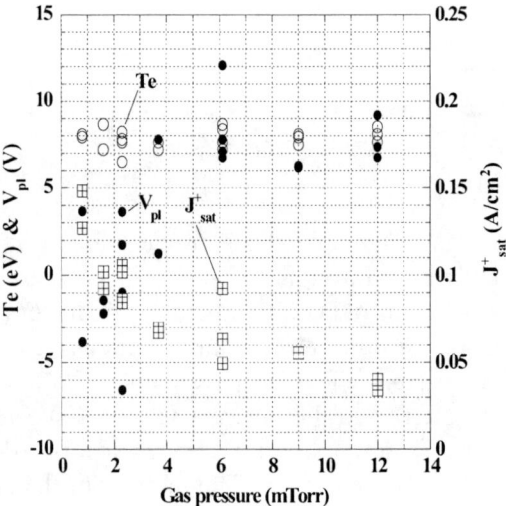

FIGURE 5. Dependence of the plasma parameters (T_e, J^+_{sat}, V_{pl}) on the gas pressure at 160-170kW.

J^+_{sat} is 0.148A/cm^2 at 174kW of net power with 3.7mTorr. This power dependence behaves similar to the result of an extracted beam of positive H$^+$ ions with a small extractor [6].

The electron temperature (by the probe P2) lies in the range of 7 to 10eV and the power dependence is small. The value of T_e seems to be a little high, and is similar to that in the filament-driven NIS in the driver region, rather than in the extraction region. The reason for a fairly high T_e in this NIS is understood to be the fact that the filter strength at a measurable position of probe P2 is an insufficiently weak filter intensity of 692Gcm from FS to P2 position, as evaluated from the field distribution in Fig.3. T_e should reduce further after crossing the LMF filter existing only in the local vicinity of the PG. The increase of J^+_{sat} with RF power is mostly due to the increase in the plasma density because T_e is almost constant.

The plasma potential (V_{pl}) ranges from 13V to 1V. The data points at a given RF power seem to be scattered over many operations. The reason for this is not clear. Experience suggests that only a small part of the net power is absorbed effectively into the produced plasmas. It might be that scattering sensitively affects plasma production and/or J^+_{sat}, especially the threshold power for high density plasma production.

When the gas pressure in hydrogen is changed at 160-170kW of net power, plasma parameters are shown in Fig.5. J^+_{sat} takes a high value at a low gas pressure until ignition pressure. This dependence of J^+_{sat} on gas pressure results in a similar dependence of the extracted H$^+$ ion beam current [6]. T_e is around 7-9eV in this pressure range.

V_{pl} ranges -6V to12V with varying gas pressure, as can be seen also from Fig.5. At a low pressure, a negative plasma potential seems to occur. It was observed during negative ion beam extraction in volume produced plasmas that when V_{pl} became more negative (for example, plasma produced without Faraday shield at high RF power), the extractable negative ion beam current appeared to decrease (to possibly nothing). This might be attributed to deterioration in the ion optics at the meniscus as well as increasing losses of the negative ions to the plasma source wall.

By estimate, the power efficiency of plasma production with the present MATISE source was considered to be fairly low compared to that of the filament-driven NIS in the same chamber. We need to improve considerably the efficiency as will be discussed in calculation section, despite the accomplishment of high positive ion current density.

Z-axis distribution

The J^+_{sat} distribution along the z-axis is shown in Fig.6. It appears to be a smooth distribution within a region of external magnetic field. There seems to be no abnormal localization of plasma production, leading to a decrease in the efficiency of volume plasma production. Approaching the PG, J^+_{sat} decreases by a factor of about 3 with respect to that around the center. It would come from cross diffusion across the magnetic filter field. If the probe P3 can deeply scan a region about the LMF magnetic field near the PG, as similarly discussed about T_e in Fig.4, J^+_{sat} should decrease further.

The power dependence of J^+_{sat} under two configurations of the external filter related to the current conductor orientation and the magnetic field lines was checked through the conditioning shots. From the view point of a considerable improvement required on the order of a tenfold power efficiency, it was concluded that changing the field configuration did not lead to an improvement (or change) of the discharge operation, nor a maximization of J^+_{sat}, although the data points in each set were scattered. As a result of this, the existing error field of 44-32G at the Faraday shielded multi-antenna system in the NIS arrangement would be of no effect concerning plasma production and efficiency.

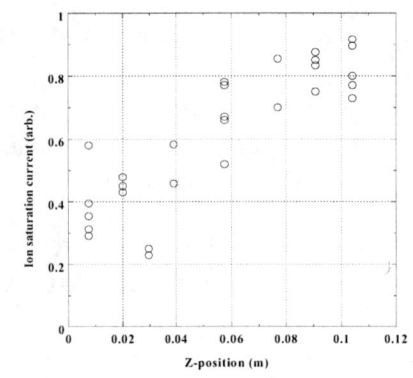

FIGURE 6. Distribution of J^+_{sat} along the z-axis.

CALCULATION OF N_{PL} AND T_e

N_{pl} and T_e are calculated by a simple analytical model. The results are compared with the experimental ones, and then the adopted approach to an improvement in plasma production is discussed.

The model uses a plasma particle balance equation (1) and the power balance equation (2), for which details are written elsewhere [11].

$$0.6 n_{pl} u_b A = K n_{pl} n_o V \qquad (1)$$

$$P_{ab} = 0.6 e n_{pl} u_b A E_t \qquad (2)$$

where in Eq.(1), n_{pl} is the plasma density at the center, u_b is the Bohm velocity, A is the loss area of the plasma source, n_o is the gas density in hydrogen, V is the volume of the plasma source, and K is the ionization rate constant at a given T_e and is taken from the data base of IAEA [12]. In Eq.(2), P_{ab} is the absorbed RF power into the plasma, e is the elementary charge, E_t is the total energy lost per ion as calculated in the following equation (3).

$$E_t = E_c + 2T_e + E_i = E_{ioniz} + 5.3 T_e \qquad (3)$$

where E_c is the collisional energy loss per e-i pair created and it assumes 1~2 times of the ionization potential of 13.6eV for a hydrogen atom, $2T_e$ is the mean kinetic energy lost per electron, and E_i is the mean kinetic energy lost per ion using $3.3T_e$.

Figure 7 shows the dependence of the calculated T_e, n_{pl} and J^+_{sat} on the absorbed RF power at 3.7mTorr. When the RF power increases, n_{pl} and J^+_{sat} are increased. T_e maintains a constant 7.2eV which approximates the experimental one. Figure 8 shows a dependence of the calculated parameters on the gas pressure at 180kW of the absorbed RF power. When the gas pressure increases, n_{pl} and J^+_{sat} are increased, while T_e is decreased inversely.

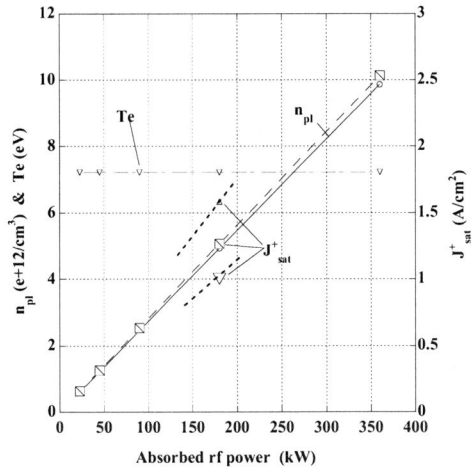

FIGURE 7. A dependence of calculated T_e, n_{pl}, and J^+_{sat} on the absorbed rf power at 3.7mTorr. As a reference, J^+_{sat} (marked by small triangles) is the case with 2 time the reaction rate of K assumed, and J^+_{sat} (marked by inverse larger triangle) is the case with 2 times the ionization potential assumed.

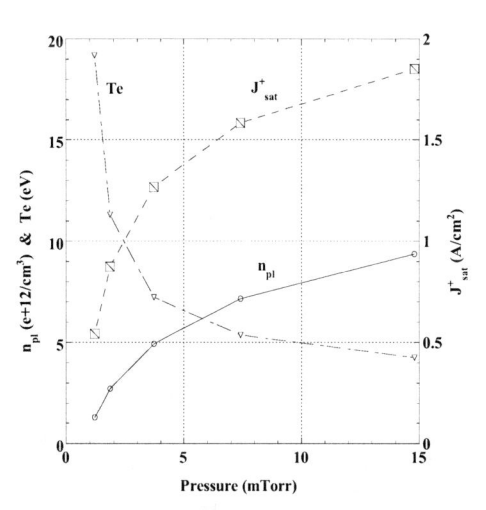

FIGURE 8. A dependence of the calculated T_e, n_{pl}, and J^+_{sat} on the gas pressure at the absorbed power of 180kW.

By calculation, J^+_{sat} is roughly 1.3A/cm² at 180kW of given absorbed power depending considerably on the pressure as well as an assumed calculation parameters, while by experiment it is ~0.15A/cm² at ~180kW of net power as described in the previous section of experiment. It means that the power efficiency in the MATISE RF source is low by a factor of about 9 compared to that of the calculation. It is also known empirically that the power efficiency by this MATISE RF plasma source is quite low compared to that of the filament-driven plasma source, as mentioned in the previous section. This low efficiency was considered to come from the fact that only a fraction of the net RF power in the experiment could be absorbed usefully into plasmas as the absorbed RF power used by the calculation. In order to increase the efficiency of the plasma production, it is required efficiently to increase the absorbed power fraction and improve the charged-particle confinement in the discharge. Our first priority is the improvement in the fraction of the absorbed power, and we will design a newly improved antenna system in the next section.

The gas pressure dependence of J^+_{sat} and probably T_e by experiment (Fig.5) and calculation (Fig.8) behaves oppositely. In Eq.(1) from the model, an increasing ionization in the volume of the plasma source by the increase of gas pressure (which results in an increased n_{pl}) balances a decreasing T_e in terms of the collision rate constant. In the experiment, energy gain of electrons by RF near field through e-n collisions including inelastic collision such as ionization ($\omega < \nu_e$ in this experiment) decreases with gas pressure. This may leads to the decrease of plasma production with the pressure (which is above ~5mTorr).

EXPERIMENTS WITH AN IMPROVED MULTI-ANTENNA

New antenna

A new antenna design to improve the fraction of the absorbed power aims to increase a coupling efficiency of the RF power to plasmas. A guiding principle of the design is to increase the RF inductive field at the plasma edge. Figure 9 shows a photograph of the exterior of the new Faraday-shielded multi-antenna system. A cross-sectional sketch of the antenna is shown in Fig.10.

FIGURE 9. A photograph of the new multi-antenna system.

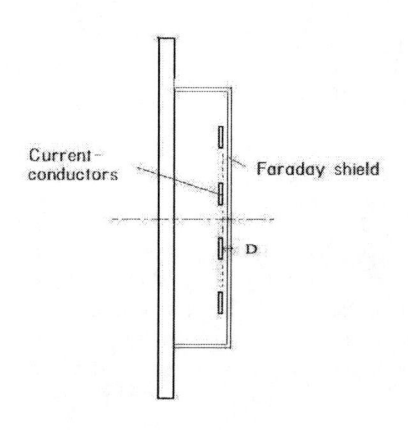

FIGURE 10. A cross-sectional sketch of the new antenna with a close coupling configuration.

Current conductor of the multi-antenna is composed of a stainless steel plate which has a rectangular cross section of 2.5cm in width and 20cm in length, instead of pipe conductors in the existing antenna (in Fig.2). Four conductors are electrically connected in parallel, again. FS is composed of a metal housing box having 9 slots perpendicular to the RF current. The surface of the FS is flat, while the previous one has a concave plane (as can be seen from Fig.2).

According to the two above-mentioned points, along with the large area current conductor and the short gap to the plasma, the new antenna could realize a close proximity of the RF field to the plasma i.e., a close coupling antenna system. A standard of the distance, D in Fig.10 is 0.65cm while in the existing antenna the distance D is ~2cm. The residual space inside of the FS is filled with ceramic dielectrics. The antenna components are not water-cooled for 10msec pulsed operation. The location of current conductor (i.e., corresponding gap D) is changeable, if needed.

Plasma production with the new antenna and discussion

Plasma production is tested with the new antenna system. Unfortunately an air leak occurred in the assembly at the back plate/flange (not the antenna). The base pressure with the air leak was 6.2mTorr, and the partial pressure at hydrogen introduced was 1.6mTorr, in preliminary experiments. When the net RF power increases, J^+_{sat} by probe P3 at z~2.2cm increases, as can be seen from Fig.11.

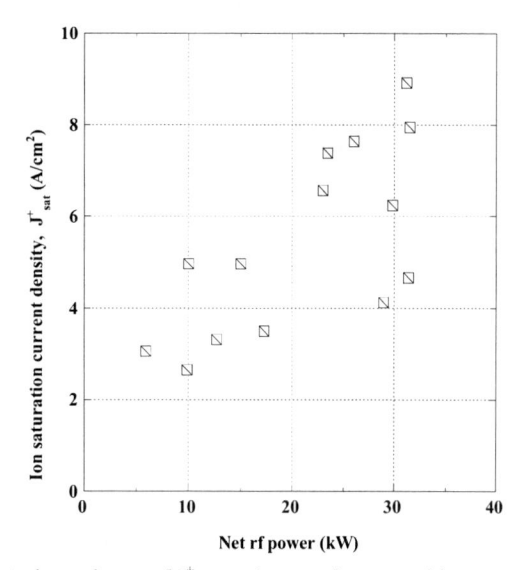

FIGURE 11. A dependence of J^+_{sat} on the net rf power with new antenna in first test.

Dense plasma (by visual observation, too) seems to be produced efficiently. It comes from the fact that the large surface area (correspondingly large RF field area) of the current conductor plate which can face the FS / plasma was realized, and furthermore it is possible for it to be moved closer to the FS / plasma edge compared to the pipe conductors. A maximum value of J^+_{sat} reaches several amperes more per cm^2 with the net power of ~30kW. The power efficiency of the plasma production in terms of current density improves over a factor of ten from a simple model calculation. As a result of this test, it is considered that an increasing fraction of the absorbed RF power into plasmas with respect to the given net power was realized. It is found that the antenna system consisting of a close coupling configuration makes a considerable advancement in the plasma production with Faraday-shielded multi-antenna plasma source for the MATISE RF ion source.

CONCLUSION

To improve the plasma production especially the power efficiency in a Faraday-shielded multi-antenna RF ion source of MATISE, the plasma parameters were obtained by Langmuir probe measurement and the calculation which used a simple analytical model. Also, effects of plasma production on the filter magnetic field were studied. A comparison of J^+_{sat} and T_e by experiment and calculation deduced that high density plasma with an improved power efficiency could be produced with a large fraction of the absorbed RF power into plasmas with respect to the net RF power for the MATISE negative ion source. To make the rf field coupling intensify, i.e., resulting in a close coupling multi-antenna system, a short gap between the current conductor and the plasma is realized. A newly designed Faraday shielded multi-antenna system consisted of 4 plate-conductors having a large plasma-facing area with a close proximity of the current conductor to the FS and plasma edges. A high density plasma of J^+_{sat} of several amperes per cm^2 at ~30kW of net power is generated in the preliminary test with a considerable improvement of the power efficiency by a factor of over 10 compared to that of the existing antenna system.

ACKNOWLEDGMENTS

This work was supported by the National Institute for Fusion Science collaborating research programs (NIFS10KCBR009 and NIFS10KCBR007). The authors acknowledge continuous supports and encouragement from Drs. O. Kaneko and Y. Takeiri in NIFS.

REFERENCES

1. O. Kaneko, this conference.
2. M. Kuriyama, N. Akino, T. Aoyagi, N. Ebisawa, N. Isozaki, A. Honda, T. Inoue, T. Itoh, M. Kawai, M. Kazawa, J. Koizumi, K. Mogaki, Y. Ohara, T. Ohga, Y. Okumura, H. Oohara, K. Oshima, F. Satoh, T. Takenouchi, Y. Toyokawa, K.Usui, K. Watanabe, M. Yamamoto, T. Yamazaki, and C. Zhou, *Fusion Engineering and Design*, **39-40**, pp115-121(1998).
3. Y. Oka, T. Shoji, T. Kuroda, O. Kaneko, and A. Ando, *Rev. Sci. Instrum.* 61, pp398-400(1990).
4. W. Kraus, H.D. Falter, U. Fantz, P. Franzen, B. Heinemann, P. McNeely, R. Riedl, and E. Speth, *Rev. Sci. Instrum.* **79**, 02C108 (2008).
5. P. Franzen et al., this conference.
6. Y. Oka, T. Shoji, O. Kaneko, Y. Takeiri, K. Tsumori, M. Osakabe, K. Ikeda, K. Nagaoka, E. Asano, M. Sato, T. Kondo, M. Shibuya, and S. Komada, *AIP Conference Proceedings* **1097**, p282(2008)
7. Y. Oka, T. Shoji, M. Hamabe, Y. Sakawa, C. Suzuki, K. Ikeda, O. Kaneko, K. Nagaoka, M. Osakabe, Y. Takeiri, K. Tsumori, E. Asano, T. Kawamoto, T. Kondo, and M. Sato, *Rev. Sci. Instrum.* **75**, pp1841-1843 (2004).
8. T. Shoji, Y. Oka, and LHD NBI Group, *Rev. Sci. Instrum.* **77**, 03B513 (2006).
9. Y. Oka, T. Shoji, K. Ikeda, O. Kaneko, K. Nagaoka, M. Osakabe, Y. Takeiri, K. Tsumori, E. Asano, T. Kondo, M. Sato, and M. Shibuya, *Rev. Sci. Instrum.* **77**, 03B506 (2006).
10. Y.Oka, A. Ando, O. Kaneko, Y. Takeiri, K. Tsumori, R. Akiyama, and T. Kawamoto, *J. Vac. Sci. Technol.* A 12(6), pp3109-3114(1994).
11. For example, in the book by M. A. Lieberman and A. J. Lichtenberg, *"Principles of plasma discharges and materials processing"*, John Wiley & Sons, Inc. New York, 1994
12. R.K. Janev, W. D. Langer, K. Evans Jr, D. E. Post Jr., H-He Plasma (1987).

Analysis of Discharge Initiation in a RF Hydrogen Negative Ion Source

T. Hayami[a], S. Yoshinari[a], R. Terasaki[a], A. Hatayama[a], A. Fukano[b]

[a]*Graduate School of Science and Technology, Keio University,*
3-14-1 Hiyoshi, Kouhoku-ku, Yokohama 223-8522, Japan
[b]*Monozukuri Department, Tokyo Metropolitan College of Industrial Technology,*
Higashioi, Shinagawa, Tokyo 140-0011, Japan

Abstract. The maintenance free RF ion source is expected to be one of the most promising candidates for the negative ion sources of plasma heating for fusion reactors. In order to make clear the condition for the discharge initiation of the RF source, we are developing an electromagnetic PIC model. The numerical result shows that a positive potential built-up with respect to the wall. As a result, the electron wall loss decreases and the electron density increases. The positive potential plays a key role for the suppression of wall loss and the electron confinement. The electromagnetic-PIC model developed is useful for the analysis of discharge initiation of the RF source.

Keywords: RF negative ion source, Initial discharge, Positive potential, PIC model
PACS: 29.25.Ni, 52.80.Pi,

INTRODUCTION

Negative hydrogen ion sources play a key role for neutral particle beam injection which is an efficient method for heating fusion plasma. The arc sources suffer from regular maintenance intervals due to the limited lifetime of the filaments[1]. As an alternative to the arc sources, the RF sources have been developed for H⁻ production. The RF source is basically maintenance-free in operation which is quite beneficial for the remote handling requirements of ITER. The initiation condition of the RF discharge in H⁻ negative ion sources, however, has not been well understood.

The purpose of this study is to make clear and obtain the knowledge of the physics of discharge initiation and the condition for the discharge initiation by electromagnetic PIC modeling of the RF H⁻ source.

Our PIC model mainly consists of two parts. One is the 2D model of electromagnetic field due to the RF-coil and the plasma current in the cylindrical geometry, in which Maxwell equations are directly solved by the FDTD method[2]. The other part is the analysis of the charged particle energy distribution function in RF-field with Monte Carlo Method[3], where the equations of motions for the charged particles are numerically solved. The important collision processes for the discharge initiation such as hydrogen molecule ionization are taken into account. Our self-consistent model including not only the electron dynamics but also the ion dynamics has been developed and electron and ion dynamics has been analyzed.

In this paper we will discuss the importance of the positive potential with respect to the wall which is created by electron and ion dynamics for discharge initiation in RF negative ion source.

SIMULATION MODEL

2D model of electromagnetic field

Electromagnetic field in the negative ion source is numerically calculated by the following Maxwell equations,

$$\nabla \times \boldsymbol{E} = -\frac{\partial \boldsymbol{B}}{\partial t} \qquad (1)$$

$$\nabla \times \boldsymbol{B} = \mu_0 \boldsymbol{j} + \mu_0 \varepsilon_0 \frac{\partial \boldsymbol{E}}{\partial t} \qquad (2)$$

where $\boldsymbol{E}$ and $\boldsymbol{B}$ are the electric field and the magnetic field, respectively. The current density $\boldsymbol{j}$ consists of the RF-coil current $\boldsymbol{j}_{rf}$ and the plasma current $\boldsymbol{j}_{plasma}$. In order to solve Eqs. (1) and (2), the FDTD method has been applied. The axial symmetry ($\partial/\partial \theta = 0$) has been assumed and two dimensional model in the r and z directions (r, z) is used in analysis as shown in Fig. 1. A simple model is used for the RF-coil as shown in Fig. 1. Five independent co-axial coils are set and their coil-centers are located at the z-axis. Not only the electron contribution but also the ion contribution is included in the plasma current. Furthermore, three components (r, θ, z) of the plasma current is taken into account in Eq. (2).

The calculation region in the r-z plane in Fig.1 is divided into $N_r \times N_z$ cells with $N_r = 48$ and $N_z = 84$, in the r and z directions, respectively. The plasma current at the numerical grid points is obtained by the linear interpolation from the electron and ion current at each space point. The following boundary conditions are used for Eqs. (1) and (2) at each boundary: i) $E_\theta = 0$ at $r = 0$, ii) the radiation boundary condition[4] by Bayliss and Turkel is used at the boundary $r = r_{max}$ in Fig. 1, and iii) the absorbing boundary condition[5] by Mur at the boundary $z = z_{min}$ and $z = z_{max}$ in Fig.1.

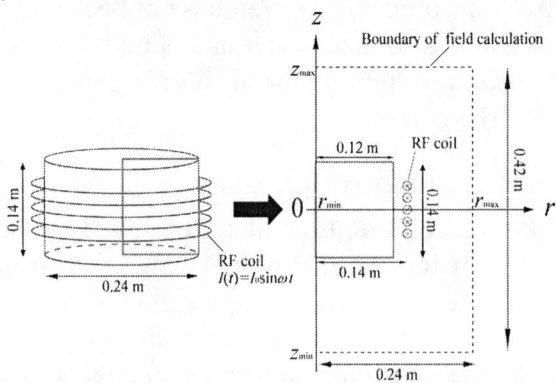

FIGURE 1. Model geometry of the numerical simulation.

Monte-Carlo model of the electron energy distribution function

Electron/ion trajectories in the electromagnetic field as described previously is followed by equation of motion,

$$m_j \frac{d\mathbf{v}_j}{dt} = q_j(\mathbf{E} + \mathbf{v} \times \mathbf{B}) + \text{collision} \quad (j = \text{ion, electron}) \quad (3)$$

where m_j, v_j and q_j are the electron and ion mass, velocity and charge, respectively. The Boris-Buneman version of the leap-flog method[6] is used to solve Eq. (3). The electric and magnetic fields at the electron/ion position in Eq. (3) are interpolated from the values at the four neighbor grid points calculated by the FDTD method as mentioned before.

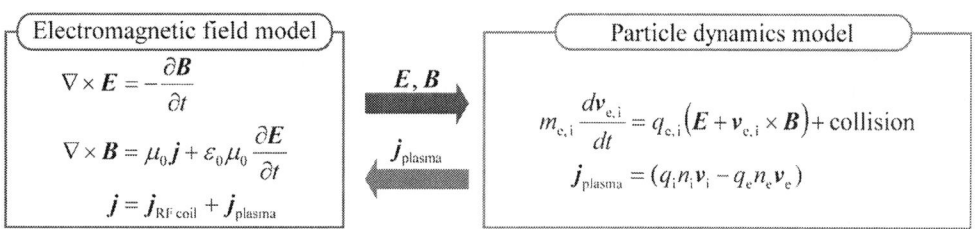

FIGURE 2. Basic concept of numerical calculation.

In our present model, the collision process for ion is not considered for the simplification. This process need to be taken into account in the future.

As for collision processes for electron, i) inelastic and elastic collisions with hydrogen atoms, molecules, ions and ii) elastic collision (Coulomb collision) with electrons are taken into account. The former i) is modeled by the Null-Collision (N.C.) method[7], while the latter ii) is by the Binary Collision (B.C.) method[8]. Table 1 summarizes the collision processes with hydrogen particles included in the analysis.

The particle reached at the wall is removed from the calculation. Trajectories of secondary electrons produced by each ionization event are also followed.

TABLE 1. Reactions taken into account in the simulation code.

Collision Species	Before	After	Reference
Elastic	e + H	e + H	[9]
	e + H_2	e + H_2	[10]
Ionization	e_1 + H(1s)	e_1 + e_2 + H^+	[11]
	e_1 + $H_2(X^1\Sigma_g^+)$	e_1 + $H_2(v)^+$ + e_2	[11]
Dissociative Ionization	e_1 + $H_2(X^1\Sigma_g^+)$	e_1 + H^+ + H(1s) + e_2	[11]
	e_1 + H_2^+ (0≤v≤9)	e_1 + H^+ + H^+ + e_2	[11]
Dissociation	e_1 + $H_2(X^1\Sigma_g^+)$	e + H(1s) + H(1s)	[11]

Model geometry and simulation condition

The time steps Δt_{EM} and Δt_e, respectively for the electromagnetic field and the electron trajectories, are taken to be the same, and $\Delta t_{EM} = \Delta t_e = 1.0 \times 10^{-12}$s mainly due to courant condition[12]. Collision processes with hydrogen species are judged at every 1.0×10^{-10}s time-step in the null-collision method, while 1.0×10^{-10}s is used for e-e Coulomb collision in the binary collision method.

The dimensions of the cylindrical discharge chamber in Fig.1 are as follows: the radius of 0.12m ($0 < r < 0.12$m), the height of 0.14 m (-0.07m $< z < 0.07$m). In addition, the dimensions of the calculation domain for the electromagnetic calculation by the FDTD method are taken to be $0 < r < 0.28$m and -0.21m $< z < 0.21$m in order to avoid the large reflection at the calculation boundary. The each coil radius is set to be 0.14m. Table 2 summarizes the calculation conditions and parameters for the calculations of electron/ion trajectories, background hydrogen species, and RF-coil.

The RF-coil parameters (such as frequency, current wave form, power), the initial gas pressure, the initial electron density has been parametrically changed case-by-case. For simplicity, the initial energy distribution of electrons is assumed to be Maxwellian distribution with the room temperature (300K). Their spatial distribution is assumed to be uniform. The density of hydrogen molecular ions H_2^+ has been changed in the self-consistent manner as ionization and recombination events. Detailed simulation parameters are shown in Table 2.

TABLE 2. Simulation parameters.

RF frequency	f_{RF}	[MHz]	variable
Coil current amplitude	I_0	[A]	variable
H_2 gas pressure	P_{H_2}	[Pa]	variable
H_2 temperature	T_{H_2}	[K]	300
Dissociation rate of H_2	α_{H_2}	-	10%
H temperature	T_H	[K]	300
Initial electron temperature	T_{e0}	[K]	300
Initial electron density	n_{e0}	[m^{-3}]	1.58×10^{11}
Initial H_2^+ temperature	T_{i0}	[K]	300
Initial H_2^+ density	n_{i0}	[m^{-3}]	1.58×10^{11}
Cell size (r – direction)	Δr	[m]	0.005
Cell size (z – direction)	Δz	[m]	0.005
Grid point number	(N_r, N_z)	-	(48, 84)
Time step	Δt_{EM}	[s]	1.0×10^{-12}

RESULT AND DISCUSSION

We have calculated the time evolution of the electron density in the discharge chamber in two cases. The density is the averaged density over the whole discharge volume. First, the case without the effect of the plasma current has been studied. In this case (Case A), we have calculated the electron dynamics and density in the electromagnetic field produced only by the external RF-coil current j_{rf}, without including the plasma current j_{plasma}. Secondly, the case with the effect of the plasma current has been studied. In this case (Case B), we have calculated the electron/ion dynamics and density in the electromagnetic field produced both by the external RF-coil current j_{rf} and the plasma current j_{plasma}. Table 3 summaries the difference between Case A and Case B.

TABLE 3. Difference between case A and case B.

Case A	· Calculate the electromagnetic field produced only by the external RF-coil current j_{rf}. · Calculate only the electron dynamics and density.
Case B	· Calculate the electromagnetic field produced by both the external RF-coil current j_{rf} and the plasma current j_{plasma} ($j_{plasma} = n_i q_i v_i - n_e q_e v_e$). · Self-consistent model including not only the electron dynamics but also the ion dynamics has analyzed in order to calculate j_{plasma} accurately.

With or Without j_{plasma}

Figure 2 shows the time evolution of the electron density under the realistic discharge parameters (peak coil current I_{peak}=100A, RF-frequency f = 1MHz, initial gas pressure P_0=0.3Pa, initial electron density n_{e0}=1.58×10^{11}m^{-3}).

Under this condition in Case A, the electron density gradually decreases in time. Ionizations of hydrogen molecules and atoms leads to the increase in the electron density, while the wall loss reduce it. The latter, especially, the wall loss is dominant and the electron density decreases with time as shown in Fig. 3 (W/O j_{plasma}). We have also calculated the case where the parameters such as the amplitude of the RF coil, hydrogen gas pressure have been changed[13]. However, the result has been similar as shown in Fig. 3 (W/O j_{plasma}).

It is expected that the electrons are lost easily, while ions remain in the discharge chamber due to large mass difference. This might lead to a positive potential built-up with respect to the wall. In order to take into account self-consistently this potential built-up, not only the electron dynamics, but also ion dynamics should be taken into account in the model.

For the reason discussed above, we have calculated the time evolution of the electron density with the consideration of both j_{rf} and j_{plasma} (Case B). In Case B, the electron density increases as shown in Fig. 3 (With j_{plasma}). Whether the electron density increases or not depends on the balance between the ionization and the wall loss. In order to understand the increase in the electron density for Case B, we will discuss the electron production (ionization) and loss mechanism in detail below.

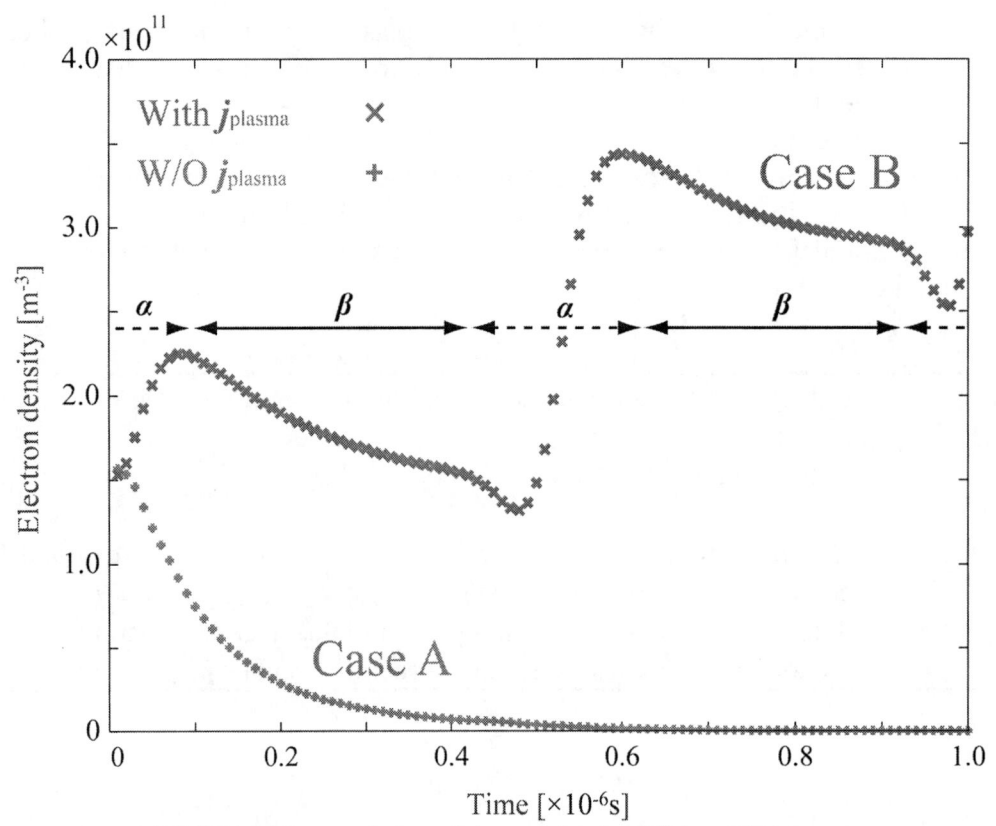

FIGURE 3. Time evolution of electron density (With or W/O j_{plasma}).

Ionization and Loss Mechanism of Electron

As for the ionization mechanism, the amount of the ionization per unit time (ionization rate) increases at each half-cycle of the RF period (Phase α). This increase in the ionization rate is mainly due to the increase in the electron energy, the acceleration trough the RF-electric field $\tilde{E}_\theta$. This ionization-increasing period corresponds to the Phase α around which the azimuthal amplitude of the RF electric field $|\tilde{E}_\theta|$ reaches its maximum $|\tilde{E}_\theta| = |\tilde{E}_\theta|_{MAX}$, while the RF magnetic field $|\tilde{B}_Z|$ becomes the minimum ($|\tilde{B}_Z| = |\tilde{B}_Z|_{min} = 0$). In this phase, the electron energy rapidly increases due to the large amplitude of the RF electric field $|\tilde{E}_\theta|$ and easily exceeds the threshold energy of the hydrogen ionization (13.6eV). As a result, the ionization rate rapidly and significantly increases. In this phase, however, the axial RF magnetic field becomes almost zero $|\tilde{B}_Z| \approx 0$ as mentioned above. Therefore, the electron loss towards the side wall, i.e., the radial loss of the electrons also rapidly increases as will be discussed. Only if the increase in the ionization rate due to electron acceleration by the RF electric field $|\tilde{E}_\theta|$ overcomes the increase in the radial loss without the confinement-effect of the axial RF magnetic field $|\tilde{B}_z|$, it is possible for the number of electrons in the system to increase.

As mentioned above, during the Phase α, the radial electron loss rapidly increasing and dominates the electron particle confinement. This is partially due to the loss of the radial confinement by the axial RF magnetic field $|\tilde{B}_z|$, because $|\tilde{B}_z| \approx 0$ as mentioned above. In addition, it should be noted that during this phase, the average electron energy rapidly increases by the $|\tilde{E}_\theta|$-acceleration. The characteristic loss time for such energetic electrons becomes very short. This is also the reason why the electron loss rapidly increases during the Phase α.

On the other hand, Phase β, i.e., during the period when the axial RF magnetic filed becomes relatively large and the RF electric field $|\tilde{E}_\theta|$ becomes relatively small, the electron loss towards the top and bottom wall dominates the radial loss. The axial loss of electrons becomes the main loss mechanism of electrons during this phase, because electrons can be magnetized/trapped by the axial RF-magnetic field $|\tilde{B}_z|$ and are mainly lost along the field line. It should be noted that the average electron energy is considerably small during this Phase β compared with that during the Phase α. Therefore, the ionization hardly takes place and the electron loss time along the field line becomes relatively longer compared with that during the $|\tilde{E}_\theta|_{MAX} - |\tilde{B}_z|_{MIN}$ (Phase α).

The Effect of Positive Potential

As mentioned above, the electron radial loss rapidly increases initially in Phase α. Almost all the ions, however, still remain in the discharge chamber due to their large mass. This difference in dynamics between electrons and ions leads to the positive potential built-up with respect to the wall. Figure. 4 shows the potential profile on (r, z) plane at t = 1.0×10^{-6}s. This potential built-up suppresses further electron radial loss in Phase α, then the electron loss rapidly decreases in the last period on the Phase α. As a result, the ionization overcomes the radial electron loss. This is the reason why the electron density in Case B increases while the electron density in Case A monotonically decreases.

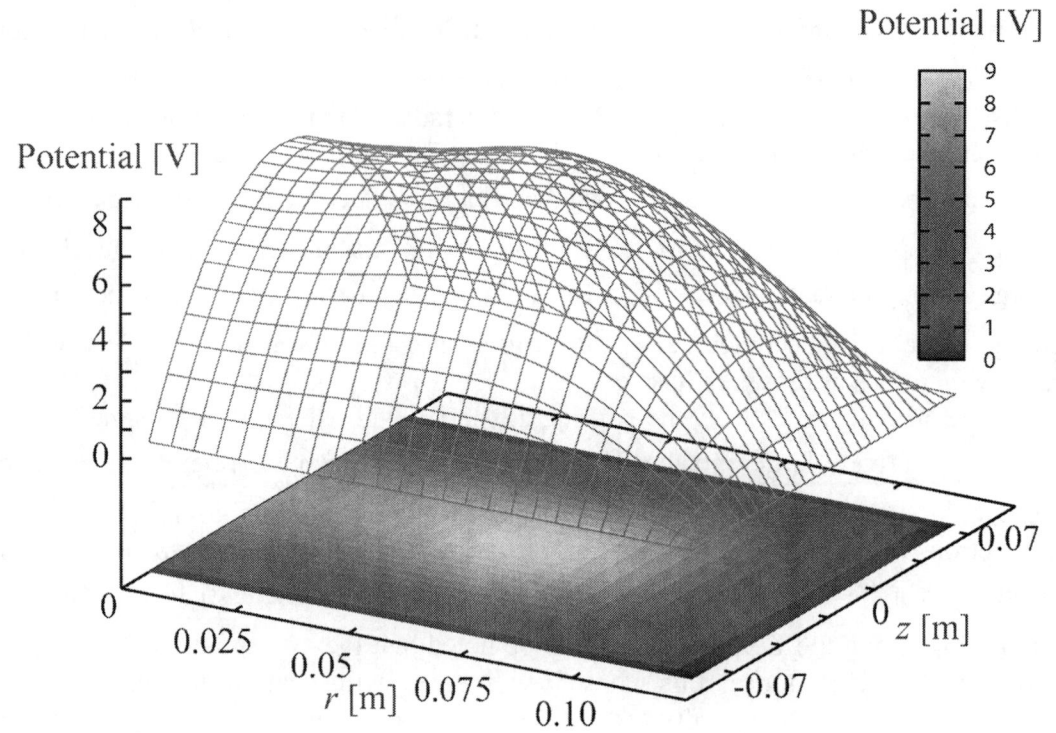

FIGURE 4. Potential profile on (r,z) plane at $t = 1.0 \times 10^{-6}$ s

CONCLUSION

Self-consistent model including not only the electron dynamics but also the ion dynamics has been developed. The model has been applied to the analysis of the RF-discharge initiation. The electron density increases during the RF-Phase where the azimutal RF electric field $\tilde{E}_\theta$ becomes maximum (Phase α in Fig. 3). At the initial stage in this phase, not only the ionization rate, but also the radial loss of the electrons rapidly and significantly increases. However due to this electron loss, the positive potential is built up with respect to the wall. This potential built up suppresses further electron radial loss and finally the ionization rate overcomes the radial electron loss. This is the mechanism for the electron density increases in Phase α, i.e., the successful discharge initiation in RF negative ion sources. The positive potential plays a key role for the discharge initiation in RF negative ion sources.

REFERENCES

1. U. Fantz, P. Franzen, W. Kraus, M. Berger, S. Christ-Koch, M. Fröschle, R. Gutser, B. Heinemann, C. Martens, P. McNeely, R.Riedl, E. Speth, and D. Wünderlich, Plasma Phys. Control. Fusion **49**, B563 (2007).
2. K.S. Yee, IEEE Trans. Ant. Prop. **14**, 302 (1966).
3. I. Fujino, A. Hatayama, and N. Takado, Rev. Sci. Instrum., **79**, (2008).
4. A. Bayliss and E. Turkel, Commun. Pure Appl. Math. **33**, 707, (1980).
5. G. Mur, IEEE Trans. Electromagn. Compat. **23**, 377, (1981).
6. C. K. Birdsall and A. B. Langdon, *Plasma Physics via Computer Simulation* (McGraw-Hill, New York, 1985).
7. K. Nanbu, IEEE Trans. Plasma Sci. **28**, 971 (2000).
8. T. Takizuka and H. Abe, J. Comput. Phys. **25**, 205 (1977).
9. S.J. Buckman *et al.*, *Interactions of photons and electrons with atoms* (Berlin, Springer-Verlag, 2000).
10. S.J. Buckman *et al.*, *Interactions of photons and electrons with atoms* (Berlin, Springer-Verlag, 2003).
11. R.K. Janev *et al.*, *Elementary Processes in Hydrogen-Helium Plasmas* (New York, Springer-Verlag, 1987).
12. R. Courant, K. Friedrichs and H. Lewy, Mathematische Annalen, **100**, 32 (1928)
13. S. Yoshinari, T. Hayami, R. Terasaki, A. Hatayama, A. Fukano, Rev. Sci. Instrum., **81**, (2010).

Quantification Of Cesium In Negative Hydrogen Ion Sources By Laser Absorption Spectroscopy

U. Fantz[a,b], Ch. Wimmer[b], and the NNBI Team[a]

[a] Max-Planck-Institut fuer Plasmaphysik, EURATOM Association, D-85748 Garching, Germany
[b] Lst. f. Experimentelle Plasmaphysik, Universitaet Augsburg, D-86135 Augsburg, Germany

Abstract. The use of cesium in negative hydrogen ion sources and the resulting cesium dynamics caused by the evaporation and redistribution in the vacuum and plasma phase makes a reliable and on-line monitoring of the cesium amount in the source highly desirable. For that purpose, a robust and compact laser absorption setup suitable for the ion source environment has been developed utilizing the Cs D_2 resonance line at 852.1 nm. First measurements are taken in a small laboratory plasma chamber with cesium evaporation. A detection limit of $\approx 5\times10^{13}$ m^{-3} at a typical path length of 15 cm has been obtained with a dynamic range of more than three orders of magnitude, limited by line saturation at high densities. For on-line monitoring an automatic data analysis is established achieving a temporal resolution of 100 ms. The setup has then been applied to the ITER prototype ion sources developed at IPP. It is been shown that the method is well suited for routine measurements revealing a new insight into the cesium dynamics during source operation and cesium conditioning.

Keywords: negative hydrogen ion source, cesium absorption, cesium density, cesium resonance line, neutral beam injection
PACS: 29.25.Ni, 52.25.Os , 52.70.Kz

INTRODUCTION

The formation of negative hydrogen ions depends of the ion source application and is based either on the volume process or on the surface process [1]. Volume production relies on the dissociative attachment requiring a reasonable amount of high vibrationally excited hydrogen molecules and electrons. As a consequence, such sources operate typically in the pressure range around 1 Pa and the extracted negative ion current is usually accompanied by a high co-extracted electron current. In the case of the surface process, hydrogen atoms and positive ions are converted on a surface into negative ions using preferentially the extraction electrode as converter surface. The conversion yield depends on the work function of the surface, which is lowered in the ion sources by cesium deposition on metallic surfaces [2]. This is achieved by cesium evaporation from a dispenser or an oven. These sources can operate at low pressures and are characterized by a reduced amount of co-extracted electrons and enhanced negative ion current densities compared to sources operating in the volume regime [3]. Typical applications are sources for accelerator systems [4,5] and neutral beam injection (NBI) systems at fusion devices [6,7].

Ion sources for fusion applications operate in hydrogen and in deuterium. In their next generation they have to deliver accelerated negative ion currents of 60 A and 40 A, respectively for up to one hour. Further requirements of the international fusion experiment ITER are the source pressure of 0.3 Pa, and an amount of co-extracted electrons which must be below one [8]. The extraction area is 0.2 m^2 in a source dimension of 1.9×0.9 m^2. Key issues of the development are the size scaling of the source and the long pulse stability. In order to extract a uniform and stable beam from such a large source for one hour, uniform and stable formation of negative hydrogen ions at the converter surface is essential. This requires a plasma illumination of the whole area with a uniform atomic and positive ion flux, as well as uniform and stable cesium coverage of the first grid of the extraction system, the plasma grid.

A manifold of diagnostic methods such as Langmuir probes and optical emission spectroscopy exists to measure and control the plasma parameters [9,10], but in-situ diagnostics for the monitoring and control of the cesium coverage is not available for such sources. Therefore indirect methods have to be used, for example the cesium emission spectroscopy which measures the cesium density in plasma phase [11].

Since the cesium vapor pressure and the work function is very sensitive on impurities [12, 13] monitoring of cesium evaporation and knowledge of the amount of cesium in the source during plasma-on phase and plasma-off (vacuum) phase is mandatory for optimizing the source performance. In particular, deeper insight in the cesium dynamics, i.e. redistribution, is expected.

The paper describes the development and the application of the laser absorption technique to the ITER prototype source at the Max-Planck-Institut für Plasmaphysik, (IPP). Results obtained in the vacuum phase, such as the cesium evaporation from the oven and the cesium redistribution after the end of the source operation will be shown. Time traces of cesium source conditioning with plasma operation will be discussed, demonstrating the capability of this technique as a sensitive on-line diagnostic tool.

CESIUM LASER ABSORPTION TECHNIQUE

The line absorption technique is an active diagnostic method for measuring particle densities using intensive light sources such as stabilized high-pressure lamps which emit a broad white light emission spectrum, or wavelength tunable lasers. Since the intensity of the signal depends on the Einstein coefficient for absorption B_{ki} for a transition from the lower excited state k into the higher excited state i resonance lines, with k as ground state, are well suited for this diagnostics. The ground state density is directly accessible by

$$n_k = \frac{8\pi c}{\lambda_0^4} \frac{g_k}{g_i} \frac{1}{A_{ik} l} \int_{line} \ln\left(\frac{I(\lambda,l)}{I(\lambda,0)}\right) d\lambda , \qquad (1)$$

in which the transition probability for emission A_{ik} is used instead of B_{ki} [14]. g_k and g_i denote the respective statistical weights, λ_0 the central wavelength of the line, l the path lengt, and. $I(\lambda,0)$ and $I(\lambda,l)$ the light intensity at position x = 0 and l, respectively.

Absorption spectroscopy has several advantages in comparison to emission spectroscopy: (1) the system does not need an intensity calibration since a relative signal is used; (2) except the well known transition probability and the path length no other parameters need to be known for the quantification; (3) measurements can be taken in the vacuum and in the plasma phase. The usage of a laser instead of a white-light source improves the signal-to-noise ratio. The combination of a laser with an optical fiber coupling and a photodiode with an interference filter as detector offers a robust and compact setup which is suitable to be used at the complex ion sources with limited access being in a harsh RF and high voltage environment.

For the detection of the cesium density, the D_2 resonance line ($6\,^2P_{3/2} - 6\,^2S_{1/2}$ transition, 852.11 nm) is used (figure 1). Due to the nuclear spin $I = 7/2$ of cesium the $6\,^2S_{1/2}$ state is split into two ($F = 3,4$) and the $6\,^2P_{3/2}$ state into four hyperfine levels ($F = 2,3,4,5$). F denotes the total angular momentum quantum number. As a consequence of the selection rule $\Delta F = -1/0/1$, the D_2 line consists of six hyperfine lines. However, due to the Doppler broadening only two peaks can be resolved as demonstrated in the calculated spectrum for a gas temperature of 1000 K using the wavelengths of the hyperfine lines and their relative intensities from refs. [15] and [16], respectively.

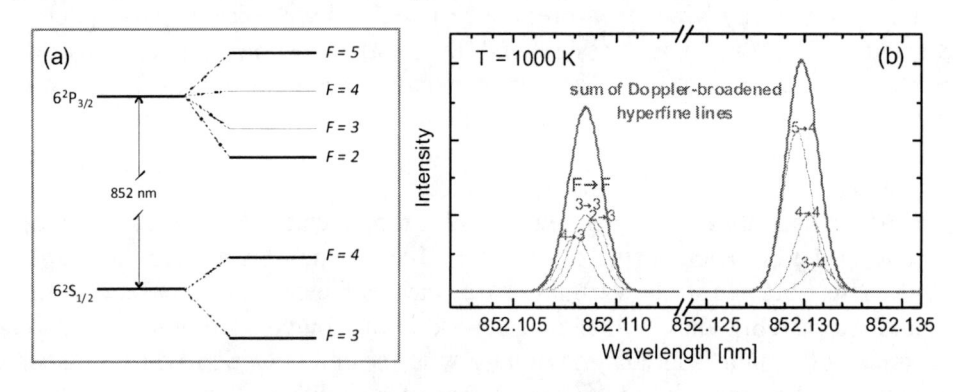

FIGURE 1. (a) Schematic energy level diagram of the hyperfine structure of the cesium D_2 resonance line. (b) Calculated spectrum using a Doppler-profile at 1000 K.

As light source a tunable mode hop free, single mode distributed feedback (DFB) diode laser system with a total tunable range of 851 nm – 853 nm, a spectral line width of 0.01 pm and a maximum output power of 150 mW is used. The wavelength of the laser is set into the region of the cesium D_2 line by temperature modulation, whereas the current modulation is applied for the tuning.

As shown in figure 2, the light of the diode laser is coupled into an optical fiber with a core diameter of 105 µm. A lens with a focal length of 25.4 mm is used to create a beam diameter of 11.5 mm. The neutral density filter with a transmission of 1.8% at 852 nm in front of the absorption medium attenuates the intensity of the laser beam. Behind the absorption medium the light is coupled by a lens with the same focal length into an optical fiber ending at a photo diode as detector. A wider fiber with a core diameter of 200 µm is used to enhance the coupling efficiency decreased by the optical aberration of the lens. The interference filter ($\lambda_0 = 852$ nm, $\Delta\lambda_{FWHM} = 10$ nm) in front of the lens reduces the offset by the plasma emission to zero. The current of the photo diode is amplified by a transimpedance amplifier and the voltage is recorded together with the corresponding modulation voltage.

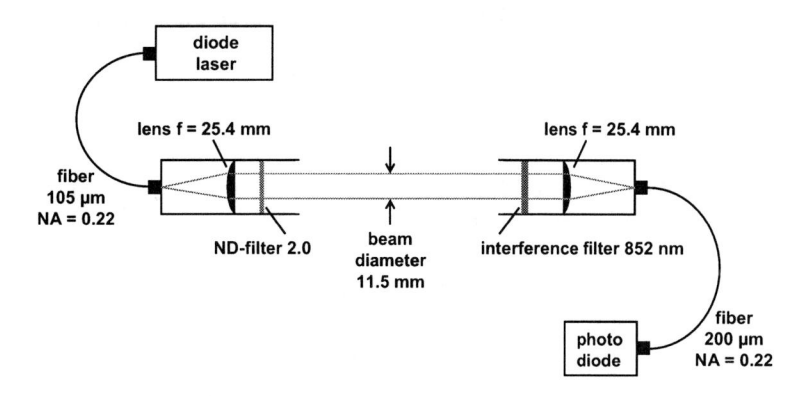

FIGURE 2. Setup for the laser absorption spectroscopy.

In order to minimize the effect of depopulation of the ground state by the intense laser light a large beam diameter and a neutral density filter is chosen. Since the speed of the photo diode amplifier depends on the amplification factor, it is not possible to further increase the attenuation without decreasing the temporal resolution of the system. With the setup 25 absorption spectra per second can be recorded.

Figure 3 shows the signals of laser absorption and white light absorption (apparatus profile of $\Delta\lambda_{App,FWHM}$ = 15 nm). Clearly a much better spectral resolution and signal-to-noise ratio is obtained. The detection limit is improved significantly: from $n_{Cs} > 10^{15}$ m^{-3} for white light absorption to $n_{Cs} > 4 \times 10^{13}$ m^{-3} for laser absorption at an absorption length of 15 cm. The wavelength positions of the two Doppler-broadened peaks are for the wavelength calibration assuming a linear wavelength modulation.

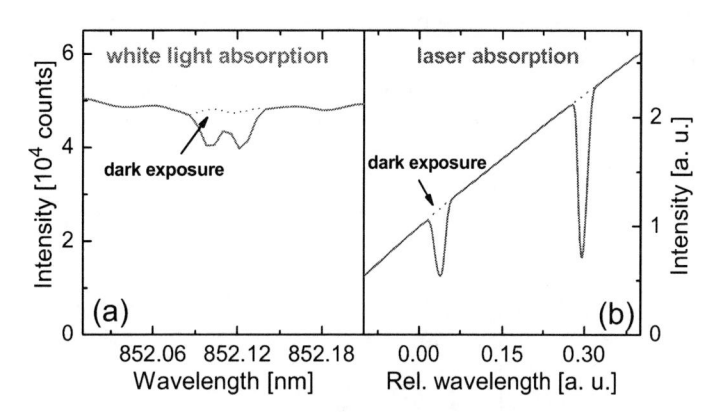

FIGURE 3. Absorption signals measured with white light absorption (a) and laser absorption (b).

Details of the setup and the analysis method are given in ref. [17] which describes also the effect of line saturation. The latter is due to a high absorption along the path length. Thus, the profiles flatten and have to be fitted by a Gaussian profile to obtain the true absorption signal. Depending on the density, or to be more precise on density × path length, typical correction factors are below 2. Both, line saturation and ground state depopulation result in an underestimation of the density.

TESTS AT AN INDUCTEVELY COUPLED PLASMA SOURCE

First tests are carried out at an inductively coupled plasma source which operates at similar vacuum condition and plasma parameters as the ion sources but offers a much better accessibility and flexibility [12]. In this case however, the fiber coupling was not available and a direct line of sight after the laser widening has been used instead.

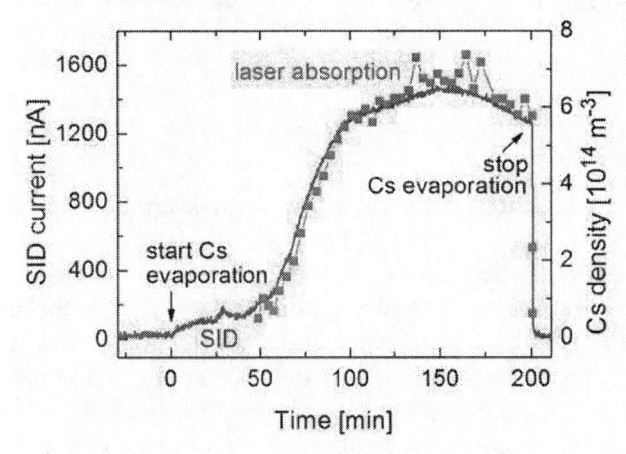

FIGURE 4. Signal of the Surface Ionization Detector (SID) in comparison with the cesium density measured by laser absorption during cesium evaporation in the vacuum phase

Figure 4 shows the evolution of the cesium density in the vacuum chamber of the ICP (background pressure of 10^{-5} mbar) during cesium evaporation from a dispenser (100 mg, Alvatec) with a typical evaporation rate of less than 10 mg/h. The results from the absorption technique are compared with the signal of the surface ionization detector, which measures the ionized cesium at a tungsten surface by applying a voltage between the wires [12]. A very good agreement between the two curves is obtained keeping in mind that they measure at different locations and different detection geometries.

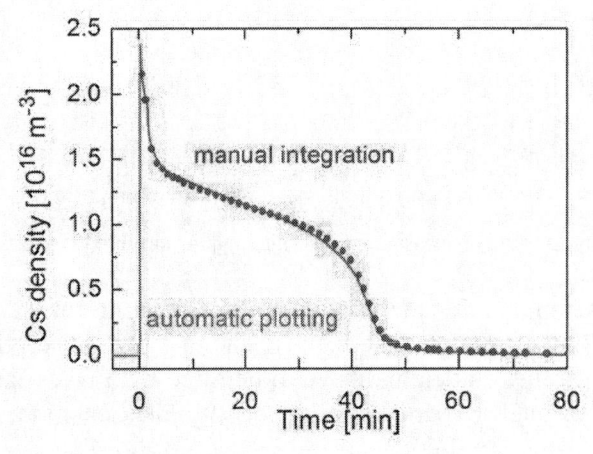

FIGURE 5. Cesium density measured by laser absorption during cesium evaporation from a Cs dispenser in vacuum. Comparison of manually evaluated data with automatic (on-line) data evaluation.

So far the absorption signal has been evaluated manually after the measurements. However, for application to ion sources on-line data evaluation is highly desirable. This has been carried out in LabView utilizing the maximum temporal resolution [17]. The automatic line integration is compared to the manual integration in figure 5 using the evaporation of an SAES dispenser (10 mg amount) as Cs source. A high intensity dynamic and a good time resolution are obtained.

Similar measurements have been performed during plasma operation. In this case the signal has been compared to the emission spectroscopy from which also the absolute density can be obtained. However, emission spectroscopy needs plasma density and temperature for data evaluation which were both known only from measurements in the plasma without cesium. Nevertheless, a satisfying agreement has been observed [12], the more reliable method being clearly the absorption technique. The accessible density range is exactly in the expected density range for the ions sources which is known from emission spectroscopy [11].

After some further investigations on reliability and reproducibility of the laser absorption in vacuum and during plasma operation the setup has been transferred to the negative ion source at the IPP.

APPLICATION TO THE NEGATIVE HYDROGEN ION SOURCE

Figure 6 shows the IPP prototype source for ITER. As described in more detail in ref. [3], the plasma is generated in the driver using inductively coupling with a generator frequency of 1 MHz and a typical RF power of 100 kW and expands into the expansion chamber. Close to the extraction system a magnetic field is installed to cool the electrons down and to reduce the amount of the co-extracted electrons. The plasma in front of the extraction system is characterized by electron densities in the range of

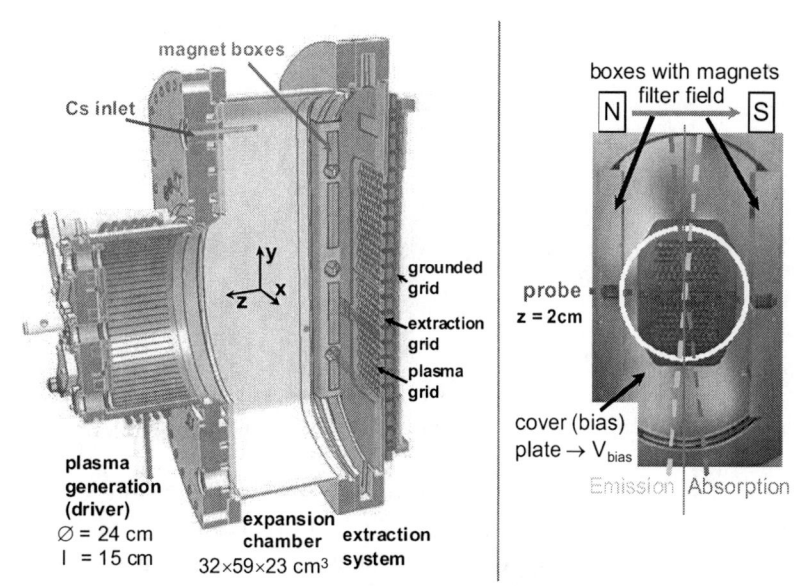

FIGURE 6. Technical drawing of the IPP RF prototype source and a view on the plasma grid of the ion source at the BATMAN test facility with the line-of-sight for emission and absorption spectroscopy.

10^{17} m^{-3} and temperatures of 1 eV [11]. Diagnostic access is provided by a diagnostic flange mounted above the extraction system. Usually this flange houses also the magnets, for the present investigations however, they have been removed and a magnetic frame has been installed outside the chamber to provide more flexibility for ion source operation at the test facility BATMAN.

The lines of sight for emission and absorption spectroscopy cross the source in a similar manner at a distance of approx. 2 cm to the plasma grid (figure 6). For laser absorption, the setup with the optical fibers has been used together with an interlock system for safety purposes. The system is coupled to the trigger system of BATMAN.

Cesium is evaporated by a cesium oven which is mounted at the upper half of the backplate housing 3 ampoules with 1g Cs each. Based on experimental experience, Cs conditioning is improved if the walls of the source chamber are kept at temperatures of typically 45 °C and the plasma grid at 150 °C. Due to these elevated temperature strong Cs accumulation is avoided and Cs redistribution enhanced.

Results in the Vacuum Phase

The measured cesium density in the vacuum phase is shown in figure 7 for different temperatures of the cesium oven and in two different stages of the cesium conditioning period. The lower curve shows the density obtained at the third day of Cs-conditioning after opening the first cesium ampoule, the upper curve refers to a later stage, i.e. the third day after opening the second cesium ampoule. As expected, the cesium density increases with higher cesium evaporation due to a higher oven temperature. In the later stage, a higher cesium density is reached for the same oven temperature.

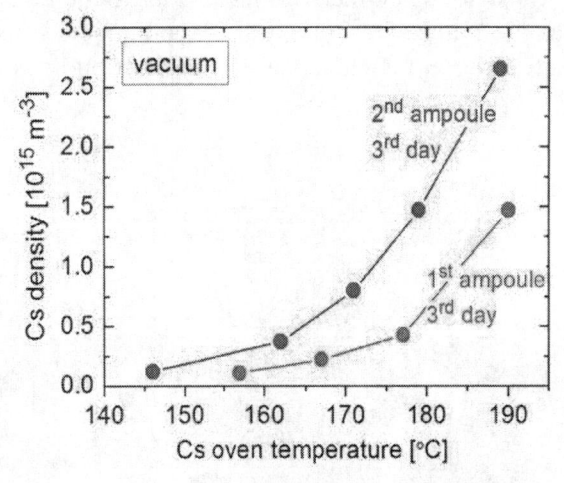

FIGURE 7. Cs density in the vacuum chamber (without plasma operation) as a function of the cesium evaporation rate controlled by the oven temperature.

This might indicate problems in reproducibility of the cesium evaporation rate of the oven caused by cesium contamination which reduces the vapor pressure. Another much more favored explanation is the sticking of the evaporated cesium at the chamber walls which is higher for the beginning at an experimental campaign. At a certain Cs level sticking is in balance with desorption.

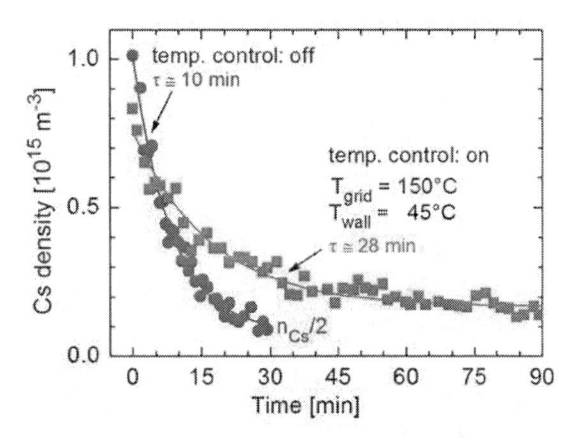

FIGURE 8. Cs density in the vacuum phase after the test facility has been switched off.

The temporal evolution of the cesium density after turning off the heating of the cesium oven after a full day of source operation is shown in figure 8 for two cases. In the first case, the temperature control of the plasma grid and the chamber walls has been switched off together with the cesium oven. The cesium density shows an exponential decay with a time constant of 10 min. In the second case, the temperature control is still operating. An increased time constant, namely 28 min, is obtained, demonstrating reduced sticking and enhanced desorption of cesium at the surfaces.

Results during Source Operation: Cesium Conditioning

In the Cs conditioning period of the ion source, the temporal behavior of the cesium density during plasma-on and plasma-off phase changes strongly in the first discharges of a day as shown in figure 9. From the first shot on, a sequence of time traces is being recorded. Pressure and power of the source has been kept constant. The oven has been turned on and Cs is being evaporated continuously.

First, the cesium signal and thus Cs density increases in successive discharges in the plasma phase, whereas in the vacuum phase almost no changes occur. A slight increase after the discharges is to be seen which is vanished before the next discharge starts. Next, after 9 plasma discharges a peak in the cesium density after the end of the plasma operation appears accompanied by a decay time. The density during the plasma, however, decreases. The peak height increases further with the following discharges, until saturation in the growth of the peak is reached after 18 discharges. This temporal behavior does not change significantly during the following discharges of the day, if continuous operation with similar source parameters takes place.

The peak after the discharge is caused by the recombination of ionized cesium, a species being not directly accessible by laser absorption. As the cesium is to more than 90 % ionized during the discharge one would expect also a decrease of the signal with starting the discharge. However, a similar neutral density is obtained as in the vacuum phase shortly before the discharge. This means a strong Cs distribution and release from the surfaces takes place during the plasma-on phase most probably due to the thermal load. As a consequence, the amount of the recombined cesium after the

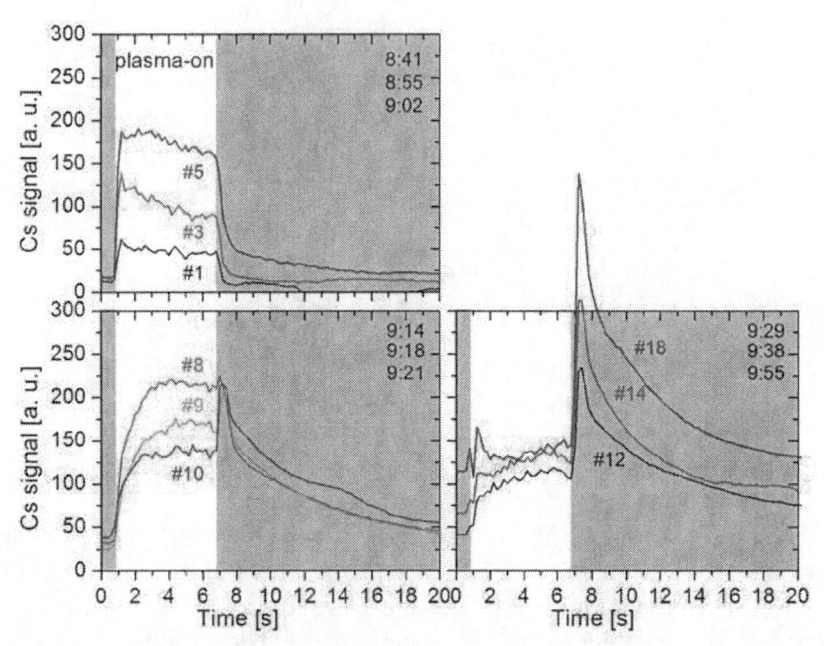

FIGURE 9. Time traces of the laser absorption signal before, during and after consecutive discharges of 6s starting from shot number #68543 (#1) to #68562 (#18).

discharge is also much higher. Then, the high cesium density decreases by transport mechanisms to the wall until an equilibrium with the evaporation of the oven is reached. The time constant of this transport mechanism strongly depends on the sticking coefficient. At the beginning of source operation, i.e. in the morning, only a small amount of cesium is evaporated. This leads to a high sticking coefficient of cesium and hence to a fast transport of cesium to the walls after the end of a plasma pulse. The temporal resolution of the absorption system is not fast enough to detect the rise and drop of the peak. In a later stage, after evaporating more cesium and thus a lower sticking coefficient, the transport mechanism is getting slower and thus the peak can be detected with the temporal resolution of the laser absorption system.

In the following, the evolution of the cesium density in the vacuum and plasma phase and the emission of the cesium 852 nm line during plasma operation, measured by a diode with an interference filter, will be discussed. The signals in the plasma are averaged during the second half of the extraction phase (4 s extraction between 1.5 s and 5.5 s of the 6 s discharge). An exponential decay $n_{Cs}(t) = n_{Cs,vac} + C \exp(-t/\tau)$ has been fitted on the decrease of the signal of the laser absorption to determine a value for the cesium density in the vacuum $n_{Cs,vac}$ and a value τ as the decay time. The parameter C is used as fitting parameter. A ratio $\Delta Cs/Cs^{peak}_{plasma} = (n_{peak} - n_{plasma}) / n_{peak}$ out of the last value during the plasma phase n_{plasma} and the value at the maximum of the peak n_{peak} is introduced, which indicates a lower limit for the ionization degree of Cs, if the transport mechanism of cesium to the walls is slow enough to be temporally resolved by laser absorption. The ratio of zero corresponds to a non-resolved peak.

The results are shown in figure 10, together with the temperature of the cesium oven and the injected RF-power. As in figure 7, the cesium density in the vacuum is in good correlation with the temperature of the cesium oven. The signal of the Cs diode rises accordingly keeping in mind that the step-like increase at the end is caused by an increase of the RF power. However, the cesium density detected by the laser absorption in the plasma reveals a maximum after a few discharges. In contrast to the emission signal which relies on the plasma length, absorption detects also signals from volumes without plasma.

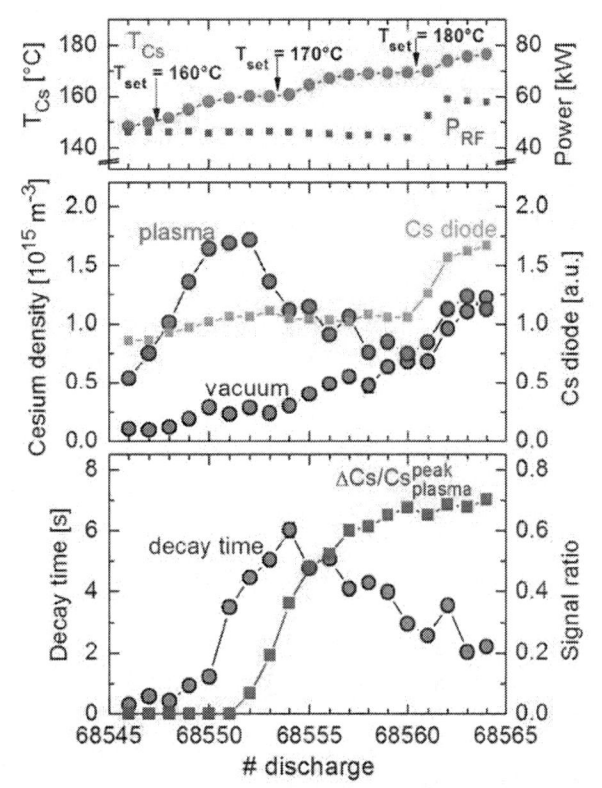

FIGURE 10. Temporal evolution of the laser absorption signals and the Cs emission line (Cs diode signal).

The bottom part of figure 10 shows that decay time reaches a maximum followed by a slight decrease. The peak-to-plasma ratio however shows a transition between two levels: no peak and maximum value of the peak; the latter taken as a measure for the lower limit of the ionization degree, i.e. $n_{Cs+}/n_{Cs} \geq 0.7$. Both behaviours indicate a quite complex Cs dynamics in the ion source during the vacuum phase, the plasma phase and explicitly after the discharge. More insight can be expected from further systematic measurements which will follow these first measurements.

CONCLUSION

The laser absorption method is been applied to the Cesium D_2 resonance line at 852.1 nm using a wavelength tunable diode laser (851-853 nm). A robust and easy-to-use setup has been developed for being suitable to be mounted at the complex negative hydrogen ion sources in RF and high voltage conditions. With an automatic control and data acquisition the method is used for on-line monitoring of the cesium densities allowing for a temporal resolution of 100 ms. A density range of typically 5×10^{13} m^{-3} to 10^{17} m^{-3} (limited by the line saturation effect) is accessible, being perfectly in the range in which the ion source operate.

The proof of principle has been demonstrated in a small laboratory experiment using cesium dispensers as evaporation source. Effects of line saturation lead to an underestimation of the cesium density at high densities, whereas the effect of depopulation of the ground state density by the intense laser light has been quantified to be below a factor of two by using a neutral density filter for an attenuation of two orders of magnitude.

Experiments at the ITER prototype source at IPP gave already a first very promising insight in the cesium dynamics: (1) in the vacuum the cesium evaporation from the oven and its reproducibility can be measured, (2) time traces before, during and after plasma pulse show the redistribution and the ionization of the cesium as well as the pumping of the cesium by the surfaces which has been expressed in decay times, (3) the influence of surface temperatures on the cesium sticking (adsorption and desorption) is clearly to be seen in the cesium amount, (4) a comparison of the vacuum density and the plasma density revealed the astonishing result that the neutral densities are in the same range and that the redistribution during the plasma phase dominates over the cesium evaporation. Further promising results are expected when the method is being applied as routine diagnostics.

ACKNOWLEDGMENTS

The work was partly supported by a grant from Fusion for Energy (F4E-2008-GRT-07). The opinions expressed herein are those of the authors only and do not represent the Fusion for Energy's official position.

REFERENCES

1. M. Bacal, *Nucl. Fusion* **46**, S250 (2006).
2. B.S. Lee and M. Seidl, *Appl. Phys. Letters* **61**, 2857 (1992).
3. E. Speth et al., *Nucl. Fusion* **46**, S220 (2006).
4. J. Peters, *Rev. Sci. Instrum.* **71**, 1069 (2000).
4. M. Kronberger et al., *Rev. Sci. Instrum.* **81**, 02A708 (2010).
5. R. Welton et al., *Rev. Sci. Instrum.* **8**, 02A727 (2010).
6. Y. Ikeda et al., *Nucl. Fusion* **46** S211 (2006).
7. Y. Takeiri et al., *Nucl. Fusion* **46** S199 (2006).
8. R. Hemsworth et al., Nucl. Fusion **49**, 045006 (2009).
9. P. McNeely et al., *Plasma Sources Sci. Technol.* **18**, 014011 (2009).
10. U. Fantz et al., *Nucl. Fusion* **46,** S297 (2006).
11. U. Fantz et al., *Fusion Eng. Design.* **74**, 299 (2005).
12. U. Fantz, R. Gutser, Ch. Wimmer, *Rev. Sci. Instrum.* **81**, 02B102 (2010).
13. R. Schletti, P. Wurz and T. Fröhlich, *Rev. Sci. Instrum.* **71**, 499 (2000).
14. A.J.M. Buuron, et al., *Rev. Sci. Instrum.* **66**, 968 (1995).
15. D.A. Steck, "Cesium D Line Data", available online at http://steck.us/alkalidata
16. H.E. White and A.Y. Eliason, *Phys. Rev.* **44**, 753 (1933).
17. U. Fantz, Ch. Wimmer, in preparation (2010).

Cavity Ring-Down System for Density Measurement of Negative Hydrogen Ion on Negative Ion Source

Haruhisa Nakano[a], Katsuyoshi Tsumori[a], Kenichi Nagaoka[a], Masayuki Shibuya[a], Ursel Fantz[b], Masashi Kisaki[a], Katsunori Ikeda[a], Masaki Osakabe[a], Osamu Kaneko[a], Eiji Asano[a], Tomoki Kondo[a], Mamoru Sato[a], Seiji Komada[a], Haruo Sekiguchi[a], and Yasuhiko Takeiri[a]

[a]*National Institute for Fusion Science, 322-6, Oroshi-cho, Toki, 5095292, Japan*
[b]*Max-Planck-Institut fur Plasmaphysik, EURATOM Association, Boltzmannstrasse 2, D-85748 Garching, Germany*

Abstract. A Cavity Ring-Down (CRD) system was applied to measure the density of negative hydrogen ion (H⁻) in vicinity of extraction surface in the H⁻ source for the development of neutral beam injector on Large Helical Device (LHD). The density measurement with sampling time of 50 ms was carried out. The measured density with the CRD system is relatively good agreement with the density evaluated from extracted beam-current with applying a similar relation of positive ion sources. In cesium seeded into ion-source plasma, the linearity between an arc power of the discharge and the measured density with the CRD system was observed. Additionally, the measured density was proportional to the extracted beam current. These characteristics indicate the CRD system worked well for H⁻ density measurement in the region of H⁻ and extraction.

Keywords: cavity ring-down, negative hydrogen ion, ion source, neutral beam
PACS: 29.25.Ni , 52.27.Cm, 52.50.Dg, 52.50.Gj, 52.80.Mg,, 52.70.Kz, 52.55.Fa, 52.55.Hc

INTRODUCTION

Negative-hydrogen-ion sources have been applied to Neutral Beam Injectors (NBIs) with the energy range of more than a few hundreds keV [1, 2]. Negative-ion-based NBI systems are scheduled to be utilized for fusion-experimental devices such as Japan Torus 60 SA (JT-60SA) [3] and International Thermonuclear Experimental Reactor (ITER) [4]. One of the most important points to enhance performance of the negative-ion source is the increase of current density of negative-hydrogen-ion (H⁻).

We have developed the H⁻ source of NBI for Large Helical Device (LHD) [1] at National Institute for Fusion Science (NIFS). The current density of the H⁻ source has been enhanced by cesium (Cs) seeding, and has achieved more than ~300 A/m². In Cs seeded into H⁻ source, the H⁻ production on Plasma Grid (PG) surface is the dominant process. Although, the H⁻ ions are generated on the PG direct to inward of the ion source, i.e. opposite to extracted beam direction, the mechanism of H⁻ transport from the generation to extraction are not fully understood. The H⁻ dynamics in generation

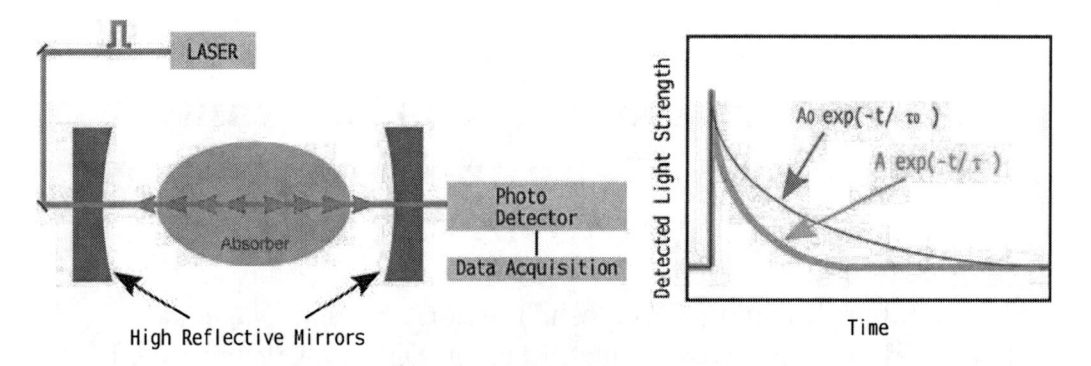

FIGURE 1. Conceptual scheme of the cavity ring down method.

and extraction region is one of the most important issues to the H⁻ source, and was discussed by numerical simulations [5, 6, 7, 8]. Common physics for the dynamics can exist in H⁻ sources, regardless the use of filament arc or radio frequency as the plasma discharge driver. In experiments, to study the dynamics using the H⁻ source for LHD-NBI development [9], we adopted the Cavity Ring-Down (CRD) method for H⁻ density measurement. The profits of the CRD method are an optical and absolute density measurement, and comparatively easily to measure the density in the device on high-voltage stage because there are any components on the high-voltage stage.

In following sections, we introduce the principle of CRD method, and describe the CRD system on H⁻ source for LHD-NBI development. And, the experimental observation results of the H⁻ density measurements and relations between H⁻ density, input power of discharge and extracted H⁻ beam current are shown.

CAVITY RING-DOWN METHOD

The CRD method is an enhanced laser-absorption-spectroscopy with high sensitivity to measure line-integrated density [10]. The CRD method is applied to wide fields [11], including H⁻ source [12, 13, 14, 15]. The CRD system basically consists of a pulse laser, an optical ring cavity with two high reflective mirrors installed both sides of absorber, an optical detector, and a data acquisition system (Fig. 1). The principle of CRD method is briefly introduced. A single laser pulse is injected from one side of cavity. The laser intensities in the cavity and passing through both mirrors gradually weaken by photons reacting to the absorber and slightly passing through both mirrors. The laser pulse passing through from another side of cavity is observed by the detector. The output signal of detector transmits to the data acquisition system. The signal forms the exponential decay curve (ring-down signal). The ring-down signals are obtained without and with the absorber. The absorber density is evaluated by comparing to the decay times without and with the absorber.

The analytical form of CRD method is derived as followings. The intensity of initial pulse at the detector (I_0) is represented by considering Lambert-Beer's law.

$$I_0 = I_{Laser}(1-R)^2 \exp(-n\sigma L), \quad (1)$$

where I_{Laser} is the intensity of input laser pulse to the cavity, the R is the reflectance of mirrors, the L is the length of the absorber along cavity axis, the n is the absorber density, and σ is the absorption reaction cross-section between the laser and the absorber. The I is the intensity of laser at the detector after N round trips of the cavity. The absorption rate I/I_0 is represented by

$$I/I_0 = R^{2N} \exp(-2N n \sigma L). \quad (2)$$

The discrete formula of Eq. (2) changes to continuous one by time variable $t = 2Nd/c$, where d is length of the cavity and c is light speed, as,

$$I/I_0 = \exp\left(-\frac{t}{\tau}\right), \quad (3)$$

Where

$$\frac{1}{\tau} = \frac{1}{\tau_0} + n\sigma \frac{L}{d} c, \quad (4)$$

$$\frac{1}{\tau_0} = \frac{-c \ln R}{d}. \quad (5)$$

The τ_0 and τ are called ring-down times without and with the absorber, respectively. There are some unknown optical losses in the cavity in actual experiments. The unknown optical losses can be include in Equation (5) as

$$\frac{1}{\tau_0} = \frac{-c \ln R}{d} + X, \quad (6)$$

where the X is the term of unknown optical losses. The absorber density is evaluated as

$$n = \frac{1}{c}\left(\frac{1}{\tau} - \frac{1}{\tau_0}\right)\frac{d}{L}\frac{1}{\sigma}. \quad (7)$$

The absolute density of absorber can be calculated from only measured values of τ_0 and τ without the need of knowing optical losses term X in Eq. (6).

EXPERIMENTAL SETUP

Negative Ion Source for LHD-NBI Development

The experiment was performed with the H⁻ source [9] for LHD-NBI development on test beam line at NIFS. The ion source plasma of the H⁻ source is generated by filament-arc discharge. The size of arc discharge chamber is 0.35 m wide x 0.7 m long x 0.23 m high, and is a half volume of the actual negative-ion source of the LHD-NBI. Two type magnetic fields are formed in the chamber. One is the cusp field formed around inner wall. The magnetic strength of cusp field is set to 0.2 T on the inner surface of arc chamber. Another is the filter field formed by the sets of magnets recessed in the opposite side of the chamber near the PG. The magnetic strength of the filter field is set to ~ 6 mT at the center of the PG. The filter field plays the role of lowering T_e near the PG. Cesium injection ports are arranged on the back plate of the chamber. Cesium storage tanks were heated around 170 to 200 centigrade. There are vertically 28 by horizontally 17 apertures with 12 mm in diameter at the PG. Additionally, extracted H⁻ beam current can be evaluated with calorimeter array installed downstream the H⁻ source.

Cavity Ring-down System

A schematic view of CRD system on the negative-ion source is shown in Fig. 2. We adopted a fundamental wave of Nd-YAG laser of which wavelength, pulse width and repetition time are 1064 nm, 5 ns and 50 ms, respectively, as a probe laser pulse. A photo-detachment cross section of H⁻ for wavelength of 1064 nm is 3.5×10^{-21} m² [16]. The energy of laser pulse is utilized less than about 50 mJ. An optical ring cavity

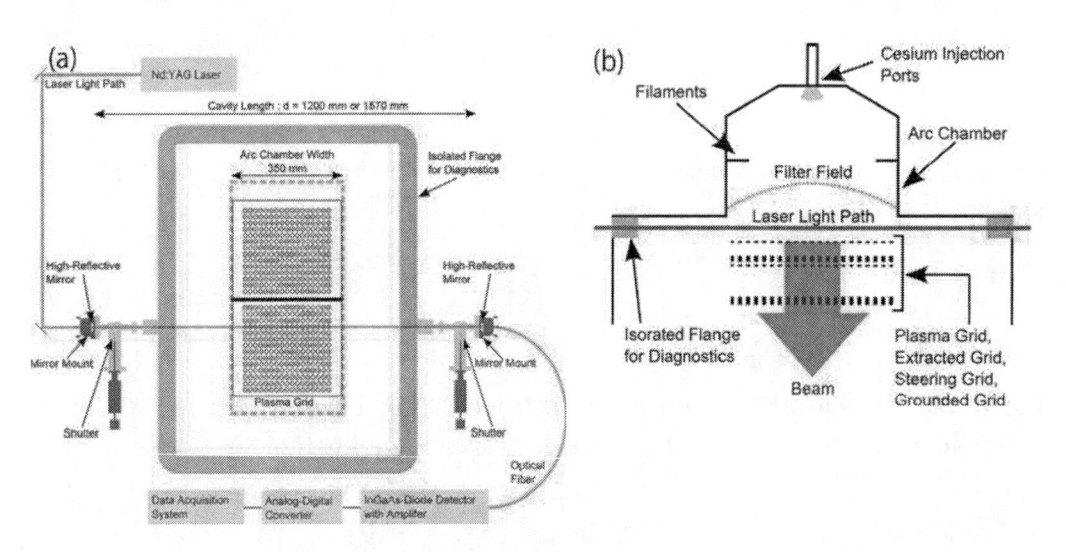

FIGURE 2. The schematic view of CRD system on the negative-ion source for LHD-NBI development. (a) Cross-section perpendicular to beam extraction in the vicinity of the PG in the arc discharge chamber side. (b) Cross-section of the ion source parallel to the beam extraction. The laser path of CRD is aligned 9 mm from the PG surface and its apertures.

is equipped on an insulator flange between the arc chamber and flange of PG, shown as blue in Fig. 2. Although the width of the insulator flange was lengthened from 20 mm to 40 mm for this experiment, property of the ion source, such as arc efficiency, change little. A cavity axis is aligned at 9 mm above PG surface and the center of apertures located at 72 mm from the vertical center of arc chamber. This axis is in H⁻ generation and extraction region. The effective plasma length along the laser axis is 0.18 m, which evaluated from a movable Langmuir probe measurement. The distance between high-reflective mirrors of cavity is $L = 1.20$ m (or 1.57 m). The reflectance and curvature of high-reflective mirrors are more than 0.99996 and 1 m, respectively. Movable shutters are installed between the mirrors and the plasma for avoiding the attachment of impurities to the mirror surface when the CRD system is not operated. The output laser pulse from the cavity is transmitted to a photo detector by an optical fiber. The output voltage signal of detector is converted to digital data by analog-digital-converter. The digital data are preserved at a digital-data-storage server. The data is acquired every 50 ms during 20 s including the discharge duration. The time evolution of H⁻ can be displayed in the shot interval of 2 minutes typically.

RESULTS

In pure hydrogen plasmas (without Cs seeding into the ion source plasma), the H⁻ density was measured with the CRD system. Figure 3 shows ring-down signals without and with the plasma. The dots and lines represent raw signals and their fittings, respectively. The practically measured ring-down time τ_0 without the plasma is 28.1 μs, while the ring-down time τ_0 calculated by Eq. (5) ($L = 1.2$ m and $R = 0.99996$) is 100.0 μs. The difference of the both ring-down times is caused by the optical losses due to the absorption and scattering of laser on the mirrors, walls and laser transmission line, which are included X in Eq. (6). With the plasma (the arc power of discharge is 72 kW and the gas pressure of introduced H_2 is 0.34 Pa), the ring-down time τ is 21.9 μs. The H⁻ density is evaluated 6.4×10^{16} m⁻³ by Eq. (7).

The coincidence between the H⁻ density $n(H^-)$ and extracted H⁻ current $I(H^-)$ is tried

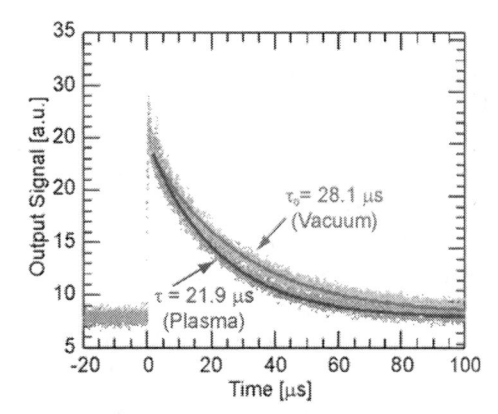

FIGURE 3. Ring-down signals without and with a pure hydrogen plasma. The dots and lines show raw data and their fittings, respectively. The cavity length $L = 1.2$ m, and the mirror reflectance $R = 0.99996$.

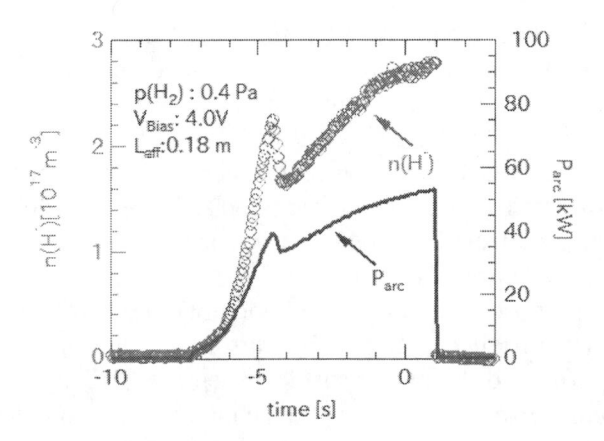

FIGURE 4. Time evolutions of H⁻ density ($n(H^-)$) and arc power of discharge (P_{arc}) in the Cs seeded plasma. Open circles and lines represent H⁻ density and arc power of discharge, respectively. The time resolution of H⁻ density is 50 ms.

to evaluate by assuming the similarity of positive-ion beam extractions. When an optimal beam is extracted from a positive ion source, the positive-ion density in the vicinity of the PG can be estimated from the extracted positive-ion-beam current by a formula of ion saturation current. Here, we assume with utilizing following equation,

$$I(H^-) = e\, n(H^-)\, c_{s,H-}\, S\, \exp(-1/2), \tag{8}$$

where $c_{s,H-} = \sqrt{T_{H+}/M_{H-}}$, the S is the summation of cross-sections of beam extraction apertures at the PG, the T_{H+} is positive hydrogen (H⁺) temperature, and M_{H-} is H⁻ mass. The H⁻ beam current in the Fig. 3 case is evaluated 0.94 A. Assuming the H⁺ temperatures are 0.1 eV and 1 eV, the H⁻ densities are 5.8×10^{16} m⁻³ and 1.8×10^{16} m⁻³, respectively, calculated by Eq. (8). The density measured with the CRD system is relatively good agreement with the density evaluated from the extracted H⁻ current by assuming the similarity of positive-ion beam extractions.

The time evolutions of H⁻ density and the arc power of discharge in the Cs seeded plasma without beam-extraction are shown in Fig. 4. The Cs vapor is seeded for 10 hours by keeping Cs oven temperature of 180 – 190 centigrade, and Cs amount is sufficient and optimized for the beam extraction. The peaks at -5 sec are caused by insufficiency of the stability of arc power supply, and it has no influence to experimental data. The H⁻ density was obtained every 50 ms.

The relations between the H⁻ density, the arc power of discharge and extracted H⁻ beam current in the Cs seeded plasma are shown in Fig. 5. The H⁻ density is indicated the value just before the beam extraction. The gas pressure of introduced H⁻ is 0.27 Pa, and the bias voltage of between the arc chamber and the plasma grid is 1 V. Extraction and acceleration voltages are 1.6 kV to 2.2 kV and 48 kV to 70 kV, respectively.

The linearity between the arc power of discharge and H⁻ density was observed. A possible explanation of linearity is that the parent particles (positive hydrogen ions and hydrogen atoms) of the surface production arriving at the PG increase with the arc power of discharge.

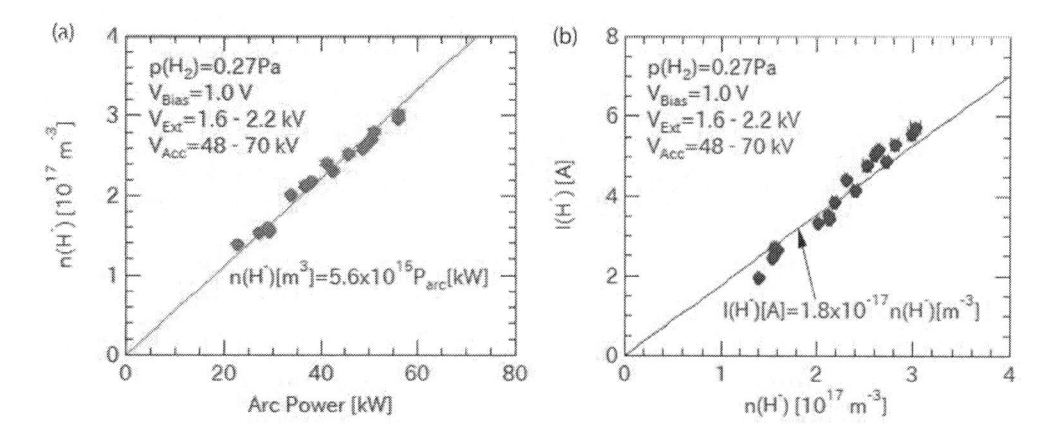

FIGURE 5. The relation between the arc power of discharge (P_{arc}), H⁻ density (n(H⁻)) and H⁻ current I(H⁻) in the Cs seeded plasma. The solid lines show fittings of plots.

The H⁻ density was proportional to the extracted H⁻ beam current. The proportionality coefficient is 1.8×10^{17} A/m⁻³. The proportionality coefficient calculated by Eq. (8) is 1.7×10^{17} A/m⁻³ when the positive ion temperature of 0.1 eV. The proportionality coefficient of experimental result is comparable to the proportionality coefficient by assuming the similarity of positive-ion beam extractions. Although the relation of absolute values between the H⁻ density and extracted H⁻ current is not much inconsistent with Eq. (8) in not only pure hydrogen but also Cs seeded plasmas, it needs furthermore consider the consistency of the relation and Eq. (8).

It is well known that the extracted H⁻ beam current is proportional to the arc power of discharge in applicable ion-source plasma condition. The both relations shown in Fig. 5 provide that the H⁻ density is intermediate parameter between the arc power of discharge and the extracted H⁻ beam current, and, simultaneously, the CRD system to measure H⁻ density works well.

SUMMARY

We have applied the Cavity Ring-Down system to the negative-ion source for LHD-NBI development to measure the negative-hydrogen-ion density in the vicinity of the plasma grid. The negative-hydrogen-ion densities were acquired in the pure hydrogen and Cs seeded plasmas, and the time evolution of negative-hydrogen-ion was obtained every 50 ms during the discharge. In the Cs seeded plasma, the proportionality relations were observed between the arc power of discharge and negative-hydrogen-ion density, and between the negative-hydrogen-ion density and the extracted negative-hydrogen-ion beam current. By these characteristics, it was confirmed that negative-hydrogen-ion density is intermediate parameter between the arc power of the discharge and the extracted negative-hydrogen-ion beam current, and the Cavity Ring-Down system worked well. Using the Cavity Ring-Down system, we will investigate the dynamics of the negative-hydrogen-ion generation and extraction region.

ACKNOWLEDGMENTS

One of the author (H N) thanks Prof. T. Minami, Dr. T. Akiyama, Prof. I. Yamada, Prof. A. Ando, and Prof. M. Sasao for fruitful discussions. This study was supported by NIFS (NIFS10ULRR008 and NIFS10ULRR702).

REFERENCES

1. Y. Takeiri, K. Tsumori, K. Ikeda, M. Osakabe, K. Nagaoka, Y. Oka, E. Asano, T. Kondo, M. Sato, M. Shibuya, S. Komada, and O. Kaneko, AIP Conf. Proc. 1097, pp. 470-479, 2009.
2. Ikeda, Y. , Umeda, N. , Akino, N. , Ebisawa, N. , Grisham, L. , Hanada, M. , Honda, A. , Inoue, T. , Kawai, M. , Kazawa, M. , Kikuchi, K. , Komata, M. , Mogaki, K. , Noto, K. , Okano, F. , Ohga, T. , Oshima, K. , Takenouchi, T. , Tanai, Y. , Usui, K. , Yamazaki, H. , and Yamamoto, T., *Nucl. Fusion* 46, pp. S211-S219, 2006.
3. Ikeda, Y., Akino, N., Ebisawa, N., Hanada, M., Inoue, T., Honda, A., Kamada, M., Kawai, M., Kazawa, M., Kikuchi, K., Kikuchi, M., Komata, M., Matsukawa, M., Mogaki, K., Noto, K., Okano, F., Ohga, T., Oshima, K., Takenouchi, T., Tamai, H., Tanai, Y., Umeda, N., Usui, K., Watanabe, K., and Yamazaki, H., *Fusion Eng. and Des.* pp. 791-797, 2007.
4. Hemsworth R., Decamps H., Graceffa J., Schunke B., Tanaka M., Dremel M. , Tanga A., De Esch H.P.L., Geli F., Milnes J., Inoue T., Marcuzzi D., Sonato P., and Zaccaria P. *Nucl. Fusion* 49, p. 045006 (15pp), 2009.
5. F. Taccogna, R. Schneider, and S. Longo, Phys. Plasmas 15, pp.10352(10p), 2008, F. Taccogna, P. Minelli, S. Longo, Capitelli, and R. Schneider, *Phys. Plasmas* 17, pp. 063502 (8pp), 2010.
6. R Gutser, D Wunderlich, U Fantz and the NNBI-Team, *Plasma Phys. Control. Fusion* 51, 045005 (14pp), 2009.
7. O. Fukumasa and R. Nishida, *Nucl. Fusion* 46, pp. S275 – 280, 2006.
8. A. Hatayama, *Rev. Sci. Instrum.* 79, 02B901(10pp), 2008.
9. K. Tsumori, Y. Takeiri, O. Kaneko, M. Osakabe, A. Ando, K. Ikeda, K. Nagaoka, H. Nakano, E. Asano, M. Shibuya, M. Sato, T. Kondo and M. Komada, *J. Fusion Sci. Technol.* 58, pp. 482-488, 2010.
10. O'Keefe A. and Deacon D.A.G., *Rev. Sci. Instrum.* 59, pp. 2544-2551, 1988.
11. G. Berden and R. Engeln eds., *Cavity Ring-Down Spectroscopy: Techniques and Applications*, John Wiley and Sons Ltd., 2009.
12. E. Quandt, I. Kraemer and H. F. Döbele, Europhysics Lett. 45, pp. 32 -37, 1998
13. F. Grangeon, C. Monard, J.L. Dorier, A.A. Howling, Ch. Hollenstein, D. Romanini and N. Sadeghi, *Plasma Sources Sci. Technol.* 8, pp. 448-456, 1999.
14. M. Berger, U. Fantz, S. Christ-Koch and NNBI Team, *Plasma Sources Sci. Technol.* 18, 025004 (8pp), 2009.
15. R. Pasqualotto, A. Alfier and L. Lotto, *Rev. Sci. Instrum.* 81, 10D710(5pp), 2010.
16. M. Bacal, *Rev. Sci. Instrum.* 71, pp. 3981-4006, 2000.

Comparison of Optical Emission Spectroscopy and Cavity Ring-Down Spectroscopy in Large-Scaled Negative-Ion Source

K. Ikeda[a], H. Nakano[a], K. Tsumori[a], U. Fantz[b], O. Kaneko[a], M. Kisaki[a], K. Nagaoka[a], M. Osakabe[a] and Y. Takeiri[a]

[a]*National Institute for Fusion Science, 322-6 Oroshi-cho, Toki 509-5292, Japan*
[b]*Max-Planck-Institut fuer Plasmaphysik, Boltzmannstr. 2, D-85748 Garching, Germany*

Abstract. Optical emission spectroscopy (OES) and cavity ring-down spectroscopy (CRDS) systems are installed in a 1/3-scaled negative hydrogen-ion source at the National Institute for Fusion Science testbed to investigate the dynamics of H$^-$ ions in the extraction region near the plasma grid. The signal form of the H$^-$ ion density rapidly drops after beam extraction on applying a low-bias voltage. A similar signal drop appears in the intensity of the hydrogen Balmer-line emission measured by OES and is caused by decreasing atomic hydrogen produced by mutual neutralization effects between H$^-$ and H$^+$. Shot trend of the beam currents are similar to the H$^-$ density and H$_\alpha$/H$_\beta$ in the extraction region, which increases twice as large immediately after Cs seeding. We observe a linear correlation between the H$^-$ density and the inclination of H$_\alpha$/H$_\beta$ which allows for experimentally benchmarking the OES measurement with that of CRDS. Thus, this approach is used for estimating the H$^-$ density by OES in negative-ion sources for high-energy neutral beam injector.

Keywords: Hydrogen, Negative ion source, Spectroscopy.
PACS: 52.70.Kz, 29.25.Ni

INTRODUCTION

A cesium (Cs) assisted negative-hydrogen ion source for a high-energy neutral beam injector (NBI) has been successfully operated for more than ten years at the Large Helical Device (LHD) of the National Institute for Fusion Science (NIFS). The maximum beam injection power of 16 MW uses three injectors with six ion sources [1], which is the same as the design power of one NBI for the International Thermonuclear Experimental Reactor (ITER) [2]. Production of H$^-$ ions and their dynamics in the extraction region near the surface of the plasma grid (PG) are important subjects in H$^-$ or D$^-$ ion sources for high power and stable operation of the present NBI [3,4] and development of future NBI for ITER [2,5].

Electrical and calorimetric diagnostics are standard techniques to optimize beam condition for extracted beam currents and beam divergence, respectively, for the NBI system. Because measuring the beam condition for optimization of H$^-$ density at the extraction region in the H$^-$ ion source is insufficient, optical emission spectroscopy (OES) is a useful diagnostic method [6]. Hydrogen dynamics, Cs consumption, and impurity trend can be simultaneously observed with simple optical components such as an optical fiber and spectrometer. The analysis of OES is quite complex because the excited state of the hydrogen population involves numerous processes. Mutual neutralization of H⁻ and H⁺ ions is an important process in the H⁻ source. According to the numerical calculation of the emission rate coefficient of hydrogen Balmer lines, an emission-line ratio H_α/H_β relates to the H⁻ density in a negative-ion source [7]. The H⁻ density can be estimated by the condition $T_e > 1$ eV. With the present plasma parameters near the PG ($T_e < 1$ eV), it is difficult to obtain the H⁻ density from the ratios of Balmer-line intensities solely using OES analysis because dissociative recombination of molecular hydrogen ions significantly contributes to the Balmer-line radiation. Therefore the calibration approach is employed in the OES measurement for H⁻ estimation used in cavity ring-down spectroscopy (CRDS). The CRDS is a powerful diagnostic tool utilized for obtaining the absolute value of H⁻ density using a decay time of laser absorption [8]. Because its complex system configuration and calibration laser alignment makes installation of a high-energy NBI system difficult, we have installed the CRDS system in a 1/3-scaled NIFS H⁻ ion source in the NIFS-testbed. Our approach has worked well for acquiring the H⁻ density [9].

In this paper, we show the optical configuration of OES and CRDS systems and the result of the time trend of H⁻ density and the optical emission. We then compare the CRDS and OES measurements in applying high-energy NBI systems.

NIFS TEST-ION SOURCE AND DIAGNOSTICS

Figure 1 shows a cross section of the 1/3-scaled NIFS H⁻ source. The inner size of the arc chamber was 700 mm in height (y-axis), 350 mm in width (x-axis), and 250 mm in depth (z-axis). The grid system consisted of the grounded grid (GG), extraction grid (EG), and PG. Each grid was divided into two segments along the y-axis. Hydrogen plasma was produced by arc discharge and confined by a magnetic field produced by cusp magnets set in the arc chamber. To reduce electron

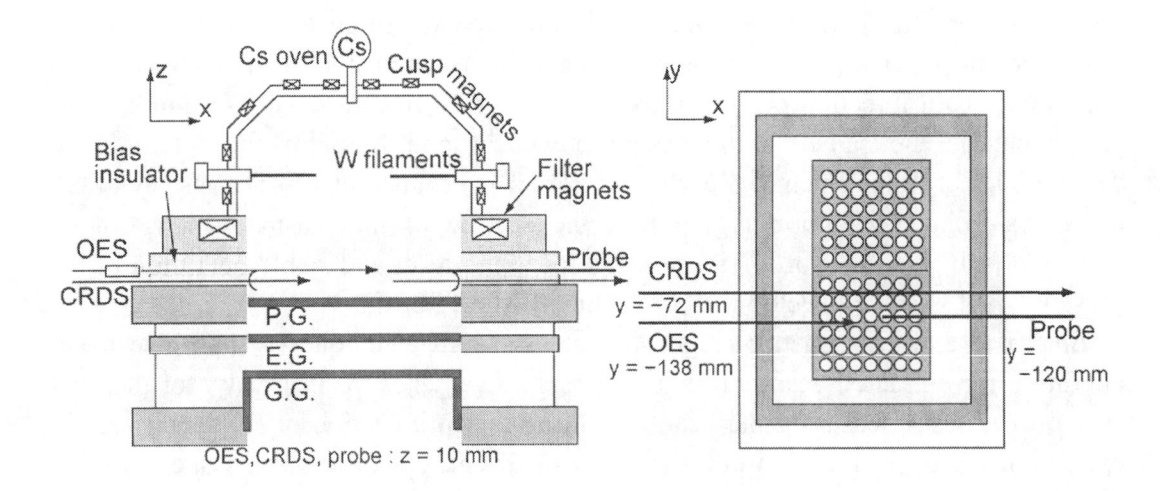

FIGURE 1. Schematic cross section of the 1/3-scaled NIFS test ion source. OES and CRDS diagnostics are installed near PG.

temperature, a magnetic filter field was created between the discharge area and PG by filter magnets. Two Cs ovens set on the back plate of the arc chamber supplied Cs vapor.

A bias insulator was installed to increase the thickness from 16 mm to 40 mm for installing quartz windows for OES, optical mirrors for CRDS, and Langmuir probe ports. All diagnostic positions were 10 mm apart from PG and parallel to its plane. The lines of sight for CRDS and OES were located near the center position at y = −72 mm and y = −138 mm, respectively. Hydrogen Balmer spectra (H_α, H_β, H_γ) and atomic Cs spectrum (Cs I: 852 nm) were observed in the extraction region by the OES system using a wide-range survey spectrometer (PLASUS EmiCon system) connected by an optical fiber. Although Cs^+ lines (Cs II: 460 nm) were intense in the driver region in a previous experiment [10], the intensity of the lines was below the measurable limit in the extraction region. The maximum electron temperature measured by the Langmuir probe was 1 eV. The OES system permitted wavelengths from 200 nm to 870 nm and was calibrated using an Ulbricht sphere with a standard lamp.

The CRDS system consisted of a pulse laser with a wavelength of 1064 nm, highly reflective cavity mirrors, and an optical diode detector. The laser pulse was absorbed by H^- ions during the round trip between the cavity mirrors, causing a decrease in intensity. Line-integrated H^- density near the PG was estimated by the decay function of the penetration laser pulse. Dr. H. Nakano discussed this CRDS measurement in detail in his research [9].

COMPARISON OF OES AND CRDS MEASUREMENT

Hydrogen spectra and H^- density in the extraction region were observed during arc discharge with a beam extraction of 1 s. Figure 2(a) shows a typical example of time evolution of arc discharge current (I_{arc}) with various bias voltages. Here we used Cs vapor to increase the H^- current. I_{arc} was set to the same growth with different bias voltages. The average arc power was 50 kW during beam extraction when I_{arc} increased from additional heat loading of back streaming H^+ ions.

For a low bias voltage of 0.1V, large extraction current (I_{ex}) was observed by an electrical diagnostic in the power supply system as shown in Fig. 2(b). Extracted electrons and secondary electrons were suppressed by the strong magnetic field at the EG. The acceleration beam current (I_{acc}) comprised H^- current and a small leaking electron current. When high-bias voltage was applied, the electron beam component was sufficiently suppressed. The H^- beam current measured on the beam calorimeter located at 4 m past the ion source and I_{acc} somewhat decreased by increasing the bias voltage as denoted by circles in Fig. 2(b). I_{acc}/I_{ex} improved from 0.64 to 0.77 with the high-bias voltage of 6.9 V.

At the extraction region in the arc chamber, the H^- density measured by CRDS grew to 1.6×10^{17} m^{-3} and I_{arc} increased with an applied bias voltage of 0.1 V (Fig. 3 (a)). The H^- density decreased to 1.1×10^{17} m^{-3} when the applied bias voltage was 6.9 V. The density of H^- ions in the lower-bias voltage was much better only in the arc discharge condition. When applying the beam extraction voltage, the H^- density suddenly dropped to 1.1×10^{17} m^{-3} for a 0.1 V bias. The H$^-$ densities of different bias voltages somewhat increased in the same manner as the increasing arc discharge during beam extraction, and the values were nearly the same. Different values in the H$^-$ density before and after beam extraction decreased as bias voltage increased. This gap was not observed with a high-bias voltage of 6.9 V. From this result, the H$^-$ profile along the z-axis was considered to be shift in the PG direction on beam extraction.

The time evolution of H$_\alpha$ emission intensity measured by OES is shown in Fig. 3 (b). The H$_\alpha$ intensity also increased in the same manner as H$^-$ density by increasing

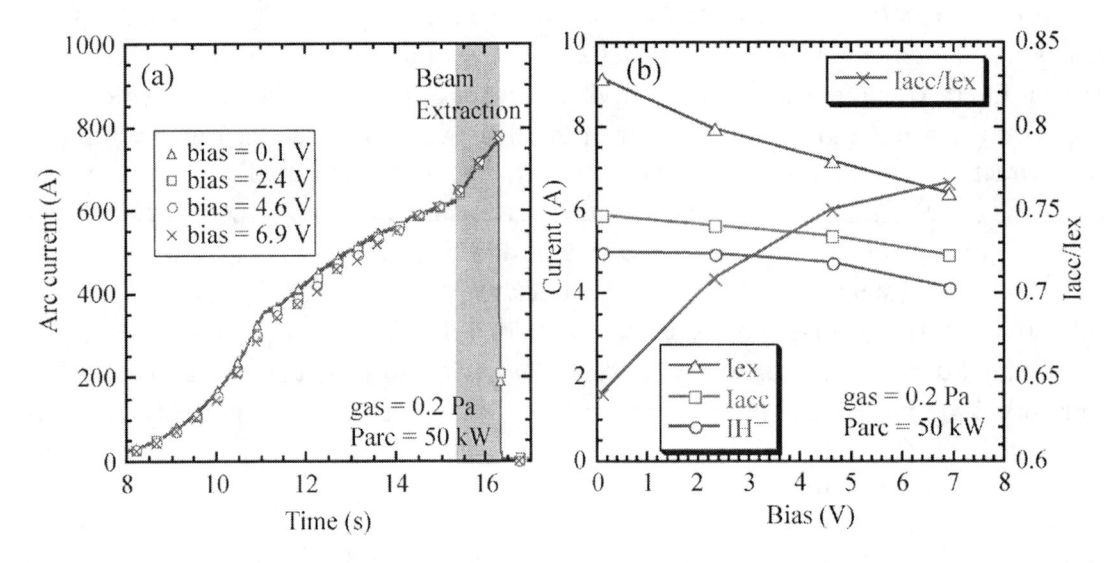

FIGURE 2. (a) Time evolution of arc discharge current with a beam extraction of 1 s. (b)Triangles and squares denote the extraction and acceleration beam current in the power supply, respectively. Crosses represent I_{acc}/I_{ex}. Circles represent H$^-$ current on the beam calorimeter.

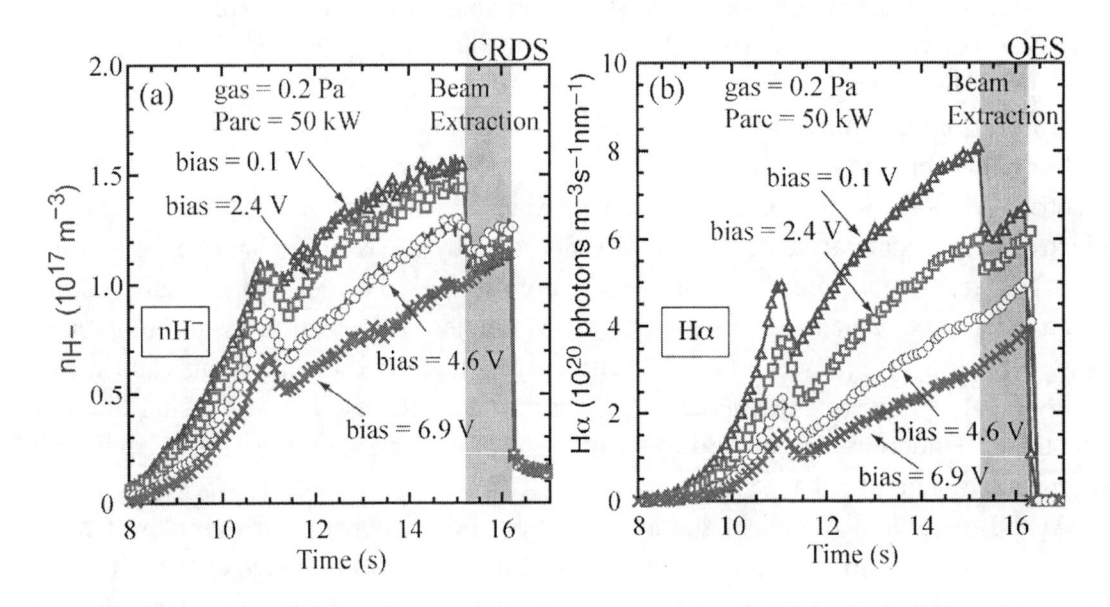

FIGURE 3. Time evolution of H$^-$ density measured by CRDS (a) and H$_\alpha$ emission intensity measured by OES (b), during arc discharge with beam extraction. Bias voltages of 0.1 V, 2.4 V, 4.6 V and 6.9 V are plotted.

the arc discharge power before beam extraction. Similar signal drops appeared in the low-bias voltage condition during beam extraction. The difference in H_α intensity before and after beam extraction also decreased with an increasing bias voltage. From the result of CRDS measurement, the H^- densities with different bias voltages were nearly the same during beam extraction. Thus, neutral hydrogen produced by the effect of mutual neutralization was the same value between the 0.1 V and 6.9 V bias voltage during beam extraction. According to the OES measurement, the H_α intensity in the 6.9 V high-bias voltage was half as large as that in the 0.1 V low-bias voltage during beam extraction. Intensity of H_α emission was proportional to the emission rate coefficient, electron density, and neutral hydrogen density. Therefore, this result suggests that the electron density is suppressed by half in the high-bias voltage condition.

Figure 4 shows the shot trend of I_{acc} at the grid system and the I_{H-} on the beam calorimeter with a constant 50 kW arc discharge condition at the beginning of Cs conditioning. In the pure volume operation, it detected a negative ion current of 1 A. Both currents jumped up to twice the size immediately after Cs seeding with an oven temperature of 160 °C. The same signal jump appeared in the negative ion density by CRDS and H_α/H_β by OES in the extraction region as shown in Fig. 5. The H^- density doubled in size without beam extraction, and a similar increase was observed in the H_α/H_β line ratio. The beam currents and H^- density gradually increased after raising the oven temperature to 185 °C. The H^- beam current reached 3 A with the H^- density of 0.9×10^{17} m^{-3} in the extraction region after 340 shots from Cs conditioning. According to this experiment with constant discharge conditions (arc power, bias voltage, and gas pressure), we observed a linear correlation between the inclination of H_α/H_β by OES and H^- density by CRDS, as shown in Fig. 6. A linear correlation also existed between the H^- beam current and the source H^- density near the PG surface. Thus, H_α/H_β measurement by OES with CRDS calibration can be applied for the

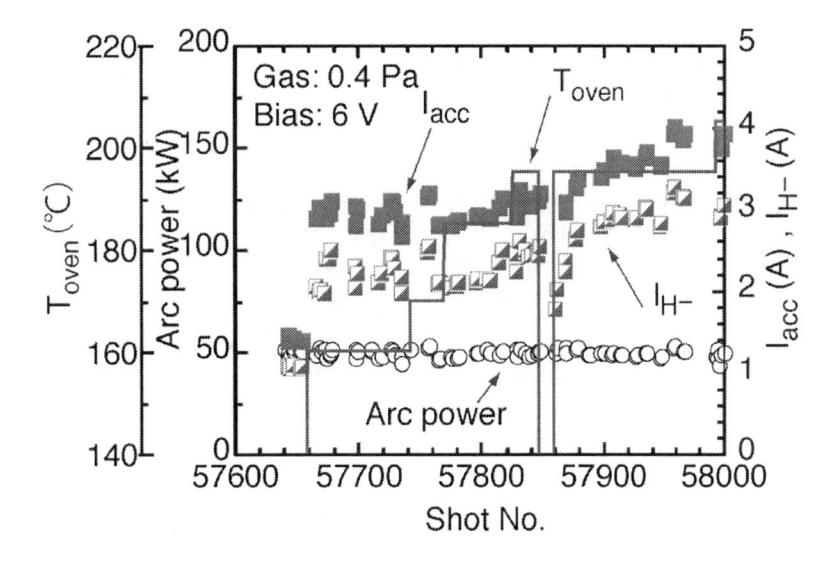

FIGURE 4. Shot trend of acceleration beam current (I_{acc}) at the grid system and negative hydrogen current (I_{H-}) at the beam calorimeter with a constant arc discharge condition (50 kW) with Cs vapor. Cs oven temperature (T_{oven}) was applied at 160 °C − 200 °C.

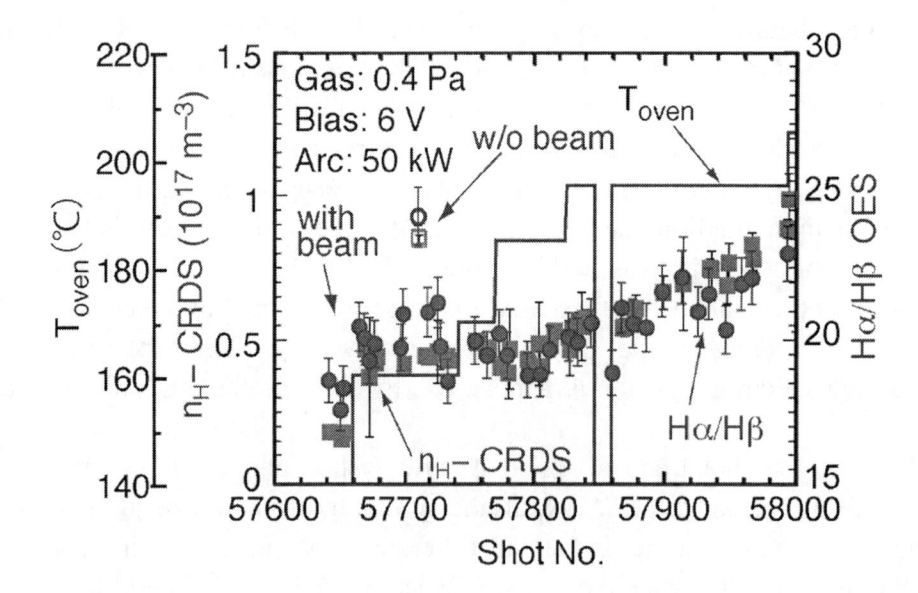

FIGURE 5. Shot trend of the H⁻ ion density obtained by CRDS and line ratio of H_α/H_β measured by OES. Closed marks denote the operation with beam extraction and open marks denote arc discharge only with the discharge power of 50 kW.

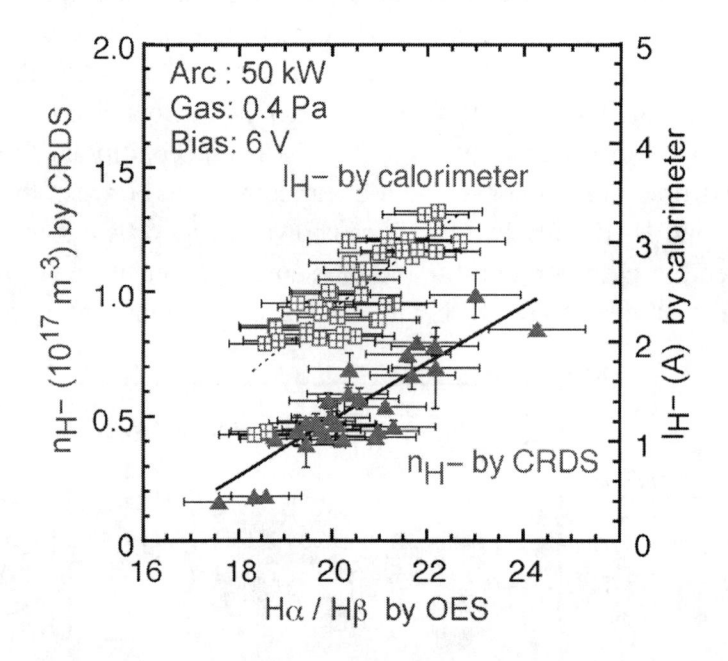

FIGURE 6. Correlations of H_α/H_β measured by OES with H⁻ ion density measured by CRDS in the extraction region and with the H⁻ beam current measured by the beam calorimeter. The condition of arc discharge is constant with different Cs conditions.

estimation of H⁻ density in the negative-ion source under severe conditions such as a high-energy NBI system in a fusion reactor.

SUMMARY

Hydrogen spectra and H⁻ density were measured by OES and CRDS diagnostics, respectively, at an extraction region of 1/3-scaled NIFS negative ion source. Intensity of H_α emission decreased after beam extraction with a low-bias voltage, which is a trend similar to that of the H⁻ density obtained by CRDS. It is suggested that atomic hydrogen strongly reflects mutual neutralization of H⁻ and H⁺ near the PG surface. A linear correlation was observed between the inclination of H_α/H_β and H⁻ density, which allows for experimentally benchmarking the OES measurement with that of CRDS.

ACKNOWLEDGMENTS

The authors thank the NBI staff for their operational support. This work was supported by the budget for NIFS10ULRR702 and NIFS10ULRR009.

REFERENCES

1. K. Ikeda, K. Nagaoka, Y. Takeiri, M. Osakabe, K. Tsumori, and O. Kaneko, *Plasma Science and Technology*, **11**, No. 4, 452-455 (2009).
2. R. S. Hemsworth, A. Tanga, and V. Antoni, *Rev. Sci. Instrum.*, **79**, 02C109 (2008).
3. Y. Takeiri, O. Kaneko, K. Tsumori, Y. Oka, K. Ikeda, M. Osakabe, K. Nagaoka, E. Asano, T. Kondo, M. Sato and M. Shibuya, *Nucl. Fusion*, **46**, S199-S210 (2006).
4. M. Kuriyama, N. Akino, N. Ebisawa, L. Grisham, H. Liqen, A. Honda, T. Itoh, M. Kwai, M. Kazawa, K. Mogaki, Y. Ohara, T. Ohga, K. Ohmori, Y. Okumura, H. Oohara, K. Usui, and K. Watanabe, *Journal of Nuclear Science and Technology*, **35**, No. 11, 739-749 (1998).
5. E. Speth, H.D. Falter, P. Franzen, U. Fantz, M. Bandyopadhyay, S. Christ, A. Encheva, M. Fröschle, D. Holtum, B. Heinemann, W. Kraus, A. Lorenz, Ch. Martens, P. McNeely, S. Obermayer, R. Riedl, R. Süss, A. Tanga, R. Wilhelm, and D. Wünderlich, *Nucl. Fusion*, **46**, S220-S238 (2006).
6. U. Fantz, H. Falter, P. Franzen, D. Wünderlich, M. Berger, A. Lorentz, W. Kraus, P. McNeely, R. Riedl, and E. Speth, *Nucl. Fusion*, **46**, S297-S306 (2006).
7. U. Fantz and D. Wünderlich, *New Journal of Physics*, **8**, 301 (2006).
8. E. Quandt, I. Kraemer, and H. F. Döbele, *Europhys. Lett.*, **45** No. 1, pp. 32-37 (1998).
9. H. Nakano, K. Tsumori, K. Ikeda, K. Nagaoka, M. Kisaki, U. FANTZa, M. Osakabe, O. Kaneko E. Asano, T. Kondo, M. Sato, M. Shibuya, S. Komada, H. Seki, and Y. Takeiri, "Negative-Hydrogen-Ion Density Characteristics in Large-Negative-Ion-Source with Arc Discharge" AIP Conference Proceeding of 2nd International Symposium on Negative Ions, Beams and Sources, Takayama, (2010), session No. O-32.
10. K. Ikeda, K. Nagaoka, Y. Takeiri, U. Fantz, O. Kaneko, M. Osakabe, Y. Oka, K. Tsumori, *Rev. Sci. Instrum.*, **79**, 02A518 (2008).

Measurement of Electron Density near Plasma Grid of Large-scaled Negative Ion Source by Means of Millimeter-Wave Interferometer

K. Nagaoka, T. Tokuzawa, K. Tsumori, H. Nakano, Y. Ito, M. Osakabe,
K. Ikeda, M. Kisaki, M. Shibuya, M. Sato, S. Komada, T. Kondo,
H. Hayashi, E. Asano, Y. Takeiri, O. Kaneko and NBI group

National Institute for Fusion Science, 322-6 Oroshi-cho, Toki 509-5292, Japan

Abstract. A millimeter-wave interferometer with the frequency of 39 GHz (λ=7.7 mm) was newly installed to a large-scaled negative ion source. The measurable line-integrated electron density ($n_e \cdot l$) is from 2×10^{16} to 7×10^{18} m^{-2}, where n_e and l represent an electron density and the plasma length along the millimeter-wave path, respectively. Our interest in this study is behavior of negative ions and reduction of electron density in the beam extraction region near the plasma grid. The first results show the possibility of the electron density measurement by the millimeter-wave interferometer in this region. The line-averaged electron density increases proportional to the arc power under the condition without cesium seeding. The significant decrease of the electron density and significant increase of the negative ion density were observed just after the cesium seeding. The electron density measured with the interferometer agrees well with that observed with a Langmuir probe. The very high negative ion ratio of $n_{H^-}/(n_e + n_{H^-}) = 0.85$ was achieved within 400 min. after the cesium seeding.

Keywords: .negative ion source, electron density, interferometer, negative ion ratio
PACS: 52.27.Cm, 52.50.Dg, 52.50.Gj, 52.70.Gw

INTRODUCTION

Negative ions are utilized in various fields such as high-energy particle accelerator, magnetically confined fusion and industrial application. Intensive efforts have been made to increase negative ion current density[1-4]. In the fusion field, the high power negative ion beams with the order of 10^7 W are being successfully operated[5]. In the near future, larger device (ITER) will start the operation, in which the beam power becomes the order of 10^8 W. In such case, the reduction of co-extracted electrons becomes more important. That is because accelerated electrons could damage the beam line components including accelerator and consume the electric power severely. Therefore, the suppression of electron beam components is also a key issue of negative ion sources for the fusion application.

Recently, a millimeter-wave interferometer was temporary installed in a 1/3-scaled ion source for LHD-NBI system to test the possibility of electron density measurement near the plasma grid. In this paper, the preliminary results of the electron density measurement by the interferometer are reported. In the session 2, the experimental setup and interferometer system used in this experiment are described. In the session 3, the wide range of electron density was measured by the interferometer. The comparison with Langmuir probe measurement shows good agreement. The estimation of negative ion ratio in the condition with cesium seeding is also discussed.

EXPERIMENTAL

Negative Ion Source

The experiments have been performed in the NBI test stand at National Institute for Fusion Science applying the 1/3-scaled negative ion source for LHD-NBI system[1], of which schematic is shown in fig. 1. The ion source is a cesium-seeded negative ion source. An arc chamber has dimensions of 700 mm x 350 mm in cross section and 220 mm in depth, and is surrounded by a strong cusp magnetic field of 0.2 T on the inner walls. The external magnetic filter is generated a transversal magnetic field in front of a plasma grid by a pair of permanent magnet rows facing each other with a separation of 350 mm. The magnetic field strength is around 6 mT at the center. Two cesium ovens are attached to a back plate of the arc chamber. The most important region for negative ion production is in front of the plasma grid. In order to observe the plasma and their characteristics, a diagnostic flange was installed between the arc chamber and the plasma grid flange. The center of the diagnostic flange is 12 mm apart from the plasma grid surface and various diagnostic tools are installed on this flange. A movable Langmuir probe can measure the local values of electron temperature, density, electrostatic potential, and their profiles in three directions. A cavity ring down (CRD) system measure the line-averaged negative ion density[6]. An optical emission spectroscopy system is also installed and monitors the plasma[7]. An interferometer was also installed through the diagnostic flange. The lines of sight of the interferometer and the cavity ring down spectrometer are chosen to be geometrically identical in this experiment. The plasma length along the line of sights is 180 mm, which is experimentally measured using the movable Langmuir probe.

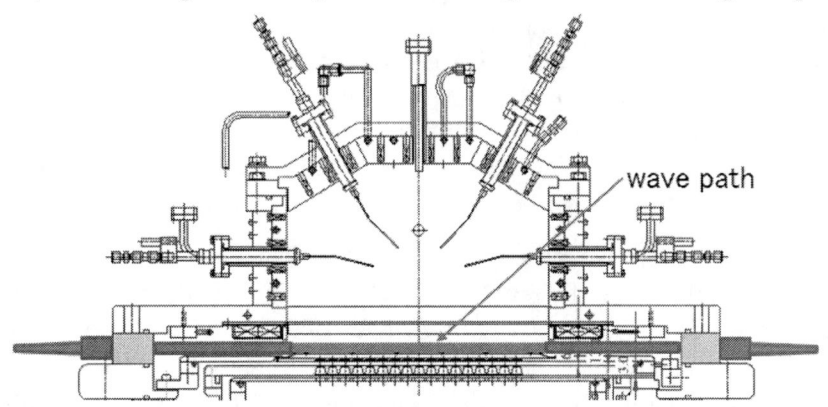

FIGURE 1. Schematic of cross section of arc chamber. The distance of millimeter-wave path is 1.2 cm from the plasma grid surface.

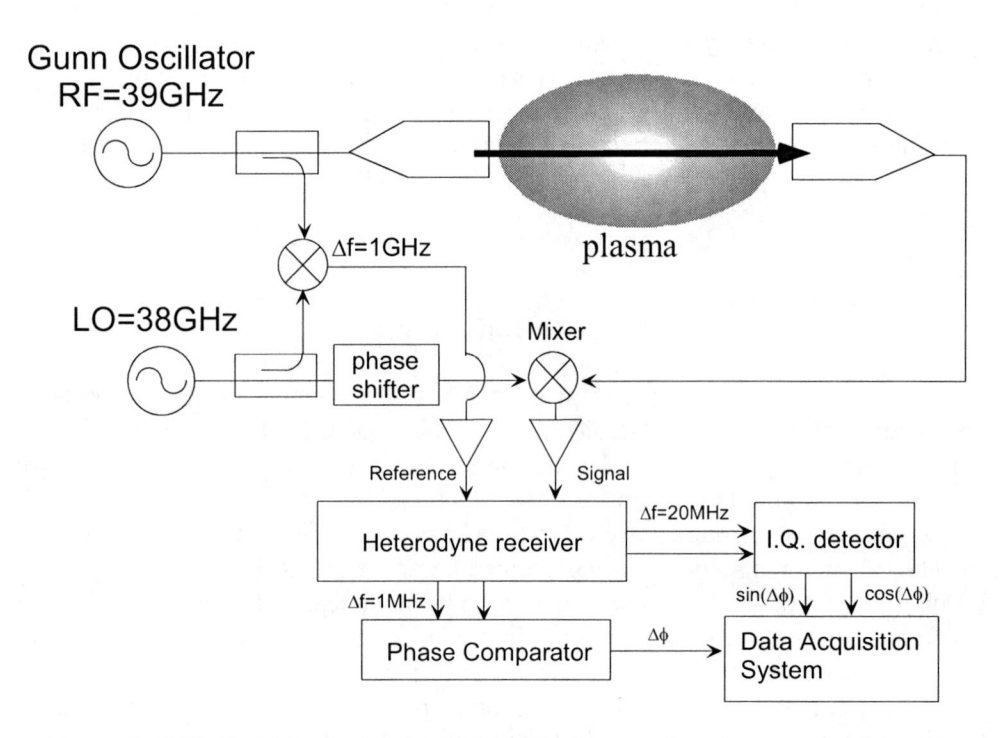

FIGURE 2. Schematic drawing of the millimeter-wave interferometer system.

Millimeter-Wave Interferometer

The heterodyne millimeter-wave interferometer system for the electron density measurements in the negative ion source is shown in Fig. 2. The system is categorized in Mach-Zehnder type interferometer. A Gunn oscillator with the frequency of 39 GHz and the power of about 100 mW is used as a probe source (RF). The output millimeter wave is transmitted through the $\phi 16$ circular waveguide and is launched to the plasma. The beam width at the center of the ion source is about 20 mm and the electric field of the millimeter wave is oriented perpendicular to the plasma grid. The phase modulated millimeter wave passed through the plasma is detected and mixed with the local wave. The frequency of local oscillator (LO) is exactly 1 GHz away from the probing frequency. Using a mixer, millimeter wave is down-converted to 1 GHz beat signal with original phase information. Also, another mixer is applied to generate a reference signal that transmits a path without the plasma. The phase difference ($\Delta\phi$) between both signals is expressed by $\Delta\phi = k_0(1-\bar{\mu})L \approx k_0 \bar{n}_e L / 2n_c$, where $\bar{\mu}$ is the line-averaged refractive index, $\bar{n}_e$ is the line-averaged electron density, L is the line length of sight, and n_c is the cutoff density of the probing frequency. In this experiment ($L = 0.18$ m), the line-averaged density $\bar{n}_e = 1.7 \times 10^{17}$ m^{-3}/fringe. Both the mixer outputs are led to the heterodyne receiver system which is adopting a super heterodyne technique in order to compensate the drift of the beat frequency. The heterodyne receiver system, which is the similar system in Ref. 8, outputs two pairs of frequency components. One pair is down-converted 1 MHz signals and they are fed to the analog phase comparator

[9]. The phase resolution of the phase comparator in this experiment is around 1/100 fringes. Another pair is 20 MHz signals which use the IQ detection. The IQ detector produces sine and cosine signals of the phase difference. The output signals of the phase comparator and the IQ detector are collected by a data acquisition system, of which sampling rate is 50 kHz.

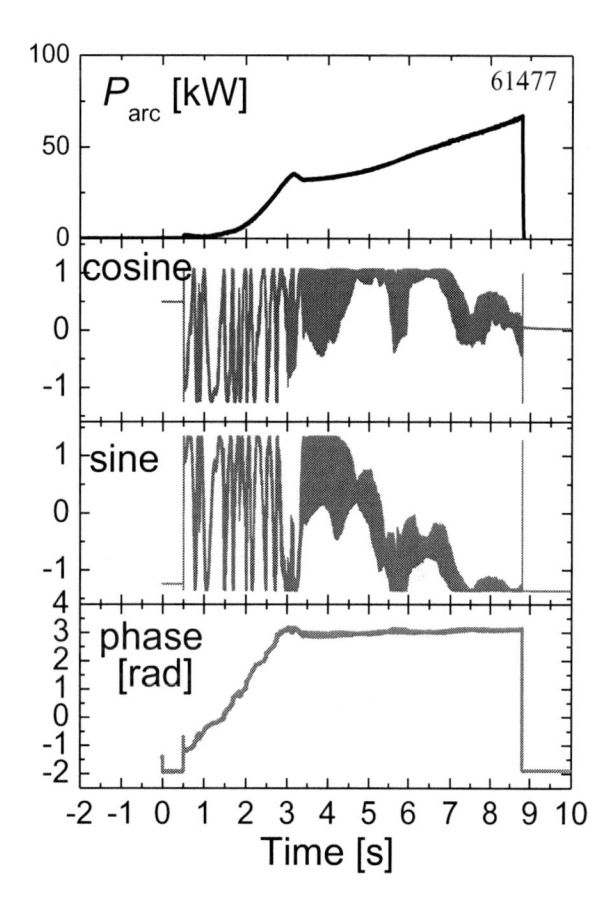

FIGURE 3. Typical wave forms of arc power, cosine and sine components of interference signal and phase during an arc discharge.

RESULTS AND DISCUSSIONS

The interferometer was operated to measure the line-averaged electron density during the discharges. The millimeter-wave can pass through the plasma and the interference wave was successfully detected. Figure 3 shows the typical waveform of the arc power, the cosine and sine components of the IQ signal, and the phase. At $t <$ 3.0 sec, the arc power increases quickly and the interference signals, sine and cosine components, also change dynamically. The phase signal corresponding to the line-averaged electron density increases similarly to the arc power. At $t > 3.0$ sec, the arc power increases gradually, while the phase keeps constant. The reason of the different behaviors between the arc power and the electron density is not understood yet. One candidate is the change of the spatial distribution of the electron density. In order to remove uncertainties, the typical line averaged density of each discharge is estimated by the average over 0.5 sec just before the plasma termination.

The arc power was scanned based on shot by shot basis without cesium seeding, which is shown in fig. 4. The observed electron density increases linearly with arc power, indicating that the interferometer works well in the wide range of electron density.

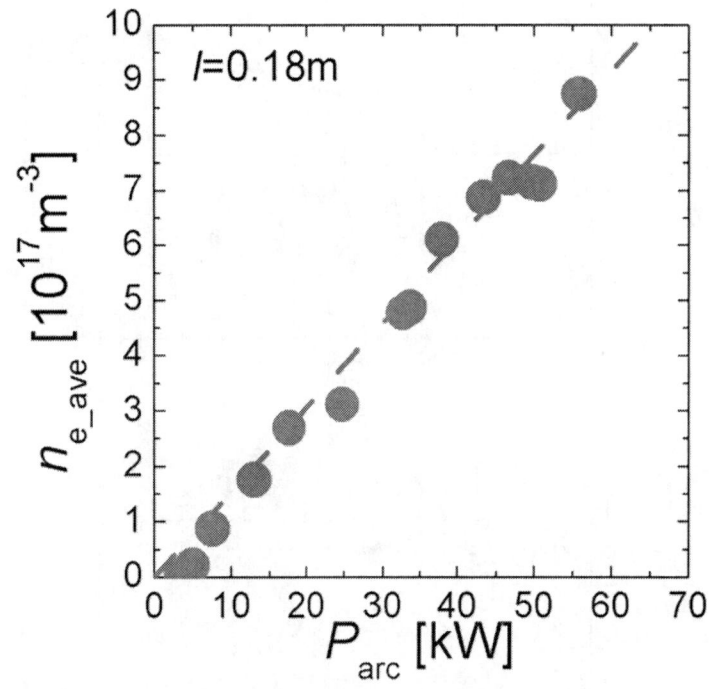

FIGURE 4. Line-averaged electron density measured by interferometer as a function arc power without cesium seeding.

In order to investigate the cesium seeding effects on electron suppression and negative ion production, the sequential operation of arc discharge in every two minutes has been carried out. Figure 5 (a) shows the time evolution of the electron density measured with the interferometer. The electron density decreased with a factor of around 5 is observed in the initial phase within 50 min after the start of cesium seeding. The quite similar result was obtained with the Langmuir probe, which is shown in Fig. 5 (b). The absolute value of the electron density also agrees well between the interferometer and the Langmuir probe. The difference is only within a factor of 1.6. In the second phase ($t > 150$ min.), the cesium seeding effect has a much longer time scale than that in the initial phase. The jump of the electron density was observed around $t = 100$ min., indicating the change of discharge state, which is considered as a mode-change like phenomenon. The significant increase of the negative hydrogen ion density in the initial phase is observed by the cavity ring down spectroscopy, which is shown in Fig. 5 (c). The interferometer and cavity ring down spectroscopy can evaluate the absolute value of line-integrated densities with geometrically identical line of sight. Therefore, the negative ion density ratio R (= $n_{H^-}/(n_e + n_{H^-})$) can be determined with very high accuracy. The negative ion density ratio without cesium seeding is $R = 0.17$, and it becomes $R = 0.73$ within 50 min. after the start of cesium seeding. Finally $R = 0.85$ is achieved at 400 min. from the start of cesium seeding.

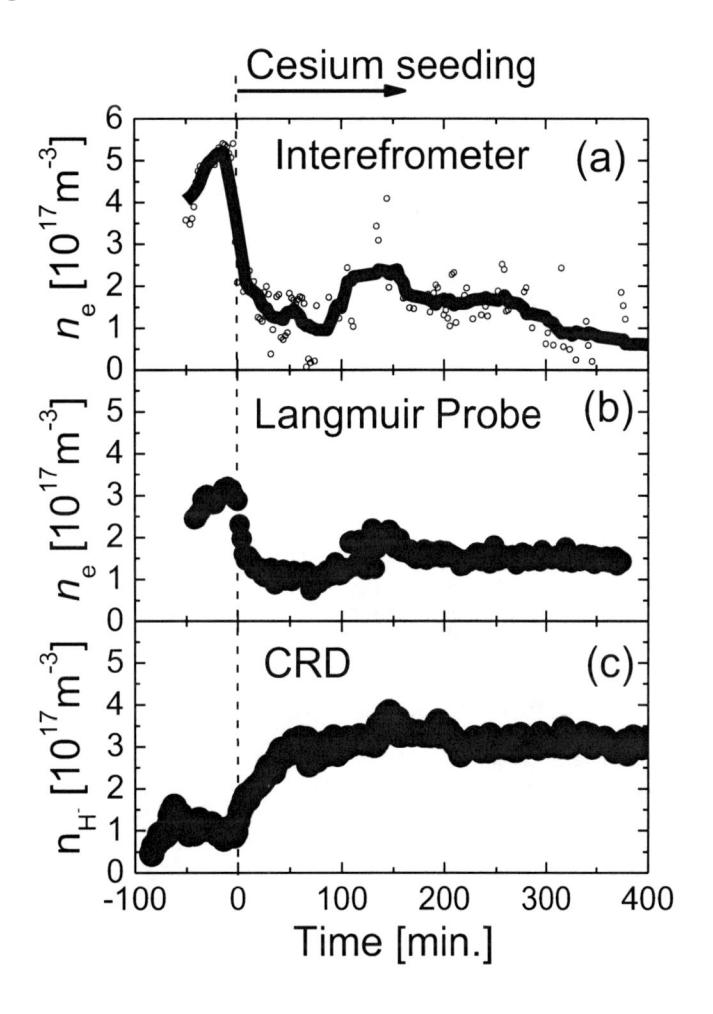

FIGURE 5. Time evolution of (a) electron density measured by interferometer, (b) that by Langmuir probe and (c) negative hydrogen ion density. The cesium seeding starts at $t = 0$ min.

SUMMARY

A millimeter-wave interferometer was temporary installed in a 1/3-scaled negative ion source of LHD-NBI system to investigate the possibility of the electron density measurement near the plasma grid. The interference signal can be observed and the electron density measurement was successfully done in the wide range of the electron density. The density agrees well with that obtained with a Langmuir probe. The effects of cesium seeding on the electron density and negative ion density were observed. The negative ion density ratio increases from 0.17 to 0.73 within 50 min. just after the start of the cesium seeding. Finally, the large negative ion density ratio of 0.85 was achieved. Phase jump phenomena were often observed in the initial phase of the plasma discharge. The optimization of interferometer system is necessary to improve stability of the system and the accuracy of the density measurement.

ACKNOWLEDGEMENTS

One of the author (K.N.) thanks Prof. U. Fantz and A. Ando for their fruitful discussions, LHD experiment group for kind supports and technical staff of LHD-NBI group for excellent operation of NBI test stand. This study was supported by NIFS (NIFS10ULRR702).

REFERENCES

1. K. Tsumori, Y. Takeiri, O. Kaneko, et al., *Fusion Science and Technology,* **58**, 489, (2010).
2. H. Tobari, et al., *Plasma Fusion Research* **2**, 22 (2007).
3. K. Ikeda, et al., *Rev. Sci. Instrum,* **75**, 1744 (2004).
4. M. Hanada, et al., *Nucl. Fusion*, **46**, S318 (2006).
5. Y. Takeiri, O. kaneko, K. Tsumori, et al., *Fusion Science and Technology,* **58**, 482, (2010).
6. H. Nakano, presented in this symposium, O-32 (2010).
7. K. Ikeda, K. Nagaoka, Y. Takeiri, et al., *Rev. Sci. Instrum.* **79**, 02A518 (2008).
8. K. Kawahata et al., *Rev. Sci. Instrum* **70**, 695 (1999).
9. Y. Ito et al., *Fusion Engineering and Design*, **74**, 847 (2005).

Experimental Mapping and Benchmarking of Magnetic Field Codes on the LHD Ion Accelerator

G. Chitarin[a,b], P. Agostinetti[a], A. Gallo[a], N. Marconato[a], H. Nakano[c], G. Serianni[a], Y. Takeiri[c] and K. Tsumori[c]

[a]*Consorzio RFX, Associazione EURATOM-ENEA, Corso Stati Uniti 4, 35127 Padova, Italy*
[b]*University of Padova, Dept. of Management and Engineering, strad. S. Nicola, 36100 Vicenza, Italy*
[c]*National Institute for Fusion Science, Particle Beam Heated Research Div, 322-6, Oroshi-cho, Toki, Gifu, 509-5292, Japan*

Abstract. For the validation of the numerical models used for the design of the Neutral Beam Test Facility for ITER in Padua [1], an experimental benchmark against a full-size device has been sought. The LHD BL2 injector [2] has been chosen as a first benchmark, because the BL2 Negative Ion Source and Beam Accelerator are geometrically similar to SPIDER, even though BL2 does not include current bars and ferromagnetic materials. A comprehensive 3D magnetic field model of the LHD BL2 device has been developed based on the same assumptions used for SPIDER. In parallel, a detailed experimental magnetic map of the BL2 device has been obtained using a suitably designed 3D adjustable structure for the fine positioning of the magnetic sensors inside 27 of the 770 beamlet apertures. The calculated values have been compared to the experimental data. The work has confirmed the quality of the numerical model, and has also provided useful information on the magnetic non-uniformities due to the edge effects and to the tolerance on permanent magnet remanence.

Keywords: Electrostatic accelerators, Negative-ion beams, Magnetostatics
PACS: 29.20.Ba, 41.75.Cn, 41.20.Gz

INTRODUCTION

A Neutral Beam test-bed facility comprising a RF-driven Negative Ion Source with 100 kV Accelerator (SPIDER) and a complete 1 MV 40 A Neutral Beam System (MITICA) is presently under construction in Padova, as a support to the realization of the Heating Neutral Beam (HNB) system for the ITER experimental Fusion Reactor [1]. The final design optimization of SPIDER and MITICA in order to assure the achievement of the ITER HNB target performances in all operating conditions has necessarily included the improvement of the beam optics by optimization of three magnetic field sources:

• "Filter" field, produced by current flowing in the Plasma Grid and in suitable busbars, reduces the number of co-extracted electrons and also affects the Negative Ion production efficiency;

- "Suppression" field, produced by permanent magnets in the Extraction grid, deflects the co-extracted and secondary electrons;
- "Compensation" field, produced by permanent magnets and ferromagnetic material in the Grounded Grid, compensates the deflection of the accelerated ions due to the electron Suppression field.

For the numerical magnetic field models of the SPIDER accelerator, ANSYS with a "mixed potential" formulation has been adopted in order to describe efficiently both the details of the single beamlet apertures on the grids and the 3D edge effects on the global scale [8]. For the validation of these magnetic models, the LHD BL2 injector [2,3] has been chosen as a first important step, because its negative ion source and beam accelerator are similar to SPIDER, even though BL2 does not include current bars and ferromagnetic materials. The simulations and the experimental mapping of LHD BL2 have also been useful for establishing a sensible requirements on magnetic field uniformity in SPIDER and MITICA, which shall be of the order of 2%.

GEOMETRY OF LHD BL2 AND FINITE ELEMENT MODEL DESCRIPTION

The horizontal section of the LHD BL2 Ion Source and Accelerator is schematized in Fig. 1. The Filter Field is produced by a large stack of permanent magnets located on both sides of the plasma grid (PG). The Filter Magnets are mechanically aligned to the PG plates and magnetized along the X direction in order to produce a rather uniform horizontal magnetic field across the ion source near the PG.

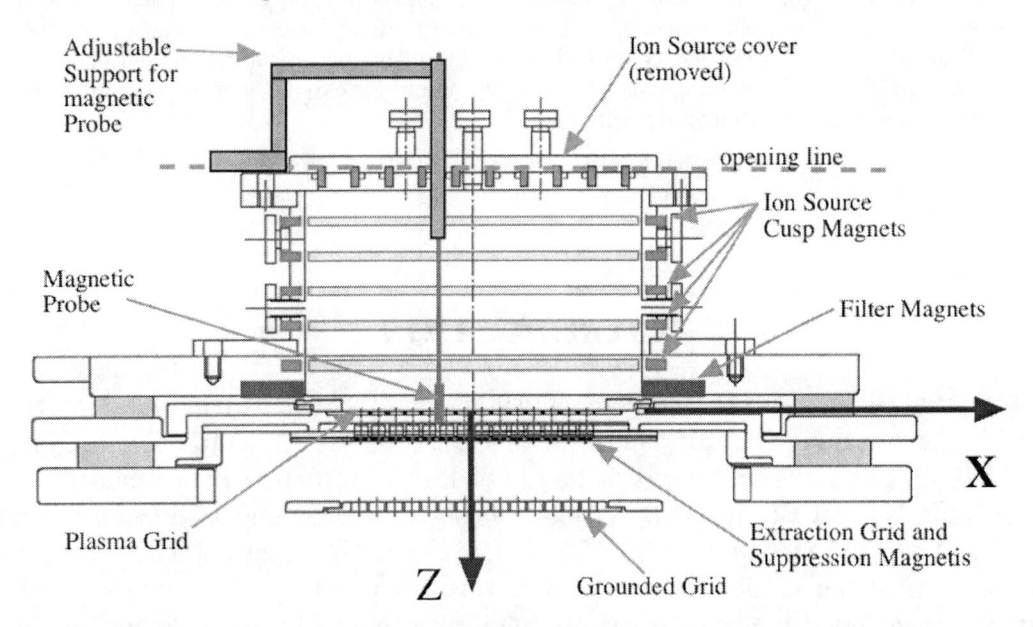

FIGURE 1. Horizontal cross-section of the BL2 source showing the measurement set-up. Z is the longitudinal axis of the accelerator and X is the horizontal transverse axis. The Cusp Magnets, the Filter magnets and the Extraction Grids containing the Suppression magnets are shown.

The Cusp Field in the Ion Source is produced by five stacks of "Cusp" magnets located on the right and left side of the Source. Within each stack, the Cusp magnets have uniform magnetization along X with opposite polarity with respect to those of the adjacent stacks. During the operation of the device Cusp Field is also produced by the Cusp Magnets located on the Ion Source Cover. However, these magnets have not been taken into account because the cover had to be removed during the magnetic measurements.

The Suppression Field is produced by 62 horizontal arrays of magnets embedded in the Extraction Grid (EG). The magnets are located inside the copper grids immediately above and below the grid apertures. Within each array, the magnetization is oriented along the beam direction (Z) but has opposite polarity with respect to the adjacent array (above or below), in order to produce a field (mainly) along Y (vertical) on the front and on the back of each EG aperture.

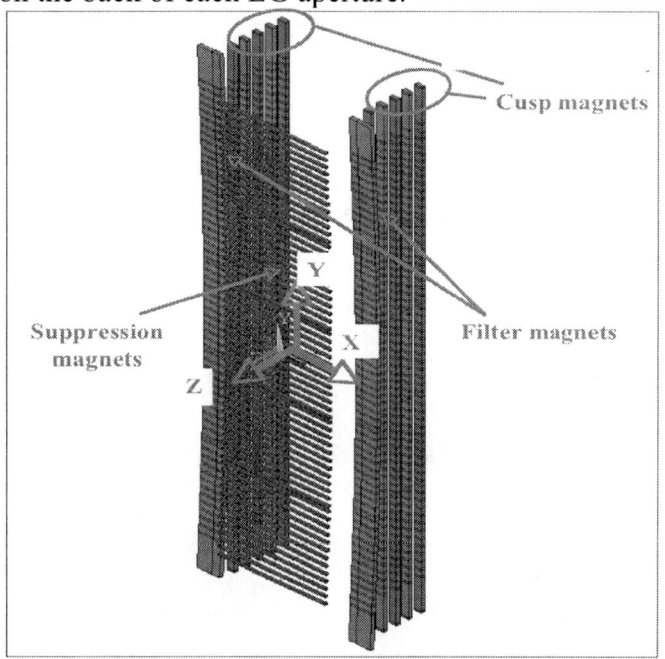

FIGURE 2. view of the 3D FEM model of LHD BL2, the accelerator grids, the vacuum region and the external boundary are not shown for simplicity.

A numerical model of the LHD BL2 magnetic configuration has been set-up using ANSYS [4], with the aim of describing both the magnetic configuration on the beamlet aperture scale and the global non-uniformities of the entire BL2 device, including edge effects. In order to provide a useful benchmark, the model has been based on the same geometrical simplifications and was essentially similar to the 3D model used for the optimization and verification of the design of SPIDER.

In spite of the evident geometrical symmetry of the device, because of the existence of the different magnetic field sources, neither an "odd" nor an "even" symmetry plane could be exploited in BL2 for reducing to ¼ or to ½ the complexity of the model and the number of unknowns.

In order to avoid errors due to the model boundary conditions, the region surrounding the device has been modeled as parallelepiped whose dimensions are more than two times the maximum dimension of the magnetic field sources. Moreover

"Infinite solid elements" have been used on the whole exterior of the model, which reproduces the effect of far-field decay in magnetic analysis.

The Cusp magnets located on the short sides of the Source (top and bottom) have been neglected for simplicity. All Permanent magnets have been described by a linear material characteristic B(H), with a coercive force of 732 kA/m and a relative permeability of 1.04, corresponding to a remanence of 0.96 T.

The model mesh size varies from about 1 mm (elements in the region of the suppression magnets as shown in the particular of Fig. 14) to 50 mm (on the boundary of the air region, excluding the elements which simulate the far-field decay), for a total of about 8.5 million nodes.

Local simplified models have also been set up using OPERA and FEMM [6,7] in order to obtain an independent validation of specific simplification assumptions.

EXPERIMENTAL MAGNETIC FIELD MAPPING

An adjustable supporting structure has been designed and built for accurate positioning the magnetic field probes inside the BL2 Ion Source and Accelerator. The conceptual scheme of the structure is shown in Fig. 1. The structure is formed by three aluminium linear sliding guides, respectively aligned along the X, Y, and Z axes of the device and allows the insertion of the magnetic probes inside the grid apertures.

FIGURE 3. View of the LHD BL2 Ion source with the 3D Supporting Structure for the positioning of the magnetic probe, the reference axes are also shown

The measurements were actually carried out on the so called BL2 "spare" Source and Accelerator, which had already operated on LHD in the past years and is strictly identical to the "original" one. Fig. 3 shows the supporting structure installed on the device.

The magnetic distortion due to the guide screws (made of carbon steel) has been found to be negligible because of the limited size and large distance of the screws from the measurement probe. The positioning tolerance of the structure, including the odometer tolerance, is ≈ 0.2 mm. For the orientation, the tolerance is 0.5 deg.

It is worth noting that the measurements of the three field components (Bx, By and Bz) inside each aperture have been taken in different moments, because of the need to change, and so re-zeroing, the probe and its orientation. For this reason, alignment errors observed in one of the field components are in principle independent on the alignment errors in the other 2 components.

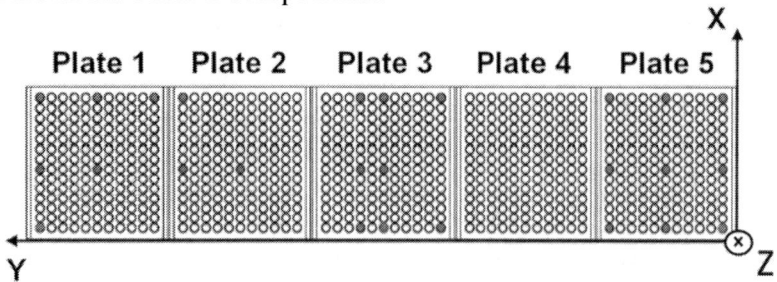

FIGURE 4. scheme of the measurement locations on the five sections (plates) of the BL2 accelerator grids. The reference directions are also shown, Z is the beam axis direction, X is the horizontal transverse direction, and Y is the vertical transverse direction.

Each Accelerator Grid Plate has 14 apertures along the x direction and 11 along the y one, the total number of apertures is therefore 770. The following convention has been assumed in order to identify the measurements related to the different apertures, as described in Fig. 4: each aperture is identified by a string in the form "p i x j y k", where i is the number of the plate, j is the number of the aperture in the x direction, and k is the number of the aperture in the y direction.

However, being the BL2 Plasma Grid plates slightly converging, the Z axis of the supporting structure and the probe itself had to be rotated by ≈1 deg. for measurements on plates 2 and 4 and by ≈2 deg. for measurements on plates 1 and 5 in order to obtain measurements along the beamlet axes (see Fig. 5).

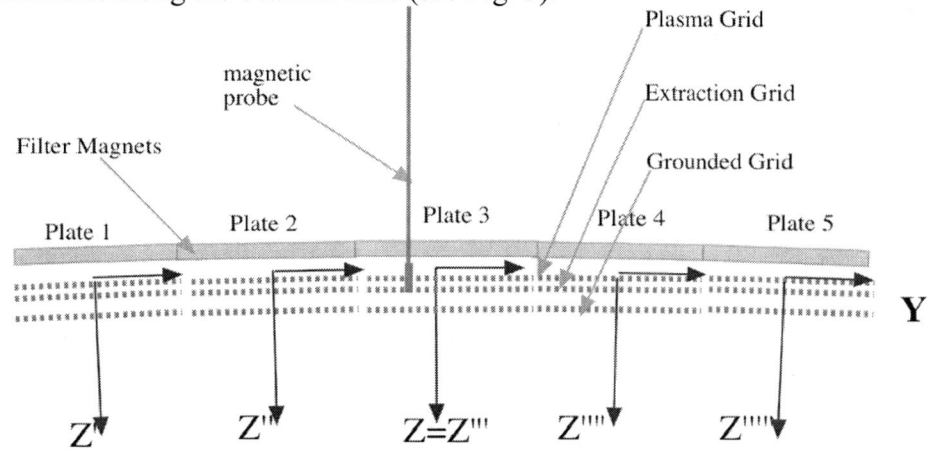

FIGURE 5. scheme of the BL2 accelerator grids showing the orientation of the Plasma Grid Plates, plates 2 and 4 are rotated by 1.03 deg and plates 1 and 5 are rotated by 2.06 deg.

The magnetic field inside 27 of the 770 apertures of the accelerator has been measured. About 150 measurement points were taken on each aperture, with a sampling step of 1 mm, starting from 50 mm upstream the plasma grid inside the ion source, to 100 mm inside the accelerator region in the z direction.

The instrumentation for the magnetic measurements consisted of a FW Bell 6010 Gaussmeter with a transverse Hall probe with rectangular cross section for measuring the X and Y magnetic field components, and an axial probe with circular cross section for the Z component (small size 3D Hall probes were not commercially available). Both probes consist of an aluminium stick with the Hall sensor located at the extremity. The certified measurement accuracy of the system is ±0.25% in the range from 10 mT to 3 Tesla, with automatic temperature compensation.

COMPARISON OF MEASURED AND CALCULATED VALUES

The values of the three components of the magnetic field obtained using the FEM model have been compared to the measured data. The calculation points are defined along paths centred on the axis of the beamlet apertures. As an example, Fig. 6 shows the results concerning six beamlet apertures located on the central plate of the plasma grid (P03), the values obtained on beamlets located on other plasma grid plates being similar.

The vertical component of the magnetic field By is mainly produced by the Suppression magnets and is by far the major one in LHD BL2 (about 58 mT).

As shown in Fig. 6, the measured spatial distribution of By is smooth and consistent with the calculated one.

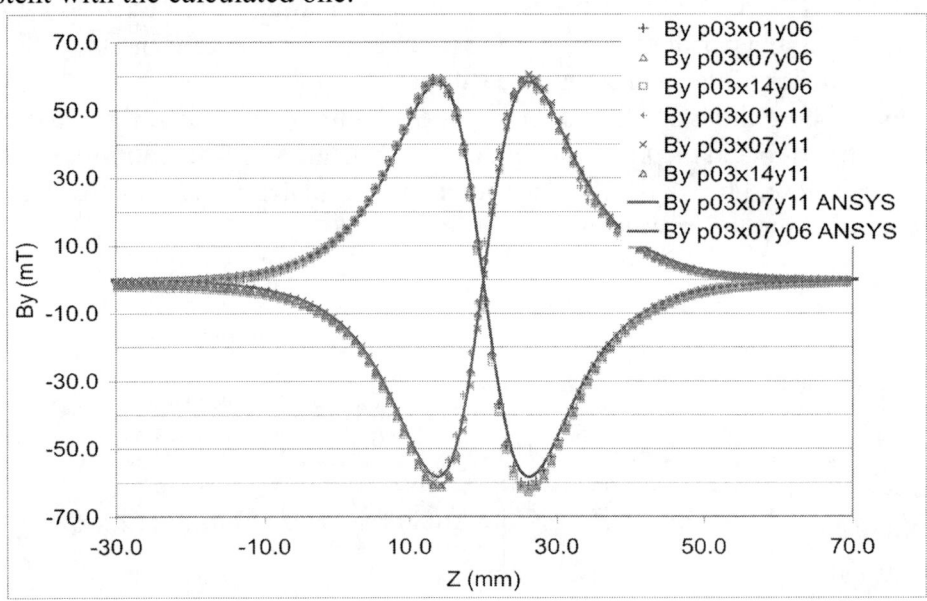

FIGURE 6. Comparison of the calculated (solid lines) and measured (symbols) By magnetic field distribution along 6 beamlets inside the LHD BL2 source and accelerator.

Although a good agreement has been found between the calculated and the measured By field component, on some beamlet apertures the measured peak By was found to be up to 3 mT larger than the calculated value. This mismatch is about 5% of the expected value and cannot be explained as due only to a probe positioning error (offset or rotation).

In order to distinguish among these sources, it was decided to evaluate numerically the effect of probe tilting of 1 deg and probe offset of 0.5-1 mm, which are the maximum plausible probe positioning errors.

In fact, a probe pure positioning error (1 or 2 mm offset) would produce a symmetrical modification both of the positive and negative peak By. As can be seen in Fig. 7 and 8, in the cases where the measured By values are larger than expected, the increase is not symmetric. In order to produce such an asymmetry of the measured values, a probe orientation error of at least 3-5 degrees in addition to the offset would be necessary, which does not seem to consistent with the actual positioning tolerance of the probe.

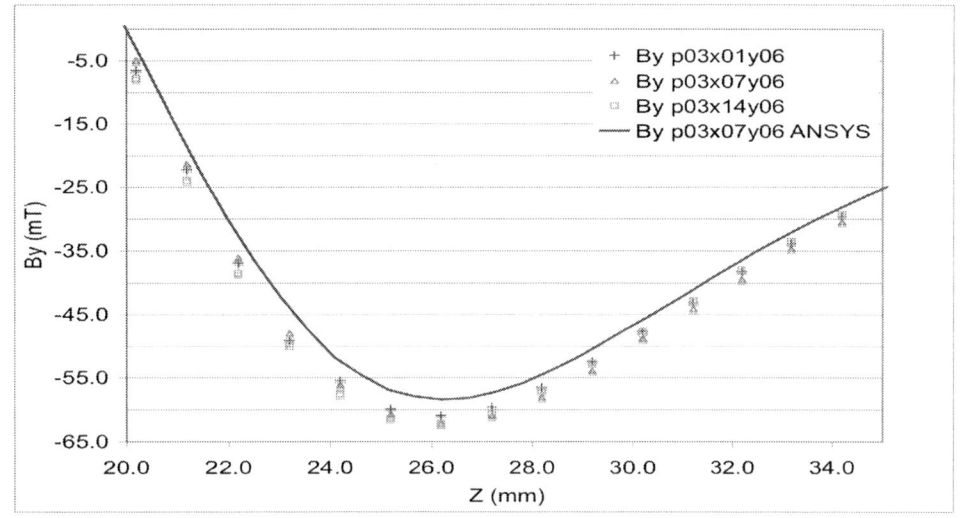

FIGURE 7. Details of the calculated (solid lines) and measured (symbols) By magnetic field distribution along 3 beamlets inside the LHD BL2 source and accelerator.

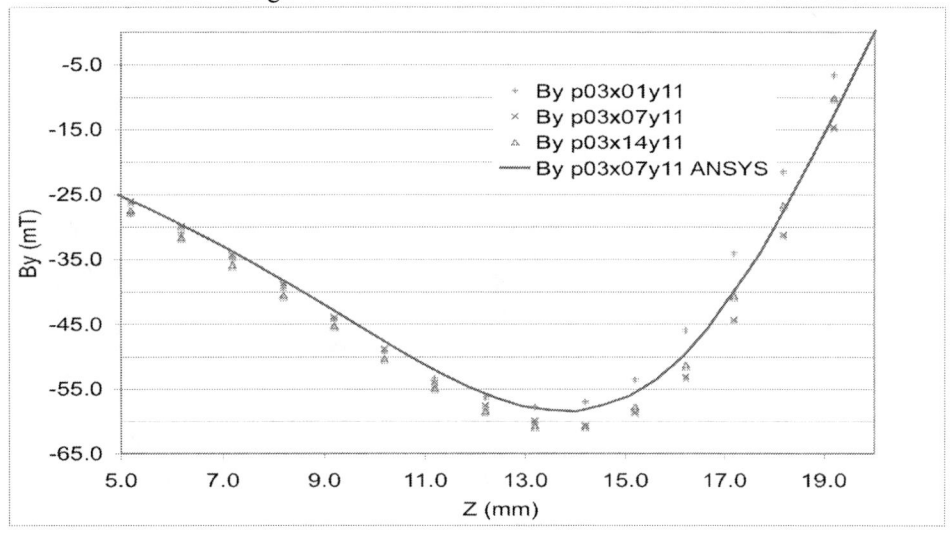

FIGURE 8. Details of the calculated (solid lines) and measured (symbols) By magnetic field distribution along 3 beamlets inside the LHD BL2 source and accelerator.

On the other hand, on the basis of the same 3D model, it has been found that a 15% local non-uniformity of magnetization of Suppression Magnets could produce a 5% non-uniformity of By, with similar shape as that evidenced in the experimental results.

Regarding the Bx component of the magnetic field, which is mainly produced by the Filter Field magnets and is relatively small inside the Accelerator (<10 mT), the FEM model was initially inconsistent with the measurements. The discrepancy was of the order of 30%, concerning both the peak value and the spatial distribution of Bx in the Ion Source region (see Fig. 9). This inconsistency certainly could be explained neither as a simple probe positioning and orientation error, nor as a numerical model error.

As a first step, two independent simplified models based on different FEM codes (OPERA and FEMM), have been developed, confirming the validity of the results of the Bx field calculated with the 3D ANSYS model. Having excluded these two possibilities, the problem appeared to be most likely related to an inconsistency of the Filter Magnet geometry assumed in the numerical models, with respect to the geometry of the real device on which the measurements had been carried out. It is worth noticing that two nominally "identical" LHD-BL2 sources and accelerators exist and that they have been exchanged and overhauled several times by NIFS during the recent years. The one which has been used for the measurements is officially defined "spare" BL2 source and accelerator.

As a matter of fact, after the thickness of the Filter Magnet in the model was increased from 10 to 14 mm, a Bx configuration much better consistent with the measurements was obtained. A further improvement was obtained by shifting the Filter Magnets model 1 mm towards the PG (Fig. 9). The verification of the actual Filter magnet geometry is in progress.

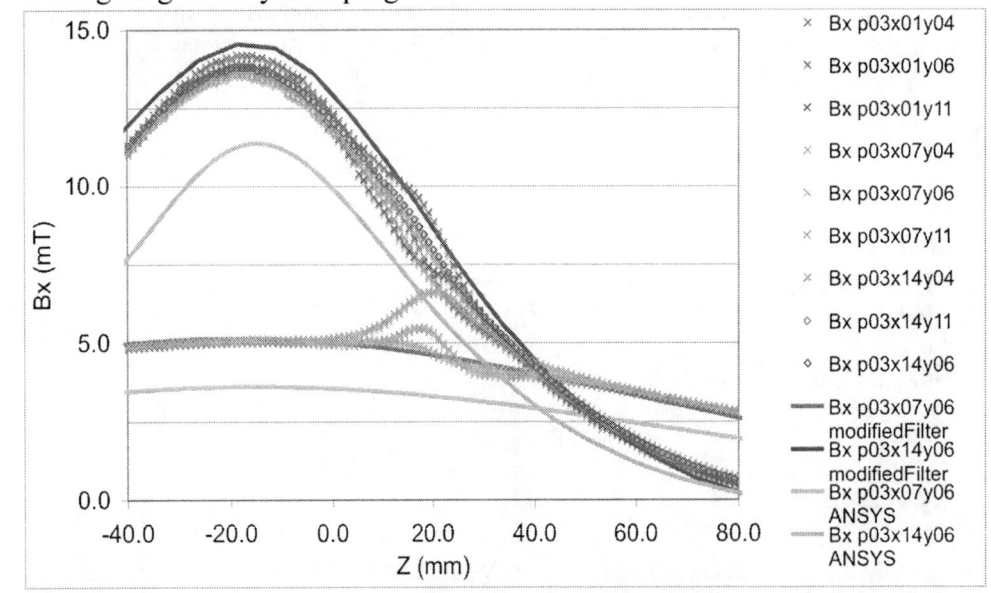

FIGURE 9. Measured (symbols) and calculated (solid lines) Bx magnetic field distribution along 9 beamlets. The calculated results are shown both for the original ANSYS configuration (yellow and light blue solid lines) and for the modified Filter Magnets (red and dark blue solid lines).

A second issue regards the "bumps" which are clearly visible in the Bx measurements and certainly indicate a "cross-talk" effect of the Suppression Magnets on Bx. In principle such bumps can be due both to a probe positioning error and/or to a non-uniform remanence of Suppression Magnets.

However, a probe positioning/orientation error of ~ 4mm + 4 deg. would be necessary in order to justify the measured Bx bumps whose amplitude is ~2 mT. Since such positioning errors appear to be unrealistic with respect to the actual probe positioning tolerance of the supporting structure, the presence of these bumps gives a further evidence of the presence of a non-uniformity in the Suppression Field magnets. A specific model describing this effect is being developed.

The Bz component of the magnetic field is much smaller than By and Bx in absolute value, but is generally very much sensitive to local field non-uniformities and also to possible probe misalignments. In any case Bz, being parallel to the beam, practically no effect on the particle trajectories.

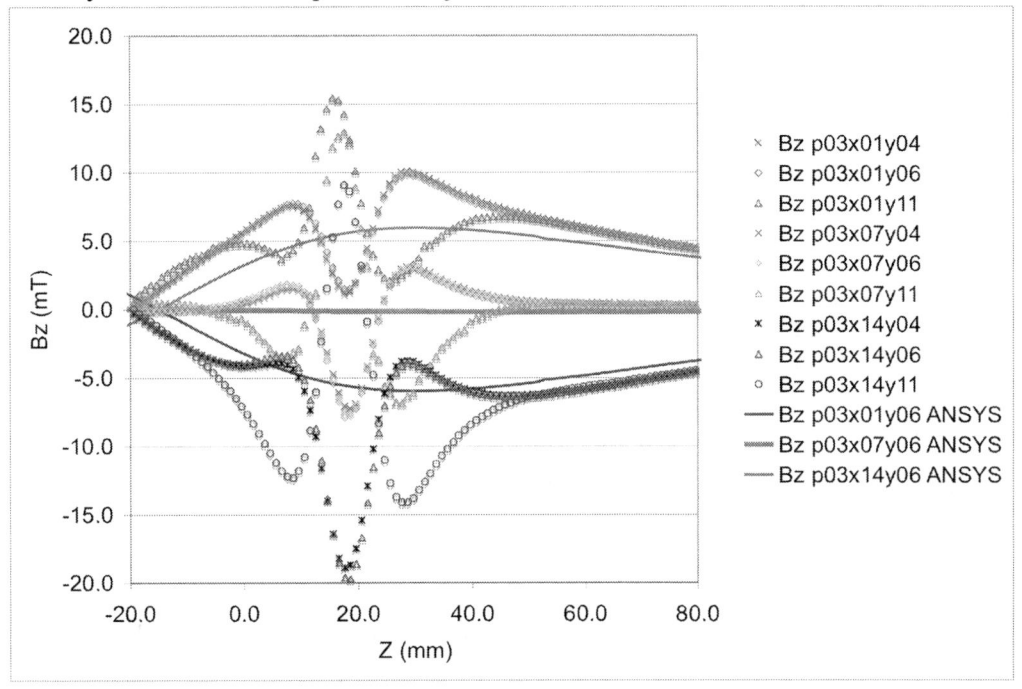

FIGURE 10. Details of the calculated (solid lines) and measured (symbols) Bz magnetic field distribution along 9 beamlets inside the LHD BL2 source and accelerator.

Due to the presence of the Filter magnets on the sides of the Plasma Grid, the value of Bz is expected to be positive on the left side of the grid and negative on the right side. This is very well confirmed by the numerical model and by the measurements shown in Fig. 10.

In addition to the effect of the Filter Magnets, the measured Bz values also show a considerable "cross-talk" effect due to the Suppression Magnets, which is clearly visible in Fig. 10. As for Bx, this effect could be ascribed either to positioning error of the probe or to lack of uniformity in the characteristics of the Suppression magnets. Being the amplitude of the cross-talk less than 15 mT, in this case it is not possible to distinguish between a positioning error (of the order of 0.5 mm or 0.5 deg) and a non-uniformity of the magnets. However, on the basis of the non-uniformity of By and Bx, it seems reasonable to interpret the results as mainly due to a non-uniformity of the magnet property, which can be roughly estimated of the order of 15%. Therefore, in order to obtain 2% magnetic field uniformity on SPIDER beamlets, a tolerance of 5% on magnet remanence seems advisable.

CONCLUSIONS

As a benchmark for the 3D numerical model used in the magnetic design of the future SPIDER facility, a comprehensive 3D numerical model of the LHD BL2 device has been set-up, based on the same assumptions and simplifications. An accurate experimental mapping of the LHD BL2 Ion Source and Accelerator has also been carried out. The numerical simulations are generally in good agreement with the measured values. The measured suppression field By confirms a peak value of 58 mT in good agreement with 3D calculations, however with 5% non-uniformity. The Measured Filter field Bx is 14 - 16 mT on edges and about 5 mT at center, FEM simulation are in reasonable agreement after correction of the Filter magnet thickness. In some cases, the maximum plausible error due to probe tilting and offset was also evaluated in order to distinguish it from the effects of non-uniformity of permanent magnet material.

These results indicate that the numerical FEM model is suitable for describing accurately the LHD BL2 geometry, which includes a large number of magnet arrays and whose complexity is not much smaller than that of SPIDER. The detailed 3D experimental mapping of BL2 will be used also for particle trajectory simulations and for the interpretation of the beam experimental data. The work has also provided a tolerance criteria for the magnetic material characteristics in order to ensure sufficient uniformity of the magnetic field in the SPIDER Accelerator.

AKNOWLEDGEMENTS

This work was set up with the mobility financial support of EURATOM.

REFERENCES

1. P. Sonato, et al.: The ITER full size plasma source device design, *Fusion Eng. Des.* 84 (2009), 269-274.
2. H. Yamada et al.: "10 years of engineering and physics achievements by the Large Helical Device project", F *Fusion Eng. Des.* 84 (2009) 186–193.
3. Y. Takeiri et al.: "High-power and long-pulse injection with negative-ion-based neutral beam injectors in the Large Helical Device", *Nucl. Fusion* 46 (2006) S199–S210.
4. P. Agostinetti et al. : "Modeling activities on the negative-ion-based Neutral Beam Injectors of the Large Helical Device", Proc. of the 2nd NIBS conference, Takayama, Japan, Nov. 2010.
5. ANSYS manual *www.ansys.com*.
6. PerMag Manual, private communication from D. Ciric (2007), UKAEA Fusion/Euratom Association, Culham Science Centre, Abingdon, United Kingdom.
7. OPERA-3D, Vector Fields Co.Ltd., *www.vectorfields.com*.
8. FEMM Finite Element Method Magnetics *www.femm.info*.
9. N. Marconato et al.: Simulation, Code Benchmarking and Optimization of the Magnetic Field Configuration in a Negative Ion Accelerator, Proc. of 26th SOFT conference, Sept 2010, Porto Portugal.

OTHER ION SOURCES

Production of a high brightness H⁻ beam by charge exchange of a hydrogen atom beam in a sodium jet

V. Davydenko[a], A. Ivanov[a], A. Kolmogorov[a], A. Zelenski[b]

[a]*Budker Institute of Nuclear Physics, 630090, Novosibirsk, Russia*
[b]*Brookhaven National Laboratory, 11973, Upton, NY, USA*

Abstract. Production of H⁻ beams for applications in accelerators via charge exchange of a high brightness hydrogen neutral beam in a sodium jet cell is experimentally studied in a joint experiment by the Brookhaven National Laboratory (BNL) and the Budker Institute of Nuclear Physcis (BINP). In the experiment, a hydrogen atomic beam with an energy 3-6 keV, an equivalent current up to 5 A and a pulse duration 200 microseconds is used. Initial results demonstrate that an H⁻ beam with a current 36 mA, an energy 5 keV and ~0.15 cm·mrad normalized emittance was obtained. The recirculating sodium jet target with an entrance aperture 2 cm, which was developed for the BNL optically pumped polarized ion source, is used in the experiment. To increase the H⁻ beam current, geometric focusing of the hydrogen neutral beam will be used in the future experiments. In addition, the effects of H⁻ beam space-charge and sodium-jet stability will be studied to determine the basic limitations of this approach. The atomic beam is produced by charge exchange of a proton beam in a pulsed hydrogen target. The proton beam is formed by an ion source with a four-electrode multiaperture ion-optical system with small-size apertures. The plasma emission surface is formed by the plasma expansion from an arc plasma generator. The transverse ion temperature at the emission surface is 0.2 eV.

Keywords: H⁻ beam, charge exchange, sodium jet target
PACS: 29.25.Ni, 29.25.-t, 41.75.i

INTRODUCTION

Intense beams of hydrogen negative ions can be efficiently obtained by charge exchange of protons and hydrogen atoms in supersonic jets of alkaline and alkaline-earth metal vapors. The cross sections of charge exchange processes and the equilibrium fractions of hydrogen negative ions in the targets are well known. For sodium, the maximum H⁻ yield is 9% at the beam energy (8.4% at 3 keV).

The Budker Institute of Nuclear Physics (BINP) and the Brookhaven National Laboratory (BNL) are developing an intense, optically pumped, polarized ion source based on a high-brightness atomic beam injector. As a logical continuation of that project, it was decided to study the production of high brightness H⁻ beams for accelerator applications. A recirculating sodium jet target, developed by the BNL, and an intense high-brightness atomic beam injector with the beam energy 3-5 keV, developed by BINP are used in the experiment. The experiment is carried out at a well-equipped experimental test stand at BNL.

In this paper we present a description of the sodium jet target and the results of initial experiments on H⁻ beam production. In addition, a design of the atomic beam injector and discussion of interaction of intense hydrogen beam with the target jet are presented.

SODIUM JET TARGET

A transverse recirculating sodium charge-exchange jet cell with a 2 cm diameter aperture [1] is used for charge exchange ionization of a hydrogen beam. A schematics of the target is shown in Fig.1.

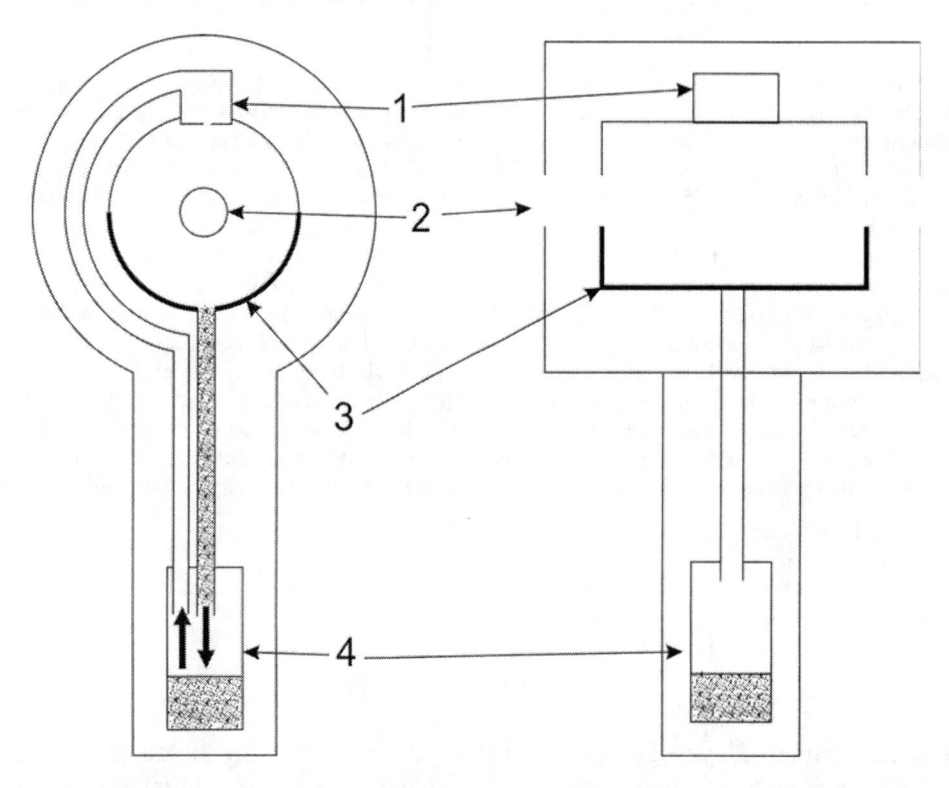

FIGURE 1. A schematic of the sodium jet target. The front and side views are shown. 1-jet nozzle, 2- beam aperture, 3- collector, 4 - reservoir.

The target reservoir is loaded with 150 g of sodium metal, and both the reservoir and the jet nozzle operate at a temperature 505 °C. At this temperature the sodium vapor density is $\sim 10^{17}$ atoms/cm^3 resulting in a vapor jet with an effective thickness $\sim 5 \cdot 10^{14}$ atoms/cm^2 which is sufficient to maximize the H⁻ yield. The sodium vapor condenses on the collector walls, which are cooled down to 120 °C (above the sodium melting point) by hot water circulation. The liquid sodium flows down the return tube and back to the reservoir. The return tube temperature is kept at 150 °C by an attached hot water line. The backstreaming vapor flow through the return line is negligible due to a low conductance at 150 °C. The sodium in the jet cell circulates along the reservoir–nozzle–collector–return line–reservoir path and the system provides continuous and stable operation for hundreds of hours with the initial 150 g load of sodium.

INITIAL EXPERIMENTS ON H⁻ BEAM PRODUCTION

The schematic of the initial experiment on H⁻ beam production is shown in Fig.2. A proton beam is extracted by a small-size four-electrode multi-slit ion optical system of acceleration-deceleration type from the surface of the plasma emitter with a low transverse ion temperature. The plasma emission surface is formed as a result of the expansion of the plasma jet from an arc plasma generator [2]. The formed proton beam with an energy 3-6 keV, current 3-5 A and a 200 microsecond pulse duration is first focused by a magnetic lens and after that is neutralized in a pulsed hydrogen target. The obtained beam of hydrogen atoms passes through the Na target located at 75 cm distance from the ion-optical system. The H⁻ beam, produced as a result of charge-exchange is deflected by a bending magnet and measured by a Faraday cup.

In Fig.3, the dependence of the measured H⁻ current on the beam energy is presented. The maximum current of the H⁻ beam 36 mA is achieved at 4.5-5.5 keV beam energy. The total current of hydrogen particles passing through the target was calculated as the ratio of the H⁻ beam current and the equilibrium H⁻ yield. The dependence of the total current of hydrogen particles through the target on the beam energy is shown in Fig.4.

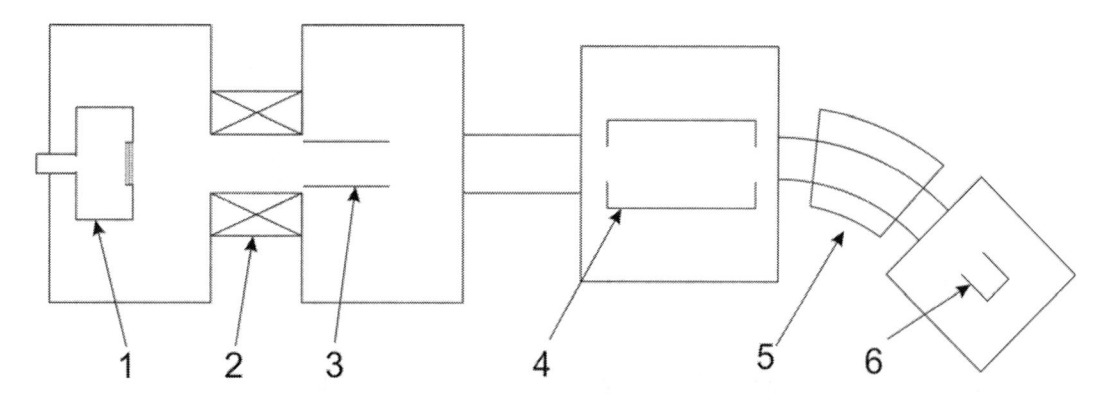

FIGURE 2. A schematic of the experiment on H⁻ beam production. 1- proton source, 2 - magnetic lens, 3 - neutralizer, 4 - sodium target, 5 – bending magnet, 6 - Faraday cup.

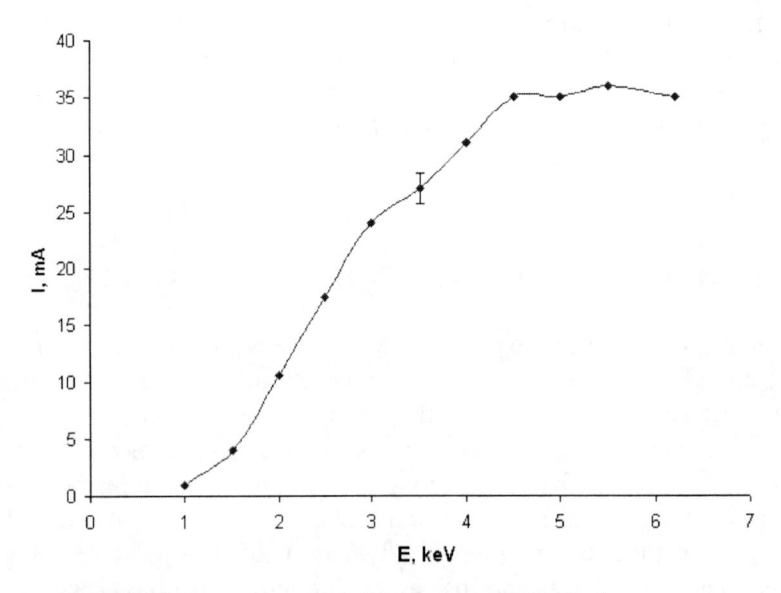

FIGURE 3. Dependence of the H⁻ beam current on the beam energy.

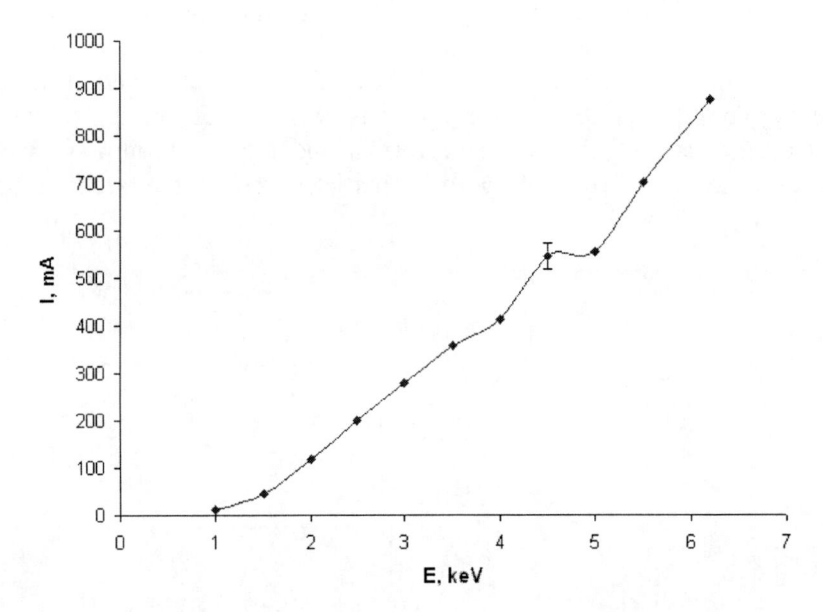

FIGURE 4. Equivalent current of hydrogen particles beam through the target versus beam energy.

The angular divergence of the formed atomic beam is ~$1.5 \cdot 10^{-2}$ rad and the normalized emittance of the produced H⁻ beam is estimated as 0.15 cm·mrad.

A dependence of H⁻ current on the energy with and without magnetic focusing of proton beam at a distance between the ion optical system and the sodium target 175 cm is shown in Fig.5. The maximum H⁻ beam current with focusing is 9.5 mA, compared to 5.5 mA without focusing.

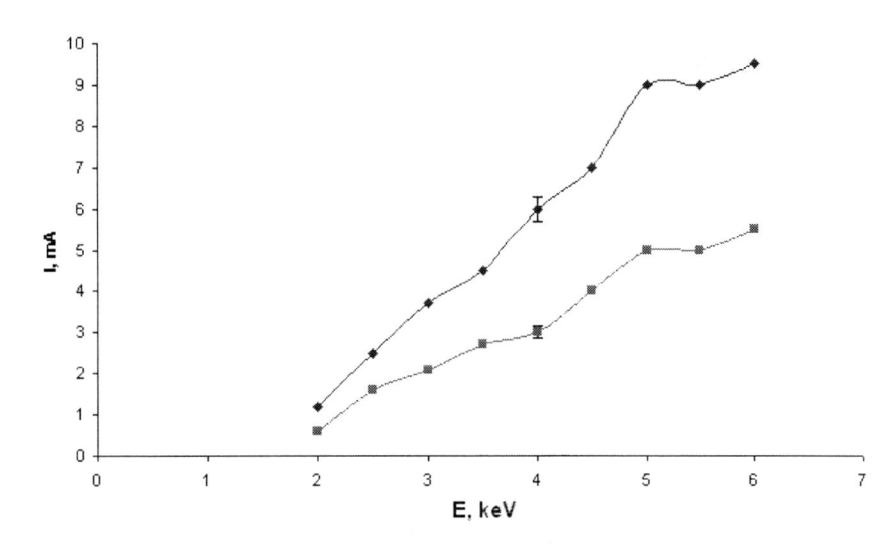

FIGURE 5. Dependence of H⁻ beam current on the beam energy at 175 cm distance to the sodium jet. The upper curve - with magnetic beam focusing, the lower curve – without focusing.

The magnetic beam focusing at low beam energies was studied in BINP and TRIUMF experiments [3,4]. Those experiments have shown that at energies below 4 keV the magnetic focusing is weak due to insufficient space charge compensation in the magnetic lens by the secondary electrons.

ATOMIC BEAM INJECTOR WITH GEOMETRICAL FOCUSING

To avoid the problems associated with the proton beam compensation at low energies, ballistic or geometrical focusing of atomic beam will be applied instead of the magnetic focusing. In this case the proton beam is focused by spherically formed grids of the ion optical system and neutralized in the target at a short distance downstream from the grids [2,5]. The ion optical system of the atomic injector should be operated at low energies. In this case thin electrodes with small diameter apertures should be used.

The elementary cell of a small-size four-electrode ion optical system of the acceleration-deceleration type was optimized by the PBGUNS code [6]. The ion trajectories in the optimized version of the elementary cell with 0.2 mm electrode thickness and 0.4 mm apertures are shown in Fig.6. The initial ion temperature is 0.2 eV. The emittance diagram of the formed elementary beam is presented in Fig.7. At the emission current density of 470 mA/cm^2 the angular divergence of the formed beam is 15 mrad. The electric field in the accelerating gap is about 50 kV/cm, which is lower than the 70 kV/cm breakdown limit.

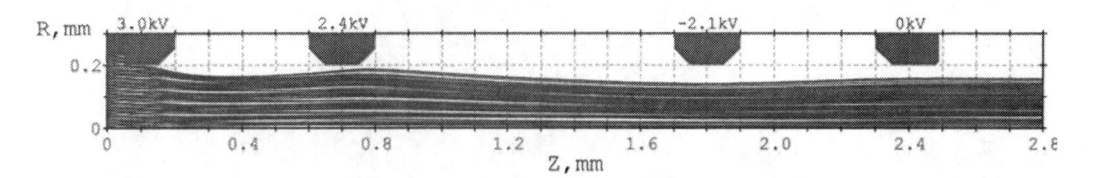

FIGURE 6. Ion trajectories for 3 keV ion beam formation.

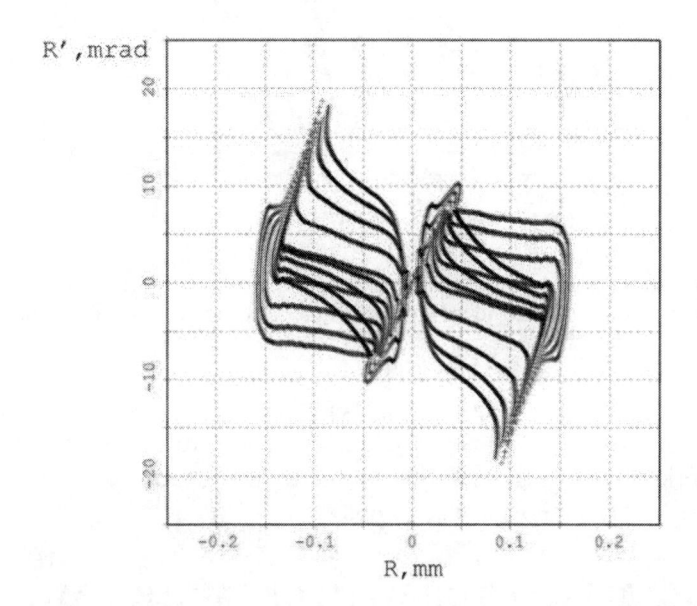

FIGURE 7. The emittance diagram of formed 3 keV elementary beam.

The ion beam formation with 4.0 A current is executed by 7466 circular apertures forming a hexagonal structure with the period of 0.55 mm and the outer diameter of 5 cm. The grids are made off 0.2 mm thick pure molybdenum. The apertures were produced by photo-etching. The grids were subsequently shaped by stress recrystallization at a high temperature. The grids were laser welded to stainless steel holders using an industrial CO_2 laser.

The atomic beam injector with the geometrical beam focusing is shown in Fig.8. The focal distance of the formed atomic beam is 60 cm. A sodium jet cell will be installed in the beam focusing region. The atomic intensity of the beam passing through the Na cell with a 2 cm aperture is expected be 3 A which would result in a 360 mA current of H⁻ ions.

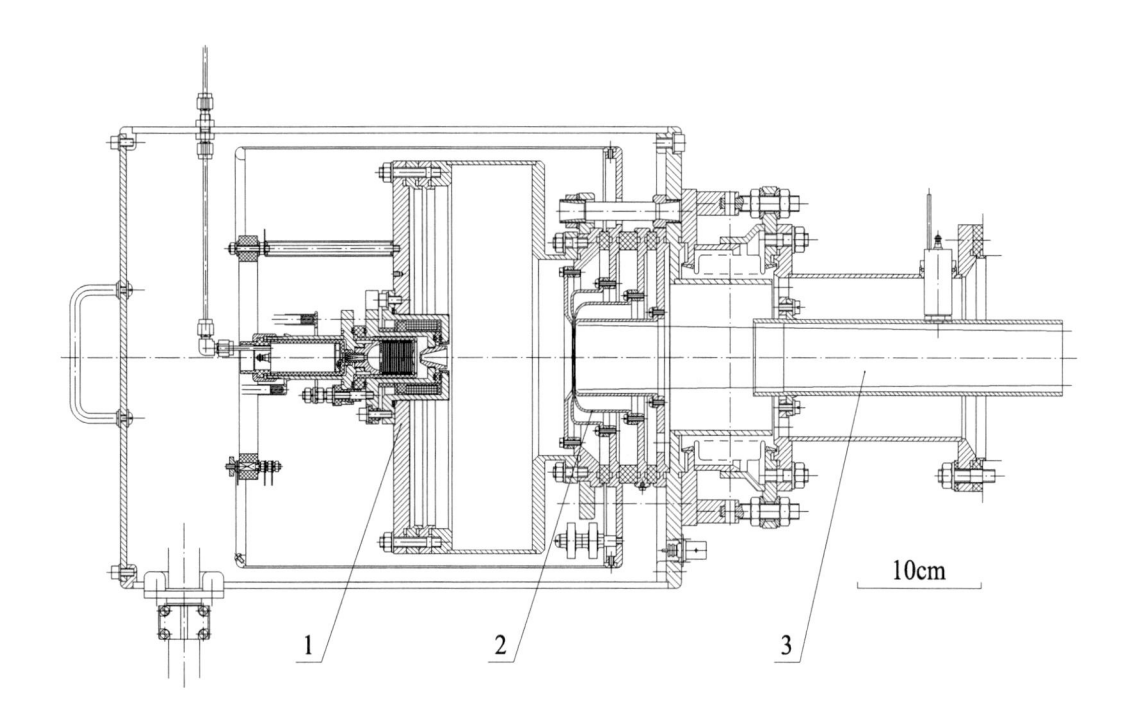

FIGURE 8. An atomic beam injector with geometrical focusing. 1- arc plasma generator, 2- ion optical system, 3- neutralizer.

INTERACTION OF A HIGH-INTENSITY HYDROGEN BEAM WITH a Na-JET CELL

A critical problem of the charge exchange method is the interaction of an intense beam with the target jet. The secondary plasma produced as a result of the interaction causes slowing down and heating of the jet which leads to increase in the jet material losses. According to paper [7], the influence of the secondary plasma on the stream is determined by a parameter $\alpha = J\sigma_i/(ev_0)$, where J is the beam current per unit of the target length, σ_i – is the target atoms ionization cross-section, v_0 – is the speed of the stream. The parameter α is the ratio of the sodium ions flux density generated by hydrogen beam to that of the initial jet. A direct ionization of sodium atoms by hydrogen particles with energy 3-4 keV is small (the ionization cross section is less than 10^{-16} cm^2) so the bulk of the sodium ionization in the jet is caused by the electron transfer from the sodium atom to the hydrogen particles. The cross section of charge exchange of protons and sodium atom is $1.2 \cdot 10^{-14}$ cm^2, the electron capture cross section of hydrogen atoms in sodium is $5 \cdot 10^{-16}$ cm^2. At the beam linear current density $J \approx 1.5$ A/cm and the beam proton fraction ~ 5% the parameter α is ~0.04.

According to results of [7] the critical value of α is ~ 0.06 so secondary plasma can have an impact on the jet material losses. To reduce Na ionization in the cell, it is desirable to remove protons by magnetic field after the beam exit from the neutralizer target.

In the future experiments, the effects of interaction between the hydrogen beam and the target will be studied. The linear density of the incoming beam ion current is supposed to be varied through the change of the hydrogen beam neutralization degree. The beam proton current to 2 cm Na cell aperture at the switched-off hydrogen neutralizer target will be ~ 2-3 A and the linear density of the incoming beam ion current will be varied up to a maximal value of about 1-1.5 A/cm.

ACKNOWLEDGMENTS

The authors would like to thank A.Sorokin for numerical simulations of the ion optical system and V.Mishagin and N.Stupishin for the design of the atomic beam injector.

REFERENCES

1. A.Zelenski. et al. *Rev. Sci. Instrum.*, **73**, 888-891(2002).
2. V.I.Davydenko. *Nucl. Instrum. Meth. A*, 427, 230-234 (1999).
3. A.Zelenski. et al. *Rev. Sci. Instrum.*, **67**, 1359-1361(1996).
4. A.Zelenski. et al. AIP Conference Proceedings, **421**, 372-380(1998).
5. V.I Davydenko, A.A.Ivanov. *Rev. Sci. Instrum.*, **72**, 1809-1811(2004).
6. J.E.Boers, Proceedings of the IEEE Particle Accelerator Conference. American Physical Society, New York, 1995, p.1312.
7. A.I.Krylov, V.V.Kuznetsov. *Fizika Plasmy*, **11**, 1508-1516 (1985), in Russian.

Study of Fluctuations in the CW Penning Surface-Plasma Source of Negative Ions

Yuri Belchenko, Andrey Sanin and Valery Savkin

Budker Institute of Nuclear Physics, av. Lavrentieva, 11, 630090, Novosibirsk

Abstract. Study of current fluctuations for cw Penning SPS with hollow cathode drive was done. The noiseproof measurements of negative ion beam current, current in extracted electrode circuit, discharge current and voltage were carried out by the low-inductive probes in wide frequency range. Spectrum and intensity of fluctuations at various operation modes, parameters and electrode geometry were recorded for two versions of cw Penning SPS. H⁻ beam current and the extracted electrode circuit current had the level of ripples higher, than the ripples in discharge current and voltage signals. Frequency spectrum of beam and discharge fluctuations displayed stable peaks. The main peak had location in the range 0.1÷1.5 MHz and FWHM of about 0.1 MHz. For the basic operational mode the main peak in frequency spectrum was in the range 0.3-0.4 MHz. The fluctuations of current in extracted electrode circuit and in accelerated electrode circuit had the similar structure and correlated with beam current fluctuations. The obtained data show that plasma density oscillations are responsible for the beam current fluctuations. The 0.1÷1.5 MHz fluctuations of plasma density could be produced by oscillations of cathode emissivity and of discharge current distribution between the specific cathode regions.

Keywords: Negative ion source, Emissivity, Fluctuation, Penning discharge, Hollow cathode.
PACS: 29.25.Ni.

INTRODUCTION

Surface-Plasma Sources (SPS) of negative ions for accelerators with Penning or magnetron geometry exploit the hydrogen-cesium discharge in external magnetic field and deliver pulsed H⁻ beams with intensity of about 100 mA [1] or the cw beams with current level 10-15 mA [2]. In these sources negative hydrogen ions are produced by conversion of atoms from plasma on the cesiated anode surface. Both sources are intensively used on accelerators of ISIS [3] and in BINP.

The beams, extracted from these sources show a noticeable level of fluctuations. The detailed study of oscillations and noise in the pulsed Penning SPS was done earlier [4]. The fluctuation level could be decreased in discharge "noiseless" mode with cesium coverage conditioning [4] or by adding of nitrogen to discharge. In the cw Penning SPS with Hollow Cathode (HC) relatively high level of discharge and beam fluctuations is recorded in the most cases, including the discharge with mixing of hydrogen with nitrogen and the case with cesium coverage, well-conditioned by long term discharge. This difference in discharge and beam noise characteristics for the pulsed and cw Penning sources is caused by very dissimilar conditions of the discharge, listed in Table 1. Penning source with HC operates steady state at 3 times lower hydrogen filling pressure, and it has about 25 times lower average discharge current density as compared with the pulsed source.

TABLE 1. Parameters of pulsed Penning SPS and cw SPS with hollow cathode.

Parameter	Pulsed Source [3]	CW Source with HC [2]
Beam, mA	50 ÷ 100	10 ÷ 15
Pulse Duration, ms	1 ÷ 0.2	cw
Discharge Voltage, V	150 ÷ 200	60 ÷ 85
Discharge Current, A	100 ÷ 180	10
Cathode Current Density, A/cm^2	150 ÷ 250	10
Hydrogen Pressure, mTorr	>100	30
Emittance 90% norm., π mm mrad	0.7/1.1	0.3
Emission Aperture, mm	Slit 10x0.6 mm	Ø 3 мм
H- Production Power Efficiency, mA/kW	5	18

An increased value of emittance was recorded in the cw Penning source with HC. Since the emittance value is depended on discharge fluctuations, the detailed study of discharge and beam current fluctuations for two variants of cw Penning SPS with HC were done in order to determine the properties and the origin of oscillations.

EXPERIMENTAL FACILITY

Versions of cw Penning Source with Hollow Cathode

The principles of operation and geometry of cw Penning SPS with hollow cathode were described earlier [2, 5]. The cross sectional view of the cw source with HC, developed for tandem accelerator at BINP, is shown in Fig.1. It consists of a massive gas-discharge chamber with the cathode inside. Cathode is preliminary heated by embedded ohmic heaters for cesium seed and discharge start. Hydrogen and cesium are delivered to discharge through the HC inserts.

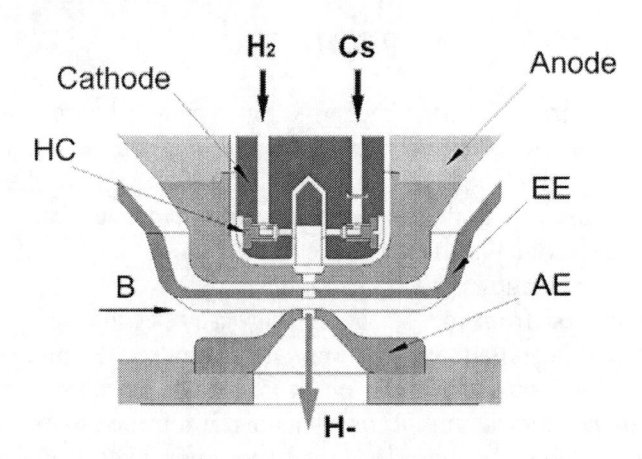

FIGURE 1. Cross sectional view of cw Penning source. HC- hollow cathode insert, EE- extraction electrode, AE- acceleration electrode, B- direction of external magnetic field.

Negative ion beam extraction and post-acceleration were produced with the help of 3-electrode ion optical system, including the anode, extraction and acceleration electrodes (EE and AE in Fig.1). Magnetic field for PIG discharge support and for electron current suppression was produced by constant magnets or by electromagnetic coils. Magnetic field direction is shown by arrow B in Fig.1.

Two variants of sources with special construction and power supply circuits, but having the similar ion optical system, discharge geometry and electrode sizes were studied. The principal differences in these sources are listed in Table 2.

TABLE 2. Principal differences in experimental and tandem sources.

	Experimental	Source for tandem
Location	Vacuum-immersed	Attached with flange
Discharge power supply	Rectifier 50 Hz	Inverter 20 kHz
Magnetic field	0.2 ÷ 1.2 kGs	0.8 kGs
Hollow cathode insert	Vertical	Longitudinal
Distance of Beam transport to FC	7 cm	40 cm

The experimental source, described in [5], is equipped by external electromagnetic coils. The measurements with variable magnetic field in the range 0.2÷1.2 kGs were produced in this case. This source is immersed into vacuum chamber, while the extracted and accelerated beam of negative ions propagates distance of about 7 cm before entering to the collector. The power supply for discharge feed of the experimental source uses a 50 Hz rectifier.

The second source, an upgraded version of the tandem source [2] has a magnetic system with permanent NIB magnets and with magnetic field of 0.8 kGs at the discharge zone. This source was installed on the external flange of the test stand, and the negative ion beam propagates 40 cm distance before measuring by Faraday cup. Power supply for discharge current feed of this source uses a 20 kHz inverter. Discharge current change is produced by width modulation techniques, while current stabilization is supplied by an embedded inductivity.

The wide range of discharge and beam extraction parameters was tested in both sources. Various HC inserts materials and geometries were studied. In order to check the negative ion production with beam extraction from peripheral PIG zone, the source with HC, shifted to the periphery of discharge window, was tested.

Scheme of measurements

Scheme of measurements is shown in Fig.2. The following signals were measured by digital oscilloscope with 300 MHz bandwidth: discharge current I_d, discharge voltage U_d, current to extraction electrode I_{ee}, current in the circuit of extraction power supply I_{ext} and current of negative ion beam I_{H^-}.

Discharge current I_d was measured by low inductive shunt in the cathode circuit. Discharge voltage U_d was recorded by probe, directly connected to the discharge electrodes. Negative ion beam current I_{H^-} was measured by shielded Faraday cup, equipped with secondary electron suppression grid. The impedance of low-inductive, low capacitive circuit of I_{H^-} measuring was matched with the FC ohmic load.

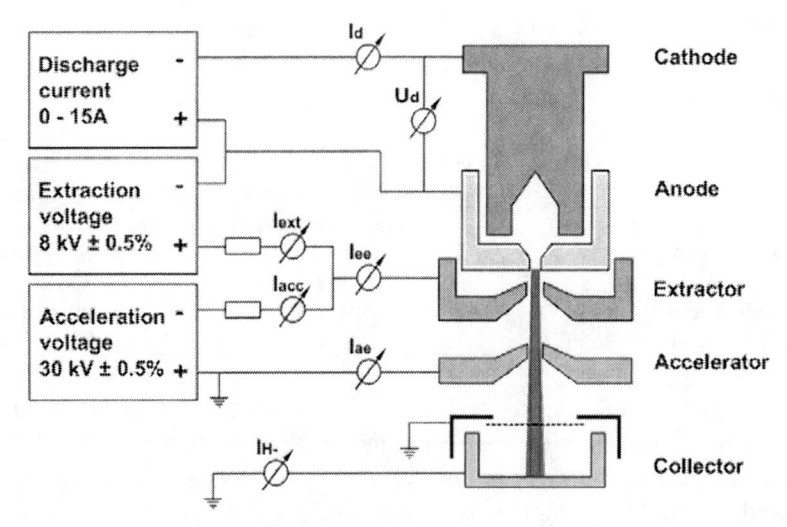

FIGURE 2. Scheme of measurements.

Signal I_{ext} was recorded by probe, biased to HV anode potential and was transmitted to ground via optical line. Discharge current I_d and its correlation with other signals were also measured during source HV operation. Biased signal I_d was transmitted to ground potential via optical line. Temperature of electrodes was monitored by embedded thermoresistors. Varied slowly average values of indicated voltages and currents were controlled by biased meters of computer control system [6].

EXPERIMENTAL DATA

Discharge Fluctuations

The basic features of discharge fluctuations were similar for both sources. Typical oscillogrammes of discharge current (top track) and of discharge voltage (lower track) are shown in Fig.3.

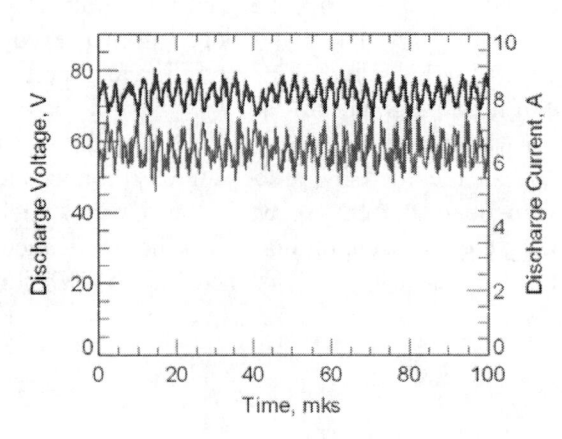

FIGURE 3. Typical oscillogrammes of discharge current (top track, right scale) and of discharge voltage (lower track, left scale).

Fluctuations of discharge current and of discharge voltage have the regular structure. Discharge current fluctuations were in antiphase with discharge voltage fluctuations. An average period of discharge fluctuations was about 2.5-3.5 µs for the standard mode of source operation. Fourier spectrum of discharge fluctuations has the main peak with frequency of 0.2-0.4 MHz. Oscillations with frequency 20÷40 MHz with low contribution to overall level of fluctuations were also recorded. Discharge current fluctuations could be suppressed (with normalized RMS <0.7 %) by introducing an additional inductive choke in the discharge current circuit. It leads to a corresponding increase of discharge voltage fluctuations, and has small influence on the beam fluctuations.

The fluctuations of discharge current have lower value at the increased discharge current and at increased cesium seed. Fourier spectrum of discharge current fluctuations and its evolution with cesium feed change are shown in Fig. 4. Discharge current fluctuations in standard mode U_d = 70-80 V (case b in Fig.4) has the peak in spectrum at frequency of 0.3 MHz. For discharge with low cathode emissivity and an average voltage U_d = 90-100 V the main peak in spectrum had frequency of about 0.2 MHz (case a in Fig. 4). Fluctuations with no distinctive peaks in Fourier spectra were displayed for cesiated, well conditioned discharge with an average discharge voltage U_d = 55-60 V (case c in Fig. 4).

The influence of cesium feed on level of discharge current fluctuations is shown in Fig. 5. The level of discharge current fluctuations was gradually increased from 1.5% to 5% (normalized RMS) with the decrease of cesium seed to discharge and corresponding increase in average discharge voltage. It is important to note, that discharge voltage depends on various parameters: on cesium seed, on magnetic field, on electrode temperature, on cesium coverage activation by discharge etc. So the spectra and dependences of Fig. 4 and 5 illustrate the influence of gradual cesium seed change, while the other source parameters were unchanged. In contrast, an increased level of fluctuations in Fig. 1 (with an average U_d = 60 V) was recorded for the case with lower electrode conditioning as compared with that for Fig. 4 and 5.

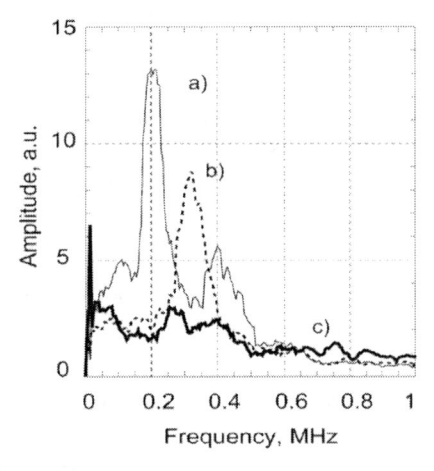

FIGURE 4. Peaks in 0÷1 MHz part of Fourier spectra of discharge current fluctuations at various U_d. a) U_d =98 V; b) 79 V, c) 60 V.

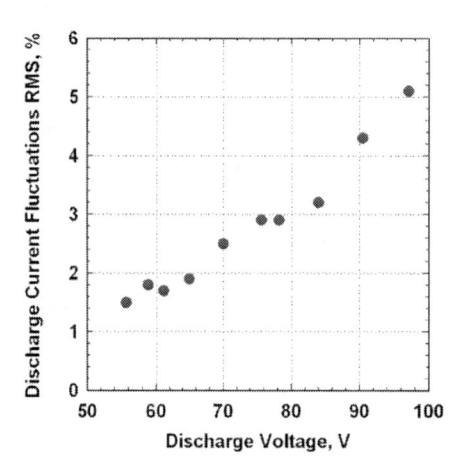

FIGURE 5. Level of discharge current fluctuations (normalized RMS) vs discharge voltage. Discharge voltage was changed by gradual decrease of cesium seed.

Fluctuations of discharge were not depended on hydrogen feed and on extraction voltage applied (for emission apertures with diameter 3 mm). The shift of peak in Fourier spectrum and RMS level of discharge current fluctuations vs magnetic field for experimental source are shown in Fig. 6 and 7. No change of the peak frequency in discharge current fluctuations and of the level of discharge current fluctuations was recorded in the standard range of source magnetic field 0.6 ÷ 1.1 kGs (Fig. 6, 7). At the low magnetic field 0.2 ÷ 0.4 kGs the 3x jump in discharge current fluctuations value and in peak frequency was displayed due to discharge rearrangement.

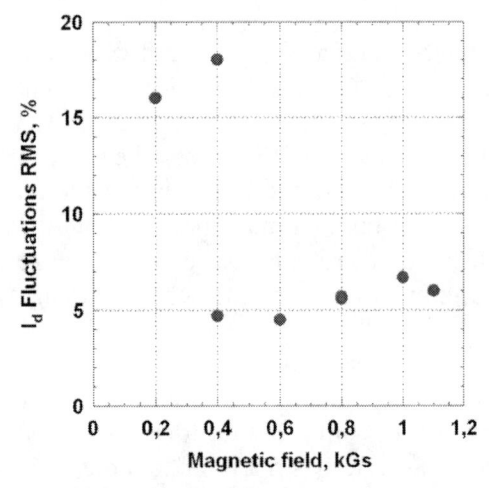

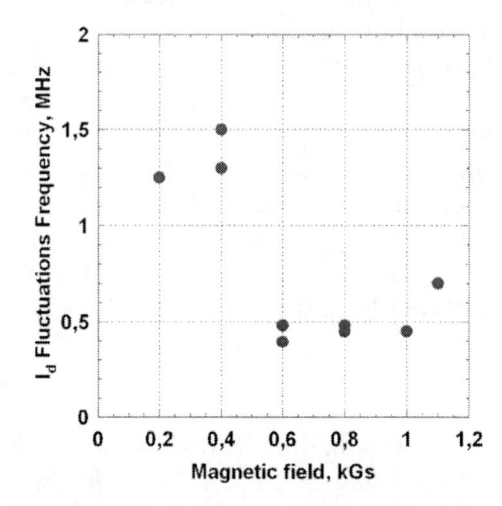

FIGURE 6. Level of discharge current fluctuations (normalized RMS) vs magnetic field.

FIGURE 7. Peak frequency in Fourier spectrum of discharge current fluctuations vs magnetic field.

Discharge fluctuations had similar structure at different tested variants of HC insert and had lower level in the case of ferromagnetic HC insert.

Extracted Current and Beam Current Fluctuations

Typical oscillogram of cw H- beam current, measured at distant Faraday Cup with the help of matched circuit is shown in Fig.8. Beam current displays regular fluctuations with amplitude up to ±25% and normalized RMS deviation 6%. Fourier Spectrum of oscillogram Fig. 8 is shown in Fig. 9. It has the peak at frequency 0.4 MHz with FWHM of about 0.1 MHz. An average period of beam current fluctuations of Fig. 8 is about 2.5 µs and it was changed in the range 0.7 - 10 µs with the cesium coverage change, similar to those for discharge oscillations. No change in frequency and level of beam current fluctuations was recorded with the change of magnetic field in the working range 0.6 ÷ 1.1 kGs. Beam current fluctuations were not depended on hydrogen feed.

Correlation of beam current fluctuations with the discharge current fluctuations was studied. Cases with correlation of these fluctuations were realized at high discharge voltage of about 90-100 V with the small cathode emissivity. On other hand, no correlation of beam current and of discharge current fluctuations was recorded in the cases of discharge with voltage 60-70 V with good cathode emissivity.

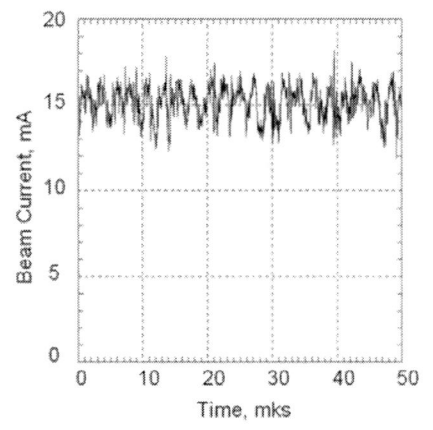

FIGURE 8. Oscillogram of cw H- beam current. Fluctuations amplitude up to ±25%, RMS=6%

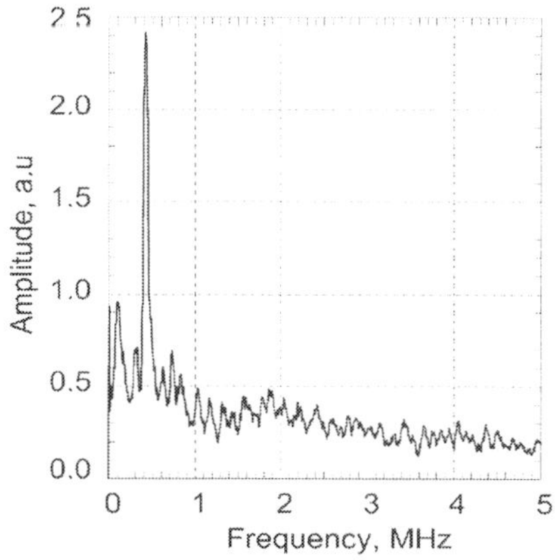

FIGURE 9. Fourier spectrum of left oscillogram.

Dependencies of beam current (triangles, right scale) and of the level of H- beam current fluctuations (circles, left scale) vs extraction voltage are shown in Fig. 10. Beam current value is increased with the growth of extraction voltage, while the level of beam current fluctuations (normalized RMS) has a minimal value at optimal extraction voltage 2.5 kV. Frequency of H- beam fluctuations does not depend on extraction voltage. Extraction voltage, which minimizes the fluctuation amplitude, is most likely corresponds to the optimal configuration of the electric field and of plasma boundary position, providing the smaller change of beam current with the extraction voltage change.

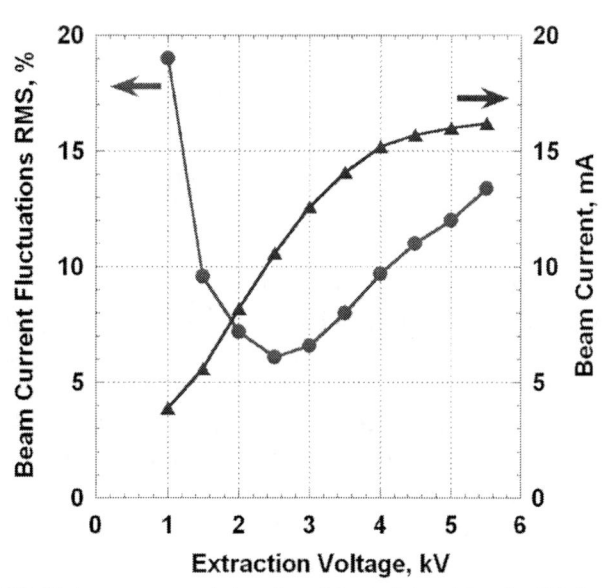

FIGURE 10. H- beam current (triangles, right scale) and level of H- beam current fluctuations (circles, left scale, normalized RMS) vs extraction voltage.

Different geometries and materials of HC insert unit were tested. The minimal value of H- beam current fluctuations was recorded for the source with ferromagnetic HC inserts, located close to the emission aperture. Namely the level of fluctuations of about 4% (normalized RMS) was obtained with ferromagnetic inserts at extraction voltage 2.5 kV.

Fluctuations of current in the circuit of extraction power supply current I_{ext} and of current to extraction electrode I_{ec} had the structure and frequency, similar to those of beam current. Both, I_{ext} and I_{ec} currents correlated in phase with beam current fluctuations. The I_{ext} current consists of the current of negative ions, of co-extracted electrons and of back-streaming positive ions, while the current, intercepted to extraction electrode I_{ec} is mostly consists of co-extracted electrons. Both, I_{ext} and I_{ec} currents are governed by plasma density near emission aperture. Correlation of fluctuations of I_{ext} and I_{ec} with the beam current fluctuations confirms, that plasma density near emission aperture is important for H- ion production and extraction in the Penning surface-plasma sources.

Shift of Hollow Cathode Unit to discharge periphery

Plasma density in the peripheral PIG zones could be significantly decreased in the case of increased hollow cathode emissivity (for the common voltage of combined PIG+HC discharges). The source with hollow cathode unit shifted to the periphery of discharge window was tested in order to check the negative ion production in area, close to peripheral PIG zone of discharge.

Discharge current and voltage in the case of cathode with shifted HC were similar to those of discharge with standard position of HC insert. The extracted H- beam current was about 2.5 times lower in the case of source with shifted HC. Fluctuations of extracted beam current were 3-5 times higher in the case of shifted HC. The recorded data demonstrate the importance of plasma from hollow cathode insert for H- ion production and extraction, and confirm the assumption on discharge current redistribution between the specific cathode regions in the combined PIG+HC discharge.

FLUCTUATIONS ORIGIN

Cesium deposition to Penning SPS provides a multiple increase in surface negative ion production and in secondary electron emission. The last one is essential for supporting the low-voltage Penning discharge under low hydrogen filling pressure. Small variation in cesium coverage could essentially change the electrode work function, discharge current value and distribution over the cathode.

The recorded oscillations of discharge and extracted currents could be explained by fluctuations of the cathode cesium coverage and emissivity. This conjecture is confirmed by high sensitivity of discharge and beam fluctuations to the cesium coverage and by their low dependence on magnetic field and hydrogen feed. The emissive nature of discharge and beam fluctuations is also confirmed by the identical basic frequency of discharge fluctuations and of extracted currents fluctuations, by their correlation in discharges with the decreased cesium seed and by identical features of fluctuations for both source versions.

The possible reason of emissivity fluctuations is the redistribution of cesium due to impact desorption. Desorption of cesium from the emissive zone of HC channel decreases the local plasma density, which is important for cesium blocking by dense plasma [7]. The drop in local plasma density increases the cesium seed from HC volume and decreases the desorption rate as well. As a result, the cesium coverage of HC emitting parts and an increased discharge current density are restored with some delay.

The fast depletion of cesium coverage could be produced by impact desorption with oblique incidence of bombarding particles to the HC walls. For example, the 10% depletion of half-monolayer Cs coverage could be produced in characteristic time 1 μs for bombarding flux with current density 3 A/cm^2 and desorption yield 1 atom/ion. The time of Cs coverage restoration is determined by cesium transport from the HC volume to HC emitting parts. For HC with temperature ~500 K the thermal cesium flux could provide the 10% gain of half-monolayer Cs coverage in 1 μs. The ratio of processes of cesium adsorption, blocking and impact desorption defines an equilibrium value of HC coverage, the period and level of coverage fluctuations. Since all the cesium is fed to discharge via HC cavity, the indicated cesium desorption /restoration cycle could results in pulsing of cesium feed and in fluctuations of the total discharge current as well.

The recorded period of the regular fluctuations of discharge and extracted currents had value of about 5-10 μs for discharge with low cesium seed, and it was decreased down to 2-4 μs with the cesium coverage increase (Fig. 7, 9). The decrease in level of discharge and beam fluctuations, recorded with cesium seed growth is explained by lower change of cathode work function for desorption / restoration cycle of enriched cesium coverage. It is known, that work function of W+Cs emitter [8] and the secondary electron emission of Mo+Cs emitter, bombarded by 50 eV ions [9] have small change with the depletion or growth of enriched cesium coverage. A decreased level of discharge and beam fluctuations was measured in discharge with higher current (i.e. higher emissivity) as well.

The change of local plasma density in the vicinity of emission aperture affects on efficiency of production and extractions of negative ions and electrons. It is clearly displayed in discharges with high emissivity, where the fluctuations of beam current correlate with oscillations of extracted I_{ext}, I_{ec} currents, but have no correlation with the oscillations of total discharge current. The importance of dense HC plasma for negative ion production and extraction is also confirmed by a 2.5 times lower level of H- beam, extracted from the "peripheral" PIG discharge region in the case of shifted hollow cathode unit.

A decreased level of fluctuations, recorded in the case of ferromagnetic HC insert could be explained by an increase of electron emitting area of HC channel due to shielding of magnetic field inside the ferromagnetic bush.

Long-term conditioning of electrodes by discharge makes the cesium coverage more uniform and stable. It reduces the fluctuations in discharge and beam.

CONCLUSIONS

Plasma and cesium injection from the hollow cathode is essential for cw source operation, but it makes the non-uniform distribution of plasma density in discharge. Changes in cesium coverage due to impact desorption and delayed adsorption from the incoming flux could be a reason of the recorded fluctuations of discharge and extracted currents.

For reducing the discharge and beam fluctuations it is necessary to operate at decreased extraction voltage, at higher discharge currents and to have the enriched cesium coverage. It is important to decrease the sensitivity of hollow cathode emission to cesium redistribution by increasing of HC emitting area, like those in the case of HC ferromagnetic screening. Long term conditioning of electrodes by discharge induces less noisy discharge and beam.

REFERENCES

1. P. Allison and J. Sherman, AIP Conference Proceedings **111**, American Institute of Physics, NY, 1983, pp.511-519.
2. Yu. Belchenko, A. Sanin, and A. Ivanov. AIP Conference Proceedings **1097**, American Institute of Physics, Melville, NY, 2009, pp. 214-222.
3. D. C. Faircloth, M. O. Whitehead and T. Wood, *Rev. Sci. Instrum.*, **77**, 03A520 (2006).
4. G. E. Derevyankin and V. G. Dudnikov, AIP Conference Proceedings **111**, American Institute of Physics, NY, 1983, pp.376-397.
5. Yu. I. Belchenko and V. Savkin, *Rev. Sci. Instrum.*, **75**, pp.1704-1706 (2004).
6. P.V.Zubarev, A.D.Khilchenko, A.N.Kvashnin, D.V.Moiseev, E.A.Puriga, A.L.Sanin, V.Y. Savkin. Computer System for Unattended Control of Negative Ion Source. These proceedings.
7. Yu. I. Belchenko and V. Davydenko, *Rev. Sci. Instrum.*, **77**, 03B702 (2006).
8. C.A. Papageogopoulos and J. Chen, *Surface Science*, **39**, pp.283-312 (1973).
9. A. Ortykov and R. Rakhimov, Influence of Cs adsorption on potential and kinetic electron emission of molybdenum. Proceedings of XIV USSR' conf. on emissive electronics, Makhachkala, 1976, Vol.2, pp.175-176. (in Russian).

Surface Plasma Source Electrode Activation by Surface Impurities

Vadim Dudnikov[a], B.Han[b], Rolland P. Johnson[a], S.N. Murray[b],
T.R. Pennisi[b], M. Santana[b], Martin P. Stockli[b], R.F. Welton[b]

[a] *Muons, Inc., 552 N. Batavia Ave., Batavia, IL, 60510, USA*
[b] *Oak Ridge National Laboratory, Oak Ridge, TN, 37831, USA*

Abstract. In experiments with RF saddle antenna surface plasma sources (SPS), the efficiency of H⁻ ion generation was increased by up to a factor of 5 by plasma electrode "activation", without supplying additional Cs, by heating the collar to high temperature for several hours using hot air flow and plasma discharge. Without cracking or heating the cesium ampoule, but likely with Cs recovery from impurities, the achieved energy efficiency was comparable to that of conventionally cesiated SNS RF sources with an external or internal Cs supply. In the experiments, optimum cesiation was produced (without additional Cs) by the collection and trapping of traces of remnant cesium compounds from SPS surfaces. Such activation by accumulation of impurities on electrode surfaces can be a reason for H⁻ emission enhancement in other so-called "volume" negative ion sources.

Keywords: cesiation, surface plasma sources, negative ion, RF discharge.
PACS: 32.80.Gc, 52.27.Cm, 29.25.Ni, 52.80.Pi

INTRODUCTION

The "cesiation effect", in which negative ion emission from gas discharges is increased while the coextracted electron current is decreased below the negative ion current, was discovered by placing a compound with about 1 mg of cesium [1] into a gas discharge chamber. In subsequent experiments [2-4], it was demonstrated that cesium adsorption decreases the surface work function, which enhances secondary emission of negative ions caused by the interaction of the plasma with the electrode surface and thereby enhances surface plasma generation of negative ions (SPG). Ion sources based on this process have been named Surface-Plasma Sources (SPS).

A small admixture of cesium or other impurity with low ionization potential (ILIP) in the gas discharge significantly improves H⁻ production [1]. When done correctly, a cesiated SPS works well [2-8]. However, improper cesiation can complicate ion source operation. For example, injection of too much cesium can cause the discharge to become unstable and sparking occurs in the extractor with loss of stable ion source operation. With low cesium concentration the efficiency of negative ion production is too low. With "proper" cesiation the efficiency of negative ion production is high and extended ion source operation is stable.

Some attempts to develop H⁻ sources with acceptable H⁻ beam parameters but without cesiation are considered in reviews and books [5-8]. The hope to eliminate the use of Cs was supported by improvements of volume generation (VG) in volume sources (VS) where dissociative electron attachment with resulting H⁻ formation increases significantly in collisions of electrons with rovibrationally excited molecules. In older VS, the highest

H⁻ intensity was below 5 mA and the co-extracted electron current was up to hundred times larger [9]. In one practical VS for cyclotrons described in [10] a relatively high efficiency of 4 mA/kW was produced with very large 13 mm diameter aperture, low emission current density of 10 mA/cm^2, and high hydrogen gas flow of about 30 sccm.

This diameter and this flow rate are each several times larger than for an SPS at the same current. For example, in a cesiated SPS, the H⁻ emission current density has been as high as 8 A/cm^2 with current up to 11 A in ms pulses and up to 1 A/cm^2 in DC mode [3,11,12]. The current of co-extracted electrons can be suppressed to be below that of the H⁻ current. In recent experiments with RF (radio frequency) saddle antenna surface plasma sources (SPS) [13], the efficiency of H⁻ ion generation was increased by up to a factor of 5 by four hours of plasma electrode activation, without adding Cs from a Cs supply, but with Cs recovery from impurities (as in other cesium free sources).

ELECTRODE ACTIVATION IN GAS DISCHARGES

By empirical electrode activation in some cesium-free sources, H⁻ current with acceptable emittance has been increased to 40 or 80 mA (emission current density from 0.1 to 1 A/cm^2) and the ratio of electron current to H⁻ current has been decreased down to e/H⁻ ~ 4 without cesium injection. This enhancement of H⁻ generation was smaller than the "cesiation effect" but significant and important for further ion source development. Understanding the reasons for these improvements is important for upgrading H⁻ sources. The enhancement of SPG in discharges without cesium has been considered in reports [3,4,11,12,14,15,19]. Possible reasons for these enhancements are discussed below.

A simple explanation of this enhancement is that impurities with low ionization potential (ILIP) are accumulated on the electrode with corresponding decrease of surface work function and corresponding increase of secondary emission of negative ions similar to "cesiation". Potassium decreases the surface work function almost the same as cesium and increases negative ion emission significantly. For optimal work function reduction, a very small quantity of the ILIP is enough (e.g., a part of a monolayer or ~10^{14} particles per cm^2), which is difficult to detect by sensitive methods including spectroscopy. For efficient accumulation and trapping of the ILIP on the emission surface it is useful to keep this surface at negative potential relative to the plasma. Particles of ILIP are ionized in the plasma and are transported by the electric field as ions to the surface with a negative potential. After sputtering by ions or evaporation, these particles are ionized again and are returned to the negative surface. Small amounts of ILIP exist in many metals, ceramics, glasses, enamels, and cathode materials. A sub-monolayer of barium, lanthanum or other lanthanides can decrease the work function significantly and can increase secondary negative ion emission.

The new DESY (Deutsches Electronen Synchrotron) RF-driven multicusp H⁻ ion source is considered by its developers as an RF-volume source [16]. However, it was concluded that generation of 80 mA of H⁻ beam current during this source operation with internal antenna could be due to an admixture of potassium from its antenna insulation (containing ~15% of K) which should enhance SPG. In source versions with an external antenna, a ceramic chamber comprising 0.01% of K$_2$O is used, and the developers claim that with this low concentration of K the volume generation must be dominating because the addition of Macor ceramic with 10% K$_2$O concentration and using an Al collar with

20% of K$_2$MnO$_4$ admixture do not enhance the negative ion emission. However for the enhancement of SPG it is enough to have a very low surface concentration of alkali atoms such that a small contamination of electrodes or ceramic such as 0.01% of K$_2$O can be enough for domination of the SPG of H$^-$ above the other generation processes in so called "volume" sources. Namely, once the optimum has been achieved, a further increase of K cannot improve H$^-$ generation.

In the J-PARC (Japan Proton Accelerator Research Complex) tandem multicusp sources [17,18], an increase of H$^-$ current from 14 to 38 mA was reached by high temperature activation by discharge of the negatively biased plasma electrode and by optimization of the plasma electrode configuration for anode surface plasma H$^-$ generation. The plasma was generated by a pulsed discharge with a directly heated LaB$_6$ cathode. A deposit of La was observed on the collar surface. This case clearly demonstrated the enhancement of an anode SPG in a discharge without cesium addition (the initial emission of ~10 mA can be interpreted as Volume Generated).

An H$^-$ emission current density up to 0.75 A/cm^2 was initially produced without cesium in a flat magnetron/planotron ion source by electrode activation in a high density discharge [3,11,19]. After cesium was added, the source emission current density was increased to 3.7 A/cm^2. As discussed in [14,15], enhanced anode SPG has also been observed in some other negative ion sources with high density discharges in crossed fields: in the Ehlers type source [20] and more clearly in the Ehlers source with insulated plasma electrode [21]; in the Dudnikov type source with LaB$_6$ cathodes [22]; in the source with a tubular magnetron discharge in a longitudinal magnetic field [23] and in a duoplasmatron with tubular discharge [9].

Cesium adsorption lowers the surface work function, and this increases the probability of sputtered and reflected particles escaping in the form of negative ions. This probability increases as the work function decreases and the velocity of particles moving away from the electrode surface increases because the probability for electron tunneling back to the

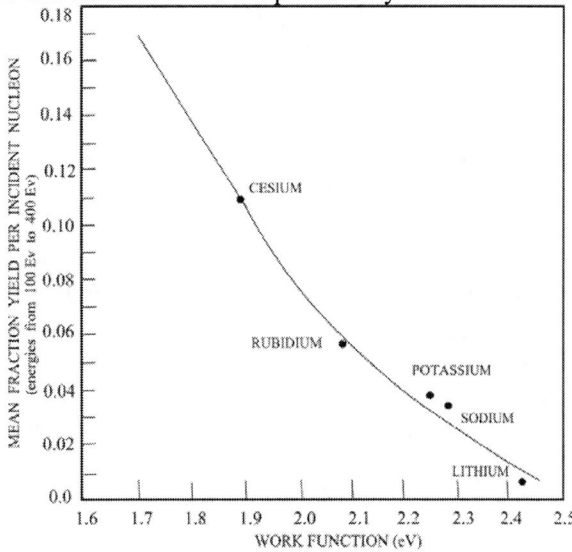

FIGURE 1. Experimental secondary emission yield for surfaces with different work functions (deposition of different LIP particles) from [14, 25]. The hand drawn curve is meant to guide the eye toward the minimal work function at 1.4 eV [25].

solid is smaller in a shorter escaping time. The theory of negative ionization of particles interacting with an electrode surface has been presented in publications [24]. The calculated probabilities of H⁻ formation as a function of initial particle velocity for different work functions and as a function of work function for different initial velocities are available in [14,15,24].

The experimental dependence of H⁻ formation on the surface work function is shown on Fig.1 from [14, 15, and 25]. As seen on the plot, secondary emission of H⁻ can be enhanced by lowering the surface work functions to 2.5 eV, corresponding to the deposition of Li, Mg, Ba or La.

The conditions for volume and surface-plasma generation of H⁻ in gas discharges coexist. In discharges with high emission current density the free path of cold H⁻ is very short (~ mm). In these conditions, the volume density of H⁻ is determined by VG. But the production of emitted H⁻ is determined by SPG on the plasma electrode surface around the emission aperture. To enhance the efficiency of emitted H⁻ production, it is necessary to optimize SPG on the plasma electrode. This can be achieved by the efficient transformation of gas molecules to a flux of hyperthermal atoms with the energy of several eV [4, 11, 12, 14]. It is necessary to use a strong magnetic field which is perpendicular to the H⁻ beam direction for strong suppression of electron diffusion to the emission surface and to steer the H⁻ beam to the emission aperture. It is useful to use a slit extraction system. It is necessary to optimize the shape and thermal properties of the plasma electrode for the efficient formation and extraction of generated H⁻ ions. It is essential to activate the emission surface by discharge and by thermal processing. After such optimization, it is possible to produce a high H⁻ density with much lower electron density on the emission surface, and to suppress the ratio of electron current to H⁻ current γ down to less than 5-10 in discharges without cesium. High optimization efficiency was demonstrated in works [4,6,11-14,16-19]. The emission current density of H⁻ ions was increased up to ~1 A/cm² in discharges without cesium. We believe that the most probable reason for this emission enhancement is the diffusion and accumulation of ILIP from the bulk of the electrodes to the surface and sputtering from the RF antenna coating or ceramic.

Refractory metals such as Mo, W, Ta have relatively high concentration of alkaline metals, mainly sodium and potassium, because compounds such as Na_2WO_4, K_2TaF_7, and Na_2MoO_4 are used to produce these metals. In experimental measurements, presented in [25-27] the secondary emission of D⁻ was increased up to 50 times by the deposition of sodium and potassium on a copper surface (the work function φ was decreased from 4.8 eV down to 2.25 eV) and up to 100 times by deposition of cesium (φ decreased from 4.8 eV down to 1.9 eV). A further decrease of the work function of the plasma electrode emission surface by an optimal deposition of cesium from a special cesium vapor supply and activation (down to $\varphi\sim1.6$ eV) increases H⁻/D⁻ emission 4 to 5 times, relative to the discharge with well activated electrodes without cesium, and decreases electron emission below H⁻ emission.

An alternative explanation of H⁻ generation enhancement through electrode activation was expressed in [28]. The hypothesis was that the activation enhances "recombinative desorption" comprising desorption of rovibrationaly excited molecules initiated by plasma particles with further VG of H⁻ by dissociative attachment of electrons. This complex mechanism of surface plasma and volume generation of negative ions is interesting and must be analyzed for possible confirmation and possible application. However, H⁻ generation enhancement by lowering the work function through accumulation of ILIP on the negative electrode surface is more probable and really unavoidable with ordinary vacuum cleaning because of widespread alkali metal impurities in ion source materials.

LV SPS SADDLE ANTENNA ELECTRODE ACTIVATION

H⁻ ion generation in a large volume (LV) SPS with a saddle RF antenna was tested in Ref. [13]. The diagram of the LV SPS with saddle antenna and plasma gun igniter is shown in Fig. 2. In this experiment an RF discharge with saddle antenna and longitudinal magnetic field is used for plasma generation. In SPS, the plasma flux is needed only inside the collar with 2 cm diameter. (In ordinary RF discharges with a solenoid antenna and cusp magnetic field the plasma flux is uniform over the 7 cm diameter discharge chamber and only small part of the plasma flux is used for negative ion production.) With the longitudinal magnetic field and saddle antenna, the plasma is concentrated near the axis and the plasma flux increases with magnetic field. Simultaneously, the transverse magnetic field created by the permanent magnet in the collar suppresses the flux of co-extracted electrons as in all SPS.

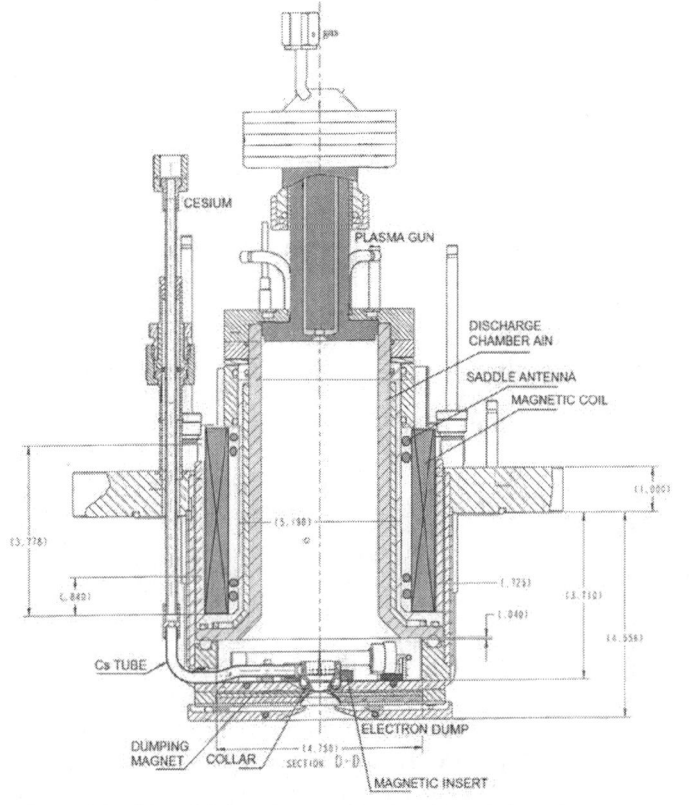

FIGURE 2. Diagram of RF SPS with saddle antenna, magnetic coil, and glow discharge plasma gun igniter (dimensions are in mm, where the collar emission aperture is 7 mm diameter).

This SPS was assembled after a regular cleaning procedure with SPS parts washed in alcohol, which should either remove alkali impurities present from previous operations or transform their remains into compounds. A strong transverse magnetic field was created by the dumping permanent magnet inside and below the collar for filtering and deflecting electrons.

The cesium transportation line was heated but the Cs ampoule was not cracked for the first operation described here. A 2 MHz RF power supply was used for these experiments. The beam current on the SNS test stand was measured using a toroidal Beam Current Monitor (BCM) and a calibrated Faraday cup (fc), both located at the exit of the LEBT (Low Energy Beam Transport) [29], which was similar to that used to develop basic SNS (Spallation Neutron Source) RF sources.

After starting the RF discharge with RF power Prf=24 kW, 10 Hz repetition rate, 0.3-0.4 ms pulse length, and magnetic field B=250 G, the initial H⁻ beam current Ifc=8 mA started growing and increased up to 42 mA after 4 hours without cracking the cesium ampoule as shown by graph 1 (blue) in Fig. 3. Over the same period, the extracted current, which includes H⁻ ions, back accelerated positive ions, and electrons, decreased significantly. The initial current density J~20 mA/cm^2 generated with cleaned collar can be interpreted as volume generation but the increase of J up to 100 mA/cm^2 with the same RF power must be interpreted as produced by SPG though electrode activation by the accumulation on the collar ionization surface of a very small amount of ILIP (Cs in this case), which remained even after alcohol cleaning.

The H⁻ current decreased linearly down to 12 mA with decrease of magnetic field B from 250 to 50 G.

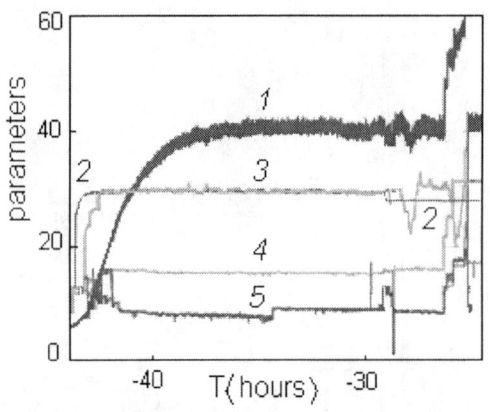

FIGURE 3. Evolution of H⁻ beam intensity of the saddle antenna RF LV SPS; 1- ion beam current; 2- gas flow; 3- collar temperature; 4- forward RF power; 5- reflected RF power.

We believe that this saturating "activation" was produced (without additional Cs) by the collection and trapping of traces of cesium compound remnants from SPS surfaces. Long conditioning is necessary because cesium is only slowly recovered from remnants by Hydrogen discharge. This slow accumulation demonstrates that the lifetime of these impurities on the collar can be very long. Nanograms of impurities are enough for enhancement of secondary emission of negative ions from the collar surface.

Stable operation of the tested prototype of the RF H⁻ SPS with the saddle antenna and longitudinal magnetic field up to highest available RF power of 56 kW, 0.3 ms, 10 Hz was successfully demonstrated. With lower magnetic field the RF discharge become unstable at gas flow Q~19 sccm. Beam current up to Ifc = 67 mA was observed with RF power of Prf = 56 kW.

The collar temperature (curve 3, green) can influence the H⁻ beam current (curve 1, blue) as seen in Fig. 3, where the second bump in the blue curve after ~30 hours is related to a decrease of gas flow (curve 2, red) and to variations of collar temperature 3 (green).

A decrease of the collar temperature below optimal (~270°C) initially increases H⁻ current by increasing cesium adsorption but then decreases after the maximum is attained, which can be interpreted as passing through the minimum of the work function dependence on the cesium surface concentration (see Fig. 4). H⁻ generation is also sensitive to hydrogen flow variations (graph 2).

The ratio of Ifc/Prf ~ 1.6 mA/kW that was produced before cracking the cesium ampoule is higher than in other SNS RF discharges using either an internal antenna or an external antenna and multicusp magnets in similar ("Cs free") conditions [29,30].

This activation from Cs contamination in this experiment was so good that the efficiency of H⁻ generation was not improved significantly by standard cesiation after cracking the cesium ampoule. However, the Cs ampoule is needed for fast start and for very long operation with optimal cesiation. We therefore believe that nanograms of ILIP are enough for enhancement of secondary emission of negative ions from a collar surface. This process should also work for other ILIP such as K, Na, and Ba [11, 14]. For example, a small contamination of ceramic such as 0.01% of K_2O can be enough for domination of the SPG above volume generation in the other "volume" RF sources [16].

PRODUCTION OF SURFACES WITH LOW WORK FUNCTIONS

For stable and reproducible Surface Plasma Generation (SPG) of H⁻ ions in gas discharges with electrode activation by cesiation, specific cleaning and processing of electrode surfaces are required. For reproducible production of surfaces with low work function it is necessary to heat them to high temperatures for hours to remove surface and volume impurities. Fortunately, from the beginning it has been possible to heat the SPS to high temperature (>1000 K) by plasma discharges.

The dependences of work function on surface cesium concentration for different tungsten crystalline surfaces at low temperatures are shown in Fig. 4. Here we show W data, which are more detailed than for Mo, because Mo with W deposition has been used for cesiation testing and because collars of with W ionization surfaces are planned to be studied. The curves demonstrate the diversity of conditions for H⁻ generation with cesiation.

The estimated relative yield Y of H⁻ secondary emission dependence on cesium surface density is shown by curve 5 in Fig. 4. Here Y = (H⁻ secondary emission current)/(flux of bombarding plasma particles, ions and atoms). The cesium surface density is a parameter determined by the surface work function and the probability of sputtered and scattered particles escaping as negative ions. The dependence of work functions and probabilities of negative ionization were determined in separate experiments with slow deposition of cesium to a cold surface.

On very clean surfaces with work function above 4 eV, the yield of secondary emission of H⁻ ions is very low (<10^{-4} for fast bombarding particles) and volume generating of H⁻ can be dominant in these conditions. With increase of ILIP particle surface concentration up to N~ 10^{14} p/cm² (~10 ngram/cm²), the work function drops below 2 eV and Y increases significantly, reaching maximum Y~0.2 at minimum work function ~1.5 eV with ILIP particle surface concentration up to N~ 2-4 10^{14} p/cm² (~30 ngram/cm²).

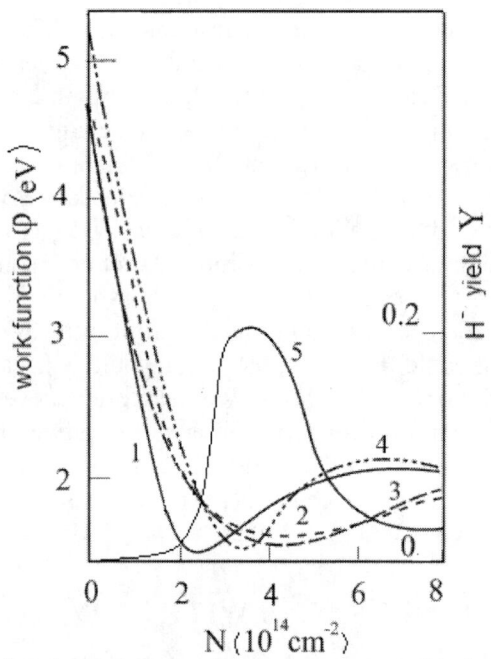

FIGURE 4. Dependence of work function on surface cesium concentration for different W crystalline surfaces: 1-(001); 2-(110); 3-(111); 4-(112) (left scale) and 5-estimated relative yield Y of H⁻ secondary emission for surface index (111) and particle energy ~3 eV (right scale).

The desorption energy H of cesium decreases almost linearly with surface cesium concentration from H ~ 3 eV for a clean surface to H < 1 eV for LIP particle surface concentration N ~ 4 to 6 10^{14} p/cm² (~50 ng/cm²), approximately consisting of one monolayer.

The thermal desorption time τ has an exponential dependence on H and temperature [32]:

$$\tau(N) = \tau_0 \cdot \exp(H(N)/k \cdot T) = 5 \cdot 10^{-13} \cdot \exp(H(N)/k \cdot T) \, [s].$$

For low T~<500 K and H~1.5 eV, which corresponds to a T of about 15,000 K, the desorption time τ is greater than several seconds and LIP impurities can be accumulated over a long time and the optimal concentration can be supported by very small cesium flux.

The equilibrium cesium concentration on the surface depends on the cesium particle flux to the surface from the volume and on a rate of desorption determined by τ. In systems with a cesium supply, such as thermionic energy converters or those of SPS, the cesium flux to the surface from the volume has an exponential dependence on the cesium reservoir temperature Tr, and the equilibrium cesium concentration on the surface depends on the ratio T/Tr [32]. The optimum cesium concentration delivering the minimal work function corresponds to T/Tr <~2, as shown in Fig. 5 and described in Ref. [32].

In SPS this equilibrium is complicated by fast particle sputtering and Cs ion trapping by the plasma. However, Cs adsorbate sputtering by hydrogen ions is low up to energies near 50 eV because the ions are much less massive than the Cs atoms. But desorption by heavier fast particles can be more important. An additional complication is due to impurity deposition which is strong in the SPS from hot cathode discharge, which necessitates Cs redeposition. Electrode sputtering by back accelerated positive ions should be suppressed in order to minimize cesium consumption. Recent experience in the SPS with RF plasma generation shows such effects to be smaller, which is favorable for operation with very low cesium consumption as was demonstrated in [13,29,30] and in this report. With an internal Cs supply, proper cesiation can be done in about 20 minutes [1-4,30].

The desorption energies of other ILIP such as Rb, K, Na, Ba, La are larger and optimal concentration of these ILIP substances can be supported by lower flux from the volume or by one time deposition.

The increase of H$^-$ generation with decreasing collar temperature below T~540 K observed in Fig. 3 following the decrease in the second bump in curve 1 after -30 hours confirms the existence of a Cs flux to the collar surface from a parasitic Cs supply corresponding to Tr~340 K. At lower collar temperature, the optimal Cs concentration N can be supported at lower Tr with a lower Cs flux.

However, the minimization of the work function is not always successful. In Cs deposition to polycrystalline molybdenum in Ref. [31] the minimum work function was not reached but only a decrease of work function from 4.4 eV to 2.8 eV was observed. Many conditions must be satisfied for minimization of the work function. For example, fast deposition of the cesium film can leave a film thicker than the optimal 0.6 monolayer necessary for minimizing the work function. The presence of some impurities including other ILIP can also interfere with optimum cesiation.

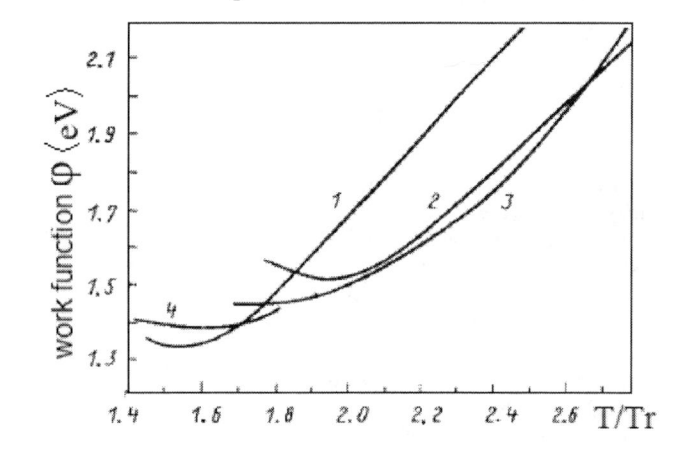

FIGURE 5. The work function dependence of cesium adsorption on the ratio of sample temperature to cesium-tank temperature for collectors of 1) a molybdenum polycrystalline with a tungsten layer on the surface, 2) (110) molybdenum, 3) a molybdenum polycrystalline, and 4) an LaB$_6$ polycrystalline.

Variations of accessible work functions very much influence the efficiency of H⁻ ion production. Production of reproducible low work functions has been developed in photocathode manufacturing and in thermionic convertors [32]. Similar procedures can be used for electrode activation in SPS. Fig. 5 shows dependences of work functions for molybdenum and LaB$_6$ surfaces as a function of the ratio of surface temperature to container temperature from [32]. Tungsten deposition does not prevent production of work functions below 1.5 eV, which is very favorable for highly efficient H⁻ generation.

Laser diagnostics of cesium dynamic in the SPS proposed in ref. [33] can be used for improving of cesiation.

Secondary emission of H⁻/D⁻ from graphite surfaces [34] should have a strong dependence of LIP impurities as was observed in [35].

Gas pulsing by fast valve [36] was important for high efficiency of H- generation in the pulsed SPS [1-4].

CONCLUSIONS

Advanced H⁻ sources need to have high temperature electrode cleaning for reproducible production of low work functions and for stable generation of H⁻ ions with high efficiency. In a clean hydrogen RF discharge the Cs lifetime on the surface can be very long and very low cesium consumption can be achieved. Small traces of impurities of low ionization potential are enough for optimal surface plasma generation on the electrode surface. With a separate internal Cs supply similar to that used in [4] it is possible to heat the plasma electrode to very high temperature (>1000 K) before cesiation and then have fast cesiation with long Cs lifetime on the plasma electrode surface.

ACKNOWLEDGMENTS

This work has been supported in part by US DOE Contract DEAC05-00OR22725 and by STTR grant DE-SC0002690.

REFERENCES

1. V. Dudnikov, The Method of Negative Ion Production, SU Author Certificate, C1.H01 3/04, No. 411542, Application, 10 March,1972.
2. V. Dudnikov, Surface-Plasma Method of Negative Ion Production, Doctor Thesis, INP, Novosibirsk, 1977(partly published in 3,4,6,8,11,12,19,24).
3. V. Dudnikov, Rev. Sci. Instrum. 63(4),2660 (1992). V. Dudnikov, Rev. Sci. Instr. 73(2), 992 (2002).
4. Yu. I. Belchenko, G. I. Dimov, and V. G. Dudnikov, BNL report BNL 50727, 79 (1977); V. Dudnikov, BNL 51304,137 (1980).
5. D. P. Moehs, J. Peters, and J. Sherman, IEEE Transactions on Plasma Science, 33, 6, 1786 (2005). Joseph Sherman and Gary Rouleau, AIP Conference Proc. 680, 1038, 2002.
6. J.Peters, EPAC'2000; J.Peters, LINAC' 98, Chicago, 1998; J.Peters, Rev. Sci. Inst., 71(2), (2000).
7. *The Physics and Technology of Ion Sources*, Edited by Ian Brown, Wiley-VCH, 2004.
8. Zhang Hua Shun, *Ion Sources, Springer*, 1999.
9. K. Prelec, and Th. Sluyters, Rev. Sci. Instrum., 44(10), (1973).
10. T. Kuo, P. Schmor, G. Dutto, Rev. Sci. Instrum., Vol 69 (2), 959 (1998).
11. Yu. Belchenko, V. Dudnikov, European Symposium Negative Ion Production, 47-66,Belfast, 1991; Yu. Belchenko, Rev. Sci. Instrum, 64, 1385 (1993).

12. Yu. Belchenko, G. Dimov, V. Dudnikov and A. Kupriyanov; Revue Phys. Appl. 23, 1847-58 (1988). hal.archives-ouvertes.fr/docs/00/.../ajp-physap_1988_23_11_1847_0.pdf.
13. V. Dudnikov et al., IPAC 2010, THPEC073, Kyoto,Japan, 2010.
14. V.Dudnikov, 10th Int. Symp. PNNIB, AIP CP 763, J. Sherman and Yu. Belchenko, p.122 (2005).
15. V. Dudnikov and R. P. Johnson, Rev.Sci.Instrum., 81, 02A711, 2010.
16. J.Peters, AIP CP 1097, edited by Surrey and Simonin, p. 171 (2009).
17. A.Ueno, K.Ikegami, Ya. Kondo, Rev.Sci.Instrum., 75 (5), 1714 (2004); H. Oguri, A.Ueno, K. Ikegami, Rev.Sci.Instrum., 79 (2), 02A506 (2008).
18. A. Ueno, H. Oguri, K. Ikegami, Y. Namekawa, and K. Ohkoshi, Rev.Sci.Instrum., 81 (2), **81**, 02A720-1 (2010)
19. Yu. Belchenko, G. Dimov, V. Dudnikov, Bulletin of the Akademy of Sciences of the USSR Physical Series, v.37, no.12, p.91-5. 1974; NUCLEAR FUSION,14(1),113-114 (1974).
20. V.Goretsky, I.Soloshenko, Rev.Sci.Instrum., 73 (3),1157 (2002).
21. K.Jimbo, K.Ehlers, K.Leung, R.Pyle, Nucl.Inst.Methods, A 248, 282 (1986).
22. K. Leung et al., AIP Conf. Proc.158, p.356, 1986. and Rev.Sci.Instrum., 58 (2), 235 (1987).
23. Yu. Kursanov, P..Litvinov, V. Baturin, 10th PNNIB, AIP Conf. Proc No.763, 229 (2005).
24. M. Kishinevsky, Sov. Phys. Tech. Phys, 45, no.6, 1281 (1975), and 48, no.4, 773 (1978). Later these calculation were repeated by Cui [H.L Cui, J. Vac. Sci. Technol.,A, 9 (3), 1823 (1991)].
25. P.J. Schneider et al., BNL 50727, 63 (1977); Phys. Rev. B, 23(3), 941 (1981); M. Wada et al., J. Appl. Phys. 67 (10), 6334 (1990).
26. F.N.Hoffman, P.E.Oettinger, BNL 51304,UC-34a, 119, (1980).
27. M. Seidl, H. Cui,J.Isenberg, H.Kwon, Appl. Phys.,79(6), 2896 (1996).
28. M.Bacal, Nucl. Fusion, 46, p. 250-259 (2006).
29. R.F. Welton, M. P. Stockli, S.N. Murray, Rev. Sci. Instrum. 75 1793 (2004).
30. Martin P. Stockli et al., AIP CP 1097, edited by Surrey and Simonin, p. 223 (2009).
31. U. Fantz, R. Gutser, and C. Wimmer, Rev. Sci. Instrum. **81**, 02B102 (2010)
32. V.Z.Kaibyshev,V.A.Koryukin and V.P.Obrezumov, Atom Energiya, Vol. 69, No.3, p. 196-197, 1990.
33. V. Dudnikov, P. Chapovsky and A. Dudnikov, Rev. Sci. Instrum. **81**, 02A714 (2010).
34. L. Schiesco et al., Plasma Sources Sci. Tech. 19 (2010) 045016 ; Appl. Phys. Lett., 95, 191502 (2009).
35. M. Gleeson and A. Kleyn, Nucl.Instrum. Mhetods., B 157, 48 (1999).
36. G. E. Derevyankin, V.G. Dudnikov, P.A. Zhuravlev, Pribory i Tekhnika Eksperimenta, **5**, 168 (1975).

Negative hydrogen ion beam extracted from a Bernas-type ion source

N. Miyamoto and M. Wada

Graduate School of Engineering, Doshisha University, Kyotanabe, Kyoto 610-0321, JAPAN

Abstract. Negative hydrogen (H⁻) ion beam was produced without cesium seeding by a Bernas-type ion source with a coaxial hot cathode. The amount of H⁻ ion beam current extracted from an original Bernas-type ion source using a hairpin shape filament as a hot cathode was 1 μA with the 0.4 A arc current, while that 300 eV beam energy. In the other hand, H⁻ ion beam current using the Bernas-type ion source with a coaxial hot cathode reached 4 μA under the same condition. Production efficiency was enhanced by the focused plasma produced by a coaxial hot cathode.

Keywords: negative ion source, Bernas ion source, hot cathode, coaxial cathode.
PACS: 41.75.Cn, 07.77.Ka, 79.40.+z, 52.27.Cm.

INTRODUCTION

High current negative hydrogen/deuterium (H⁻/D⁻) ion sources have been developed for a neutral beam injection (NBI) system to heat up a plasma to thermonuclear fusion condition [1, 2]. These ion sources have large size plasma generators to produce high current negative ion beams. The principal mechanism of the high current density H⁻/D⁻ ion beam formation is believed to be the so called volume production process [3]. However, the present sources are operated with cesium seeding, which should alter the interaction between the source plasma and the surface of plasma grids.

The volume production process consists of two steps: (i) production of highly excited states of vibrationally excited molecules and (ii) negative ions are produced by dissociative attachment of low energy electrons to vibrationally excited molecules. The production of vibrationally exited molecules needs high energy electrons. The reaction rate for negative ion production by dissociative electron attachment takes the maximum at low electron energy, while high energy electrons destroy H⁻ in the plasma. Therefore, a high density high temperature plasma should be separated from a plasma electrode of an ion extractor.

Recently, a small size Bernas-type ion source equipped with a coaxial thermionic cathode has been developed [4, 5]. The plasma produced by a coaxial cathode in a Bernas-type ion source had indicated clearer and focused plasma shape along the external magnetic field [6]. A high density plasma column has been sustained straight from the coaxial cathode along the external magnetic field. Namely, two layers necessary to form negative ions exist in the arc chamber: a layer to form vibrationally excited molecules, and a layer to produce H⁻ by dissociative electron attachment. The Bernas-type ion source with a coaxial cathode can produce large current H⁻ ions as the plasma geometry can formed two regions necessary form to H⁻ in the arc chamber.

The details of the Bernas-type ion source and the experimental results will be described in this paper.

EXPERIMENTAL SETUP

A high temperature cathode having a coaxial electrical current path for heating was produced with the structure similar to the BEAR (Bernas-type Electron Active Reflection) ion source developed by Nissin Ion Equipment Co. Ltd [7]. A schematic diagram of the coaxial cathode is shown in Fig. 1. The coaxial cathode consisted of two thin-wall tubes. The dimensions of the inner tube were 4.0 mm in diameter and 0.3 mm in thickness. The outer tube was 5.0 mm in diameter and 0.2 mm in thickness. The length of the cathode was about 50 mm. The inner tube was made of tungsten and the outer tube was made of tantalum. A pair of tips of both tubes was connected so that the cathode heating current can run through the tubes. At the other end, the cathode terminals were connected to molybdenum current feedthroughs inserted from the end flange supporting the arc chamber of the ion source.

To compare negative ion beam production efficiencies, hairpin shape filament cathode for BEAR ion source was also tested. The dimension of the filament was 2 mm in diameter and 150 mm in length. Both cathode positions along the arc chamber were the same. The arc chamber structure including the cathode position was optimized for higher positive ion production. According to previous research, thermionic electron emission is known to localize near the region of the negative end of the filament [8, 9]. The emitted electrons were trapped near the cathode by magnetic field produced by the cathode heating current, and this effect should broaden the energy distribution function of the emitted electrons. A coaxial cathode structure producing only small magnetic field at the surface of the external conductor can emit electrons directly from the tip of the coaxial cathode, forming a high energy electron beam aligned along an external magnetic field. Figure 2 shows a schematic diagram of the BEAR ion source arc chamber equipped with a coaxial hot cathode. The dimensions of an arc chamber are 90 mm in length, 36 mm in width and 28 mm in depth. Two electron reflectors biased negatively at several tens of volts with respect to the anode are located at both ends of the arc chamber under the normal operation of the BEAR ion source. A proper electrical potential of the reflector enhance the

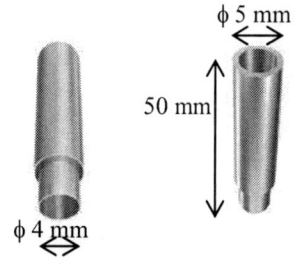

FIGURE 1. Schematic diagram of a coaxial cathode.

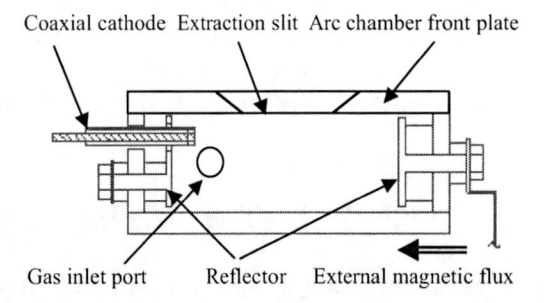

FIGURE 2. Schematic diagram of an arc chamber for BEAR ion source equipped with a co-axial cathode.

ionization efficiency of the ion source. However, in these experiments, the potential of the reflector was not optimized, but the potential was set to the same negative potential as the cathode heating power supply.

An external magnetic field is applied to the region inside the vacuum chamber where the ion source is inserted. The field is produced by a coil and an iron yoke which are set outside the vacuum chamber. The direction of the external field is perpendicular to the direction of beam extraction. An ion beam was extracted by a single stage extractor. The top plate of the arc chamber has a perpendicular slit along the external magnetic field. The dimensions of the slit are 3 mm in width and 30 mm in length. A grounded electrode made of graphite was located 3 mm downstream from the top plate of the arc chamber. An extraction potential was supplied to the extractor up to 300 V. The extracted negative ion beam current was measured by a Faraday cup located the other end of the vacuum chamber. A schematic diagram of the vacuum chamber for performance tests is shown in Fig. 3. The vacuum chamber was cylindrical in shape with 250 mm in diameter and 400 mm in length. A turbo

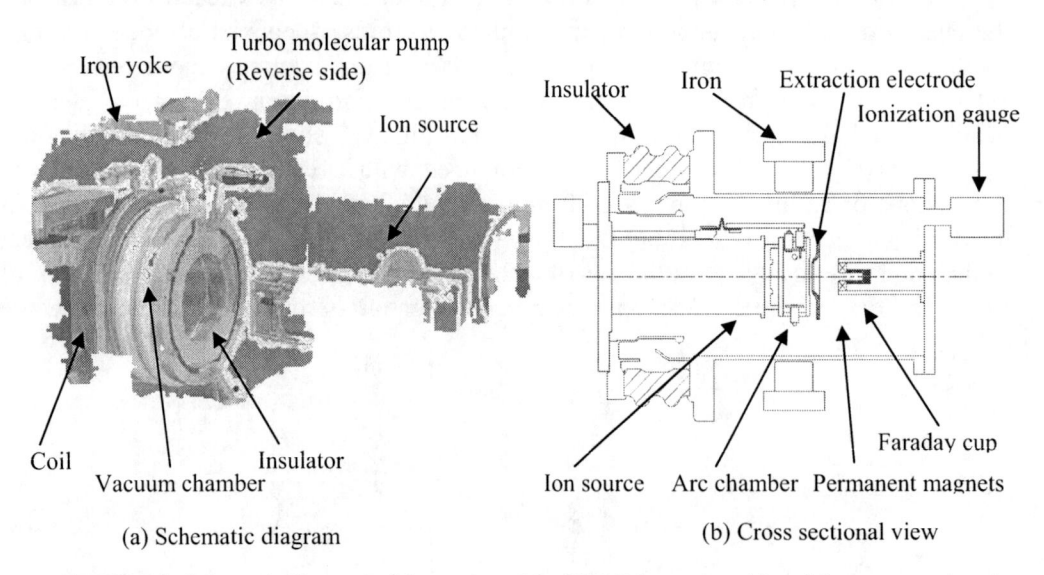

FIGURE 3. Schematic diagram of the test stand for BEAR ion source (a) and the cross sectional view of the vacuum chamber in the test stand.

molecular pump and an oil rotary pump were connected to the vacuum chamber for evacuation. The pumping speed of the turbo molecular pump was 250 ℓ/min. An ionization gauge was installed onto the vacuum chamber to measure the gas pressure.

The Faraday cup was located about 35 mm downstream from the ground electrode to measure the negative ion beam current. The Faraday cup was made of graphite. The Faraday cup has a transverse magnetic field produced by a pair of permanent magnets to eliminate secondary electron emission. The strength of the transverse magnetic field was 80 mT·cm. The transverse magnetic field was aligned along the external magnetic field direction. The dimensions of the Faraday cup were 10 mm in diameter and 22 mm in depth. Hydrogen gas was supplied by a hydrogen electric generator. The gas was fed into one of the side walls of the arc chamber. Gas flow was controlled by a needle valve.

EXPERIMENTAL RESULTS AND DISCUSSIONS

Plasma Production Test

Hydrogen plasma production test was carried out at the test stand. The arc currents as a function of arc voltage in various hydrogen gas pressures using the BEAR ion source with a hairpin cathode and that with a coaxial cathode are shown in Fig. 4. These data were obtained under the conditions of 155 A cathode heating current and 10 mT external magnetic field. The hydrogen gas pressure was adjusted from 0.1 Pa to 0.2 Pa. The arc current characteristics were similar under this range of hydrogen gas pressure. The arc current produced with a coaxial cathode reached about 0.5 A. On the other hand, the arc current with a hairpin cathode was about 1.2 A. The

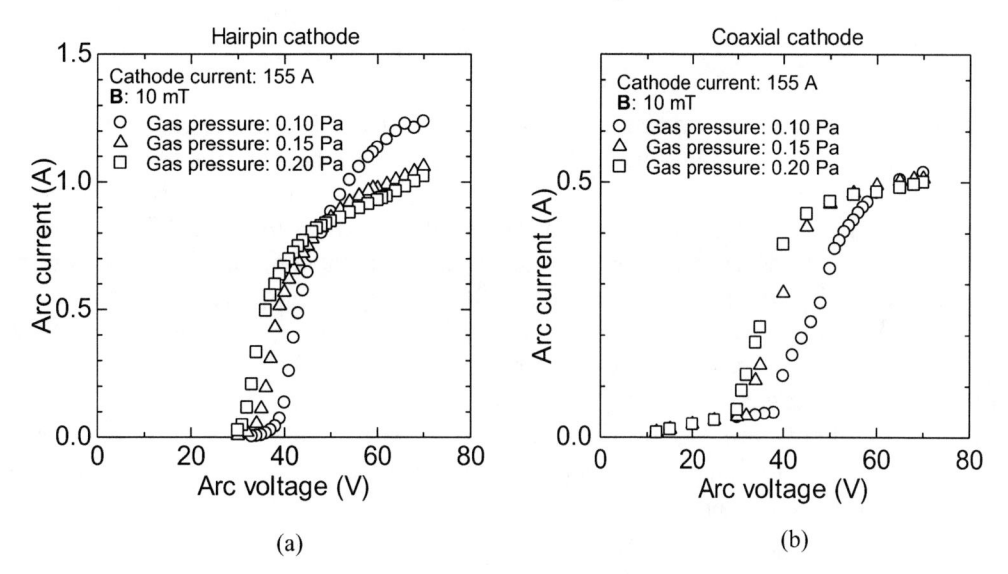

FIGURE 4. Arc discharge characteristics in various hydrogen gas pressure using with (a) a hairpin cathode and (b) a coaxial cathode for the BEAR ion source.

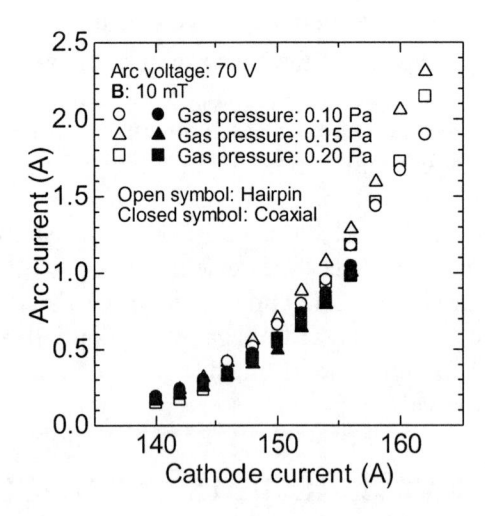

FIGURE 5. Arc current as a function of cathode current in various hydrogen gas pressure using with a hairpin cathode and a coaxial cathode for the BEAR ion source.

difference suggests that the temperature of a coaxial cathode was lower than a hairpin cathode under the same heating current.

The arc current as a function of cathode heating current in various gas pressures is shown in Fig. 5. These data were obtained under the conditions of 70 V arc voltage and 10 mT external magnetic field. Using both of cathodes, the arc currents were independent of hydrogen gas pressure. The increase in arc current by the cathode heating current was steeper for a hairpin cathode than a coaxial cathode. It was thought that a hairpin cathode was able to heat up easier than a coaxial cathode.

Negative Ion Beam Extraction Test

Negative hydrogen ion beam was extracted from the BEAR ion source using a coaxial and a hairpin cathode to compare the negative ion beam current. Figure 6 shows the negative ion beam current as a function of arc current in various external magnetic fields. In the figure, panel (a) shows the result using a BEAR ion source with a hairpin cathode and (b) shows the one with a coaxial cathode. These data were obtained under the conditions of 70 V arc voltage, 0.05 Pa hydrogen gas pressure and 300 V extraction potential. Using hairpin cathode, the negative ion current increased lineally with increasing arc current except at 8 mT external magnetic field. At the condition of 8 mT an external magnetic field, the negative ion beam current saturated to more than 400 mA arc current. The maximum negative ion beam current was about 0.8 µA. On the other hand, the negative ion beam produced with a coaxial cathode increased exponentially against increasing an arc current. The negative hydrogen ion current reached 2 µA under the same conditions using a hairpin cathode. The maximum beam current produced by a coaxial cathode was twice larger than that by a hairpin cathode. The structure and configurations of the arc chamber including the cathode position were adjusted to produce a large current density positive ion beam.

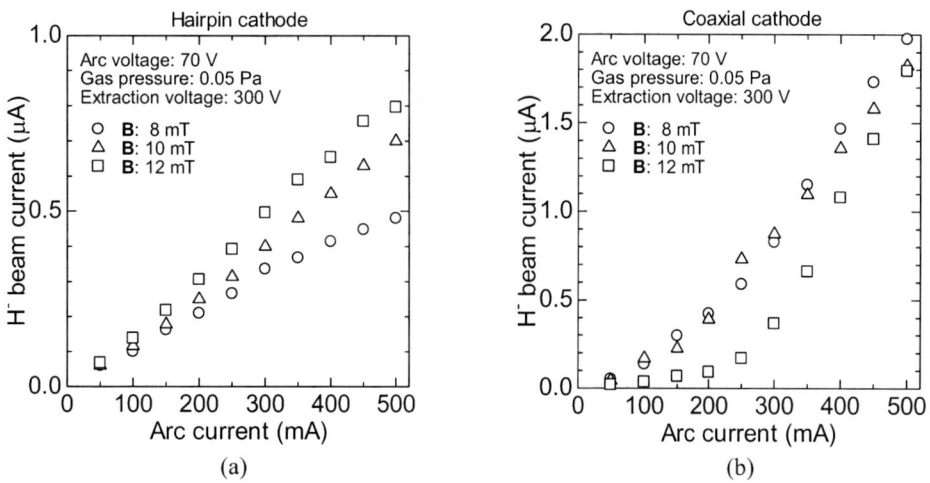

FIGURE 6. Negative hydrogen ion beam current as a function of arc current in various external magnetic field using with (a) a hairpin cathode and (b) a coaxial cathode for the BEAR ion source.

Therefore, the negative hydrogen ion beam was not large enough against an ion source which was optimized to produce a high current negative ion beam.

The gas pressure characteristics of the negative ion beam current in various arc currents are shown in Fig. 7. These data were obtained under the conditions 300 V extraction potential, 70 V arc voltage and 10 mT external magnetic field. The minimum ion beams were observed at 0.08 Pa hydrogen gas pressure for the BEAR ion source with a hairpin cathode and 0.06 Pa for the BEAR source with a coaxial cathode. The negative ion beam current extracted from the BEAR ion source with a coaxial cathode reached about 3 μA. The negative ion beam from the BEAR source with a coaxial cathode was three times larger than that with a hairpin cathode.

The negative ion beam current as a function of arc voltage is shown in Fig. 8. These data were obtained under the conditions of 10 mT external magnetic field, 0.05

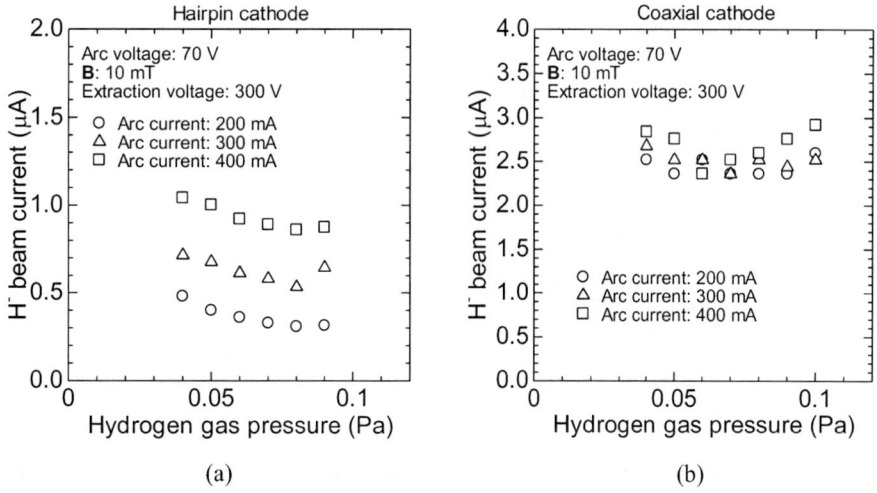

FIGURE 7. Negative hydrogen ion beam current as a function of gas pressure in various arc current using with (a) a hairpin cathode and (b) a coaxial cathode for the BEAR ion source.

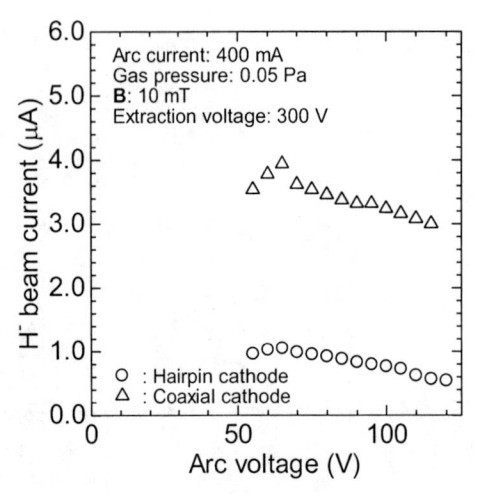

FIGURE 8. Negative hydrogen ion beam current as a function of arc voltage in various arc current using with a hairpin cathode and a coaxial cathode for the BEAR ion source.

Pa hydrogen gas pressure, 400 mA arc current and 300 V beam extraction voltage. Both configurations exhibited similar beam current behavior. The arc voltage which produced maximum negative ion beam current for both configuration was about 65 V. The result suggests that the optimum arc voltage for BEAR ion source to produce vibrationally excited molecules was about 65 V. The negative ion beam current produced by BEAR ion source with a coaxial cathode reached 4 µA. This value was four times larger than the negative ion beam produced by BEAR ion source with a hairpin cathode.

Based on the experimental results, the negative ion current for identical discharge condition using a coaxial cathode was higher than that using a hairpin cathode. Figure 9 shows that the schematic diagrams of the positive column produced with a hairpin and a coaxial cathode. According to the previous research [6], the positive column produced a hairpin cathode distributed widely. In contrast, the positive column with a coaxial cathode was focused. It was thought that the coaxial cathode was able to produce double layer plasma which was required high efficiency negative ion production. This means a coaxial cathode can perform better as a volume production negative ion source.

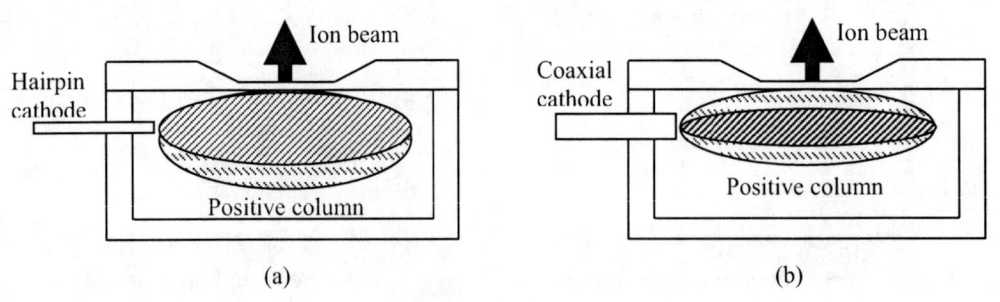

FIGURE 9. Schematics of the positive column in the arc chamber using with (a) a hairpin cathode and (b) a coaxial cathode.

CONCLUSIONS

Negative ion beam production was carried out using a Bernas-type ion source designed for ion implantations based on positive ion beam extraction. Two types of hot cathodes were adapted to a Bernas-type ion source to test their suitability for cathodes of a negative ion source. The arc discharge current produced with a coaxial were smaller than that using a hairpin cathode. However, the negative hydrogen ion beam current produced by a coaxial cathode was four times higher than that by a hairpin cathode under the same conditions. The results suggest, a coaxial cathode was able to produce double layer plasma column in the arc chamber because a hot cathode having a coaxial heating current path can produce clearer and focused plasma column without magnetic field distortion compared to a hairpin cathode.

REFERENCES

1. Y. Okumura *et al.*, *Rev. Sci. Instrum.* **67** (3), 1018 (1996).
2. Y. Takeiri *et al.*, *Journal of plasma and fusion research* **74**, 1434 (1998).
3. M. Bacal and G. W. Hamilton, *Phys. Rev. Lett.* **42**, 1538 (1979).
4. N. Miyamoto *et al.*, *AIP Conference Proceedings* **1066**, 304 (2008).
5. N. Miyamoto *et al.*, *Journal of Plasma and Fusion Research Series* **8**, 1542 (2009).
6. S. Imakita *et al.*, *Journal of Plasma and Fusion Research Series* **8**, 764 (2009).
7. N. Miyamoto *et al.*, *Rev. Sci. Instrum.* **73**, 819 (2003).
8. K. W. Ehlers and K. N. Leung, *Rev. Sci. Instrum.* **50**, 356 (1979).
9. M. E. Arciaga *et al.*, *The Science and Engineering Review of Doshisha University* **44**, 185 (2003).

Development of Hydrogen Pair-Ion Source on the Basis of Catalytic Ionization

W. Oohara, T. Maeda, and T. Higuchi

Department of Electronic Device Engineering, Yamaguchi University, Ube 755-8611, Japan

Abstract. To develop a hydrogen pair-ion source comprising only H^+ and H^- ions, the efficient production of pair ions is required. When discharged hydrogen plasma is used for positive-ion irradiation to a Ni catalyst, the pair ions are produced from the back of the irradiation plane. The number of pair ions produced increases proportionally with the irradiation flux.

Keywords: hydrogen pair ions, catalytic ionization, porous catalyst
PACS: 52.50.Dg, 52.25.Jm, 52.40.Hf

INTRODUCTION

Negative-ion production mechanisms can be classified in terms of the electron source [1]. In the surface effect, an electron at the Fermi level in the conduction band of a metal shifts by tunneling to the electron affinity level of an atom or a molecule approaching the metal surface. The probability of the electron shift is enhanced as the effective work function of the metal surface decreases. The electron shift occurs in particle reflection and sputtering phenomena. Cesium has the lowest work function of all elements. A small admixture of cesium vapor in a hydrogen discharge significantly improves H^- production and decreases the current of coextracted electrons [2-4]. However, the use of cesium complicates ion source operation and requires a careful stabilization of cesium injection and discharge parameters.

There have been many attempts to develop H^- sources with acceptable H^- beam emittance but without a cesium admixture. Plasma electrons are the source of electrons in volume production [5,6]. A highly vibrationally-excited hydrogen molecule effectively captures a low-energy plasma electron to form a H^- ion through dissociative electron attachment. However, there is the problem of a relatively low current density of H^- ions in volume production.

Hydrogen atomic pair ions, H^+ and H^- ions, are the lightest ions and have high response frequencies to electromagnetic fields. To generate a hydrogen pair-ion plasma consisting of only the pair ions [7-9], the production of equal quantities of H^+ and H^- ions and the absence of impurities such as electrons and other ions are required [10]. It is difficult to satisfy these requirements in surface production with a cesium admixture or in volume production. In order to solve this difficulty, we have suggested a catalytic ionization method with plasma assistance [11].

In our previous work, a Penning ionization gauge (PIG) discharge plasma was used for positive-ion irradiation to a porous catalyst [11]. The both irradiation energy and flux of positive ions depend on the discharge power (anode voltage). The irradiation flux should be controlled independent of the irradiation energy. In this paper, initial results of the operation of an dc arc discharge source with a nickel (Ni) porous catalyst are discussed.

EXPERIMENTAL APPARATUS

To produce the hydrogen pair ions, a hydrogen plasma is generated by a dc arc discharge in a rectangular chamber of 25 cm×25 cm in cross section and 19 cm in length. A schematic diagram of the experimental setup is shown in Fig. 1 (a). The apparatus for producing the pair ions comprises mainly a discharge section for the irradiation and a pair-ion production section. Four horseshoe tungsten filaments of 0.7 mm in diameter and 15 cm in length are installed on the side walls of the chamber. The filaments are the thermal cathode biased at a discharge voltage V_d and the chamber walls are the grounded anode. The plasma discharged is efficiently confined by line-cusp magnetic fields adjacent to the chamber walls which are applied by permanent magnets. A commercially available Ni porous plate (Celmet, Sumitomo Electric Toyama Co., Ltd.) acts a catalyst with a porous body of 56-60 cells/inch, a pore size of 0.45 mm, a thickness of 1.4 mm, a specific surface area of 5,800 m^2/m^3, and a porosity of 96.6 %. The porous catalyst is biased at $V_{pc} = -600$ V here and the irradiation current I_{ir} is measured. The circular irradiation area is 19.6 cm^2 (diameter of 5 cm) and the other electrode is covered with a mica plate. The catalyst is located at $z = 0$ cm, the discharge section is in $z < 0$ cm, and the pair-ion production section is in $z > 0$ cm. Plasma parameters are measured using Langmuir probes at $z = -7$ cm and 3 cm. The operating pressure in the source is about 0.2 Pa.

The surface production mechanism of the pair ions on the porous catalyst is considered to be completely different from the conventional mechanism of surface production with a cesium admixture which is based on the electronic transition of metals with low work functions. Hydrogen atoms, produced by dissociative adsorption, are covalently bound with the surface-metal atoms, and can easily move along the surface, i.e., surface migration occurs. Surface migration on catalysts is a well-known fundamental phenomenon. Hydrogen atoms migrate along the pore surface in the porous catalyst to the back of the irradiation plane. An electronic transition occurs between the surface-metal atoms and hydrogen atoms during desorption from the surface if hydrogen atoms are provided a sufficient energy. Desorption ionization from solid surfaces in the gaseous phase is known to occur in laser desorption/ionization. In our previous experiment, the pair ions are clear not to be produced without the irradiation even though the porous catalyst is additionally heated. The irradiation supplies the energy required for desorption ionization; the thermal energy is too small for desorption ionization to occur. A part of kinetic energy of positive ions irradiated will be transferred to the back of the irradiation plane owing to the migration of hydrogen atoms. If two shared electrons transfer to the surface atoms during desorption, H^+ ions are produced. H^- ions, on the other hand, are produced if the electrons transfer to hydrogen atoms. The work function of Ni (5.15 eV) is relatively high, the contact-ionization probability of H⁻ ions is infinitely close to zero. The catalytic activity for hydrogen is important here, whereas the work function does not affect the ionization. It is known that negative ions are produced from the irradiation plane in the conventional converter-type production, but fast electrons and fast positive ions promote the collisional detachment. It is highly beneficial that the pair ions are produced from the back of the irradiation plane, because the negative-ion production from the back can prevent the collisional detachment. We refer to this process involving dissociative adsorption, surface migration, and desorption ionization as catalytic ionization, which is schematically illustrated in Fig. 1 (b).

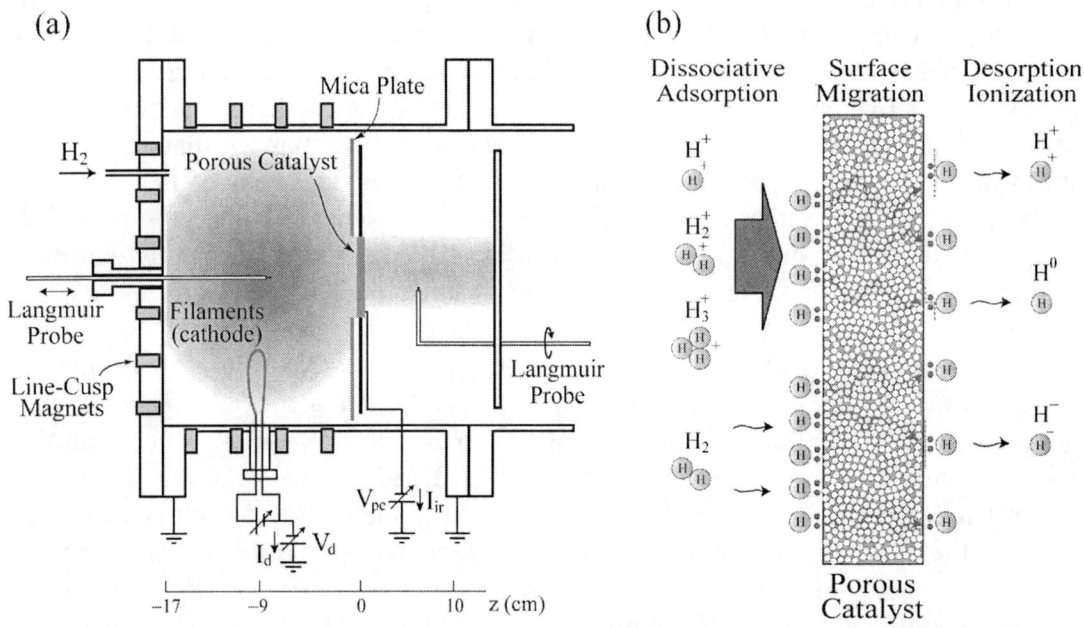

FIGURE 1. (a) Diagram of experimental setup for producing hydrogen pair ions on the basis of catalytic ionization. (b) Process of catalytic ionization under the positive-ion irradiation, which comprises dissociative adsorption, surface migration of hydrogen atoms, and desorption ionization.

RESULTS

The discharged plasma in $z < 0$ cm is used to irradiate positive ions to the Ni porous catalyst. Electron density n_e and plasma potential ϕ_s at $z = -7$ cm are measured. The dependence of n_e and the irradiation energy of positive ions $e(\phi_s - V_{pc})$ on the discharge power of $P_d = V_d \times I_d$ is shown in Fig. 2, where I_d is the discharge current. Positive ions are accelerated up to $e(\phi_s - V_{pc})$ (eV) in the sheath formed in front of the porous plate. V_{pc} is fixed at -600 V here and ϕ_s is varied in $+10\sim20$ V with P_d, but the variation width is narrow with respect to V_{pc}. The irradiation energy is almost constant independent of P_d. The irradiation current (flux) of positive ions I_{ir}, not shown here, increases proportionally with n_e. Therefore, the irradiation flux can be controlled by P_d carrying on the constant irradiation energy.

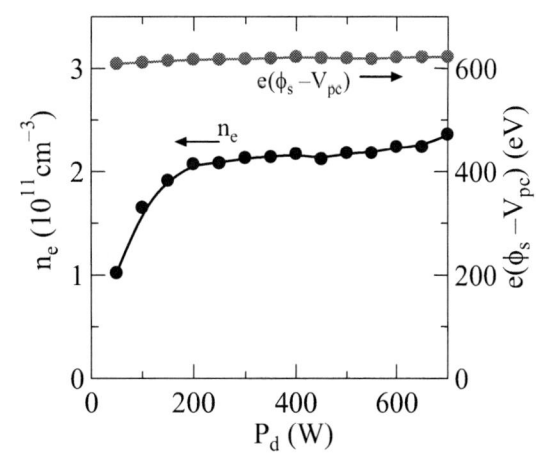

FIGURE 2. Dependence of the electron density and the irradiation energy of positive ions on the discharge power in the discharge region.

Under the irradiation, the pair ions are produced and an ionic plasma is generated in $z > 0$ cm. Typical current (I_p)-voltage (V_p) characteristics of the single probe at $z = 3$ cm are shown in Fig. 3. The I_p-V_p characteristics are symmetrical with respect to the ground voltage and zero current. The plasma potential indicated by a voltage of inflexion point almost coincides with the floating potential. The plasma potential is almost 0 V which is equal to the chamber wall. The positive- and negative-saturation currents of the characteristics are shown to be approximately equal. If electrons measurably exist in the plasma, the negative current would be higher than the positive current. Thus the existence of electrons can be negligible.

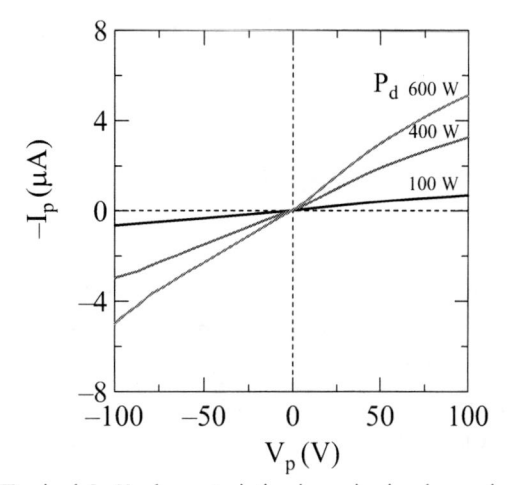

FIGURE 3. Typical I_p-V_p characteristics in an ionic plasma in $z > 0$ cm.

The positive- and negative-saturation currents of the probe, I_+ and I_-, are obtained at probe bias voltages of $V_p = -120$ V and $+120$ V, respectively. $I_\pm$ and the irradiation current I_{ir} depending on P_d are shown in Fig. 4. Since the porous catalyst and the thermal cathode are biased at $V_{pc} = -600$ V and $V_d = -70$ V, respectively, electrons cannot reach to the porous catalyst and positive ions are only irradiated. I_- appears to be the only negative-ion current. The current ratio I_-/I_+ is close to 1, indicating an ionic plasma comprising only H^+ and H^- ions without electrons. Both $I_\pm$ increase proportionally with I_{ir}; that is, the number of pair ions produced increases proportionally with the irradiation flux. The production quantity of H^+ ions from the catalyst surface will be bigger than that of H^- ions in the case of using the Ni porous catalyst, but the obtained $I_\pm$ are almost the same. The flow of H^+ ions toward downstream appears to be restricted to maintain quasi-neutrality in the plasma by the potential structure formed in front of the catalyst. The production balance between H^+ and H^- ions is mainly attributable to the catalyst material. In solid state properties, electronegativity describes the ability of an atom to attract electrons toward itself, which has an influence on desorption ionization. The electronegativities of Ni and H are 1.8 and 2.2 in Pauling units, respectively. In order to produce H^- ions more, the electronegativity of catalyst had better be lower.

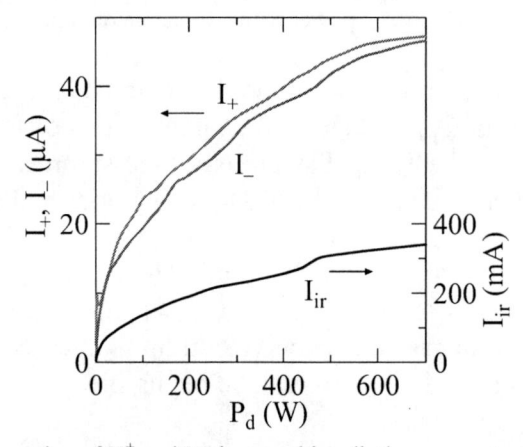

FIGURE 4. Production properties of H^+ and H^- ions and irradiation current of positive ions as function of the discharge power.

SUMMARY

A Ni porous catalyst is used to produce hydrogen atomic pair ions, i.e., H^+ and H^- ions. When hydrogen positive ions produced by a dc arc discharge are irradiated to the porous catalyst, the pair ions are produced from the back of the irradiation plane. We refer to the production mechanism as catalytic ionization. The irradiation flux is controlled by the discharge power maintaining the constant irradiation energy. The number of pair ions produced increases proportionally with the irradiation flux.

ACKNOWLEDGMENTS

The authors thank Dr. K. Tsumori and Dr. Y. Takeiri for their collaboration and Dr. O. Fukumasa for his encouragement. This work is partially supported by NIFS09KKMB004.

REFERENCES

1. J. Ishikawa, "Negative Ion Sources", in *The Physics and Technology of Ion Sources, Second Edition*, edited by Ian G. Brown, WILEY-VCH Verlag GmbH & Co. KGaA, Weinheim, 2004, pp. 285-310.
2. G. D. Alton, *Surf. Sci.* **175**, 226-240 (1986).
3. K. N. Leung and K. W. Ehlers, *Rev. Sci. Instrum.* **53**, 803-809 (1980).
4. Y. Okumura, M. Hanada, T. Inoue, H. Kojima, Y. Matsuda, Y. Ohara, Y. Oohara, M. Seki, Y. Suzuki, and K. Watanabe, *Proceedings of the 16th Symposium on Fusion Technology*, Amsterdam, 1991, **2**, p. 1026.
5. M. Bacal and G. W. Hamilton, *Phys. Rev. Lett.* **42**, 1538-1540 (1979).
6. O. Fukumasa and S. Mori, *Nucl. Fusion* **46**, S287-S296 (2006).
7. W. Oohara and R. Hatakeyama, *Phys. Rev. Lett.* **91**, 205005-1-4 (2003).
8. W. Oohara, D. Date, and R. Hatakeyama, *Phys. Rev. Lett.* **95**, 175003-1-4 (2005).
9. W. Oohara and R. Hatakeyama, *Phys. Plasmas* **14**, 055704-1-7 (2007).
10. W. Oohara and O. Fukumasa, *J. Plasma Fusion Res. SERIES* **8**, 860-864 (2009).
11. W. Oohara and O. Fukumasa, *Rev. Sci. Instrum.* **81**, 023507-1-6 (2010).

BEAM FORMATION, ACCELERATIONS, NEUTRALIZATION AND TRANSPORT

Application of 3D Code IBSimu for Designing an H⁻/D⁻ Extraction System for the Texas A&M Facility Upgrade

T. Kalvas[a], O Tarvainen[a], H. Clark[b], J. Brinkley[b] and J. Ärje[a]

[a]*Department of Physics, University of Jyväskylä, Jyväskylä, 40500, Finland*
[b]*Texas A&M University, Cyclotron Institute, College Station, TX 77843, USA*

Abstract. A three dimensional ion optical code IBSimu is being developed at the University of Jyväskylä. So far the plasma modelling of the code has been restricted to positive ion extraction systems, but now a negative ion plasma extraction model has been added. The plasma model has been successfully validated with simulations of the Spallation Neutron Source (SNS) ion source extraction both in cylindrical symmetry and in full 3D, also modelling electron beam dumping and ion beam tilt. A filament-driven multicusp ion source has been installed at the Texas A&M University Cyclotron Institute for production of H⁻ and D⁻ beams as a part of the facility upgrade. The light ion beams, produced by the ion source, are accelerated with the K150 cyclotron for production and reacceleration of rare isotopes. The extraction system for the ion source was designed with IBSimu. The extraction features a water-cooled puller electrode with a permanent magnet dipole field for dumping the co-extracted electrons and a decelerating Einzel lens for adjusting the beam focusing for further beam transport. The ion source and the puller electrode are tilted at 4 degree angle with respect to the beam line. The extraction system can handle H⁻ and D⁻ beams with final beam energies from 5 keV to 15 keV using the same geometry, only adjusting the electrode voltages. So far, 24 µA of H⁻ and 15 µA of D⁻ have been extracted from the cyclotron.

Keywords: ion source, negative hydrogen, plasma sheath, simulation, ion beam, cyclotron
PACS: 07.77.Ka, 41.20.Cv, 41.75.Cn, 52.40.Kh, 52.65.-y

INTRODUCTION

The use of computer simulations has become a standard procedure of designing ion optical systems. Many specialized codes exist with capabilities for modelling positive ion plasma extraction problems in two [1,2] and three dimensions [3]. For negative ion plasma extraction there exists fewer codes ([4] for example), most of them capable of 2D modelling only. The lack of available three dimensional codes for negative ion extraction causes difficulties because negative ion sources are often equipped with magnetic dipole filter fields, which often protrude to the extraction and furthermore the co-extracted electrons are usually deflected using magnetic fields. The application clearly calls for three dimensional simulation. Fortunately, work has been done to develop simplified models for negative ion plasma extraction [5], which can be extended to 3D.

The Ion Beam Simulator, IBSimu, is a program for three dimensional ion optics [6]. The development of IBSimu, was started at LBNL in 2004 for designing a slit-beam plasma extraction system for a neutron generator [7]. The development of the code has been continued at the University of Jyväskylä, Department of Physics (JYFL) with a drive to making the code modular and suitable for many different types of problems. The code has been published as open source [8] to be used by the community and it has been benchmarked against other codes used for extraction ion optics. IBSimu has been applied to designing an H^+ grid triode extraction, several neutron generator accelerators, slit-beam extraction for diagnostic neutral beams [9], modelling of ECR ion source extraction, designing an $E \times B$ filter for diagnostics and many other problems. The code has only had a plasma sheath model for positive ion extraction, but recently also a negative ion plasma extraction model, which is described in this article, has been developed.

The ongoing facility upgrade for the Texas A&M University Cyclotron Institute aims at extending the reasearch possibilities with stable beams and adding rare ion beam capabilities [10]. This is done by re-activating the K150 cyclotron to deliver high intensity light particle and heavy ion beams. These beams will be used for production of rare isotopes in the targets of light and heavy ion guides for reacceleration with the K500 cyclotron. As a part of the upgrade project, a filament-driven multicusp ion source has been installed for injecting H^- and D^- beams into the K150 cyclotron. The extraction system for the ion source has been designed using IBSimu. This paper reports the development work done for the H^-/D^- project.

SHEATH MODEL FOR NEGATIVE ION EXTRACTION

It is assumed that the extractable negative ions, which are either volume or surface produced, are born in the plasma electrode (wall) potential. These charges form a potential well and counteract the formation of a saddle point at the extraction. The non-existence of the saddle point is supported by the observed good emittance from H^- ion sources. The negative ion plasma extraction model in IBSimu is based on these assumptions and on the existence of an equipotential surface between the bulk plasma and the extraction, where the potential $U = U_W = 0 \text{ V}$. The potential deviates from zero going into the bulk plasma due to the plasma potential and towards the extraction due to the acceleration voltage. This potential structure causes positive ions from the bulk plasma to be accelerated towards the extraction, having energy eU_P at the zero potential. These ions propagate until they are reflected back to the plasma by the increasing potential in the extraction. The potential well acts as a trap for thermal positive ions. The negative ions and electrons are accelerated from the wall potential towards the bulk plasma and more importantly towards the extraction. Schematic view of the potential structure of the negative ion extraction is shown in fig. 1.

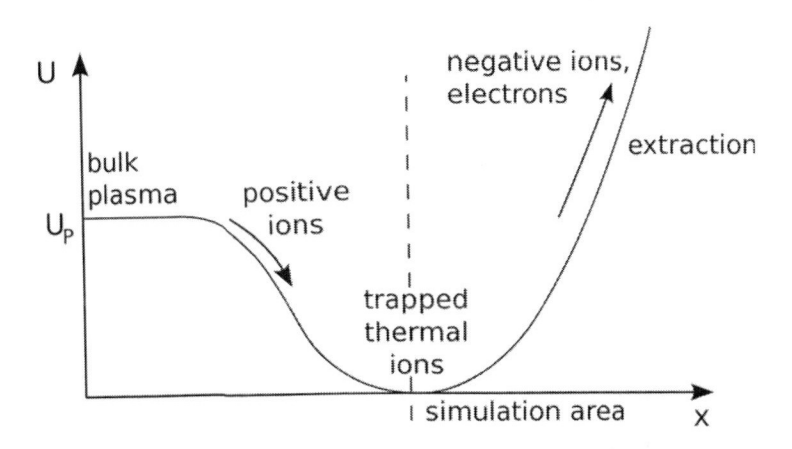

FIGURE 1. Potentials in a negative ion source: Potential drops from positive plasma potential of bulk plasma to wall potential and raises then again going towards the extraction. Simulations model the area starting from wall potential.

The negative ion plasma extraction implementation in IBSimu follows the guidelines of references [4] and [5]. The simulation starts at the plasma electrode potential, where the extracted negative ion and electron beams originate from. The volume between the bulk plasma and this boundary is not simulated, only the flux of directed positive ions is taken in account. The negative particle beams are defined by setting current density, initial drift energy and temperature values. The beams are propagated by standard ray-tracing techniques, using Runge-Kutta integration of the Lorentz force taking into account the electric and magnetic fields. The charge of the beams is deposited on a space charge density mesh during the calculation. The electric field is calculated from a solution of the Poisson equation, using the space charge of the negative beams from the previous round of the so-called Vlasov iteration. Positive space charges are taken into account using analytic formulations presented below. The resulting non-linear Poisson equation is solved using Newton-Raphson iteration. The electric field for the first iteration round of the simulation is acquired by setting zero space charge density and forcing $U=0\,\text{V}$ inside the estimated plasma volume and solving the resulting Laplacian. The Vlasov iteration described here is a standard method for producing self-consistent solutions of space charge dominated problems. The Poisson equation describing the system is

$$\nabla^2 U = \frac{-\rho}{\epsilon_0} , \qquad (1)$$

where $\rho = \rho_{rt} + \rho_f + \rho_{th}$. Here ρ_{rt} is the space charge density of negative particles from ray-tracing, ρ_f is the space charge of fast positive ions and ρ_{th} is the space charge of trapped positive thermal ions.

The model allows several different negative ion species to be extracted from the ion source and also many positive ion species to be used as background plasma. Each of the thermal ion species has a separate Maxwellian velocity distribution with the associated space charge distribution

$$\rho_{th} = \rho_{th0} \exp\left(\frac{-eU}{kT_i}\right) ,\qquad(2)$$

where ρ_{th0} is the space charge density of the thermal ion species at the wall potential and T_i is the corresponding thermal ion temperature. The fast ions are decelerated and turned back to plasma by the extraction voltage. The space charge distribution of the fast ions is defined by the virtual cathode formation and it is

$$\rho_f = \rho_{f0}\left(1 - \frac{eU}{E_i}\right) ,\qquad(3)$$

at $U < E_i$ and zero otherwise. Here ρ_{f0} is the space charge density of fast ions at the wall potential and E_i is the corresponding kinetic energy, which should be around eU_p as the particles are flowing from the bulk plasma. The amount of particles and at least one energy (thermal or directed) must be given by the user for all particle types. The quasi-neutrality of the plasma requires $\rho_{rt} + \rho_f + \rho_{th} = 0$ at $U = 0\,\text{V}$. Otherwise the parameters can be freely selected.

In case of modelling a typical negative ion extraction with a dipole magnetic field in 3D, one problem in the model needs to be addressed: the ray-traced particles will deflect already inside the plasma. This is an unphysical artefact of the simulation and occurs because the particle collisions and cross field diffusion are not being modelled. In reality the ions and electrons are highly collisional and move diffusively until they are accelerated by the extraction electric field. In the simulations this is achieved by suppressing the magnetic field at potentials less than some threshold value given by the user. This threshold defines a boundary between the plasma volume, where ions and electrons are collisional and the extraction volume, where collisions don't happen anymore. In most cases the threshold value should be around 1–20 V as this corresponds to the energy range with typical plasma densities where collisional properties become negligible. The physically correct threshold value is hard to estimate accurately, but it isn't relevant in this context as the ion optics is not very sensitive to the parameter in most cases.

PLASMA MODEL VALIDATION

The plasma model was tested using the widely simulated and thoroughly published extraction of the SNS ion source [11,12]. The simulations were made using published data about the geometry, magnetic field and electrode voltages of the SNS extraction. Therefore the simulated system might not be exactly identical to the existing extraction, but care has been taken to make it as representative as possible. Many simulations were run with different parameters both in cylindrical symmetry and full 3D to be able to compare to different published results. Examples of cylindrical simulations with different plasma densities are shown in fig. 2.

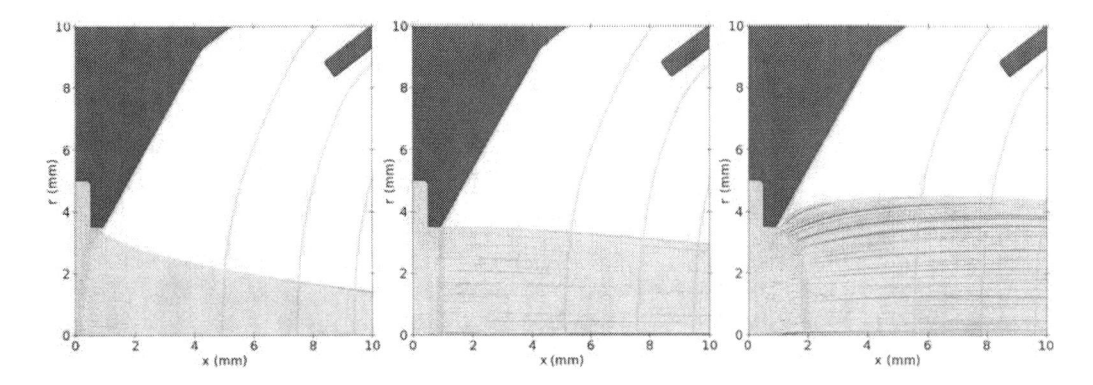

FIGURE 2. Simulations of plasma meniscus formation on the SNS ion source with different plasma densities. In the center the current density J=50 mA/cm^2, which is optimal for aberration free extraction with the electrode voltages used. On the left J=20 mA/cm^2, which is too small for optimal extraction and on the right J=100 mA/cm^2, which is too large. Electode voltages were kept at values shown in fig. 3.

The results show some deviation from previously published simulations, but this is to be expected because of slight differences in the problem definition and because the new simulations feature capabilities which don't exist in the previously used plasma odels. The most important new phenomenon modelled is the unsymmetric increase of negative space charge near the plasma meniscus resulting from the electrons, which are deflected by the electron dump magnetic field. This leads to an unsymmetric plasma meniscus, which affects the optimal tilt angle of the ion source. The biggest difference between the experimental and simulated emittances is that the simulated distributions are much more aberration-free (see [11] for experimental emittance plots). This can also be seen in the emittance value. An example of three dimensional simulations is shown in figures 3 and 4.

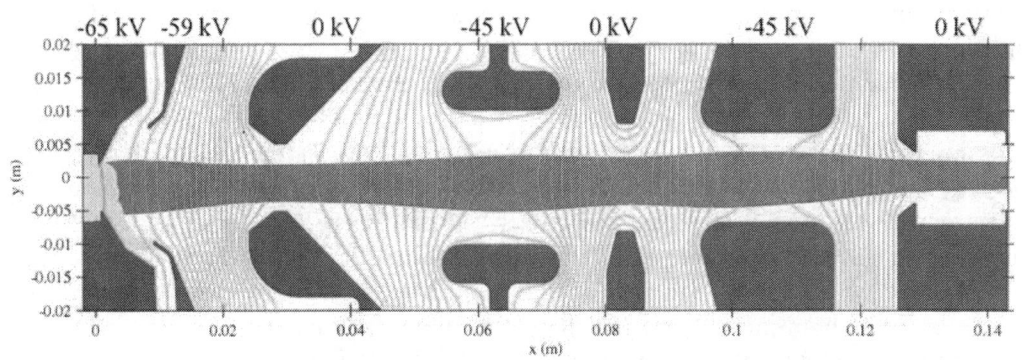

FIGURE 3. Simulation of the SNS ion source extraction in 3D with 38 mA extracted H$^-$ (red) and 230 mA e$^-$ (yellow) deflected by the magnetic filter.

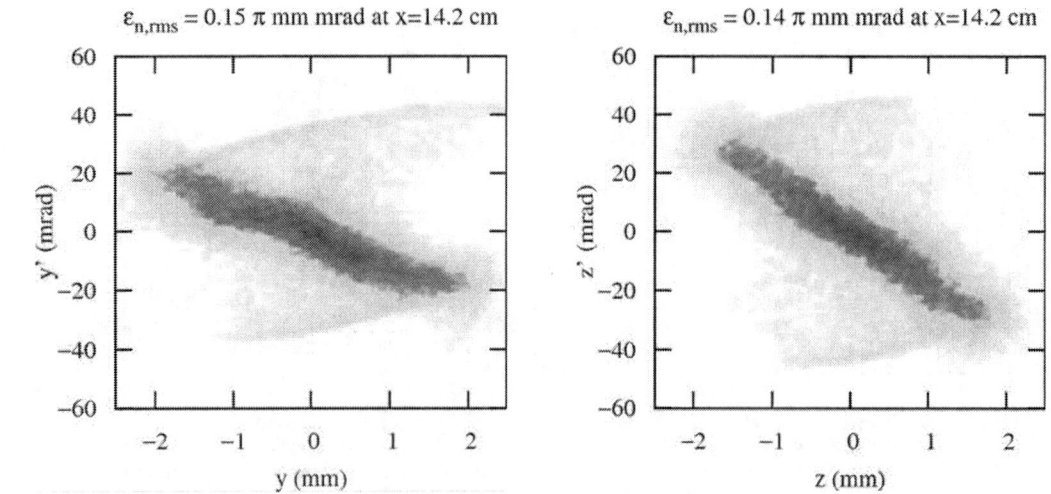

FIGURE 4. Transverse emittance plots from the 3D simulation shown in figure 3.

Overall, the results achieved with the new negative ion extraction model in IBSimu are consistent with earlier studies and experimental observations from the SNS ion source. This suggests that the plasma model is reasonable.

DESIGN OF NEGATIVE ION EXTRACTION

A filament-driven cesium-free multicusp ion source was installed at the Texas A&M University Cyclotron Institute for production of H⁻ and D⁻ beams. The source requires an extraction system capable of extracting up to 1 mA of ion beam and transporting it to the next focusing element of the beam line with low emittance growth. The energy of the ion beam has to be variable from 5 to 15 keV. The extraction system has to be also able to deflect tens of milliamps of co-extracted electron beam into an electron dump. This extraction system was designed using IBSimu.

A dominant feature of a negative ion extraction system is the dumping of the co-extracted electrons. There are several ways of dealing with the electrons. One possibility is to have a transverse magnetic field at the extraction aperture and a separate dumping electrode before the actual puller electrode for dumping the electrons at low energy for decreased power dissipation problems. The ion beam will be deflected by the magnetic field, which is corrected by having the ion source at a small angle with respect to the rest of the beam line [13]. This method is especially beneficial with high-intensity, high-voltage extraction systems.

Another possibility is to have two anti-parallel dipole fields later in the extraction. The first dipole field is used to deflect the electrons to a beam dump and behind this the second dipole field is used to correct the angle of the ion beam. This will cause an offset to the beam axis, which is dependent on the beam energy, but in many cases this is easy to correct mechanically or by using active magnetic elements [14].

In our case, the filter field of the ion source protrudes to the extraction area and thus there will be a tilt in the ion beam regardless of the dumping method. Therefore the tilted ion source method was chosen. The dumping magnetic field was oriented anti-parallel to the filter field to minimize the magnetic field strength at the extraction aperture for minimal interference to the slow particles accelerated from the plasma. The dipole field was constructed using 10 6.35 mm cube SmCo magnets for a maximally flat magnetic field in the transverse direction to minimize emittance growth. The magnetic field peak strength is 32 mT and the FWHM length of the peak is roughly 24 mm. The magnets were integrated in the water cooled puller electrode, engineered to handle the electron beam power.

For the ion beam to tilt to the same angle in the magnetic field regardless of the final energy, the beam energy at the puller electrode has to be fixed. This is done by having the power supply for the puller electrode in the ion source potential. In this case a puller to plasma electrode voltage difference of 6 kV was chosen. After the puller electrode the beam is accelerated to the final energy while the focusing is adjusted using a decelerating Einzel electrode between the puller and the ground electrode. A series of simulations was made in 2D (axially symmetric) to design the extraction electrode geometry and potentials and to check for the sensitivity of simulations to the plasma model parameters. Because of some uncertainty in the plasma parameters and the source performance, the gap between plasma and puller electrodes was made adjustable.

For finding the optimal tilt angle and the center of rotation three dimensional simulations were done. The 3D magnetic field was modelled using Radia3D [15] and the field data was imported into IBSimu. The same plasma model parameters were used as in 2D with the exception of the magnetic field suppression added for potentials $U < 8\,V$. The optimal tilt angle for hydrogen was observed to vary from $4.1°$ to $4.9°$ with changing final beam energy because the magnetic dipole field is still nonzero at the Einzel lens. The variation of the tilt is so small that the tilt was made fixed. The beam deflection can be corrected by a xy steering magnet which is installed on the beam line after the extraction. A simulation of the optimized three dimensional extraction is shown in fig. 5.

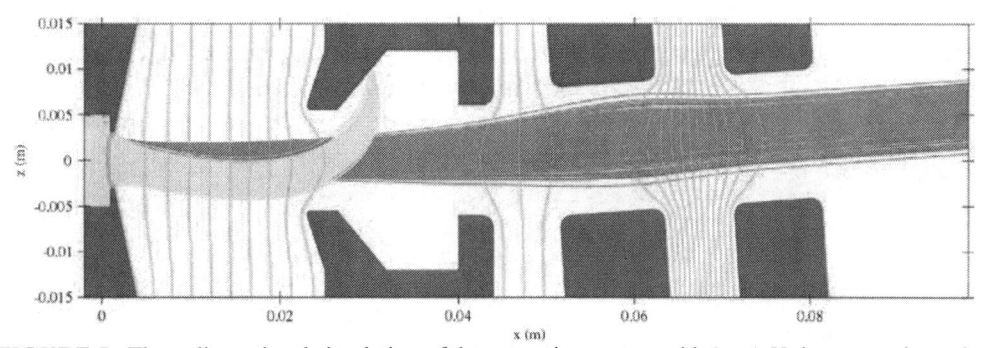

FIGURE 5. Three dimensional simulation of the extraction system with 1 mA H⁻ beam accelerated to 12 keV final energy. Co-extracted 25 mA electron beam is dumped inside the puller electrode. Ion beam exits the simulation at 74 mrad angle.

The extraction system is capable of handling both H⁻ and D⁻ beams with the same electrode configuration using roughly the same voltages. This is possible because the change in beam bending between the ion species isn't proportional to $\sqrt{2}$ as it should if only the mass would change, but it is less than half of this. There are several reasons for this. One of the reasons is that the electron beam is deflected by the residual filter field of the ion source into –z direction (see fig. 6). The electron beam makes a local negative space charge cloud, which pushes the ion beam to +z direction. This effect is magnified with deuterium extraction because the electron-to-ion ratio is higher. Also, the plasma boundary is concave with deuterium because of lower plasma density, which causes the electron beam to deflect more into –z direction amplifying the first effect. This feature makes it possible to change between H⁻ and D⁻ beams easily, minimizing the downtime of the facility. It is also a good example of features that are very difficult to model without three dimensional ion optical codes with plasma modelling capabilities.

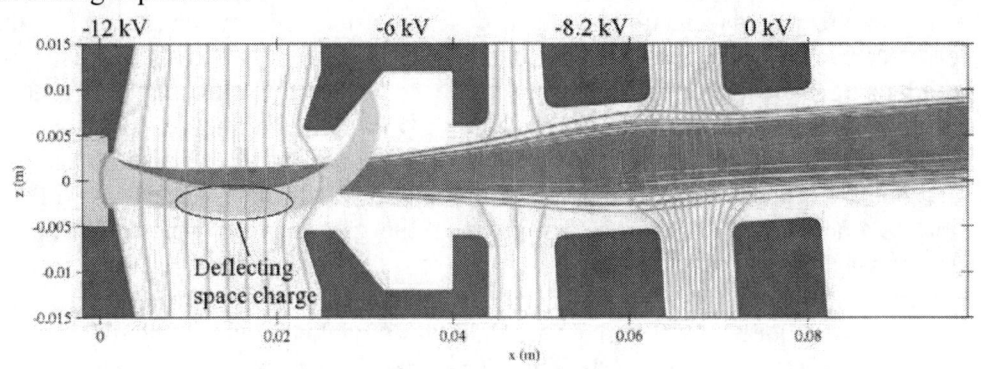

FIGURE 6. Three dimensional simulation with 0.3 mA D⁻ extracted with the same electrode voltages used as with hydrogen. Beam exits the simulation at 66 mrad angle.

EXPERIMENTS

The ion source was installed on the injection line of the K150 cyclotron at Texas A&M. The beam line was equipped with a vacuum chamber for housing the extraction, a fine tuning xy steering magnet and three 1000 l/s turbo pumps (see fig. 7). The chamber has an electron suppressed Faraday cup for measuring ion current. The electron currents are measured from the puller electrode. At the end of the chamber there is an Einzel lens for focusing the beam to the next ion optical element in the injection line, which is shared with an ECR ion source. The cyclotron inflector is located roughly 6 meters from the ion source.

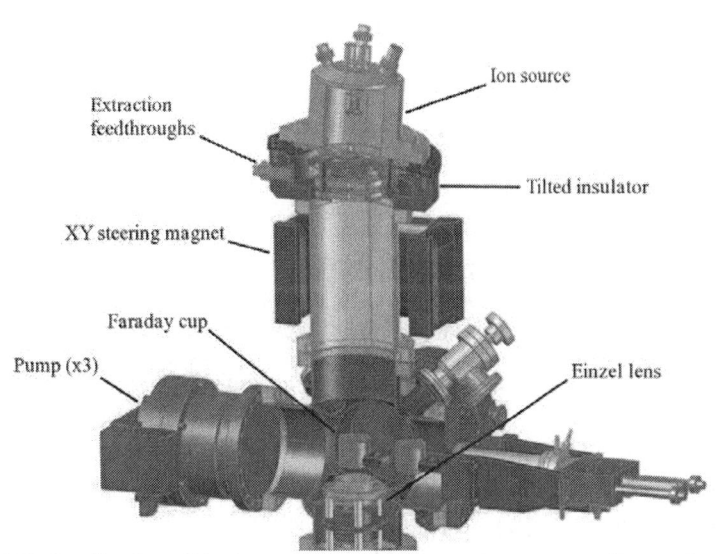

FIGURE 7. CAD visualization of the ion source, extraction system, pumps, diagnostics and the first beam tuning elements. The next Einzel lens used for focusing is roughly 1~m below the extraction chamber.

The extraction of the ion source proved to function as predicted by the simulations made for both H⁻ and D⁻. The optimal transmission to the first Faraday cup was found very close to the simulated electrode voltages. Also the deuterium beam angle was observed to behave as in the simulations. The H⁻ current of 1 mA on the first Faraday cup was reached with an arc current of 12.7 A at the arc voltage of 100 V. The e⁻/H⁻ ratio was around 25. For D⁻ 285 µA was measured on the first cup with 10 A, 100 V arc. The e⁻/D⁻ ratio was about 87. Beam currents of 25 µA of H⁺ and 15 µA of D⁺ have been extracted at energies of 30 MeV and 20 MeV respecively from the cyclotron with stripping extraction and measured from the first Faraday cup outside the cyclotron.

Overall, the experimental work done with the ion source extraction shows that the design process has been successful.

FUTURE WORK

At the University of Jyväskylä we are starting a project for building a filament driven multicusp H⁻ ion source to be used on the pelletron accelerator of the laboratory. During this project there will be possibilities for further experimental validation of the sheath model by traditional diagnostics methods including Faraday cup measurement and beam profile determination with Kapton foils. Further development of the code is planned to be done using this data.

REFERENCES

1. J. E. Boers, *International Conference on Plasma Sciences*, Vancouver, BC, 7-9 June 1993.
2. R. Becker and W. B. Herrmannsfeldt, *Rev. Sci. Instrum.* **63** pp. 2756-2761 (1992).
3. P. Spädtke, *KOBRA3-INP User Manual*, 2000.
4. R. Becker, *Rev. Sci. Instrum.* **75**, 1723 (2004).
5. R. Becker, *10th Int. Symp. on Prod. and Neutralization of Neg. Ions and Beams*, AIP Conference Proc. 763, 194 (2005).
6. T. Kalvas, et al., *Rev. Sci. Instrum.* **81**, 02B703 (2010).
7. T. Kalvas, et al., *Rev. Sci. Instrum.* **77**, 03B904 (2006).
8. T. Kalvas, *Ion Beam Simulator distribution website*, http://ibsimu.sourceforge.net/, 25 May 2011.
9. J. H. Vainionpaa, et al., *Rev. Sci. Instrum.* **79**, 02C102 (2008).
10. Texas A&M University Cyclotron Institute Upgrade Project Management Plan, Rev 2, April 30, 2009 (unpublished).
11. B. X. Han, et al., *Rev. Sci. Instrum.* **79**, 02B904 (2008).
12. R. Becker, *Rev. Sci. Instrum.* **77**, 03A504 (2006).
13. R. Thomae, et al., *Rev. Sci. Instrum.* **73**, 2016 (2002).
14. T. Kuo, et al., *Rev. Sci. Instrum.* **67**, 1314 (1996).
15. O. Chubar, P. Elleaume, J. Chavanne, *Radia3D — A computer program for calculating static magnetic fields*, http://www.esrf.eu/Accelerators/Groups/InsertionDevices/Software/Radia/.

Improvement of voltage holding and high current beam acceleration by MeV accelerator for ITER NB

M.Taniguchi, M.Kashiwagi, T.Inoue, N.Umeda, K.Watanabe, H.Tobari, M.Dairaku, H.Yamanaka, K.Tsuchida, A.Kojima, M.Hanada and K.Sakamoto

Japan Atomic Energy Agency, Naka Fusion Research Establishment, 801-1 Mukoyama, Naka, 311-0193, Japan

Abstract. Voltage holding of -1 MV is an essential issue in development of a multi-aperture multi-grid (MAMuG) negative ion accelerator, of which target is to accelerate 200 A/m^2 H$^-$ ion beam up to the energy of 1 MeV for several tens seconds. Review of voltage holding results ever obtained with various geometries of the accelerators showed that the voltage holding capability was about a half of designed value based on the experiment obtained from ideal small electrode. This is considered due to local electric field concentration in the accelerators, such as edge and steps between multi-aperture grids and its support structures. Based on the detailed investigation with electric field analysis, accelerator was modified to reduce the electric field concentration by reshaping the support structures and expanding the gap length between the grid supports. After the modifications, the accelerator succeeded in sustaining -1 MV for more than one hour in vacuum. Improvement of the voltage holding characteristics progressed the energy and current accelerated by the MeV accelerator. Up to 2010, beam parameters achieved by the MAMuG accelerator were increased to 879 keV, 0.36 A (157 A/m^2) at perveance matched condition and 937 keV, 0.33 A (144 A/m^2) slightly under perveance.

Keywords: Neutral Beam Injector (NBI), Negative ion accelerator, Vacuum insulation
PACS: 41.75.Cn

INTRODUCTION

The neutral beam (NB) injection system for ITER is required to inject 16.5 MW of D^0 beams per one injector at the energy of 1 MeV for heating and current drive of the fusion plasma [1]. To realize such a high power injector, development of negative ion accelerator which can generate 1 MeV, 200 A/m^2 D$^-$ ion beams for 3600 s is necessary. For this purpose, Japan Atomic Energy Agency (JAEA) has developed a proof-of-principle accelerator called "MeV accelerator" at MeV test facility [2-4]. A target of the MeV accelerator is to demonstrate 1 MeV H$^-$ ion acceleration at the current density level of 200 A/m^2 for several tens seconds.

The current and energy achieved in the MeV accelerator was 796 keV, 320 mA (140 A/m^2) in 2007 [5]. These parameters were limited due to the voltage holding of the accelerator during the beam acceleration. To fulfill the ITER requirement (1 MeV, 200 A/m^2), further improvement of the voltage holding capability was strongly desired.

In the present work, voltage holding characteristics of the MeV accelerator was examined. It was found that the local electric field concentration at the grid and/or grid support structure caused the breakdowns in the accelerator. According to the experimental results and electric field analysis, MeV accelerator was modified in order to improve the voltage holding capability. As the results, MeV accelerator succeeded in sustaining -1 MV in vacuum and the energy and current of H$^-$ ion beam accelerated by MeV accelerator was increased to 937 keV, 330 mA levels. This paper reports the recent progress of the voltage holding study in MeV accelerator and the modification of MeV accelerator.

MEV ACCELERATOR

Figure 1 (a) shows a cross sectional illustration of the MeV accelerator. In the ITER NBI, accelerator was designed to hold high voltage of -1 MV in vacuum to prevent the radiation induced conductivity of insulation gas [6]. To simulate this vacuum insulation in ITER, the MeV accelerator is installed inside vacuum chamber composed of a stack of five FRP (Fiber reinforced plastic) insulators. To suppress the surface flashover along the FRP surface, specially designed large stress ring has been installed [7].

Hydrogen negative ion is produced by the cesium seeded hydrogen arc discharge

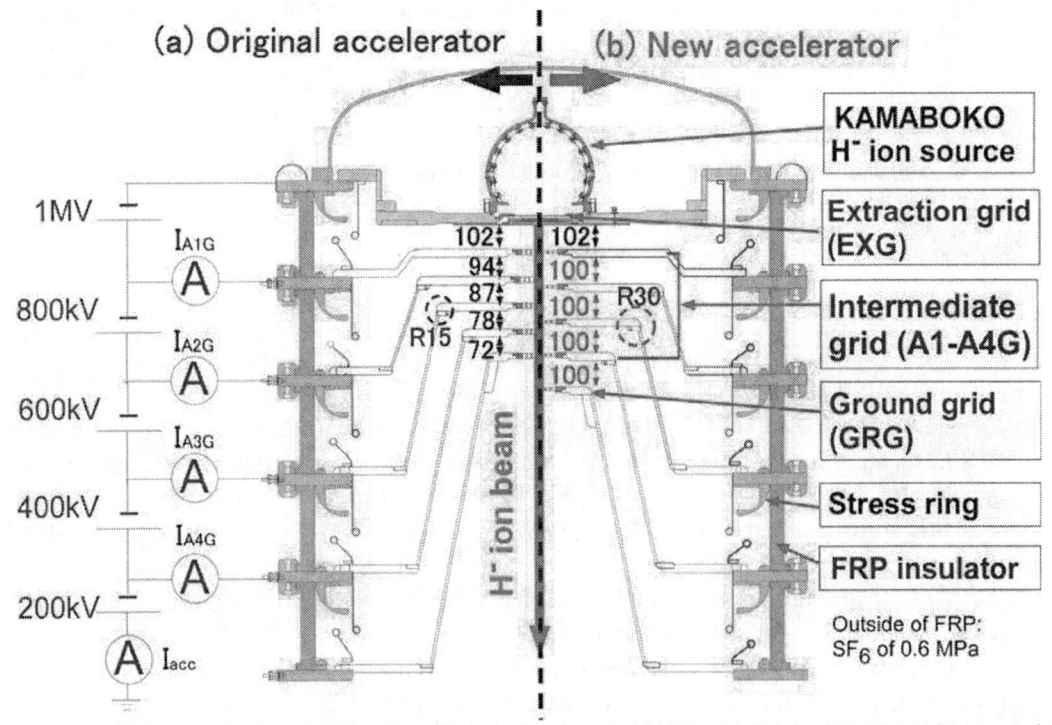

FIGURE 1. Fig.1 Schematics of the MeV accelerator developed at JAEA. (a) Original MeV accelerator and (b) New accelerator

in a KAMABOKO ion source [8], which is mounted on the top of the accelerator. The produced H⁻ ions are extracted through 3 x 5 apertures by potential difference of several kilo volts applied between the plasma grid (PG) and the extraction grid (EXG). The extracted H⁻ ions are accelerated by the five stage electrostatic accelerator composed of four intermediate grids (A1G – A4G) and the grounded grid. Between each grid, 200 kV is applied and the H⁻ ions are accelerated up to 1 MeV by five stages. The minimum gap lengths between the grid supports of each stage were 102/94/87/78/72 mm, respectively.

Accelerated negative ion current was measured by calorimetry using a copper plate equipped with thermocouples at 2.3 m downstream of the accelerator. Beam profile was estimated by using the 1D-carbon fiber composite (CFC) target located at 2.5 m downstream of the accelerator. By observing the temperature increase of the target with an IR camera, beam foot print can be observed, and the information on beam profile can be evaluated from the IR image. The residual gas in the accelerator is monitored by the quadruple mass analyzer.

VOLTAGE HOLDING OF THE MEV ACCELERATOR

In the beginning of the MeV accelerator R&D, suppression of the surface flashover along the FRP insulator columns in the vacuum was an issue. To overcome this problem, large stress ring which protects the cathode triple junction (interface of

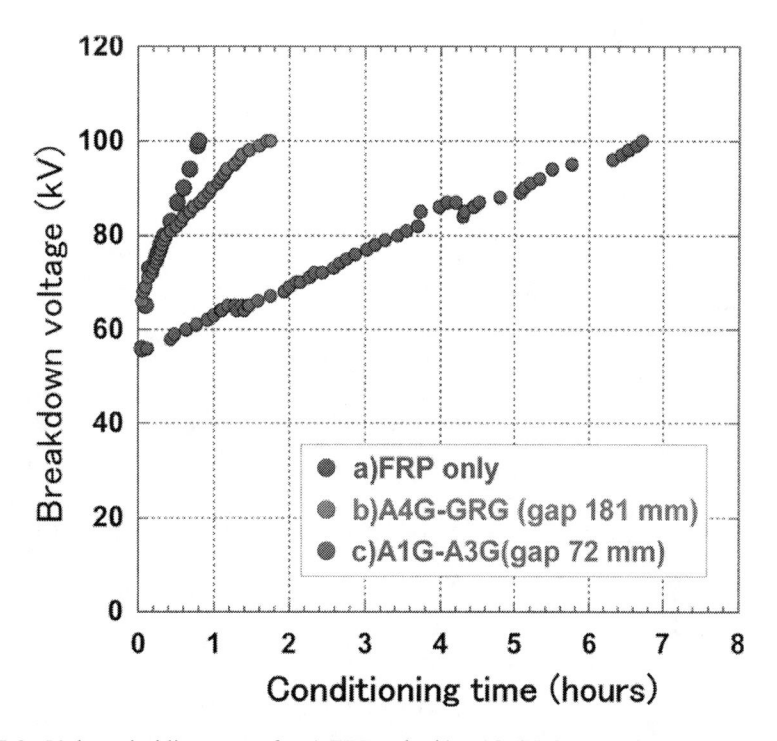

FIGURE 2. Voltage holding curve for a) FRP only, b) A4G-GRG, c) A1G – A3G. In the case of c), A2G was removed.

vacuum, FRP and metal flange) was installed to the MeV accelerator. This stress ring was effective to lower the electric field strength at the triple junction from 3.6 to 1.2 kV/mm. By installing the large stress ring, MeV accelerator succeeded in sustaining 835 kV without gas feeding [7]. Increase of the voltage holding was observed by feeding H_2 gas to the ion source, and -1 MV could be sustained in the pressure range of typical MeV accelerator operation (0.04 – 0.1 Pa). However, the voltage holding during beam acceleration was lower than that without beam acceleration. During the beam acceleration, accelerated H- ions and/or secondary electrons are impinged on the grids to cause the reduction of voltage holding capability. This fact indicates voltage holding higher than -1 MV is required for the stable beam acceleration at 1 MeV, and -1 MV holding without gas feeding was an issue for long years.

For the improvement of voltage holding capability, voltage holding tests for one stage of the accelerator was performed. The effect of gap length on maximum voltage holding was examined by removing some of the intermediated grids of the accelerator. Fig.2 shows the high voltage conditioning curve up to 100 kV for a) one stage of FRP without grid and grid support, b) A1G – A3G (A2G was removed, gap:183 mm), and c) A4G – GRG (gap: 72 mm). In the case of FRP without grid and grid support, increase of the voltage is much faster than with grid and grid support. Conditioning time to reach 100 kV increases with decreasing the gap length between the grids. During the conditioning, evolution of the gas species such as hydrocarbons (mass 44) and hydrogen (mass 2) were observed. The amount of released gas during the high voltage conditioning increases with decreasing the gap length. These facts suggest that the discharge between the grids determines the conditioning time of the accelerator.

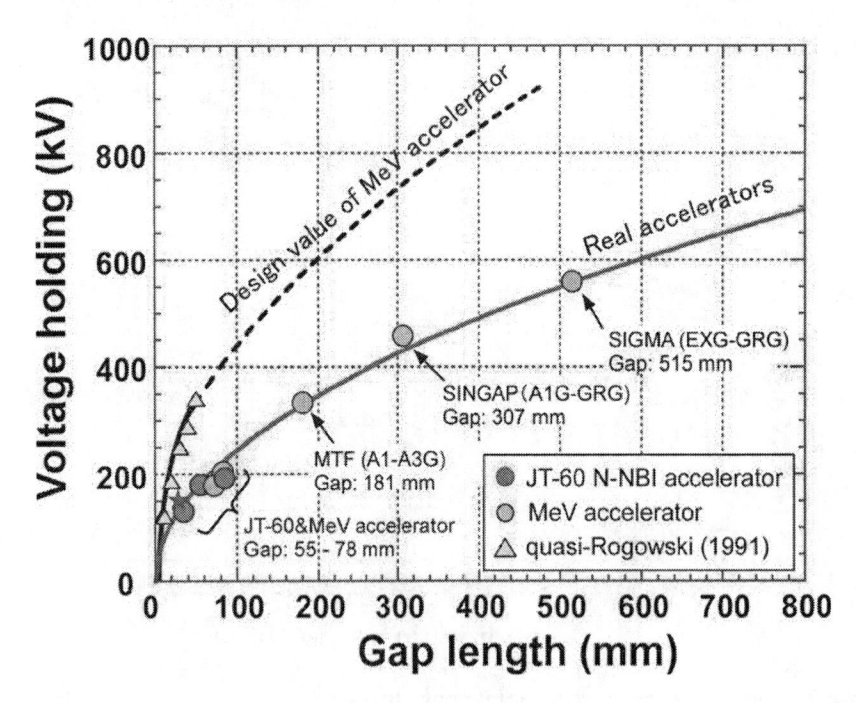

Figure 3 summarizes the dependence of voltage holding on gap length for various

FIGURE 3. Voltage holding v.s. gap length for MeV and JT-60U accelerators. Data of quasi-Rogowski grid is also shown for comparison.

accelerator geometries obtained in the MeV accelerator [9-11] and JT-60U negative ion accelerator [12]. The maximum voltage holding is plotted as a function of the minimum gap length between the grid supports. Although the accelerators have many edge and steps where the electric field concentration is anticipated, the voltage holding capability of the accelerators follows a clump theory [14] of vacuum discharge, that is, the maximum voltage holding is proportional to the square root of the gap length in wide range of 50 - 500 mm. Furthermore, the structure and area of grid and grid support are different, voltage holding capability of MeV accelerator and JT-60U accelerator shows the same dependence on the gap length. This is considered to be because 1) same quasi-Rogowski electrode data shown in fig.3 was used for a design of gap length of both accelerators, and 2) grid and/or grid supports have the same curvatures at the edges, and hence, the local electric field concentration is in the similar level of 4.9 kV/mm at the cathode side.

In fig.3, the data for quasi-Rogowski electrode [13] (160 mm in diameter) is also shown for comparison. The voltages sustained in the accelerators are much lower than those obtained with quasi-Rogowski grid. This could be attributable to the local electric field concentrations in the accelerators, such as those at edge and steps of the support structures. After the high voltage tests, various discharge marks were found between the grid supports. Most of these discharges seemed to start at the edge and the steps of the grid support. These facts suggest that the voltage holding in the accelerators is influenced by the local electric field concentration on the grid and its supports.

MODIFICATION OF THE MEV ACCELERATOR

To improve the voltage holding capability of the MeV accelerator, decrease of electric field concentration by expanding the gap is considered to be effective. The data shown in fig.2 indicates that the minimum gap length between the grid supports (72 mm) is marginal for stable -200 kV holding (-1 MV for five stages). During the beam acceleration, accelerated H⁻ ions and/or secondary electrons are impinged on the

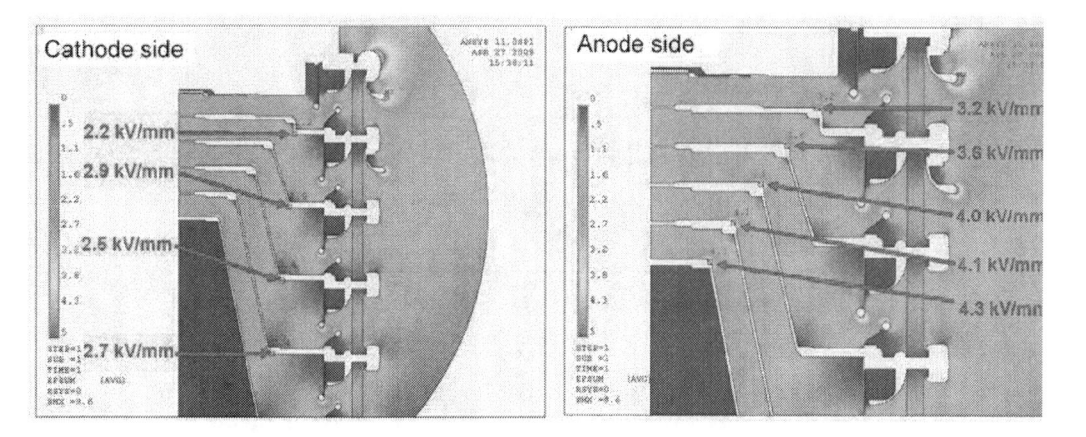

FIGURE 4. Results of electric field analysis of new MeV accelerator.

grids to cause the reduction of voltage holding capability. In fact, the voltage holding of MeV accelerator during the high current beam acceleration was -796 kV although -1 MV could be sustained without beam acceleration. This means 20 % of higher voltage holding is required for stable beam acceleration. To include this 20 % margin, the gap length between the grid supports was expanded to 100 mm (-240 kV can be sustained from the data in fig.2) so as to hold -200 kV stably during the beam acceleration. The effect of gap expansion on the beam optics was examined by the 2D beam trajectory calculation. As the result, it was found that the beam divergence was increased from 3.1 to 3.7 mrad by this gap expansion. This is within the allowable level compared with the requirement for ITER NBI (7 mrad).

Figure 1(b) shows the MeV accelerator after modification. The minimum gap between the grid supports was expanded to 100 mm. Furthermore, curvatures of the edge of the grid supports were increased from R15 to R30 to decrease the electric field concentration. Figure 4 shows the result of electric field analysis of the new MeV accelerator after the modifications. By the above modifications, local electric field at the edge of the grid support was lowered. The maximum electric field concentration at the cathode and anode side was decreased from 4.9 to 2.9 kV/mm and 6.4 to 4.2 kV/mm respectively.

Figure 5 shows the conditioning curve of the MeV accelerator before and after the modifications. The conditioning was performed in the same way before and after the modifications. During the high voltage conditioning, out gassing and subsequent pressure evolution, power supply drain current, intermediated grid current and X-ray

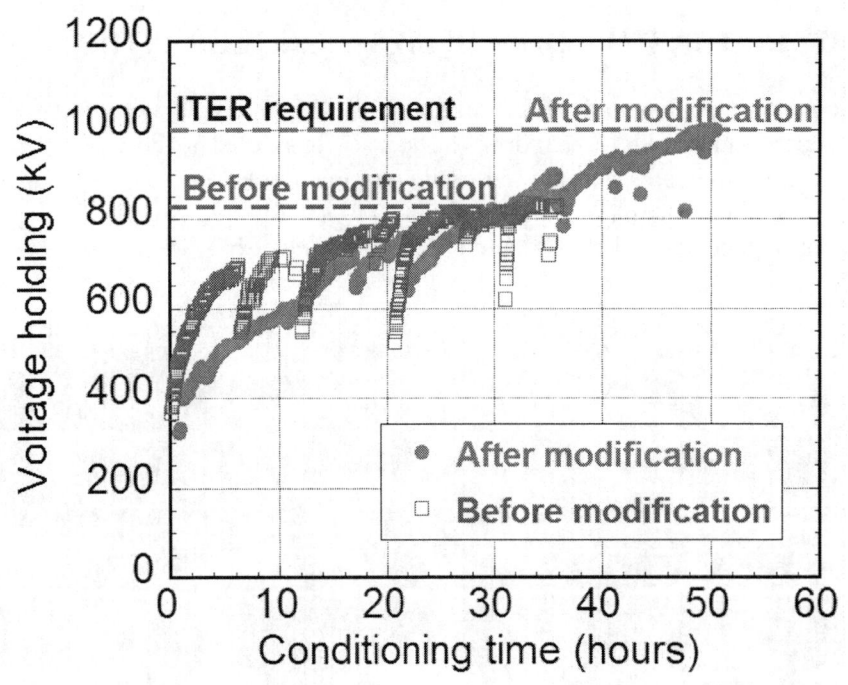

FIGURE 5. Conditioning history of MeV accelerator before and after the modification.

emission were carefully monitored. Typically, when the applied voltage was increased, increases of these values were observed. Applied voltage was increased only when these values decreased to be negligible levels. After the modifications, voltage holding capability was drastically improved. After the 50 hours of conditioning, the MeV accelerator achieved voltage holding of -1 MV, which is a limit of power supply, without gas feeding (base pressure; 2 x 10^{-4} Pa). Voltage holding was very stable and -1 MV was sustained for more than 3600 s without breakdowns.

ACCERLERATION OF HIGH CURRENT H⁻ ION BEAM WITH NEW MEV ACCELERAOTR

High current H⁻ ion beam acceleration test was performed with the new MeV accelerator with seeding Cs into the KAMABOKO negative ion source. Figure 6 shows the result of beam acceleration test with the new accelerator. The voltage holding enhanced the accelerated H⁻ ion current and energy. At the perveance match condition, an 879 keV, 360 mA (157A/m^2) beam was successfully accelerated. Slightly below the perveance but at the comparable power of the above beam, a 937keV, 330 mA (144 A/m^2) beam was obtained. The accelerated H- ion beam power was increased from 0.25 MW to 0.32 MW. Such progress of the beam acceleration performance has been obtained by improvement of voltage holding of the accelerator.

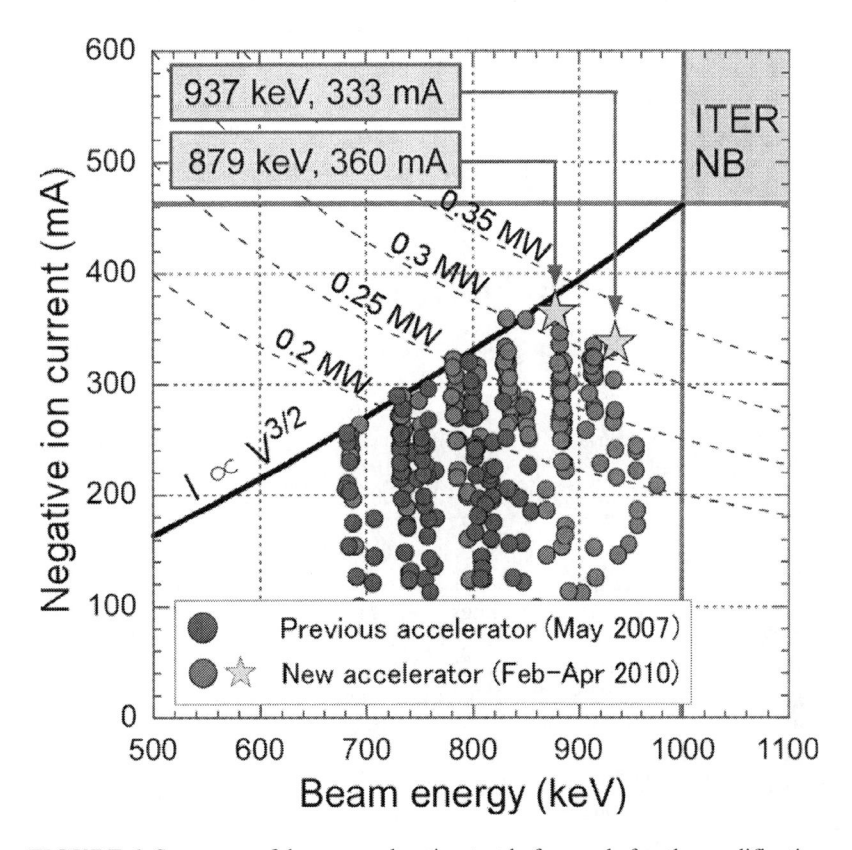

FIGURE 6. Summary of beam acceleration test before and after the modification.

However, the pulse length, as well as the energy and current were limited at these levels due to the high heat load by the direct interception of the beam at the acceleration grids as a consequence of beam deflection. A 3D multi-beamlet analysis indicated that this beam deflection is caused by magnetic field and space charge repulsion between the beamlets [15]. Study on the compensation methods against this beam deflection is now in progress to achieve the target of MeV accelerator of 1 MeV, 463 mA (200 A/m^2).

SUMMARY

To improve the voltage holding of the MeV accelerator, voltage holding test of various accelerator geometries were performed. As a result, it was found that;
1) Voltage holding or the accelerator depends on the clump theory in the wide gap range of 50 – 500 mm.
2) Voltage holding of the accelerator geometries was much lower than that of the small quasi-Rogowski electrode. This is considered to be due to the local electric field concentration at the edge or steps in the grid and its support structure.

According to the above results, MeV accelerator was modified. The gap length between the grids was expanded and the curvature of the grid support was increased to lower the local electric field in the accelerator. As a result, MeV accelerator succeeded in sustaining -1 MV without gas feeding. By the improvement of the voltage holding capability, energy and current accelerated by the MeV accelerator was increased. Up to 2010, the MeV accelerator succeeded in accelerating 879 keV, 360 mA (157A/m^2) and 937keV, 330 mA (144 A/m^2) beam, compared to 796keV, 320 mA in 2007.

REFERENCES

1. ITER EDA Final Design Report, ITER technical basis, Plant Description Document (PDD), G A0 FDR 1 01-07-13 R1.0, IAEA EDA documentation No. 24 (2002).
2. K.Watanabe et al, *Rev. Sci. Instrum.* **73 (2)**, 1090-1092 (2002).
3. T.Inoue, et al., *Nucl. Fusion* **46(6)**, S379-385 (2006).
4. M.Taniguchi et.al., *Rev. Sci. Instrum.*, **81**, 02B101 (2010).
5. M.Taniguchi et.al., *Rev. Sci. Instrum.*, **79**, 02C110 (2008).
6. E. Hodgson et al., *J. Nucl. Mater,* **256-263**, 1827 (1998).
7. T. Inoue et al., *Fusion Eng. Design,* **66-68**, 597-602 (2003).
8. T. Inoue et al., *Fusion Eng. Design,* **55**, 291-301 (2001).
9. M.Taniguchi et.al., AIP conference proceedings 1097, American Institute of Physics, Melville, NY, 2008, pp.335-343.
10. H.P.L.de Esch et.al., *Fusion Eng. Design,* **84**, 669 (2003).
11. K.Watanabe et.al., *JAEA-Tech* 2005-002, (2005) [in Japanese]
12. A. Kojima et al., *Rev. Sci. Instrum.*, **81(2)**, 02B112 (2010).
13. K.Watanabe et.al., *J.Appl.Phys.*, **72**, 3949 (1992).
14. L. Cranberg, *J.Appl. Phys*, **23**, 518(1952).
15. M.Kashiwagi et.al., In this conference

Study of beamlet deflection and its compensations in a MeV accelerator

Mieko Kashiwagi[a], Takashi Inoue[a], Masaki Taniguchi[a], Naotaka Umeda[a], Larry R. Grisham[b], Masayuki Dairaku[a], Jumpei Takemoto[a], Hiroyuki Tobari[a], Kazuki Tsuchida[a], Kazuhiro Watanabe[a], Haruhiko Yamanaka[a] and Keishi Sakamoto[a]

[a] *Japan Atomic Energy Agency (JAEA), 801-1, Mukoyama, Naka 311-0193, Japan.*
[b] *Princeton Univ. Plasma Physics Lab., P.O.Box 451, Princeton, NJ 08543, USA*

Abstract. In a five stage multi-aperture and multi-grid (MAMuG) accelerator in JAEA, beam acceleration tests are in progress toward 1 MeV, 200 A/m^2 H$^-$ ion beams for ITER. The 1 MV voltage holding has been successfully demonstrated for 4,000 s with the accelerator of expanded gap length that lowered local electric field concentrations. This led to increase of the beam energy up to 900 keV-level. However, it was found that beamlets were deflected more in long gaps and direct interceptions of the deflected beamlet caused breakdowns. The beamlet deflection and its compensation methods were studied utilizing a three-dimensional multi beamlet analysis. The analysis showed that the 1 MeV beam can be compensated by a combination of the aperture offset of 0.8 mm applied in the electron suppression (ESG) and the metal bar called a field shaping plate with a thickness of 1 mm attached beneath the ESG. The paper reports these compensation methods and analytical predictions, with experimental results of the MAMuG accelerator in which those compensation techniques have been applied.

Keywords: negative ion, accelerator, ion beam, ITER.
PACS: 29.20.Ba, 29.25.Ni, 29.27.Fh, 52.20.Dq, 52.50.Gj

INTRODUCTION

In the ITER neutral beam heating & current drive (NB H&CD) system, a negative ion accelerator is required to generate deuterium negative ion (D$^-$) beams of 1 MeV, 40 A at the current density of 200 A/m^2 for 3,600 s [1]. To fulfill the requirement, R&D of a negative ion accelerator, called the "MeV accelerator" [2, 3] have been carried out at the Japan Atomic Energy Agency (JAEA) as the Japan domestic agency (JADA) for ITER project. A target of the MeV accelerator is to accelerate 0.5 A (200 A/m^2) H$^-$ ion beam at the energy of 1 MeV for several tens of seconds.

In the ITER EDA (Engineering Design Activity) phase, two accelerator concepts were proposed and designed, one was a multi-aperture multi-grid (MAMuG) accelerator [4], and the other was a single gap, single aperture (SINGAP) accelerator [5]. In 2007-2008, direct comparison of these accelerator concepts have been carried out under collaboration among JAEA, ITER Organization and CEA Cadarache. As the test result the MAMuG showed better voltage holding capability and suppression of electron acceleration, followed by higher beam energy and current by the MAMuG. According to the result of this comparison, the MAMuG accelerator has been chosen as a baseline design for the ITER NB H&CD accelerator [6-7].

The MeV accelerator is a MAMuG type with five acceleration stages, each with 5×3 apertures in intermediate grids of each stage. In a first attempt of long pulse test of the MeV accelerator with water-cooled grids, 600 keV, 161 mA beam was accelerated for 10 s [8, 9]. By this long pulse beam acceleration tests, two issues were revealed. One was insufficient voltage holding capability in the accelerator [10]. The other was excess heat load on the intermediate grids, in particular on the grounded grids. In order to understand source of the grid heat load and probably resulting voltage breakdowns, a three dimensional model of 10×5 beamlets [11] were modified to analyze the present configuration of the MeV accelerator with 5×3 beamlets. As the result, it was shown that the beamlet was deflected 11 mrad at 600 keV and partly intercepted at the grids causing excessive heat loads to melt the grids [12]. Such large beamlet deflection was caused by superposition of the beamlet deflections due to i) magnetic field and ii) space charge repulsion among beamlets.

This paper describes the magnetic deflection and beamlet repulsion of MeV class energy negative ion beams. Three-dimensional (3D) beam analyses have been developed and utilized to simulate and characterize the beam deflections of the MeV accelerator. Aperture displacement [13] and so called "field shaping plate [14]" are introduced to compensate the beamlet deflection in the MeV accelerator. The compensation and resulting increase in the beam energy and current is also reported in the paper.

MeV ACCELERATOR

A cross sectional view of the MeV accelerator is shown in Fig.1. The H⁻ ions are produced in the KAMABOKO ion source [15, 16] directly mounted on the top of the accelerator. The extractor consists of the plasma grid (PG), the extraction grid (EXG) and the electron suppression grid (ESG). The detail of the extractor is shown in Fig.2 with the numerical model. The produced H⁻ ions are extracted through apertures (each 14 mm in diameter) drilled in a lattice pattern of 5×3 in the PG. Here coordinate of each aperture is defined as (R_1-R_5) in the row and (C_1-C_3) in the column depending on position of each aperture as shown in Fig.2. The beams emerging from the extractor are injected immediately into the MeV accelerator where aperture diameter is 16 mm each. The accelerator consists of four intermediate acceleration grids (A1G-A4G) and the grounded grid (GRG). The applied voltage between each grid is 200 kV at the maximum and 1 MV by the stack of five stages. Each acceleration grid and the GRG are supported by Fiber-Reinforced-Plastic (FRP) insulator columns. To suppress the surface flashover along the FRP surface, specially designed large stress rings have been installed [17]. It was confirmed that the voltage holding capability of the FRP insulator is more than 300 kV in one stage by installing the large stress ring. Thus it

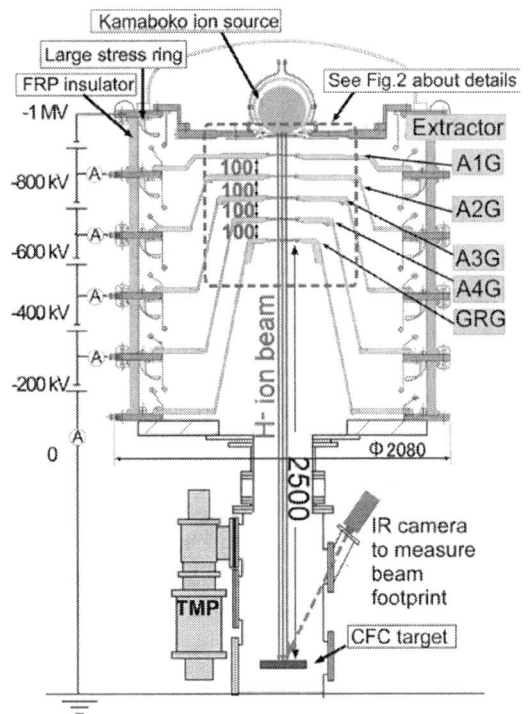

FIGURE 1. A cross sectional illustration of the MeV accelerator which hold 1 MV stably even in high vacuum with extended gap length of 100 mm in each stage.

FIGURE 2. A numerical model of the MeV accelerator in OPERA-3d. Details of the extractor are also illustrated.

should be noted that each stage of the accelerator column can sustain 1.5 times higher voltage than the rated voltage (200 kV). The grid gap length in each stage was adjusted to 100 mm to hold 240 kV in one stage including 20 % margin. This gap length was decided from past results of the MeV accelerator tested with various gap length and voltage [10]. The beam footprints were observed on a one-dimensional carbon fiber composite (CFC) plate utilizing an infrared (IR) camera. The beam current was measured by an inertia-cooled calorimeter located 2.5 m downstream of the GRG in a short pulse (≤ 0.2 s) operation. The current and energy achieved in the MeV accelerator was 796 keV, 0.32 A (140 A/m^2) in 2008 [6].

ANALYTICAL DESCRIPTION OF THE MODEL

A three-dimensional (3D) multi beamlet analysis using OPERA-3d code [18] was applied to study the beamlet deflections. Fig. 2 shows a numerical model of the MeV accelerator in OPERA-3d. The model includes the following details:

Ion emitter

Although negative ions can be extracted very easily from plasma in a similar manner to that of positive ions, physics of negative ion extraction from plasmas is not yet fully understood. And hence, many analytical efforts have been done with assumptions, such as emission of negative ions from fixed emitter. In some cases computer codes for positive ion extraction/acceleration have been utilized and the results were compared and discussed with experimental [19]. Here we have utilized a two-dimensional positive ion trajectory code "BEAMORBT [20]" to define the emitter of the negative ions. Namely, ion position, energy and directions were recorded at the exit of the aperture of PG from output result of BEAMORBT, and then utilized as a set of input data for OPERA-3d, defining a new emitter at the same position preserving the beam parameters of the BEAMORBT.

Magnetic fields

The model includes magnetic field from magnetic filter of the KAMABOKO source. The KAMABOKO source has a pair of large permanent magnets just outside of the aperture area on PG. These magnets were modeled to simulate magnetic field penetration in the extractor/accelerator. The model also included small magnetic field formed by small permanent magnets embedded in the extraction grid (second grid, EXG). As shown in top of Fig. 2, right hand side, small permanent magnets were embedded in EXG to form a transverse magnetic field above and below each aperture of EXG. It should be noted that by changing polarity of each magnet row embedded between aperture rows, the direction of magnetic field lines, and hence, direction of beam deflections changes line by line.

Space charge of beam particles

Of course, the code analyzes the beam trajectory taking into account the space charge of each beam ions. This in turn results in beamlet deflection due to space cherge repulsion between each beamlet.

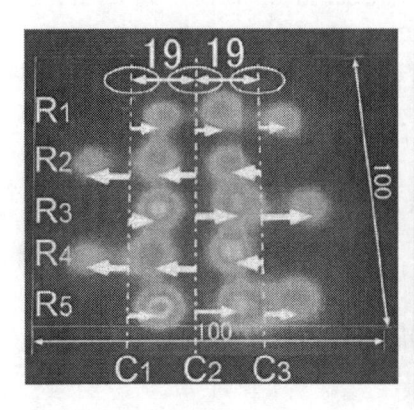

FIGURE 3. A typical measured beam footprint extracted from 5×3 aperture array and accelerated in the MeV accelerator. Beam energy: 500 keV with aperture offset of 0.5 mm at ESG and FSP of 1mm thick.

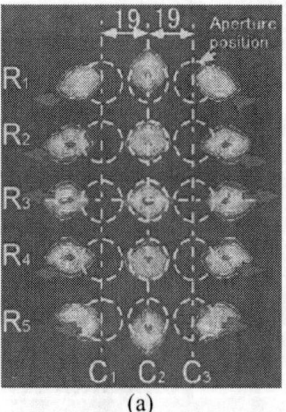

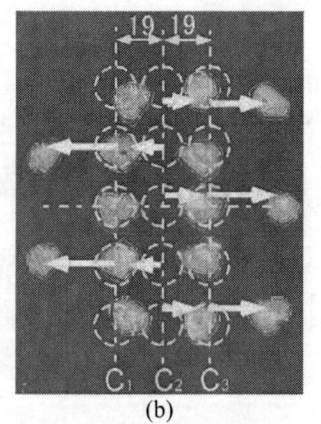

FIGURE 4. Calculated beam footprint at beam energy of 1 MeV. (a) space charge repulsion and (b) magnetic deflection.

Geometry of the accelerator

So far it has been clarified in a careful investigation of the previous results of OPERA-3d analyses that the geometries of the extractor and the accelerator, in particular, steps, edges and corners around the grid and even grid support structures could have an impact on the beam trajectory. Thus the whole geometry of the grids and grid support structures were modeled in the current OPERA analyses as shown in Fig.2, in order to simulate as much as possible the accurate electric field profile in the accelerator.

Stripping loss

The new version of OPERA-3d (version 13) allows analyzing the beam trajectory with reducing the current in beam along the beam axis. This is one of crucial function to be taken into account to simulate the space charge repulsion of beamlets with the reducing space charge properly to reflect the actual loss of space charge in the course of acceleration.

BEAM DEFLECTION AND COMPENSATION

Figure 3 shows a typical measured beam footprint. The beamlets were deflected in the ±X directions alternatively in each row due to i) the magnetic field. The beamlets were also deflected outwards of the aperture array due to ii) the space charge repulsion. As a result of the beamlet deflections, intensity of outermost beamlets, namely R_2C_1,

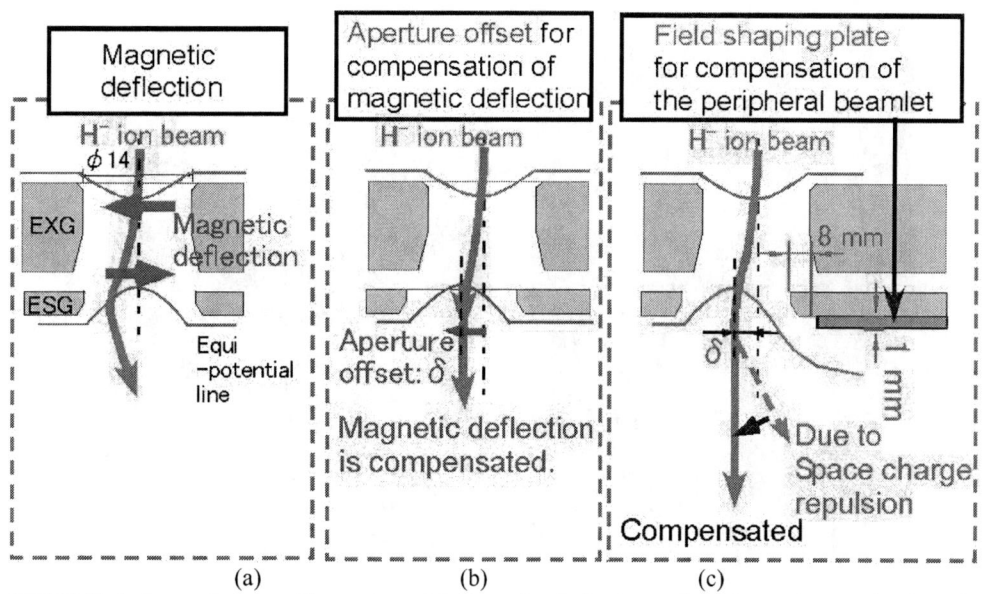

FIGURE 5. An explanatory illustration of the beamlet deflection and compensation mechanisms in a cross sectional view of the apertures of the R_3 row in the extractor. (a) magnetic deflection by magnets embedded in EXG, (b) Aperture offset to compensate the magnetic deflection, and (c) Field shaping plate to make intentional distortion of the electric field.

R_4C_1, R_1C_3, R_3C_3, R_5C_3 beamlets, became weaker. This is due to direct interception of the beamlets at grid apertures by excessive deflection angle by superposition of the deflections i) and ii).

This has been reproduced by the OPERA analysis as shown in Fig.4. Figure 4 (a) was obtained without magnetic field in the EXG. The beamlets extracted from aperture array of lattice pattern repelled each other due to ii) the space charge repulsion, and the outer most beamlets were deflected outward. As a result, outline of the beam footprint at 2.5 m downstream from the GRG looks like a barrel, with larger deflection angle at R_3C_1, R_3C_3, R_1C_2, R_5C_2 beamlets, but without deflection at the center (R_3C_2) beamlet by a balance of space charges from surrounding beamlets.

Figure 4 (b) was obtained turning on the magnetic filed in the analysis of Fig. 4 (a). Then the beamlets were deflected alternately in a horizontal direction in each row due to the magnetic field formed by the magnets in EXG. It is clearly indicated in the figure that the deflection angle of outermost beamlets, such R_3C_3 beamlet is larger (9.5 mrad) than that of center (R_3C_2) beamlet. This is because the superposition of magnetic deflection and space charge repulsion. The calculated deflection angles of the C_1, C_2 and C_3 columns in the R_3 row were 0, 4.7 and 9.5 mrad at 1 MeV, respectively.

Figure 5 shows an explanatory illustration of the beamlet deflections and proposal of compensation methods in a cross sectional view of the aperture in the R_3 row in the extractor. The center beamlet of the C_2 column is deflected only by i) the magnetic deflection as shown in Fig.5 (a). The deflection of the peripheral beamlet in the C_3 column is larger than that of the beamlet in the C_1 and C_2 columns. This is due to

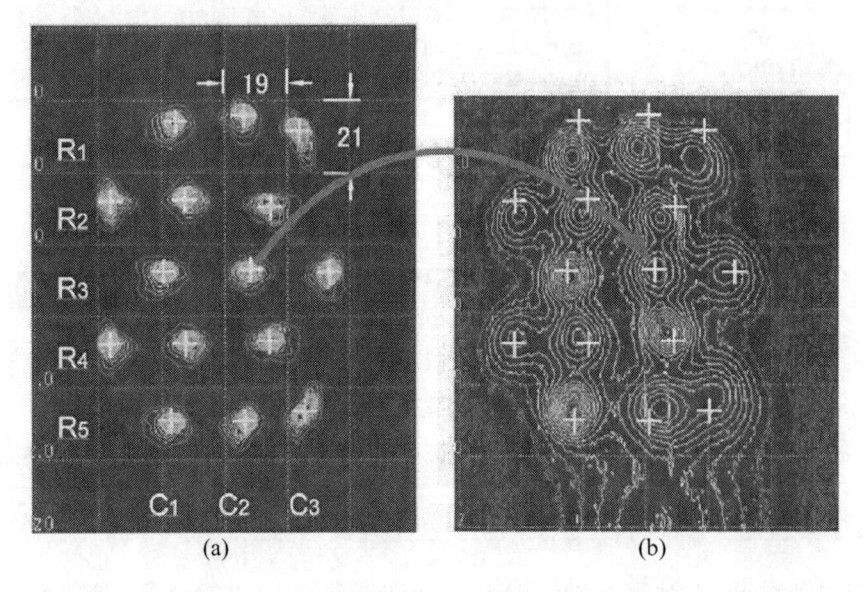

FIGURE 6. Beam footprints in (a) the 3D beam analysis and (b) the experiment. The symbol "+" expressed the beamlet center in the calculation. The aperture offset of 0.5 mm and FSP with 1mm thickness were applied.

superposition of i) and ii): In R$_3$C$_3$ beamlet, the deflections due to i) and ii) are in the same direction. On the contrary, the beamlet deflection of the C$_1$ column is the smallest because the beamlet deflections due to i) and ii) cancel out each other since they are in opposite direction.

To compensate the beamlet deflections, the aperture offset was applied to each row of apertures in alternate directions at the ESG as shown in Fig.5 (b). For the most deflected C$_3$ beamlet, another compensation method was also provided. This is the field shaping plate (FSP), which is a metal plate of 1 mm thick attached at the ESG as shown in Fig.5 (c). It generates electric field distortions, which steer the peripheral beamlets inward.

If the beamlet deflection angle as shown in Fig.4 (b) is to be lowered below 1 mrad only by the aperture offset, it has been estimated that the necessary aperture offsets in the apertures of the C$_2$ and C$_3$ columns in the R$_3$ row are 0.8 mm and 1.8 mm for the beamlet of 1 MeV. However, from two dimensional beam trajectory analysis, it has been predicted that the aperture offset of more than 1 mm would intercept the beamlet at the ESG. Therefore, a conservative aperture offset of 0.5 mm and the field shaping plate were applied in the first campaign of the compensation test in the MeV accelerator. Then in the second campaign, aperture offset of 0.8 mm was applied in the ESG apertures.

The calculated footprint by the 3D beam analysis and the measured one in the experiment are shown in Fig.6 (a) and (b), respectively for the case of beam energy at 700 keV as the typical operational conditions in the first campaign (aperture offset = 0.5 mm). Both figures (a) and (b) show the footprints 2.5 m downstream from the

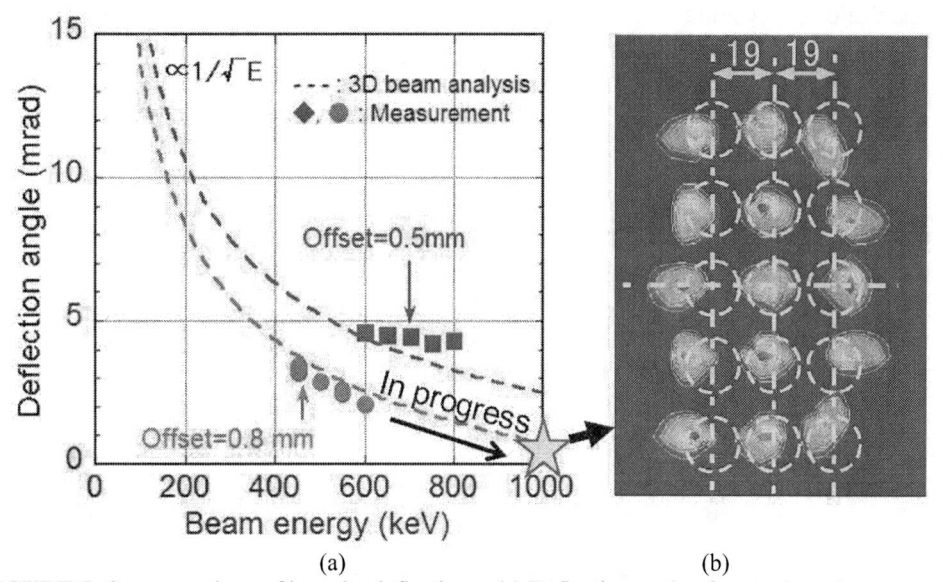

FIGURE 7. Compensations of beamlet deflections. (a) Deflection angle of center beamlet measured on the CFC target and (b) the calculated beam footprint for 1 MeV beam at 2.5 m downstream from the accelerator exit.

GRG exit. In the calculation, the extraction voltage and the current density were adjusted to be the same as the experimental conditions, that is, 4.2 kV and 120 A/m^2, respectively. In Fig.6 (a), the center of calculated beamlet is marked by "+" symbols. These "+" are superimposed on Fig.6 (b) to compare the beamlet positions directly. In horizontal direction, the positions of each beamlet are in good agreement between the calculation and the experiment.

In the second campaign the aperture offset was increased to 0.8 mm in ESG. The results of compensations of the beamlet deflections are summarized in Fig.7 for both the first and second campaigns. The beamlet deflection angles of the center aperture R_3C_2 are shown as a function of the beam energy. When the aperture offset of 0.5 mm was applied in the first campaign, it was expected that the beamlet deflection would be compensated from 4.7 mrad to 2.8 mrad at 1 MeV. The beamlet deflections were changed as expected in the analysis and the beam energy was increased gradually. The beam parameters achieved by the MAMuG accelerator were increased to 879 keV, 0.36 A (157 A/m^2) at perveance matched condition and 937 keV, 0.33 A (144 A/m^2) slightly under perveance from 800 keV-level in the previous test [10].

However, the beam energy was limited up to this level due to breakdowns. The breakdowns were caused by the direct interception of the beamlet because the beamlet deflection still remained. Then, the aperture offset was increased to the proper value of 0.8 mm in the second campaign. As shown in Fig. 7(a), deflection angle with the aperture offset of 0.8 mm was estimated to be 0 mrad at 1 MeV. Fig. 7 (b) shows a beamlet footprint analyzed by the OPERA-3d, which indicates well aligned beamlets in lattice pattern as in the aperture arrays (shown in broken lines) at the beam energy of 1 MeV. In the experiment, the beamlet deflection angle is decreasing according to increase in the beam energy as expected in the 3D beam analysis. The extraction current, which represents the direct interceptions of beams at the EXG and the ESG, was not increased in the first and second campaigns. This indicated that interceptions of beamlet at the ESG could be tolerable as expected in the 2D beam analysis. The second campaign is still in progress to achieve higher beam energy and current simultaneously with the compensation of the beamlet deflections.

SUMMARY

In order to demonstrate stable accelerations of 1 MeV, 200 A/m^2 H$^-$ ion beam, the beamlet deflections and the compensation methods were studied in experiments and three dimensional multi beamlet analyses. It was clarified that the beamlet from the peripheral aperture was most deflected about 9.5 mrad by superposition of the magnetic deflection and the space charge repulsion. As the compensation methods, the aperture offset was applied for all beamlets and the field shaping plate with the thickness of 1 mm was applied additionally for the outermost beamlet from the peripheral apertures. The results are summarized as follows.

- In the first campaign, the beam parameters were increased to 879 keV, 0.36 A (157 A/m^2) at perveance matched condition by applying the compensation

methods (aperture offset = 0. 5 mm). However, the beam energy was limited up to this level due to breakdowns caused by the direct interception of the beamlets because the aperture offset was not enough.
- In the second campaign, the aperture offset was increased to 0.8 mm. In the 3D beam analysis, it was demonstrated that the aperture offset of 0.8 mm would be enough for compensation of the deflection avoiding direct interception of the beamlets at ESG due ot the aperture offset. In the experiment, it has been demonstrated that the beamlet deflections were compensated as designed. At present, the beam acceleration test is progressed toward 1 MeV.

ACKNOWLEDGMENTS

The authors would like to acknowledge the late Dr. T. Tsunematsu, who was a leader of the ITER project in Japan for a long time.

REFERENCES

1. *ITER Technical Basis*, ITER Engineering Design Activities (EDA) Document Series No.24, IAEA Vienna (2002).
2. K. Watanabe et al, *Rev. Sci. Instrum.* 73 (2), 1090-1092 (2002).
3. T. Inoue, et al., *Nucl. Fusion* 46(6), S379-385 (2006).
4. T. Inoue, et al., Joint meeting "Production and neutralization of negative ions and beams" Seventh Int. Symp. "Production and application of light negative ions" Sixth European Workshop, Upton NY, 1995, *AIP Conf. Proc.* No.380, 397-405 (1996).
5. M. Fumelli, F Jequier, J. Pamela, A. Simonin, *Particles and fields* series 53 "Production and neutralization of negative ions and beams" Sixth Int. Symp., Upton NY, 1992, *AIP Conf. Proc.* No.287, 934-949 (1992).
6. M. Kashiwagi et al., *Nucl. Fusion* 49, 065008 (2009).
7. H.P.L. de Esch, et al., *Fusion Eng. Design*, 84, 2-6, 669-675 (2009).
8. M. Taniguchi et al., *Rev. Sci. Instrum.*, 81, 02B101 (2010).
9. N. Umeda et al., *J. Plasma Fusion Res.* SERIES, Vol. 9, 259-263 (2010).
10. M. Taniguchi et al., to be submitted in *Proc. of 2nd Int. Symp. on Negative Ions, Beams and Sources* 2010.
11. M. Kashiwagi et al., "1st Int. Symp. on Negative ions, Beams and Sources 2008", *AIP Conf. Proc.* 1097, 421-430 (2008).
12. M. Kashiwagi et al., *Plasma Fusion Research*, 5, S2097 (2010).
13. T. Inoue et al., *JAERI-Tech*.2000-051(2000).
14. M. Kamada et al., *Rev. Sci. Instrum.*, 79, 02C114 (2008).
15. T. Inoue et al., *Fusion Eng. Design* 55, 291-301 (2001).
16. R. S. Hemsworth and T. Inoue, *IEEE Trans. on Plasma Sci.* 33 (6), 1799-1813 (2005).
17. T. Inoue et al., *Fusion Eng. Design* 66-68, 597-602 (2003).
18. "OPERA-3d", Cobham Co. Ltd., Vector Fields software, http://www.cobham.com/about-cobham/avionics-and-surveillance/about-us/technical-services/kidlington.aspx.
19. T. Inoue et al., "FUSION TECHNOLIGY 1996", *Proc.19th Symp. Fusion Tech.* Lisbon, Portugal 16-20 Sep. 1996, vol.1,701-704 (1997).
20. Y. Ohara, *JAERI-M* 6757 (1976).

Acceleration of 500 keV Negative Ion Beams By Tuning Vacuum Insulation Distance On JT-60 Negative Ion Source

A. Kojima[a], M. Hanada[a], Y. Tanaka[a], M.Taniguchi[a], M.Kashiwagi[a], T.Inoue[a], N.Umeda[a], K.Watanabe[a], H.Tobari[a], S. Kobayashi[b], Y. Yamano[b] L. R. Grisham[c] and JT-60 NBI group[a]

[a] *Japan Atomic Energy Agency, 801-1, Mukoyama, Naka 311-0193, Japan.*
[b] *Saitama University, Saitama, Saitama-ken, 338-8570, Japan.*
[c] *Princeton Plasma Physics Laboratory, Princeton, NJ 08543, USA.*

Abstract. Acceleration of a 500 keV beam up to 2.8 A has been achieved on a JT-60U negative ion source with a three-stage accelerator by overcoming low voltage holding which is one of the critical issues for realization of the JT-60SA ion source. In order to improve the voltage holding, preliminary voltage holding tests with small-size grids with uniform and locally intense electric fields were carried out, and suggested that the voltage holding was degraded by both the size and local electric field effects. Therefore, the local electric field was reduced by tuning gap lengths between the large size grids and grid support structures of the accelerator. Moreover, a beam radiation shield which limited extension of the minimum gap length was also optimized so as to reduce the local electric field while maintaining the shielding effect. These modifications were based on the experiment results, and significantly increased the voltage holding from <150 kV/stage for the original configuration to 200 kV/stage. These techniques for improvement of voltage holding should also be applicable to other large ion sources accelerators such as those for ITER.

Keywords: negative ion source, voltage holding capability, vacuum insulation.
PACS: 07.77.Ka, 41.75.Cn

INTRODUCTION

A negative-ion-based neutral beam injector (N-NBI) is utilized for high-energy and high-power beam injections to advanced fusion plasmas [1-3]. Since the JT-60SA superconducting tokamak is designed to have 500 keV, 10 MW D^0 beams for 100 s, a N-NBI with two large negative ion sources has been developed, each of which accelerates 500 keV, 22 A negative ion beams with multi aperture and three-stage accelerator [4]. As a result of previous R&D, two key issues on the development of the JT-60SA ion source were found in the JT-60U ion source. One issue was a large heat load on acceleration grids, which prevented long pulse injections. However, by applying beam-steering techniques for multi-beamlets, the grid heat load was reduced to an allowable level for 100 s beam injection [5-6]. The other issue was low voltage holding capability of the three-stage accelerator.

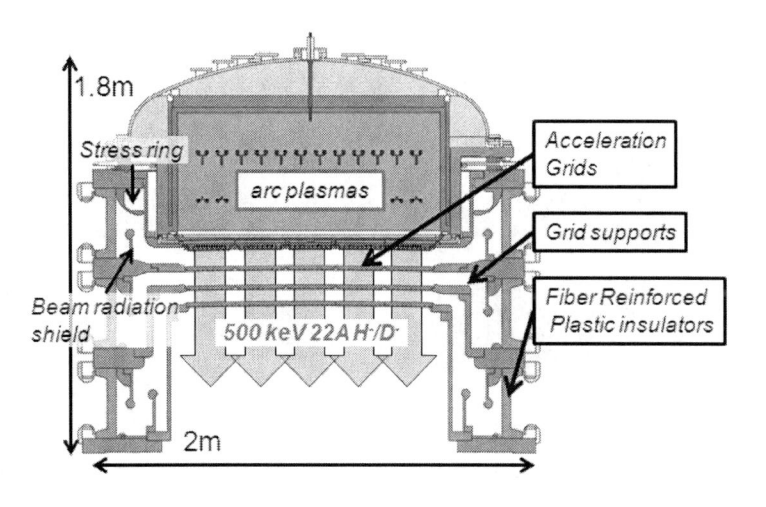

FIGURE 1. JT-60 negative ion source with a three-stage accelerator. The accelerator consists of acceleration grids, grid supports and fiber reinforced plastic insulators.

In past operations, the beam energy of the JT-60U ion source was limited to about 400 keV due to breakdowns which occurred at the accelerator, where three possibilities were considered. Up to now, some countermeasures against these possibilities have been taken as follows. As for external breakdowns in atmosphere, the breakdown voltage (V_{BD}) of external spark gaps was examined and these gap lengths were optimized in order not to constrain the beam energy. As for breakdowns at insulators; in order to suppress surface flashover on the Fiber Reinforced Plastic (FRP) insulators for the accelerator, large stress rings were developed to reduce electric field concentrations at triple junctions [7-8]. By installing these stress rings, the rated voltage for single acceleration stage was sustained by removing large-size acceleration grids in the accelerator. The remaining possibility was breakdowns in the vacuum gaps between the large-size grids (2 m^2) and between their support structures, whose gap lengths had been determined from past results using a small-size electrode (0.02 m^2) [9]. Although the small-size Rogowski electrode had the capability to sustain 400 kV with a gap length of 55 mm, the voltage holding of the accelerator was degraded to 150 kV/stage. In order to clarify the degradation mechanism and to improve the voltage holding of the accelerator, vacuum insulation of both the large-size grids and small-size electrodes having locally strong electric fields were investigated.

In this paper, the voltage holding tests by using the large-size grid for the ion source and the small-size electrode are reported. In addition, the modifications of the ion source based on the experimental results are described.

VOLTAGE HOLDING CAPABILITY ON JT-60 NEGATIVE ION SOURCE

JT-60 negative ion source is the largest ion source presently in use for nuclear fusion. The schematic view of the ion source with three-stage accelerator is shown in Fig. 1. This accelerator was developed in a design concept of a multi-aperture, multi-

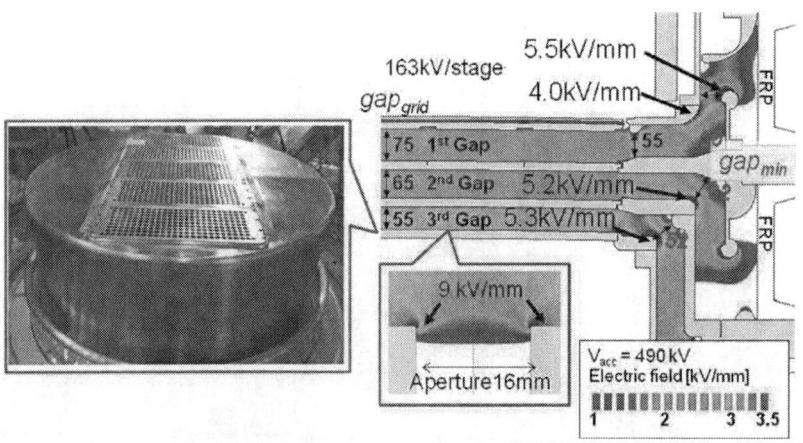

FIGURE 2. Photograph of the grounded grid and electric field analysis for the ion source. gap_{grid} denotes the gap lengths between the acceleration grids, and gap_{min} denotes the minimum gap length in each acceleration stage. The electric field around the aperture edge is enlarged.

grid (MAMuG) accelerator which was adopted as the baseline design of the ITER accelerator [10]. In the three-stage accelerator, multi beamlets are accelerated through 1080 apertures on the large acceleration grids (0.45 m x 1.1 m) which are supported by grid supports. Each acceleration stage is insulated by large FRP insulators whose diameter is 1.8 m. In addition to the stress rings, beam radiation shields were installed in each acceleration stage so as to suppress surface flashover due to photoelectron emission induced by radiation such as soft X-ray and ultraviolet light.

After the improvements of the spark gap spacings and stress rings, it was concluded that the remaining possibility was breakdown in vacuum gaps between these acceleration grids and their grid supports. Actually, traces left by discharges were observed at these regions when the accelerator was disassembled. An electric field analysis showed that locally strong electric fields of 9 and 5 kV/mm at the aperture edges on the acceleration grids and the corner of the grid supports were generated as shown in Fig. 2, where gap_{grid} were defined as distances between the acceleration

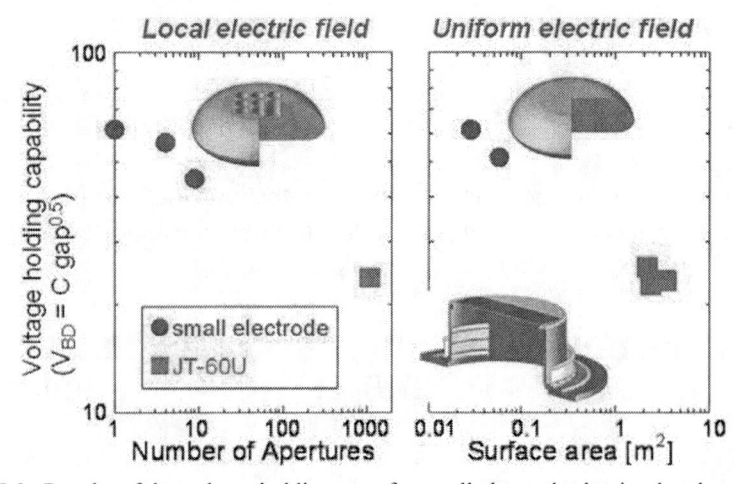

FIGURE 3. Results of the voltage holding tests for small electrodes having local and uniform electric fields. Local electric field is generated by the aperture whose size is similar to the ion source.

grids, and *gap_min* were defined as shortest distances between each acceleration stage. Although these electric fields and gap lengths were much weaker and longer than those of the small-size electrodes which had the capability to sustain 163 kV with a gap length of 8 mm, the accelerator was not able to sustain its rated voltage.

Degradation Of Voltage Holding Capability From Small Electrode

In order to investigate the large difference of the voltage holding between the large accelerator and small-size electrodes, preliminary experiments were carried out by using small-size electrodes. There are differences between the small-size electrodes and the large accelerator in terms of electric field and size. The impact of the local electric field was examined by utilizing the small-size electrodes with multi apertures whose size are identical to that of the accelerator. It appears that the voltage holding capability decreases as the aperture number increases, as shown in Fig. 3, where the voltage holding capability was defined as the coefficient of the gap dependence of the square root of the gap. The dependence with aperture number suggests that the extent of the local electric field plays an important role for the voltage holding capability. In the accelerator, local electric field concentration was observed at the edges of 1080 apertures and the corners of large grid supports, where the discharge traces had been observed. In addition to this local electric field effect, the voltage holding capability

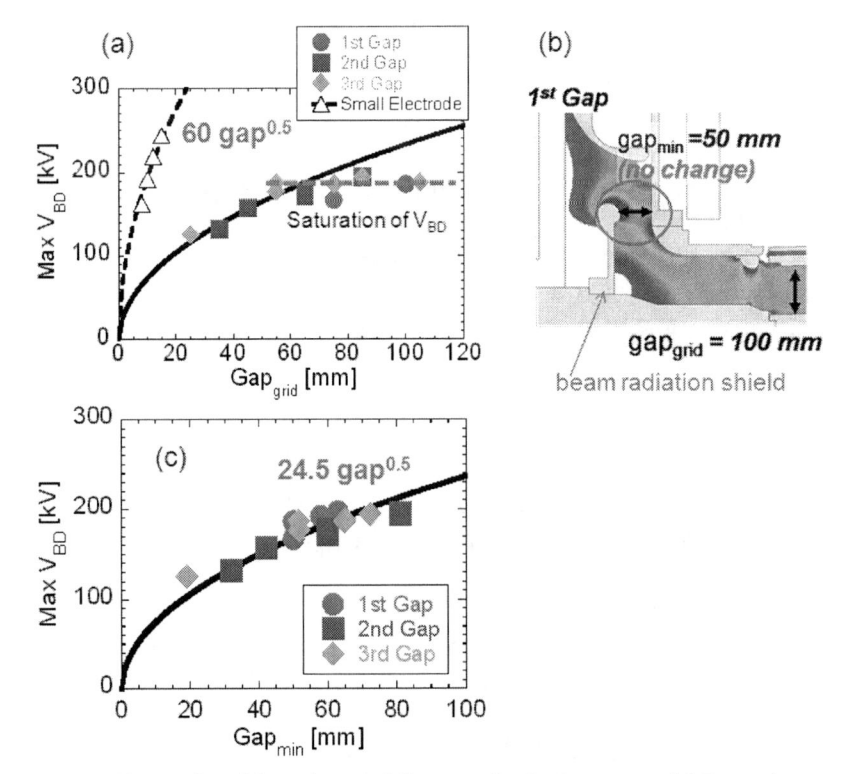

FIGURE 4. The results of the voltage holding tests for the ion source. (a) Dependence of the breakdown voltage (V_{BD}) on the gap length between the acceleration grid (gap$_{grid}$). (b) Schematic view of the saturation mechanism at the corner region of 1st acceleration stage. The minimum gap (gap$_{min}$) was not varied during the extension of gap$_{grid}$. (c) Dependence of V_{BD} on gap$_{min}$.

was also degraded with increase in size [11]. Both of these effects are connected with the product of the electric field (E) and the surface area (S),

Improvement Of Voltage Holding By Tuning Gap Lengths

From the preliminary experiments, we found that improvement of the voltage holding was required to reduce the electric field profile in terms of its strength and the area over which it was high. When the reduction of the electric field was considered, although gap extension reduced the electric field certainly, the gap extension should be minimized from the viewpoint of beam performance. However, there is no database of the voltage holding capability for the size effect including the local electric field, which is necessary to minimize the gap extension and to design the voltage holding capability of large accelerators. Therefore, the gap tuning was carried out based on the experimental results of the voltage holding capability on the large accelerator for the JT-60 negative ion source. To clarify the voltage holding capability of the large accelerator, voltage holding tests were carried out by changing gap_{grid} between the acceleration grids while keeping the original configuration. These experiments were carried out in each acceleration stage with a vacuum pressure of 1×10^{-4} Pa. A high voltage of 163 kV/stage would be required to reach a total acceleration voltage of 500 kV, including the extractor stage. The obtained relation between maximum breakdown voltages (V_{BD}) and gap_{grid} are shown in Fig. 3(a). As a result, although the voltage holding capability of the large accelerator was much lower than the small-size electrode, the gap extension was able to increase V_{BD}. However, if gap_{grid} was extended over 70~80 mm, no further increase in V_{BD} was obtained. This is explained in Fig. 3(b). In this experiment, because gap_{grid} was extended while keeping the other geometry (vertical direction in the figure) the same, gap_{min} between the corner and the beam radiation shield was not varied even if gap_{grid} was extended. Thus, as would be expected, gap_{min} is also an important parameter in improving the voltage holding capability. Therefore, V_{BD} was analyzed in terms of gap_{min} as shown in Fig. 3(c). In order to improve V_{BD} for per acceleration stage, gap_{min} was increased by removing the beam radiation shield. As a result, 200 kV/stage has been achieved in each

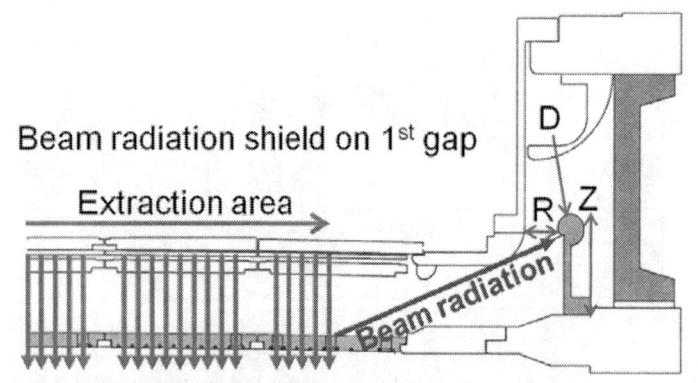

FIGURE 5. Schematic view of the shielding effect by the beam radiation shield, where R is the gap length between the corner and the shield, Z is the height of the shield and D is the diameter of the top of the shield. In this figure, the negative ion beams are extracted from 3 segments and 2 half segments. The shield prevents the photo-electron emission on FRP induced by the beam radiation.

acceleration stage according to the function of the square root of gap_{min}. These experimental results were utilized for the gap tuning of the modified accelerator. It was found that gap_{grid} affects the voltage holding with gaps shorter than 70~80 mm, however with longer gaps gap_{min} was limiting factor if the gaps in the support structure led to higher field concentrations than in the accelerator, as would be expected.

OPTIMIZATION OF BEAM RADIATION SHIELD

As a result of the voltage holding test, it was found that the beam radiation shield and its gap_{min} restricted the improvement of the voltage holding on the 1st gap. Therefore, the shield has been redesigned to ensure the sustainment of the rated voltage while keeping its shielding effects from the beam-induced radiation. Fig. 5 shows the schematic view of the shield on the 1st acceleration stage, where R is the distance between the corner (cathode) and the shield (anode), Z is the height of the shield and D is the diameter of the top of the shield. These parameters characterize the performance of the shield and also the local electric field at surrounding components. Optimization of these parameters was carried out within the existing configuration in terms of the shielding effect, surface electric field on the FRP insulator and local electric field at the corner and the shield.

From the view point of the shielding effect, a boundary condition for R and Z was determined by the beam extraction area as shown in Figs. 5 and 6(a). In Fig. 6(a), the relations of required R and Z in the cases of beam extractions from 1 to 5 segments are shown together with the contour of the surface electric field on the FRP insulator. Because the beam is extracted from 4 segments (3 segments and 2 half-segments) in the JT-60SA ion source, R and Z needed to be selected above the line for 4 segments in Fig. 6(a).

From the view point of the surface electric field on the FRP insulator, smaller R and Z are preferable. In the gap scan experiment as mentioned above, the surface

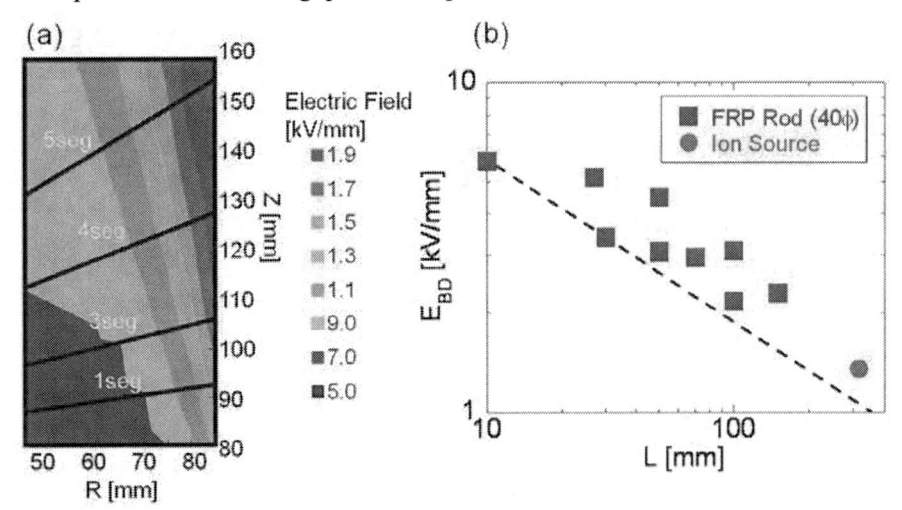

FIGURE 6. (a) Contour of the surface electric field of FRP insulator for the ion source. The boundary condition for the shielding effect is also shown. (b) The relation between the electric field at the breakdown (E_{BD}) and the length of FRP rods and the FRP insulator for the ion source.

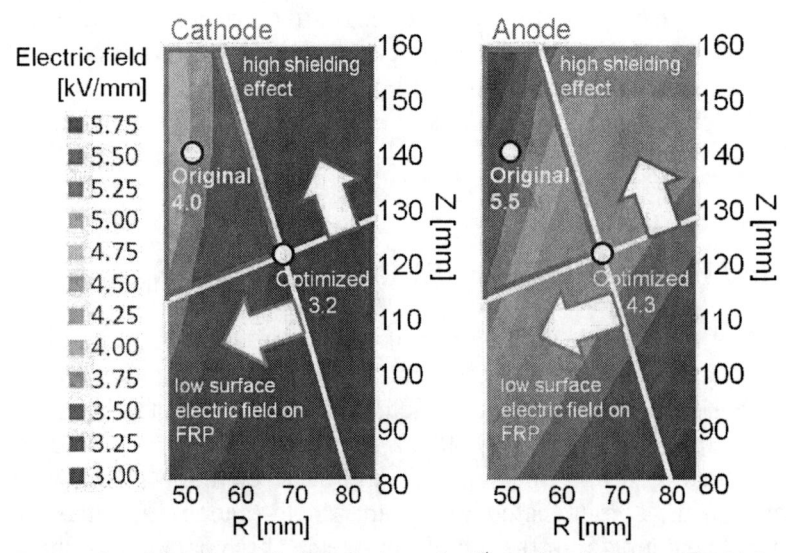

FIGURE 7. Results of the electric field analysis on the 1st acceleration stage. The highest electric field at cathode and anode are calculated, which are located at the corner and the beam radiation shield, respectively. Both boundary conditions about the shielding effect and surface electric field on FRP are shown.

electric field of 1.3 kV/mm on the FRP insulator has been experienced as shown in Fig. 6(b). In addition, voltage holding tests for FRP rods with a diameter of 40 mm showed that the allowable electric field was estimated to be 1.1 kV/mm by extrapolating the length dependence to 315 mm which corresponds to the FRP insulator for the ion source. Therefore, R and Z needed to be determined so as to satisfy the surface electric field below 1.1 kV/mm in Fig. 6(a).

From these two boundary conditions, the available range of R and Z were limited to the region shown in Fig. 7. The electric field analysis showed that local electric fields at the corner and shield on the 1st acceleration stage were decreased with larger R and smaller Z. As a result, the local electric fields of 4 and 5.5 kV/mm at cathode and anode for the original configuration were reduced to 3.2 and 4.3 kV/mm, respectively. The dependence of the local electric field on D, the diameter of the top of the shield as

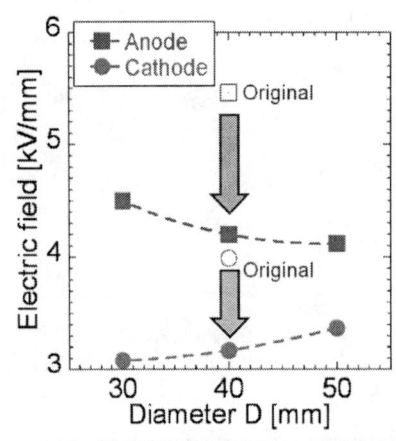

FIGURE 8. The relation between the diameter D of the top of the shield and the optimized electric fields at cathode and anode. Original shield had the diameter of 40 mm.

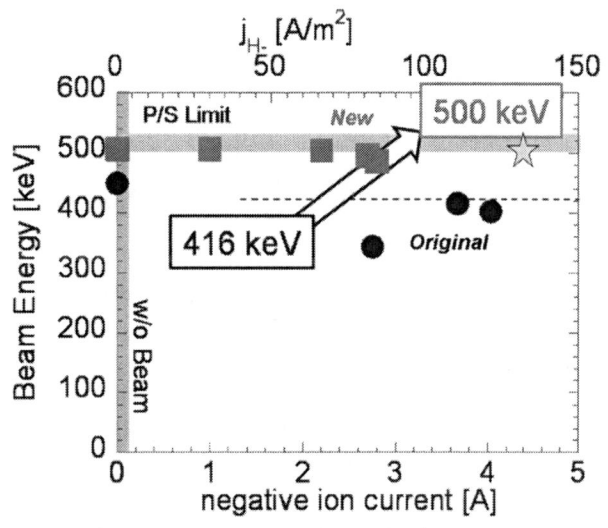

FIGURE 9. The progress of beam energy on JT-60 negative ion source. The beam energy has been improved to 500 keV which is the limit of the acceleration power supply.

shown in Fig. 8. It was found that increase of D reduced the local electric field at anode and increased that at the cathode. However, the impact of each electric field on voltage holding has not been clarified yet. Therefore, D of 40 mm was selected to minimize the product of the both electric fields. Thus, the shield was optimized to satisfy the required conditions with the shielding effect retained.

ACCELERATION OF 500 KEV BEAMS WITH MODIFIED ION SOURCE

In order to improve the beam energy, the gap lengths were tuned in addition to optimization of the beam radiation shield as shown in Fig. 8. Based on the experimental results, gap$_{min}$ was extended from 52/62/50 to 68/81/72. These new gap lengths have the possibility to sustain high voltage over 200 kV/stage. Gap$_{grid}$ was also modified to 85/85/85, which was selected so as to reuse the existing grid support. This configuration was predicted not to affect the beam performance such as the beam optics and stripping loss of negative ions by calculations. In the future, gap$_{grid}$ is going to be optimized after further investigations so that both the beam performance and the voltage holding capability satisfy the JT-60SA requirements.

As a result of the voltage holding tests by using the three-stage accelerator in vacuum at the pressure of 1×10^{-4} Pa, these modifications have improved the sustainable acceleration voltage of the three stage accelerator from 450 kV of the original ion source to 500 kV, the acceleration power supply limit. The required conditioning time to obtain the highest voltage was half of that for the original accelerator configuration. In addition, long pulse sustainment has been achieved up to 490 kV for 40 s which is also limited by the acceleration power supply. These results showed that the stability of the voltage holding in vacuum was also improved.

Because the degradation of the voltage holding with beam acceleration during cesium seeding was experienced in previous experiments, the voltage holding with beam was examined by utilizing the 20% of the extraction area due to the limited electric-capacity of the test facility during this period. Due to this limitation, the target for this experiment was the acceleration of 4.4 A (130 A/m^2) of negative ion beams of 500 keV, for 0.8 s. After the voltage conditioning up to 500 kV, a 507 keV, 1A beam was obtained without cesium seeding as shown in Fig. 9. Even with cesium seeding to enhance the negative ion production, 500 keV beams also have been attained up to negative ion current of 2.8 A without degradation of the voltage holding. In addition, it was observed that the modified gap$_{grid}$ did not affect the beam optics and stripping loss significantly [12]. From these results, the available beam energy has been improved from 416 keV to 500 keV which satisfies the JT-60SA requirement. In this experiment, because the voltage holding of the modified accelerator exceeded the capability of the acceleration power supply, the limit of the voltage holding was not determined. As for the achieved negative ion current, it was limited by the surface flashover on the feed-through for the extraction voltage which extracted the negative ions to lead to the accelerator. For the realization of the JT-60SA ion source, this feed-through is going to be upgraded to ensure stable operation.

SUMMARY

The voltage holding capabilities of the large negative ion source and small-size electrodes were investigated in order to overcome the critical issue of the low acceleration voltage. The experimental results of the small-size electrodes with local electric field showed that the voltage holding capability was degraded by both the local concentration of electric field and increasing size. In addition, the results of the gap scan with the large accelerator showed that the voltage holding is limited by the minimum gap, or more explicitly by the local electric field, whether that occurs between the acceleration grids or within the support structure and the shield. Therefore, in addition to the gap tuning, the local electric fields around gap$_{min}$ in the 1st acceleration stage were optimized by the redesign of the beam radiation shield taking into account the shielding effect and surface electric field on the FRP insulator. These modifications have much improved the beam energy of the large ion source from 416 keV to 500 keV. These techniques are useful to obtain and design the required voltage holding capability for future ion sources, especially JT-60SA ion source and the ITER accelerator because both accelerators have been developed along the same design concept of multi-aperture, multi grid accelerators.

REFERENCES

1. M. Kuriyama et al., "High energy negative-ion based neutral beam injector for JT-60U.", *Fusion Eng. Des.* **26** (1995), p. 445.
2. ITER EDA Final Design Report (2002).
3. Y. Takeiri, "Negative ion source development for fusion application", *Rev. Sci. Instrum.* **81**, 02B114 (2010).
4. M. Hanada et al., "Development and design of the negative-ion-based NBIfor JT-60 Super Advanced", *J. Plasma Fusion Res. SERIES*, Vol. **9**, p208 (2010).
5. M. Hanada et al., " Development of Long Pulse Neutral Beam Injector on JT-60U for JT-60SA", IAEA Fusion Energy Conference 2008, FT/P2-27.

6. M. Kamada et al., "Beamlet deflection due to beamlet-beamlet interaction in a large-area multiaperture negative ion source for JT-60U", Rev. Sci. Instrum. **79**, 02C114 (2008).
7. T. Inoue et al., "Accelerator R&D for JT-60U and ITER NB systems", *Fusion Eng. Des.* **66**, 597 (2003).
8. Y. Ikeda et al., "Recent R&D activities of negative ion based ion source for JT-60SA", *IEEE Trans. Plasma Sci.* **36**, 1519 (2008).
9. K.Watanabe et.al., "dc voltage holding experiments of vacuum gap for high - energy ion sources", *J. Appl. Phys.*, **72**, 3949 (1992).
10. M. Kashiwagi et al., "R&D progress of the high power negative ion accelerator for the ITER NB system at JAEA", *Nucl. Fusion* **49**, 065008 (2009).
11. J. M. Lafferty, ed.: "Vauum Arcs, Theory and Application", p43, John Wiley & Sons, New York (1980)
12. A. Kojima et al., "Demonstration of 500 keV beam acceleration on JT-60 negative-ion-based neutral beam injector", IAEA Fusion Energy Conference 2010, FTP/1-1Ra.

Magnetic Insulation for Electrostatic Accelerators

L. R. Grisham

Princeton Plasma Physics Laboratory, P. O. Box 451, Princeton, New Jersey 08543, USA

Abstract. The voltage gradient which can be sustained between electrodes without electrical breakdowns is usually one of the most important parameters in determining the performance which can be obtained in an electrostatic accelerator. We have recently proposed a technique which might permit reliable operation of electrostatic accelerators at higher electric field gradients, perhaps also with less time required for the conditioning process in such accelerators. The idea is to run an electric current through each accelerator stage so as to produce a magnetic field which envelopes each electrode and its electrically conducting support structures. Having the magnetic field everywhere parallel to the conducting surfaces in the accelerator should impede the emission of electrons, and inhibit their ability to acquire energy from the electric field, thus reducing the chance that local electron emission will initiate an arc. A relatively simple experiment to assess this technique is being planned. If successful, this technique might eventually find applicability in electrostatic accelerators for fusion and other applications.

Keywords: Magnetic insulation, electrostatic accelerator, voltage holding, electrical breakdown
PACS: 41.20.Gz, 41.75.Ak, 41.75.Cn, 52.59.Bi

INTRODUCTION

Sustaining high electric field gradients between electrodes is probably the oldest and most intractable challenge facing designers of electrostatic accelerators. The strength of the electric field which can be sustained without breakdowns is the principal determinant of the design options which are practical, including the strength of the lenses, the flux density of ions, electrons, or other charged particles which can be handled, the distances between successive accelerator stages, and the gaps required within the accelerator support structure for reliable operation. The practical design of any accelerator system constitutes a mix of tradeoffs between the need to limit electrical breakdowns, or discharges between surfaces at different potentials, to a tolerable level, and constraints imposed by other design factors, such as minimizing the accelerator length in order to reduce charge-changing reactions in ion beams.

Some electrostatic accelerators, such as those used to produce energetic beams which heat and drive current in magnetically confined plasmas used for fusion research[1] are of necessity designed near the practical limits of voltage holding and, as a consequence, require large amounts of high voltage conditioning time before they can operate at usable levels of reliability. To the extent that this conditioning time reduces the high performance operational time of the overall facility, it can be extremely expensive. Some electrostatic accelerators, such as those used for negative ions in some beam systems for magnetically confined fusion experiments, are also exposed to cesium vapor, which can lower the electron work function of electrodes and thus decrease the electric field gradient which is sustainable. Many electrostatic accelerators are in vacuum environments which periodically must be opened to air for maintenance or modifications, requiring substantial conditioning time after each opening to regain their full voltage holding capability.

Because of the large impact voltage holding considerations have upon both the design and useful operating time of electrostatic accelerators, many techniques have been developed over the past century to increase the sustainable voltage gradients among their components. These techniques were for the most part comprised of surface treatments, bakeout procedures, or modifications to high voltage conditioning methods. Due to the large potential impact of even modest improvements in voltage holding for some applications, such as magnetically confined fusion beam accelerators, it is worthwhile to consider other possible venues for improvement.

MAGNETIC FIELDS AS VACUUM INSULATION

We recently proposed[2] a possible technique to increase the sustainable electric field gradient in electrostatic accelerators by enveloping the accelerator electrodes and their electrically conducting support structures in magnetic fields which are everywhere parallel to the conducting surfaces. If successful, this would have the advantage of improving the voltage holding characteristics not only between successive accelerator stages, but also between their conducting electrical feed and support structures, where breakdown suppression is often even more of a design challenge than between accelerator stages. Since the reference in which we proposed this technique[2] gave a comprehensive description of the idea and its relation to different theories of electrical breakdown and vacuum discharges, we will not repeat that here, but will instead give a brief discussion of some salient features of the technique and the plans to experimentally test it. We also include a figure illustrating one possible generic embodiment of magnetic insulation, as the previous publication did not include one.

The magnetic insulation field will be produced by passing an electric current through each stage of an electrostatic accelerator and the conducting support structure which holds it in place. The significant feature of such a magnetic field is that it is parallel to the surfaces of all the components it flows through, so long as they are structurally smooth. If the magnetic field can prevent electrons or other charged particles from leaving the surface and moving far enough away to pick up appreciable energy from the electric field, then it might be effective in increasing the electric gradient which an accelerator can accommodate.

In this approach to magnetic insulation, magnetic field components parallel to electrode surfaces are beneficial, because they increase the vacuum impedance by inhibiting charged particle motion away from the surface, and they may also inhibit electron mobility to the surface. Magnetic field components normal to conductor surfaces, on the other hand, are most likely detrimental, since they may channel electron flows and promote breakdown if they link conducting components at different electrical potentials. The requirement that the magnetic field be everywhere parallel to the surface of the electrodes being insulated largely precludes the use of any configuration of permanent magnets being utilized to perform a similar function.

The proposed magnetic geometry has, in fact, been previously utilized for an entirely different purpose in the first stage of some negative ion beam accelerators[3,4] used in magnetically confined fusion experiments. In these systems, a current of several kiloamps is run along the length of the plasma grid, which faces an arc discharge where negative ions are formed. The purpose of the resulting magnetic field is to form a magnetic filter which inhibits the flux of energetic primary electrons from the arc, which can destroy negative hydrogen ions, and which also impedes the flow of electrons to the plasma grid, allowing the extraction of the negative ions and reducing co-extraction of electrons. This plasma grid filter field also extends forward into the accelerator.

While this configuration is basically similar to the geometry of the proposed magnetic insulation, it also differs in some important aspects. The arc chamber, which feeds plasma to the plasma grid, is surrounded by a large number of magnetic cusps produced by arrays of powerful permanent magnets which produce magnetic field components orthogonal to the plasma grid surface, which is undesirable. Additionally, the extractor grid, which is separated from the plasma grid by a narrow gap, contains hundreds of powerful permanent magnets which remove co-extracted electrons from the many beamlets comprising the negative ion beam, and which introduce orthogonal components into the plasma grid filter field. By the time the downstream component of the plasma grid filter field reaches the much longer gaps of the main accelerator the magnetic field strength is declining and, more importantly, it is developing orthogonal components towards the edges.

MAGNETIC INSULATION CONFIGURATIONS

The magnetic field will, of course, deflect the electrically charged beams which are being accelerated. In the case of electron beam accelerators, this will a significant deflection, and may preclude the usage of magnetic insulation in electron accelerators, even if the technique is found to be otherwise practical. For beams of ions, or any heavier charged stream, the net deflection can be small. Since the magnetic insulation field wraps around each acceleration stage, it deflects a charged particle beam in opposite directions on the upstream and downstream side of each acceleration stage. Thus, the net deflection is primarily due to the fact that beam velocity is higher on the downstream side, and is less than the deflection on either side. Moreover, in multistage accelerators, the currents driving the magnetic insulation can either be all run in the same direction, or in opposite directions in alternating accelerator stages, which could in principal result in an arbitrarily small total net beam deflection. The mechanical forces acting upon the accelerator and its support structure are different for these two configurations, and might play some role in determining their relative appeal depending upon the application.

Figure 1 shows a generic rendering of a magnetic insulation configuration with all the currents producing the magnetic insulation flowing in the same direction. The power supply to drive the current producing the magnetic field has to be floated at the potential of each acceleration stage. While each power supply would need to be capable of large currents, the required voltage would be very low, of the order of a few volts, since the impedance of each acceleration stage will be small. Nonetheless, room would have to be provided in the high voltage deck for each stage, and the power required would need to be supplied to it either by an isolation transformer or mechanical means. These are substantial disadvantages which would need to be weighed in considering whether to apply magnetic insulation if it proves feasible.

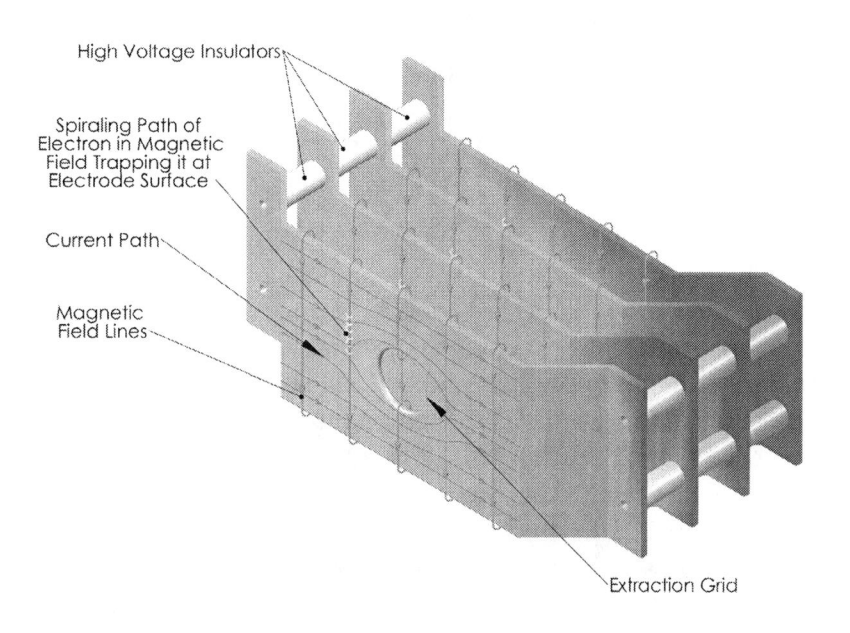

FIGURE 1. Generic figure of an electrostatic accelerator with the currents generating the magnetic insulation field flowing in the same direction in each accelerator stage. An alternative arrangement would be with alternating stages having different current flow directions. The power supplies to drive the magnetic insulation current flows have to be floated at the potential of each accelerator stage.

ELECTRICAL BREAKDOWN INSTIGATORS

Electrical breakdown in accelerators can be initiated through a number of mechanisms. Possible models for the behavior of breakdown with gap length are discussed in our earlier paper[2], and do not result in a completely satisfactory explanation. However, there are thought to be a number of processes which start the process that leads to an electrical breakdown.

By far the most ubiquitous breakdown instigator in a vacuum is field emission of electrons from microprojections on surfaces and sharp edges on components, where the electric field strength reaches levels much higher than would pertain if the surface were a perfect plane. Because these electrons should be born at an energy characterized by the temperature of the electrode,[5,6] this phenomenon should be the most amenable to suppression by magnetic insulation, since the lower the electron birth energy, the less magnetic field should be required to inhibit electrons from leaving the surface and picking up energy from the electric field.

Other sources of breakdown include the photoelectric effect and production of secondary electrons due to impacts by energetic electrons, ions, or neutrals. All of these breakdown sources can release electrons with birth energies greater than thermal, and thus will likely be less amenable to any benefits which magnetic insulation may convey. However, if it should prove beneficial just for spontaneous field emission, magnetic insulation might still find some applications, since this is the most common source of electrical breakdown under vacuum conditions.

PLANNED EXPERIMENTAL TEST

We are presently planning an experiment to verify whether magnetic insulation offers measurable improvement in voltage holding under realistic conditions. Two copper busbars will be separated by a variable distance vacuum gap, with a potential of up to 160 kV applied between them. The grounding will be arranged such that the ground of the high voltage supply is at cathode potential, and the magnetic insulation current will be passed through the cathode potential busbar. This arrangement will allow the 4 kA 6 volt supply which will be used for the magnetic insulation to be at ground potential for this test. The maximum voltage which can be sustained across the gap will be measured with no magnetic insulation field, and at various fields as different currents are passed through the cathode busbar.

If this test of efficacy against spontaneous field emission is successful, then an ultraviolet light source will be added to test effectiveness against photoelectric emission, and it will then be tried on a small ion source where effectiveness in the presence of ion beams and electrons can be checked.

It is difficult to estimate what sort of magnetic field might be required, since it is likely to be a strong function of surface conditions, but as an example, a 100 gauss field at an electrode surface would restrict an electron with 0.025 eV (room temperature) to a gyroradius of 3.8×10^{-3} cm. A gyroradius of this size might still allow an electron which escaped the surface to gain energy in a sufficiently high field, such as the 4×10^4 volts/cm in the highest field gap in the ITER accelerator design.[1] Nonetheless, since the magnetic field will also be present within the accelerator electrode, where the applied electric field is very close to zero, it may impede the motion of electrons sufficiently to reduce the field emission current. In event, if it proves necessary, it would be practical to provide surface magnetic fields significantly higher than 100 gauss.

CONCLUSION

We have discussed a novel technique for improving the voltage holding in electrostatic accelerators by enveloping the accelerator electrodes and their support structures in magnetic fields which are everywhere parallel to the conductor surfaces.

If demonstrated to be useful at practical magnetic fields, this might provide an additional design options in some applications.

ACKNOWLEDGEMENTS

It is a pleasure to acknowledge helpful discussions with Dr. Jill Foley, who also drew the figure, Al von Halle, Dr. Tom Kornack, Dr. Masaaki Kuriyama, and Dr. Ken Young. This work was supported by U.S. DOE Contract No. AC02-CH03073.

REFERENCES

1. R. S. Hemsworth, *Nucl. Fusion* **43**, 851 - 860 (2003).
2. L. R. Grisham, *Physics of Plasmas* **16**, 043111-1 – 043111-5 (2009).
3. M. Kuriyama et al, *Fus. Sci. & Tech.* **42**, 410-419 (2002).
4. O. Kaneko et al, *Nucl. Fus.* **43**, 692-696 (2003).
5. L. Nordheim, *Proc. R. Soc. London* **A121**, 626 - 634 (1928).
6. R. H. Fowler and L. Nordheim, *Proc. R. Soc. London* **A119**, 173 - 181 (1928).

Space Charge Neutralization of DEMO Relevant Negative Ion Beams at Low Gas Density

Elizabeth Surrey and Michael Porton

EURATOM/CCFE Fusion Association, Culham Science Centre, Abingdon, Oxon OX14 3DB, U.K.

Abstract. The application of neutral beams to future power plant devices (DEMO) is dependent on achieving significantly improved electrical efficiency and the most promising route to achieving this is by implementing a photoneutralizer in place of the traditional gas neutralizer. A corollary of this innovation would be a significant reduction in the background gas density through which the beam is transported between the accelerator and the neutralizer. This background gas is responsible for the space charge neutralization of the beam, enabling distances of several metres to be traversed without significant beam expansion. This work investigates the sensitivity of a D^- beam to reduced levels of space charge compensation for energies from 100keV to 1.5MeV, representative of a scaled prototype experiment, commissioning and full energy operation. A beam transport code, following the evolution of the phase space ellipse, is employed to investigate the effect of space charge on the beam optics. This shows that the higher energy beams are insensitive to large degrees of under compensation, unlike the lower energies. The probable degree of compensation at low gas density is then investigated through a simple, two component beam-plasma model that allows the potential to be negative. The degree of under-compensation is dependent on the positive plasma ion energy, one source of which is dissociation of the gas by the beam. The subsequent space charge state of the beam is shown to depend upon the relative times for equilibration of the dissociation energy and ionization by the beam ions.

Keywords: Negative ion beams, space charge, neutralization, neutral beam injection heating.
PACS: 29.27Eg, 41.75Cn, 52.20Hv, 52.40Mj

INTRODUCTION

For most Neutral Beam Injection (NBI) systems, based on either positive or negative ion precursor beams, the space charge neutralization of the beam is of no concern. This is primarily due to the large gas target presented by the gas neutralizer, which ensures that full compensation is achieved. Furthermore it is common in NBI systems for the neutralizer to immediately follow the grounded grid of the accelerator, so there is no drift distance between the two during which the beam may expand under space charge. The situation is different for the ITER beamline, where there is a 1.6m interspace distance between the grounded grid and the neutralizer channel of the heating beams and 1m for the diagnostic beam. Even in this case, there is sufficient gas density in the interspace to achieve full space charge compensation of the negative ion precursor beam [1].

For NBI on DEMO tokamaks, the 58% neutralization efficiency of gas neutralizers limits the electrical efficiency of the heating system to 34% [2] and precludes their use for steady state current drive. The most promising alternative is the photoneutralizer [3,4,5], for which neutralization efficiency of 95% has been postulated, boosting the electrical efficiency to above 60% [2]. Using this technology obviously removes the gas source due to the neutralizer and, as minimisation of plasma formation in the optical cavity is a major consideration, the imperative to operate at maximum possible vacuum is strong. The corollary is that the negative ion precursor beam may not be fully space charge compensated resulting in an increased rate of expansion in the interspace and a consequent reduction in transmission of NBI power to the plasma. This would reduce the electrical efficiency, negating the benefits of the photoneutralizer. Furthermore, during commissioning, operation at reduced beam energy, when the beam is more sensitive to space charge effects, will be necessary and increased direct interception on beamline components may result. It is therefore necessary to ascertain the probable impact of gas-free neutralizers on the beam optics before embarking upon a significant development programme.

This work considers firstly the effect of the degree of space charge neutralization on beam transport and secondly the probability that the space charge will not be fully compensated. Results are presented for three beam energies: 100keV, 750keV and 1.5MeV to represent phases from early commissioning to full energy operation and, more significantly, the operation of a scaled, proof-of-principle prototype.

BEAM TRANSPORT IN THE PRESENCE OF PARTIALLY COMPENSATED SPACE CHARGE

Space charge neutralization plays an important role in beam propagation by reducing the repulsive force between beam particles and enabling the beam to be transported significant distances without the use of focusing elements. The improvement in the efficiency of negative ion based neutral beam systems and the role of a gas free neutralizer based on photo detachment was discussed in the introduction. This device offers potentially 95% neutralization efficiency and an improvement in beam transmission due to reduced losses through stripping in the accelerator and re-ionization further downstream due to the reduction in background pressure. However, if the background gas pressure is reduced to a level where the space charge neutralization is too low, the beam will expand rapidly before entering the neutralizer. Furthermore these expanding trajectories will be maintained after neutralization and it may not be possible to transport the beam. Another route to improving the beamline efficiency would be to reduce the beam divergence but if the halo is a consequence of over-compensation of the space charge [1], this can only be achieved by operating at lower background pressure; so the balance of gas density and space charge is crucial.

Having established that space charge under-compensation may be an issue at low background pressures, the effect of this on the beam optics and consequences for transmission are considered. The effect of the self-field of an under-compensated beam can be included through modelling of the evolution of the emittance diagram along the beamline. To avoid lengthy PIC calculations, the model is formulated in terms of the Twiss parameters describing the size and orientation of the phase space ellipse. This is an economical method of computation, for which space charge effects are well supported in the literature. It also lends itself to comparison with experimental measurement and represents a technique for further investigating space charge neutralization at low gas density.

Emittance Transport Model

Considering the beam to be axi-symmetric, the beam equation of motion can be expressed as in Eq. 1 where r indicates the radial distance, the prime denotes differentiation with respect to distance z along the longitudinal beam axis, and where k represents the radial beam forces:

$$r'' + k^2 r = 0 \tag{1}$$

A simple solution to permit evolution of the beam profile with longitudinal space is offered by Eq. 2 where δz represents a small distance in longitudinal direction z, **r** represents the column vector (r r´) and r´=dr/dz:

$$\mathbf{r}(z+\delta z) = R\mathbf{r}(z) \tag{2}$$

The action of space charge can be represented as a distributed, divergent lens of strength k with the transformation matrix R_{sc}:

$$R_{sc} = \begin{bmatrix} \cosh(k\,\delta z) & \frac{1}{k}\sinh(k\,\delta z) \\ k\sinh(k\,\delta z) & \cosh(k\,\delta z) \end{bmatrix} \tag{3}$$

and k can be evaluated from consideration of the electric field, E, at some radius r within the beam as in Eq. 4:

$$E(r) = \frac{c_0 e n_b r}{2\varepsilon_0}\left(1-\beta_v^2\right)(1-h) \tag{4}$$

where β_v is the usual relativistic parameter, n_b is the beam density, h is the degree of space charge compensation and c_0 is the fraction of the beam distribution contained within radius r, allowing different distributions to be compared. Consideration of the equation of motion provides Eq. 5, and therefore that k^2 is defined as in Eq. 6, where m_b is the mass of the beam ion and v_b is its velocity:

$$eE(r) = m_b \frac{d^2 r}{dt^2} = m_b v_b^2 \frac{d^2 r}{dz^2} = m_b v_b^2 r'' \tag{5}$$

$$k^2 = \frac{ec_0 I_b (1-h)}{2\pi\varepsilon_0 m_b v_b^3 r_0^2} \tag{6}$$

At each longitudinal location, the beam properties of **r** can be visually summarised via the beam emittance ellipse described by four parameters: beam emittance ε and the Twiss parameters, α, β, γ. The former, multiplied by π, corresponds to the area of the ellipse, whilst the Twiss parameters then allow specification of the extremities and intercepts on the emittance diagram, as indicated in Fig. 1. Note that the Twiss parameters are subject to the condition that $(\gamma\beta - \alpha^2) = 1$ and that the beam envelope radius r_0 can be considered to correspond to the indicated r_{max} for the emittance contour of interest.

The emittance ellipse is defined at some initial point by specifying sufficient known parameters. Most commonly, the beam envelope radius r_{max} (corresponding to some fraction of the beam current), the beam envelope divergence (r´ at r_{max}) and the emittance are known and these are sufficient to specify all the initial Twiss parameters. From the specified beam current, c_0 and beam energy all other parameters can be derived. The effects of drift and space charge over a distance δz are evaluated simultaneously via the transformation:

$$\begin{cases} \alpha(z+\delta z) = \left(\cosh^2(k\delta z)+\sinh^2(k\delta z)\right)\widetilde{\alpha}(z)-k\cosh(k\delta z)\sinh(k\delta z)\widetilde{\beta}(z)-\sinh^2(k\delta z)\widetilde{\gamma}(z) \\ \beta(z+\delta z) = -\frac{2}{k}\cosh(k\delta z)\sinh(k\delta z)\widetilde{\alpha}(z)+\cosh^2(k\delta z)\widetilde{\beta}(z)+\frac{1}{k^2}\sinh^2(k\delta z)\widetilde{\gamma}(z) \\ \gamma(z+\delta z) = -2k\cosh(k\delta z)\sinh(k\delta z)\widetilde{\alpha}(z)+k^2\sinh^2(k\delta z)\widetilde{\beta}(z)+\cosh^2(k\delta z)\widetilde{\gamma}(z) \end{cases} \quad (7)$$

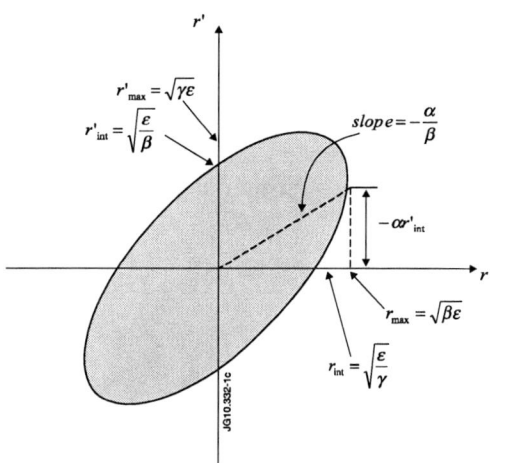

FIGURE 1. Phase space ellipse diagram showing the relationship between the Twiss parameters α, β, γ and the beam properties r, r´ and emittance ε

A convergence study showed that the beam profile was suitably convergent with respect to grid confinement for longitudinal step size smaller than the initial beam radius. The model was validated against an analysis by Humphries using the beam envelope equation and zero emittance of a 200mA, 300keV C^+ beam [6]. Agreement was within 1% for r_{max} and r'_{max} over a distance of z=0.3m.

To illustrate the use of the code consider Fig. 2, which shows, as functions of the space charge neutralization fraction, h, computed values of r_{max} and slope (=-α/β) as defined in Fig. 1 and corresponding to the Gaussian ($\sqrt{2}$rms) contour containing 63% of the beam distribution. The computation is for a 33mA, 36keV H^- beam generated at the EFDA-JET neutral beam test stand and the emittance diagnostic is one metre from the grounded accelerator grid. The initial ellipse parameters were estimated from ray tracing calculations where it is assumed a beam waist is formed at the grounded grid i.e. α_0=0. The figure clearly shows the effect of space charge on the emittance ellipse and how this technique might be used to measure h as well as study the effects on transport. At low values of h the beam envelope radius is the preferred indicator, reducing almost linearly as h increases. At higher values of h, above 0.8 where most beamlines operate, the slope gives a very strong dependency on h and so can be used to distinguish between small variations in this parameter.

From the experimental emittance ellipse, the Gaussian radius is 9±2mm and the slope is 0.9±0.2mrad/mm (the error in the radius is largely due to the resolution of the data presentation but the error in the slope mainly arises from identifying the centre of the ellipse). From Fig. 2 the beam transport code suggests that the two values of h are $h_r = 0.97^{+0.03}_{-0.02}$ and $h_s = 0.95^{+0.03}_{-0.10}$. The background gas density in this system during the experiment was approximately $1.5 \times 10^{19} m^{-3}$ and the three component model of [7] as modified in [1], gives an equivalent value of h=1.05 (slight over-compensation) and the model presented in the section below gives h=1.00 for T_i=0.025eV. Given the uncertainties in the experimental conditions (r_0, $r_0{'}$, α_0, and gas density along the beam path) the agreement between the models and the diagnostic is encouraging as a tool for measuring the degree of space charge neutralization in real beams. A similar technique was described by Baartman, et. al. [8] using the rms values of beam size and emittance in the Kapchinskij-Vladimirskij beam envelope equations to trace the beam dimensions between two emittance scanners. This technique is only strictly exact for the rms values of the beam parameters, so cannot be extended to contain larger fractions of the beam distribution which may be disadvantageous if non-linear space charge driven effects are evident.

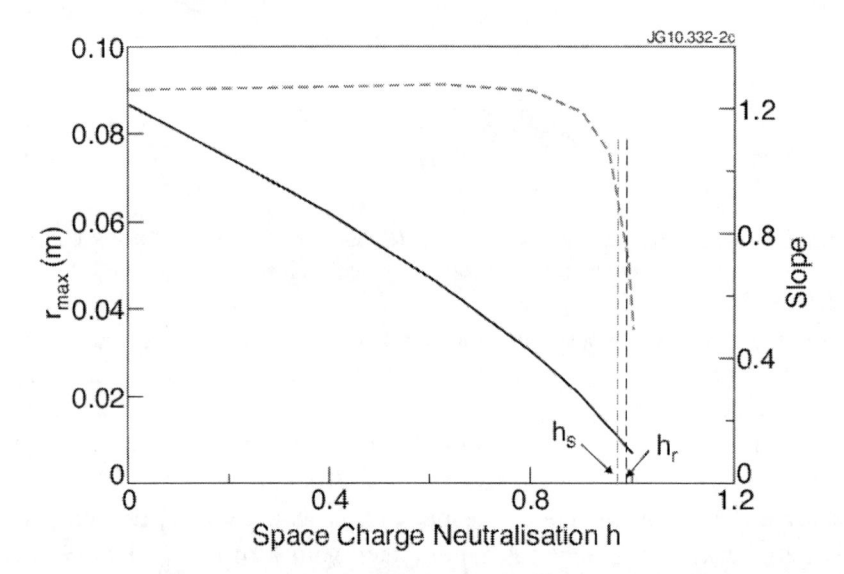

FIGURE 2. Values of beam Gaussian envelope radius ——— r_{max} and — — slope as functions of the space charge neutralization fraction h for the 33mA, 36keV H⁻ beam emittance diagram measured on the JET neutral beam test bed. Values of h determined from comparison with the measured radius, h_r, and slope, h_s are indicated by the broken lines.

Application to Beam Optics of DEMO Beams

The effect of space charge neutralization on the beam optics of the DEMO relevant beam can be examined using the same technique with an estimate of the initial properties of the beam ellipse. For this purpose the code follows the evolution of a single beamlet and there is no attempt to include the effect of transporting multiple beamlets or of merging multiple beamlets, when the beam potentials may become superimposed [1]. The initial beamlet parameters are given in Table 1 and the results of the computation over a distance of 1.5m shown in Fig. 3. The beamlet current of 45mA and envelope divergence of 3mrad correspond to the beam configuration identified in [2] as necessary to provide the desired neutral beam electrical efficiency for a power plant.

TABLE 1. Beam parameters and initiating values of the emittance space charge code for DEMO relevant beams.

Beamlet Parameters		Initial Beam Envelope Parameters (rms values)		Initial Twiss Parameters	
Energy	1.5MeV	r_{max}	3.00mm	α_0	0
Current	45mA	r'_{max}	3.00mrad	β_0	1.000
Ion	D$^-$	ε (un-normalised)	9.00πmm.mrad	γ_0	1.000

Figure 3(a) shows the ratio of the beam envelope radius at a distance of 1.5m to the initial radius for values of space charge compensation, h, from 0 to 1 and for beam energies 100keV, 750keV and 1.5MeV. It is clear from Fig. 3 that the effect of space charge on the beam optics is small at the higher energies, but considerable at lower energies. No adjustment of the current with beam energy has been made since operating at constant perveance results in currents less than 1mA at 100keV and it is difficult to envisage operating the ion source at this level.

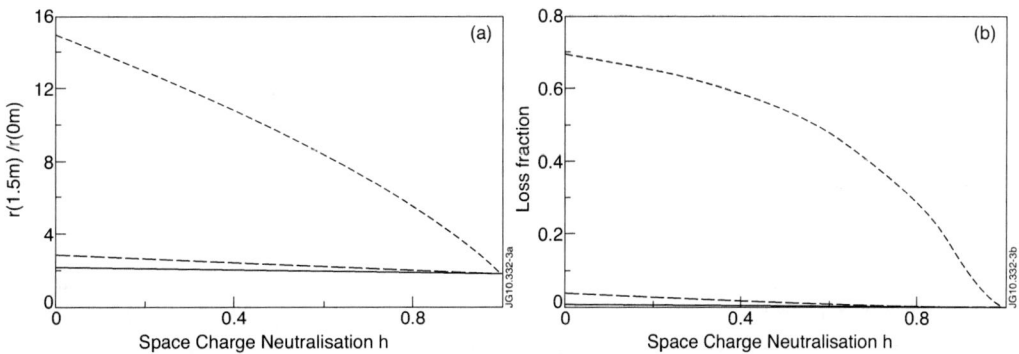

FIGURE 3. (a) Computation of beam envelope as a function of space charge neutralization factor for DEMO relevant beamlet of 45mA at energy 1.5MeV ——, 750keV − − and 100keV ·······; (b) the fractional loss on a 100mm wide aperture, representative of a neutralizer channel. The transport distance is 1.5m.

The implication for beam transmission can be appreciated by considering an aperture of total width D, such as the cavity in the photoneutralizer. The fraction, F, of the beam lost by direct interception is given by:

$$F = erfc\left(\frac{D/2}{r_{max}}\right) \qquad (8)$$

and is shown in Fig. 3(b) for an aperture of D=0.1m, such as the ITER neutralizer channel width. The laser power required by the photoneutralizer is proportional to the channel width [5], so there is incentive to minimise this quantity. From Fig. 3(b) even at reasonably high values of space charge neutralization, there is significant transmission loss at low beam energy, so commissioning such a system would not be simple. The same result would apply to any scaled prototype photoneutralizer experiment and an assessment of the factors contributing to this is presented below.

SPACE CHARGE COMPENSATION AT LOW GAS DENSITY

It may be surprising to find that very little attention has been paid to the problem of negative ion beams propagating through a rarefied background gas. This is probably due to the common proposition that the positive ions, produced by beam ionization of the gas, accumulate within the beam potential well until the beam is fully space charge compensated [8,9]. The time over which this process occurs is:

$$\tau = (N\sigma_i v_b)^{-1} \qquad (9)$$

where N is the density of the background gas and σ_i the cross section for ionization by the beam. The ionization time typically takes values of the order of microseconds depending upon the beam energy and gas density. However, this approach assumes that there is no loss channel for the positive ions created by the beam and that they effectively have zero energy so cannot escape the beam potential. At high gas density this description of the beam-plasma system is correct, at least to first order, as the ionization rate is sufficiently high to maintain a three component system [10,11] (beam ions, plasma ions and plasma electrons). However as the ionization rate is reduced through reduced gas density or increasing beam energy, the system becomes essentially two component, as the plasma electrons are expelled by the (un-compensated) beam potential [10,11]. For a given beam energy, the critical gas density, N_c, at which this condition is met is given by [11] for negative ion beams as:

$$N_c = \frac{2v_i}{v_b \sigma_i r_0}\left[1 - \frac{v_i \sigma_e}{v_e \sigma_i}\right]^{-1} \approx \frac{2v_i}{v_b \sigma_i r_0} \qquad (10)$$

where $v_{b,i,e}$ is the respective velocity of the beam ions, the plasma ions and the plasma electrons and $\sigma_{i,e}$ is the respective cross section for ion and electron production by the beam. For a typical DEMO relevant beam of 1.5MeV D$^-$ ions, $N_c \sim 10^{18} m^{-3}$ as a consequence of the low ionization cross section but, depending on the ion velocity and beam radius, can range over an order of magnitude above and below this value.

Existing models of space charge neutralization in continuous negative ion beams [1,7,11] assume that the three component system applies and that the plasma ion energy is arbitrary [7] or is determined by coulomb collisions with the beam ions [11]. In both cases, this results in a model that effectively limits the maximum negative excursion of the beam potential well to fractions of a volt as it cannot exceed the ion temperature (to avoid a singularity) and the ion heating by the beam is minimal.

To circumvent this issue a new, simple model is proposed which treats the two component system and allows an estimation of the degree of space charge compensation in a DEMO relevant beam. This is then used in conjunction with the beam transport code to estimate the probable influence on beam transmission under such circumstances.

Two Component Space Charge Model

As the D⁻ beam passes through low pressure deuterium gas it creates ion and electron pairs within its path. The beam exhibits a potential well of depth ϕ defined relative to an arbitrary beam edge. The potential difference between the beam edge and the axis is negative so that electrons are expelled radially and the beam-plasma system is comprised of beam ions and positive ions. Continuity at a beam radius r gives:

$$2\pi r L dn_i(r) v_i(\rho, r) = 2\pi \rho d\rho L\, \partial n_i / \partial t \quad (11)$$

where $dn_i(r)$ is the incremental slow ion density at r due to the production rate $\partial n_i / \partial t$ at radial position ρ. Integrating and substituting

$$\partial n_i / \partial t = N n_b v_b \sigma_i \quad (12)$$

gives:

$$N n_b v_b \sigma_i r_0 = 2 n_{i0} v_i \exp(-\phi/T_i) \quad (13)$$

where the slow ions are assumed to have a Boltzmann distribution characterised by temperature T_i and r_0 is some characteristic radius of the beam.

The potential ϕ can be expressed in terms of the beam current as:

$$\phi = \frac{c_0 I_b (1-h)}{4\pi \varepsilon_0 v_b} \quad (14)$$

where $h = n_{i0}/n_b$ is the degree of space charge neutralization and the parameter c_0 is the fraction of the beam current contained within r_0. Substitution into Eq. 13 gives:

$$ln\left[\frac{N v_b \sigma_i r_0}{2 h v_i}\right] = -\frac{c_0 I_b (1-h)}{4\pi \varepsilon_0 v_b T_i} \quad (15)$$

which can be solved numerically to give h. Figure 4 shows the degree of space charge neutralization for a 1.5MeV D⁻ beam as a function of gas density obtained from Eq. 15 for values of T_i from 0.077 to 0.77eV.

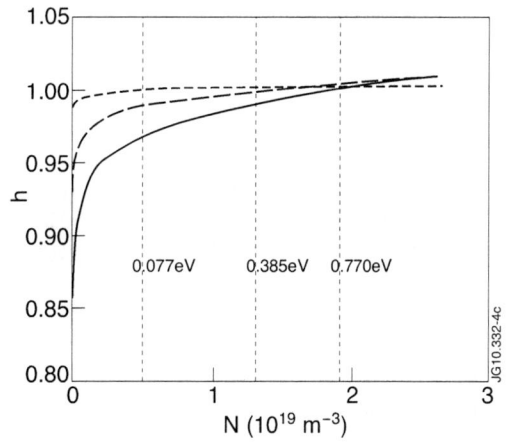

FIGURE 4. Space charge compensation, h, as a function of gas density, N, for a uniform 1.5MeV D⁻ beam of 45mA, r_0=12mm. The parameter is ion temperature T_i: —— 0.077eV, – – 0.385eV and ······ 0.77eV. The vertical broken lines indicate the value of N_c for each ion temperature.

These ion energies are fractions of the electron temperature resulting from the 1.5MeV beam as obtained from the fitting parameters given in Rudd [12]. The calculation is for a uniform distribution with r_0=12mm (i.e. 4rms of the beam in the previous section) and c=1; the cross section σ_i=4.5x10^{-21} was taken from [13] and corresponds to ionization by protons as argued in [1] and the reduction in beam space charge due to electron stripping and expulsion from the beam is taken into account.

Figure 5 shows the equivalent plot for the three different beam energies with T_i=0.77eV, this being identified as the worst case. The apparently unusual behaviour of the 750keV data results from a combination of cross section, σ_i, and beam velocity. At low beam energies the cross section for negative ions is similar to that of the neutral atom [14] but moves towards that of the positive ion with increasing velocity. The details of the transition are not clear in the literature and in this study the 100keV data was equated with the atomic cross section and the higher energies with the ionic cross section. Thus there is a discontinuity in the smooth transition between cross section and velocity which manifests itself as the apparently high ionization in the 750keV beam. Applying the atomic cross section to the 750keV case gives N_c~2.14x10^{19}m^{-3} and the curve would lie below the other two.

At the higher end of the density range, the electron population will become more important to the equilibrium of the beam-plasma system and Eq. 13 is not strictly valid, although values of h>1 obtained for N>N_c are not dissimilar to the equivalent values calculated in [1] for the ITER beams using a model based on [7] (i.e. assuming a three component system).

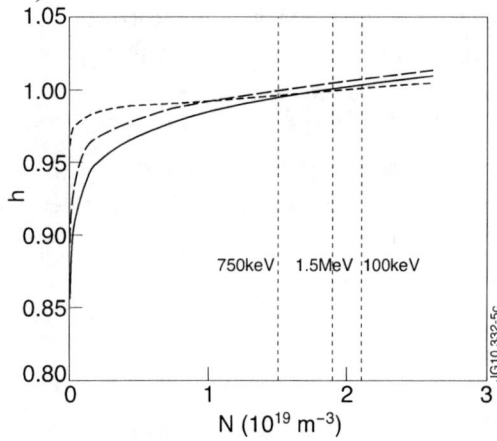

FIGURE 5. Space charge compensation, h, as a function of gas density, N, for uniform D⁻ beams of 45mA, r_0=12mm and T_i=0.77eV. The parameter is beam energy: —— 1.5MeV, — — 750keV, ······ 100keV. The vertical broken lines indicate the value of N_c for the three beam energies. For explanation of 750keV data see text.

It is clear that over most of the density range the beam will be fully compensated even for such a high ion temperature, however even the slight shortfall in compensation (~2%) for the 100keV beam could cause a doubling of the beam radius at 1.5m. It is therefore necessary to determine the probability that the plasma ion temperature could reach fractions of an electron volt.

It is obvious from Fig. 4 that the ion temperature plays an important role in determining the space charge state of the beam, although very little work exists on this subject in the literature. The choice of T_i is not entirely arbitrary as there are two possible sources of energy, other than coulomb interactions with the beam ions, available to the beam plasma, namely dissociation by the beam ions and plasma electrons and coulomb interaction with the plasma electrons as they leave the beam channel. Beam dissociation of the background deuterium to create D^+ ions can occur through two channels:

$$\underline{D}^- + D_2 \rightarrow \underline{D}^- + D^+ + D + e$$
$$\rightarrow \underline{D}^0 + D^+ + D + \underline{e} + e$$

$$\underline{D}^- + D_2^+ \rightarrow \underline{D}^- + D^+ + D$$
$$\rightarrow \underline{D}^0 + D^+ + D + \underline{e}$$
$$\rightarrow \underline{D}^- + 2D^+ + e$$
$$\rightarrow \underline{D}^0 + 2D^+ + \underline{e} + 2e$$

where the energy per dissociation product is indicated in Table 2. For electrons the favoured route is by dissociation of the molecular ion only. The degree of dissociation from each route can be estimated from a series of rate equations of the form:

$$\frac{dn_d}{dt} = n_t n_f \sigma v_t \qquad (16)$$

where the subscript t refers to test particle (beam or electron) and the subscript f refers to the field particle (molecule or ion). These rates can be normalised to the ion production rate given by Eq. 12 and the degree of dissociation relative to the ion density derived. The expressions are given in Table 2 with the relevant cross sections or rate coefficients obtained from velocity equivalence of the corresponding values for electrons given in [15]. The energy of the plasma electrons arising from beam ionization is derived from the fitting parameters in Rudd [12], which give the average electron energy as 15.4eV at 1.5MeV and 3.5eV at 100keV and the plasma electron temperature is set to half this value as described in [1].

Energy transfer between the energetic ions and electrons and the cold plasma ions depends on the collision frequency, ν [16] and the difference between the energies of the two sets of particles. For interaction between 15eV electrons and D_2^+ ions $\nu_{e,i} \sim 1 \times 10^{-15} n_i$ s^{-1}, so the electrons do not contribute to the ion energy. For 5eV D^+ ions and cold D_2^+ ions $\nu_{i,i} \sim 1 \times 10^{-13} n_i$ s^{-1}, so with ion density of the order 5×10^{14}m^{-3} the equilibration time for the dissociated ions is of the order 20ms, to be compared to an ionization time constant between 0.7µs and 650µs over the pressure range considered. It should be noted that some fraction of the energetic ions will be trapped in the beam potential and exhibit oscillatory motion about the beam axis. (It is also possible that the potential well also contributes to the ion heating through acceleration of the ions which accommodate before completing the full cycle of motion). It is therefore feasible that, at least at low background pressure the compensating ion distribution could exhibit an enhanced temperature, whilst as the pressure increases the potential well is "filled" with cold ions by the increasing ionization rate and the ion loss rate reduces. The system therefore moves towards the standard model of a continuously accumulating compensating ion density. Interestingly, a simple weighted average of the dissociated ion energies and density fractions over the whole ion population yields $T_i \sim 0.3$eV, assuming the cold ions are created at room temperature 0.025eV, implying that sufficient energy is available from the dissociation products to produce significant ion heating under suitable conditions.

TABLE 2. Dissociation channels, product energies and dissociation fractions calculated for the three beam energies.

Channel	Normalised Dissociation	Dissociation Energy per Particle (eV)	Dissociation Fraction (%) 1.5MeV	750keV	100keV
$\underline{D}^- + D_2 \to D^+ + D$	σ_1/σ_i	5	3.6	6.3	0.06
$\underline{D}^- + D_2^+ \to D^+ + D$	$n_i\sigma_2/2N\sigma_i$	5	0.26	0.03	0.23
$\underline{D}^- + D_2^+ \to 2D^+$		8			
$e + D_2^+ \to D^+ + D$	$\alpha n_i^2 \langle \sigma_3 v_e \rangle / 3 N n_b v_b \sigma_i$	5	0.01	0.00	0.00
$e + D_2^+ \to 2D^+$	$\alpha = n_e/n_i \ll 1$	8			

Estimation of Gas Density in Photoneutralizer Systems

Given that there is no design for a photoneutralizer system and that the ultimate background pressure will be dependent on pumping provision no definitive statement regarding the relevant density can be made. However, it is possible to estimate the probable density from the ITER system values [17], adjusted for the absence of the neutralizer gas flow. It is unlikely that the gas flow to the ion source will be significantly different (unless exotic technology emerges in the following decade), so assuming an operating pressure of P_S the required source flow, Q_S, can be estimated from:

$$Q_S = P_S C_{acc} + Q_b \qquad (17)$$

where C_{acc} is the total accelerator conductance and Q_b is the flow equivalent of the extracted ion beam. This latter flow does not contribute to the background pressure in the beamline, which is determined solely by the first term on the right hand side of Eq. 17. Taking the ITER MAMuG accelerator design, $C_{acc} \sim 5.1 m^3 s^{-1}$ at 500K and assuming a source pressure of 0.3Pa, Eq. 17 yields $P_S C_{acc} \sim 4 \times 10^{20}$ molecules s^{-1}. The neutralizer gas flow in the ITER four channel design is 5.5×10^{21} molecules s^{-1} which, combined with the residual source flow gives an average pressure of ~0.01Pa in the beamline between the accelerator and the neutralizer. Thus removing the neutralizer flow would result in a background pressure of $\sim 8 \times 10^{-4}$Pa, towards the lower end of the range investigated in the previous section. Of course these pressures are only indicative and are dependent on the details of the pumping system.

CONCLUSION

The effect of incomplete compensation of the space charge of a negative ion deuterium beam has been investigated over a range of energies corresponding to commissioning phases and full energy operation, motivated by the need to implement the gas-free photoneutralizer on NBI for DEMO. It is found that, fortuitously, the beam-plasma system is most likely to be fully compensated at lower energies when it is most sensitive to the effects of space charge on beam transport. Indeed, over a wide range of pressure and energy the beam will tend towards a fully compensated system so the impact on transmission will be minimal. This is primarily due to the poor rate of transfer of energy between the plasma dissociation products and bulk of the plasma ions compared to the ionization rate of the beam. As a result the beam potential well "fills" with cold ions at a faster rate than the heated ions can escape.

There is no reason, therefore to anticipate problems in transporting a DEMO relevant beam into such a neutralizer even during low energy prototype testing and commissioning phases.

ACKNOWLEDGMENTS

The authors would like to thank B Crowley and M Kovari for supplying the neutral beam test bed emittance data. This work was funded by the United Kingdom Engineering and Physical Sciences Research Council under grant EP/G003955 and the European Communities under the contract of Association between EURATOM and CCFE. The views and opinions expressed herein do not necessarily reflect those of the European Commission.

REFERENCES

1. E. Surrey *11th Int. Symp. Production and Neutralization of Negative Ions and Beams, Santa Fe, NM 2006*, AIP Conference Proceedings 925, American Institute of Physics, Melville, NY, 2007, pp. 278-289.
2. E. Surrey, D.B. King, J. Lister, M. Porton, W. Timmis and D. Ward, *Symp Fus Tech 2001, Porto, Portugal* to be published in Fus. Eng. Des.
3. J.H. Fink, *3rd Int. Symp. Production and Neutralization of Negative Ions and Beams, Brookhaven, NY 1983* AIP Conference Proceedings 111, American Institute of Physics, Melville, NY, 1984, pp. 547-560
4. T. Inoue, M. Hanada, M. Kashiwagi, et al, *Fus. Eng. Des.*, **81**, 2006, pp. 1291-1297
5. M. Kovari and B Crowley, *Fus. Eng. Des.*, **85**, 2010, pp. 745-751
6 S. Humphries, *Charged Particle Beams, Ch 5.4*, 1990, John Wiley & Sons, Inc. New York
7. A.J.T. Holmes, *Beam Transport, The Physics &Technology of Ion Sources*, Ian G Brown Ed., Ch 4 1st Edition, 1989, John Wiley & Sons, New York
8. R. Baartman and D. Yuan, *Space Charge Neutralization Studies of an H Beam*, EPAC 88, 1988, pp. 949-950
9. E.J. Horowitz, C.R. Chang and M. Reiser, *Transport of Intense, High Brightness H Beams*, SPIE 1061 Microwave and Particle Beam Sources and Directed Energy Concepts, 1989, pp. 483-488
10. M.D. Gabovich, *Sov. Phys. Usp*, **20**, 1977, pp. 134-148
11. I.A. Soloshenko, V.N. Gorshkov and A.M. Zavalov, *11th Int. Symp. Production and Neutralization of Negative Ions and Beams, Santa Fe, NM 2006*, AIP Conference Proceedings 925, American Institute of Physics, Melville, NY, 2007, pp. 262-277
12. M.E. Rudd, Y-K Kim, D.H. Madison and T.J. Gay, *Rev. Mod. Phys*, **64**, 1992, pp. 441-490
13. C.F. Barnett, J.A. Ray, E. Ricci et al. *Atomic Data for Controlled Fusion Research*, ORNL-5206, 1977, Oak Ridge National Laboratory, Oak Ridge, TN, USA
14. Ya.M. Fogel, A.G. Koval and Yu.Z. Levchenko, *Sov. Phys. JETP*, **11**, 1960, pp. 760-767; ibid. **12**, 1961, pp. 384-391
15. E.M. Jones, *Atomic Collision Processes in Plasma Physics Experiments*, CLM-R 175, 1977, Culham Laboratory, Abingdon, U.K.
16. B.A. Tubnikov, *Rev. Plasma Phys.*, **1**, 1965, pp. 105
17. ITER Design Description Document N53 DDD 29 01-07-03 R0.1, *Neutral Beam Heating & Current Drive (NBH & CD) System*, 2003, IAEA

SIPHORE: Conceptual Study of a High Efficiency Neutral Beam Injector Based on Photo-detachment for Future Fusion Reactors.

A. Simonin[a], L. Christin[a], H. de Esch[a], P. Garibaldi[a], C. Grand[a],
F. Villecroze[a], C. Blondel[b], C. Delsart[b], C. Drag[b], M. Vandevraye[b],
A. Brillet[c] and W. Chaibi[c]

[a] IRFM, CEA Cadarache, IRFM, St. Paul-lez-Durance, France.
[b] LAC : Aimé-Cotton Laboratory, Univ. Paris-sud, Orsay, France
[c] ARTEMIS Laboratory, Côte-d'azur Observatory, Nice, France.

Abstract. An innovative high efficiency neutral beam injector concept for future fusion reactors is under investigation (simulation and R&D) between several laboratories in France, the goal being to perform a feasibility study for the neutralization of intense high energy (1 MeV) negative ion (NI) beams by photo-detachment.

The objective of the proposed project is to put together the expertise of three leading groups in negative ion quantum physics, high power stabilized lasers and neutral beam injectors to perform studies of a new injector concept called SIPHORE (SIngle gap PHOto-neutralizer energy REcovery injector), based on the photo-detachment of negative ions and energy recovery of unneutralised ions; the main feature of SIPHORE being the relevance for the future Fusion reactors (DEMO), where high injector efficiency (up to 70-80%), technological simplicity and cost reduction are key issues to be addressed.

The paper presents the on-going developments and simulations around this project, such as, a new concept of ion source which would fit with this injector topology and which could solve the remaining uniformity issue of the large size ion source, and, finally, the presentation of the R&D program in the laboratories (LAC, ARTEMIS) around the photo-neutralization for Siphore.

Keywords: Negative ions, Plasma source, Fusion reactor, Plasma-heating, Photo-detachment.
PACS: 52.27.Cm, 52.50.Dg, 28.52.Cx, 52.50.-b, 32.80.Gc

INTRODUCTION

In parallel to the ITER construction is raising the question of the next step toward a real Fusion power plant (called DEMO reactor), which should be a full ignition and high power Fusion machine. Reactor simulations and debate around the main set of factors that appear to be the most important (design, size, plasma parameters, electricity production, etc..) are going on, the target of DEMO being the demonstration of electricity generation with self sufficient fuel supply at a moderate cost.

The choice of the additional plasma Heating and Current Drive (H&CD) systems and amount of additional power required are centrepieces in these studies as the H&CD systems have to provide an initial plasma heating to enter in the burn phase (ignition), and, to sustain a non inductive current drive for steady state or long pulse operations. It turns out that the H&CD power requested by the Fusion plasma amounts to about 20% of the net electrical power produced by the reactor [1,2,3]: 150 to 200MW of additional heating is required for 1GW of net electrical power.

As a consequence, the global efficiency of the reactor and electricity cost are strongly correlated to the re-circulating power necessary to supply the different auxiliary systems (H&CD systems, Helium pumping, etc...). Calculations show that a tolerable electricity cost requires a global H&CD efficiency factor higher than 60% (~ 330MW of H&CD re-circulating electrical power for the 1GW produced) while on the present H&CD devices [4], it only ranges between 15 to 25%.

Two H&CD systems are under consideration for DEMO, the NBI system mainly dedicated to plasma heating, which amounts of about 75% of the total H&CD power [3], and the electron cyclotron (ECR) heating system devoted to the plasma profile control. It is clear that NBI system remains an essential part of a reactor device, with very stringent conditions such as, the realisation of powerful high energy beams (several tens of MW at 1-2MeV) with high efficiency (> 60%) and small footprint in the reactor environment. The achievement of such performances requires a considerable R&D effort in parallel to the ITER construction.

In the present N-NBI systems, negative ion beam neutralization is produced via stripping reactions in a gas cell (the neutralizer); it is a simple and reliable method, but the neutralization efficiency is modest (up to 58%) and the background gas density along the beam line is such that in the ITER-NBI system, about 28% of the negative ions are lost due to molecular collisions in the accelerating channel. Together with the beamline transmission losses this causes the overall injector efficiency to be lower than 30%. Whereas all these losses are tolerable for an experimental fusion machine like ITER, they will be unacceptable in any future fusion-based power reactor (DEMO). This paper describes a new concept of high efficiency high energy NBI system (called SIPHORE); it is based on Photo-neutralization of the 1MeV negative ions (D⁻) coupled with an energy recovery system which collects the remaining D⁻ at the neutralizer exit. A feasibility study of this concept and R&D on negative ion photo-detachment will be performed in the near term in different laboratories, to address and highlight the main issues of the concept.

In the following, in section 1 is presented the principle of the Siphore concept, in section 2, a presentation of a new negative ion source concept which would fit with a Siphore injector, and in section 3, the present R&D on photo-neutralization which is addressed in the framework of this project.

PRESENTATION OF THE SIPHORE CONCEPT

Main Issue of the ITER-NBI Concept for a DEMO Reactor

The ITER machine will be heated with two lines of 17MW of D° at 1MeV [5]; the NBI system is based on the acceleration of negative ions D⁻ which are partially neutralized in a gas cell, the so called "neutralizer" where stripping reactions occur ($D^-_f + D_2 \rightarrow D^0_f + D_2 + e^-$), and cause part of the beam (~55%) [11] to be converted to neutrals (D°). Due to the background gas density, caused by the D_2 gas required to operate the ion source and the gas neutraliser, 28% of the extracted D⁻ ions are lost by stripping reactions generating stray particles in the accelerating channel and important thermal losses on the accelerator grids [6], yielding a low overall injector conversion efficiency of ~25%. A very important technological key to be addressed either for

ITER and DEMO is the reduction of the gas injection to minimise as much as possible the beam losses and stray particles in the accelerating channel. As the beam neutralization by a gas target accounts for about 75% of the total gas load in the beam line, R&D has been underway to explore other neutralization processes with higher efficiency and lower gas load, such as Li-vapor neutraliser [7], or Optical neutraliser[8,9].

Photo-Neutralization of Intense High Energy Negative Ion Beam

The neutralization of high energy negative ions by photo-detachment (photo-neutralization) is a very seducing and challenging alternative that appears to offer at the same time a complete suppression of the gas injection in the neutraliser, with a potentially excellent beam neutralization rate. An exploratory study dedicated to the feasibility of a photo-neutralizer relevant with the NBI system has been conducted over the last years in France [9]. The output of the study shows that the photo-detachment process of negative ions requires a large amount of photon power: for the same neutralization rate as on ITER (~55%) [11], a 1MeV 10A D⁻ beam sheet of 6cm width, 1.5m height (⇔one segment of the ITER injector), requires about 15MW of photon power (λ=1064nm). Such very high Laser power could be achieved by the use of a multi-refolded multi-metric (20-30 m long) high finesse (finesse ~5000-10000) Fabry-Perot cavity (see fig. 1) fed by a kW laser.

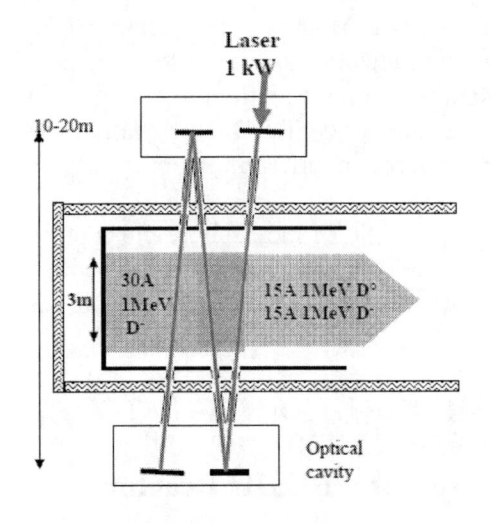

FIGURE 1. Principle of the Cavity re-folding (side view)

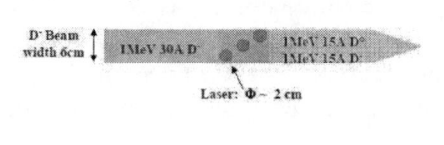

FIGURE 2. Overlap of the D⁻ beam by the Laser intra-cavity beam (Top view)

The refolding inside the cavity would allow a complete overlap of the ion beam, the number of refolding has to be as low as possible, it depends of the Laser and ion beam widths, the target being 3 refolding with a Laser width of 2cm, and an ion beam width lower than 6cm (see fig 2). For an optimized overlap, the laser beam has to keep approximately the same width over the ion beam height (along the cavity length), which compels the cavity close to a flat-flat, and therefore marginally stable configuration. Adding two extra mirrors to the cavity with appropriate curvature radius can move away the instability threshold; this concept has been applied with success in the recycling cavity of the Advanced Virgo project [10].

Additional major issues have also been identified in this study: the use of high quality optical components in a "polluted" environment (power plant + NBI), where radiations (hard X-rays, neutrons), mechanical vibrations and plasma-gas-metallic sputtering co-exist may considerably reduce the life time of the optical components. This issue could be overcome by a long (multi-metric) Fabry-Perot cavity, with optical components located far away from the injector and outside the nuclear island of the beam line and Tokamak; intermediate active pumping cells and absorbing materials would contribute to keep the optical cells under high and clean vacuum conditions.

An other issue outlined is related to the lack of an available high power (~kW range) highly stabilized laser ($\delta\nu/\nu \sim 2.\ 10^{-14}$) required to supply the FP cavity.

A dedicated Laser development has to be addressed, such performance could be achieved by the use of "Fiber Laser" (see paragraph 3).

Principle of the Siphore: SIngle gap PHOto-Neutralizer Energy REcovery Injector Concept (see Fig 3)

The photo-neutralization process has the double advantage to considerably reduce the amount of gas injected along the beam line (low stripping rate) and to avoid the formation of a secondary plasma in the neutralizer chamber. In conventional injector (with gas target), this plasma diffuses along the beam and leaks at both ends of the neutralizer [11]. Released from these constraining issues, a photo-neutralizer based injector can advantageously operate with a Singap accelerator [12] where the pre-accelerated ion beamlets, free of high level of stray electrons, are merged in a single beam and a post-accelerated to 1MeV in a single gap; moreover, this post-acceleration allows a lateral beam compression with a minimum waist in the photo-detachment area. An energy recovery system (ERS) [13] can be implemented at the neutralizer exit (free of secondary plasma) to decelerate and collect the remaining un-neutralized 1MeV D⁻ beam at low energy (~100keV); the ERS has a major consequence on the injector electrical set-up as the ion source has to be grounded, and the neutralizer at the high voltage (1MV). The beam line (source, accelerator) has to be designed in such a way to make the conditions for photo-neutralization and energy recovery as favourable as possible; both concepts requires a narrow D⁻ beam sheet: the injector topology would be vertically extruded, 30A of D⁻ could be extracted from a 3m height ion source (10cm width) coupled with a high transparency (40%) pre-accelerator. The neutralizer being held at 1MV, the stripped electrons released by the photo-detachment are trapped in the 1 MV potential well; on the contrary, a very efficient secondary electron trapping has to be implemented on the recovery electrode surface in order to avoid secondary electrons generated by impinging ions to be back accelerated towards the neutraliser (+1 MV).

In case of high voltage breakdowns or troubles along the beam line (or optical cavity), a very fast switch on the pre-accelerator stage (100kV) can interrupt the ion beam in the µs range.

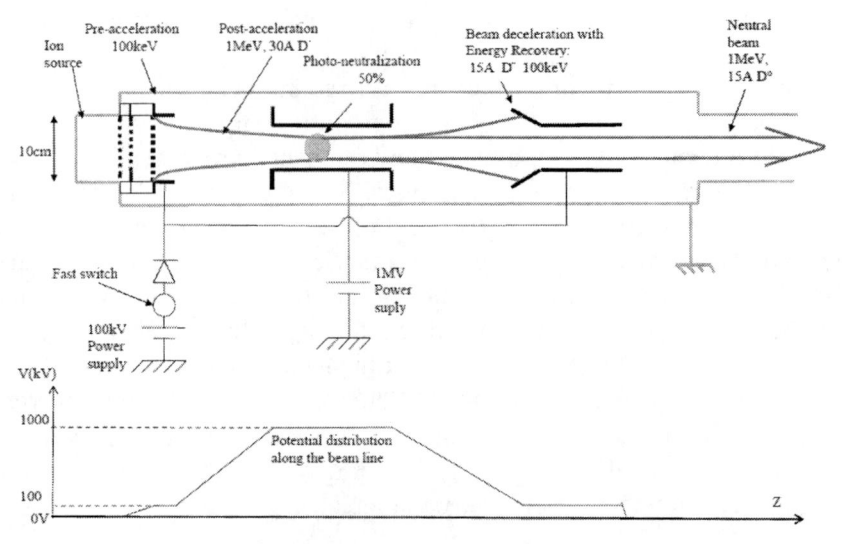

FIGURE 3. Principle of the Siphore injector

Figure 4 shows a side view of the injector topology as it could be implemented on a reactor: to reduce the damage of the optical components (mirrors) by particles, radiation or neutrons, the optical cells are located far away from the beam and injector, outside of the reactor nuclear island; the cavity pipes must be instrumented with intermediate pumping cavities, insulating valves (or fast shutter), and neutron absorbing materials.

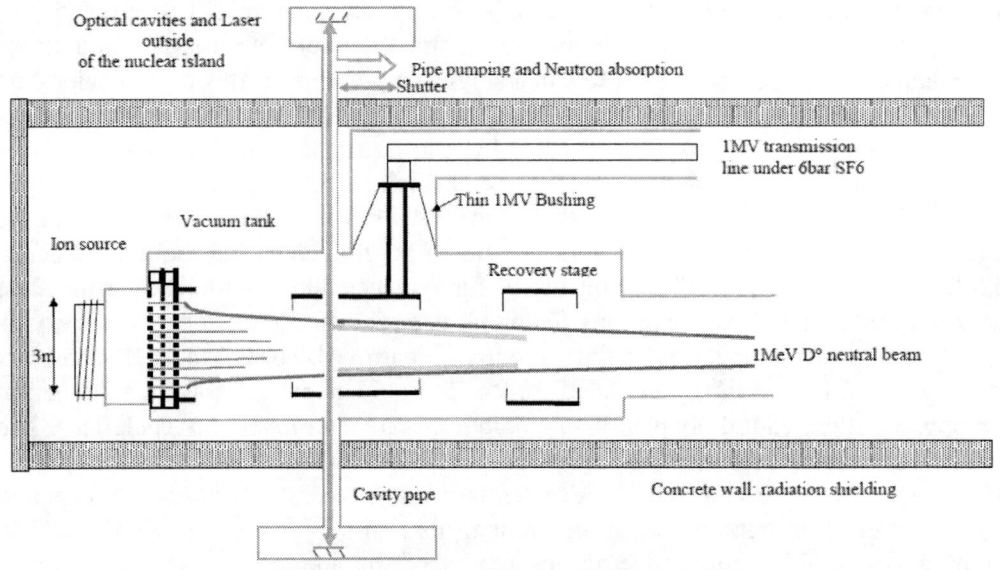

FIGURE 4. Transverse view of the Siphore Injector

Efficiency of a Siphore Injector

The global efficiency of the injector is the ratio of the neutral power at the entrance of the Tokamak chamber to the total electric power consumed (ion source + accelerator + recovery electrode + auxiliary systems). If one assumes a photo-neutralization rate of 50% (~15MW of intra-cavity power) on a 30A 1MeV D⁻ beam sheet, the injector would generate 15 MW of D° beam, while the powers spent being the power fraction of the recovered ion beam:~ 1.5 MW (15 A of D⁻ at 100 keV), 10% of losses (due to beam re-ionization and beam interception in the duct .5 MW) during the neutral beam transport between the injector and the plasma, about 4 MW of electrical power for the source, pre-accelerator, stripping losses, and sub-systems (pumping, etc.), 95% efficiency of the 1MV power supply [17], the laser power being in the range of a few kW (for 15MW of photon power intra-cavity) is negligible; in these conditions, the overall Siphore conversion factor (efficiency) is close to 65%.

Noticeable Advantages of the Siphore Injector

-) A grounded ion source yields a substantial simplification and cost reduction of the overall injector: no high voltage platform with the ion source power supplies at 1MV, simplification of the transmission line and Bushing, an important cost reduction, and an easy access to the source for remote handling

-) 1MeV powerful Tritium beams could be envisaged with Siphore for plasma heating and/or fuelling, in order to maintain high ion temperature and optimum D:T mix in the plasma core. The central Tritium gas would fill the grounded ion source to the required pressure, while the 1MeV T⁻ would be photo-neutralized (no Tritium gas injection at 1MV) or recovered at low energy.

Pending Developments around Siphore

A feasibility study of the Siphore concept has been launched this year in France, it should last for the next 3-4 years, the main objectives are :
-1) Ion beam optics simulations of the injector
-2) Development/modeling of a Siphore ion source
-3) Development of a high transparency pre-accelerator
-4) High voltage conditioning studies and 1MV Thin Bushing development
-5) Development of high power, high finesse Fabry-Perot cavity
-6) Development of high power highly stabilized laser to supply the FP cavity
-7) Basic physics in the field of negative ion photo-detachment.

ION SOURCE DEVELOPMENT FOR SIPHORE

Plasma Uniformity Issue in Large Size Negative Ion Sources

The Siphore ion source has to produce a long (~3m), thin (~10cm), intense (J_D~250A/m^2) and uniform negative ion beam sheet, outlining the unresolved plasma uniformity issue observed in large size negative ion sources [14,15].

This recurring difficulty faced by conventional negative ion sources results from the transverse magnetic filter (filter field) which crosses the source front face (the extraction region) to prevent negative ion destruction from hot electrons; this filter field generates a vertical plasma drift (**Bx∇B**), caused by the magnetic field gradient which exists between the back (or centre) of the source (free of B-field) and the extraction region (with the filter field).

An attempt to reduce such effect has been performed at IRFM on a large size ion source [16] (see fig 5b), where additional transverse magnetic filters (magnetic barriers) subdivide vertically the source in 5 modules. These magnetic barriers confine the primary electrons emitted by filamented cathodes in each module centre (free of magnetic field) (see fig 5a) and prevent them to drift vertically.

Figure 5c shows the ion density distribution on the extraction surface for different arc power; these experimental results show an important plasma in-homogeneity between top and bottom of the source, which increases with the arc power. At low power (50-90kW), a plasma uniformity of ± 10% is achieved by an over-polarization of the cathodes in low density regions, while, at higher arc power (P> 90kW), the plasma drift is dominant over the cathode over-polarization: the ion density on the top of the source is 7 times higher than in the bottom (10 times higher for electrons).

A 3D trajectory code which is a combination of Runge-Kutta method for the 3D particle trajectories with Monte-Carlo methods for particle kinetics has been adapted to simulate plasma particle trajectories (e-, D+) in the source configuration of figure 5a (only 3 source modules); the code does not simulate any plasma effect. It shows that the primary electrons emitted in the module centre (B< 3Gauss) have a mean free path (mfp) lower than 15cm, they are confined in the magnetic well of the module, as expected, where they undergo inelastic collisions (ionization). On the other hand, ions are not collisionnal in the plasma, there mfp is in the range of 1.5m (for a source pressure 0.2 Pa); the code shows a tendency of ions to propagate upwards due to the filter field, where they recombine to neutrals by collisions with the source walls. Moreover, these neutrals with a mfp of the cm range, contribute to increase locally the ionization rate and plasma density. This ion drift could act as neutral pumping from bottom to top involving plasma in-homogeneity; a confirmation of this effect would require a dedicated plasma code including either plasma-wall interactions, neutral & particle transport and plasma diagmagnetic effects.

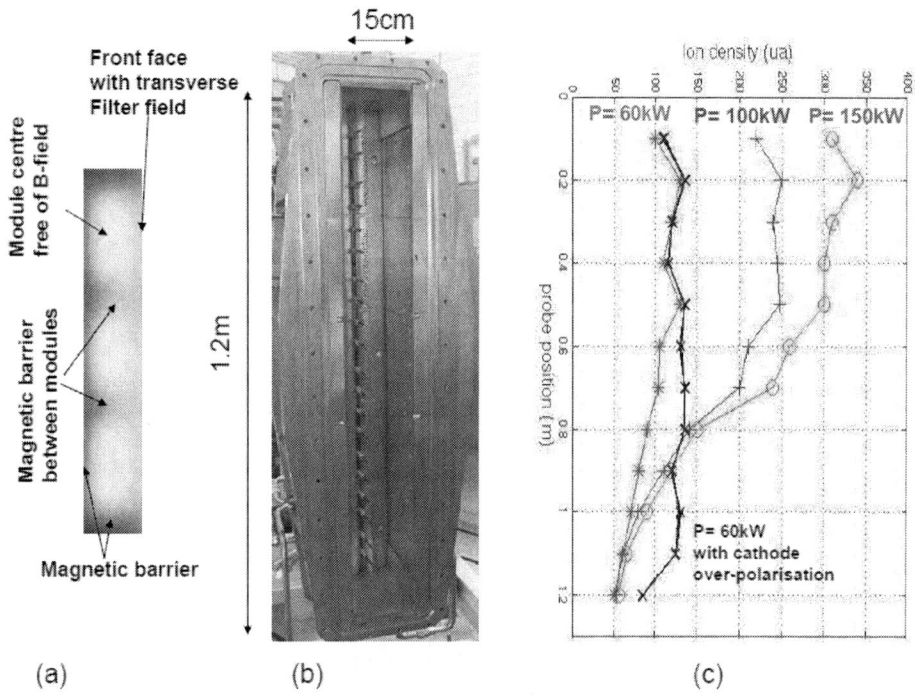

FIGURE 5. (a) Magnetic field distribution with 3 modules on the vertical plan
(b) The Ion source with 5 modules
(c) Source uniformity on the vertical axis for different arc power

Principle of the Siphore Ion Source

It was shown above that the Cybele magnetic confinement (with transverse filter field) cannot be considered for a long and thin ion source; we propose for Siphore, a magnetic confinement based on a vertical magnetic field generated by lateral coils with iron core (see fig 6a, 6b). The current in the coils is 20A/cm for about 60 Gauss in the plasma volume (see fig 6b); we note that the magnetic field is homogenous in the plasma area and drops to zero on the accelerator side with an additional transverse current flowing in the plasma grid (~10A/cm). Moreover, this topology can be vertically extruded without any limit (3m height) keeping an uniform B-field along the source.

Simulation of the 3D particle trajectories with this magnetic field topology shows that primary electrons are trapped by the magnetic field lines and do not drift in the source. On the other side, figure 6c shows the ion trajectories in the source: they are magnetized and rotate around the B-field lines, and leak on both source ends (top and bottom); further simulations would require the development of a dedicated "magnetized" plasma fluid code.

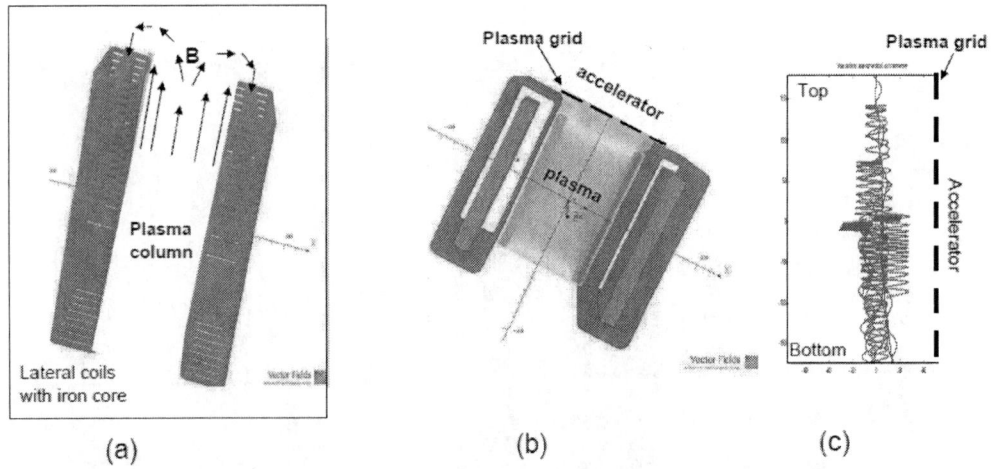

FIGURE 6. (a) Source magnetic confinement: 2 lateral coils (red) with iron core (blue)
(b): Horizontal cross section of a source with B-field distribution
(c): Side view of the ion trajectories in the source (source height 1.2m)

NEGATIVE ION PHOTO-DETACHMENT, OPTICAL CAVITY AND LASER DEVELOPMENTS

The Optical Cavity Configuration

The optical cavity is a key element of the Siphore concept. Indeed, it provides the opportunity to neutralize high energy negative ions beam without secondary plasma generation. This allows non-neutralized ions energy to be recycled which enhance the overall injector efficiency. Nevertheless, the optimum cavity configuration is still unknown. The number of refolding in one hand and the mirrors specifications (curvature, reflectivity…) in the other hand are to be determined in order to reach an optimum coupling between the laser and the Fabry-Pérot cavity.

Actually, the cavity geometry and the mirrors specifications are interconnected matters, and shall be chosen following the discussion below:

-) The finesse of the cavity, i.e. the mirrors reflectivities (mirrors with average scattering and absorption <5ppm), can be chosen as low as possible (provided the laser power is high enough to reach the needed intracavity power.

-) The number of mirrors increases with the number of refolding which enhance the overall cavity losses and limits the cavity finesse.

-) Decreasing the number of refolding increases the laser beam width needed to cover the whole negative ion column and decreases the cavity length

-) Increasing the laser beam width increases the light scattering losses on each mirror.

-) The system is less sensitive to wave front thermal distortions while increasing the cavity length and decreasing the laser beam width.

-) The thermal compensation systems are likely to control wave front thermal distortions and allow the cavity to be close to the instability threshold.

in collisions with a proper buffer gas and the ion trajectories can be focused to a small region near the longitudinal axis of the device. As a result, beams with significantly reduced energy spread and emittance can be obtained. This maximizes the effective isobar suppression of the magnetic isotope separator and thus improves the purity of RIBs for HRIBF research. In this paper we report the latest development and performance of the RFQ cooler for cooling and improving the beam quality of negative ions.

There are still many cases in which the isobaric contaminants in the RIBs cannot be removed effectively by magnetic separation even after beam cooling. Consequently, additional effective and efficient beam purification techniques are necessary. A highly efficient isobar suppression method based on photodetachment in a RFQ cooler has been developed [2]. The use of the RFQ cooler makes it possible to achieve near 100% photodetachment efficiency and thus almost complete suppression of the isobar contaminants in certain negative RIBs [3-5]. Recently, the feasibility of depleting the excited populations in atomic negative ions by photodetachment in the RFQ cooler has been demonstrated [6]. This implies that pure ground-state negative ion beams could be obtained for high-precision studies of negative ions with enhanced selectivity and accuracy. The experimental results obtained at HRIBF on improving the purity of negative ions by photodetachment in the RFQ cooler are presented.

DESCRIPTION OF THE RFQ ION COOLER

A RF-only quadrupole acts as a high-pass mass filter or ion guide [7]. Ions with mass larger than a certain value can be radially confined within the quadrupole while proceeding in the axial direction. When a buffer gas is introduced into the quadrupole, the ions will encounter collisions with the buffer gas. If the gas atoms or molecules are significantly lighter than the ions, the net effect is that ions will lose energy [8]. With sufficient buffer gas pressure, ion energies can be reduced to approximately the thermal energy of the buffer gas and ion transverse motions will also be damped to a small region near the quadrupole axis. As a result, ion beams extracted from the cooler will have very small emittance and high brightness. Thus, gas-filled RFQ ion guides have been extensively used for beam cooling and bunching for high-precision measurements. More details of the principle of RFQ ion cooler and buffer gas cooling can be found in [1].

Figure 1 is a schematic view of the RFQ cooler at HRIBF. It is designed for cooling light negative RIBs such as ^{17,18}F$^-$, ^{33,34}Cl$^-$, and ^{56}Ni$^-$. The RF quadrupole consists of four parallel cylindrical rods of 8-mm diameter and 40-cm length, equally spaced with an inscribed circle of 3.5 mm radius. Equal and opposite-polarity RF voltages are applied to adjacent rods. The quadrupole operates at frequencies around 2.75 MHz with RF amplitudes up to 500 V. The quadrupole rod structure is mounted inside a Cu cylindrical enclosure of 3 cm inner diameter, with a 3-mm diameter entrance aperture and a 2-mm diameter exit aperture. He is used as the buffer gas and is introduced into the quadrupole enclosure through an orifice located near the middle of the Cu cylinder.

After the ions are cooled to near the thermal energy of the buffer gas, the ion motions along the longitudinal axis are essentially governed by diffusion. It is

To sum up, the cavity configuration depends on the light power the laser system can provide and the efficiency of the thermal compensation systems we can design.

The Laser-Cavity Coupling

In order to couple the laser into the optical cavity, two conditions are to be verified. The laser transverse mode has to match the cavity natural mode, which is simply done by using a sequence of lenses. Moreover, the laser is to be kept resonant with the optical cavity. It is much easier to act on the laser frequency than changing the cavity length for a wide frequency range. Therefore, the laser has to be stabilized on the cavity: its frequency noise has to be reduced with respect to the cavity resonance frequency which is related to its optical length noise by $\delta v/v = \delta(nL)/nL$, where n represents the optical index if the intracavity area, L, the cavity length and v, the laser frequency.

The relation $\delta(nL)/nL = \delta n/n + \delta L/L$ displays two different noise sources to the cavity optical length. The seismic noise is responsible for the mechanical fluctuation of the cavity length, with $L \simeq 100\text{m}$, $\delta L/L \simeq 10^{-8}$ at maximum. Whereas, the optical index variation is due to the presence of the ion beam within the cavity. As a first approximation, we neglect the laser-ion interaction, and consider the ion beam as a neutral gas crossed by the laser. For 40A at 1MV, the ion beam density is $2.5\ 10^{13}\text{m}^{-3}$ which gives the optical index of the ion beam with an acceptable approximation $n \simeq 1 + 3\ 10^{-8}$. Hence, $\delta n/n \simeq 3\ 10^{-8}$ represents a higher limit for the contribution of the ion beam index fluctuation to the optical length noise. These are common values in the stabilized lasers domain and should not represent any difficulties which could not be overcome.

Photodetachment R&D Program

The research program for the photo-neutralizer can be subdivided into two subjects:
-i) A small scale experiment is being built in Aimé Cotton laboratory where a 1nA-1μA ion beam shall be fully neutralized through an optical cavity which should be able to focus 40kW of light power out of a 10W laser system. The aim is to confirm the optical cavity based photo-neutralizer. Moreover, fundamental studies on the ion-laser interaction shall be carried out in order to enhance the photodetachment cross section.
ii) A metric scale high power cavity is under construction in the ARTEMIS laboratory. The Artemis group is working on a fiber laser system which shall provide 200W light power. The same concept should be able to reach 1kW in the near future. This laser will be injected in a high finesse cavity (F=30000) which is designed in such a way that the intracavity beam has a constant cross section on a metric distance. Thermal compensation systems shall be added to the mirrors in order to optimize the laser-cavity coupling efficiency. The intracavity power should reach 4MW.
Following this R&D program, all conceptual and technical issues shall be addressed in order to indentify the optimum photo-neutralizer configuration from an optical point of view. At the end of this program, an injector scale experiment shall be designed in

which a high energy ion beam will be photo-neutralized within an optical cavity that fits neutral beam injector's constraints.

ACKNOWLEDGMENTS

This work, supported by the European Communities under the contract of Association between EURATOM and CEA, was carried out within the framework of the European Fusion Development Agreement and "Federation de Recherche " in France. The views and opinions expressed herein do not necessarily reflect those of the European Commission.

REFERENCES

1. LI-Puma et al., "*Consistent integration in preparing the helium cooled lithium lead DEMO 2007 reactor*"; Fusion Eng. And design 84, 2009, pp 1197-1205
2. H. Zohm," *On the minimum size of DEMO*"; Fusion science and technology, vol 58 Oct. 2010, pp 613
3. J. Garcia et al.; "*Analysis of DEMO scenarios with the CRONOS suite codes*"; Nucl. Fusion 48 (2008), 075007
4. J. Pamela et al.; "*Efficiency and availability driven R&D issues for DEMO*"; Fusion Eng. And design 84,2009, pp 194-204
5. RS. Hemsworth et al.;"*Status of the ITER neutral beam injection system*";Review of scientific instrument; vol 79, issue 2, pp02C109 - 02C109-5, 2008
6. G. Fubiani, et al.; "*Modeling of secondary emission processes of the negative ion based electrostatic accelerator of ITER*"; Physical review special topics Accelerator and beams; vol 11 Issue 1; n° 014202; 2008
7. L.R. Grisham; " *Lithium JET Neutralizer to improve Negative Ion Neutral Beam performances*"; AIP Conference proceeding 1097, pp 364, Negative Ion Beam and Source (NIBS) Conference 2008
8. J.H. Fink, "*Photodetachement now*"; 12th Symposium on Fusion Engineering, Monterey, CA 12-16 October 1987
9. W. Chaibi et al.," *Photo-Neutralization of negative Ion beams for Future Fusion reactor*", AIP Conference Proceedings 1097, pp385; Negative Ion Beam and Source (NIBS) Conference 2008
10. A. Brillet et al., "VIRGO : Proposal for the construction of a large interferometric detector of gravitational waves", Unpublished (1989)."
11. T. Minea et al., "*Simulation code for the D- beam transport through the ITER neutralizer* Journal of Optoelectronics and Advanced Materials, 10(8), 2008, pp. 1899-1903
12. H.P.L. De-Esch et al.; "*Updated physics design ITER-SINGAP accelerator*"; Fusion Engineering and Design Volume 73, Issues 2-4, Oct. 2005, pp 329-341
13. J. Pamela et al. "*Energy Recovery experiments with a powerful 100keV D- based neutral beam injector*"; Nuclear Instr. And methods in Physics research B73 (1993), pp 296-302
14. M. Hanada et al.; "*Improvement of beam uniformity by magnetic filter optimization in a Cs-seeded large negative-ion source*"; Review of Scientific Instrument; March 2006, Vol 77, issue n°3, pp 03A515
15. A. Simonin et al. " The DRIFT source : a negative ion source module for direct current multi-ampere ion beams"; Review of Scientific Instruments, vol.70, n°12 (1999) p.4542-4544 (1999)
16. A. Simonin et al.; "Cybele: A large size ion source of modular construction for the Tore-Supra diagnostic injector",Review of scientific Instrument, vol77, issue 3, pp 03A525 - 03A525-3, 2006
17. Private communication from the ITER NBI team

Improving Negative Ion Beam Quality and Purity with a RF Quadrupole Cooler

Y. Liu

Physics Division, Oak Ridge National Laboratory, Oak Ridge, Tennessee 37831, USA

Abstract. Recent progress in the development of a gas-filled RF quadrupole ion cooler for cooling negative ions is reported. Experiments demonstrate that negative ion beams can be cooled to 2 eV FWHM energy spread with more than 50% transmission through the cooler. The RFQ cooler can potentially improve the purity of radioactive ion beams by magnetic mass separation. New developments on purifying negative ion beams by photodetachment in the RFQ cooler are presented. With a laser of proper photon energy, nearly 100% suppression of the unwanted negative ions in the RFQ cooler has been observed, while the desired ions remain mostly intact. A recent experimental study demonstrates that pure ground state negative ion beams can be obtained by state-selective photodetachment in the RFQ cooler.

Keywords: RF quadrupole ion guide, ion cooling, photodetachment, negative ion beam
PACS: 29.27.Eg, 37.10.Rs, 32.80.Gc, 41.75.Cn

INTRODUCTION

Ion beam quality and purity are of crucial importance to many basic and applied areas of research. The Holifield Radioactive Ion Beam Facility (HRIBF) at Oak Ridge National Laboratory (ORNL) provides accelerated radioactive ion beams (RIBs) for experimental research in nuclear physics and nuclear astrophysics. The RIBs are produced using the isotope separator on-line (ISOL) method. In the ISOL process, ions of neighboring nuclei having the same mass number are produced simultaneously, usually in quantities exceeding the isotopes of interest by orders of magnitudes. The isobaric contaminants complicate and often compromise experiments. Therefore, they must be either eliminated or reduced to tolerable levels. A two-stage magnetic isotope separator with a nominal mass-resolving power of $M/\Delta M \cong 20000$ is used at HRIBF for isobaric purification. However, such resolving power can only be achieved with very high-quality ion beams (i.e., beams with very small emittance and energy spread). A 25-MV tandem electrostatic accelerator is used at HRIBF to accelerate the mass-selected RIBs to required energies. The tandem accelerator requires negative ions as input. Negative ion beams are usually generated with a Cs-sputter negative-ion source or indirectly by positive-to-negative charge exchange. Such beams often have inherently large emittance and large energy spread, which limit the degree of isobaric purification that can be achieved by the isotope separator.

A Gas-filled RF quadrupole (RFQ) ion cooler has been developed for improving the quality of negative RIBs at HRIBF [1]. In the RFQ ion cooler, ions can be cooled necessary to provide an electric field in the axial direction to push the ions out. This is done with four inclined DC electrodes which are placed between the quadrupole rods. When the DC electrodes are negatively biased at appropriate potentials, a weak longitudinal field is created along the axis of the quadrupole that gently pushes the negative ions toward the exit aperture.

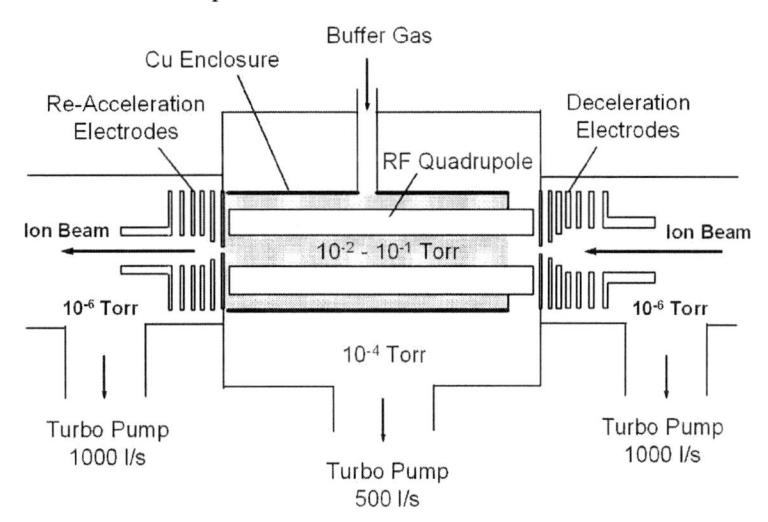

FIGURE 1. Schematic view of the RFQ cooler with deceleration and re-acceleration electrodes.

Ions are extracted from ion sources at high energies. They are decelerated to low energies to enter the cooler and re-accelerated to high energies upon exiting the cooler. Negative ions are fragile since their electron affinity is less than 3.7 eV. Electron detachment can take place in collisions with the buffer gas. The threshold energy for electron detachment in collisions with He has been estimated to be about 36 eV and 20 eV for F^- and O^-, respectively [1]. Thus, negative ions must be decelerated to very small energies, typically less than 40 eV, before entering the quadrupole in order to eliminate collisional detachment losses. The deceleration and re-acceleration electrodes are coupled with the cooler as illustrated in Fig. 1.

EXPERIMENTAL SETUP

The performance of the RFQ cooler has been measured at one of the off-line ion source test facilities at HRIBF using stable negative ions. The negative ions are produced using Cs-sputter or plasma-sputter negative ion sources. Ions extracted from the ion source are mass-separated in a 90 degree dipole magnet and the selected ions are then focused into the RFQ cooler by an Einzel lens. The ion intensity injected into the cooler is measured with a Faraday cup before the Einzel lens. As shown in Fig. 2, ions re-accelerated out of the cooler are sent to a scattering chamber where they are directed by an electrostatic deflector to a Faraday cup for intensity measurement or to an electrostatic energy analyzer for energy spread measurement. Ion-transmission efficiencies are determined as the ratio between the ion-beam intensities measured in Faraday cups before and after the cooler. The energy spread in the ions prior to the cooling process is also measured by the electrostatic energy analyzer with the cooler grounded and evacuated. Due to limitations of the experimental apparatus, all the measurements are performed with ions extracted from the ion source at about 5 keV energy.

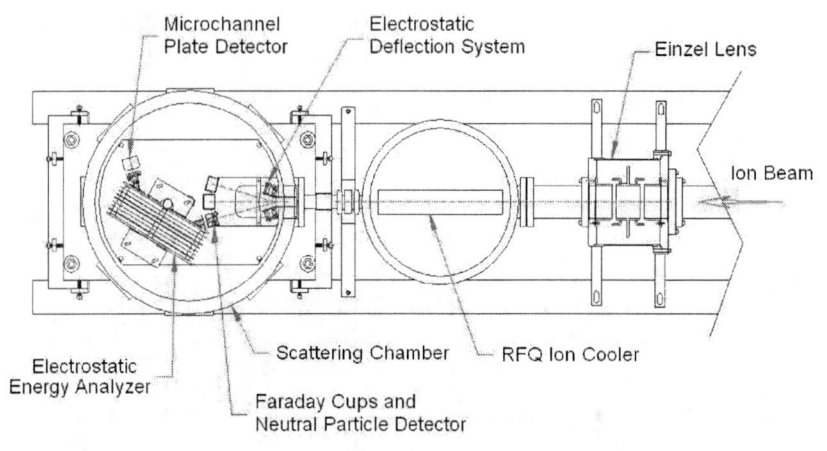

FIGURE 2. Schematic view of the setup for beam cooling and energy spread measurements.

COOLING OF NEGATIVE ION BEAMS

The negative ions produced with the Cs-sputter or plasma-sputter ion sources at HRIBF have emittances of 50-70 π mm mrad for 20 keV beams and energy spreads of about 5 eV full width at half maximum (FWHM) with long tails extending more than 10 eV. Fig. 3(a) shows a typical energy spread of ^{19}F$^-$ beams produced in a RF-plasma sputter negative ion source.

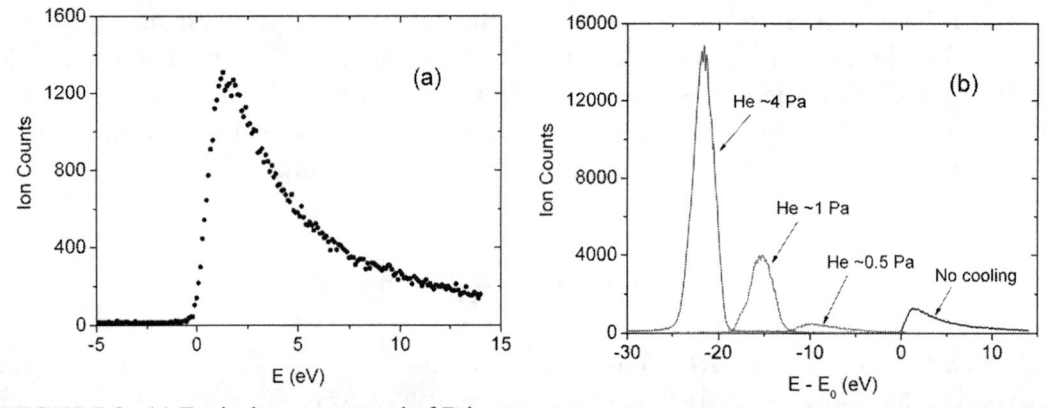

FIGURE 3. (a) Typical energy spread of F$^-$ beams produced in a RF-plasma negative ion source. (b) Measured energy distributions of F$^-$ beams before (no cooling) and after the RFQ cooler at different He buffer gas pressures. The RFQ was operated at 2.75 MHz with RF amplitude of ~ 80 V.

The effects of cooling the ion beams by the RFQ cooler can be seen in Fig. 3(b), which shows the measured energy distributions of the F$^-$ ion beams passing through the RFQ cooler at different He buffer pressures. The F$^-$ beams were injected into the RFQ cooler after being decelerated to less than 25 eV. With increasing He gas pressure, the energy spread of the ions became smaller and the mean energies of the ions after cooler also shifted to lower values. This indicates that the initial kinetic energy of the ions was dissipated in collisions with the buffer gas and, consequently, the ions at cooler exit had lower energies than the ions at the point of injection. It is also seen that the measured ion counts increased rapidly with increasing He pressure, suggesting higher transmission through the cooler for colder ions. The measured ion energy spread was about 2.2 eV FWHM after cooling with about 4 Pa He buffer gas.

Figure 4 shows an example of cooling ^{16}O$^-$ beams produced with a Cs-sputter negative ion source. The initial energy spreads of the O$^-$ beams were larger than 7 eV FWHM. After cooling in the RFQ cooler at a He pressure of 2.4 Pa, the energy spread was reduced to about 2 eV FWHM with near Gaussian distribution. Similar cooling results have been obtained with other negative ions studied.

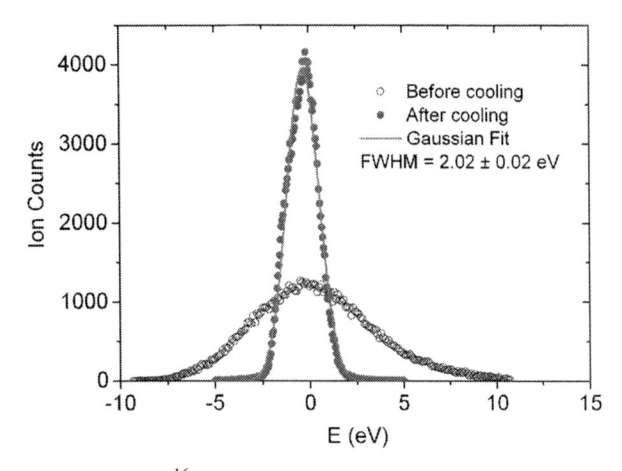

FIGURE 4. Energy distributions of ^{16}O$^-$ beams before and after cooling. The O$^-$ ions were produced in a Cs-sputter negative ion source. The RFQ cooler was operated at 2.76 MHz, RF amplitude of ~ 60 V, and about 2.4 Pa He pressure.

Figure 5 shows the calculated mass resolutions of the HRIBF isotope separator for ^{17}F + ^{17}O RIBs. If the ^{17}F + ^{17}O beams have an emittance of 70 π mm mrad at 20 keV and energy spread of 5 eV FWHM, which correspond to real beam properties without cooling, it is impossible to suppress the ^{17}O contaminants without substantially scarifying the ^{17}F intensity [Fig. 5(a)]. If the beam emittance is reduced to 10 π mm mrad at 20 keV and the energy spread is reduced to 1 eV FWHM, ^{17}O can be well separated from ^{17}F and pure ^{17}F beam can be obtained, as seen in Fig. 5(b). Moreover, the ^{17}F beam intensity passing the isobar separator will be significantly higher.

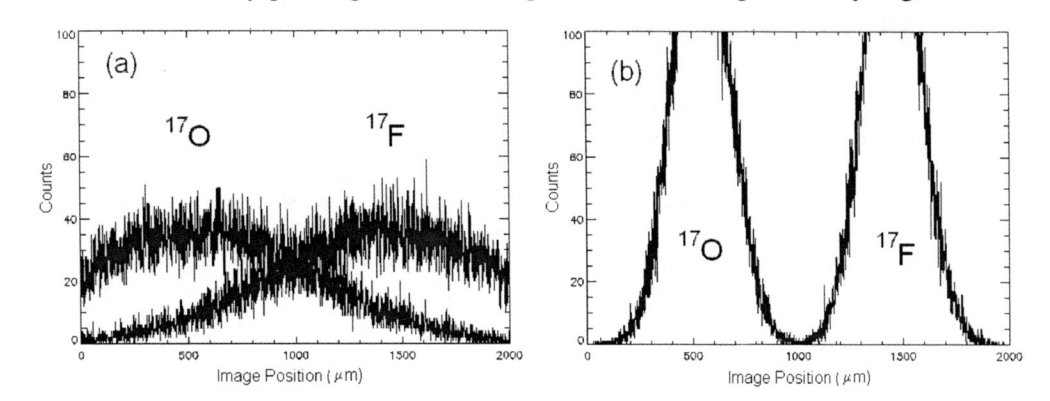

FIGURE 5. Calculated mass resolutions of the HRIBF isotope separator with different input beam qualities. (a) Beam emittance 70 π mm mrad and 5 eV FWHM energy spread. (b). Beam emittance 10 π mm mrad and 1 eV FWHM energy spread.

The smallest energy spread that has been obtained with the RFQ cooler is 2 eV at FWHM. We are not able to measure the ion beam emittance after the cooler. However, the emittance of the ion beams from existing RFQ coolers, with similar or larger energy spreads, has been reported to be well below 10 π mm mrad for beam energies of 5 kV - 60 kV [9-11]. Therefore, the output beam emittance of our RFQ cooler is expected to be less than 10 π mm mrad at 20 keV beam energy. This would be a substantial improvement in beam quality. According to the calculations in Fig. 5, the present RFQ cooler can significantly enhance the isobar suppression capability of the HRIBF isobar separator for RIBs. Further reduction in energy spread is possible by improving the scheme for re-accelerating the cooled ions and gas pumping in the re-acceleration region to minimize ion collisions with residual buffer gas during re-acceleration.

Improving Transmission Efficiency for Negative Ions

A key requirement for the use of the RFQ cooler in HRIBF research is high transmission efficiency. The first transmission efficiencies measured were only 10-14% for F⁻ and 5-6% for O⁻ ion beams. It was found that significant detachment loss occurred near the entrance aperture where the ions encountered many collisions with the outgoing buffer gas while their energy was still too high. To minimize the detachment loss, it was necessary to remove the buffer gas in the injection region so that negative ions will not encounter collisions before their energy is reduced to the desired value. Thus, a section of the Cu cylindrical enclosure, about 1 cm long, at the entrance end was removed, as illustrated in Fig. 1. The quadrupole section outside the Cu enclosure acted as a differential pumping region for quickly removing the buffer gas. It minimized the residual gas reaching the entrance aperture and escaping to the deceleration region. This modification increased the cooler transmission efficiency by a factor of more than three [12].

Table 1 lists the transmission efficiency obtained for the negative ions studied. It can be seen in Table 1 that for ions of similar masses the transmission efficiency is higher for the ions of larger electron affinity. This suggests that there are still some detachment losses, most likely in the deceleration and re-acceleration regions. Therefore, the transmission efficiency for these negative ions can be even higher if the cooler structure and operation parameters are further optimized.

TABLE 1. Overall Transmission Efficiencies of the RFQ Cooler

Negative Ion	EA (eV)	Mass (amu)	Transmission (%)
C	1.263	12	17 ± 2
O	1.461	16	30 ± 4
F	3.399	19	35 ± 3
Si	1.386	28	44 ± 2
S	2.077	32	49 ± 2
Cl	3.617	35	52 ± 4
Ni	1.156	58	56 ± 3
Co	0.661	59	52 ± 2
Cu	1.228	63	56 ± 3

BEAM PURIFICATION BY PHOTODETACHMENT

For many negative RIBs, magnetic separation is not sufficient to clean the isobar contaminants even with cooled beams. Additional effective beam purification means is necessary. Isobar contaminants are usually the adjacent-Z species of the isotope of interest. If the electron affinity (EA) of the isobaric contaminant (EA$_1$) is lower than the EA of the desired radioactive isotope (EA$_2$), it is possible to selectively remove the isobar contaminant by photodetachment with photons of energy EA$_1$ < $h\nu$ < EA$_2$. This method was proposed by Berkevits et al [13, 14]. However, the overall degree of isobar suppression reported was far from practically useful.

For a given laser power and photon energy, the fraction of negative ions not neutralized by the laser radiation is given by

$$\frac{n}{n_0} = \exp(-\sigma\phi t), \quad (1)$$

where n_0 is the initial number of negative ions before interaction with the laser radiation, n is number of ions not neutralized by the laser at time t, ϕ is the laser flux (photons cm^{-2} s^{-1}), and σ is the photodetachment cross-section (cm^2). To obtain high photodetachment efficiency and thus a high degree of isobar suppression, one needs either a large photon flux or a long interaction time.

We have developed a novel technique that can substantially enhance the efficiency of photodetachment by increasing laser-ion interaction time. The technique is based on directing laser radiation into the RFQ cooler and performing photodetachment in the cooler [2]. As discussed in the previous section, in such a RFQ cooler, ions can be cooled to approximately the thermal energy of the buffer gas and move many times slower through the RFQ. Thus, much longer laser-ion interaction time can be obtained. Simulations and experiments show that ion transit times can be on the order of milliseconds [4]. In addition, the ion trajectories will be confined to a small region near the axis, which improves the spatial overlap between the laser beam and the ions. Therefore, the RFQ cooler provides ideal conditions for laser-ion interactions.

Isobar Suppression by Photodetachment

A number of important RIBs, ^{17,18}F, ^{33,34}Cl, and ^{56}Ni, needed for studies in nuclear structure and nuclear astrophysics at HRIBF, are often dominated by isobaric contaminants ^{17,18}O$^-$, ^{33,34}S$^-$, and ^{56}Co$^-$, respectively. The electron affinities of these elements are F (3.41 eV), O (1.43 eV), Cl (3.62 eV), S (2.08 eV), Ni (1.156 eV) and Co (0.661 eV). As noted, the electron affinity of the contaminants is smaller than that of the corresponding radioactive ions of interest. We have investigated the efficiency of suppressing these isobars by photodetachment in our RFQ ion cooler. The experimental setup has been described in detail in previous publications [2,3].

A Nd:YAG laser at 1064 nm = 1.165 eV) was used to selectively detach the Co$^-$ ions. The photon energy is well above the electron affinity of Co$^-$ (0.661 eV), but also slightly larger than the electron affinity of Ni$^-$ (1.156 eV). Fig. 6(a) shows the ^{59}Co$^-$ current measured at the off-axis Faraday cup after the cooler. The RFQ cooler was operated at 2.76 MHz and ~ 6 Pa He pressure. When a CW 1064 nm laser beam of 5.4

W (measured at the laser output) was sent into the cooler, the Co⁻ current reaching the Faraday cup instantly dropped to less that 2%. When the laser was blocked, the Co⁻ current immediately returned to its original value. Under the same conditions, only about 20% of the ^{58}Ni⁻ ions were detached. For comparison, we also measured the efficiency of photodetachment without the cooler. In this case, the laser had to interact with 5 keV ^{59}Co⁻ ions and only about 3% of the ^{59}Co⁻ ions were detached with 6.8 W CW laser, as shown in Fig. 6(b). That is, photodetachment efficiency was more than 30 times higher when laser-ion interaction was made in the RFQ cooler.

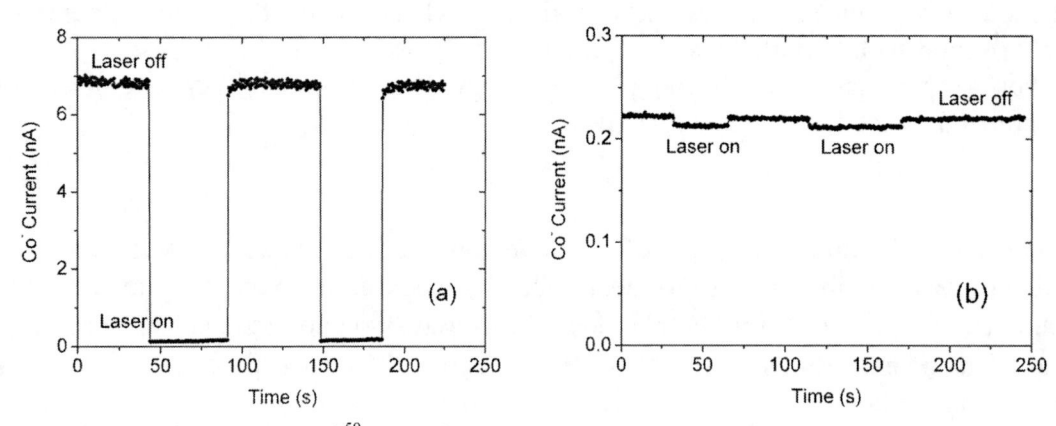

FIGURE 6. Photodetachment of ^{59}Co⁻ ion beams using a CW laser at 1064 nm. (a) Photodetachment in the RFQ cooler at a He pressure of ~ 6 Pa. More than 98% of the ^{59}Co⁻ ions were removed in the cooler. (b) Without the RFQ cooler, only 3% of the ^{59}Co⁻ ions were photodetached.

Similar results were obtained for photodetachment of O⁻ and S⁻ ions. Fig. 7 shows the measured intensities of ^{16}O⁻ and ^{32}S⁻ beams accelerated from the RFQ cooler. A pulsed Nd:YAG laser beam at 527 nm ($h\nu$ = 2.35 eV) was sent into the RFQ, with average laser power of 2.1 W. The cooler was operated at a He pressure of ~6 Pa. As seen, more than 99% of the ^{16}O⁻ and ^{32}S⁻ ions were removed when the laser beam was on. Under the same conditions, no photodetachment of ^{19}F⁻ and ^{35}Cl⁻ in the cooler was observed. Without the RFQ cooler, only 1% detachment of ^{32}S⁻ was obtained.

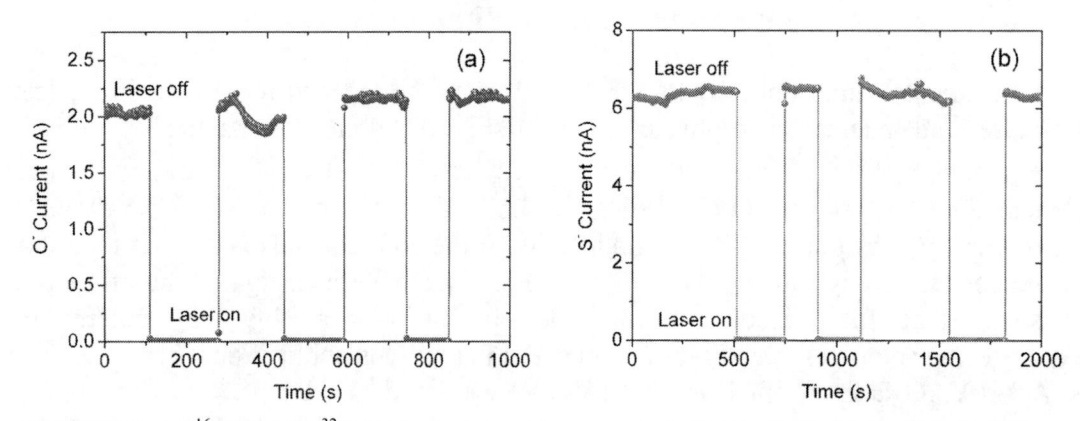

FIGURE 7. (a) ^{16}O⁻ and (b) ^{32}S⁻ ion beam intensities measured after the RFQ cooler with the 527 nm photodetachment laser beam on and off. The cooler was operated at 2.76 MHz, He pressure ~ 6 Pa. More than 99% of the ^{16}O⁻ and ^{32}S⁻ ions were detached when the laser was on.

The data show that nearly 100% photodetachment efficiency is possible in the RFQ cooler. To date, photodetachment of 99.97% of ^{32}S$^-$ and 99.9% of ^{16}O$^-$ ions in the RFQ cooler has been observed with about 2.5 W average power of the 527 nm laser, while under the same conditions, the desired Cl$^-$ and F$^-$ ions were not affected [4]. For the (Co, Ni) pair, 99.99% suppression of ^{59}Co$^-$ ions was demonstrated with less than 3 W average power from a pulsed 1064 nm Nd:YAG laser at 10 kHz repetition rate. Under identical conditions, only about 20% of ^{58}Ni$^-$ ions were depleted [5].

The laser power needed to achieve nearly 100% photodetachment is readily available from existing commercial lasers. Therefore, this technique is very promising for real applications. It is being developed for use at the HRIBF for on-line purification of RIBs such as ^{56}Ni. It can also be applied in accelerator mass spectrometry to suppress the ^{36}S contaminants in ^{36}Cl beams [15].

Production of Pure Ground-State Negative Ions

Negative ions are usually produced in sputter ion sources or plasma ion sources and the resulting anions are internally hot with all bound excited-states populated. For fundamental studies of negative ions, beams of pure ground-state anions may be desired. For example, the presence of excited populations may limit the precision and accuracy of laser photodetachment threshold spectroscopy or complicate the studies of collisions between electrons and atomic or molecular anions. There are no ion sources that can produce pure ground-state negative ions.

The feasibility of depleting the excited populations in negative ions by photodetachment in a RFQ cooler has been recently investigated. In collaboration with University of Gothenburg and other institutes, the study was conducted at HRIBF with C and Si negative ions [6]. C$^-$ has a loosely bound excited ^{2}D state at 0.033 eV below the continuum, while Si$^-$ has two bound excited states: a ^{2}P state of binding energy of 0.029 eV and a ^{2}D state that has a binding energy of 0.527 eV. A 1064 nm Nd:YAG laser (1.165 eV) was used to remove the ions in excited states. The photon energy is large enough to detach the excited ions in both C$^-$ and Si$^-$ but not sufficient to affect the ground-state ions. Fig. 8 shows the energy levels of C$^-$ and Si$^-$ as well as the relative 1064 nm photon energy from the ground and excited states.

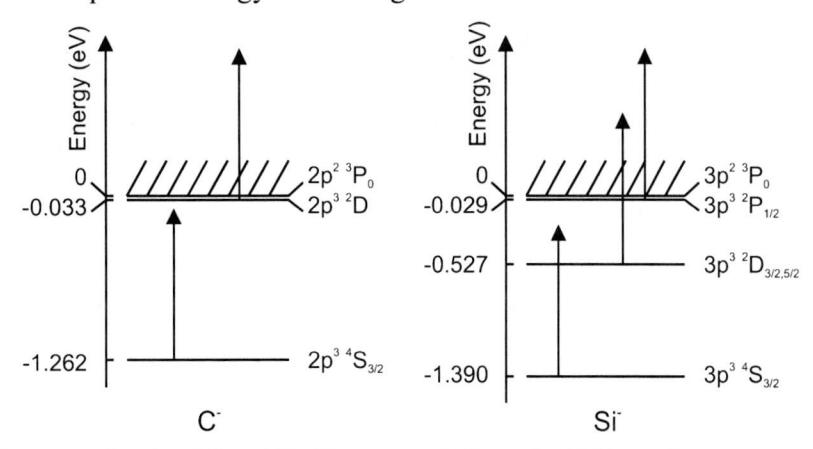

FIGURE 8. Energy levels of C$^-$ and Si$^-$. The arrows indicate the 1064 nm photon energy (1.165 eV).

The experimental setup has been reported in Ref. 6. It was observed that the ^{2}D excited state in C$^-$ was completely depleted in the RFQ cooler through collisional detachment alone. Similarly, the loosely bound ^{2}P state in Si$^-$ was expected to be depopulated through collisions in the cooler. With a CW 1064 nm laser of 2 W, about 98% of the remaining ^{2}D population in Si$^-$ was removed by photodetachment inside the cooler. The total reduction of the excited populations in Si$^-$, including collisional detachment and photodetachment was estimated to be 99 ± 1 %.

This work demonstrates the possibility of obtaining pure ground-state atomic negative ion beams by collisional detachment and state-selective photodetachment in a RFQ cooler. Such beams can improve fundamental studies on negative ions, such as enhancing the selectivity, as well as accuracy, in high-precision experiments on many atomic and molecular negative ions. This highly efficient and state-selective technique can be applied to any atomic or molecular negative ions that possess bound excited states.

ACKNOWLEDGMENTS

I would like to thank Dr. Paul Mueller for providing the mass resolution calculations for ^{17}O and ^{17}F beams shown in Figure 5. This work has been supported by the Office of Nuclear Physics, U.S. Department of Energy.

REFERENCES

1. Y. Liu, J. F. Liang. G. D. Alton, J. R. Beene, Z. Zhou, H. Wollnik, *Nucl. Instr. and Meth.* B **187**, 117 (2002).
2. Y. Liu, , J. R. Beene, C. C. Havener, J. F. Liang, *Appl. Phys. Letters* **87**, 113504 (2005).
3. C. C. Havener, Y. Liu, J. F. Liang, H. Wollnik, J. R. Beene, *AIP Conf. Proc.* **925**, 346 (2007).
4. Y. Liu, J. R. Beene, C. C. Havener, A. Galindo-Uribarri, T. L. Lewis, *AIP Conf. Proc.* **1097**, 431 (2009).
5. P. Andersson, A. O. Lindahl, D. Hanstorp, C. C. Havener, Yun Liu, Yuan Liu, *J. Appl. Phys.* **107**, 26102 (2010).
6. A. O. Lindahl, D. Hanstorp, O. Forstner, N. Gibson, T. Gottwald, K. Wendt, C. C. Havener, Y. Liu, *J. Phys. B* **43**, 115008 (2010).
7. H. Dawson, *Quadrupole Mass Spectrometry and Its Applications*, Amsterdam: Elsevier, 1976.
8. F. G. Major, H. G. Dehmelt, *Phys. Rev.* **170**, 91 (1968).
9. J. Dilling, et al., *Int. J. Mass Spectrom.* **251**, 198 (2006).
10. F. Herfurth, J. Dilling, A. Kellerbauer, G. Bollen, S. Henry, H.-J. Kluge, E. Lamour, D. Lunney, R.B. Moore, C. Scheidenberger, S. Schwarz, G. Sikler, J. Szerypo, *Nucl. Instr. and Meth.* A **469**, 254 (2001).
11. J. Äystö, A. Jokinen and the EXOTRAPS Collaboration, *J. Phys. B: At. Mol. Opt. Phys.* **36**, 573 (2003).
12. Y. Liu, J. F. Liang, J. R. Beene, *Nucl. Instr. and Meth.* B **255**, 416 (2007).
13. D. Berkovits, E. Boaretto, G. Hollos, W. Kutschera, R. Naaman, M. Paul, Z. Vager, *Nucl. Instr. and Meth.* A **281**, 663 (1989).
14. D. Berkovits, E. Boaretto, G. Hollos, W. Kutschera, R. Naaman, M. Paul, Z. Vager, *Nucl. Instr. and Meth.* B **52**, 378 (1990).
15. A. Galindo-Uribarri, C. C. Havener, T. L. Lewis, Y. Liu, *Nucl. Instr. and Meth.* B **268**, 834 (2010).

BEAMLINES AND FACILITIES

Stability of High Power Beam Injection in Negative-Ion-Based LHD-NBI

K. Tsumori, O. Kaneko, Y. Takeiri, M. Osakabe, K. Ikeda, K. Nagaoka, H. Nakano, M. Shibuya, E. Asano, T. Kondo, M. Sato, S. Komada and H. Sekiguchi

National Institute for Fusion Science, 322-6 Oroshi Toki Gifu 509-5292, Japan

Abstract. We describe the characteristic of stable beam injection in a neutral beam injector (NBI) for Large Helical Device (LHD) in high injection power of more than 6 MW. In the NBI, it takes a week after starting Cs seeding to finish the pre-injection conditioning. The injection starts with the beam power ~ 6.2 MW, and the maximum power reaches ~ 7 MW. The Cs-seeding rate affects the beam stability in such high power injections. By optimizing the rate to 0.65 mg / shot, the success ratio, which is defined as a ratio of actual pulse duration to setting one, increases to 85 - 90 % in the power and energy range of more than 6.2 MW and 185 keV, respectively. The weights of Cs adsorbed on several surfaces in the ion sources of the NBI are measured by means of Inductively-Coupled Plasma Atomic Emission Spectroscopy (ICP-AES), and averaged surface densities are calculated by dividing with the several surface areas. The seeded Cs of 99.5 % is condensed in the plasma generator, and very tiny amount of Cs reaches the surfaces of the accelerator grids. This very low amount of Cs on the grids is interpreted that most of the Cs atom evaporated from the inner walls is ionized during the arc discharges, and repelled to the source plasmas by the electrostatic field for H- extraction.

Keywords: NBI, hydrogen negative ion, cesium, negative ion source.
PACS: 52.27.Cm, 52.50.Dg, 52.50.Gj

INTRODUCTION

Hydrogen/deuterium negative ion sources are adopted for neutral beam injection (NBI) for plasma heating, production of high-β plasmas and neutral-beam current drive in large-scaled fusion experimental devices [1-7]. High beam powers of several mega-watts are required to generate high-performance plasmas confined in those devices. In order to enhance the H$^-$/D$^-$ currents, small amounts of cesium (Cs) are introduced in the plasmas produced inside the negative ion sources.

Negative ion sources seeded Cs have been researched and developed for the LHD from 1989 [8-12], and the LHD experiments with beam injections have started with the negative-ion-based NBI (negative NBI) systems since 1998 [13-18]. The NBI systems have been continuously improved their performances, and the total injection power obtained with three NBI systems has exceeded 15 MW of the nominal power of LHD-NBI since 2007. The negative NBI systems have been the main plasma-heating device in LHD, and the pulse duration of each beam line is important to complete the scenario of the LHD plasmas in many cases. Hence, the injection success ratio defined as the actual pulse length to setting one is here introduced to evaluate the performance of NBI. The success ratio in negative NBI is lower than that in positive one. Especially, the ratio goes down near maximum injection power of each beamline. Assuming that the probability of a voltage breakdown in the accelerator is independent of the other beamlines, the overall success ratio becomes a product of the ratios of three beamlines; for instance, the overall ratio decreases 12.5 %, when each NBI has the success ratio of 50 %. To improve the ratio, it is necessary to understand the causes of the breakdown in the accelerators of negative ion sources.

Comparing to positive ion sources, a negative ion source has large differences in the beam acceleration. Those are (1) co-extracted and accelerated electrons, which are extracted from source plasma and produced via stripping process, respectively, (2) ambiguity of beam perveance depending on the Cs condition in the plasma generator, and (3) Cs adsorption on the surfaces of accelerator grids, which cause the emissions of secondary Cs ions and electrons. Those items are relevant to the Cs condition inside the plasma generators of the ion source. Supplying Cs less than the optimum condition, the beamlet divergence is not minimized and the tail part of beamlets strike accelerator grids. Furthermore, the tail part of the beam superposing all the beamlets hits the internal structures of beamline. Consequently, breakdowns occur in the accelerator and the port-through efficiency becomes lower. Excess Cs seeding also induces breakdowns due to the Cs adsorption on the accelerator grids. Therefore, Cs monitoring is important and can be a clue to reduce breakdowns in negative ion source.

Results of Cs monitoring in the periods of pre-injection conditioning and of beam injection are reported here. We describe in following sections about the conditioning and injection statuses of LHD-NBI #1 in 2009, the comparison of the status to the other years and the surface density of Cs evaluated with Inductively-Coupled Plasma Atomic Emission Spectroscopy (ICP-AES) after the experimental campaign in 2009.

BEAM CHARCTERISTICS AND CS CONDITION IN CONDITIONING AND INJECTION

Details of the LHD-NBI #1 and its negative ion source are described elsewhere [16-18], and the digest of the structures is indicated here. Two ion sources are installed at each beamline of the NBI system. Each ion source has three Cs lines, and every line consists of a Cs reservoir, two valves and a Cs-guiding tube. At the exit of the guiding tube, a nozzle made of molybdenum is attached to diffuse the Cs vapor on the sidewalls of plasma generator. The sidewalls and the back-plate wall of the plasma generator are cooled by water. The beam accelerator is composed by four electrode grids named plasma grid (PG), extraction grid (EG), steering grid (SG) and multi-slot grounded grid (MSGG). In those grids, the PG is thermally insulated to keep the temperature at 200 – 350 °C, and the SG is indirectly cooled by attaching to the EG cooled by water. Inside the MSGG, water-cooling channels are also installed to remove the heat load caused by stripped electrons and H^o beams. To avoid the concentration of local electric field, the MSGG is rounded at the edges on the beam-upstream side.

The beam-accelerating situation in the period of the pre-injection conditioning is a suitable sample of beam acceleration with insufficient Cs seeding. The conditioning history of the beam energy and the current ratio (*Ri*), which is defined as a ratio of the acceleration current (*Iacc*) to the extraction one (*Iext*), are respectively shown in Fig. 1(a) and (b). Roughly speaking, the current ratio corresponds to that of the H⁻ current

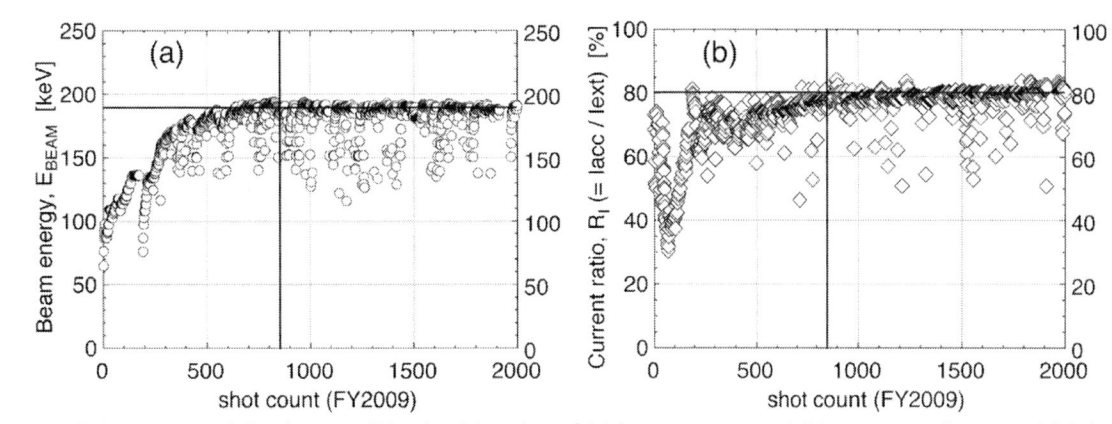

FIGURE 1. Pre-injection conditioning histories of (a) beam energy and (b) current ration *Ri*, which is defined as a ratio of acceleration to extraction currents (*Iacc* / *Iext*).

passing into the acceleration gap to the extraction current, since the electron current and neutralized H° beam in the acceleration gap is mainly originated from the stripping process. At a certain extraction voltage, the ratio is one of the guiding parameter of a ratio of the H⁻ to the total extracted currents, equivalent to extraction current, even if excess extraction voltage is applied for the perveance matching. The H⁻ current involved in the extraction current is a function of the amount of seeded Cs. Therefore, the ratio of *Ri* is a guiding parameter to control the Cs seeding rate. High-voltage grid conditioning without beam extraction is done before the Cs seeding. In the case of the pure hydrogen discharge, extracted H⁻ current saturates input arc power, and the frequency of voltage breakdowns in the accelerator increases because of the mismatching of the perveance. On the other hand, the saturation is removed by seeding Cs in the plasma generator, and the beam energy increases smoothly with keeping the perveance matching. The current ratio *Ri* decreases initially due to rapid

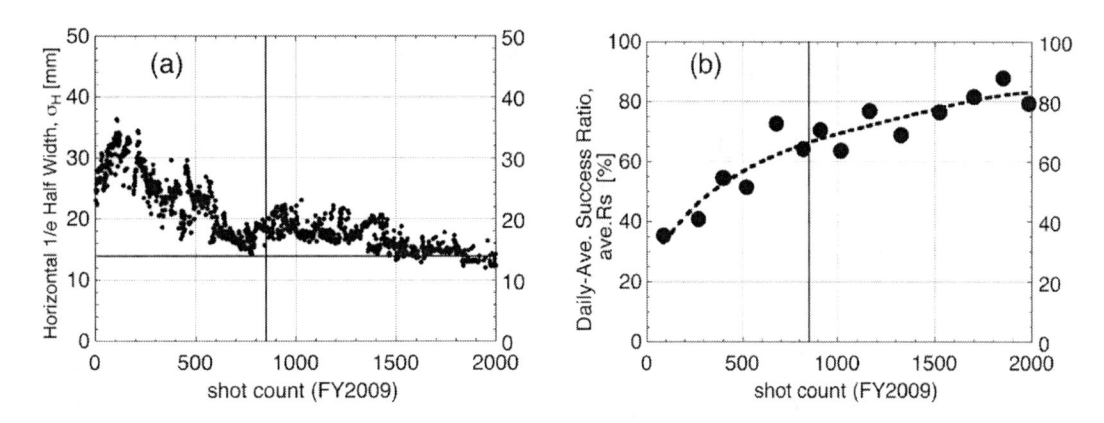

FIGURE 2. Pre-injection conditioning histories of (a) beam energy and (b) current ration Rs, which is defined as a ratio of acceleration to extraction currents (*Iacc* / *Iext*).

raise of the beam energy, and then recovers with accumulating the Cs in the plasma generator. Perveance matching is tried to start to adjust as much as possible near the sharp valley at ~100 shots. The conditioning has been finished at the shot count of 1000, and the beam energy reaches 190 keV at the shot count of ~850, and the timing is indicated as a vertical lines in the figure. According to the increase of the amount of the seeded Cs, the beam divergence decreases, since the perveance approaches to the matching condition. In the beam power more than 6 MW, it is not available to increase the pulse length more than 0.8 sec, since the calorimeter array would be damaged due to the power concentration.

Figure 2 shows (a) the changes of e-folding half width of the beam profile on the calorimeter array and (b) the success ratio averaged by day. After saturating the beam energy and the current ratio at the maximum values near the end of conditioning, the beam profile at the calorimeter array on neutral beam dump becomes sharper as shown in Fig. 2. This is interpreted the beam divergence decrease by approaching the Cs amount to optimal value.

The whole history of the injection power in 2009 is shown in Fig. 3, in which the injection power is indicated with open diamonds. The origin of the shot count indicates the stating point of the Cs seeding. Through the conditioning and injection, the temperature of the Cs reservoir has been kept constant at 150 °C, and the Cs valve has been opened during daily operation of the ion source. The power is estimated with calorimeter arrays installed in the neutral-beam dump and beam amour on the LHD inner wall in conditioning and injection. The injection power separated to two groups near 6 MW and 3MW. Those groups correspond to the beam injections with two ion sources and single one to adjust the injection power. The maximum injection power reaches ~7 MW at an energy of 187 keV for 1.5 sec.

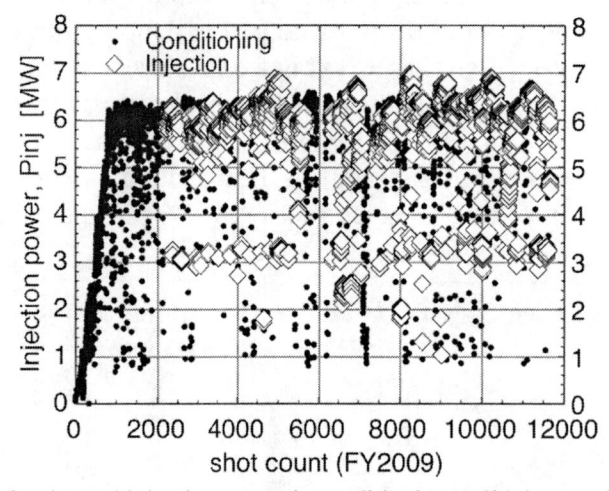

FIGURE 3. History of estimated injection power in conditioning, solid dots, and LHD-Injection, open diamonds, in 2009.

Since the Cs is constantly seeded by keeping the reservoir temperature, the H⁻ current depends not on the total Cs amount but the seeding rate; the shot interval is fixed for 3 min with the constant temperature of the Cs reservoir at 150 °C. The seeding and evaporating rate of Cs is considered to have equilibrium inside the plasma generator under this condition. In our Cs line, the optimal range of the Cs reservoir is within 150 ± 5 °C, less and excess Cs supply causes breakdowns inside the accelerator after holding the temperature for more than 1 hours. In the optimal temperature range, the current ratio of Iacc / Iext has been almost constant within the range of 80 - 82 % during the beam-injection phase. The H current is not so sensitive to the PG temperature within the rage of 200 – 350 °C. Once the current ratio becomes stable, no change is observed within the time range of several-tens minutes in the case of the pulse duration of a few seconds.

INFLUENCE OF CESIUM SEEDING

Cesium Seeding Rate and Its Total Amount

As wrote in the previous section, the current ratio of Ri ($Iacc / Iext$) is a guiding parameter to monitor the ratio of H to total extracted current, and is a parameter to optimize the Cs seeding rate. However, no system to monitor the dynamic Cs vapor pressure has been installed in our ion source yet. To understand the Cs recycling in the plasma generator, it is necessary to estimate the pressure and amount of Cs. According to J. B. Taylar and I. Langmuir [20], the Cs pressure (P_L [in Torr]) as a function of temperature (T [in K]) is indicated in following equation:

$$\log_{10} P_L = 11.0531 - 1.35 \log_{10} T - \frac{4041}{T} \qquad (1)$$

Assuming the seeded amount of Cs is proportional to the vapor pressure above, the time evolution of the Cs amount inside the plasma generator can be evaluated by integrating the Cs pressure with the reservoir temperature and opening duration of the Cs valve. To make the pressure difference between the reservoir and introducing tube

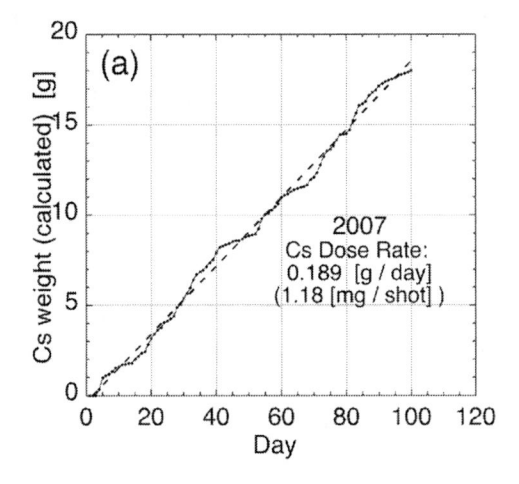

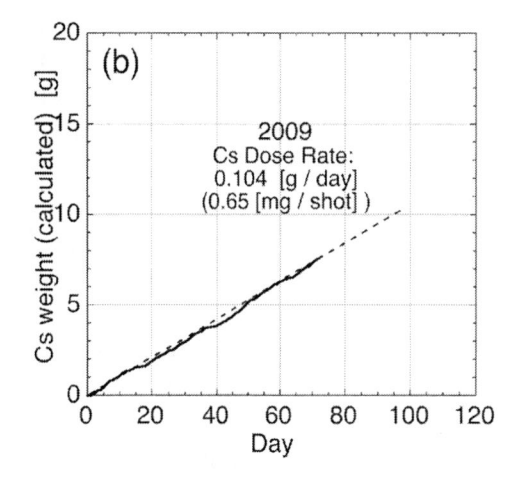

FIGURE 4. Histories of calculated Cs weights seeded in the plasma generator in (a) 2007 and (b) 2009. The weights are integrated with time using Cs vapor pressure in eq. (1) as a function of the reservoir temperature. The dotted lines indicate the linearly fitting line of the integrated Cs weights.

as low as possible, the bellow valve with a large orifice diameter of 7.92 mm (Swagelok SS-8UW-TW) is adopted for the Cs valve. All the tubes in the Cs system have an inner diameter of 8 mm to avoid local Cs condensation in the tubes.

Figure 4(a) and (b) show the evolutions of estimated Cs weights in plasma generators in 2007 and 2009, respectively. The Cs-seeding rate in 2009 is 0.104 g / day, which corresponding to 0.65 mg / shot. The rate in 2009 is about 55 % of that in 2007. In the latter case, a trouble with water leakage at the calorimeter array occurred, and the higher seeding rate was required to obtain the injection power more than 5 MW. Due to the high rate, the frequency of the voltage breakdowns in the accelerator increases. The success ratio with respect to the injection power and the power distribution in 2007 and 2009 are shown in Fig. 5(a) and (b), respectively. As indicated in Fig. 5(a), the success ratio is ~60 % and the power distribution becomes wider. The success ratio in 2009 reaches 85 - 90 % in the shaded zones, which

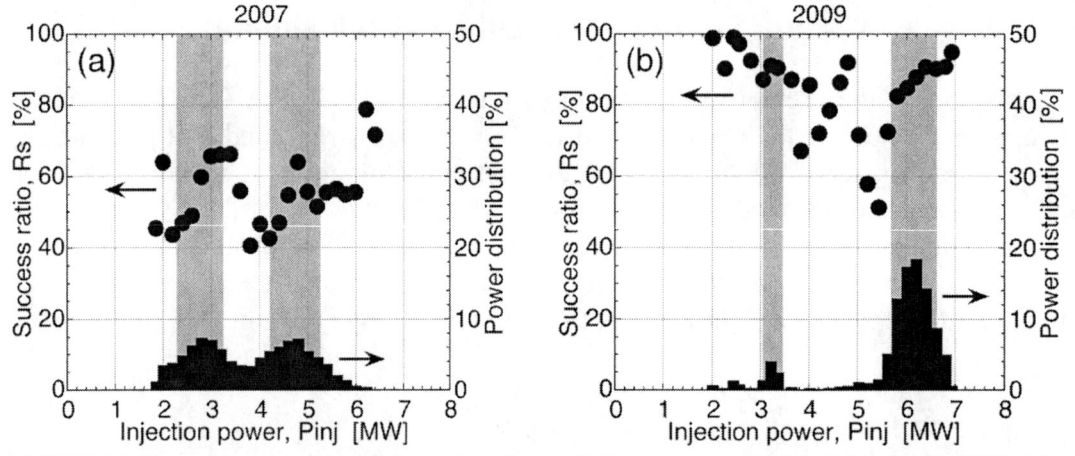

FIGURE 5. Success and number ratios as functions of injection power in (a) 2007 and (b) 2009. The success ratio and number ratio are indicated with solid circle and solid bar, respectively. The shaded zones show the injections with single (lower power) and full ion sources (higher power).

indicate the optimal injection powers with the single or full of the two ion sources in the beamline. The power distribution concentrates at the peak position of 6.2 MW. The beam energy is more than 185 keV in those high power injections. In the unstable case, the beam energy is changed to reduce the frequency of accelerator breakdown, and the power distribution becomes broad consequently.

Cesium Distribution

The Cs weight and its vapor pressure are estimated in the previous section. In this estimation, the spatial distribution of Cs is unknown. The distribution is difficult to measure during ion-source operation. That is because the contamination of Cs by oxygen and water molecular causes di-cesium oxide (Cs_2O) and cesium hydro-oxide (CsOH), and those compounds affect the surface process to produce H$^-$ ions and as well as the arc discharges. In order to measure very tiny weight of Cs, the Inductively-Coupled-Plasma Atomic Emission Spectroscopy (ICP-AES) is applied to measure the Cs concentration on the walls. The method has been adopted by Y.

Okumura and K. Watanabe [20]. After finishing the LHD experimental period in 2009, the ion sources have been detached from the NBI #1. By opening the ion sources, adsorbed Cs on the inner walls of two ion sources is thoroughly wiped with BEMCOT[TM] (Asahi Kasei Co). The material of the BEMCOT is made of 100 % cellulose, and water-wet BEMCOT is applied to wipe adsorbed Cs efficiently. After this procedure, the BEMCOT absorbing Cs is soaked and the Cs is dissolved in one litter of distilled water. The Cs-dissolved water is analyzed by means of the ICP-AES to measure the Cs concentration, and the Cs concentration is converted to the Cs weight. The inner walls of the ion sources are separated to 8 surfaces indicating in Table 1.

TABLE 1. Position index and corresponding regions

Position Index	Distance from Cs-nozzle exit, L [mm]	Regions in ion source
1	125.3	Sidewalls (SW)
2	7.6	Back plate (BP)
3	53.2	Inside filament ports
4	104.3	Filament assemblies
5	212.3	Surfaces between plasma-generator and PG flanges
6	212.3	PG surface facing to plasma
7	236.0	EG surface on upstream surface
8	326.0	GG surface on upstream surface

The distribution of the Cs-surface density to the distance from the position of the Cs-nozzle exit is shown in Fig. 6(a). About 99.5 % of the seeded Cs is localized on the inner surface of the plasma generator, the filament assemblies and inside surface of filter flange. Although the flow rates of cooling water and density of cooling tubes are similar, the Cs weight on the back plate is ~50 % of that on the sidewall. The difference is caused that (1) the direction of Cs-nozzle exit is directed to sidewall and (2) the back plate is exposed by back streaming positive ions during beam extraction. Watching the back plate, less condensation of Cs is observed on the footprints caused

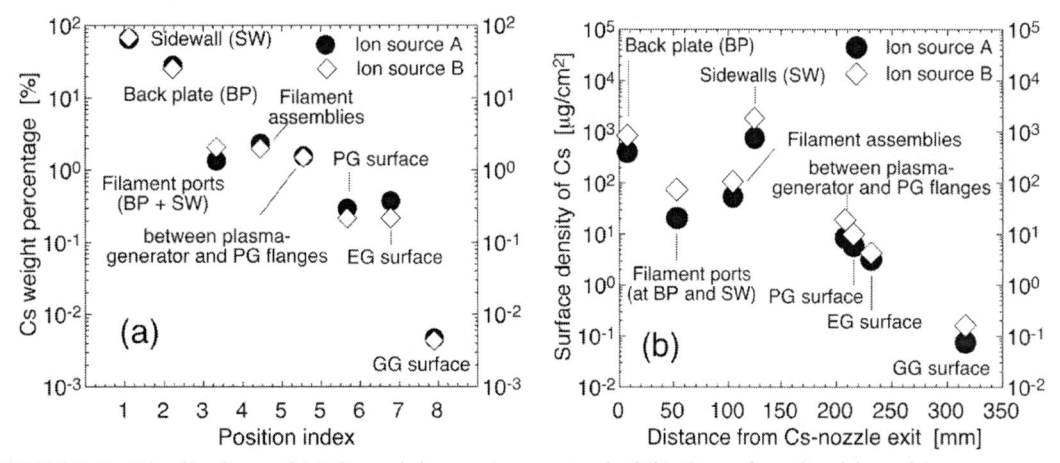

FIGURE 6. Distributions of (a) Cs weight percentages and of (b) Cs surface densities with respect to the distance from the exit position of Cs nozzle. Two plot marks of the solid circle and the open diamond indicate the data obtained with two ion sources of LHD-NBI #1.

by the back streaming, and higher Cs condensation is observed on the other part of the back plate. Therefore, it is interpreted that the Cs on the back plate is heated and evaporated by back-streaming positive ions. The cusp lines is heated with the heat load carried dominantly by the electrons, and condensed Cs on the sidewalls evaporates mainly from the lines. The condensed Cs weight on the filament-port walls is comparable to that at the filament assembly. The coverage is one order lower than that on the back plate and the sidewalls, because the effective surfaces of those filament units are small to the exposure of the Cs vapor. On the PG surface facing to the plasma, the Cs amount is 0.01 – 0.1 times lower than those on the other walls inside the plasma generator, and this low condensation can be attributed to the PG temperature of 200 – 350 °C. Although the total surface area at a region between the plasma-generator flange and the PG flange surrounding the PG is cooled by water, the Cs weight on the surface is comparable to the PG surface. That is because a narrow gap of 2 mm between the region and PG prevents Cs vapor to reaches the region freely.

As described previously, total Cs weight of 99.5 % localizes in the plasma generator, and the residual Cs of 0.5 % reaches the surfaces of accelerator grids. The Cs amount on the EG surface on the beam-upstream side is dominant in the accelerator surfaces, since the EG is cooled by water and is the nearest grid to the plasma generator. Comparing the other surfaces, very small amount of Cs accumulates on the GG surface on the beam-upstream side. Simulating the Cs evaporation from the plasma generator in the atomic state, the Cs weight on the surface of GG is much higher than the measured weight. Most plausible explanation is as follows: the Cs atom is immediately ionized after evaporating from the chamber walls because of its low ionization potential. The Cs^+ ion near the PG aperture, furthermore, is repelled by the extraction field and confined in the source plasma. The Cs surface density on each surface area is shown in Fig. 6(b), the origin of the horizontal axis is the Cs nozzle exit. The Cs densities on the EG and the GG surfaces are less than ~30 µg / cm^2 and ~1 x 10^{-1} µg / cm^2, respectively. The latter density is considerably lower than the critical density of several tens of µg / cm^2 leading to the discharge breakdown in the acceleration gap [20]. Additionally, the transparency of this GG is twice as large as that of usual multi-rounded aperture GG, and is considered to have an advantage to reduce the total weight of stray Cs on the GG.

The combination of optimized Cs seeding rate, the GG structure with high beam transparency and significantly low Cs adsorption on the GG is considered to contribute the stability with the success ratio of 85 – 90 % in high injection power more than 6.2 MW.

ACKNOWLEDGMENTS

The authors would like to appreciate his thanks to LHD experiment group for kind supports and to technical staffs of LHD-NBI group for excellent operation of NBI test stand. The authors also express their appreciation to Prof. Dr. A. Komori, the Director general of NIFS, for his encouragement for this experiment. This research is supported by NIFS (NIFS10ULRR702).

REFERENCES

1. O. Kaneko, A. Komori, H. Yamada, et al., *Phys. Plasmas* **9**, 2020 (2002).
2. Y. Takeiri, S. Morita, K. Ikeda, et al., *Nucl. Fusion* **47** 1078 (2007).
3. T. Oikawa, Y. Kamada, A. Isayama, et al.,*Nucl. Fusion* **41** 1575 (2001).
4. A. Isayama, Y. Kamada, T. Ozeki, et al., *Nucl. Fusion* **41** 761 (2001).
5. S. Inagaki, H. Takenaga, K. Ida, et al., *Nucl. Fusion* **46** 133 (2006).
6. T. Inoue, R. Hemsworth, V. Kulygin, Y. Okumura, *Fusion Eng. Design*, **55**, 291 (2001).
7. M. Hanada, N. Akino, Y. Endo, et al., *J. Plasma Fusion Res.* SERIES, **9**, 208 (2010).
8. O. Kaneko, A. Ando, Y. Oka, et al., *Proc. 17th Symp. Fusio Technology* (Rome, 1992) P.544.
9. A. Ando, K .Tsumori, Y. Oka, et al., *Phys. Plasmas*, **1**, 2813 (1994).
10. Y. Takeiri, M. Osakabe, Y. Oka, et al., *Rev. Sci. Instrum.* **68**, 2012 (1997).
11. K. Tsumori, T. Takanashi, S. Asano, et al., *Proc. Joint Meeting of 8th International Symp. on Production and Neutralization of Negatibe Ions and Beams* (Villagium de Giens, France, 1997), AIP Conf. Proc, No.439, p. 93.
12. Y. Takeiri, O. Kaneko, K. Tsumori, et al., *Rev. Sci. instum*, **71** 1225 (2000).
13. O. Kaneko, Y. Takeiri, K. Tsumori, et al., *Nucl. Fusion* **43**, 692, (2003).
14. Y. Takeiri, K. Ikeda, M. Hamabe, et al., *Rev. Sci. instum*, **73** 1087 (2002).
15. Y. Oka, K. Tsumori, Y. Takeiri, et al., *Rev. Sci. Instrum.* **75**, 1803 (2004).
16. K. Tsumori, O. Kaneko, Y. Takeiri, et al., *Proc. 10th International Symposium on Production and Neutralization of Negative Ions and Beams*, (Kiev, Russia, 2005), AIP Conf. Proc, No.**763**, p.35.
17. K. Tsumori, K. Nagaoka, M. Osakabe, et al., *Rev. Sci. Instrum.* **75**, 1847 (2004).
18. K. Tsumori, M. Osakabe, O. Kaneko, et al., *Rev. Sci. Instrum.* **79**, 02C107 (2008).
19. J. B. Taylor and I. Langmuir, Phys. Rev, 51, 753-760 (1937) .
20. Y. Okumura and K. Watanabe, *Extra Issue on progress of ITER Engineering R&D Japanese J. Society of Plasma and Fusion,* **75**, p.66 (1999) (in Japanese).

Modeling activities on the negative-ion-based Neutral Beam Injectors of the Large Helical Device

P. Agostinetti[a], V. Antoni[a], M. Cavenago[b], G. Chitarin[a], H. Nakano[c], N. Pilan[a], G. Serianni[a], P. Veltri[a], Y. Takeiri[c] and K. Tsumori[c]

[a] *Consorzio RFX, Associazione Euratom-ENEA sulla Fusione, Corso Stati Uniti 4, 35127 Padova, Italy*
[b] *INFN, Laboratori Nazionali di Legnaro, V.le dell'Università 2, 35020 Legnaro, Italy*
[c] *National Institute for Fusion Science, 322-6 Oroshi-cho, Toki 509-5292, Japan*

Abstract. At the National Institute for Fusion Science (NIFS) large-scaled negative ion sources have been widely used for the Neutral Beam Injectors (NBIs) mounted on the Large Helical Device (LHD), which is the world-largest superconducting helical system. These injectors have achieved outstanding performances in terms of beam energy, negative-ion current and optics, and represent a reference for the development of heating and current drive NBIs for ITER.

In the framework of the support activities for the ITER NBIs, the PRIMA test facility, which includes a RF-drive ion source with 100 keV accelerator (SPIDER) and a complete 1 MeV Neutral Beam system (MITICA) is under construction at Consorzio RFX in Padova.

An experimental validation of the codes has been undertaken in order to prove the accuracy of the simulations and the soundness of the SPIDER and MITICA design. To this purpose, the whole set of codes have been applied to the LHD NBIs in a joint activity between Consorzio RFX and NIFS, with the goal of comparing and benchmarking the codes with the experimental data. A description of these modeling activities and a discussion of the main results obtained are reported in this paper.

Keywords: Neutral Beam Injector, Large Helical Device, beam, modeling
PACS: 41.75.Cn, 52.50.Gj

INTRODUCTION

A comprehensive set of numerical and analytical codes is currently being used and developed at Consorzio RFX, aiming at simulating the most important aspects inside the negative ion electrostatic accelerator, the electrical and magnetic fields, beam aiming and optics, pressure inside the accelerator, stripping reactions, transmitted and dumped power, operating temperature, stresses and deformations of the accelerator grids. These codes provide a description of the physical phenomena taking place in the different components of the injector system (source, accelerator, neutralizer, electron and ion dumps, calorimeter) and therefore are necessary for the engineering design optimization as well as for the prediction of the operating conditions and the interpretation of the experimental results.

This paper describes a set of optics and physics analyses on the accelerator of the LHD Neutral Beam Injector [1, 2]. The codes used are some of the ones currently considered

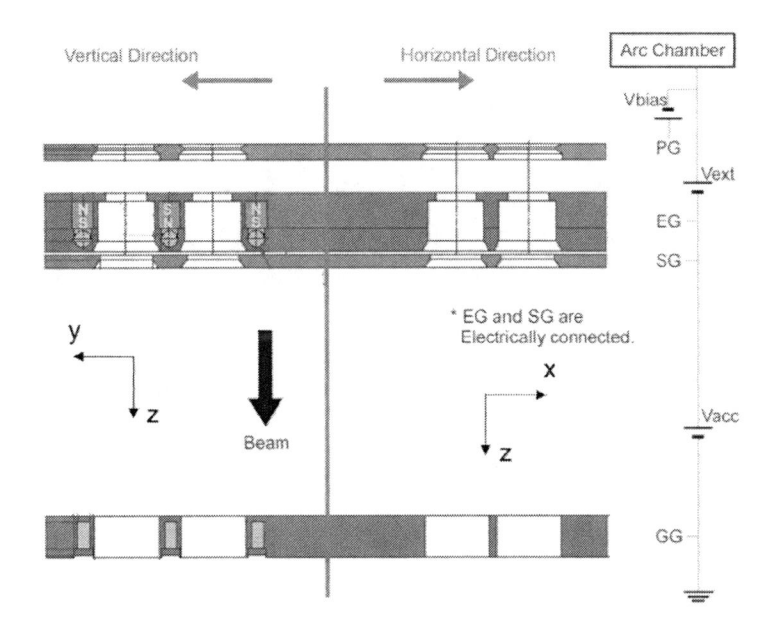

FIGURE 1. LHD Neutral Beam Injector, Beam Line 2: aperture geometry and electrical connections.

for the design of the SPIDER accelerator [3, 4]:

- CONDUCT to estimate the conductance through the grids of a gas in low pressure conditions;
- STRIP for the pressure and density profiles inside the accelerator and the stripped losses;
- SLACCAD for the beam optics;
- OPERA, ANSYS and PERMAG for the magnetic fields;
- EAMCC for the estimation of the heat loads on the grid, transmitted beam and amount of backstreaming ions.

ACCELERATOR GEOMETRY AND BOUNDARY CONDITIONS

The geometry considered for the first set of simulations on LHD NBI is that of the Beam Line 2 (BL2), shown in Fig. 1. This is a negative-ion-based Neutral Beam Injector. The extraction/acceleration system is made of a Plasma Grid (PG), an Extraction Grid (EG), a Steering Grid (SG) and a Grounded Grid (GG). EG and SG are electrically connected, so they are biased at the same potential.

The EG features embedded magnets with alternated polarities. The magnetic remanence is taken equal to 0.96 T, which is a typical value for Samarium Cobalt permanent magnets [5]. EG and GG feature cooling channels to remove the heat loads by co-extracted and secondary electrons. The number of apertures is 1540, calculated considering 2 sources, 5 grid sections per source, 14x11 apertures per section. The source

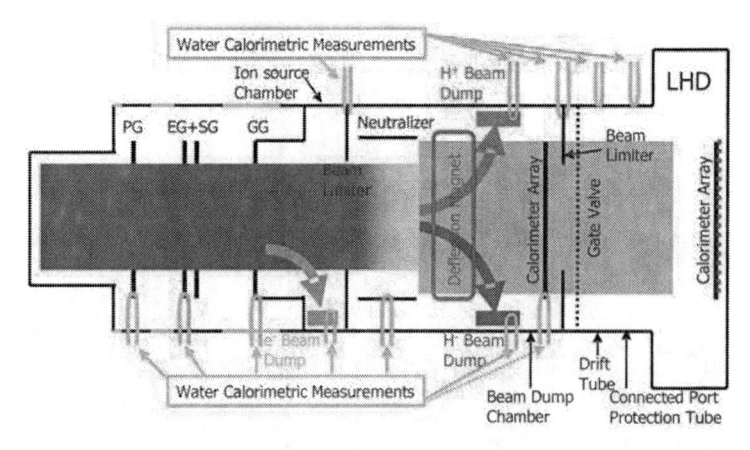

FIGURE 2. LHD Neutral Beam Injector, Beam Line 2: schematic view of the calorimetric measurement set up.

and tank pressures, which are respectively the pressure at the entrance and at the exit of the accelerator, were not measured during the considered campaign.

As a reference value we have considered the pressure distribution inside the 1/3 scaled negative ion source at the NIFS testbed [1], where the filling pressure is about 0.43 Pa and the tank pressure is about 0.06 Pa. One should bear in mind that these data are referred to a different system, with smaller dimensions and two stages of acceleration (there is an acceleration grid between EG and GG), and the pressure boundary conditions could be different in the case of BL2.

For all the considered models, the z coordinate is considered along the beam direction, while x and y are considered respectively along the transverse horizontal and transverse vertical directions (see Fig. 1).

VOLTAGE, CURRENT AND CALORIMETRIC MEASUREMENTS

In order to carry out the comparison between the results of the models and the experimental data, typical pulses of the LHD BL2 injector have been considered. Fig. 2 schematically shows the calorimetric measurement set up adopted in the BL2 injector. The total energy absorbed by each component can be estimated by integrating the power carried away by the cooling water during and after the pulse, according to the formula:

$$E = \int \dot{m} C \Delta T \, dt \qquad (1)$$

where $\dot{m}$ is the water flow, C is the water heat capacity and ΔT is the difference between inlet and outlet temperature of the cooling water. The power deposited on the components can be then evaluated by dividing the corresponding energy by the pulse length:

$$P = \frac{E}{\Delta t_{pulse}} \qquad (2)$$

TABLE 1. Voltage, current and calorimetric measurements.

LHD shot number		98006	98007	98013
Pulse length	ΔT_{pulse} [s]	1.96	1.94	2.96
Extraction voltage	$V_{PS,ext}$ [kV]	9.7	9.8	8.5
Acceleration voltage	$V_{PS,acc}$ [kV]	163	163	143
Total extraction current (sources A+B)	$I_{PS,ext}$ [A]	91.9	92.4	82.9
Total acceleration current (sources A+B)	$I_{PS,acc}$ [A]	70.5	70.8	64.5
Average power absorbed by the EG (sources A+B)	P_{EG} [kW]	312	308	244
Average power absorbed by the GG (sources A+B)	P_{GG} [kW]	1189	1113	875

where Δt_{pulse} is the pulse-on time. This is to be considered as an average power deposited on the components during the pulse duration. Voltage, current and calorimetric measurements recorded during three typical pulse in LHD BL2 accelerator are reported in Tab. 1. In particular, the data from pulse 98006 will be here considered.

OPTICS ANALYSES

The SLACCAD code has been used to estimate the electric field inside the accelerator by integration of the Poisson's equation, with cylindrical geometry conditions [6]. This is a modified version of the SLAC Electron Trajectory Program [7], adapted to include ions, a free plasma boundary and a stripping loss module [8].

The CONDUCT and STRIP codes [9] have been used to generate the boundary conditions for SLACCAD. CONDUCT has been used to estimate the conductance of a gas in low pressure conditions across the aperture, following the classical approach for molecular flow [10]. The temperature of the background gas inside the ion source is estimated to be in the range between 1000 and 5000 K (depending on the operating conditions of the arc chamber ion source), while it is expected to be close to the room temperature in the acceleration gaps.

Gas density profiles should be calculated for this temperature range. With a fixed flow rate and molecular flow conditions, the velocity of the molecules V is foreseen to increase as the gas temperature increases. In other words, the gas density in the ion source and the accelerator decreases as the gas temperature increases, i.e. the conductance increases with temperature. Regarding this topic, Krylov and Hemsworth propose the relation $V \propto T_{gas,source}^{0.5}$ [11]. The STRIP code requires a source pressure (for the region upstream of the accelerator) and a tank pressure (for the region downstream of the accelerator) to calculate the pressure profile inside an electrostatic accelerator. The source pressure to be given as input to this code should take into account that the gas in the source is at high temperature. To make this, the following formula is used in the calculations:

$$p_{eff} = p_{fill} \cdot \left(\frac{T_{room}}{T_{gas,source}} \right)^{0.5} \quad (3)$$

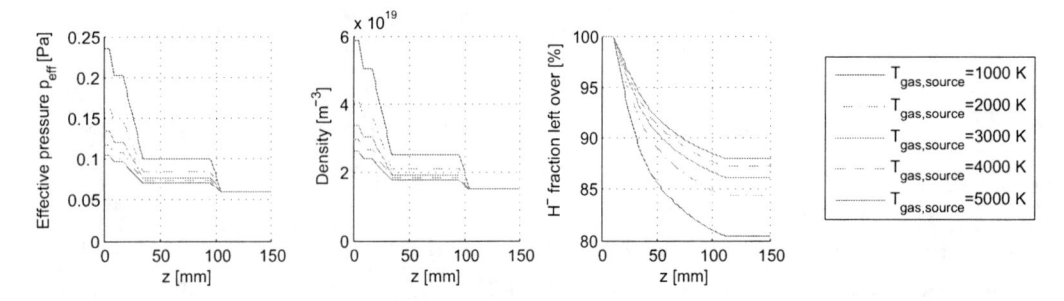

FIGURE 3. Pressure profile, density profile and stripping percentage on beam axis, calculated with STRIP.

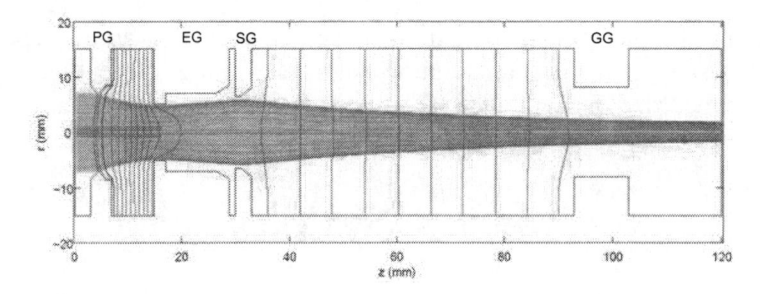

FIGURE 4. Trajectories and equipotential lines (2D) calculated with SLACCAD.

where:

- p_{eff} is the effective source pressure, to be used for the stripping calculations with the STRIP code;
- p_{fill} is the filling source pressure, intended with no source operation and the system at room temperature. This is taken equal to 0.43 Pa according to [1]
- T_{room} is the room absolute temperature, taken equal to 300 K;
- $T_{gas,source}$ is the gas absolute temperature inside the source, assumed to be in the range 1000-5000 K.

The potentials of the grids are respectively 0 kV (PG), 9.7 kV (EG/SG) and 172.7 kV (GG), according to the data recorded during the pulse 98006 of Tab. 1, considered for these analyses.

With these boundary conditions, STRIP has been used to estimate the pressure profile and stripping percentage on beam axis (considering the cross sections reported in the ORNL Redbooks [12]). The stripping and charge exchange reactions have been then taken into account by SLACCAD for the space charge evaluations.

The pressure, density and stripping profiles calculated with STRIP are plotted in Fig. 3, while Fig. 4 shows the trajectories of the negative ions and equipotential lines calculated with SLACCAD. The beamlet meniscus corresponds to the first equipotential line (0 kV). The grids are acting as converging and diverging electrostatic lenses.

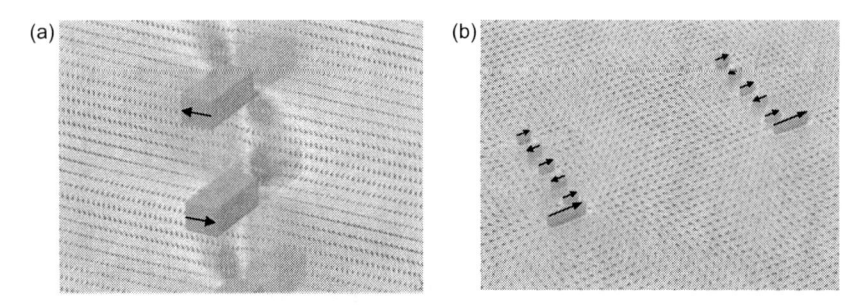

FIGURE 5. Evaluation of the magnetic field with OPERA: (a) Electron suppression field (vertical section); (b) Filter and cusp field (horizontal section).

MAGNETIC FIELD ANALYSES

The magnetic field distribution inside the accelerator has a considerable influence on the operating behaviour of this component. In particular, it affects the beam deflection, the amount of heat load on the grids and the electron filtering capability.

The suppression field (given by the suppression magnets embedded in the EG) is the most important magnetic field inside the accelerator, because it provides the electron filtering capability. This field, which is mostly along the transverse vertical direction y, can horizontally deflect the charged particles inside the accelerator. It has been calculated with three different codes: OPERA [13] (see Fig. 5a), ANSYS [14] and PERMAG [15] (being the first one and the second one finite element method codes and the third one a semi-analytical code). An optimal agreement among the three simulations was found. The data has also been compared and checked with measurements of the magnetic field on the LHD BL2 source [16].

The effect of filter and cusp magnetic fields inside the accelerator (see a simulation with OPERA in Fig. 5b) is much smaller than the suppression field. Nevertheless, these fields, which are mostly along the transverse horizontal direction x, can slightly deflect vertically the charged particles inside the accelerator. The results of the simulations show that the filter and cusp magnetic fields are of the order of 10^{-4} T in the accelerator area. This is about two orders of magnitude lower than the suppression field. On the other hand, filter and cusp fields are present also inside the arc chamber, with the functions of enhancing the plasma generation and filtering the electrons before they are extracted from the plasma source.

ESTIMATION OF THE EXTRACTED CURRENT AND HEAT LOADS

The collisions among particles inside the accelerator, as well as secondary particle production processes, were analysed with the code EAMCC. This is a 3-dimensional relativistic particle tracking code where macroparticle trajectories, in prescribed electrostatic and magnetostatic fields, are calculated inside the accelerator [17]. In the code, each macroparticle represents an ensemble of rays considering the time-independent

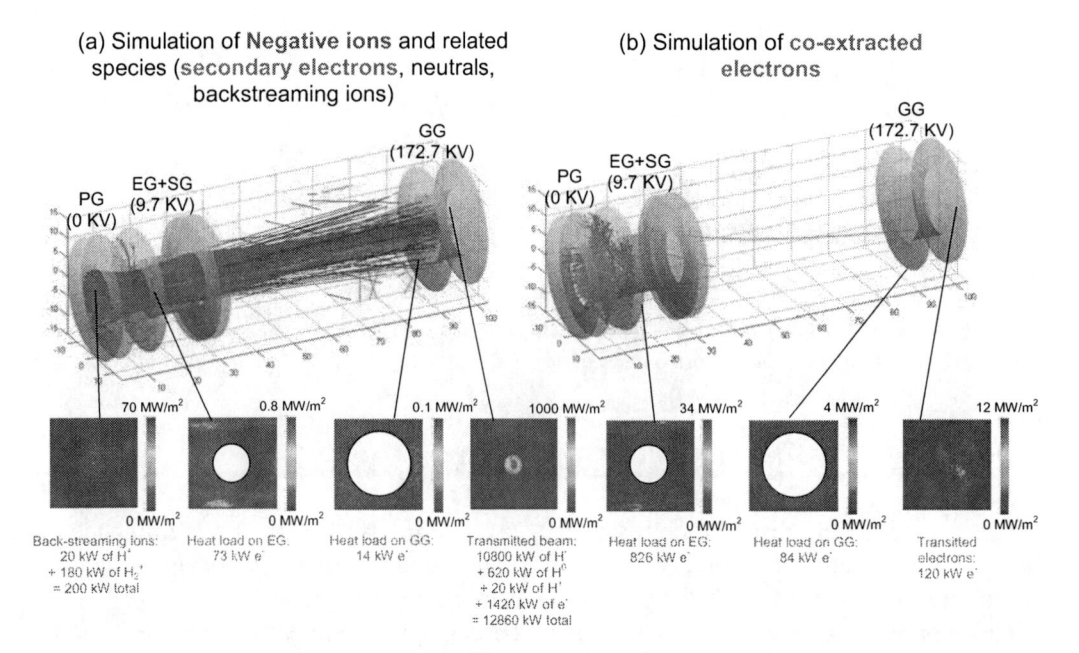

FIGURE 6. EAMCC simulation (a) of a negative ion beamlet and (b) of the co-extracted electrons in the LHD BL2 accelerator: the particle trajectories and stripping reaction are simulated with a Monte Carlo approach in a domain with electrical and magnetic fields.

physical characteristics of the system. This code needs as inputs the electric and magnetic fields inside the accelerator. The former was calculated with SLACCAD, the latter with OPERA (magnetic fields from suppression, filter and cusp magnets are included in the simulations).

Collisions are described using a Monte-Carlo method. The collisions considered in the code are the ones between the particles (electrons, ions and neutrals) and the grids, the negative ion single and double stripping reactions and the ionization of background gas. The main output of an analysis with EAMCC are summarized in Fig. 6. In this case, as a first hypothesis, a 342 A/m^2 current density have been assumed both for the negative ions (H^-) and for the co-extracted electrons (e^-), with the power loads and heat density referred to these values.

In order to compare the simulations with the experimental data (pulse 98006), the H^- and e^- extracted current densities have been calculated as a function of the currents measured at the two power supplies ($I_{PS,ext}$ and $I_{PS,acc}$). The power corresponding to these extracted current densities have been then compared with the calorimetric measurements of the grids.

To make this, the currents measured at the power supplies can be written as a function of the currents (hydrogen ions and electrons) extracted from the plasma source:

$$\begin{bmatrix} I_{PS,ext} - I_{PS,acc} \\ I_{PS,acc} \end{bmatrix} = K \cdot \begin{bmatrix} I_{H^-,ext} \\ I_{e^-,coext} \end{bmatrix} \quad (4)$$

The matrix K can be evaluated as:

$$K = \begin{bmatrix} \dfrac{I_{e^-,str,EG}+I_{e^-,str,SG}}{I_{H^-,ext}} & \dfrac{I_{e^-,coext,EG}+I_{e^-,coext,SG}}{I_{e^-,coext}} \\ \dfrac{I_{H^-,exit}+I_{e^-,str,exit}+I_{e^-,str,GG}-I_{H^+,exit}-I_{H_2^+,backstr}-I_{H^+,backstr}}{I_{H^-,ext}} & \dfrac{I_{e^-,coext,exit}+I_{e^-,coext,GG}}{I_{e^-,coext}} \end{bmatrix} \quad (5)$$

The following parameters can be obtained from EAMCC simulations where hydrogen ions are launched from the emitter (all the parameters are calculated as a percentage of the extracted H^- ions):

- $\dfrac{I_{H^-,exit}}{I_{H^-,ext}}$ is the relative amount of hydrogen ions exiting the accelerator;

- $\dfrac{I_{e^-,str,EG}}{I_{H^-,ext}}$ is the relative amount of electrons produced by stripping and charge exchange reactions impinging on the EG;

- $\dfrac{I_{e^-,str,SG}}{I_{H^-,ext}}$ is the relative amount of electrons produced by stripping and charge exchange reactions impinging on the SG;

- $\dfrac{I_{e^-,str,GG}}{I_{H^-,ext}}$ is the relative amount of electrons produced by stripping and charge exchange reactions impinging on the GG;

- $\dfrac{I_{e^-,str,exit}}{I_{H^-,ext}}$ is the relative amount of electrons produced by stripping and charge exchange reactions exiting from the accelerator;

- $\dfrac{I_{H^+,exit}}{I_{H^-,ext}}$ is the relative amount of H^+ ions exiting from the accelerator;

- $\dfrac{I_{H^+,backstr}}{I_{H^-,ext}}$ is the relative amount of H^+ ions backstreaming to the plasma source;

- $\dfrac{I_{H_2^+,backstr}}{I_{H^-,ext}}$ is the relative amount of H_2^+ ions backstreaming to the plasma source.

The following parameters can be obtained from EAMCC simulations where electrons are launched from the emitter (all the parameters are calculated as a percentage of the co-extracted e^-):

- $\dfrac{I_{e^-,coext,EG}}{I_{e^-,coext}}$ is the relative amount of co-extracted electrons impinging on the EG;

- $\dfrac{I_{e^-,coext,SG}}{I_{e^-,coext}}$ is the relative amount of co-extracted electrons impinging on the SG;

- $\dfrac{I_{e^-,coext,GG}}{I_{e^-,coext}}$ is the relative amount of co-extracted electrons impinging on the GG;

- $\dfrac{I_{e^-,coext,exit}}{I_{e^-,coext}}$ is the relative amount of co-extracted electrons exiting the accelerator.

Hence, the ion and electron extracted currents can be estimated with the formula:

$$\begin{bmatrix} I_{H^-,ext} \\ I_{e^-,coext} \end{bmatrix} = K^{-1} \cdot \begin{bmatrix} I_{PS,ext} - I_{PS,acc} \\ I_{PS,acc} \end{bmatrix} \quad (6)$$

The main results of the EAMCC analyses are plotted in Fig. 7, where different values for the source gas temperature have been considered, as this quantity is not well known.

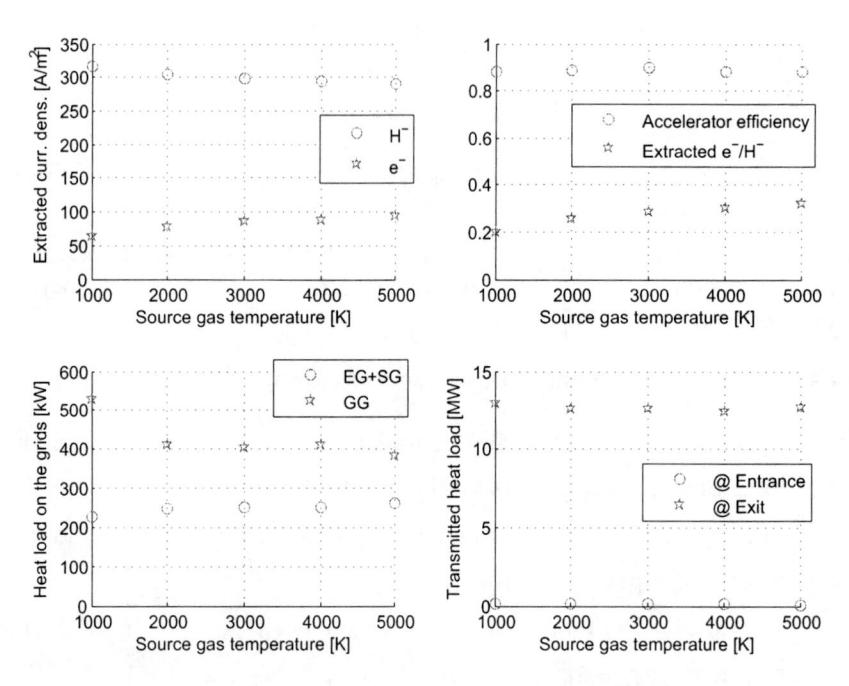

FIGURE 7. Main results of the EAMCC analyses, with a sensitivity analysis on the effect of the source gas temperature. The acceleration efficiency is defined as the ratio between the H^- power at the accelerator exit and the total power from the power supplies.

CONCLUSIONS AND OPEN ISSUES

The main outputs of the analyses carried out on the LHD BL2 accelerator with the CONDUCT, STRIP, SLACCAD, OPERA and EAMCC codes are:

- H^- extracted current: 69-75 A;
- H^- extracted current density: 290-320 A/m^2;
- e^- extracted current: 15-22 A;
- e^- extracted current density: 60-90 A/m^2;
- extracted e^-/H^- ratio: 0.2-0.33;
- stripping losses: 12-20%;
- Total power deposited on the extraction and steering grids: 220-271 kW (close to the calorimetric evaluations, see Tab. 1);
- Total power deposited on the grounded grid: 384-527 kW (not in agreement with the calorimetric evaluations, see Tab. 1);
- H^- power at accelerator exit: 10900-11900 kW;
- e^- power at accelerator exit: 1100-1400 kW;
- H^0 power at accelerator exit: 270-460 kW;
- H^+ power at accelerator exit: 10-40 kW;
- H_2^+ power at accelerator entrance: 70-150 kW;

- H^+ power at accelerator entrance: 20-40 kW.

It can be observed that the calorimetric evaluations carried out for the extraction and steering grids confirm the results of the simulations, while this is not true for the grounded grid. The reason for this will be investigated in future, by performing further measurements (for example by means of calorimetry, pressure sensors, thermo-cameras, spectroscopy etc.) on the accelerator. In this way, more precise comparisons could be carried out between experimental measurements and results of the simulations. In particular, it would be interesting to compare the power from the various species exiting the accelerator (evaluated with the described models) with the calorimetric measurements on the components located downstream of the accelerator (electron dump, H^+ beam dump, H^- beam dump, and calorimeter).

ACKNOWLEDGEMENTS

This work was set up in collaboration and financial support of Euratom.

REFERENCES

1. Y. Takeiri, et al., High-energy acceleration of an intense negative ion beam, J. Plasma Fusion Res. 71 (1995), 605-614.
2. Y. Takeiri, et al., Development of a High-Current Hydrogen-Negative Ion Source for LHD-NBI system, J. Plasma Fusion Res. 74 (1998), 1434-1443.
3. P. Agostinetti, et al., Design of a low voltage, high current extraction system for the ITER Ion Source, AIP Conf. Proc. 1097 (2009) 325.
4. P. Sonato, et al., The ITER full size plasma source device design, Fusion Eng. Des. 84 (2009), 269-274.
5. P. Agostinetti, et al., Choice of the permanent magnets for the SPIDER and MITICA beam sources, Technical note RFX-SPIDER-TN-121 (2011).
6. J. Pamela, A model for negative ion extraction and comparison of negative ion optics calculations to experimental results. Rev. Sci. Inst. 62 (1991), 1163-1172.
7. W.B. Hermannsfeld, Electron Trajectory Program, SLAC report, Stanford Linear Accelerator Center, SLAC-226 (1979).
8. H.P.L. de Esch, R.S. Hemsworth and P. Massmann. Updated physics design ITER-SINGAP accelerator. Fusion Eng. Des. 73 (2005) 329-341.
9. H.P.L. De Esch, Personal communication and support (2007), Association EURATOM-CEA, CEA-Cadarache, St. Paul-Lez-Durance, France.
10. K. Jousten, *Wutz Handbuch Vakuumtechnik - theorie und praxis*, 9th edition, Vieweg und Teubner Verlag, Wiesbaden (2006), ISBN-13 978-3-8348-0133-3
11. A. Krylov and R.S. Hemsworth, Gas flow and related beam losses in the ITER neutral beam injector, Fusion Eng. Des. 81 (2006), 2239-2248.
12. E.W. Thomas, Oak Ridge National Laboratory, Technical Report No. ORNL-6088, 1985.
13. OPERA-3D, Vector Fields Co.Ltd., http://www.vectorfields.com/
14. ANSYS Manual, ANSYS Inc., http://www.ansys.com/
15. PerMag Manual, private communication from D. Ciric (2007), UKAEA Fusion/Euratom Association, Culham Science Centre, Abingdon, United Kingdom.
16. G. Chitarin et al., Experimental mapping and benchmarking of magnetic field codes on the LHD Ion Accelerator, Proceedings of the 2nd International Symposium on Negative Ions, Beams and Sources, Takayama, Japan, 2010.
17. G. Fubiani et al., Modeling of secondary emission processes in the negative ion based electrostatic accelerator of the International Thermonuclear Experimental Reactor. Phys. Rev. Special Topics Accelerators and Beams 11, 014202 (2008).

Development of the JT-60SA Neutral Beam Injectors

M. Hanada[a], A.Kojima[a], T.Inoue[a], K.Watanabe[a], M.Taniguchi[a],
M.Kashiwagi[a], H.Tobari[a], N.Umeda[a], N.Akino[a], M.Kazawa[a], K.Oasa[a],
M.Komata[a], K.Usui[a], K.Mogaki[a], S.Sasaki[a], K.Kikuchi[a], S.Nemoto[a],
K.Ohshima[a], Y.Endo[a], T.Simizu[a], N.Kubo[a], M.Kawai[b] and L.R.Grisham[c]

[a] *Japan Atomic Energy Agency, 801-1 Mukoyama, Naka311-0193, Japan*
[b] *Nippon Adovanced Technology Co.Ltd, 3129-45 Hibara Muramatsu, Tokai 319-1112, Japan*
[c] *Princeton Univ., Plasma Physics Lab, P.O.Box451, Princeton, NJ08543, USA*

Abstract. This paper describes the development of the neutral beam (NB) systems on JT-60SA, where 30-34 MW D^0 beams are required to be injected for 100 s. A 30 s operation of the NB injectors suggests that existing beamline components and positive ion sources on JT-60U can be reused without the modifications on JT-60 SA. The JT-60 negative ion source was modified to improve the voltage holding capability, which leads to a successful acceleration of 2.8 A H^- ion beam up to 500 keV of the rated acceleration energy for JT-60SA.

Keywords: negative ion source, voltage holding capability, vacuum insulation.
PACS: 07.77.Ka, 41.75.Cn

INTRODUCTION

The JT-60SA (JT-60 Super Advanced) project is a combined project of JAEA's program for national use and JA-EU Satellite Tokamak Program collaborating with Japan and EU fusion community [1]. The main mission of the JT-60SA project is to contribute to early realization of fusion energy by supporting the exploitation of ITER [2] and by complementing ITER in resolving key physics and engineering issues for DEMO reactors [3].

Neutral beam (NB) injection is main plasma heating and current drive system on JT-60SA, where existing NB injectors on JT-60U is required to be upgraded and reused. The upgrade of the NB injector will be made in two steps. In the first step, the power supplies will be upgraded to extend the pulse duration time from 10 s of the present rated value to 100 s of a design value for JT-60SA. In addition, a part of the magnetic shield in the positive-ion-based NB (P-NB) injector, which is composed of cancelling coils and passive magnetic shield installed outside of an NB vacuum vessel, will be upgraded to reduce the stray field in the beam path since the stray field on JT-60SA is three-times larger than that on JT-60U. However, the upgrade of the magnetic shield is insufficient for guiding residual ions to residual ion dump. To avoid the residual ions out of the RID, a part of the grid apertures, where the improper residual ions are emerged, will be masked. This results in a reduction of 15% of a total ion

TABLE 1. Specifications of the NB injectors on JT-60SA

	Positive-ion-based NB injector	Negative-ion-based NB injector
Injection power per injector	1.7-2.0 MW	10 MW
Pulse length	100 s	100 s
Number of units	12	1
Number of ion sources per unit	2	2
Beam energy	80-85 keV	500 keV
Ion beam current per ion source	24.9-27.5 A	22 A
Injection positions	4 units: tangential injection to equatorial plane 8 units : perpendicular inj.	tangential injection to off-axis plane (55cm downward from equatorial plane)

extraction area, and hence a reduction of injection power to 1.7 MW from one injector although the achieved injection power is 2 MW. From 12 P-NB injectors, 20 MW D0 will be injected at 85 keV. One negative-ion-based NB (N-NB) injector will be upgraded to inject 10 MW, 500 keV D^0 beams for 100 s. The upgrade of these injectors allows a total injection power of 30 MW for 100 s.

In the second step, each of the P-NB injector will be upgraded to increase the injection power from 1.7 MW to 2.0 MW, for which cancelling coils inside the NB vacuum vessel will be upgraded to guide the residual ions to residual ion dump. By this modification, a total injection power will be increased up to 34 MW.

The pulse duration time of the NB system for JT-60SA is > 5 times longer than that in existing NB systems. The key issues to produce such as high powerful neutral beams for long pulse duration time are a heat removal of beamline components and a long pulse production of the high power ion beams. The heat removal capability of existing beamline components on JT-60U is examined in P-NB and N-NB injectors. The feasibility of long pulse beam production was examined in the JT-60 positive ion source at Korea Atomic Energy Research Institute under the agreement between Japan and Korea in fusion collaboration program. The long pulse production of the high energy negative ion beams has been tested in the JT-60 negative ion source, which has been modified to reduce the power loading of the acceleration grids and improve voltage holding capability. In order to reduce the grid power loading, the beam steering angle of multiple beamlets of the negative ion beam was tuned by suppressing outward deflection of outermost beamlets through multi-aperture grids with electric field shaping plates. In order to improve the voltage holding capability, vacuum insulation for large acceleration grids was experimentally examined. Based on this result, the ion source was modified to sustain 500 kV of the rated acceleration voltage for JT-60SA.

In this paper, the R&D results on long pulse productions of the powerful positive and negative ion beams are reported with the assessment on the heat removal capability of existing beamline components on JT-60U after a brief description of the specifications for the NB system on JT-60SA.

PLAN OF JT-60SA

The schedule of the JT-60 SA project was approved by the Broader Approach

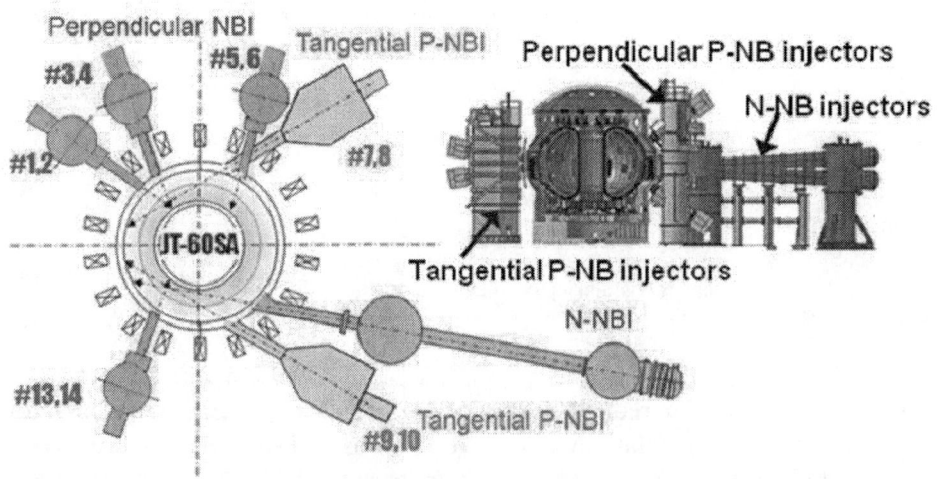

FIGURE 1. Layout of the NB injectors on JT-60SA. 12 positive-ion-based NB injectors and one negative-ion-based NB injector are allocated around JT-60SA.

Steering Committee in late 2008. By taking into account a more detailed analysis of the manufacturing and assembly schedule, the milestone of "First Plasma" is foreseen in 2016. Towards the first plasma of JT-60SA, the design and development of the NB system for JT-60SA is in progress.

On JT-60 SA, twelve positive-ion-based NB and one negative-ion-based NB (N-NB) injectors are allocated to inject the D^0 beams for 100 s. A total injection power from the positive-ion-based NB injectors is 20-24 MW, and the remaining power of 10 MW is to be injected from the N-NB injector. As shown in Fig.1, eight and four positive-ion-based NB injectors are perpendicularly and tangentially allocated to the plasma, respectively. Each injector is designed to inject 1.7 MW-2.0 D^0 beams from two ion sources, each of which produces 24.9-27.5 A, 80-85 keV D^+ ion beams. The N-NB injector is designed to be tangentially allocated to plasma in order to inject 10 MW, 500keV D^0 beams. The N-NB injector has two large negative ion sources, each of which is designed to produce 500 keV, 22 A D^- ion beams. The negative ion source has been developed to fulfill the requirement for JT-60SA. The detailed specifications of the NB injectors are summarized in Table 1.

DEVELOPMENT OF THE POSITIVE ION SOURCE

The P-NB system, which consists of 14 beamline units and has beam energy of 70 keV to 100 keV, started operation in 1986 with hydrogen beams and injected a neutral beam power of 27 MW at 75 keV into the JT-60 plasma. In 1991, the P-NB system was modified to be able to handle deuterium beams and to increase the beam energy. At the same time, the beam injection angle to the JT-60U plasma was changed from a perpendicular injection to a tangential one on four beamlines out of 14 units. After executing some research and developments, deuterium beams of 40 MW at 95 keV were injected in 1996 [4-6]. Since 2003, the long pulse injection has been executed to study long pulse production of the high β plasma. By taking account of the limit of existing power supply on JT-60U, the P-NB system was modified to extend the pulse duration time from 10 s to 30 s. Non-water cooled limiter for covering inner surface of the beam drift duct was modified to increase the heat receiving area. The cooling capability of the power supplies was also enlarged by replacing with the larger heat capacitance electric parts and by increasing number of fans. Both of the modifications were devoted to four tangential P-NB injectors, and only power supply modification to four perpendicular P-NB injectors. The pulse duration time of 2 MW, 85 keV D^0 beams was increased step by step with monitoring the power loading due to re-ionization of the D^0 beams in the drift duct. Finally, four tangential P-NB injectors injected 2 MW D^0 beams for 30 s. On the modified perpendicular P-NB injections, the pulse duration time was increased to 17-25 s at the injection power of 2 MW. From these operations during 30s, the water cooling capability was assessed. Figure 2 shows the JT-60 positive ion source [7] and the time evolution of the water temperature rises in the acceleration grids. The ion source has four acceleration grids of a plasma grid (PG), a gradient grid (GG), a suppression grid (SG) and a grounded grid (GRG) which are well cooled by water. Each of the temperature rises in the acceleration grids was measured when the D^0 beam of 2 MW was injected via neutralization of 85keV, 27.5 A D^+ ion beams for 30 s. The temperature rise rapidly increases at < 15 s, and then saturated. The highest temperature was saturated to be 8 °C in the PG, which is the allowable level for JT-60SA. From this result, the cooling capability of the ion source was suggested to be sufficient for the 100 s injection. The water temperature rise of the beamline components such as residual ion dumps was also saturated to be less than 10 °C with a short thermal time constants of < 30 s.

The ion source was operated with the longer pulse length to examine the feasibility of the longer pulse beam production since the thermal time constant of the arc chamber, where permanent magnets are distributed for plasma confinement, was longer than 30 s. The experiment was carried out at Korea Atomic Energy Research Institute under the agreement between Japan and Korea in fusion collaboration program. The ion source was stably operated for 100 s at an arc power of 28 kW (68V, 430A) required for the 2 MW injection. The water temperature rise of the arc chamber

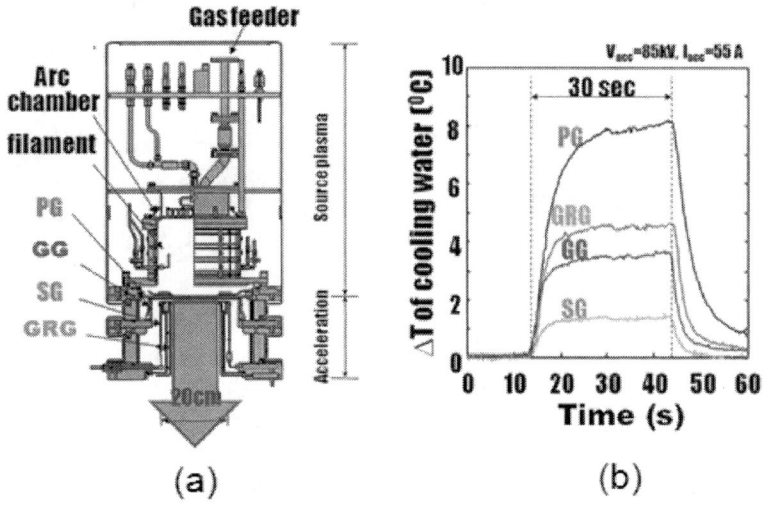

FIGURE 2. JT-60 positive ion source (left schematic) and time evolutions of acceleration grids during 30 s at the injection power of 2 MW from one injector with two ion sources in each of which 85 keV, 55 A D^+ ion beam are produced (right graph).

during 100 s. was as low as 6 °C, which is much lower than an allowable level for JT-60 SA. At the slightly lower arc discharge power of 23 kW, 60 keV, 18 A H$^+$ ion beams were stably produced during 200 s. These results suggest that existing JT-60 positive ion source will be reused without modification in JT-60SA.

DEVELOPMENT OF THE NEGATIVE ION SOURCE

Production of the high-energy negative ion beam

The N-NB system started the operation in 1996. In the period of 1996-2003, the N-NB injector had been modified to increase the injection power. The achieved maximum power injected from two ion sources was 5.8 MW for 0.87 s via the neutralization of 400keV, 49.2 A(drain current of power supply) D$^-$ ion beams. In this high power operation, a poor voltage holding capability of the ion source with three acceleration stages was found to be critical issue to achieve the injection power of 10 MW for JT-60SA[8]. Although the voltage holding capability can be readily improved by the extension of the gap length, an excess extension of the gap length causes the degradation of beam optics and the negative ion losses in the accelerator. Therefore, the increase of the gap length should be minimized. However, database on the vacuum gap insulation for the large grids is insufficient. The breakdown voltage was examined by varying the gap length in the JT-60 negative ion source. The breakdown voltages obtained in all three stages were plotted as a function of the shortest cathode-anode distance in each of gaps as shown in Fig.3. In this figure, the breakdown voltages obtained in the small sample electrode utilized for the design of the JT-60 negative ion source are also shown for comparison. Each of the breakdown voltages were obtained after sufficient conditioning. With the square root of gap length, the breakdown voltages increased. This shows that the breakdown voltage obeys Clump theory [9] even in the JT-60 negative ion source. However breakdown voltage in the JT-60 negative ion source is a half of that in the small electrodes. Based on this result, the gap lengths for each acceleration stage were extended from 55-75mm to 85mm in

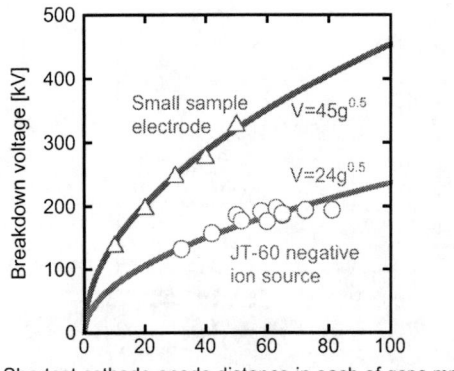

FIGURE 3. Breakdown voltages in all three stages as a function of the shortest cathode-anode distance in each of gaps.

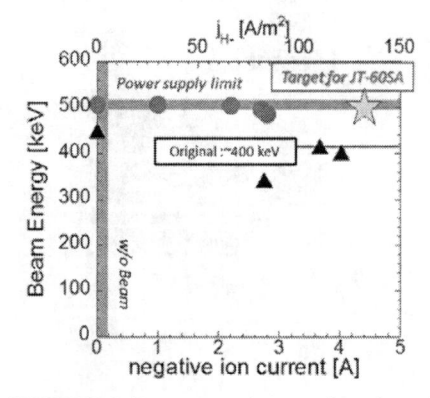

FIGURE 4. Beam energies as a function of beam current before and after modification of the ion source.

order to target 200 kV for each of the acceleration stage. No significant degradations of the beam optics and the beam losses are indicated by computational calculations [10].

After the modification of the ion source, voltage holding capability was drastically improved from 400 kV before modification to 500 kV of the power supply limitation. The conditioning time was as short as 1 hour. High voltage of 490 kV is stably sustained without breakdowns for 40 s of the power supply limitation. No degradation of the voltage holding capability during the long pulse has been observed. This suggests the feasibility of long pulse injection in JT-60SA.

After the voltage holding capability was confirmed to be improved, the beam production was tested on the N-NB injector on JT-60U. Only through 20% of an ion extraction area, the hydrogen negative ion beams were produced. Since available input electric power to the acceleration power supplies was restricted, the pulse length and the accelerated beam current were limited to be < 1s and < 10 A, respectively. The ion source was first conditioned with relatively low current beam of 1 A, and then the beam current was gradually increased to 4 A with seeding the cesium into the ion source. No significant degradation of voltage holding capability was observed even for the higher beam current and Cs seeding.

Figure 4 shows beam energies as a function of beam current before and after modification of the ion source. The beam energy was successfully improved from 400 keV before modification to 500 keV of the requirement for JT-60SA. The negative ion beams of 2.8A, 490 keV and 1A, 510 keV have been stably produced. These are the first demonstration of a high-energy negative ion acceleration of more than one-ampere to the beam energy of 500 keV in the world. The power loading of the acceleration grids and beamline components showed no significant degradations of the beam optics and the beam losses. The beam current density is 85A/m^2 and 65% of the design value of 130 A/m^2. This relatively low current density is caused by a poor ion extraction voltage of ~5 kV due to a bad setting of coaxial feed-through for the extraction voltage. New coaxial feed-through with the larger gap between inner and outer feeders is designed for JT-60SA, and could increase the rated current density up to the rated value for JT-60SA.

Long pulse operation of the negative ion source

The feasibility of the long pulse injections was examined to realize the N-NB system on JT-60SA. In the previous operation, the critical issue for the long pulse injection was high power loading of the acceleration grids. The highest grid power loadings in the two ion sources, called as the U and L ion source, were 9% and 7% of the accelerated beam power on the grounded grids in the L and U ion sources [11], respectively. These grid power loadings were higher than an allowable level of 5% for the long pulse injection.

To clarify the origin of the grid power loading, the ion trajectory was calculated using a 3 D calculation code [12]. The grid power loading on the grounded grid was found to be caused by direct interception of outmost beamlets due to the un-optimized field shaping (FS) plates which are equipped with the extraction grid (EXG) to suppress an outward deflection of the outmost beamlets by space charge of the inner beamlets [13]. To suppress the direct interception of outermost beamlets, new FS plates were designed using the 3 D calculation code. The distances between the FS plates and the outmost apertures and the geometry of the FS plates were optimized.

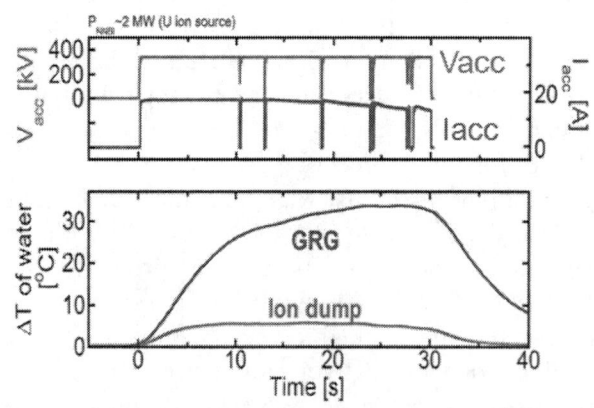

FIGURE 5. Time evolutions of acceleration voltage (Vacc), accelerated current (Iacc), temperature rises of the grounded grid (GRG) and residual ion dump (Ion dump) during 30 s.

The new FS plates properly steer the outermost beamlets as predicted by the simulation, resulting in the reduction of the grid power loadings. At a typical operation pressure of 0.3 Pa, the grid power loadings of the grounded grid in the U and L ion sources have been reduced from 9 % to 7% and 7% to 5%, respectively. The power loading of the grounded grid in the L ion source was successfully reduced to an allowable level for JT-60 SA. The difference of the grid power loading between in the U and L ion sources is due to the beam uniformity. Namely, the uniformity of the beam produced in the U ion source was worse than that in the L ion source because of 1.5 times broader ion extraction area, resulting in the higher grid power loading due to the direct interception of the negative ions by the acceleration grids.

The power loading in the U ion source could be reduced to the allowable level by improving the beam uniformity with the tent-shaped filter [14]. After the power loadings on the acceleration grids were confirmed to be suppressed sufficiently, the injection pulse length was extended from the rated value of 10 s to 30 s of the acceleration power supply limit. Via neutralization of 12.5A and 9.8A D⁻ ion beams at 340 keV from U and L ion sources, ~2.0 MW and ~1.0 MW D^0 beam were injected for 30 s and 20 s, respectively. A total injection energy from two ion sources, defined as the product of the injection power and the pulse duration time, was as high as 80 MJ. This is the highest injection energy of the N-NB in the world since operating N-NB injection at NIFS is lower than 40 MJ [15].

From the results of the 30s injection, the feasibility of water cooling capability in the acceleration grids for JT-60SA was studied. Figure 5 shows the time evolution of the water temperature rise of the grounded grid during the 30 s. The injection D^0 power and the power loading of the grounded grid were 2MW and 470 kW, respectively. The water temperature rise was saturated at ~33°C within ~25s. The saturation time was much shorter than the pulse length required for JT-60SA. Since injection power is required to be 5 MW for one ion source in JT-60SA, the water temperature rise is estimated to be ~83°C for full power injection for JT-60SA. This temperature could be reduced to an allowable level of < 60°C by improving the beam uniformity with the tent-shaped filter since the grid power loading for the more uniform beam obtained in the L ion source was ~70% of that obtained in the U ion source. The water temperature rise of the residual ion dump, where 40% of the

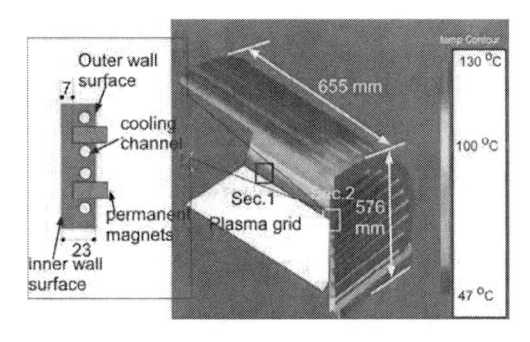

FIGURE 6-a. Surface temperature distribution on the inner wall of the arc chamber just after 100s arc discharge.

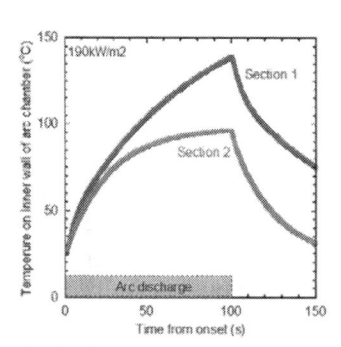

FIGURE 6-b. Calculated time evolutions of the surface temperatures at Sec.1 and 2.

accelerated ion beam is dissipated, was estimated to be < 10 °C for the full power injection on JT-60SA. These estimations suggest that water cooling capability of the acceleration grids and beamline components is sufficient for JT-60SA.

The water cooling capability of the JT-60 negative ion source was examined by the thermal analyses of the arc chamber with a three dimensional code [16] since the thermal time constant of the arc chamber was longer than 30 s. The arc discharge power and filament powers, giving the beam current density of 130 A/m2 required for JT-60SA, are assumed to be 300 kW and 130 kW from the extrapolation of the measurements in the 30s operation, respectively. These powers are assumed to be uniformly dissipated on the surface of the inner wall. The 84 water cooling channels with inner diameter of 6.6 mm are allocated around the arc chamber. A total water flow rate is assumed to be 300 L/min. Figure 6-a shows surface temperature distribution on the inner wall surface of the arc chamber just after 100s arc discharge. There is a temperature distribution. The highest temperature is 140 °C at sec1 close to the plasma grid, and lowest temperature is 90 °C at sec2 close to the permanent magnets. The time evolutions of the temperature at sec1 and sec2 are shown in Fig.6-b. The temperature at sec2 was saturated to be 90 °C after 70s, and is lower than temperature range of use in the permanent magnets (250 °C) for plasma confinement around the arc chamber. Water cooling capability of arc chamber in JT-60 negative ion source is suggested to be sufficient for the operation on JT-60 SA.

SUMMARY AND FUTURE PLAN

At Japan Atomic Energy Agency, high power and long pulse NB injectors are being developed to realize the neutral beam injectors on JT-60SA where a total injection power of 30-34 MW for 100 s. As for the positive-ion-based NBI, 2 MW D^0 beams of the rated value for JT-60SA has injected for 30s from one of existing injectors. From the measurements of water temperature rise during 30 s, water cooling capability of existing beamline components is sufficient for JT-60 SA although the power supplies and non-water cooling components is to be upgraded. The long pulse operation during 100-200 s were demonstrated in collaboration test between JAEA and KAERI. As for the negative-ion-based NBI, a 30 s injection was demonstrated at an injection power of 3 MW. From the measurements of water temperature during 30 s, water cooling capability of existing ion source and the beamline components is confirmed to be sufficient for JT-60SA. The key issue of the N-NB injector for JT-60SA, i.e., voltage holding capability of the ion source was also improved to be 500 kV of the rated acceleration voltage for JT-60SA, leading to a stable production of 500 keV, 2.8 A H⁻ ion beam.

Remaining issues in the N-NB injector for JT-60 SA are the improvement of the beam uniformity and the long pulse production of the D⁻ ion beams with high current density. To improve the beam uniformity, the arrangement of the permanent magnets around the arc chamber will be modified to form the tent-shaped filter, and tested in the JT-60 negative ion source. To produce the D⁻ ions stably for 100 s, the plasma grid will be modified to keep its temperature in a range of 200-250 ^{o}C to produce high current density beams of > 100 A/m^2. The plasma grid is being designed to be actively controlled at 200-250 ^{o}C by flowing GALDEN 270 fluid [17] that is fluorine compound. To test the modified JT-60 negative ion source, a test stand is planned to be fabricated. Using the power supplies on JT-60 U, negative ions will be produced and extracted in the JT-60 negative ion source.

ACKNOWLEDGMENTS

The authors would like to express their appreciation to Dr.D.H.Chang, B.H.Oh, S.H.Jeong for their valuable discussions and experimental supports. They are also grateful to Dr. Y. Ikeda, Dr. K.Kurihara, Dr. H.Ninomiya and the late Dr. T.Tsunematsu for their continuous encouragement and support.

REFERENCES

1. S. Ishida et al., Proc. of the 23rd Fusion Energy Conf., 2010, OV/P-4.
2. K.Ikeda,'Progress in the ITER Physics Basis', *Nucl. Fusion 47* (2007) S1.
3. K. Tobita et al., *Nucl. Fusion* 49 (2009) 075029.
4. S.Matsuda et al., *Fusion Eng. Des.*, 5, 85(1987).
5. Y.Okumura et al., *Proc. 10th Int. Conf. Plasma Physics and Controlled Nuclear Fusion Research*, London, Unitted Kingdom, September 12 1984, p.329, International Atomic Energy Agency (1985).
6. M.Kuriyama et al., *FUSION SCIENCE AND TECHNOLOGY*, 2002, VOL.42.
7. Y.Okumura et al., *Rev. Sci. Instruments.***55**,1(1984).
8. M.Hanada et al., *J.Plasma Fusion Res.SERIES*, Vol.9(2010)208
9. L. Cranberg, *J. of Appl. Phys.*, 23, 518, 1952.
10. A.Kojima, Proc. of the 23rd Fusion Energy Conf., 2010, FTP/1-1Ra.
11. M.Kuriyama et al., *Fusion Eng.Des.*, 39-40, 115 (1998).
12. *http://www.* fieldp.com.
13. M.Kamada et al., *Rev. Sci. Instruments.*79.02c114(2008).
14. M. Hanada et al., *Rev. Sci. Instruments*, vol.77, No.3(2006) 03A515-1-3.
15. Private communication with K. Tsumori.
16. http://www. ad-tech.co.jp.
17. *http://www*.solvaysolexis.com

Status of the 1 MeV Accelerator Design for ITER NBI

M.Kuriyama[a], D.Boilson[a], R.Hemsworth[a], L.Svensson[a], J.Graceffa[a], B.Schunke[a], H.Decamps[a], M.Tanaka[a], T.Bonicelli[b], A.Masiello[b], M.Bigi[c], G.Chitarin[c], A.Luchetta[c], D.Marcuzzi[c], R.Pasqualotto[c], N.Pomaro[c], G.Serianni[c], P.Sonato[c], V.Toigo[c], P.Zaccaria[c], W.Kraus[d], P.Franzen[d], B.Heinemann[d], T.Inoue[e], K.Watanabe[e], M.Kashiwagi[e], M.Taniguchi[e], H.Tobari[e] and H.De Esch[f]

[a] *ITER Organization, 13067 Saint-Paul-lez-Durance Cede, France*
[b] *Fusion for Energy, C/ Josep Pla 2, 08019 Barcelona, Spain*
[c] *Consorzio RFX. Corso Stati Uniti 4 35127 Padova, Italy.*
[d] *Max-Planck-Institut für Plasmaphysik, PO Box 1533, 85740 Garching, Germany*
[e] *Japan Atomic Energy Agency, 801-1, Mukoyama, Naka 311-0193, Japan*
[f] *CEA-Cadarache, IRFM, F-13108 Saint-Paul-lez-Durance, France*

Abstract: The beam source of neutral beam heating/current drive system for ITER is needed to accelerate the negative ion beam of 40A with D- at 1MeV for 3600 sec. In order to realize the beam source, design and R&D works are being developed in many institutions under the coordination of ITER organization. The development of the key issues of the ion source including source plasma uniformity, suppression of co-extracted electron in D beam operation and also after the long beam duration time of over a few 100 sec, is progressed mainly in IPP with the facilities of BATMAN, MANITU and RADI. In the near future, ELISE, that will be tested the half size of the ITER ion source, will start the operation in 2011, and then SPIDER, which demonstrates negative ion production and extraction with the same size and same structure as the ITER ion source, will start the operation in 2014 as part of the NBTF. The development of the accelerator is progressed mainly in JAEA with the MeV test facility, and also the computer simulation of beam optics also developed in JAEA, CEA and RFX. The full ITER heating and current drive beam performance will be demonstrated in MITICA, which will start operation in 2016 as part of the NBTF.

Keywords: ITER, NBTF, SPIDER, MITICA, beam source, ion source, accelerator
PACS: 52.57.Cm, 52.50.Gj, 52.55.Fa, 52.59.Bi, 52.65.Cc

INTRODUCTION

The design and R&D activities toward the construction of the ITER Heating/Current drive NB system (HNB)[1], which is designed to inject 16.5 MW of 1 MeV D^0 or

0.87 MeV H^0 into plasma from each beamline for 3600 s, is being developed in collaboration with JAEA, IPP, RFX, CCFE, CEA and CIEMAT under the coordination of the ITER Organization. The required performance for the beam source (BS) is 40 A of D$^-$ at 1 MeV with a current density of 200 A/m^2 at a source filling pressure of ≤0.3 Pa. In order to obtain these high performance beams stably and reliably several issues need to be solved. These include uniform source plasma of the entire extraction region (≤ ±10%) with D$^-$ or H$^-$ in order to produce low divergence beamlets from the entire array, reliable source start-up, reliable source operation at ≤0.3 Pa in order to reduce the stripping losses and power loading of the accelerator grids to acceptable levels, temperature control of the plasma grid, low co-extracted electron currents, and stable high voltage holding.

To attain the specified beam performance, many R&D and design activities are underway in the above mentioned institutions. In parallel, construction has started of the Neutral Beam Test Facility (NBTF), including MITICA which is essentially 1MeV beam injector just same as the HNB with a full size, full power and full pulse length, and SPIDER which is an ion source test bed with a full ITER size, full power and full pulse length but its accelerated beam energy is 100keV, but not 1MeV. The SPIDER will be used to carry out the final development of the ITER ion source, whilst the MITICA will finalize the development of the 1 MV accelerator, carry out integrated testing of the HNB, demonstrate the beam performance required in ITER and develop all the conditioning and commissioning protocols and procedures for the ITER HNBs. This paper provides an overview of these research activities and the status of the design of the injectors for the ITER-NBI.

REQUIRED BEAM PERFORMANCE AND GENERAL DECRIPTION OF THE BEAM SOURCE FOR HNB

The major specifications of the BS for HNB [1] are listed in Table 1. These BS parameters will allow a neutral beam power of 16.7MA per beamline into tokomak plasma.

TABLE 1. Specification of the beam source for the HNB

	D$^-$	H$^-$
Beam Energy	1MeV	870keV
Extraction current	48A	51.6A
Acceleration current	40A	46A
Current density accelerated	200A/cm2	230A/m2
Source filling pressure	0.3Pa	0.3Pa
Extracted electrons ratio to ions	<1	<1
Beam divergence (core) *)	<7mrad	<7mrad

*) Core part is 85% of the total beam power, and the other of 15% is halo part (typically beam divergence of 10-15mrad)

The BS for HNB consists of a RF type negative ion source (RF source) [2-3] and the Multi Aperture Multi Gap (MAMuG) accelerator.[4] The RF ion source, which is mainly developed in IPP, has been chosen in the HNB since it requires in principle no maintenance except refilling of Cs into Cs oven, unlike arc-type ion source which requires replacement of filaments every year using a remote handling system. The chosen accelerator is the MAMuG developed in JAEA. The alternative accelerator, the so called SINGAP developed at CEA [5], showed a lower voltage holding capability and more co-accelerated electrons, so this is not any longer an alternative.

The outline figure of the BS is shown in Fig.1. The BS consists of ion source, extractor and accelerator, and has the following approximate dimensions; height=3m, width=2.7m and the length=2.6m. The total weight of the BS is roughly 25 tons. The ion source as shown in Fig.2, which produced negative ions with a uniform current density of less than ±10% under an operation pressure of <0.3Pa, and with a current density of 240 A/m^2 with D$^-$ at the extractor for 3600 sec, is composed of eight RF drivers connected with four independent RF power supplies, expansion chamber, starter filaments and Cs oven.

The extractor consists of plasma grid (PG) and extraction grid (EXG), electron suppression grid (ESG) and bias plate (BP). All the grids are supported mechanically from the plasma grid support structure. The total extraction area of beam is roughly 0.6m in width and 1.6m in height, and is divided into four groups in vertical, and those are sub-divided into four segments in horizontal. Each segment is the size of 8cm^{W}x 33cm^H, having 5x16 apertures. The total amount of apertures in the extractor is therefore 4x4x5x16=1280. All the beamlets from the grids are faced to a focal point through tilting individually the group of segments in vertical direction and through aperture offsets in horizontal direction to maximize neutral beam injection power and to minimize heat load onto the beamline components.

In the PG, a dc current up to 5kA flows toward vertical direction to form magnetic filter field in horizontal direction (Bx) to enhance negative ions and to suppress co-extracted electrons.

The EXG, which is applied extraction voltage, is insulated electrically from the PG support structure by ceramic spacers. Sm-Co permanent magnets are embedded in the EXG to make a dipole magnetic field for deflecting co-extracted electrons onto the ESG.

The MAMuG accelerator has five stages acceleration grids. Each stage of the accelerator is divided into four groups, each group is subdivided into four segments, each group consisting of an array of 5x16 circular apertures (16-mm diameter) same as the EXG, and roughly size of the acceleration area is 0.6m x1.6 m^2 shown in Fig.3. An array of post insulators made of ceramic is adopted for the electrical insulation and mechanical connection between flanges of each potential.

PRESENT STATUS OF DESIGN FOR REALIZING THE BS

The key issues, for realizing the higher performance beam described in section 2, of the ion source and accelerator for the BS are summarized in Table 2.

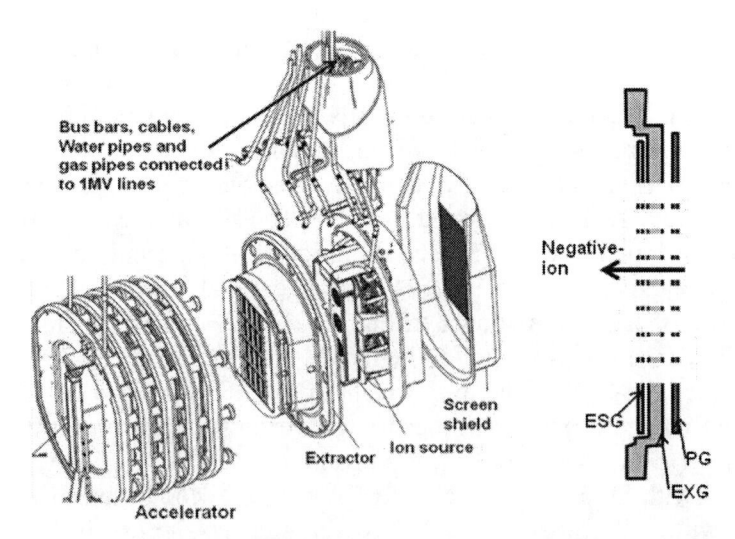

(a) Diagram of the BS for HNB (b) Diagram of the Extractor
FIGURE 1. Schematic view of the Beam source and Extractor for the HNB

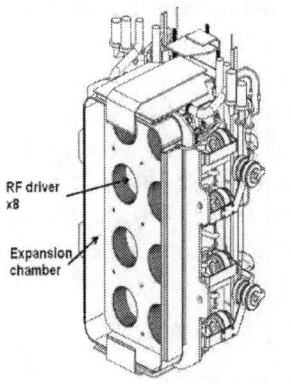

FIGURE 2. Schematic vie of RF ion source

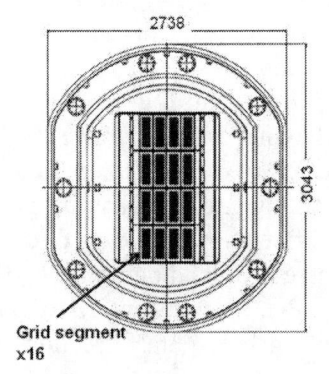

FIGURE 3. Plane view of acceleration grid and its support

Issues and their solutions of the ion source

The key issues of the ion source are discussed here. Among them, the more critical issues are to make uniform source plasma within ±10% and to operate at low filling pressure with low co-extracted electron current.

a) Asymmetry of the source plasma has been remarkably observed in large ion source such as JT-60[6] and LHD[7]. The cause of the asymmetry has not been clearly understood yet, though some of physical phenomena are speculated as the driving force, such as ExB drift, JxB drift and/or grad B drift or a combination of these effects caused by PG filter field and/or bias current. The influence of the PG magnetic filter field for the asymmetry has been investigated in IPP, JAEA and NIFS.[8-11]
The asymmetry is largely affected by the filter field, B_x, at the PG surface, and also somewhat by $\int B_x dz$.

TABLE 2. Issues for constructing the BS for ITER

Issue	Prospects for the solution	step to the solution
1. Ion source		
To produce uniform source plasma less than ±10%	> Optimization of PG filter current and Bias current > RF power regulation of each driver so as to minimize the asymmetry > Making uniform magnetic field in the expansion chamber for low Bx and high Bxdl is being investigated	BATMAN, MANITU, RADI and ELISE (IPP), (also JAEA and NIFS)
To keep constant Cs effect for 3600 sec	Optimize Cs supply, etc.	As above
To operate stable at high RF power	Optimize RF driver size, etc.	As above
To keep constant PG temperature	Adding exclusive water tube for PG	JAEA, RFX
To reduce the magnetic field at the driver region	Addition of a counter magnetic field using the return busbar of the PG filter current is being investigated.	IPP, RFX
2. Accelerator		
Compensation of deflected beamlets caused by the magnetic field for bending of co-extracted electrons and also by space charge beam repulsion	Optimization of aperture offset of ESG and field shaping plate through simulations and experiments	MTF/JAEA Simulation; JAEA, CEA
Focusing the beam at the exit of drift duct by aperture offset in accelerator grids and also tilting grids	Optimization of aperture offset of acceleration grids	MTF/JAEA Simulation; JAEA, CEA
Optimization of accelerator grid gaps from the points of view of both beam optics and voltage holding	Optimization of the gap by simulations and experiments, and the optimization of electric field relaxation rings	JAEA, CEA

A new PG filter field configuration has been proposed by IPP and RFX.[12-13] Its configuration is to make a uniform magnetic field inside the expansion chamber by the combination of the current of PG and that of return feeder lines. This magnetic field configuration seems to be anticipated to give the following results; i) minimizing the asymmetry through low Bx at PG and also almost no grad B drift, ii) minimizing electron current ratio to the ion through high ∫Bxdl, iii) reducing the penetration of PG filter field into the RF driver which causes instability of RF discharge.

Bias current also affects the plasma asymmetry, and the asymmetry increases with proportion to the bias current.[3] Therefore the bias current needs to minimize as low as possible with devising bias plate and/or optimizing PG filter field.

The distribution of Cs vapour affects the asymmetry of the source plasma, too. The optimization of the position of Cs injection nozzle is being developed through experiments and Cs vapour flow calculation at IPP.[14]

b)   The high filling pressure, which is typical characteristics of the RF source, is a serious disadvantage in the negative ion source owing to enhance stripping loss in the accelerator, though it was not serious issues in the positive ion source. Last several years many efforts to reduce the filling pressure have been done mainly in IPP [3], and as the result the filling pressure has been lowered to 0.4 Pa at optimum source condition. Furthermore, recently, the source has been possible to operate at 0.3 Pa though it is off from an optimum condition. However the ion source for HNB is needed to operate in filling pressure of less than 0.3 Pa with the optimum condition. In IPP Garching, Further efforts to reduce the filling pressure are being developed by applying a vertical magnetic field or by using Helicon type discharge at the driver region.

c) In long pulse operation more than a few 100 sec with MANITU in IPP,[14] the electron current co-extracted with ions is increased gradually though ion current does not change and is finally more than that of ions after a few 100 sec. According to the IPP data, the Cs content at near PG is decreased gradually as the time of beam duration in spite of constant feeding of Cs vapour. At present time this phenomena does not clear. The issue is now being investigated in IPP, and one of countermeasure proposed is to regulate the amount of Cs vapour with the time of beam duration.

In D⁻ beam operation, co-extracted electron ratio is considerably large in compared to H⁻ beam.[3] The issue for suppressing the co-extracted electron with D beam is also under investigation in IPP.

d) PG temperature affects negative ion production and co-extracted electron. In particular co-extracted electron current ratio is more sensitive to the PG temperature in a range of 120–200°C, though ion current does not change so much.[14-15] In order to control the PG temperature with water, an exclusive water line for the PG in addition to common water feeder line on 1MV potential is being considered in JAEA and RFX. A heat exchanger to control the water temperature is equipped on the ground potential and an additional water line for the PG to the common water feeder line is installed from ground potential to the BS on the 1MV potential. However the technical issues including setting space of the exclusive water line in the high voltage transmission line, and the influence of by-product gas which is generated by chemical reaction of SF6 gas exposed to the high temperature pipe of 150-200°C are being considered.

e) The operation of the RF driver is affected by a magnetic field.[16] In particular the PG filter magnetic field roughly more than 4mT at the driver position makes unstable RF discharge in the experiment with BATMAN in IPP. A countermeasure for decreasing the PG filter field at the RF driver is required for stable discharge. The countermeasure to make another magnetic field at the roof of expansion chamber with the return line of PG current feeder, above described, is being confirmed experimentally in IPP.

Issues and their solutions of accelerator

a) Beamlet deflection by dipole magnetic field at the EXG is needed to compensate by aperture offset at the ESG for getting good beam optics. Furthermore beamlets at the edge of the bundle of those are deflected outward by space charge repulsion force. Field shaping plate attached at the backside of the ESG to compensate the deflection has been proposed [17], and it has been applied the beam source of JT-60, and it is also being tested in the MTF.[18]   In the recent experiments with the MTF, a H⁻ current density of 144A/m^2 at 937keV and also 156A/m^2 at 879keV have achieved.[18] The beam performance, however, was restricted by the heat load onto the accelerator grids because of the ion beam deflection caused by the dipole magnets at the EXG and space charge beam repulsion for beamlets at the edge of a beam bundle. In this experiment, aperture offset of 0.5mm and field shaping plate of 1mm thick was taken the measures in order to compensate the ion beam deflection. However those measures were not sufficient for compensating the deflection. The JAEA is now being considered a new configuration of aperture offset and field shaping plate using the computer simulation with OPERA-3D code [19] by considering more detailed boundary condition, and after that it will be confirmed experimentally with the MTF.

b)   Gap length of each step in the accelerator is needed to decide so as to optimize beam optics and to keep easily the voltage holding between gaps. In the MTF and JT-60U, improvement of the voltage holding has been required. In the voltage holding test utilizing these accelerators in 2009-2010, it was found that the voltage holding capability is about a half of that of the qusi-Rogowski electrode used to design of these accelerators. To sustain 200 kV in each acceleration stage at the MTF, the local electric field concentrations were lowered more by expansion of the minimum gap length between metals from 72 mm to 100 mm including the margin of 20 % and rounding edges of grid support. After that, the voltage of 1 MV was sustained in a vacuum for 4000 s. In JT-60U, the minimum gap length between metals was expanded from 50 mm to 68 mm to hold 163 kV in one stage. [20]   After that, the voltage was increased from 400 kV to 500 kV. In ITER, the minimum gap length of the original design of the accelerator was 50mm. It was obvious that this gap length was not enough from a point of view of the voltage holding. The gap length to satisfy for both beam optics and voltage holding is needed to decide. In the ITER task of the physics design of the accelerator, JAEA reported that all acceleration gaps can be expanded to about 90 mm for improvement of voltage holding, maintaining better beam optics.[21]. The CEA has proposed the equal gap length of 84mm from a point of view of the beam optics. This still has the margin of 10 % to sustain 200 kV. We are being examined not only gap length but also thickness and structure of grid and its support considering beam optics, electric field reduction and mechanical strength.

c)   Beam aiming of each beamlet in the segments and also each grid segment are necessary for maximizing injection power and minimizing the heat load on the beamline components. Each group of the segments for vertical arrangement is tilted toward to the focal point near exit of the drift duct. Meanwhile for horizontal direction, aperture offset has been adopted.   The optimum condition, however, is under examination with the computer code of OPERA-3D in JAEA and in CEA concerning which grid of aperture offset or which combination aperture offset is reasonable.

d)   Backstreaming by positive ions, generated in the accelerator through charge exchange by collision between beam accelerated and gas is one of the issues. The energy distribution and the power density have been calculated in IPP using the

EAMCC code, and its localized spot power density reaches at more than a few tens MW/m2. As a countermeasure against the back-streaming ions, 3mm Mo coating is foreseen at back plates of expansion chamber and RF driver. The final confirmation will be done in MITICA.

Issues of 1 MV Insulation Against BS Vessel

The vacuum voltage holding between the BS at 1MV potential in maximum and BS vessel at ground potential is another issue. In the mechanical design of the BS and BS vessel, it is essential to keep the insulation distance between them as large as possible. Since, however, the space for the BS vessel is restricted by the floor and wall of the ITER Tokamak building, a careful design is needed to provide the necessary distance. In the present design the vacuum insulation up to 1MV is based on a vacuum insulation criteria proposed by IO.[22] However the data of the 1 MV is extrapolated from 500kV data in accordance with square-root law of gap. The vacuum insulation data near 1MV is necessary for ensuring the design of the BS and BS vessel. A new experimental device of vacuum voltage holding up to 800kV is being constructed at the Padua University in Italy. It will be provided more rigid data for the vacuum voltage holding.

ACHIEVED BS OPERATION PARAMETERS AND FUTURE PLANS FOR DEVELOPING THE BS

Achieved values of BS operation parameters

The achievement of acceleration current density and beam energy, which are key parameters of the BS, are shown in Fig.4, and also the achieved values the operation parameters for negative ion source are summarized in Table 3.
Each value for the beam energy for H⁻, the extracted and accelerated current density, the beam duration time, the co-extracted electron and the filling pressure has been achieved independently to the ITER parameters. Since, however those have not been obtained simultaneously, integrated development for the beam performance required to the HNB is essential.

Future Plans Toward realizing the BS for HNB

A new facility called "ELISE" in IPP will start the operation in 2011 with a half size ion source of that of ITER-NB in addition to BATMAN. Many of the ion source issues described in section 3 are anticipated to make clear. The full size ion source test facility, called "SPIDER", for ITRT-NB is being constructed at Padova in Italy as a part of the NBTF, and it will start the operation in the beginning of 2014 to demonstrate every characteristics of the ion source required to ITER-NB. In the accelerator area, in addition to the R&D works for the accelerator with the MTF, the full scale device of just same as the HNB, so called MITICA, is being constructed in also Padova as another part of the NBTF. The whole beam performance manufacturing feasibility of the DNB bushing. The design is validated to electrical and mechanical stability for different load cases and reported Since HVB faces ITER environment, all conventions of safety, voltage vacuum compatibility, selection of material, manufacturing feasibility, tests etc. are folded in to consideration. Figure 1 shows DNB bushing with ces.

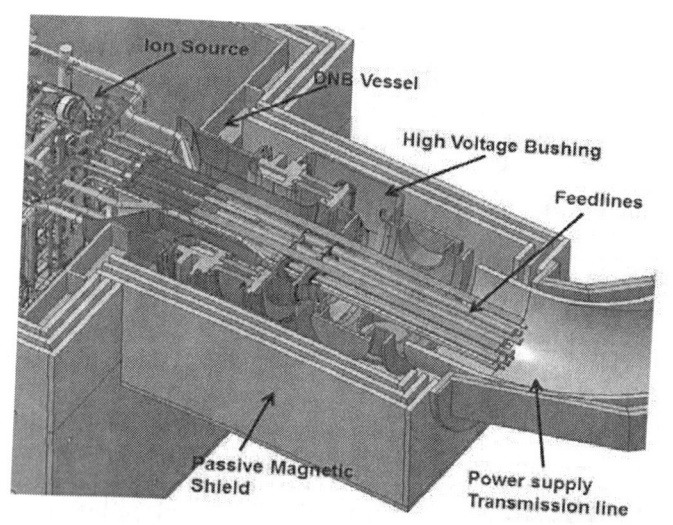

FIGURE 1. DNB bushing with its interfaces

brief details related to the classification, subcomponents of HVB, design requirements, design validation by FEA analysis and tentative manufacturing assembly sequence is presented in subsequent sections.

CLASSIFICATION OF HVB

HVB is classified according to its position in NB cell. Based on this ITER specific classification of vacuum, quality, safety, seismic and RH, design criteria are defined. According to ITER criteria, all the components forming primary vacuum confinement boundary to prevent leakage of radiological substances are classified as Safety Important Components (SIC) which comes under quality class 1 (QC-1). Further, SICs are designed to withstand the effects of earthquake without loss of capability to perform their safety functions in the event of an SL-2 earthquake according to ITER criteria [8]. Additionally, the components which form part of primary vacuum boundary come under VQC-1A category and those which are facing ITER vacuum but not forming vacuum boundary are classified as VQC 1B. The components which do not directly face ITER vacuum are considered as Non-SIC and come under quality class 2 (QC-2). Based on these definitions, different components of the HVB are classified in Table-1. As the HVB is considered to be a component which requires least maintenance during operation period of ITER, it is considered as RH3 as far as remote handling (RH) is concerned.

TABLE 3. Achieved operation parameters on beam source

		IPP (RF)	MTF (Arc)	JT-60 (Arc)	LHD (Arc)	ITER (RF)
Iacc (A)/Vacc (kV)	H	-	0.36/879 **0.33/937**	3/500 20/370	**35/190**	46/870
	D	-	-	17.4/400	-	40/1000
Extraction current density (A/m2)	H	**400**	-	-	-	280
	D	**280**	-	-	-	240
Acceleration current density (A/m2)	H	**330**	157	150	**318**	230
	D	**230**	-	130	-	200
filling pressure (Pa)		**0.3**	**0.3**	**0.2**	**0.3**	0.3
Je-/JH-	H	< 1	-	0.4	-	<1
	D	1.5-2.0	-	0.5	-	<1
Beam Duration (sec)		**3600**	10	30	128	3600

Note; Bold figures show achieved values or closed to the ITER operation parameters

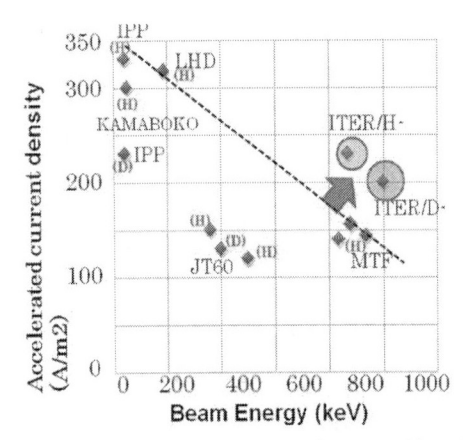

FIGURE 4. Accelerated current density vs. achieved beam energy

required to the HNB will be demonstrated in the MITICA including the neutral beam power of 16.7MA at 1MeV for 3600 sec with the beam divergence of <7mrad.

SUMMARY

The design and R&D works of the ion source and accelerator for realizing the BS of the HNB is being progressed vigorously in many institutions under the coordination of ITER Organization, and furthermore, in the near future, a new test facility, ELISE, will start the operation in IPP and it will make clear many of the issues of the ion source. A few years later, the full size ion source test facility, SPIDER, will start the operation in RFX to demonstrate the ion source performance required to ITER, and after that the full size BS test facility, MITICA, will start the operation also in RFX to confirm the beam performance of 40A/D- at 1MeV for 3600sec.

ACKNOWLEDGMENTS

This report is based on work undertaken within the framework of the ITER Project and supported by the ITER Organization and/or its Members, i.e. China, European Union, India, Japan, Korea, Russia and the United States of America. Dissemination of information contained in this paper is governed by the applicable terms of the ITER agreement.

DISCLAIMER

The views and opinions expressed herein do not necessarily reflect those of the ITER Organization.

REFERENCES

1. R Hemsworth, et al., "Status of the ITER heating neutral beam system", Nucl.Fusion, 49(2009) 045006
2. R Hemsworth et al., *Review of Scientific Instruments*, vol.79, no.2, Feb. 2008, pp. 02C109-1-5
3. P. Franzen et al., *Nuclear Fusion* 47 (2007) 264
4. T. Inoue, et al., "1 MeV, ampere class accelerator R&D for ITER", Nucl.Fusion 46(6), S379-385 (2006).
5. L. Svensson, et al., "Experimental results from the Cadarache 1 MV test bed with SINGAP accelerators," *Nucl. Fusion*, vol.46, p. S369.
6. M Kuriyama, et al., "Operation and Development on the 500keV Negative-Ion Based neutral Beam Injection System for JT-60", *J. Nucl. Science and Technol.*, 35, 739 (2002)739
7. O. Kaneko, et al., "Engineering Prospects of Negative-Ion-Based Neutral Beam Injection System from High Power Operation for the Large Helical Device," *Nucl. Fusion*, 43, 692 (2003).
8. P. Franzen, et al., "Physical And Experimental Background Of The Design Of The ELISE Test Facility", *Proc. the 1st Int. Symp. of NIBS*, 1097 2009) 451
9. M.Hanada, et al, "Improvement of beam uniformity by magnetic filter optimization in a Cs-seeded large negative-ion source", *Review of Scientific Instruments* 77 (2006), 03A515.
10. H.Tobari, et al., "Negative ion production in high electron temperature plasma", *Plasma Fusion Research*, 2, 022(2007).
11. H.Tobari, et al., "Uniform H⁻ ion beam extraction in a large negative ion source with a tent-shaped magnetic filter", *Rev. Sci. Instrum.* 79 (2008), 02C111
12. B.Heinemann, et al., to be published
13. P. Agostinetti,"Optimisation of the magnetic field configuration for the negative ion source of ITER neutral beam injectors ", *Proc. of ITC18*, 328 (2008).
14. W.Kraus, et at., "Long pulse H- beam extraction with an RF driven ion source on a high power level", *Rev. Sci. Instrum.*, 02B110, 81,pp3(2010).
15. E. Speth, *Nucl. Fusion*, 46 (2006) S220–S238
16. P.Franzen, et al., to be published.
17. M. Kamada et al., "Beamlet deflection due to beamlet-beamlet interaction in a large-area multiaperture negative ion source for JT-60U", *Rev. Sci. Instrum.*, 79, 02C114 (2008).
18. M.Kashiwagi, et al., "1 MV Holding and Beam Optics in a Multi-aperture Multi-grid Accelerator for ITER NBI", *Proc. Of the 23rd IAEA Fusion Energy Conference*, ITR/2-4Rb, Oct.11-16, 2010, Daejeon, Korea
19. "OPERA-3d", Vector Fields Co. Ltd., http://www.vectorfields.com/.
20. A. Kojima, et al.,"Demonstration of 500 keV beam acceleration on JT-60 negative-ion-based neutral beam injector", *Proc. of the 23rd IAEA Fusion Energy Conference*, FTP/1-1Ra, Oct.11-16, 2010, Daejeon, Korea
21. *Intermediate report 1 of task No.C53TD48FJ*
22. M.Tanaka, et al, "1 MV vacuum insulation for the ITER neutral beam injectors", *Proc. of the 24th Int.Symp. on Discharge and Electrical Insulation in Vacuum*, Brounschweig (2010) p536

TABLE 1. Quality classification of HVB components

HVB components	Safety Class	Quality Class	Vacuum Class
Ceramic ring with Kovar	SIC	1	VQC 1A
Dished head	SIC	1	VQC 1A
Bushing vessel	Non-SIC	2	N/A
Bus bars, pipes, tubes and cables, screen shields	Non-SIC	1	VQC 1B
Feedthroughs at the dished head for busbars, cables, pipes and coaxial RF tubes	SIC	1	VQC 1A

DESIGN OF HVB

The design DNB bushing was initiated by taking ITER reference document (DDD) design as a baseline. Following points list some of the major modifications (as compare to DDD) carried out to accommodate design requirement and manufacturing feasibility of the bushing.

- DDD bushing is based on filament based source whereas RF source is adopted by ITER. Hence all the feedlines for RF source has been incorporated in DNB bushing.
- DDD design uses SF_6 for insulation, whereas dry air with a dielectric strength of ~ 40 kV/cm is sufficient for 100 kV voltage holding requirement of DNB bushing. Therefore HVB vessel is filled up by dry air.
- The positioning of the HVB was considered vertical on DNB vessel according to DDD. Present design incorporates optimizations considering its horizontal position at rear of DNB vessel.
- Further localized changes have been incorporated which include shape and size optimization of electrostatic shields to reduce electric stress value to acceptable limit and to reduce breakdown possibilities at various places of HV bushing. The shape of SS auxiliary flange is also changed in the interspace between Alumina and FRP for the same reason.
- Some minor modifications are done to ensure manufacturing assembly.

Figure 2 shows the exploded view of the HVB where each subcomponents are shown with its position. The optimized dimension of HV bushing is 1.98 m (height) x 1.79 m (width).

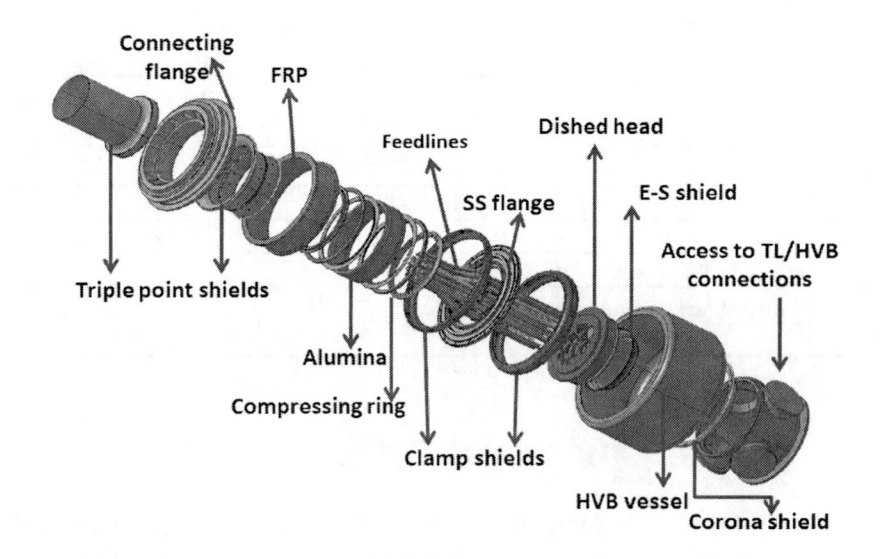

FIGURE 2. Exploded view of HVB

Different components of the HVB with their functional requirement and specifications have been explained in the following sub-sections.

Dished Head

All the electric and cooling connections passes through a metallic plate, called 'dished head' as shown in Fig. 3. Dished head is made up of two metallic plates to ensure double layer geometry. The bottom plate of the dished head is connected with auxiliary flange. The overall geometry is kept at -100 kV.

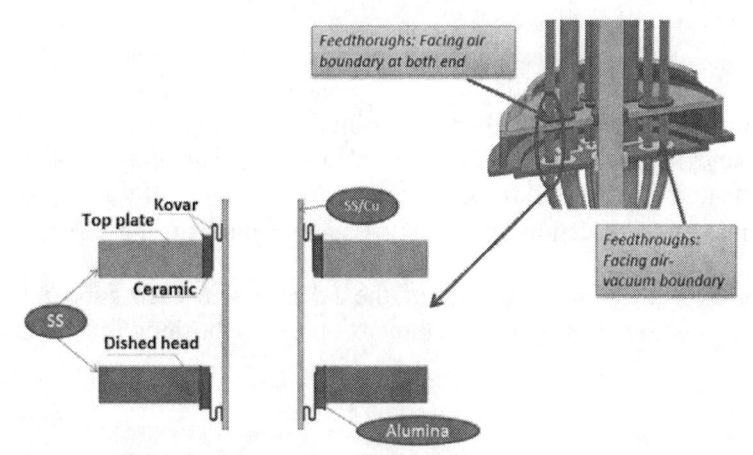

FIGURE 3. Isometric CAD view of metallic dished head

It provides mechanical support to the penetrations. Double wall protection at interspace (at dished head and at insulator) provides the double confinement for any tritium in the NB injector. Dry air at 0.07 MPa will be filled in interspace. Isolation to each busbars is given by ceramic feedthroughs. The design of the feedthrough is shown in Fig. 3. One set of feedthroughs faces slight pressure difference (~0.03 MPa) at both sides but the other group of feedthroughs is foreseen to form interface between the 0.07 MPa dry air and high vacuum. These feedthroughs (forming stress free design) are connected to dished head and pipe via kovar transition as shown in Fig. 3. Material for the insulated feedthroughs shall be high purity alumina, since the injector will be subjected to ionizing radiations [9]. Welding transitions are foreseen for cooling and diagnostic lines.

Feedlines

The number and types of feedlines have been decided according to the needs of RF based ion source. All electrical and cooling services required for beam source is summarized in Table 2 with their specifications.

TABLE 2. Specifications of lines passing through HV bushing

Name	Type	No. of line	Current requirement(A)	Voltage w.r.t GG (kV)	Diameter (mm)	Operating voltage (V)
Arc/Bias	Bus bar	1	< 10	-100	25	100
Bias plate	Bus bar	1	600	-100	80x20	
PG filter	Bus bar	1	6000	-100	70x40	15
PG filter+bias return	Bus bar	1	6600	-100	70x40	15
Cs oven	Cable & multipin feedthrough	3	2	-100	8	230
Starter filaments	Cable & multipin feedthrough	16 (1)	60	-100	8	20
Gas line	Gas	1	---	-100	12	
Fiber optics	pipe	3		-100	13.5	
Thermocouple	Pipe	4	---	-100	38(3) 50(1)	
RF co-axial tr. lines	RF	4	---	-100	79	
RF coils	Coolant	2	---	-100	27	
Faraday shields	Coolant	2	---	-100	60.3	
Chamber cooling	Coolant	2	---	-100	60.3	
Plasma Grid	Coolant	2	---	-100	27	
Extraction	Coolant/elect	2	140	-90	60.3	12000
		Total:45				

All the busbars have been designed with the current density of 1A/mm². To check the cooling requirement of busbars, the Joule heating for each busbar was calculated for selected diameter and pulse length of 3S on/20 s OFF. Temperature rise for this pulse length in bushing is negligible hence additional cooling is not envisaged for the busbars.

Separate cooling provisions for Plasma Grid (PG), Radio Frequency (RF) coil, Faraday shield and chamber wall are provided and separate monitoring system to measure the temperature of each cooling line is supplied. For extraction grid (EG), a common line is used for electric and cooling purpose. The coolant inlet pressure is fixed to 2 MPa [10]. Pressure drop calculation in coolant circuit is carried out from source up to HV deck. Total pressure drop was calculated to be 0.64 MPa.

Near BS-HVB interface, flexible connection will be provided to compensate the thermal expansion and mechanical stress.

Insulating Rings

Two SS flanges (one at -100 kV and the other grounded) will be isolated by cylindrical insulator rings made up of alumina and epoxy. Interspace between the insulators is filled up with dry air at the pressure of ~ 0.07 MPa.
(i) Inner Alumina ring: Alumina has minimum degradation of its electrical and mechanical properties upon irradiation. The main purpose of this insulator is to provide electrical insulation and good vacuum compatibility. Alumina ring is connected to kovar plate which in turn is welded to SS flange to establish the seal. Typical clearance of 1 mm is left between the flange and the ceramic insulator to ensure that no vertical load and bending moment is transferred to it. The design of both insulating rings (Alumina & FRP) is adopted from DDD 5.3 with slight modification of Alumina chamfering angle. At both ends the surfaces are chamfered by 10° angle, because the small angle helps to avoid discharges on the insulator surface which can occur during fast transient due to the possible charge accumulation close to the triple points [9].

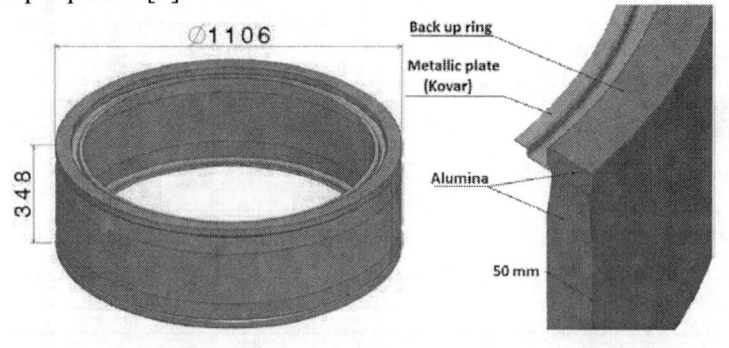

FIGURE 4. Isometric CAD view of inner insulator ring (Alumina)

The kovar/alumina brazed joint geometry is based on the use of two alumina "back up rings" which is beneficial to reduce the stress concentration at the brazed joint. Figure 4 shows the isometric view of alumina ring and its transition with kovar and back up rings. The thickness of Kovar plate is considered to be 1 mm. The kovar plate is welded to the stainless steel flange using conventional TIG process.

(ii) Outer FRP ring: Figure 5 shows the 3D CAD image of outer insulating ring made up of Fiber reinforced polymer (FRP) with height of 350 mm.

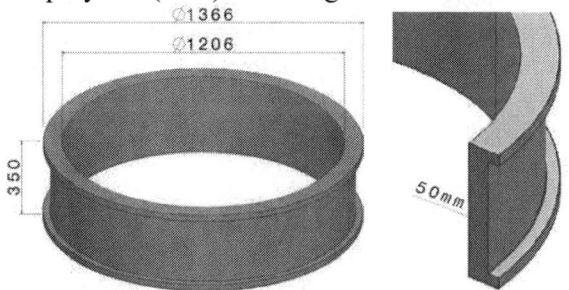

FIGURE 5. Isometric CAD view of Outer insulator ring (FRP)

In particular, the compressive load from the external pressure is taken by FRP insulator. The outside pressure i.e. the pressure between FRP and HVB vessel is 0.1 MPa. Clamp connects the glass-reinforced epoxy insulator at the top and bottom SS flanges. A possibility of casting of FRP with SS flange is under consideration. This if implemented, would preclude the possibility of elastomer seal.

Electrostatic Shields

The main purpose of these shields is to avoid any sharp edges and to provide equal stress distribution throughout the structure. Electrostatic shield at -100 kV is welded with dished head to maintain uniform potential gradient.

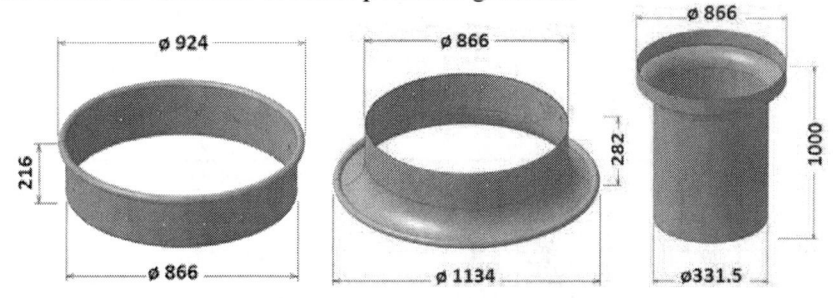

FIGURE 6. Triple point shields

To protect triple point (the point where insulators, metal and vacuum meets) from breakdown possibility, a set of electrostatic shields (triple point shield) has been incorporated which is known as triple point shields as shown in Fig. 6. All the dimensions are in mm. Electrostatic analysis is done to optimize shape and size of shield.

External clamp shields are integrated to screen the edges of the SS flange in dry air. Height and curvature of the clamp shield has been selected to minimize electrostatic field. Due to assembling reasons they are divided in three pieces, each spanning 120° angle. Possibility of overlapping joints will be carried out to avoid relatively large gap between the rings and favor an intimate contact [9].

Compressing Rings and Connecting Flange

Upper and lower compressing rings are used for better connection/compression of kovar plate with auxiliary flange.

Connecting flange connects HVB with DNB vessel. The shape of SS flange which acts as a connecting flange at the interspace between two insulating rings has been optimized to provide shield at interspace triple point.

DESIGN REQUIREMENTS

Vacuum

As HV bushing forms primary vacuum barrier and facing the vacuum of Torus, it comes under VQC 1A class. Therefore the parts of HVB, facing vacuum should be vacuum compatible and approved by ITER vacuum requirements. Table 3 lists the approved grade material which can be used for HVB in vacuum environment particularly for VQC-1A component.

TABLE 3. ITER Approved Material

Sr. No.	Material	Grade
1	SS for HVB vessel, Electrostatic shields, Dished head, clamp shield	SS 304 L
2	Alumina	Kyocera A479 grade ceramic
3	Kovar	ASTM F15 KV-1~9

All the joining process to connect different components of HVB should be vacuum compatible and ITER approved for VQC 1A and are represented in Table 4

TABLE 4. ITER Approved joining process

Sr. No.	Material	Joining process
1	SS –SS	Welding
2	SS-Alumina	Brazing
3	Kovar-Alumina	Brazing
4	Kovar-SS	Welding

Vacuum sealed electrical feedthrough at the dished head SS plate is used for vacuum tightness. Double containment is applied at dished head where all penetrations are facing vacuum boundary. According to ITER vacuum handbook, VQC 1A components that are considered to be vulnerable shall be doubly vacuum contained with a monitored interspace connected to the service vacuum system. Hence present configuration of the HVB consists of double containment at each vacuum facing surface with continuous pressure monitoring system. Further, to avoid radiation induced conductivity for stagnant air, gas circulation is provided in the interspace. Interspace between two insulators and two plates of dished head is merged and common connection is provided. Inlet and outlet for the interspace gas is connected to ground flange of HVB.

Electrostatic

Electrostatic shields are incorporated in design to protect some critical areas of HVB from high electrostatic field generation which may induce electrical breakdown. The shape and size of the electrostatic shields and the gap between two conductors are selected in order to reduce the electric stress value within acceptable limit as stated in Table 5 [8] and to avoid sharp edges.

TABLE 5. Acceptable limits for electrostatic analysis

Position	Triple points	Metal (Cathode) areas in Vacuum	Insulator Surface in Vacuum	Cathode areas in Dry air
Electric field limit (kV/cm)	< 5	< 26	< 15	< 14

Structural

The structural requirement of the HVB is to provide barrier between ITER primary vacuum and atmospheric dry air. It provides stiff cantilever support to its own weight. It is also designed to withstand the over-pressurization accident due to large coolant leak in the vacuum vessel. Structural analysis is done for the load cases listed in Table 6 and presented in [8].

TABLE 6. Load cases for structural analysis

Load case	Internal Pressure (MPa)	External Pressure (MPa)
Normal Operation	0	0.084
Loss of Coolant Accident (LOCA)	0.18	0.084

Seismic

As HVB is a SIC component, it is required to perform safety functions during or after an SL-2 earthquake. The collapse, falling, dislodgement or any other response of a component as a result of an earthquake shall not affect the functioning of other components providing a safety function. The model analysis of HVB has been carried out to satisfy SL-2 condition [11].

ANALYSIS OF THE HVB

The design of HVB is validated by electrostatic and structural analysis [8]. The structure is optimized to obtain electric stress below the mentioned limits in Table-5. Maximum stress obtained on the triple point shield in vacuum is 20.6 kV/cm, the stress at triple point is 3.2 kV/cm and the maximum stress on insulator surface is 10.5 kV/cm. All these stresses are in acceptable limit. The mechanical design incorporates a mechanical stress free mounting of the ceramic rings, aided by ~1.3 m diameter load bearing FRP cylinder which is joined to connecting flange by bolting. Casting possibilities of FRP with SS flange is under discussion. A structural analysis is carried out to assess the stress distribution in the FRP and alumina insulator and the integrated structure of HVB during commissioning, operation and accidental conditions. Furthermore the compatibility with the seismic classification required for ITER has been confirmed.

ASSEMBLY SEQUENCE

Preliminary manufacturing assembly sequence has been studied. Bushing structure is subdivided in to three sub-assemblies. All the subassemblies will be manufactured individually and merged to form final configuration of bushing. This section presents tentative manufacturing assembly sequence for DNB bushing.

Subassembly-I

This subassembly is made up of two SS flanges (one at -100 kV and the other at grounded) and two insulator rings (Alumina and FRP). Foreseen manufacturing assembly sequence of these parts is shown in Fig. 7. Kovar connected alumina ring is placed on the SS connecting flange followed by welding of kovar to the SS flange at the periphery. The compression rings are then placed and bolted with the SS flange to support kovar. FRP is then placed and bolted to the assembly followed by screwing of Triple point shields. This assembly is connected with the top SS flange (at -100 kV). Triple point shield is bolted with this flange. Clamp shields are attached with both SS flanges to avoid sharp edges.

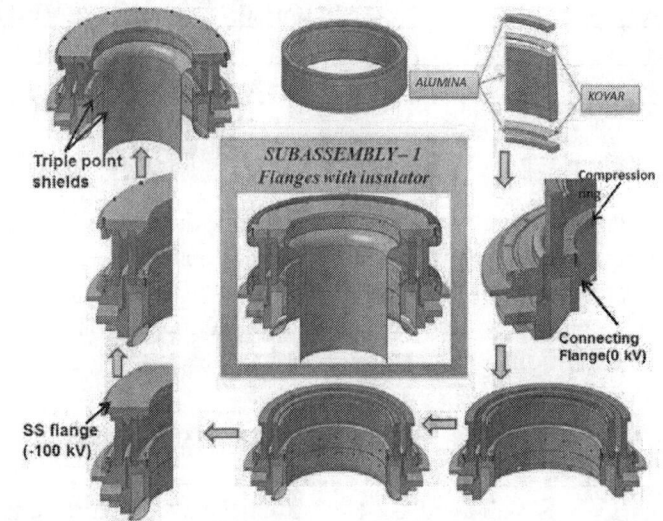

FIGURE 7. Manufacturing assembly sequence of subassembly-I of ITER DNB

Subassembly-II

This is made up of dished head, feedthroughs and all feedlines. Off the shelf or custom made feedthroughs prepared by kovar-ceramic-kovar brazing are used for isolation. Grouping of the pipes in a common feedthrough is also proposed.

Feedthroughs will be first welded to feedline and this integrated part will be welded to dished head plates. The sequence for this assembly is shown in Fig. 8.

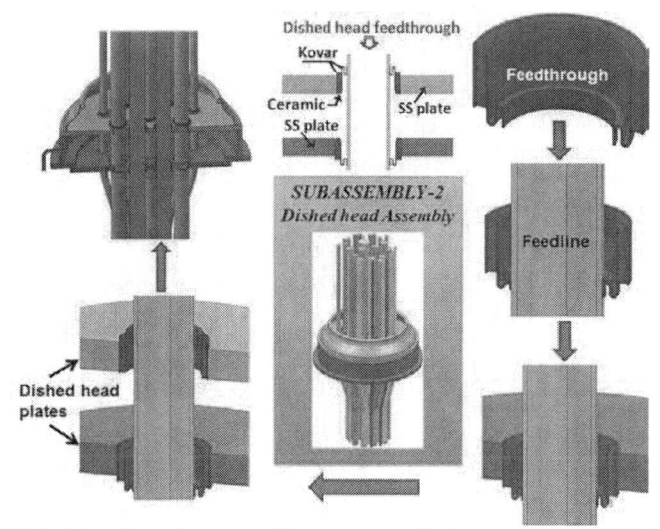

FIGURE 8. Manufacturing assembly sequence of subassembly-II of ITER DNB

Subassembly-III

HVB Vessel is manufactured in two parts and these parts are welded together. Stress shield is connected to avoid sharp edges and provide equal potential distribution. This assembly sequence is shown in Figure 9.

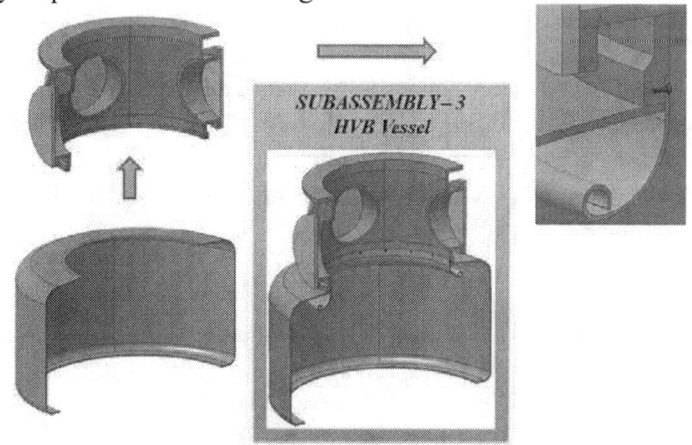

FIGURE 9. Manufacturing assembly sequence of subassembly-III of ITER DNB

Final assembly sequence

Figure 10 shows the roadmap to achieve final HVB configuration using above assembly sequences. Subassembly 1 & 2 will be welded and this assembly will be connected to subassembly 3 which forms entire HVB. Further discussions are foreseen with vendors to validate this concept.

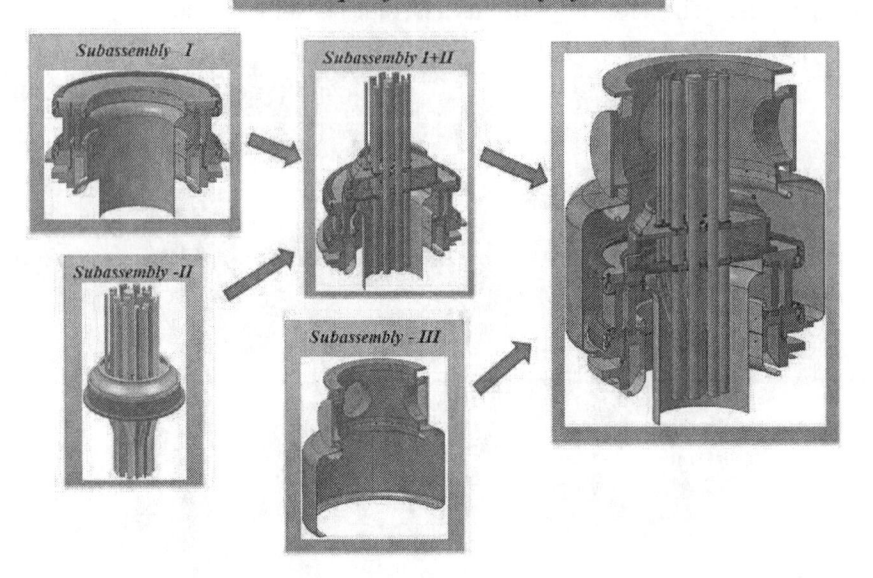

FIGURE 10. Roadmap to final assembly of HVB

SUMMARY

The 100 kV HVB is designed considering DDD design as a baseline and incorporating modifications for RF source requirements, orientation, shaping of the shields to reduce electrostatic stresses and provision for gas circulation. All conventions of vacuum requirement, voltage holding requirement, joining processes, material requirement have been taken in to account. Electrostatic and structural analysis of the model is carried out in ANSYS and the stresses are under acceptable limit. Preliminary concept of assembly sequence has been studied and presented. The design is under discussion with vendors for assessing manufacturing feasibility.

REFERENCES

1. R. Hemswroth et. al., "ITER neutral beam injector design", Fusion Energy 1996 (Proc. 16th Int. Conf. Montreal, 1996), Vol. 2, IAEA, Vienna (1997) 927.
2. A.K. Chakraborty et. al., *IEEE Transaction on Plasma Science* **38** 248 (2010).
3. D. Marcuzzi et al. *Fusion Engineering and Design* **82** 798-805 (2007).
4. M. J. Singh & H. P. L. De Esch *Rev. of Sci. Instruments* **81** 13305-13313 (2010).
5. A. Masiello *Nucl. Fusion* **46** S340-S351 (2006).
6. N. Umeda et al., *Fusion Eng. & Design* **84** 1875-1880 (2009).
7. E. Di. Pietro *Fusion Eng. & Design* **66** 603-608 (2003).
8. Sejal Shah et al., Fusion Eng. and Design-Accepted.
9. A. Masiello, Design and electrostatic analysis of the high voltage bushing for SINGAP configuration, RFX_NBTF_TN_006, 2006.
10. M. Dalla Palma, S. Dal Bello, F. Fellin, P. Zaccaria, *Fusion Engineering and Design*, **84** 1460-1464 (2009).
11. Sorin V. M., Seismic Analysis of the Tokamak Building, Vol. 1: Main Results (2DKWMG v1.0), Final report, 2009.

RAMI Analyses of Heating Neutral Beam and Diagnostic Neutral Beam Systems for ITER

D. H. Chang[a], S. Lee[b], R. Hemsworth[c], D. van Houtte[c], K. Okayama[c], F. Sagot[c], B. Schunke[c] and L. Svensson[c]

[a]*Korea Atomic Energy Research Institute(KAERI), Daejeon 305-353, Republic of KOREA*
[b]*ITER-Korea Domestic Agency, National Fusion Research Institute(NFRI), Daejeon 305-333, Republic of KOREA*
[c]*ITER Organization, 13067 St. Paul-lez-Durance Cedex, FRANCE*

Abstract. A RAMI (Reliability, Availability, Maintainability, Inspectability) analysis has been performed for the heating (& current drive) neutral beam (HNB) and diagnostic neutral beam (DNB) systems of the ITER device [1-3]. The objective of these analyses is to implement RAMI engineering requirements for design and testing to prepare a reliability-centred plan for commissioning, operation, and maintenance of the system in the framework of technical risk control to support the overall ITER Project. These RAMI requirements will correspond to the RAMI targets for the ITER project and the compensating provisions to reach them as deduced from the necessary actions to decrease the risk level of the function failure modes. The RAMI analyses results have to match with the procurement plan of the systems.
Keywords: RAMI, ITER, heating neutral beam, diagnostic neutral beam, risk level, failure mode
PACS: 51.90.+r, 51.90.+z

INTRODUCTION

The objective of these analyses is to implement RAMI (Reliability, Availability, Maintainability, Inspectability) engineering requirements for design and testing to prepare a reliability-centred plan for commissioning, operation, and maintenance of ITER neutral beam heating and current drive (NB H&CD, HNB) system and diagnostic neutral beam (DNB) system in the framework of technical risk control to support the ITER Project.

Initial and expected reliability and availability have been calculated by using commercial BlockSim7 software [4], before and after the suggested actions, respectively. A total 232 of failure modes for HNB system [5] and a total of 169 failure modes for DNB system have been postulated, respectively. Proper actions have been suggested for 25 failure modes in HNB system and 24 failure modes in DNB system to reduce the criticality of failure modes from the major risk region to the minor risk region in the operation of ITER machine.

ANALYSIS PROCESSES

The methodology, which has been defined in an ITER RAMI analysis program, includes four main steps which have to be performed in close relationship with the RAMI RO (Responsible Officer) and NB RO in ITER organization (IO).

Input Data Collection for Functional Breakdown Analysis

The whole ITER plant was considered as an assembly of systems connected between themselves and utilities. The first step of input data collection was to define a clear perimeter of the system to be RAMI analyzed and interfaces of the system with the other ITER systems, and with utilities including power supplies, water cooling, compressed air, and CODAC infrastructure. Then, it was necessary to inventory the functions performed by the system using the IDEFØ methodology to break the functions down to their elementary components, with regard to the other systems and within the system itself, for every main operational states of the ITER plant. This step corresponded to a top-down process.

RAMI Data Base

It was translated into function breakdowns as a functional description of the system by creating a project in the BlockSim toolkitTM of Reliasoft [1]. Then, it was made into a preliminary bottom-up reliability analysis by using the reliability block diagram (RBD) created in the BlockSim7 software.

FMEC Analysis

A list of all the function failure modes was made, and the list indicated their causes and their effects on the function itself, the system and the overall ITER machine. For each function failure mode, the criticality level of the failures was assessed by giving figures for the severity of the effects, the occurrence of the causes and the non-detection level. The results were translated the results by the "ITER Reliasoft XFMEA software toolTM" making it possible to fill a "Function Failure Mode Identification & Analysis Card". Then, the analysis proceeded to the initial criticality evaluation for each failure mode by using the criticality diagram provided by the RAMI software and then the risk level for each function failure was terminated.

Actions to Decrease the Risk Level

These results were used to propose the compensating provisions in terms of design optimizations, tests, operation procedures and maintenance actions to lower the risk level. The analysis proceeded to the revised (estimated) criticality evaluation, taking into account the compensating provisions for each failure mode by using the criticality diagram (matrix) provided by the software and as used to determine whether the revised risk level had became acceptable or not.

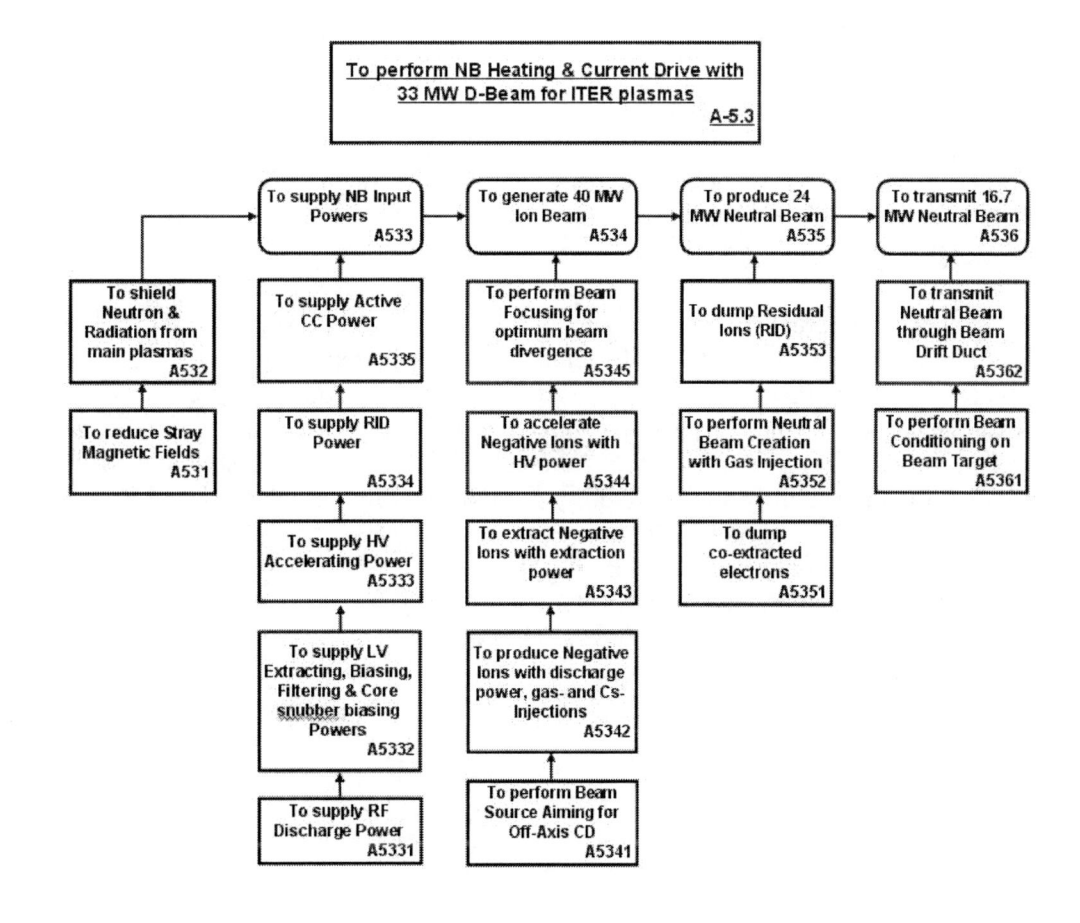

ANALYSIS RESULTS

Functional breakdown analysis

An NB H&CD system shall provide neutral beam heating and current drive to the ITER plasmas. A beam energy of 1 MeV will be required to obtain, primarily, deep enough penetration of the neutral beam into the ITER plasmas and to the heat and drive current, and finally to maximize the power per injector [6-8]. In each injector, a D- ion beam of 40 A will be neutralized to form a D^0 neutral beam, which will deliver 16.7 MW to the plasma. Thus, the NB system will provide H&CD power of 33 MW from two injectors, and can be upgraded to 50 MW in total with a third injector. The NB system will be able to operate for long pulses of up to 3,600 s. The specific function of the NB system will be singular and described as the following: The NB H&CD shall provide a heating & current drive of the ITER DT, D, H and He plasmas, accessing the H-mode and achieving Q>10. The current drive power shall provide steady-state current drive capability through on- and off-axis NB injection, which shall be carried out simultaneously with the heating function, achieving Q>5 for the central current drive.

FIGURE 1. Top-down levels of a functional breakdown structure for the NB H&CD system.

The diagnostic neutral beam (DNB) injector shall provide a probe beam of 100 keV H^0 to be used by the CXRS (charge exchange recombination spectroscopy) diagnostic system. The main purpose shall be to allow a measurement of the local density of thermal alpha-particles (helium ash). The diagnostic neutral beam (DNB) injector may also be used for other diagnostic measurements such as:
- Local density of light impurities (Be, C, O, Ne)
- Plasma rotation velocity
- Ion temperature
- MSE (motional stark emission) measurements

The DNB has a configuration which is very similar (almost all the same) to the NB H&CD (HNB) injector. The characteristics of DNB system compared to HNB system can be summarized, as the following:
- Ion beam species : H-   [H-, D- for HNB]
- Beam energy : 100 keV/60 A   [1 MeV/40 A for HNB]
- Current density (Ion beam) : 300 A/m2   [200 A/m2 for HNB]
- Beam size (Total hole area) : 0.125 m2   [0.2 m2 for HNB]
- Pulse length : 5 Hz-Modulation (for 3 s after 20 s dwell time)
- Accelerator type : Single-stage acceleration (PG + EG + GG)
- One injector system   [Two injectors for HNB]
- No source angle manipulation
- Reduction of flexible connections (for coolant, etc.)
  - Reduced heatload on beam source and beamline components
- Minor reduction of supporting dimension
  - Reduced mechanical weight

This functional breakdown was then transferred into Microsoft Visio following guidelines derived from the IDEFØ methodology. The next step of the functional breakdown was to implement this approved functional breakdown in the BlockSim7 software using RBD structures, and to complete it with the components' reliability and maintenance data in order to estimate the resulting reliability and availability of the functions. Initial reliability and availability of HNB system were 91.6 % and 93.6 %, respectively, and 97.1 % and 97.9 % of DNB system, respectively. In these results, two NB systems would need to be reliable for at least for the duration of an experimental day, i.e. 16 hours between routine maintenance shifts. For an availability calculation, it is expected to have around 1000 hours of cumulated plasma pulse time over the ITER lifetime of about 20 years.

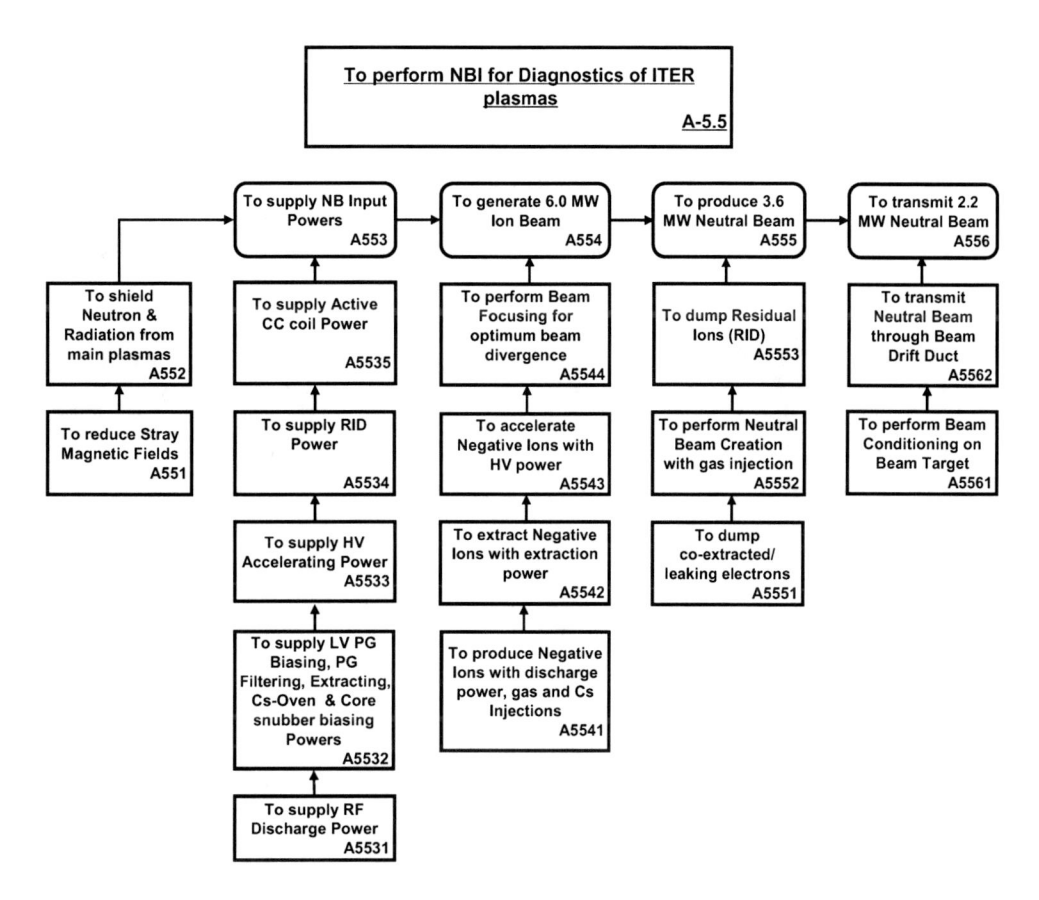

FIGURE 2. Top-down levels of a functional breakdown structure for the DNB system.

Failure mode analysis and RAMI data base

After the functional breakdown analysis, the failure mode analysis was carried out for the NB H&CD (HNB) and DNB systems in XFMEA_4 software. For the failure mode
analysis, postulated failure modes, causes, and effects were listed by a discussion with RAMI RO and NB RO in the ITER organization (IO). Then, these postulated failures were saved as the input data of XFMEA_4. Each failure cause needed to be given weight by a value of 1 to 6 (and 5 for detection) for its severity, occurrence and detection scales. A total of 232 failures were listed for the HNB system and a total of 169 failure modes for DNB system in the ITER machine. The first analysis was to give initial severity, occurrence and detection values assuming the system was installed and commissioned for the first time with mean time between failure (MTBF) values of the components and without considering any actions on the design change, tests between system operations, critical operation stage and maintenance between system operations. The second analysis was to suggest actions on design change, tests between system operations, critical operation stage and maintenance between system operations according to the revised (expected) severity, occurrence and detection values. After the suggested actions, the revised reliability and availability of HNB

system were 91.7 % and 97.2 %, respectively, and 97.1 % and 98.8 % of DNB system, respectively.

FMEC analysis results

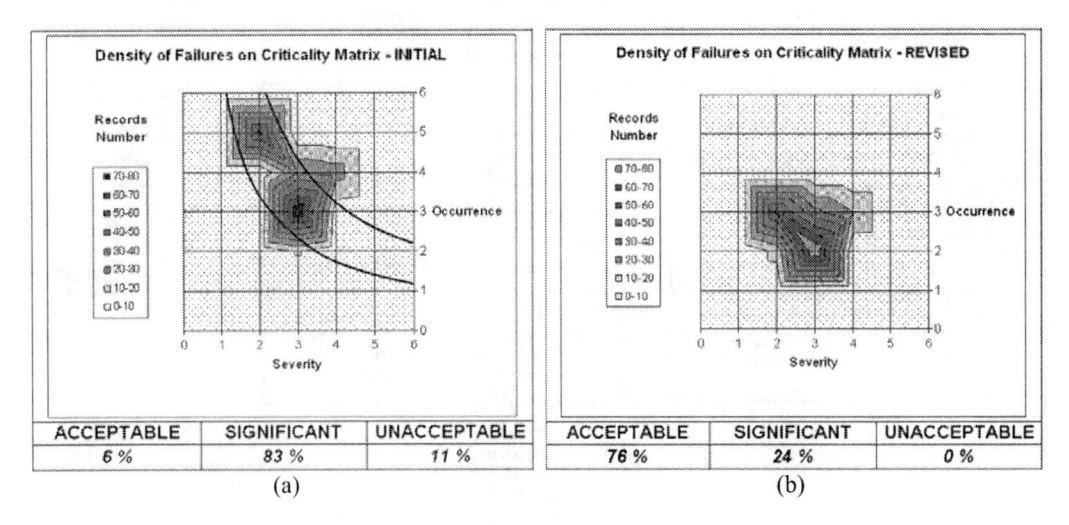

ACCEPTABLE	SIGNIFICANT	UNACCEPTABLE	ACCEPTABLE	SIGNIFICANT	UNACCEPTABLE
6 %	83 %	11 %	76 %	24 %	0 %
(a)			(b)		

FIGURE 3. Initial (a) and revised (b) criticality charts of the HNB system for the ITER machine.

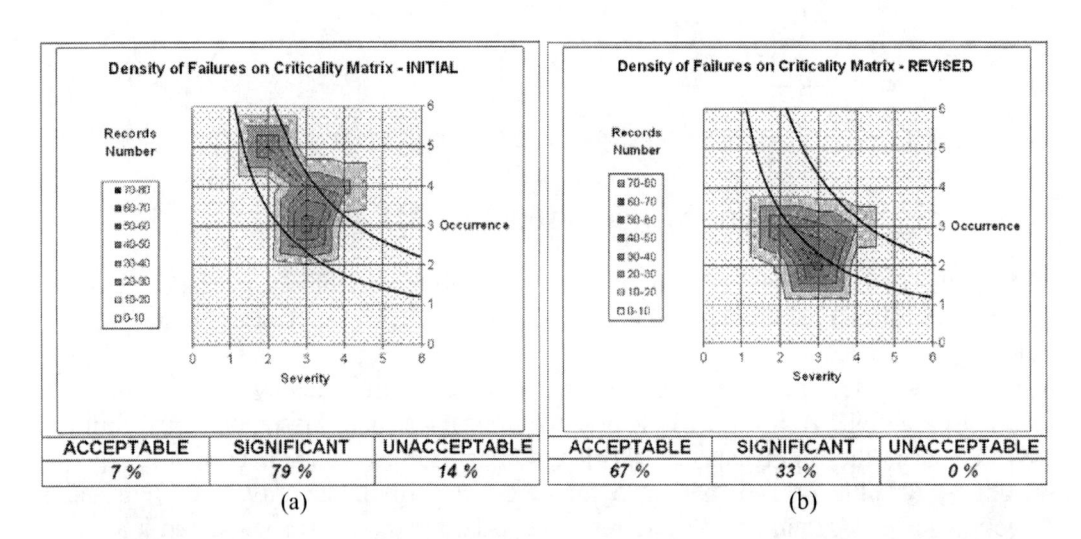

ACCEPTABLE	SIGNIFICANT	UNACCEPTABLE	ACCEPTABLE	SIGNIFICANT	UNACCEPTABLE
7 %	79 %	14 %	67 %	33 %	0 %
(a)			(b)		

FIGURE 4. Initial (a) and revised (b) criticality charts of the DNB system for the ITER machine.

Preliminary operation and maintenance plan

XFMEA analysis suggests major and minor actions to bring down the criticality level of failure modes. These actions are related to design, operation, testing and maintenance. The operation states of the NB H&CD and DNB systems are supported by a system requirement document (SRD) in addition to actions supported by the XFMEA [8]. Operation plans cover testing actions and detection methods for the failure mode of basic functions by checking interlock systems before the start of operations. Long term and short term maintenance plans are also proposed for all basic function of the HNB and DNB systems. The spare parts required to maintain the systems are also covered in the maintenance plan. In order to improve the maintainability of the HNB and DNB systems and to reduce the overall cost of the project, some components shall be considered for standardization so that the other systems may use the standardized components.

CONCLUSION

The RAMI analyses of the ITER HNB and DNB systems has been performed, and the actions were suggested for mitigation of the major risks of failures. Spare parts and the proposed standardization have been requested for the preliminary operations and maintenance plans of the system. There was no-major risk of failures remaining in the HNB and DNB systems after an implementation of the actions. The basis of the RAMI requirements for the system has been prepared for the first time and will be included in a system requirement document (SRD).

ACKNOWLEDGMENTS

This work was supported by RAMI Program through the ITER-Korea Domestic Agency (DA) at the Fusion Research Institute of Korea (NFRI) funded by the ITER Organization (IO) funds. Authors thank for the great efforts of research staffs at the IO and the Korea Atomic Energy Research Institute (KAERI).

REFERENCES

1. BLOCKSIM and XFMEA softwares (in http://www.reliasoft.com).
2. M. Boldrin, A. De Lorenzi et al., Potential failure mode and effects analysis for the ITER NB injector, *Fusion Eng. Des.*(2009), doi:10.1016/j.fusengdes.2009.02.010.
3. R. Hemsworth, H. Decamps et al., Status of ITER heating neutral beam system, *Nuclear Fusion* **49** 1-15 (2009).
4. *Software Training Guide* for ReliaSoft's BlockSim 7 (in http://BlockSim.Reliasoft.com).
5. Doo-Hee Chang and Sangil Lee, RAMI Analysis of Neutral Beam Heating and Current Drive System for ITER, *Transactions of the Korean Nuclear Society Autumn Meeting*, Gyeongju, Korea, October 29-30, 2009.
6. PBS 5.3 (PBS 53-01,-02,-03) in ITER IDM.
7. DDD 2001 (DDD 53) : NB H&CD in ITER IDM.
8. SRD-42, SRD-42C, SRD-53 in ITER IDM.

Physics and engineering studies on the MITICA accelerator: comparison among possible design solutions

P. Agostinetti[a], V. Antoni[a], M. Cavenago[b], G. Chitarin[a], N. Pilan[a], D. Marcuzzi[a], G. Serianni[a] and P. Veltri[a]

[a] Consorzio RFX, Associazione Euratom-ENEA sulla fusione, Padova, Italy
[b] INFN-LNL, Legnaro, Italy

Abstract. Consorzio RFX in Padova is currently using a comprehensive set of numerical and analytical codes, for the physics and engineering design of the SPIDER (Source for Production of Ion of Deuterium Extracted from RF plasma) and MITICA (Megavolt ITER Injector Concept Advancement) experiments, planned to be built at Consorzio RFX.

This paper presents a set of studies on different possible geometries for the MITICA accelerator, with the objective to compare different design concepts and choose the most suitable one (or ones) to be further developed and possibly adopted in the experiment. Different design solutions have been discussed and compared, taking into account their advantages and drawbacks by both the physics and engineering points of view.

Keywords: Neutral Beam Injector, MITICA, beam, modeling
PACS: 41.75.Cn, 52.50.Gj

INTRODUCTION

In the framework of the activities aimed at developing and optimizing the Heating and Current Drive Neutral Beam Injectors for ITER [1, 2], the SPIDER and MITICA experiments are planned to be built at Consorzio RFX in the PRIMA facility [3]. Several physics and engineering analyses for the conceptual design of the MITICA accelerator (see Fig. 1) have been carried out in order to choose the most suitable concept or concepts to be further developed.

The geometries considered as starting points for the development of the MITICA accelerator (see Fig. 2) are the one described on the ITER document DDD 5.3 [4] and the one with 90 mm acceleration gaps and 10 mm thickness of the acceleration grid, described in [5] and considered for physics evaluations by H.P.L. de Esch [6].

A comprehensive set of numerical and analytical codes is currently being used and developed at Consorzio RFX, aiming at simulating the most important aspects inside the negative ion electrostatic accelerator, the electrical and magnetic fields, beam aiming and optics, pressure inside the accelerator, stripping reactions, transmitted and dumped power, operating temperature, stresses and deformations of the accelerator grids. These codes, which provide a description of the physical phenomena taking place in the different components of the injector system (source, accelerator, neutralizer, electron and ion dumps, calorimeter) are necessary for the engineering design optimization as well as for the prediction of the operating conditions and the interpretation of the experimental

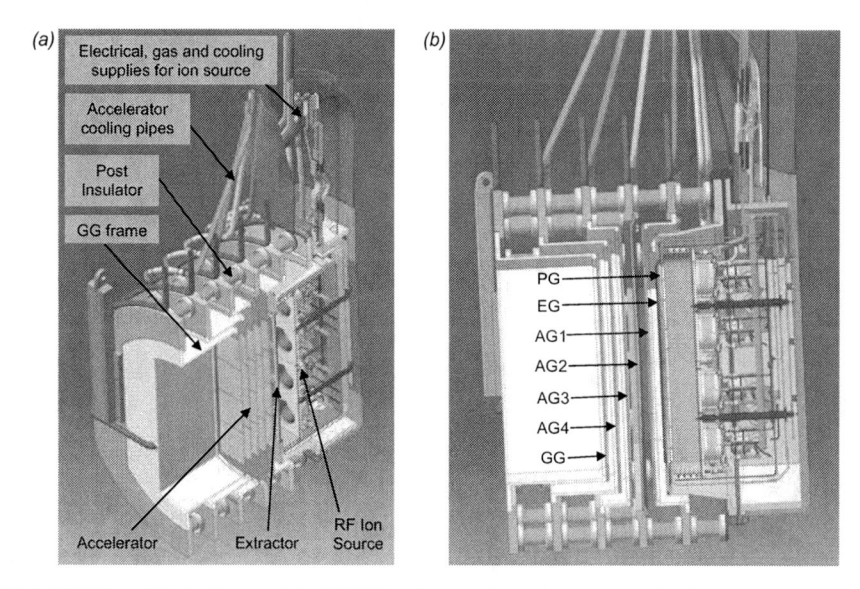

FIGURE 1. CAD views of the MITICA beam source: (a) isometric view; (b) side view.

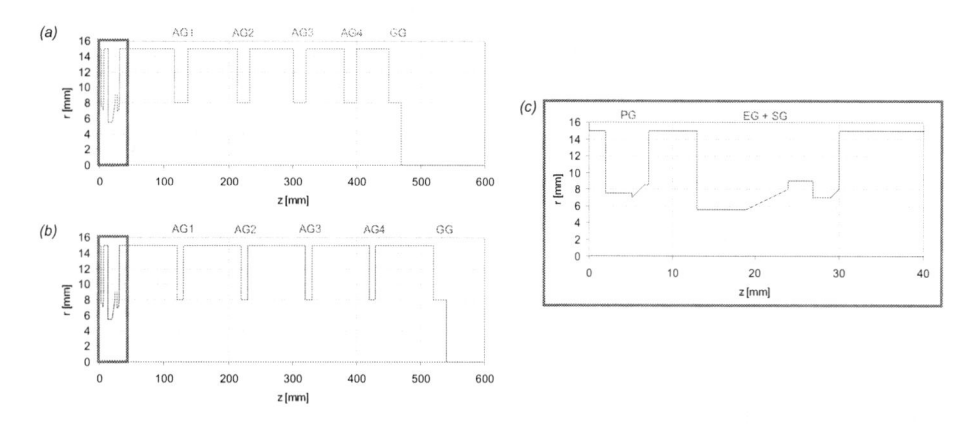

FIGURE 2. Accelerator geometries considered as starting points for the analyses: (a) DIFF_GAPS_01: reference MAMuG geometry from ITER document DDD 5.3 [4], with gaps of 6, 86, 77, 68, 59, 50 mm and grid thicknesses of 6, 17, 20, 20, 20, 20, 20 mm; (b) EQ_GAPS_01: updated geometry with equal gaps, proposed in [5], with gaps of 6, 90, 90, 90, 90, 90 mm and grid thicknesses of 6, 17, 10, 10, 10, 10, 20 mm; (c) detail of the extraction region (extraction grid and steering grid are at the same potential and considered as a single grid).

results.

An integrated approach, taking into consideration at the same time physics and engineering aspects, is adopted for the conceptual design process here described. Particular care has been taken in investigating the interactions between physics and engineering aspects of the experiments. The codes used are some of the ones currently considered also for the design of the SPIDER accelerator [7, 8]:

- CONDUCT to estimate the conductance through the grids of a gas in low pressure conditions;
- STRIP for the pressure and density profiles inside the accelerator and the stripped losses;
- SLACCAD for the beam optics investigations;
- OPERA for the evaluations of magnetic fields [9];
- EAMCC for the estimation of the heat loads on the grid [10], transmitted beam and amount of backstreaming ions.
- ANSYS for the thermo-mechanical and magnetic analyses [11].

OPTICS ANALYSES

The SLACCAD code has been used to estimate the electric field inside the accelerator by integration of the Poisson's equation, with cylindrical geometry conditions [12]. This is a modified version of the SLAC Electron Trajectory Program [13], adapted to include ions, a free plasma boundary and a stripping loss module [14].

The CONDUCT and STRIP codes [6] have been used to generate the boundary conditions for SLACCAD. CONDUCT has been used to estimate the conductance of a gas in low pressure conditions across the aperture, following the classical approach for molecular flow [15]. The molecular temperature during discharge inside the ion source is assumed to be 1000 K, while the background gas in the acceleration gaps is expected to be close to the room temperature. The gas density profile, which is important (together with the cross sections of the stripping and charge exchange reactions) for the estimation of the stripping losses, should be calculated for these temperatures. With a fixed flow rate and molecular flow conditions, the velocity of the molecules V increases as the gas temperature increases. In other words, the gas density in the ion source and the accelerator decreases as the gas temperature increases, i.e. the conductance increases with temperature. Regarding this topic, Krylov and Hemsworth propose the relation $V \propto T_{gas,source}^{0.5}$ [16]. The STRIP code requires as input a source pressure (for the region upstream of the accelerator) and a tank pressure (for the region downstream of the accelerator) to calculate the pressure profile inside an electrostatic accelerator. The code also assumes a uniform temperature of 300 K everywhere. However, the temperature in the actual MITICA ion source is foreseen to be about 1000 K. As it is not easy to modify the source temperature in the code, a virtual source pressure p_{virt} is given to the code instead of the filling pressure p_{fill} in order to compensate for the non-uniform temperature. To make this, the following formula is used in the calculations:

$$p_{virt} = p_{fill} \cdot \left(\frac{T_{room}}{T_{gas,source}} \right)^{0.5} \quad (1)$$

where:

- p_{virt} is the virtual source pressure, to be used as input for the stripping calculations with the STRIP code;

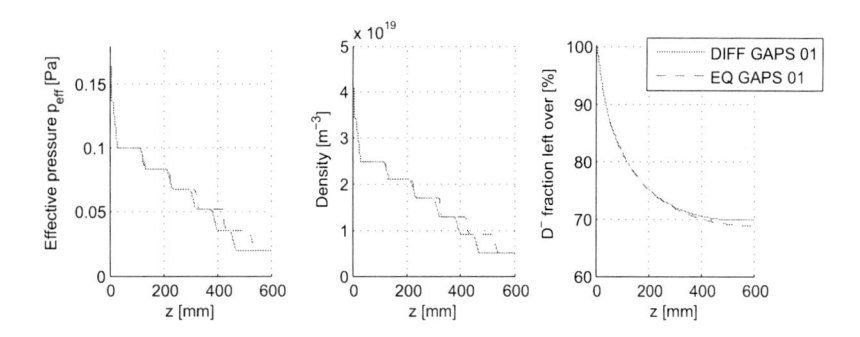

FIGURE 3. Pressure profile, density profile and stripping percentage on beam axis, calculated with STRIP with the DIFF_GAPS 01_and EQ_GAPS_01 geometry.

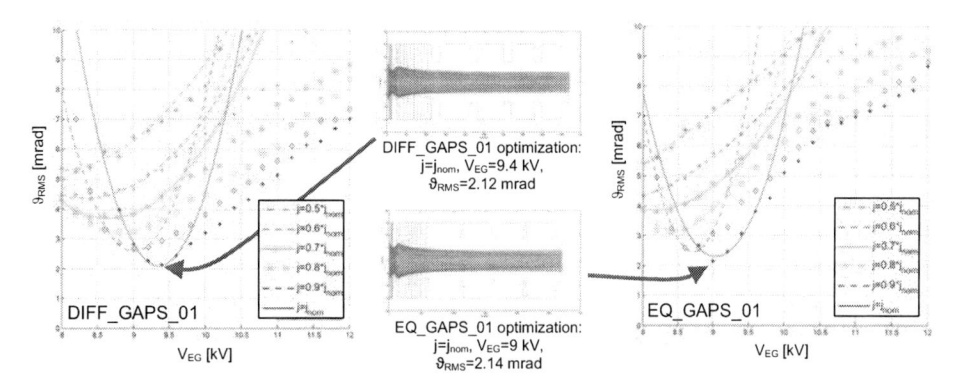

FIGURE 4. Average divergence angle of the beamlet at the accelerator exit with the DIFF_GAPS_01 and EQ_GAPS_01 geometries as a function of the Extraction Grid voltage and extracted current density. The extracted current density of 50% 60%, 70%, 80%, 90% and 100% of the nominal value are considered. The PG, AG1, AG2, AG3, AG4 and GG voltages are fixed respectively at 0, 209, 409, 609, 809 and 1009 kV, while the EG voltage is variable between 8 and 12 kV. The points close to the minimum are interpolated with second degree polynomial curves, to estimate the minimum θ_{RMS}, corresponding to the optimized configuration. D$^-$ trajectories and θ_{RMS} in the optimized configuration are reported for the two geometries.

- p_{fill} is the filling source pressure, intended with no source operation and the system at room temperature. This is taken equal to 0.3 Pa to the ITER requirements;
- T_{room} is the room absolute temperature (300 K);
- $T_{gas,source}$ is the gas absolute temperature inside the source, assumed to be 1000 K.

The ITER DDD5.3 reference geometry (DIFF_GAPS_01) and the updated geometry (EQ_GAPS_01) have been considered. The profiles of the effective pressure, density and stripping losses calculated with STRIP are plotted in Fig. 3, while Fig. 4 shows the sensitivity analysis on the extracted current and EG voltage calculated with SLACCAD. It can be observed that:

- stripping losses are slightly higher with EQ_GAPS_01 geometry; this is due to the fact that the accelerator is 70 mm longer;

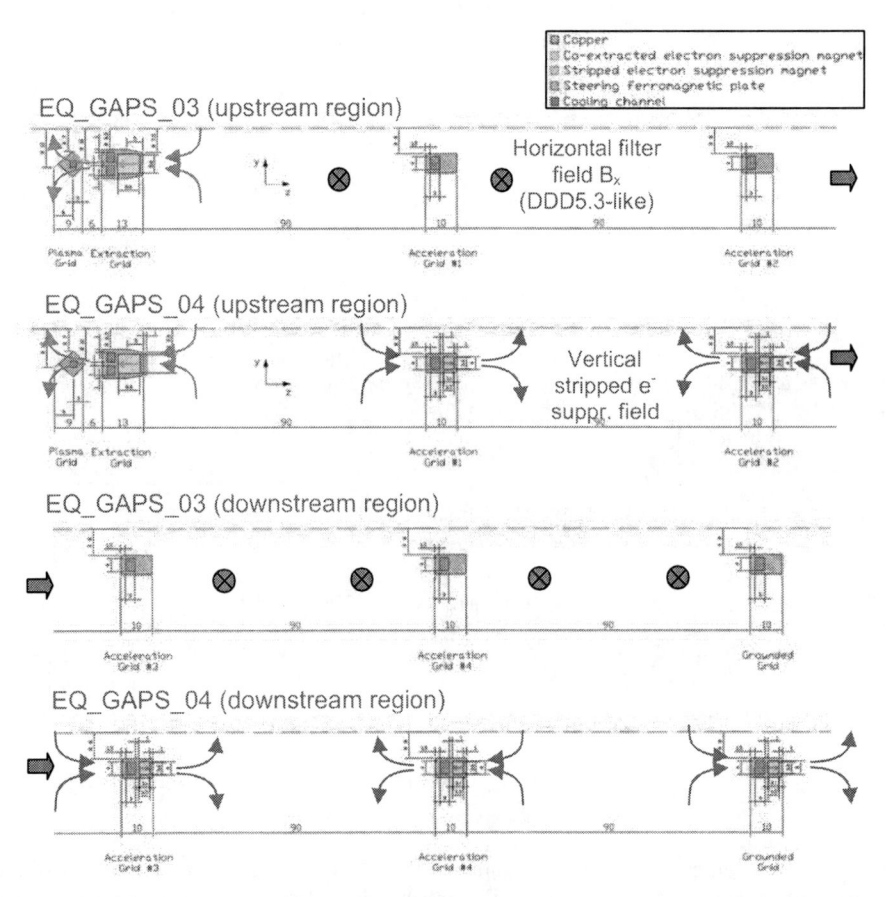

FIGURE 5. EQ_GAPS_03 and EQ_GAPS_04 geometries. All the dimensions are in mm.

- with both the considered geometries, the values of the average divergence angle (θ_{RMS}) are rather low (< 2.5 mrad) in the optimized current/voltage combination, satisfying with some margin the ITER requirement $\theta_{RMS} < 7$ mrad [17];
- the minimum values for the θ_{RMS} are similar with the two geometries and reached at slightly different V_{EG}; both geometries are quite sensitive to changes on extracted D^- current density and grid potentials; these plots can give some trends that could help the tuning of the accelerator operating parameters during the operations in the MITICA testbed;
- the DIFF_GAPS_01 is slightly better performing when the perveance match is not reached, especially when the EG voltage is higher than the optimized one (this can be seen by comparing the two sensitivity plots on the region with V_{EG} between 10 and 12 kV).

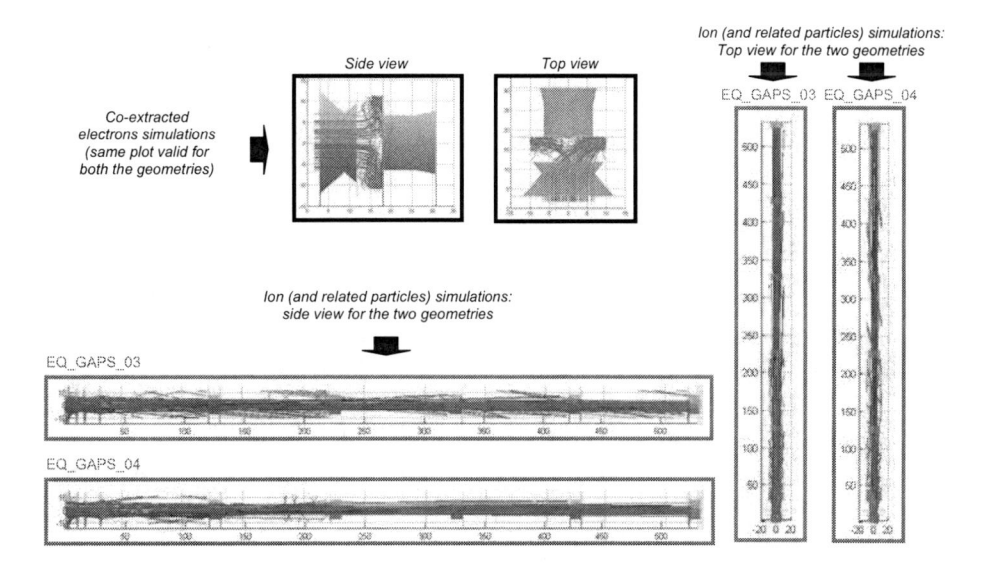

FIGURE 6. EAMCC simulations with EQ_GAPS_03 and EQ_GAPS_04 geometries: trajectories of particles. Boundary conditions: Electrical field from SLACCAD (taking into account space charge and stripping losses), density profile along the accelerator from STRIP, magnetic fields by OPERA, 10^6 macro-particles.

DEFLECTION OF ELECTRONS: COMPARISON BETWEEN TWO POSSIBLE APPROACHES

Though the EQ_GAPS_01 geometry is found to have some drawbacks regarding optics (slightly higher sensitivity to out-of-reference operating conditions), it is considered better performing than DIFF_GAPS_01 for voltage holding [5], and is here considered for further developments.

For the design of the SPIDER accelerator, a plasma grid optimized for the RF ion source (based on the IPP experience) and an extraction grid optimized for long pulses (with enhanced cooling system suitable for long pulses) were developed [7]. For MITICA, a similar design for these two grids is foreseen, with some small modifications regarding the EG thickness (that has been increased by 2 mm) and the Co-extracted Electron Suppression Magnets (CESMs) dimensions (increased from 4.6 mm x 5.6 mm to 6.6 mm x 5.6 mm).

Regarding the acceleration grids, two magnetic configurations, shown in Fig. 5, have been proposed and compared. In fact, while the co-extracted electrons are deflected in both the options by the CESMs embedded in the EG, two different approaches are chosen for the suppression of the electrons generated by stripping and charge exchange reactions (lately referred to as "stripped electrons"). In one approach, named EQ_GAPS_03, the stripped electrons are suppressed by means of an horizontal filter field (with an average value for B_x along the accelerator of 1 mT), generated by two filter permanent magnets at the sides of the plasma source and a current flowing along the PG (analogously to the case considered in the DDD5.3 [4]). In another approach, named EQ_GAPS_04, permanent magnets (named Stripped Electrons Suppression Magnets, SESMs) are embedded in all the acceleration grids and grounded grid. The SESMs are chosen with the same properties as the co-extracted electron suppression magnets (Sm$_2$Co$_7$ with 1.1 T magnetic remanence). These magnets give a B$_y$ field in the acceleration region, which should horizontally deflect the stripped electrons.

The CONDUCT, STRIP, SLACCAD and EAMCC codes are used to compare the operating behaviour foreseen with each option.

The main results of these analyses are:

- The pressure, density and stripping profiles are identical for the EQ_GAPS_03 and EQ_GAPS_04 geometries. The stripping losses, with the same hypotheses assumed for the DIFF_GAPS_01 and EQ_GAPS_01 geometries, are estimated around 34% (slightly higher than in Fig. 3 because the pressure in the extraction gap is higher).
- The minimum θ_{RMS} (average divergence angle) is calculated as 1.7 mrad and reached with the nominal extracted D$^-$ current density (293 A/m^2) and V_{EG} =9.4 kV. The curves are very similar to the ones of EQ_GAPS_01 in Fig. 4.
- The trajectories of the co-extracted electrons are almost identical in the EQ_GAPS_03 and EQ_GAPS_04: nearly all of these electrons are foreseen to impinge on the EG after a horizontal deflection due to the CESMs; the consequent heat load on the EG is about 0.5 MW for both the geometries.
- The trajectories of the stripped electrons are different for the EQ_GAPS_03 and EQ_GAPS_04 geometries, with consequent differences on the power loads on the grids and transmitted electrons (see Fig. 6):
    - in the EQ_GAPS_03 case, they are vertically deflected by the horizontal filter field onto the acceleration grids; the consequent heat loads are quite high (up to 2.5 MW for a single acceleration grid); the stripped electrons that exit from the accelerator together with the negative ions have a power of about 1.5 MW;
    - in the EQ_GAPS_04 case, they are horizontally deflected by the SESMs; the heat loads foreseen on the grids are in this case lower (up to 1.5 MW on a single grid), but the power of the electrons at the accelerator exit is much higher (about 7 MW).
- The heat loads foreseen on the grids with EQ_GAPS_03 are about 15% higher than the ones foreseen in [10], where a similar magnetic configuration was simulated. This could be due to the fact that in that case also the lateral pumping inside the accelerator was considered, hence the density profile was lower and the so the stripping losses. Moreover, a geometry similar to the DIFF_GAPS_01 was there considered, hence the accelerator was about 70 mm shorter.
- The EQ_GAPS_03 geometry is better performing than EQ_GAPS_04 regarding the capability to filter the stripped electrons, producing better results in terms of suppression of stripped electrons inside the accelerator. Nevertheless, this option has the important drawback that the vertical deflection of the negative D$^-$ ions is not uniform on the beam source cross section, because of the non-uniformities on the filter field (due to the edge effect). In fact, until now it has been found difficult to have an acceptable uniformity for a horizontal filter field generated by two filter permanent magnets at the sides of the plasma source and/or a current flow along the PG. Hence, a local approach for the suppression of the stripped

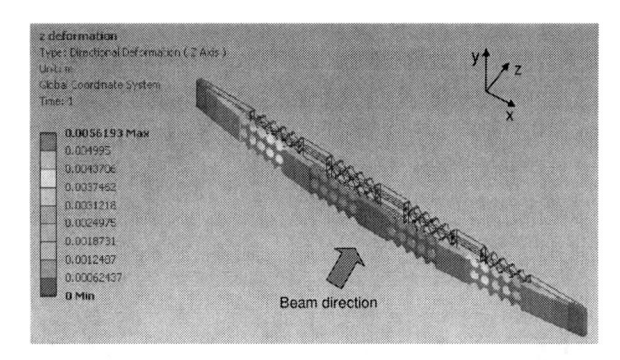

FIGURE 7. Out of plane deformation contour plot [m] of the AG2 with the EQ_GAPS_03 geometry, cooling channels dimensions 4 mm x 4 mm and cooling water velocity 10 m/s.

electrons, like EQ_GAPS_04, is at the moment preferred to a global approach, like EQ_GAPS_03.
- The power of the stripped electrons at exit is considered excessive in both cases; hence, further developments are foreseen to keep this power under acceptable values.

PRELIMINARY THERMO-MECHANICAL ANALYSES

The most critical grid by the thermo-structural point of view is the second acceleration grid (AG2) in the case of EQ_GAPS_03 geometry. The thermo-structural behaviour of this grid has been investigated by means of a tri-dimensional model in the ANSYS finite element code. The model (see Fig. 7) represents a slab region of the AG2 with full width and reduced height (two cooling channels and two aperture rows).

A first thermal analysis is carried out to evaluate the temperature in every point. Periodic boundary conditions are applied on the upper and lower surfaces. The expected heat load is applied on the surfaces and the convective heat flux to the channels walls. The convective heat transfer coefficients are calculated along the cooling channels [18], following the Sieder-Tate formula. Then, a structural analysis is carried out considering the temperatures calculated in the first step. For this analysis, the water pressure is applied and the following boundary conditions: z displacement blocked at the four corners, x displacements blocked at the left side corners, y displacements blocked at the horizontal midplane. These boundary conditions simulate the way the grids are mounted on the frame.

In the case of the EQ_GAPS_03 geometry, the AG2 grid is subject to a total power load of 2590 MW (2140 MW on the front surface, 420 MW on the aperture internal surfaces and 30 kW on the back surface).

An elasto-plastic model is used for the material, with the properties of a specific type of copper obtained by galvanic electrodeposition and analogous to the one to be used for the grids manufacturing, measured during a dedicated thermo-mechanical test campaign [19].

The cooling water velocity in the channel is 10 m/s (considered as a limit due to

TABLE 1. Sensitivity analysis on the thickness of the acceleration grids cooling channels: main results. In all cases, the water velocity along the cooling channels is 10 m s^{-1}, considered as a limit for vibrations and erosion.

Cooling channels dimensions [mm]	4 x 4	5 x 4	6 x 4	7 x 4
Cooling water inlet temperature [° C]	35	35	35	35
Cooling water outlet temperature [° C]	93	81	73	68
Maximum temperature on the grid [° C]	263	253	247	244
Maximum Von Mises stress on the grid [MPa]	65	70	71	69
Maximum Von Mises elastic strain on the grid [%]	0.058	0.062	0.063	0.060
Maximum Von Mises plastic strain on the grid [%]	0.071	0.070	0.069	0.067
In plane deformation [mm]	1.4	1.3	1.2	1.1
Out of plane deformation [mm]	5.6	6.9	8.0	8.9
Total water flow on the grid [kg/s]	10.7	13.4	16.0	18.7
Pressure drop along the channel [MPa]	0.275	0.243	0.221	0.216

the erosion processes). As the flow rate changes with the cross section, also the water temperature increase is different with different channel geometries.

Tab. 1 reports the main results of the preliminary thermo-mechanical simulations, while Fig. 7 shows the grid deformation contour plot in the case of a 4 mm x 4 mm cooling channel of the grid.

It can be observed that:

- The estimated operating temperatures, equivalent Von Mises stress and equivalent Von Mises strain (plastic and elastic) are rather high, due to the considerable amount of power deposited on the grid. These values are to be lowered by improving the design and to be checked regarding fatigue life;
- The in-plane deformations of the grids are acceptable and coherent with the ones of the PG and GG, while the out-of-plane deformations are too high and not acceptable for the usage in MITICA; in fact the beam optics and aiming would be spoiled by such high deformations. Further concepts are being considered in order to solve this problem, like increasing the grid thickness, changing the cooling channel characteristics (dimensions, flow, number of channels, position) and changing the magnetic configuration.

CONCLUSIONS

A thorough investigation of the existing conceptual designs of MITICA accelerator [5, 4] is being carried out using a comprehensive set of codes (STRIP, SLACCAD, OPERA, EAMCC and ANSYS).

Improved solutions are currently being developed, aiming at better operating behaviour in terms of voltage holding, compatibility with a Radio Frequency ion source, long pulse operation, satisfaction of the requirements on beam optics (in terms of divergence and deflection), suppression of co-extracted and stripped electrons inside the accelerator, reduction of secondary electrons inside the accelerator and better thermo-structural behaviour.

After selecting one or more suitable design concepts, further investigations will be carried out regarding the complete thermo-mechanical assessment of the grids (stress, strain, deformation, fatigue life), strategies on overall beam aiming, beamlet-beamlet and beamlet-wall interactions (to be studied with multi-beamlet models) and capability to absorb the stripped electrons exiting the accelerator with a suitable electron dump.

ACKNOWLEDGEMENTS

This work was set up in collaboration and financial support of F4E.

REFERENCES

1. R. Hemsworth, et al., Status of the ITER heating neutral beam system, Nucl. Fusion 49 (2009) 045006.
2. T. Inoue, et al., Design of neutral beam system for ITER-FEAT, Fus. Eng. and Design 56-57 (2001) 517-521.
3. P. Sonato, et al., The neutral beam test facility in Padova: the necessary step to develop the neutral beam injectors for ITER, presented at 26th Symposium on Fusion Technology, to be published in Fusion Eng. Des., Porto, 2010.
4. ITER DDD5.3, Neutral Beam Heating and Current Drive (NBH&CD) System, N 53 DDD 29 01-07-03 R 0.1.
5. T. Inoue, et al., Physics design of 1 MeV D^- MAMuG accelerator for H&CD NB, 1st Intermediate Report, ITER Task No. C53TD48FJ.
6. H.P.L. De Esch, Personal communication and support (2010), Association EURATOM-CEA, CEA-Cadarache, St. Paul-Lez-Durance, France.
7. P. Agostinetti, et al., Design of a low voltage, high current extraction system for the ITER Ion Source, AIP Conf. Proc. 1097 (2009) 325.
8. P. Sonato, et al., The ITER full size plasma source device design, Fusion Eng. Des. 84 (2009), 269-274.
9. OPERA-3D, Vector Fields Co.Ltd., http://www.vectorfields.com/
10. G. Fubiani, et al., Modeling of secondary emission processes in the negative ion based electrostatic accelerator of the International Thermonuclear Experimental Reactor. Phys. Rev. Special Topics Accelerators and Beams 11, 014202 (2008).
11. ANSYS Manual, ANSYS Inc., http://www.ansys.com/
12. J. Pamela, A model for negative ion extraction and comparison of negative ion optics calculations to experimental results. Rev. Sci. Inst. 62 (1991) 1163.
13. W.B. Hermannsfeld, Electron Trajectory Program, SLAC report, Stanford Linear Accelerator Center, SLAC-226 (1979).
14. H.P.L. de Esch, et al., Updated physics design ITER-SINGAP accelerator. Fusion Eng. Des. 73 (2005) 329.
15. K. Jousten, Wutz Handbuch Vakuumtechnik - theorie und praxis, 9th edition, Vieweg und Teubner Verlag, Wiesbaden (2006), ISBN-13 978-3-8348-0133-3
16. A. Krylov and R.S. Hemsworth, Gas flow and related beam losses in the ITER neutral beam injector, Fusion Eng. Des. 81 (2006), p. 2239
17. L. Svensson, SRD-53-PR, -MI, -MP, -SI, -SP (NEUTRAL BEAM TEST FACILITY), Version 1.2, IDM UID 2WCCSG, October 2009.
18. P. Agostinetti, et al., PCCE - A Predictive Code for Calorimetric Estimates in actively cooled components interested by pulsed power loads, 26th Symposium on Fusion Technology, 2010, Porto, Portugal.
19. P. Agostinetti, et al., Investigation of the Thermo-mechanical Properties of Electro-deposited Copper for ITER, 14th International Conference on Fusion Reactor Materials, Sapporo, Japan, 2009 (accepted for publishing in the Journal of Nuclear Materials).

Numerical Assessment of the Diagnostic Capabilities of the Instrumented Calorimeter for SPIDER (STRIKE)

M. Dalla Palma[a], M. De Muri[a,b], R. Pasqualotto[a],
A. Rizzolo[a], G. Serianni[a], P. Veltri[a]

[a] *Consorzio RFX, Euratom-ENEA association, Padova, Italy)*
corso Stati Uniti, 4- 35127 – Padova - Italy
[b] *Dipartimento di Ingegneria Elettrica, Padova University, Italy*
via Gradenigo, 6/A - 35131 - Padova - Italy

Abstract. An important feature of the ITER project is represented by additional heating via injection of neutral beams from accelerated negative ions. To study and optimise their production, the SPIDER test facility is under construction in Padova, with the aim of testing beam characteristics and to verify the source proper operation.

STRIKE (Short-Time Retractable Instrumented Kalorimeter Experiment) is a diagnostic to characterise the SPIDER negative ion beam during short operation (several seconds). During long pulse operations, STRIKE is parked off-beam in the vacuum vessel. The most important measurements are beam uniformity, beamlet divergence and stripping losses. STRIKE is directly exposed to the beam and is formed of 16 tiles, one for each beamlet groups. The measurements are provided by thermal cameras, current sensors, thermocouples and electrostatic sensors.

This paper presents the investigation of the influence on the response of STRIKE of: thermal characteristics of the tile material, exposure angle, features of some dedicated diagnostics.

The uniformity of the beam will be studied by measurements of the current flowing through each tile and by thermal cameras. Simulations show that it will be possible to verify experimentally whether the beam meets the ITER requirement about the maximum allowed beam non-uniformity (below ±10%). In the simulations also the influence of the beam halo has been included; the effect of off-perveance conditions has been studied. To estimate the beamlet divergence, STRIKE can be moved along the beam direction at two different distances from the accelerator. The optimal positions have been defined taking into account design constraints.

The effect of stripping on the comparison between currents and heat loads has been assessed; this will allow to obtain an experimental estimate of stripping. Electrostatic simulations have provided the suitable tile biasing voltage in order to reabsorb secondary particles into the same tile as the one where they were emitted from.

Keywords: Negative beams, Diagnostics, Instrumented Calorimeter, SPIDER, Numerical Simulations, Thermal Cameras
PACS: 52.50.Gj, 52.70.Nc, 52.59.Fn, 29.25.Ni, 07.20.Fw, 07.20.Ka

INTRODUCTION

The thermonuclear experimental reactor ITER will require additional heating via injection of at least two 1 MeV, 17 MW neutral beam injectors [1] in order to fulfil its mission. Since the occurrence of such conditions has never been experimentally verified, a test facility is under construction in Padova, Italy, to optimise the injector operation in view of ITER. As several beam features depend on the ion source parameters, the SPIDER test facility, entering the tender stage for construction, aims at improving the beam characteristics at 100 kV acceleration voltage [2].

One of the most important beam features to be verified is the uniformity of the beam power, which, according to ITER requirements, shall not exceed the interval ±10% with respect to the average over the beam. A very fine spatial resolution (few mm) is required of the measurement; this can only be provided by dumping the beam onto tiles made of suitable material and observed by thermal cameras.

Diagnostic calorimeters are used in several neutral beam injectors: some of them are surfaces with a clever machining and a suitable arrangement of thermocouples so as to provide some local measurements of the energy flux [3, 4, 5]. In other cases thermal cameras observe graphite plates [6] or carbon fibre composite plates, with fibres arranged along the beam direction, so as to preserve the contrast of the thermal image despite perpendicular heat transport [7, 8, 3, 9]; in the latter two cases also thermocouples are added for absolute calibration of the thermal cameras.

The STRIKE (Short-Time Retractable Instrumented Kalorimeter Experiment) diagnostic, whose design phase is completed, is dedicated to characterise the SPIDER beam during short operations (several seconds), in terms of beam uniformity, stripping losses and beam divergence.

In order to reproduce the geometry of the SPIDER beam, which is made of 4x4 groups of 80 beamlets each, STRIKE is divided into 16 tiles directly exposed to the beam, the main measurements being provided by thermal cameras and current sensors.

The present paper describes some numerical analyses of the behaviour of STRIKE when exposed to the beam. STRIKE tiles are made of unidirectional carbon-fibre-carbon composite, subjected to huge energy fluxes (peaking at 20 MW/m^2) and radiating towards surfaces at room temperature (namely plasma source and vacuum vessel). Thermal cameras observe the rear surface of the tiles, due to radiation expected on the front due to the interaction between the beam and the background gas. A first mechanical design for STRIKE is found in [10].

In the present paper, simulations concerning the operation of STRIKE are presented. First of all the dependence of the STRIKE response on the properties of the material is given; then the capabilities of STRIKE to investigate beam uniformity and beamlet divergence are studied, including the effect of beam halo; finally the factors affecting the current measurements are discussed.

In the simulations, several operational cases have been considered, not all together:
- heat radiation: no radiation; radiation only from the back side of the tile, radiation both from the back and the front side of the tiles
- angle between the beam and the perpendicular to the tile: 0 deg and 60 deg
- distance of STRIKE from the SPIDER grounded grid: 0.5 m and 1 m.

Results from all these conditions will be presented.

EFFECT OF MATERIAL PROPERTIES

To make the proper selection for the calorimeter, various materials have been considered and compared.

To study the beam feature it is important that heat transmission from the front to the rear side of the calorimeter occurs with a small systematic error. The best thermal response is exhibited by the MFC-1A carbon fibre composite produced by Mitsubishi, which is used in JET instrumented calorimeter [3]. In the following results are presented for this material. Figure 1 shows the dependence of the specific heat and the parallel and orthogonal thermal conductivities on temperature.

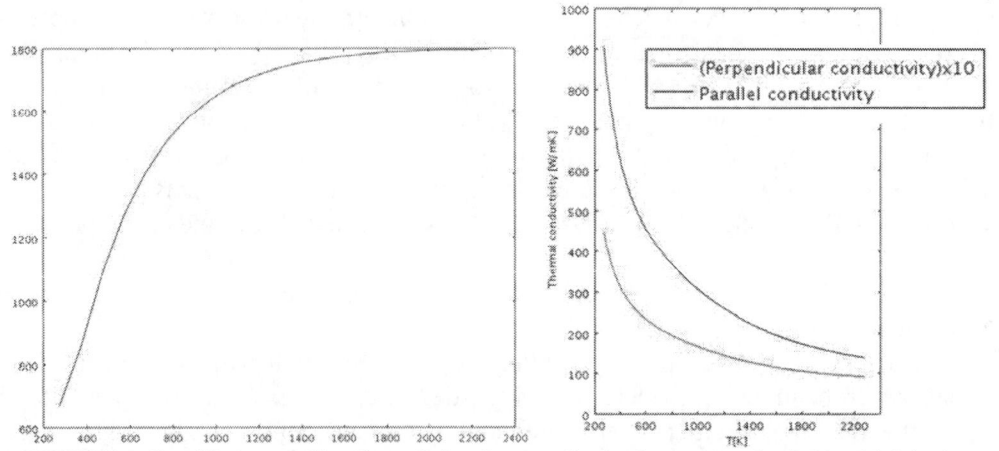

FIGURE 1. Specific heat (left) and parallel and perpendicular thermal conductivities (right).of Mitsubishi MFC-1A as a function of temperature.

An important issue that shall be taken into account is the temperature reached on the front side of the calorimeter. The model consists of a parallelepiped that represents only a part of a sector around the area hit by one beamlet.

The incidence angle of the beam on the tiles is supposed to be 90°. Surface dimension is chosen depending on the transverse conductivity coefficient: heat should be transferred to the rear side without significant lateral losses. The simulations have been performed with several thicknesses: 5 mm, 20 mm and 30 mm; different pulse durations: 0.1 s, 1 s and 2 s and different heat fluxes: 10 MW/m^2, 20 MW/m^2 and 100 MW/m^2. Every combination has been checked for all materials under consideration.

The selected material is MFC1-A: the maximum temperature on the front surface is comparable to the other studied materials but the thermal conductivity ratio is the highest.

Figure 2 presents the temperature distribution over front and rear side of the simulated sample, 30 mm thickness (a and d), 20 mm thickness (b and e) with beam power density of 20 MW/m^2 and 1 s pulse duration. The images are taken when the heat load ends. It can be noticed, from Figure 3 a and d, that the pattern on the back side is only about 1 mm larger than on the front side: heat has been essentially transmitted in the beam direction towards the rear side.

case is compared to two other cases, in which the heat load has been raised by 5% or reduced by 15% with respect to the reference case. The thermal patterns are not shown, as they do not look too different. The maximum temperatures on both sides of the tiles are described in the following table, along with the percent variation of the temperature and the temperature increase per a variation of the flux by 1%:

TABLE 1. Comparison between: reference case; case with heat load raised by 5%; case with heat load reduced by 15%.

reference case		+5% wrt reference		-15% wrt reference	
front temp. [K]	rear temp. [K]	front temp. [K]	rear temp. [K]	front temp. [K]	rear temp. [K]
2657	1733	2765	1783	2310	1572
percent variation		+4.6%	3.5%	-14.7%	-11.2%
temperature increment per 1%		21 K/%	10 K/%	-23 K/%	-10.7 K/%

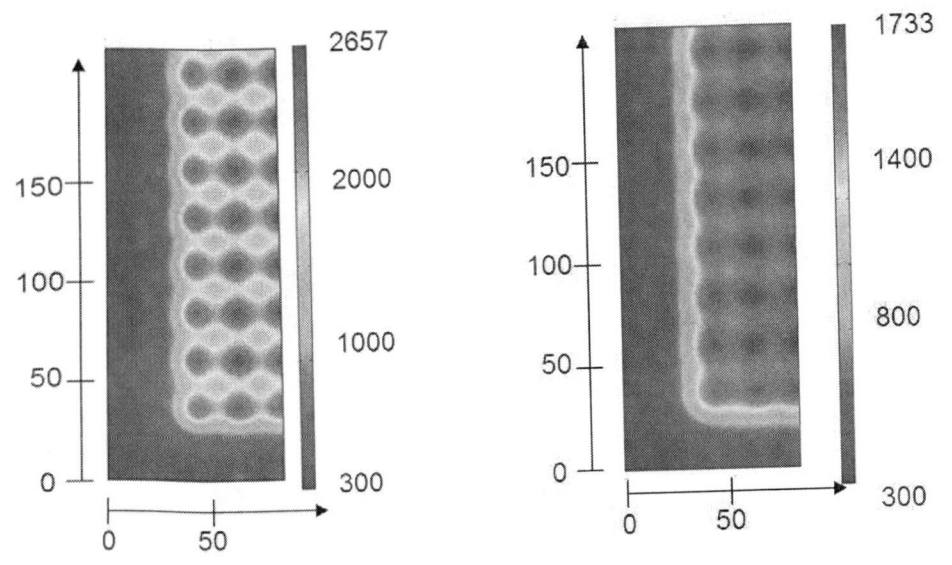

FIGURE 5. Thermal pattern on the front (left) and the rear (right) side of the tile in the reference case.

It can be deduced that a temperature variation around 10 K is expected whenever the impinging flux changes by 1%. This is marginally feasible with real thermal cameras, which have an error of 1-2%.

Previous analyses, without radiation and with angled calorimeter gave 20 K per 5% variation of the energy flux. For images a and c no halo included, b and d included 15%, 30 mrad halo.

Beamlets can be described as a superposition of a core, carrying most of the energy, and a halo. Assuming that the halo carries 15% of the total energy and possesses a 30 mrad divergence (Gauss curve), the resulting thermal pattern for 1 s beam pulse with halo (Fig. 6, b and d) is compared with the normal conditions (Fig. 6, a and c), finding that the peak temperature decreases and that the thermal pattern is slightly more blurred.

The effect of radiation is shown in STRIKE panels, which are inside SPIDER vacuum vessel. Radiation towards surfaces at room temperature can be applied to the

Figure 2 c and f show the effect of the heat flow after the load ends: observation performed 1 s after heat load ends. The front temperature decreases whereas the back temperature increases as heat distributes in the material.

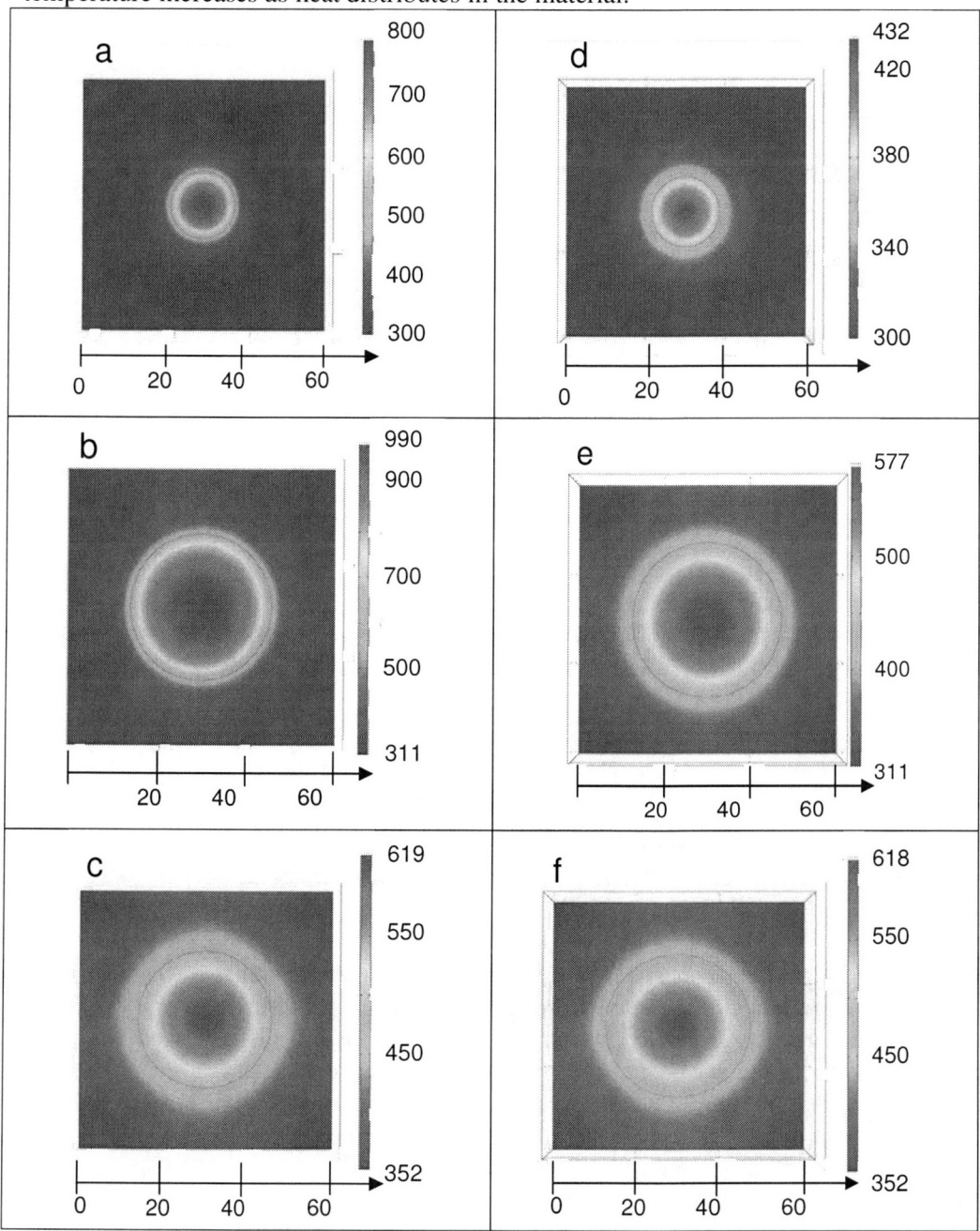

FIGURE 2. Temperature distribution over front and back side of the simulated sample for MFC-1A. For comparison, the area where heat is deposited is indicated (thin black line).

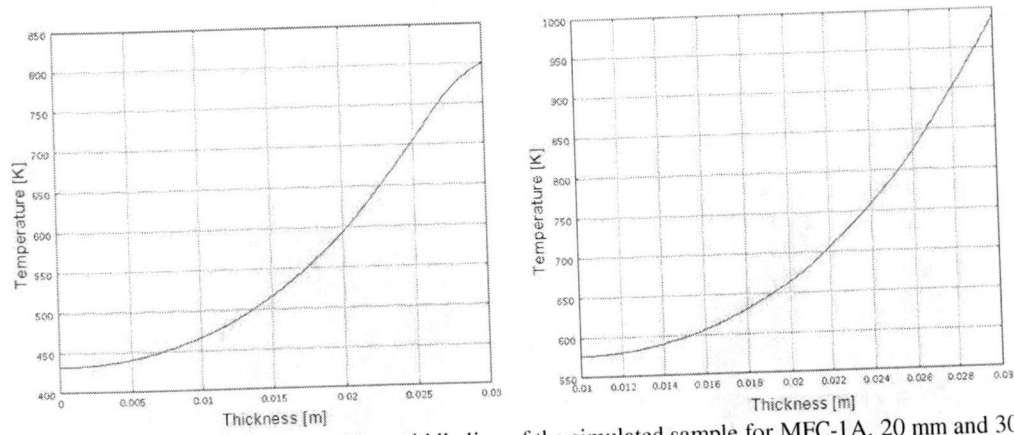

FIGURE 3. Temperature profile along middle line of the simulated sample for MFC-1A, 20 mm and 30 mm thickness, 20 MW/m^2 and 1 s pulse duration; observation performed when heat load ends.

Figure 3 shows the temperature profile along the thickness related to Fig. 2a, b and Fig. 2d, e.

INVESTIGATION OF BEAMLET DIVERGENCE

The main scope of STRIKE is the assessment of the uniformity of the SPIDER beam. At full parameters, in the simulations the beam is made of 1280 beamlets for a total of 50 A and 100 kV. In the present paper, beamlets are assumed to exhibit a Gauss curve profile in two dimensions with a width given by a 3 mm intrinsic width (that is at the GG) and a contribution due to a divergence $\delta = 3$ mrad (see eq. 1); so, in a generic position in the plane (x,y) perpendicular to the beam, the amplitude, G_g, of the energy flux associated to each beamlet is represented as follows:

$$G_g(x,y) = A_g \exp\left(-\frac{(x-x_{0g})^2}{2\sigma_{xg}^2} - \frac{(y-y_{0g})^2}{2\sigma_{yg}^2}\right) \qquad \sigma_g = \sigma_{0g} + \delta z \qquad (1)$$

where σ_{xg} and σ_{yg} are respectively the beamlet half-widths in the x and y directions and A_g is the amplitude of the energy flux at the maximum position (x_{0g}, y_{0g}).

One of the scopes of STRIKE is the measurement of the beamlet divergence. Given the previous representation of the energy flux associated to the beamlets, in each direction, the beamlet width can be written as the superposition of an intrinsic width, σ_{0g} (which is assumed = 3 mm in the examples) and the increase of the beamlet width due to the increase of the angular aperture of the beamlet as a function of the distance along z from the GG (divergence), as indicated in eq. 1.

Due to the intrinsic symmetries, only one fourth of the whole tile is simulated. The value of the beamlet divergence affects the thermal pattern on STRIKE tiles. This can be verified by comparing the simulated cases, not reported herein, with 3 mrad divergence to 5 mrad: the maximum temperature decreases and the temperature peaks stand out less clearly.

Measuring the beamlet divergence calls for two measurements of the beamlet width in different positions along the beam, in order to deduce the value of δ. Only when the effect of beamlet divergence is overwhelming with respect to the intrinsic width of the beamlet, can a single measurement be sufficient. This condition is more and more justified as the distance of the measurement position from the GG increases.

The results which can be expected from these measurements are shown in Fig. 4 for the case of the angled design of STRIKE, 3 mm beamlet divergence and radiation only on the rear surface. Least mean squares fitting is applied to the measurements at different distances from the GG. The reconstructed average width of the beamlets changes by about 3 mm between the two positions ($\sigma = 11$ mm and $\sigma = 14$ mm respectively); considering the distances from the GG and the angled exposure, the correct value of divergence is reproduced.

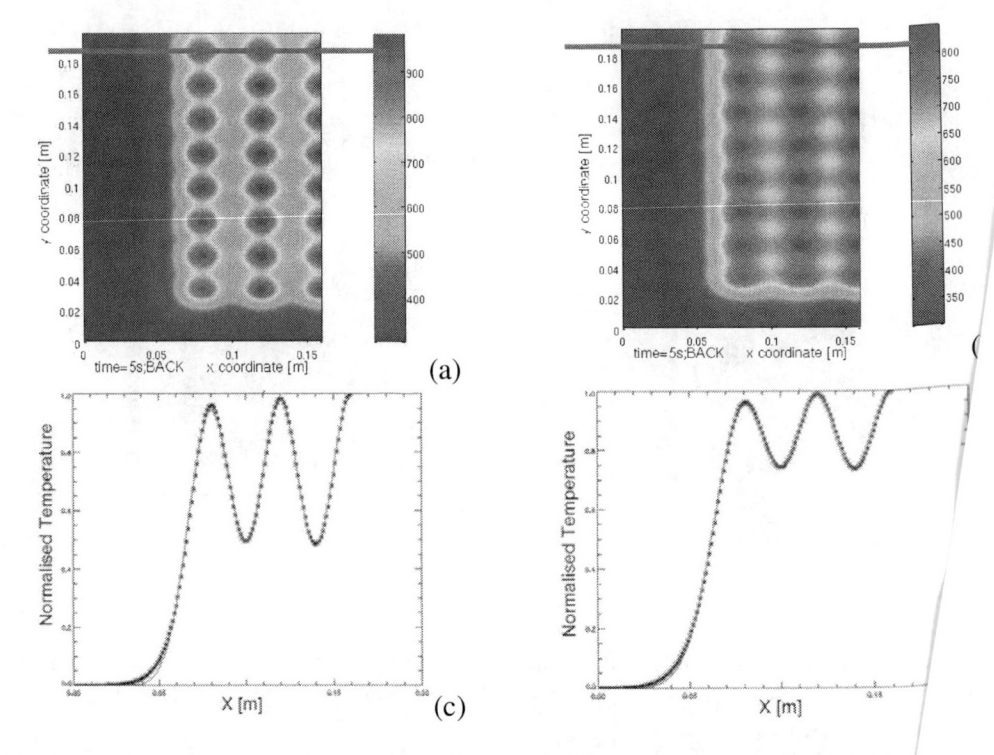

FIGURE 4. Simulation of measurements performed at different distances from the GG: 0.5 m GG (left) and 1 m from the GG (right); thermal pattern on the rear side (top) and temperature along the line shown (bottom); the best fitting curve is superposed to the temperature p

INVESTIGATION OF BEAM UNIFORMITY

Thermal radiation towards surfaces at 300 K (GG and vacuum vessel) account.

The thermal pattern for the reference case is given in Fig. 5 after 1 beginning of the application of the thermal load. A maximum temperature is found on the front side, and of 1733 K on the rear side of the tile.

model. After a transient period, radiation results in the oscillation of the temperatures mainly between two values (figure not reported herein).

Sensitivity to beamlet deflection has already been shown and is not reproduced [11]

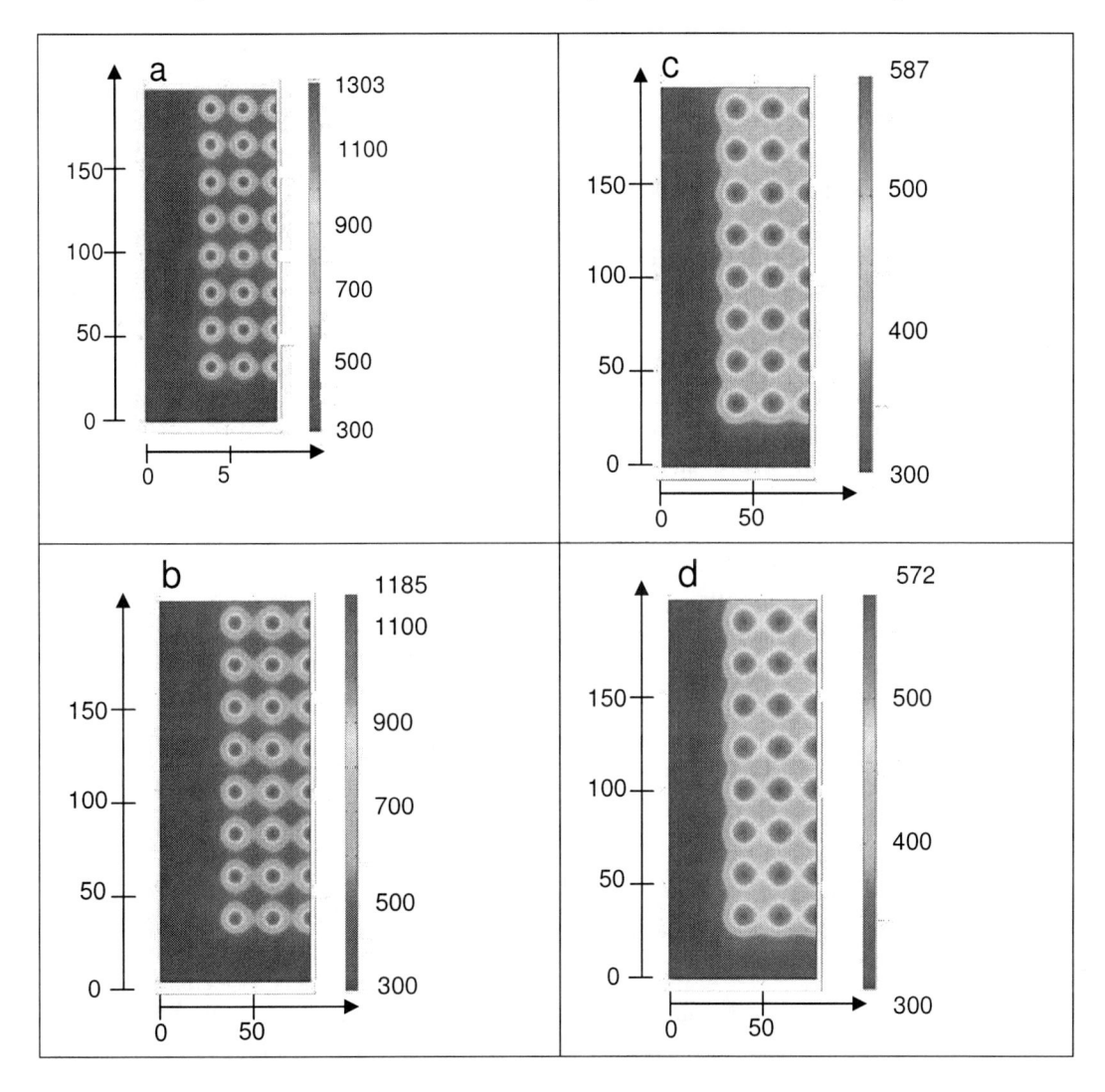

FIGURE 6. Temperature distribution over front (a and b) and back (c and d) side of STRIKE tile in the reference case, 20 MW/m² and 1 s pulse duration; radiation on both sides; observation performed when heat load ends. Top row, no halo; bottom row halo (see text).

MEASUREMENT OF CURRENT

Numerical simulations of the SPIDER accelerator [12] provide the beam composition at the GG: the beam is primarily made of negative hydrogen ions, but it also includes positive ions, neutrals and electrons. Positive ions and neutrals are produced by the process of loss of electrons from the negative ions (stripping); electrons are generated by stripping and by ionisation of the background gas.

Electrons are expected to be mainly deflected onto a dedicated device, the electron dump, just outside of the accelerator; anyway deflection of electrons makes negligible their contribution to the secondary emission of electrons from the tile surfaces.

Though all particles contribute to the emission of electrons form the tile surface [13], however, the sum of the energy fluxes associated to negative hydrogen outbalances all other contributions. Consequently, the current reaching the tiles can be assumed to be associated to negative ions.

When hydrogenic particles hit the STRIKE surface, secondary electron emission occurs; this affects the current measurements but not the thermal measurements, since the ion/atom energy is anyway deposited inside the crystalline lattice even if electrons are simultaneously emitted. Considering the data for graphite, in the case of normal incidence, the coefficient of secondary electron emission is slightly larger than one, meaning that more than one negative charge leaves the tile surface per incident particle. Emitted electrons possess the energy of a few electron volts [13].

Consequently, biasing will be applied to the tiles with respect to ground, in order to pull the secondary electrons back to the tile surface.

Finite element analyses have been performed to assess the maximum range of electrons emitted with 3 eV initial energy [14]. Analyses have adopted a two-dimensional approximation with a 50 mm thick STRIKE tile (including the supporting frame), since the distances from the various conductors (at least 0.5 m from GG and vacuum vessel) are quite large, so that it is assumed that the 3D geometry cannot change qualitatively the results on the relevant space scale (few tens of mm). The correctness of this assumption has been checked numerically.

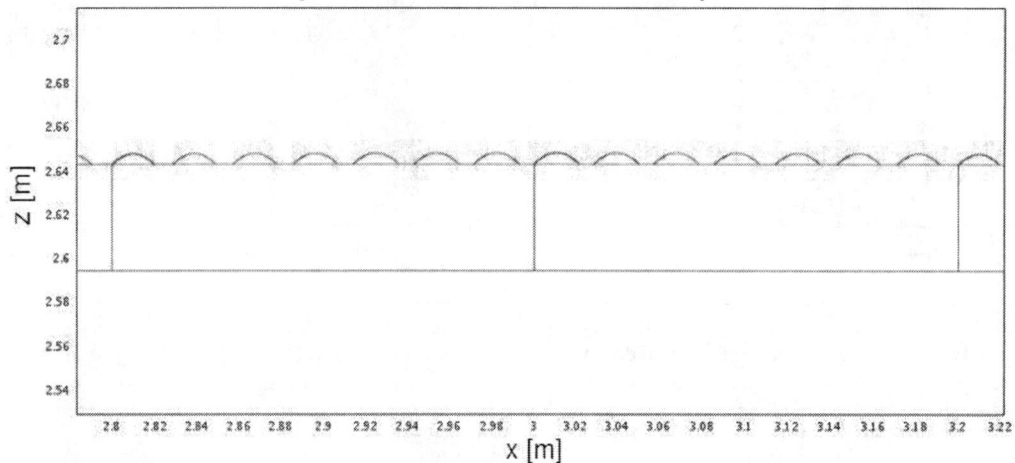

FIGURE 7. Zoom of electron trajectories for 45° and 200 V in the central region of the calorimeter.

A preliminary computation with 60 V biasing shows (non reported here) that the surface at about 1.5 V from the tile voltage is located at a distance of about 20 mm. Hence it is deduced that a voltage of about 200 V should be sufficient to restrain electrons from hopping from one tile to the neighbouring ones, thus preserving the integrity of current measurements. The electron trajectories are displayed in Fig. 7 and it is indeed found that the maximum range of electrons is about 20 mm, so that essentially all secondary electrons should return to the tile they were emitted from.

Simultaneous measurements of current and deposited power provide a way to estimate the stripping losses inside the accelerator. Indeed, by measuring the negative ion current in two different positions along the beam, the stripping outside the accelerator can be deduced and the density of negative ions at the GG can be computed. By reconstructing the negative ion energy flux at the GG and using the measured energy flux, an estimate of the energy flux associated to negative ions stripped inside the accelerator is finally obtained.

CONCLUSIONS

The present paper describes some simulations devoted to the realisation of the design of the instrumented calorimeter for the SPIDER experiment. In particular, the capabilities of STRIKE as a diagnostic of beam uniformity and of beamlet divergence have been displayed and a discussion of the possibility of quantifying stripping has been given. Some work is still ahead, e.g. concerning the analysis of the effect of off-normal conditions on STRIKE diagnostic capabilities and the expected signals of the measurement systems will be assessed.

ACKNOWLEDGEMENTS

This work was set up in collaboration and financial support of F4E.

REFERENCES

1. ITER Physics Basis Editors et al, *Nucl. Fusion* **39**, 2495 (1999).
2. P. Sonato et al., *Fusion Eng. Des.* **84**, 269 (2009).
3. D. Ciric et al., *Fusion Technol.* **2**, 827 (1994).
4. Y. Takeiri et al., *Rev. Sci. Instrum.* **71**, 1225 (2000).
5. A. Encheva, IPP Report 4/283 (2004).
6. K. Tsumori et al., *Rev. Sci. Instrum.* **81**, 02B117 (2010).
7. L. Svensson et al., *Fusion Eng. Des.* **66-68**, 627 (2003).
8. H. P. L. DeEsch et al., *AIP Conf. Proc.* **1097**, 353 (2009).
9. C. Fuentes et al., *Rev. Sci. Instrum.* **77**, 10E519 (2006).
10. A. Rizzolo, et al., "Design and analyses of a one-dimensional CFC calorimeter for SPIDER beam characterisation", *Fusion Eng. Des.* (2010), doi:10.1016/j.fusengdes.2010.09.003.
11. M. De Muri et al., "Thermal and Electrostatic Analyses of One Dimensional CFC Diagnostic Calorimeter for SPIDER Beam Characterisation", Proc. COMSOL Conference Milan 2009 poster no. 32, edited by COMSOL, Milano, Italy, 2009.
12. P. Agostinetti et al., "Physics and engineering design of the Accelerator and Electron Dump for SPIDER", *Nucl. Fusion* **51**, 063004 (2011).
13. E. W. Thomas, "Secondary electron emission" in *Nuclear Fusion Special Issue "Data Compendium for Plasma-Surface Interactions"*, edited by IAEA, Vienna, Austria, 1984, pag. 94, Chap. 9.
14. G. Fubiani et al., *Phys. Rev. STAB* **11**, 014202 (2008).

Sensitivity Analysis of the Off-Normal Conditions of the SPIDER Accelerator

P. Veltri[a], P. Agostinetti[a], V. Antoni[a], M. Cavenago[b], G. Chitarin[a], N. Marconato[a], N. Pilan[a], E. Sartori[a] and G. Serianni[a]

[a]*Consorzio RFX, Euratom-ENEA Association, Corso Stati Uniti 4, 35127 Padova, Italy*
[b]*Laboratori Nazionali di Legnaro-Istituto Nazionale di Fisica Nucleare (LNL-INFN), Viale dell'Università 2, 35020 Legnaro (PD), Italy*

Abstract. In the context of the development of the 1 MV neutral beam injector for the ITER tokamak [1], the study on beam formation and acceleration has considerable importance. This effort includes the ion source and accelerator SPIDER (Source for Production of Ions of Deuterium Extracted from an Rf plasma) ion source, planned to be built in Padova, and designed to extract and accelerate a 355 A/m^2 current of H$^-$ (or 285 A/m^2 D$^-$) up to 100 kV [2][3]. Exhaustive simulations were already carried out during the accelerator optimization leading to the present design [4]. However, as it is expected that the accelerator shall operate also in case of pre-programmed or undesired off-normal conditions, the investigation of a large set of off-normal scenarios is necessary. These analyses will also be useful for the evaluation of the real performances of the machine, and should help in interpreting experimental results, or in identifying dangerous operating conditions.

The present contribution offers an overview of the results obtained during the investigation of these off-normal conditions, by means of different modeling tools and codes. The results, showed a good flexibility of the device in different operating conditions. Where the consequences of the abnormalities appeared to be problematic further analysis were addressed.

Keywords: Neutral Beam Injector, SPIDER, accelerator modeling
PACS: 29.25.Ni

INTRODUCTION

During the first phase of the SPIDER accelerator project, a large set of modeling tools was used to optimize the critical parameters of the device, and the reference design is now in agreement with the ITER requirement about extracted current, beam divergence, etc. A second step, presently in progress, is based on the individuation and characterization of the operational scenarios that the device could experience in case of unexpected deviations from the standard conditions. These analyses are the subject of the present paper, and will be presented in the following pages.

The first section describes the major modification made on existing codes to better fit the characteristics of our scenarios. The other sections are divided according to the different causes that could affect the performances. The second section treats magnetic and electric field dishomogeneities (due to voltage and magnetic field ripple, field uniformity, etc.) The third section deal with the possibility of deviations of some operational parameters from their reference values, while section the fourth covers the event of mechanical modifications (caused by manufacturing tolerances or, with the beam on, thermal expansion or vibrations).

CODE ADAPTATION

The simulations were performed by means of commercial modeling tools (OPERA[5], ANSYS[6], COMSOL[7]) or dedicated codes (SLACCAD[8], BYPO [9], EAMCC[10], EDAC and BACKSCAT[11]). Sometimes modifications of the codes appeared necessary to include new effects. In particular the code SLACCAD needs a set of sub-codes to calculate the pressure distribution inside the accelerator (CONDUCT), or the stripping rate (STRIP). The code "CONDUCT" was substituted by a new code which calculates the grid conductances with a better approximation for tapered holes and takes into account the effects of using gases other than D_2.

Moreover, in order to handle the interaction of SLACCAD with the accompanying codes and with the post-processing routines, a suitable graphical user interface was developed, having the possibility of handling multiple analyses and allowing initializing hundreds of simulations at once.

Finally some modifications in the BYPO code were performed in order to consider the complex magnetic configuration of SPIDER. BYPO is indeed capable of calculating the magnetic field of permanent magnets, but still does not account for the effects of a ferromagnetic material layer and of permanent magnets, which are present in the grounded grid (GG) of SPIDER, to compensate for the horizontal deflection that the beamlets experience in crossing the suppression magnets located in the extraction grid (EG); moreover the ferromagnetic material reduces the filter field downstream of the GG, in order to decrease the vertical deflection of ions. To overcome the problem an interpolation of the magnetic field calculated with OPERA is fed to BYPO.

INFLUENCE OF FIELD DISHOMOGENEITIES

The extraction grid voltage is the most influential parameter on the electric field configuration inside the accelerator, since also small changes in its value have strong impact on the beamlets optic. In order to investigate the consequences that the device could experience in case of voltage ripples and, above all, to individuate trends of interest helpful in optimizing the accelerator divergence, some sensitivity analysis on the EG voltage were performed on the latest version of SPIDER, by means of the SLACCAD code, considering as reference parameter the value of θ_{RMS}. This is defined as:

$$\theta_{RMS} = \sqrt{\frac{\sum |r_i| \vartheta_i^2}{\sum |r_i|}} \quad (1)$$

i.e. the root mean value of the average divergence angle of single particles, weighed over the radius, to respect the 2D axial-symmetry of the model.

The results are plotted in fig. 1. The points found by SLACCAD numerical simulations are fitted with a second degree polynomial curve to better highlight the trend. The beam divergence results fairly constant for small variations of the voltage around the reference value of 9.4 kV, corresponding to the best quality of the optics.

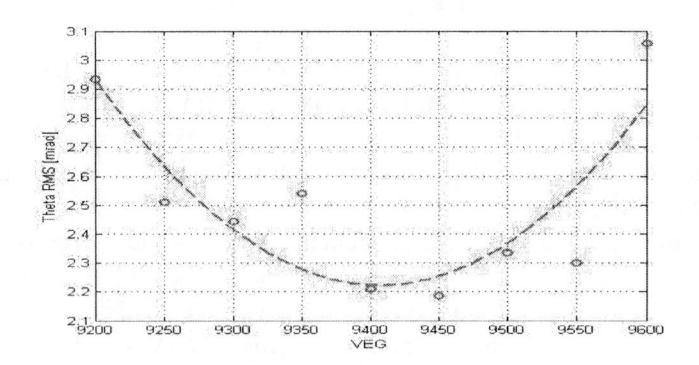

FIGURE 1. SLACCAD analysis of θ_{RMS} as a function of the EG voltage.

Regarding the magnetic field, one case of interest is the lack of homogeneity that can be caused by the inefficiency of one of the permanent magnets embedded in the grids, due to manufacture inaccuracy, or, more likely, to loss of the magnetic properties in case of magnet overheating. In fact the extended exposition to heat loads could statistically cause losses of magnetization also with temperatures well below the Curie temperature of the magnet, which is about 800 °C. The worst case has been tested, namely a total failure of the magnet in the EG, where the largest fraction of electrons should be filtered.

To model this effect, a 3D OPERA analysis was performed using an array of vertical beamlets, with one of the magnets turned off. Figure 2 shows the consequent magnetic field behavior with respect to the case of a nearby functioning magnet. The resulting magnetic field basically preserves its distribution along the z direction, but the field intensity is decreased roughly by 50%. The field of surrounding beamlets results unperturbed. Both the beam optic and the heat loads are affected by this failure. As a consequence, under the effect of the compensating magnetic field of the GG, the beamlet suffers a strong over deflection, estimated with OPERA itself in term of the angle α=-3.35 mrad, with respect to α=0.02 mrad in the reference case. As a consequence, under the effect of the compensating magnetic field of the GG, the beamlet suffers a strong over deflection, estimated with OPERA itself in term of the angle α=-3.35 mrad, with respect to α=0.02 mrad in the reference case.

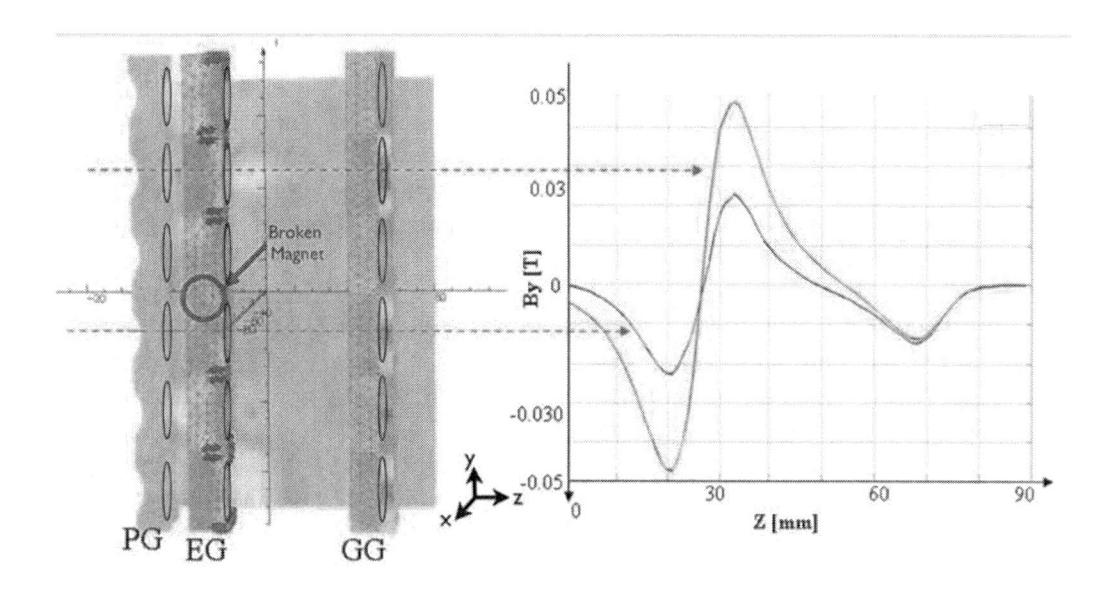

FIGURE 2. OPERA Analysis of a vertical array of SPIDER beamlets. One of the magnet embedded in the EG, in the circle, is turned off. The consequent modification in the field behavior is visible in the right side plot, together with the behavior of the reference case field corresponding to a nearby hole on the top line.

To account for the variations in terms of power deposition of co-extracted electrons and the secondary they produce, some analysis in EAMCC were carried out to account for the relative heat loads.

The total power deposition on EG is quite similar in both cases, but in case of broken magnets, the smaller magnetic field causes more electrons to impinge in the inner cylindrical part of the EG, aperture where they produce additional secondaries or being backscattered toward the accelerator exit, gaining the full acceleration. This results in a significant increasing of the load on the GG, which passes from about 300 W per aperture in the reference case to more than 430 W; this values, however, are still well manageable by the cooling system.

INFLUENCE OF OPERATIONAL PARAMETERS

The performances of the source are affected by a large set of physical operational parameters which in many cases are difficult to be controlled. The following subsections contain some samples of the investigations performed about the effects of the variation of operational parameters with respect to the values usually assumed in reference cases.

Off-normal Pressure Scenarios

The pressure in the source is strictly related with the needs of containing the stripping losses in the accelerator and the investigation of its distribution in the whole NBI was the subject of several studies [12]. The values usually assumed in standard modeling, are about 0.3 Pa for the source pressure, and 0.05 Pa for tank pressure. The uncertainty on this parameter, due to difficulties to directly control it in the source, justifies the following series of analyses.

A first test was performed with SLACCAD, setting a fixed value of 0.05 Pa for the tank pressure, and by varying the source pressure around the reference value. It should be noted that since SLACCAD accounts for the pressure difference between source and tank only with the module "STRIP", by calculating the stripping loss profile inside the accelerator. Therefore such variation affects the space charge deposited by the rays representing the macro particles along their path, which, in turn, can cause small differences in the beamlet divergence. The results are plotted in fig. 3a. The trend clearly exhibits a minimum close to the reference value. This result is not surprising, since the whole accelerator was optimized for this pressure. Changing the value of the tank pressure has an even smaller effect on the optics, since it affects the pressure profile only in the final part of the accelerator, where the largest part of the stripping has already occurred. However, a similar trend could be identified, showing an increased θ_{RMS} for increasing pressure (fig. 3b).

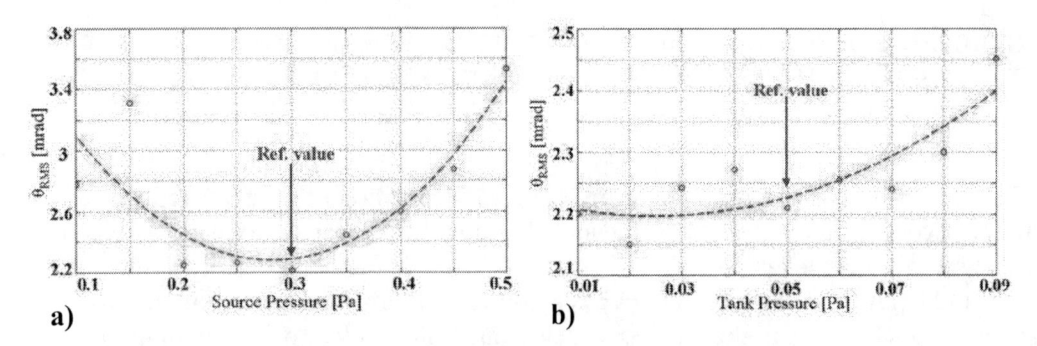

FIGURE 3. SLACCAD Pressure sensitivity analysis. a) Off- normal source pressure scenarios. b) off- normal tank pressure scenario.

In case of high pressure, a larger amount of secondary particles is produced and accelerated by the electric field. In order to estimate the consequences on the device components in terms of thermal stresses, some analyses were done with the EAMCC code, capable of quantifying these secondary fluxes and the power load they deposit on the accelerator grids. The same quantities were then estimated in the electron dump with the code BACKSCAT code, which includes the effect of electrons backscattering between ED pipes using 2 different backscattering models [13] [14].

To simulate the worst condition of loads, the value of 0.5 Pa of source pressure was imposed in both codes, whose results are resumed in table 1, compared with the loads for the reference pressure value. Even if the thermal stress of the grids and pipes increases, they are still manageable by the cooling system, which is dimensioned to tolerate load 100% greater than the reference case; the increased flux of back-streaming ions is acceptable too. The BACKSCAT code also allows evaluating the power deposition on the rear of GG due to backscattering electrons, and, since its value is not negligible in both cases, adding a dedicated cooling system in the GG was necessary to protect the ferromagnetic material.

TABLE 1. Result from the EAMCC and BACKSCAT code for the transmitted particles and heat loads on the SPIDER components, for the 0.3 and 0.5 Pascal source pressure scenarios.

Pressure [Pa]	EAMCC Trans. Particles [kW] H-	e	H0	Thermal Loads [kW] EG	GG	BACKSCAT Thermal Loads [kW] GG$_{rear}$	P-1	P-2	P-3
0.3	5720	400	330	666.4	470	50.5	177	119	87
0.5	5150	480	420	685	584	65.4	228	150	112
Variation %	-9	+17	+21	+3	+20	+23	+22	+21	+22

Influence Of The Starting Energy Of Ions

The energy of the negative ions before being extracted depends on several phenomena (like plasma temperature in the source or the production mechanism) and usually a value of some eV is assumed in standard modeling. Fig. 4 reports the results of a sensitivity analysis on this parameter by means of SLACCAD. A slightly worse beam optics is found for increasing particles energy. Note that below initial energies of 2 eV the rays hit the EG and the resulting divergence was unreliable.

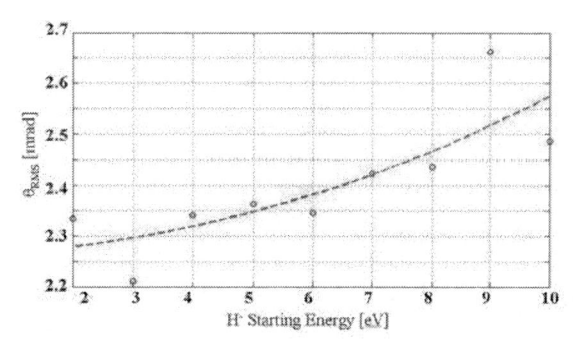

FIGURE 4. SLACCAD sensitivity analysis of the initial energy of ions before being extracted.

EXTRACTED CURRENT RATIO SENSITIVITY ANALYSIS

The ratio between the extracted current of negatively charged particles, defined as $R_j = J_e/J_{H^-}$ is another parameter difficult to be controlled, which can have consequences on the performances of the device. Since the electron space charge is distributed asymmetrically in the accelerator tube, following the suppression magnetic field, the negative ions beam optics could be influenced in case of high value of R_j. The unique code which accounts for the space charge of both electrons and ions together is BYPO [3]. The results in fig. 5 show the behavior of θ_{RMS} with respect to R_j. Here only slight variations around the values of R_j suggested by the IPP experience ($0.5 < R_j < 1.5$) were performed. Note that the values of θ_{RMS} of BYPO are not comparable with those given by SLACCAD, due to the different symmetries of the models (2D Cartesian and 2D axially symmetric respectively). A clear trend is however found, and the fit shows increasing beam divergence for increasing extracted electron current.

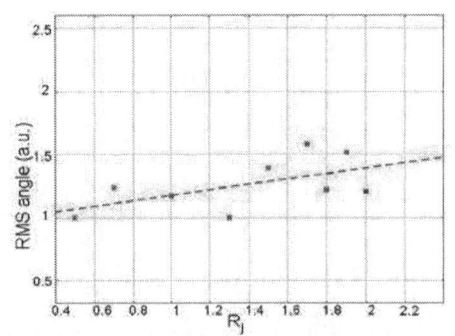

FIGURE 5. BYPO sensitivity analysis on the R_j parameter. Numerically found points and their trend. Note the rms value is normalized to the minimum value.

From the point of view of the electron thermal load on the EG, its value is simply expected to increase linearly with R_j, with small differences in its distribution.

MECHANICAL MODIFICATIONS

The thermal expansions and the bowing of the grids, due to the load of impinging electrons, the electrostatic pressure of nearby grids and the vibrations of the whole device, could all have influence on beamlet optics. The net effect is, in all cases, a variation in the accelerator geometry, resulting in an offset of the accelerating grid apertures in the horizontal (x-axis) and vertical (y-axis) directions or in a shift of the grid along the beam direction (z-axis).

The in-plane grid offset, corresponding to a misalignment which causes an electrostatic deflection of the beamlets, is due to thermal loads on the grid; these phenomena do not affect the entire grid in the same way but are more effective in the more external rows with respect to the positions where the grids are supported (fixed points). The effect has been minimized by an appropriate cooling: in the horizontal direction the shift should remain under 0.2 mm and in the vertical it is even negligible. An estimation of the beam deflection due to this expansion can be found by exploiting the linear relationship found in previous works [11] between the grid offset δ and the beamlet deflection angle α

$$\alpha\,[mrad] = K \cdot \delta\,[mm] \qquad (2)$$

with K=7.7 for the EG and -8.06 for the GG. With our values the deflection stays under 1.5 mrad which respects the requirement for the grid alignment.

Another contribution to the misalignment could come from the expansion of the support frame of the grid, which accounts for other 0.1-0.2 mm in the worst case. This effect is accounted at the expected steady-state temperature of the support frame in beam-off configuration, setting an offset in the fastening position of PG, EG to the support frames; notice that aperture misalignment will rise during beam operation.

In the case of a grid shift in the z-direction, the effect is a variation in the beamlet optics. The thermal expansion is expected to shift the grids in the beam direction by 0.1 mm for the EG or 0.6 mm for the GG. Also in this case, in spite of the smaller shift value, the EG deformation is the principal responsible of the beam optics variation, and the analysis were oriented towards its modification. Also in this case, the deformation of the support frames of the grids could increase the total shift, acting on the whole grid in a homogeneous way. Some preliminary analyses showed large values of the shift of the whole grid (about 0.5 mm), calling for further investigations and possibly reference configuration adjustment to the actual design.

Another factor which can change the z-position of the EG is the electrostatic pressure of nearby grids. The consequent shift can be estimated, by considering the EG and GG as two slabs of copper having length L, height h and thickness s as

$$\delta = \frac{5}{384} \frac{qL^4}{E_y \cdot I} \qquad (3)$$

where $E_y = 1.1 \cdot 10^{11}$ [N/m^2] is the elastic modulus for copper, I the moment of inertia of the slab and $q = 1/2 \cdot \varepsilon_0 E^2 \cdot h$ the load due to the electric field. The resulting δ value, representing the maximum shift which is expected in the grid centre, is about 0.1 mm. Even in the case of breakdowns of the system, when this load suddenly disappears and the grid a start vibrating around its nominal position, the amplitude of this oscillation stays below δ. Obviously a similar effect should be expected for the PG-EG interaction, but the potential difference is much smaller, so the electrostatic pressure in the backward direction is lower. The SLACCAD analysis reported in fig. 6 shows the result on the beam divergence of different positions of the EG with respect to the nearby grids, to account for all the conditions which can cause a grid movement. The trend highlighted by the fitting curve suggests that for small grid deflections the beam quality stays almost constant, but it rapidly decreases in case of larger values of the shift.

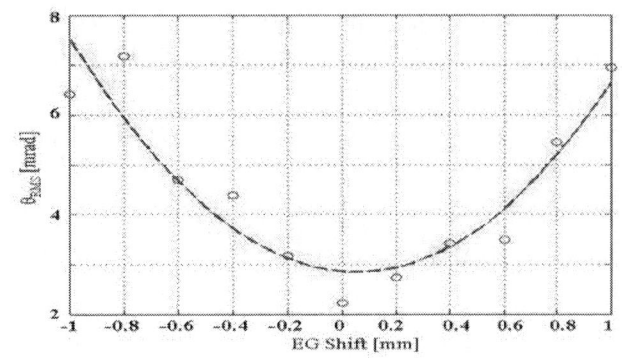

FIGURE 6. SLACCAD sensitivity analysis of the EG position along the beam direction. The 0 correspond to the design value for the EG position: 6 mm after the PG and 35 before the GG.

Finally, manufacturing dishomogeneities of part of the grids can also be listed between the mechanical modifications having similar effects on the optics. However since the percentage of manufacturing error guaranteed by the constructor is much smaller than the values of offset or misalignments due to thermal expansion, the net effect is negligible compared with previously analyzed cases.

CONCLUSIONS

The flexibility of the SPIDER source was tested in a wide set of situations in which the operational parameters differ for some reasons from the reference values. The device appeared basically stable to quite all the off-normal conditions: beam optics reasonably keeps its quality for small deviations of quite all tested parameters. Moreover, in case of a magnetic configuration failure or of increased production of secondaries, the cooling system is expected to handle the increased thermal load in the structures without any problems. In the cases where the thermal load increase was unacceptable, some modification to the cooling systems were implemented, as it was the case of the overheating of the ferromagnetic material covering the rear part of the GG. The preliminary analysis of the support frame deformation for the accelerator grids revealed higher values than expected. Since this effect could affect the beam divergence, further investigations are foreseen. Finally, the investigations reported here of some trend for the parameters of interest could be extremely useful in aiding the experimental phase of the device.

ACKNOWLEDGMENTS

This work was set up in collaboration and financial support of F4E.

REFERENCES

1. R. Hemsworth, et al., Status of the ITER heating neutral beam system, *Nucl. Fusion* **49** (2009).
2. P. Agostinetti, et al., Design of a low voltage, high current extraction system for the ITER Ion Source, *AIP Proceedings of the 1st International Conference on Negative Ions, Beams and Sources, 2008, Aix en Provence, France*
3. P. Sonato, et al., The ITER full size plasma source device design, *Fusion Eng. Des.* **84** (2009), 269-274
4. P. Agostinetti et al., Physics and engineering design of the Accelerator and Electron Dump for SPIDER, accepted for publication in *Nuclear Fusion*.
5. OPERA-3D, Vector Fields Co. Ltd., http://www.vectorfields.com/
6. ANSYS Inc, http://www.ansys.com
7. Comsol Multiphysics 3.5, (2009), http://www.comsol.eu.
8. J. Pamela, "A model for negative ion extraction and comparison of negative ion optics calculations to experimental results". *Rev. Sci. Inst.* **62** (1991).
9. M. Cavenago et al., Negative ion extraction with finite element solvers and ray maps, *IEEE Transactions on Plasma Science*, **36** (4), pp 1581-1588 (2008).
10. G. Fubiani et al., *Phys. Rev. Special Topics Accelerators and Beams* **11**, 014202 (2008)

11. P. Agostinetti et al., Optics and physics sensitivity analysis on the SPIDER accelerator, *RFX-SPIDER Technnical Note -75, 24/3/2010*.
12. A. Krylov and R.S. Hemsworth, Gas flow and related beam losses in the ITER neutral beam injector, *Fusion Eng. Des.* **81** (2006), p. 2239.
13. P.F. Staub, *J. Phys.* D 27, 1533 (1994).
14. E.Thomas, in *Nuclear Fusion Special Issue "Data compendium for plasma surface interaction"*, cap 9 (1984).

RF - Plasma Source Commissioning in Indian Negative Ion Facility

[a]M.J. Singh, [a]M. Bandyopadhyay, [b]G. Bansal, [b]A. Gahlaut, [b]J. Soni, [b]Sunil Kumar, [b]K. Pandya, [b]K.G. Parmar, [b]J. Sonara, [a]Ratnakar Yadava, [a]A.K. Chakraborty, [c]W. Kraus, [c]B. Heinemann, [c]R. Riedl, [c]S. Obermayer, [c]C. Martens, [c]P. Franzen, and [c]U. Fantz

[a]*ITER- India, Institute for Plasma Research, A-29, Sector 25, GIDC, Gandhinagar, Gujrat, India*
[b]*Institute for Plasma Research, Bhat Gandhinagar, Gujrat, India*
[c]*Max-Planck-Institut für Plasmaphysik, Boltzmannstraße 2, D-85748 Garching, Germany*

Abstract. The Indian program of the RF based negative ion source has started off with the commissioning of ROBIN, the inductively coupled RF based negative ion source facility under establishment at Institute for Plasma research (IPR), India. The facility is being developed under a technology transfer agreement with IPP Garching. It consists of a single RF driver based beam source (BATMAN replica) coupled to a 100 kW, 1 MHz RF generator with a self excited oscillator, through a matching network, for plasma production and ion extraction and acceleration. The delivery of the RF generator and the RF plasma source without the accelerator, has enabled initiation of plasma production experiments. The recent experimental campaign has established the matching circuit parameters that result in plasma production with density in the range of $0.5 - 1 \times 10^{18} / m^3$, at operational gas pressures ranging between 0.4 – 1 Pa. Various configurations of the matching network have been experimented upon to obtain a stable operation of the set up for RF powers ranging between 25 – 85 kW and pulse lengths ranging between 4-20 s. It has been observed that the range of the parameters of the matching circuit, over which the frequency of the power supply is stable, is narrow and further experiments with increased number of turns in the coil are in the pipeline to see if the range can be widened. In this paper, the description of the experimental system and the commissioning data related to the optimisation of the various parameters of the matching network, to obtain stable plasma of required density, are presented and discussed.

Keywords: RF generator, plasma source, matching network
PACS: 52.50.Dg

INTRODUCTION

The responsibility to deliver the diagnostic neutral beam (DNB) package [1] to ITER [2] and the need of high energy neutral beams for heating and current drive applications on future Tokamaks has established the basis for initiating a negative ion beam development program in the Institute for Plasma Research (IPR), India. RF based negative ion sources have been accepted as the baseline for such beam lines. Development of such a program requires not only the knowledge about manufacturing technologies involved in the fabrication of such sources but also a hands on training towards operating and maintaining such sources during the plasma and the beam production, extraction and acceleration. As a first step towards initiation of such a technology intensive program an inductively coupled RF based negative ion source facility is under establishment at (IPR), India. The facility is being developed under a technology transfer agreement with IPP Garching. Experience gained on this set up related to various aspects of the plasma and beam production by inductive coupling of RF power shall extend to the operation of large ITER sized DNB source at the Indian test facility (INTF) in IPR.

Keeping time constraints into consideration and the long lead time into the procurement of the extractor and accelerator system of the source, the program on this facility has been divided into two phases. The first phase has been activated soon after the delivery of the plasma box and related components from the manufacturer. Commissioning experiments in this phase have concentrated on optimizing various parameters of the matching network to produce stable plasma in the source through the inductive coupling of the RF power. The experimental facility and the commissioning procedure are presented in the following sections of the paper. The second phase shall be dedicated to extracting and accelerating beams after the grids become available.

THE EXPERIMENTAL FACILITY

The facility is housed in IPR and occupies a floor area of 21 m x 8 m. At present it consists of a plasma box with RF components coupled to a vacuum chamber and supported by auxiliaries. A brief description of each component is given below.

The ion source is a replica of the BATMAN [3] source working for over a decade in IPP, Garching. As shown in Fig. 1, it consists of a water cooled plasma chamber with a back plate at one end and a diagnostic flange at the other. The back plate supports the RF driver mounting. The RF driver is a 6.5 turn copper coil supported on

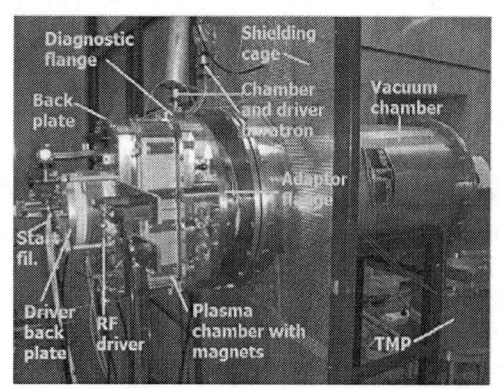

FIGURE 1. Experimental set up of the ion source in the facility

an alumina cylinder. The water cooled copper coil, a tube with 6 mm OD and 4 mm ID, is sleeved with a 10 mm OD 8 mm ID poly propylene tubing and the interspace is filled with transformer oil for better inter turn isolation during operation. Inside the ceramic cylinder is a Faraday shield made of copper. The back plate of the Faraday shield is connected to a SS flange which presses the alumina cylinder against the source back plate to provide the vacuum sealing. The SS flange also termed as the driver back plate has ports for gas feed, starter filament, driver pressure measurement and viewing. The starter filament provides the seed electrons required to ignite the plasma when RF power is applied. While the inner walls of the Faraday shield are coated with Molybdenum to prevent copper sputtering when the plasma is formed, the inner walls of the back plate and the plasma box facing the plasma have a 1 mm thick layer of electrodeposited copper for better heat conduction. The back plate of the driver, the back plate of the source and the side walls of the plasma box are lined with magnets for better plasma confinement. A diagnostic flange connected to the other end of the plasma box houses the filter magnets required during the beam operation and several ports for performing optical as well as probe diagnostics. The present set up is without the set of extractor and accelerator grids. In order to simulate plasma production under similar conditions as it would be if the source is coupled to the grid system, a dummy grid (30 apertures of 8 mm diameter) is connected to the diagnostic flange. The diagnostic flange is coupled through an adaptor flange to a cylindrical vacuum chamber 1300 mm in length and 600 mm in diameter. The other end of the cylindrical chamber is connected to a 5000 l/s turbo molecular pump through a gate valve.

The RF coil on the source is connected to the RF generator through a matching network. The 100 kW generator procured from Himmelwerk Germany is a self excited oscillator operating at 1 MHz. The matching circuit, used to match the 50 ohm of the generator to a few ohm plasma resistance, is shown in Fig. 2. As seen in the figure

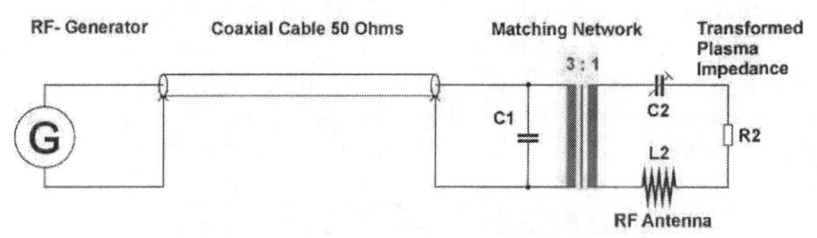

FIGURE 2. Schematic of the matching circuit

the circuit is a combination of the series (C2) and shunt (C1) capacitances and a 3:1 transformer in the shunt arm. The primary and the secondary of the transformer are made of RG 220 cables sleeved with ferrites. In this set up two types of ferrites, white coloured ones with a mu of 1500 and the grey coloured ones with a mu of 5800, have been used. Tuning C2, in the series arm, helps to ensure proper matching. The transformer helps to isolate the source from the generator in the event of source being floated at high potential during beam extraction and acceleration.

Various parts of the source and the generator are cooled using a hydraulic circuit. A total flow of 130 lpm (litres/minute) at 5.4 bar is required during the experiment, out of which 90 lpm goes to generator, 2.5 lpm to the RF coil, 5 lpm to the source back plate and 32 lpm to the plasma box. The flows to different regions of the source are diagnosed using flowmeters in each arm of the hydraulic circuit.

The gas feed system in the present set up consists of a set of two open shut valves, one for the gas puff and the other for maintaining the operational pressure in the source during the pulse. The two valves have a common gas feed line from the gas cylinder. The gas puff is essential to ignite the plasma and to avoid discharges across the coil when the RF power is applied. A typical plot showing the behaviour of the gas feed system is shown in Fig.3. Peak A in the figure relates to the trapped gas in the SS pipe between the cylinder and the valve in between two gas pulses. Peak B relates to the gas puff followed by the constant gas flow in the source during the pulse.

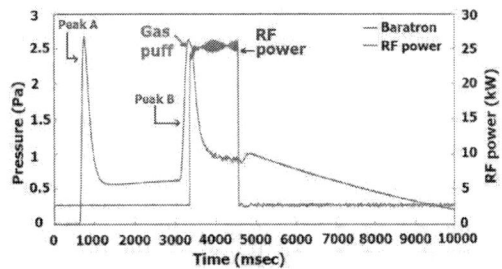

FIGURE 3. Plot of the gas pulse and the RF power during a typical shot

Sufficient delay, ~ 2500 msec, has been added between the opening of the main gas valve and the gas puff valve to differentiate between the two peaks. Further it is noticed that the gas pressure does not remain steady and tends to drop during the pulse length. One of the possible reasons could be a non stable fore pressure in the cylinder regulator. This behaviour of the gas feed system needs rectification for future experiments. The pressures are monitored at the driver and the source location using capacitance manometers isolated from the main chamber.

The entire experimental set up is run by a control system. It consists of three different Siemens PLC systems viz. (1) S-7 400 PLC as a Master Control (master), (2) S-7 300 PLC for Vacuum system control (slave) and (3) C-7 PLC for RF generator control (slave). Communication is established through Industrial Ethernet (IE). Control program and GUI are developed in Siemens Step-7 PLC programming software and WINCC SCADA software, respectively. There are approximately 150 analog and 200 digital control and monitoring signals dedicated for closed loop control of the system. Since the source floats at high potential a combination of galvanic and fiber optic isolation has been implemented. The control program has been implemented in such a way that the systems could be run independently or together. Safety interlocks prevent any damage in the event of malfunction of any component.

PXI based Data Acquisition (DAQ) system is a combination of PXI RT (Real time) system, front end signal conditioning electronics, host system and DAQ program. Fiber optic links have been used for signal conditioning, electrical isolation and better noise immunity. Ethernet cable ensures transfer of acquired data to the host system. Real time and host application programs are developed in LabVIEW. The data is stored in binary format with a facility of online display of selected parameters. Calculations and report generation occur at the end of each beam shot. The data acquisition channels for the present campaign include pressure signals, the water inlet and outlet temperatures from the generator and different source components, the voltage, current, power and the ratio of the active to the apparent power i.e. cos(phi), signals from the generator and the signals corresponding to the ON/OFF status of the gate valves. The diagnostic signals are acquired using the program supplied alongwith the spectrometer.

The filament heating (16 V, 10 A) and the discharge power supplies (100 V, 1 A) are mounted in the rack and connected to the filament.

The area housing the source and the generator and that around the tuning circuit has been shielded with perforated Al sheets to reduce the RF interference with the modules placed in the control and data acquisition racks. Shielding is also essential for capacitance manometers, the hydraulic flowmeters and the water thermocouples to avoid RF interference in measurements. As a result of the shielding a radiation level of 14 V/m was observed at 70 kW of power just outside the experimental cage. In the region where the DAC racks have been placed the value reduces to <5 V/m.

The present experimental set up uses optical diagnostics. The optical spectrometer from Plasus has 4 channels. Each channel has a spectral resolution of < 1.5 nm in the spectral range from 200 nm to 1100 nm. The optics for these measurements is mounted on the driver back plate, the source back plate and the upper and lower side ports placed diagonally on the diagnostic flange. The optical signal from driver back plate location also acts as an interlock to shut off the RF generator if the plasma does not ignite 20 ms after the RF power is coupled.

In order to ensure proper isolation and avoid formation of ground loops, the chamber and the water supply system (valves pipes etc.) are isolated from the ground using rubber pads. The grounds of all the equipment mounted inside the rack and the rack grounding cable are connected to a common copper strip. The chamber ground, the copper strips in all the data acquisition and control racks and the shielding ground are terminated at the star point. Ceramic isolators are used to isolate the vacuum gauges, gas feed connection and capacitance manometers from the source and the chamber.

EXPERIMENT

The present experimental campaign is aimed at optimizing the matching network to couple the RF power and produce the desired plasma in the source. The methodology adopted for the experiment is to establish the matching at lower powers and subsequently raise the power to higher values. This exercise has been performed at higher gas pressures, typically 2.5 Pa gas puff and the operational gas pressure of 1.5 Pa. However, keeping into consideration the fact that the ITER sources have to operate at lower pressures of 0.3 Pa the second step of the experiment is to establish the matching at lower operational gas pressures.

The following experimental parameters have been varied to obtain the best possible matching condition, high cos(phi) and resistance as close as possible to 50 ohms, for a given gas pressure,:
- Variations in the matching circuit for a given transformer configuration
    a. Series (C2) and shunt capacitance (C1) Fig. 2
- Various transformer and ferrite combinations
    a. 3:1, grey ferrites (16 ferrites per arm)
    b. 2:1, white ferrites (15 ferrites per arm)
    c. 2:1 combination of grey and white ferrites (39 ferrites per arm)

- Varying the capacitance in the tank circuit of the RF generator

RESULTS

Matching

The typical pressure pulse at the beginning of the experiment is shown in Fig. 3. The timing of the RF is adjusted to match the peak of the gas puff. As seen from the plot, the pressure inside the source during the gas puff is 2.5 Pa and the source operational pressure during the pulse varied between 0.8 -1 Pa. The filament was heated up to provide the seed electrons before the application of the RF power. With the filament heater supply at 8.3 V and 6 A, a discharge current of 2.5 mA was measured for a discharge voltage of 90 V.

The initial settings of the matching unit were evaluated using the excel sheet [4]. The Ferrites in the 3:1 isolation transformer have a coupling constant of 0.95 and the plasma resistance was taken to be 2.5 ohms. Both these values are based on the experimental inputs from IPP. The frequency of the RF generator measured on a 50 ohm dummy load was 980 kHz. The shunt arm of the matching unit had 3 nF disk capacitors from VISHAY and a 1 nF tuning capacitor from Jennings. The series arm had a 2 nF fixed capacitor and 1 nF tuning capacitor from Jennings. Both the tuning capacitors have 50 turns each with a calibration factor of 20 pF per turn. The set up was such that the coil with the cables connected to the matching network had an inductance of 15μH. With this setting the matching did not work. The cable length between the coil and the matching network was reduced to 5 m and the new arrangement had an inductance of 11.8 μH. For these settings the value of the series arm capacitance of the matching unit was calculated to be 2.3 nF, for a shunt arm capacitance of 3 nF.

The experimental matching exercise was carried out at a power level of 25 kW. Fig. 4 shows the variation of the cos(phi) and the resistance as a function of the series capacitance variation. It is seen that the observed frequency is 940 kHz and is different from the expected 980 kHz of the generator. Further with a circuit resistance of 50 ohm the coupling is poor as a low cos(phi) is observed. The resistance increases with the increasing cos(phi). It is also seen that the optimal matching, best cos(phi), is obtained for a series capacitance value of 2.62 nF. For this parameter, the power from the generator was raised in steps of 10 kW. The cos(phi) and the resistance values are not seen to change upto a RF power of 50 kW. For 60 and 70 kW RF power the series capacitance had to be increased by 2 turns each to get the same cos(phi). The resistance of the circuit was observed to be 100 ohms. The shunt capacitance of the matching circuit was varied in steps of 15 turns to study its effect on the resistance while maintaining good matching conditions. Fig. 5 shows the variation of the resistance and the cos(phi) as a function of the shunt capacitance. A minimum resistance of 78 ohm is obtained for a shunt capacitance by 3.6 nF without any change in cos(phi). No further reduction in the resistance was possible due to frequency flips at higher shunt values.

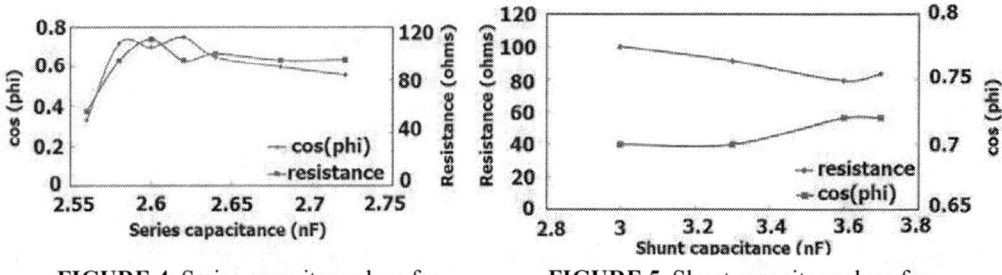

FIGURE 4. Series capacitor values for optimal cos(phi)

FIGURE 5. Shunt capacitor values for minimal resistance

No significant change in the matching parameters has been observed as the gas puff pressure is reduced to 1.6 Pa and the source pressure varies between 0.8-0.4 Pa during the shot.

Effect of transformation ratio and the ferrites on matching

In the matching exercise mentioned above the range of the series capacitance values over which the matching was sustainable was very narrow. The effect of the transformation ratio and the type of ferrites on the range over which good matching could be sustained was studied by changing the transformation ratio from 3:1 to 2:1 and using white ferrites (μ 1500) instead of the grey ones (μ 5800). It was observed that over the entire range of series capacitance values the cos(phi) never got better than 0.2. The circuit resistance varied between 28 to 40 ohms and the frequency varied between 770 to 840 kHz. Any efforts to obtain the matching by adding additional variable 1 nF capacitors in the shunt as well as series only resulted in large frequency flips between 770 to 1280 with the best cos(phi) of 0.25. Reverting back to 3:1 transformer configuration with 11 white ferrites per arm resulted in best cos(phi) of 0.4 with a circuit resistance of 83 ohms. The value of the shunt and the series capacitor were 3.5 nF and 2.6 nF respectively, for this configuration. It is worth mentioning that both configurations of ferrites have been used at IPP and cos (phi) values ~ 1 have been achieved. The point to note here is that the IPP configuration of generator is a two tetrode configuration in push pull operation which provides a better frequency stability of the tank circuit as compared to the present generator with a single tetrode tube. As no improvement in the matching conditions was observed due to either of these changes, the set up was reverted back to the original 3:1 transformer configuration with grey ferrites. The matching parameters were checked for their repeatability.

Changes in the tank circuit of the RF generator

The tank circuit of the generator, giving a frequency of 980 kHz across a 50 ohm dummy load, is equipped with seven capacitors connected in parallel with each having a 250 pF rating. This implies that the capacitance in the tank circuit of the generator is 1750 pF. A higher capacity in the tank circuit increases the reactive power of the generator and in this way the frequency stability of the generator. This was applied to similar generators coupled to the ASDEX upgrade ion sources. Another alternate to increasing the stability range of the matching network performance could be increasing the capacity in the tank circuit. In such a case

New Frequency = Old Frequency x sqrt (Old capacitance/New capacitance)

Experiments were performed for three configurations of the tank circuit and the results are listed in Table 1.

TABLE 1. Matching parameters for various configurations of the generator tank circuit

Capacitance added	New Frequency	Best cos(phi)	Range C_{series}	Frequency observed	C_{shunt} for minimum resistance	Minimum resistance
0	980	0.7	2.58-2.62	940	3.6	78
1	917	0.9	2.86-3	900	3.9	92
2	865	0.7	3.14-3.18	840	4.8	75
3	820	0.8	3.38-3.4	810	4.6	86

As seen from the above table, changing the frequency of the tank circuit with the addition of the capacitors helped to improve the cos(phi). However the range of the capacitances over which the best cos(phi) could be retained is still narrow. Fig. 6 shows the variation of the cos(phi) and frequency as a function of the variation in the series capacitance during the matching exercise.

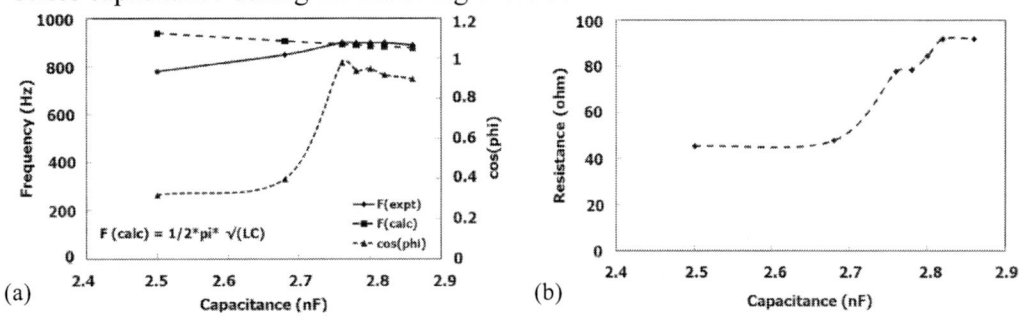

FIGURE 6 : Variation of frequency, cos(phi) (a) and resistance (b) for tank circuit frequency of 917 kHz.

Optical spectroscopy

The Plasus spectrometer is used to measure H_α, H_β and H_γ lines in the hydrogen plasma produced in the source at the measurement locations mentioned in Section 2. A typical plot of the recorded spectrum is shown in Fig. 7. The H_β and H_γ line intensities measured at the driver back plate location are seen to increase by a factor of 2.6 as the cos(phi) changed from 0.32 to 0.9. Further increase in the cos(phi) from 0.9 to 0.98 increases the line intensities by 25%. For all these measurements the set value of the power was 50 kW. The real power varied between 35 kW, low cos(phi) to 50 kW, high cos(phi). A similar increase is not observed for the H_α signal for reasons of signal saturation and needs to be improved by addition of neutral density filters in the next experimental campaign. The H_β and H_γ line intensities increase by a factor of ~2 as the applied power is increased from 50 kW to 80 kW. However the intensity of the H_α line does not increase with the applied RF power once again for the reason cited above.

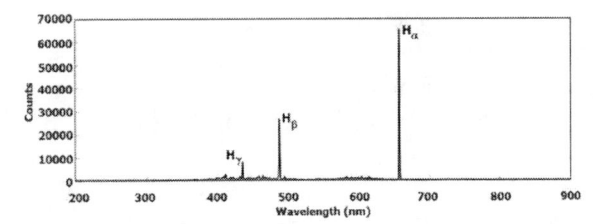

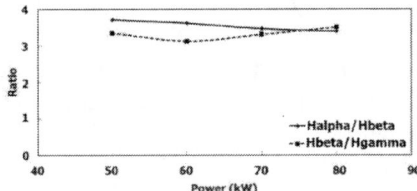

FIGURE 7 : A typical spectrum recorded during a plasma shot

FIGURE 8 : H_α/H_β and H_β/H_γ ratios as a function of RF power (diagnostic flange location)

The argument of saturation gains strength from the fact that similar measurement performed at the diagnostic flange shows the intensities for all the three lines to increase by a factor of 2 for RF power applied in the range between 50 – 80 kW. The H_α/H_β and H_β/H_γ ratios plotted as a function of the applied RF power are seen to remain the same, Fig. 8. Further the H_β/H_γ ratio at the diagnostic flange is lower 25% as compared to the one at the driver back plate. From the H_β/H_γ ratio the plasma density is estimated to be 5×10^{17} m^{-3} [5] for an assumed electron temperature (T_e) of 3 eV. The assumption is based on the experimental results performed at BATMAN under similar operating conditions [6,7].

DISCUSSION AND CONCLUSION

As shown above the best matching conditions were achieved for the case where one additional capacitor was added in the tank circuit of the RF generator. Also plotted in the figure is the frequency calculated using eq.1 where C corresponds to the series capacitance of the tuning circuit and L is the inductance of the coil and the cables connected to the matching circuit and measured as 11.8 μH.

$$F = \frac{1}{2\pi\sqrt{LC}} \qquad (1)$$

It is seen that for lower values of the series capacitance the frequency calculated using eq. 1 is much different from the frequency actually observed during the measurement. With the increase in the series capacitance, as one approaches optimal matching conditions, the observed frequency matches the calculated frequency within 0.8%. It may thus be concluded that when the tank frequency of the generator matches with the frequency due to the series capacitance of the tuning circuit and the inductance of the matching unit, the plasma resistance is transformed to a high value at the output of the generator, something necessary for the frequency stability.

The fact that a high circuit resistance is observed for best matching conditions under any generator configuration may corroborate with the maximum power transfer theorem [6]. The maximum transfer of power from the generator to the source may occur at 50 ohm but the efficiency of power transfer (ratio of the output power to the input power) attains a maximum at higher values of resistance.

Based on the above experimental observations it may further be concluded that for the present configuration of the RF generator the range over which the matching remains optimized is extremely narrow. Optimization of the series capacitance in the matching network helps to achieve the best matching while the variation of the shunt capacitance reduces the circuit resistance. Once established the matching parameters are independent of the applied RF power upto 50 kW. Higher powers require slight tuning of the matching circuit i.e adjusting C2 by ~ 40pF/10 kW. Tuning the tank circuit of the generator also helps to improve the matching conditions. The dependence of the frequency stability on the type of ferrites is still an issue. It worked in the set-up only with the grey ferrites, which have a higher µ value. But the µ value depends on the frequency, because the operation range is only 0.4 MHz. This dependence differs not only from type to type, but also from year to year of delivery, something experienced at IPP over the number of years. It should be mentioned that the poor frequency stability is a property of the used generator and not of the source. It is better with generator using two tetrodes, experience at the IPP testbeds, which are foreseen for the ITER sources. With such generators the matching is seen not to be sensitive to the type of ferrites or the turns in the coil.

SUMMARY

The single driver RF source without the extractor and accelerator system has been successfully commissioned in the IPR negative ion facility. The gas feed system is being improved to achieve a stable operating pressure in the source. The matching circuit is being moved closer to the RF coil of the source. This will help to reduce the inductance in the circuit. Measurements with this new arrangement and with a coil having larger number of turns shall be carried out in the next experimental campaign to see if the range over which a stable matching performance is achieved can be improved as compared to the present set up.

ACKNOWLEDGMENTS

The authors from IPR acknowledge the help and guidance from Mr Frank Fackert, Mr. Peter Pollner and Mr Peter Turba of IPP Garching during the training sessions at IPP. The help and support of the IPR workshop, power and water supply group in the setting up of the facility and during the experiment is duly acknowledged.

REFERENCES

1. A.K. Chakraborty et.al., *IEEE Transaction on Plasma Science* **38**, 248 (2010).
2. M B.J. Green et.al., *Plasma Physics Control Fusion* **45**, 687 (2003).
3. E. Speth et.al, *Nucl. Fusion* **46**, S220 (2006)
4. W. Kraus (private communication)
5. U. Fantz, et.al., *Nucl. Fusion* **46**, S297 (2006)
6. U. Fantz et.al. *Review of Sci. Instrum.* **77**, 03A516 (2006)
7. U. Fantz et.al. *New Jl. Of Physics* **8**, 301 (2006)

Cesium Delivery System for Negative Ion Source at IPR

G. Bansal[a], K. Pandya[a], M. Bandyopadhyay[b], A. Chakraborty[b], M.J.Singh[b], J. Soni[a], A. Gahlaut[a], K.G. Parmar[a]

[a] *Institute for Plasma Research, Bhat, Gandhinagar, Gujarat, India 382 428*
[b] *ITER- India, Institute for Plasma Research, A-29, Sector 25, GIDC, Gandhinagar, Gujarat, India*

Abstract. The technique of surface production of negative ions using cesium, Cs, has been efficiently exploited over the years for producing negative ion beams with increased current densities from negative ion sources used on neutral beam lines. Deposition of Cs on the source walls and the plasma grid lowers the work function and therefore enables a higher yield of H$^-$, when hydrogen particles (H and/or H$_x^+$) strike these surfaces.
A single driver RF based (100 kW, 1 MHz) negative ion source test bed, ROBIN, is being set up at IPR under a technical collaboration between IPR and IPP, Germany. The optimization of the Cs oven design to be used on this facility as well as multidriver sources is underway. The characterization experiments of such a Cs delivery system with a 1 g Cs inventory have been carried out using surface ionization technique. The experiments have been carried by delivering Cs into a vacuum chamber without plasma. The linear motion of the surface ionization detector, SID, attached with a linear motion feedthrough allows measuring the angular distribution of the Cs coming out of the oven. Based on the experimental results, a Cs oven for ROBIN has been proposed. The Cs oven design and experimental results of the prototype Cs oven are reported and discussed in the paper.

Keywords: Cesium, Negative Ion Source, RF Plasma
PACS: 52.50.Dg

INTRODUCTION

It is now well known that the surface production of the negative ions is enhanced in an ion source which is effected by lowering the work function of the source walls by depositing the thin layer of Cs onto them. Since the life time of negative ions is quite short in the ion source -in the range of few cm-, the negative ions produced near the extraction apertures have much higher extraction probability [1,2,3]. Therefore, the Cs is mainly deposited on the plasma grid so that once the negative ions are produced, they are immediately extracted by the extraction grids.

The optimum Cs layer thickness is 0.5 to 1 monolayer, which is about 2.4 to 4.8 × 10^{14} atoms/cm^2. The Cs deposition on the plasma grid should be as uniform as possible for the best performance of the negative ion source. During plasma off time, the Cs atoms coming out of the oven travel in the source and deposit on the source walls. Therefore, as a possibility, it is considered that if Cs is sprayed directly over the plasma grid, the plasma grid can be covered with Cs layer rather quickly and hence the source conditioning time can be reduced though it may increase the chances of the breakdowns between the grids. During direct spray of Cs onto the plasma grid, the angular distribution of Cs coming out of the oven will decide the uniformity of the Cs layer on the plasma grid.

A single driver RF based (100 kW, 1 MHz) negative ion source test bed, ROBIN, is being set up at IPR under a technical collaboration between IPR and IPP, Germany [4,5,6]. A 1MHz RF generator launches 100 kW RF power into a single driver on the plasma source to produce a plasma of density ~5 x 10^{12} cm^{-3}. Use of Cs enables delivery of a 10 A negative ion beam with a current density of ~30 mA/cm^2. The present set up shall allow for acceleration of beams to 35 kV. To characterize the distribution, a prototype configuration was built and tested to optimize the Cs delivery system for ROBIN.

PROTOYPE CESIUM OVEN DESIGN

Figure 1(a) shows a schematic of the prototype Cs oven. The Cs oven is designed by keeping the IPP's Cs oven as a baseline [7]. The oven is heated up to 150 – 250^{0}C using flexible heating elements controlled by PID temperature controllers. The Cs oven consists of three main parts: (1) liquid Cs reservoir; (2) Cs oven body; (3) delivery tube.

For testing purpose, a 1 g Cs ampoule was used. The whole Cs oven is encased in glass wool thermal insulation to avoid heat loss from the oven. The Cs reservoir is kept at the lowest temperature, which controls the vapor pressure of the Cs and therefore the Cs flux into the ion source. The other parts are kept at higher temperatures (typically 50 to 60^{0}C) to avoid any cold spots in the oven. All three parts of the oven have separate heating elements and thermocouples to precisely control the temperatures of each part.

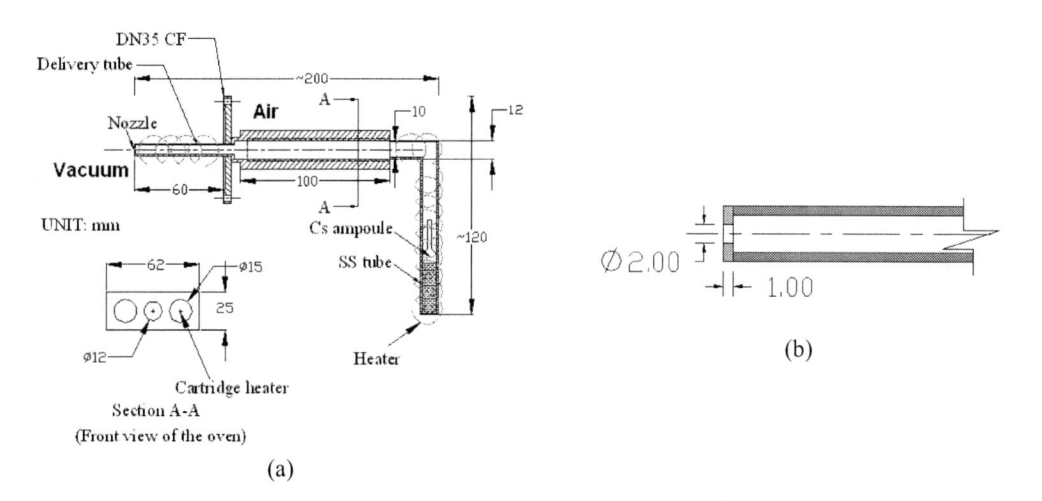

FIGURE 1. (a) Schematic of the prototype Cs oven at IPR; (b) nozzle of the Cs oven

The nozzle at the end of the delivery tube is shown in Fig. 1(b). Since the Cs oven in negative ion source does not require high collimation, a single aperture hole (2 mm diameter) nozzle is adopted.

THEORY

Cesium Flux

The intensity and collimation of the Cs atomic beam depends on the vapor pressure (which is governed by the coldest part of the Cs oven), mean free path, and size of the nozzle [8]. The vapor pressure [9] and mean free path of the Cs in the Cs oven is given by the following relation.

$$\log_{10}(P/133) = -4075/T - 1.45\log_{10} T + 11.38 \tag{1}$$

$$\lambda = \frac{1}{\sqrt{2}\pi\sigma^2 n_0} \tag{2}$$

where P, T, n_0, λ, σ are respectively, the vapor pressure, oven temperature, particle density, mean free path of Cs atom, and effective molecular diameter for Cs atoms collisions, which is taken from reference [10].

The total Cs flux coming out of the oven is given by

$$N = \frac{n_0 \bar{c} A}{4} \tag{3}$$

where A is the nozzle area and $\bar{c}$, the average particle velocity, is given by

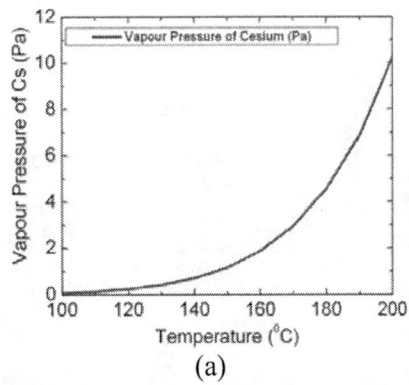

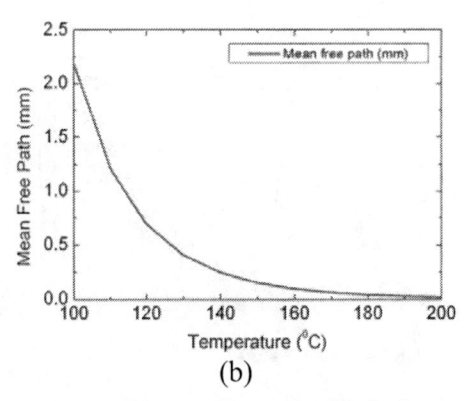

$$\bar{c} = \sqrt{2k_B T / m} \tag{4}$$

FIGURE 2. (a) vapor pressure of Cs as a function of temperature; (b) mean free path of Cs in the oven.

where k_B, m are the Boltzmann's constant and mass of a Cs atom, respectively. The n_0 is calculated by

$$n_0 = \frac{P}{k_B T} \tag{5}$$

Using above mentioned equations, the vapor pressure and mean free path of the Cs in the oven are plotted as a function of reservoir temperature. As shown in Fig. 2 (a), the vapor pressure of the Cs is a strong function of oven temperature. At 100^{0}C, it is less than 0.1 Pa and increases up to about 10 Pa at 200^{0}C. The typical operational pressure in negative ion source is about 0.3 Pa, which sometimes may go up to ~1 Pa. Hence, the operating temperature of the Cs oven should be equal to or more than 150^{0}C to keep a positive pressure in the Cs oven during the ion source operation.

Angular Distribution

The angular distribution of intensity is also calculated for the Cs oven using expressions in ref [8]. Since the mean free path of the Cs atoms at 150^{0}C or higher temperatures is smaller compared to nozzle dimensions, the Cs flow is in viscous flow regime at typical operating temperature of 150^{0}C or higher. The calculated and experimentally measured angular distribution of Cs intensity at different temperatures are compared and discussed in Section 4.

EXPERIMENT

Experimental Setup

Figure 3 shows the schematic of the experimental setup. The vacuum vessel was evacuated by a diffusion pump of a pumping speed of 1000 l/s.

The pressure in the vessel was obtained in the range of 2×10^{-6} - 5×10^{-6} mbar. At this pressure the mean free path of the particles inside the vacuum vessel is larger than its dimensions and therefore collisional effects are avoided.

In the present set up, the Cs reservoir was made of a stainless steel tube which could house a 1 g Cs ampoule.

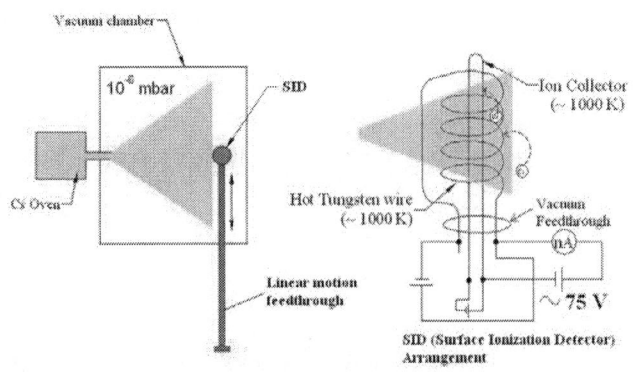

FIGURE 3. Experimental arrangement to measure Cs out flux and angular distribution.

The Cs ampoule was broken by pinching the stainless steel tube from outside. The Cs reservoir was kept at a temperature of 150 – 190^{0}C. All other parts of the oven were kept at a temperature ~50^{0}C higher than the container to avoid any cold spot in the oven. The PID temperature controllers are used to control the temperatures of the various oven parts. A surface ionization detector, SID, has been installed in the vacuum vessel to measure the Cs ion current. The SID is kept at an angle of about 40^0

from the nozzle axis and the perpendicular distance between nozzle and SID is about 50 mm. A hot tungsten filament (~1000°C) of SID ionizes the Cs atoms striking on it whereas other tungsten filament collects the Cs ions and thereby Cs ion current is measured. The Cs ion current was measured by a pico ammeter (Keithley Model no. 6487) which was in micro-ampere range. The bias voltage between ionizing and collector filaments was kept about 75 V. By measuring the Cs ion current, the Cs flux output at a particular Cs reservoir temperature is measured. Also, the SID is attached to a linear motion feedthrough which allows moving the SID without breaking the vacuum. By moving the SID, the angular distribution of the Cs flux output was measured.

Experimental Results

The Cs flux intensity was measured by measuring the Cs ion current. The SID was kept at a fixed location throughout the experiment. The Cs reservoir was kept at 150°C most of the time except for about 10% of the total time at different temperatures when temperature effects on the Cs flux intensity were measured. The 1 g Cs was exhausted in about 85 hours which gives an average Cs out flux of about 11.76 mg/hr. The calculated and experimentally measured Cs flux at 150°C are shown Fig. 4(a).

The overall size of the plasma grid in ROBIN negative ion source is ~231 mm (width) × 494 mm (height) i.e. the cross-sectional area is ~ 1141 cm^2. Therefore, the required number of Cs atoms to cover the whole area (for monolayer) of the plasma grid would be ~4.8 × 10^{14} × 1141 (or ~ 5.5 × 10^{17}). Figure 4(a) shows that the typical Cs evaporation rate is 11.76 mg/hr. at 150°C which is equivalent to 1.47×10^{16} atoms/s.

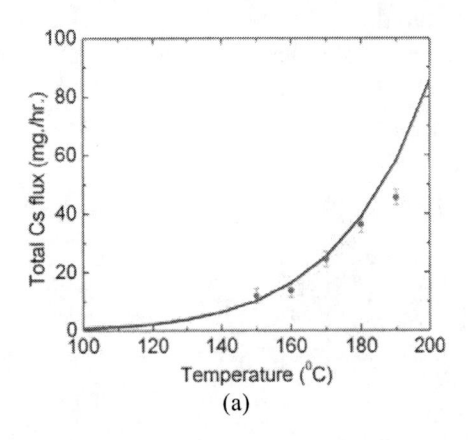

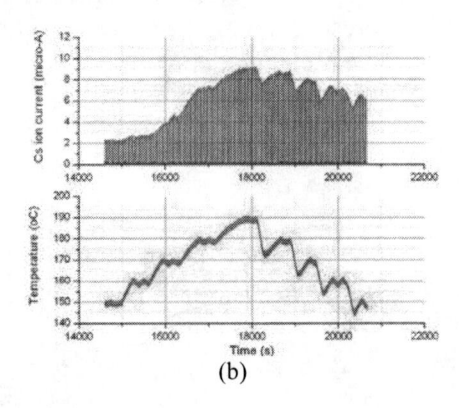

FIGURE 4. (a) Calculated Cs out flux from the oven. The solid circles are experimentally observed data.; (b) Cs flux at different reservoir temperatures.

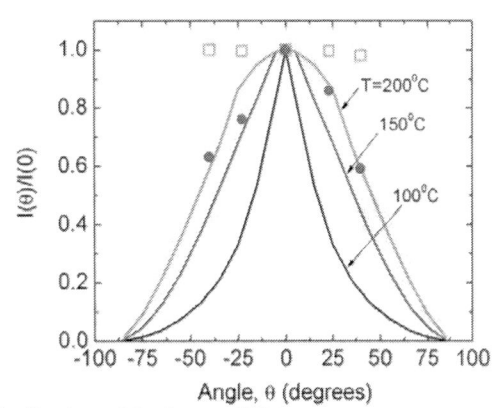

FIGURE 5. Angular distribution of Cs flux at different oven temperatures. Solid circles and hollow squares are experimental data points at 150°C and 180°C, respectively.

Therefore, if the Cs oven is operated at 150°C, the necessary time to form a monolayer of Cs atoms on the plasma grid would be 55/1.47 or ~37 seconds assuming 100% efficiency. The other data points are derived from Fig. 4(b) where Cs reservoir temperature was increased stepwise and Cs ion current was measured. The Cs flux is considered to be in the same ratio as the Cs ion currents at different temperatures. The experimental data are in good agreement with calculated Cs flux near 180°C, however, far from 180°C, there is some deviation. In the next experimental campaign, the SID is planned to be calibrated with a 10 mg. Cs dispenser so that precise Cs flux from the oven can be measured at different temperatures. It should be noted that in Fig. 4(b), a modulated Cs ion current was measured by switching ON and OFF the bias voltage and switching OFF and ON the collector filament heating to reduce the SID sensitivity effect which was found to change due to Cs coverage on the collector filament. Therefore, the peak values of the ion current should be considered.

The angular distribution of Cs flux intensity was measured by moving the SID linearly attached with a linear motion feedthrough as shown in Fig. 3. For angular distribution measurements, a miniaturized SID (~6 mm long x 6 mm diameter) was used for precise measurements. These measurements were done in second experimental campaign as the first campaign the SID was larger in size therefore precise angular distribution measurement was not possible. The experimental data are plotted in Fig. 5 along with the calculation results. The calculations show that at lower temperatures, the Cs flux is confined within a narrow band. However, at larger temperatures, the distribution starts to become like a Maxwellian distribution and Cs is spread over a wide angle. The measured angular distribution has the similar trend at 150°C as the calculated distribution. However, at 180°C or higher temperatures, it deviates from the calculated distribution. Figure 5 suggests that it might be better to operate Cs oven at higher temperatures for depositing Cs on the larger area of the plasma grid. The detailed study of the angular distribution is being carried out which will be presented elsewhere.

Other Observations

There are several other experimental observations during the Cs oven characterization.

Desorption from Adsorbed Layer in Cs Reservoir

As shown in Fig. 6, a rapid rise in Cs ion current was observed when the Cs reservoir was started heating up, and then it comes down and starts rising again gradually with temperature rise. This behavior was observed every day after the first day operation of the oven. Initially, it was considered that this might be due to the accumulation of some Cs in other parts of the oven which comes out first and shows a rapid rise in current. However, to confirm it, during next day operation, the other parts of the oven (except Cs container) were first heated up to the desired temperature of 200^0C but no Cs ion current was observed. However, once the Cs reservoir was started heating up, the same behavior was observed as shown in Fig. 6.

This might be due to desorption of Cs from the Cs reservoir wall above the liquid Cs. During Cs desorption, the ion current rises and after the desorbed Cs is exhausted quickly, the evaporation from bulk Cs contributes to the ion current.

Change in SID Sensitivity and Its Recovery

The change in SID sensitivity was also observed. As shown in Fig. 6, even though the reservoir temperature stabilizes at 150^0C, the Cs ion current keeps rising. This might be due to the Cs compounds formation on the ion collector which disturb the Cs ion current measurements. Similar behavior has been reported in previous experiments [7]. To confirm this, the ion collector filament was heated up to about 1000^0C for 5, 10, and 15 minutes to remove the Cs compounds and then Cs ion currents were measured. Every time, the collector filament was heated, the ion current was reduced which was probably due to the cleaning of the filament. For better results, the longer duration heating (typically 20 minutes) was required.

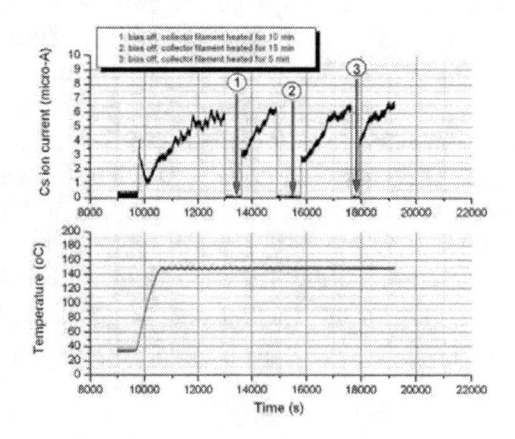

FIGURE 6. Cs ion current (or Cs flux) with increasing Cs reservoir temperature. Cs ion current changes due to the change in SID sensitivity.

Hysteresis in Cs oven

The hysteresis phenomenon in the Cs oven was also observed. Figure 7 shows the experimental results. In Fig. 7(a), the Cs reservoir was heated upto 180^{0}C and then heaters were switching off. The Cs reservoir was allowed to cool naturally. Similarly, in Fig. 7(b), the reservoir was heated upto 120^{0}C and then was cooled naturally. From Fig. 7, it is observed that higher the reservoir temperature, higher the hysteresis. This observation suggests that the reduction in Cs flux is not as quick as the temperature of the oven drops down and therefore the control of the Cs flux by the controlling the temperature may not be as good as expected.

Cs Getter Effects

During the experiments, the getter effect of Cs was also observed. For precise angular distribution measurements, before a tiny SID was used, a cylindrical envelope of stainless steel (42 mm dia. × 80 mm long) housed the larger SID. The envelope was having a hole of 8 mm diameter in the line of sight of Cs oven. The idea was to restrict the Cs reaching SID from all other unwanted directions. After getting a reasonably good vacuum in the vacuum vessel, the Cs oven was started. As soon as the oven was started, the pressure of the vessel was dropped. This pressure drop is attributed to the getter effects of the Cs. The experimental observation is shown in Fig. 8.

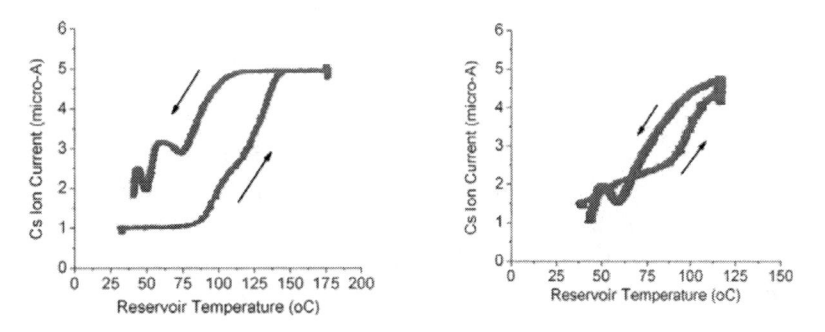

FIGURE 7. Hysteresis behavior of the Cs oven **(a)** Cs reservoir was heated upto 180^{0}C; (b) Cs reservoir was heated upto 120^{0}C.

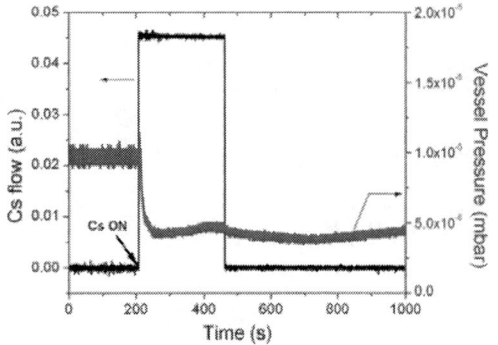

FIGURE 8. Vacuum vessel pressure with Cs flow. Vessel pressure drops as a result of getter effect of Cs.

The idea of envelope over the SID was however, abandoned due to increased outgassing from the envelope after heating from the SID filaments, probable recirculation of the Cs inside the envelope and therefore affecting the measurements, etc.

PROPOSED CS OVEN FOR ROBIN SOURCE

Based on the experimental results obtained from prototype Cs oven, a Cs oven for ROBIN has been proposed. Figure 9 shows a schematic of the proposed Cs oven. The oven shall be mounted on the back plate of the ion source and heated up to 150 – 250^{0}C using flexible heating elements controlled by PID temperature controllers. The Cs oven consists of three main parts: (1) liquid Cs reservoir; (2) Cs oven body; (3) delivery tube. A Cs inventory of 10 g in the Cs reservoir is considered to avoid frequent Cs refilling. A remotely controlled all metallic angle valve is incorporated above the Cs reservoir to avoid the contamination of the Cs in case of fault conditions or venting of the source. An all metallic manual valve (not shown in the figure) is envisaged between the Cs reservoir and angle valve which is useful during the refilling of the Cs reservoir under argon gas environment in a glove box. The whole Cs oven is encased in glass wool thermal insulation to avoid heat loss from the oven.

The Cs reservoir is kept at the lowest temperature, which controls the vapor pressure of the Cs and therefore the Cs flux into the ion source. The other parts are kept at higher temperatures (typically 50 to 60^{0}C) to avoid any cold spots in the oven. All three parts of the oven have separate heating elements and thermocouples to precisely control the temperatures of each part.

There are several advantages of the proposed Cs oven.
1. All metallic valves allow dismantling the Cs reservoir easily without contaminating the remaining Cs in the Cs container during the maintenance of the oven or refilling the reservoir.
2. The liquid Cs is stored separately into a reservoir and only Cs vapors come into the oven body and delivery tube. Therefore, there are reduced chances of delivering the Cs spurt into the source.
3. The Cs reservoir is at the lowest location and therefore during cooling down of the oven, almost all the condensed Cs goes back into the reservoir. Hence, all the Cs is used effectively.

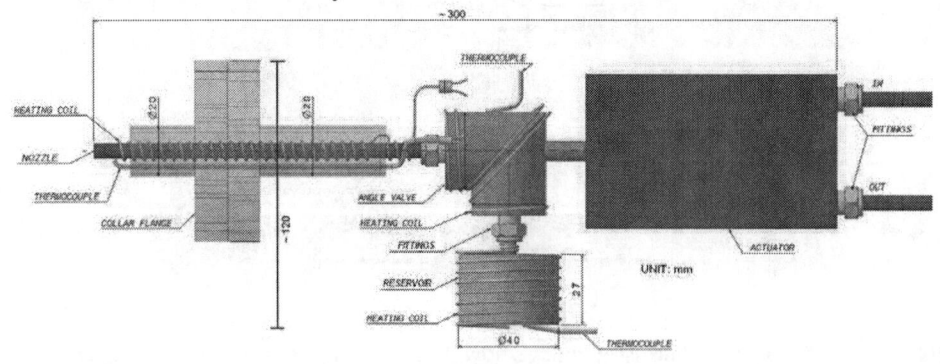

FIGURE 9. Schematic of the proposed Cs oven for negative ion source at IPR

SUMMARY

A single driver RF based negative ion source, ROBIN, is being set up at IPR under a technical collaboration between IPR and IPP, Germany. The characterization of the prototype Cs oven has been carried out. The Cs out flux and angular distribution have been measured and have been found generally in good agreement with the calculations. Other observations like change in SID sensitivity, Cs oven hysteresis, getter effects of Cs have also been made.

Based on the experimental results of prototype Cs oven, a Cs oven with 10 g Cs inventory has been proposed for ROBIN source.

As a next step, it is planned to characterize the Cs oven with source like plasma for better understanding of the Cs dynamics when source is operated.

REFERENCES

1. M. Bandyopadhyay et. al., *Rev. of Sci. Inst.* **75**, 1720 (2004).
2. O. Fukumasa et. al., *Nucl. Fusion* **46,** S275 (2006).
3. E. Speth et.al, *Nucl. Fusion* **46**, S220 (2006).
4. G. Bansal et. al., *Journal of Physics: Conf. Series* **208** 012060 (2010).
5. A. Gahlaut et. al., *Journal of Physics: Conf. Series* **208** 012030 (2010).
6. M.J. Singh et al., *this conference*.
7. M. Froschle et. al., *Fusion Engineering and Design* **84**, 788 (2009).
8. J. A. Giordmaine et al., *J. Appl. Phy.* **31**, 463 (1960).
9. T. Ikegami, *Jpn. J. Appl. Phy.* **33**, 4795 (1994).
10. I. Estermann et. al., *Physical Review* **71**, 250 (1947).

Conceptual Design, Implementation and Commissioning of Data Acquisition and Control System for Negative Ion Source at IPR

Jignesh Soni[a], Ratnakar Yadav[b], A. Gahlaut[a], G. Bansal[a], M.J.Singh[b], M. Bandyopadhyay[b], K G Parmar[a], K. Pandya[a] and A. Chakraborty[a]

[a] *Institute for Plasma Research, Bhat, Gandhinagar (Gujarat) 382 428 India.*
[b] *ITER-India, Institute for Plasma Research, Gandhinagar (Gujarat) 382 025 India*

Abstract. Negative ion Experimental facility has been setup at IPR. The facility consists of a RF based negative ion source (ROBIN) – procured under a license agreement with IPP Garching, as a replica of BATMAN, presently operating in IPP, 100 kW 1 MHz RF generators and a set of low and high voltage power supplies, vacuum system and diagnostics. 35keV 10A H- beam is expected from this setup. Automated successful operation of the system requires an advanced, rugged, time proven and flexible control system. Further the data generated in the experimental phase needs to be acquired, monitored and analyzed to verify and judge the system performance. In the present test bed, this is done using a combination of PLC based control system and a PXI based data acquisition system. The control system consists of three different Siemens PLC systems viz. (1) S-7 400 PLC as a Master Control, (2) S-7 300 PLC for Vacuum system control and (3) C-7 PLC for RF generator control. Master control PLC directly controls all the sub-systems except the Vacuum system and RF generator. The Vacuum system and RF generator have their own dedicated PLCs (S-7 300 and C-7 respectively). Further, these two PLC systems work as a slave for the Master control PLC system. Communication between PLC S-7 400, S-7 300 and central control room computer is done through Industrial Ethernet (IE). Control program and GUI are developed in Siemens Step-7 PLC programming software and Wincc SCADA software, respectively. There are approximately 150 analog and 200 digital control and monitoring signals required to perform complete closed loop control of the system. Since the source floats at high potential (~ 35kV); a combination of galvanic and fiber optic isolation has been implemented. PXI based Data Acquisition system (DAS) is a combination of PXI RT (Real time) system, front end signal conditioning electronics, host system and DAQ program. All the acquisition signals coming from various sub-systems are connected and acquired by the PXI RT system, through only fiber optics link for signal conditioning, electrical isolation and better noise immunity. Real time and Host application programs are developed in LabVIEW and the data shall be stored with a facility of online display of selected parameters. Mathematical calculations and report generation will take place at the end of each beam shot. The paper describes in detail about the design approach, implementation strategy, program development, commissioning and operational test result of ROBIN through a data acquisition and control system.

Keywords: DACS, ROBIN, Negative Ion, PXI, PLC, control, data acquisition, plasma source
PACS: 52.70.-m , 52.70.Ds

INTRODUCTION

A RF based negative ion source test bed has been setup at Institute for Plasma Research (IPR), in March 2010. The facility consist an inductively coupled RF based negative ion source, ROBIN – procured under a license agreement with IPP Garching. The ROBIN ion source is similar to BATMAN source, which is presently operating at IPP. The test bed at IPR consists of a single RF driver based beam source (ROBIN) coupled to a 100 kW, 1 MHz RF generator, with a self excited oscillator, through a matching network, for plasma production and ion extraction and acceleration. The system is supported by set of various low and high voltage power supplies, Cs Oven system, hydraulic system, Gas feed system, a vacuum chamber housing a calorimeter, a set of turbo molecular and cryo pump, optical and electrical probe based diagnostics and an in-house developed automated Data acquisition and control system (DACS).

The pictures of - ROBIN source and its various sub-systems are shown in Fig. 1 and 2 respectively. To achieve continuous and high performance operation of the source, a powerful DACS was required to be designed and constructed. A carefully

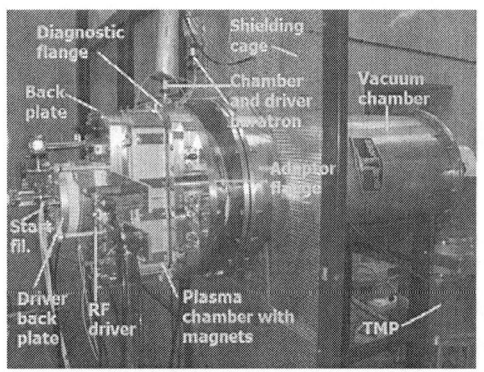

FIGURE 1. ROBIN source.

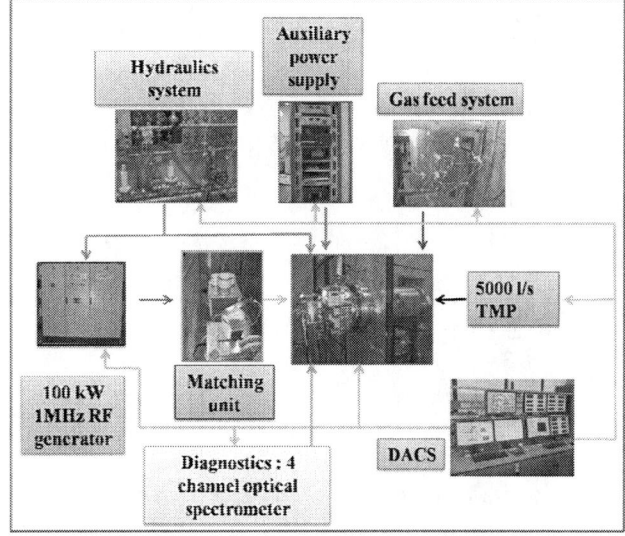

FIGURE 2. Various sub-system of ROBIN.

designed DACS will meet the experimental requirements and provide the physicist/operator with a friendly graphical-user interface.

BASIC REQUIREMENT AND SELECTION CRITERIA OF DACS FOR ROBIN

The DACS is the key for the successful operation of ROBIN experimental setup. It is essentially the brain of the system, which co-ordinates with various sub-systems, based on some pre-defined protocols. Experimental systems and their DACS requirements are shown in Table 1. The DACS will be operated in High voltage, noise prone and RF environment; and as the source essentially floats at a high (~ 35kV) potential, a proper signal isolation and communication mechanism also needs to be implemented. Moreover, the implemented control and acquisition electronics must comply with the RF radiations from the source, and hence poses another critical requirement. These decide the configuration for control and monitoring strategy.

To meet the system requirements based on the above constraints, a selected DACS should be:

(1) Advanced, rugged, flexible, reliable, scalable and optimized
(2) High precision, capable of high speed data acquisition, network communication supported
(3) Technically compatible with the existing DAC system at IPP
(4) EMI and RFI compatible

TABLE 1. Experimental systems and DACS requirements.

System	Distributions and Specification	Control signal requirements Analog	Control signal requirements Digital	Acquisition requirements Signal	Acquisition requirements Sampling rate
ROBIN	SPECIE : Hydrogen Energy : 35 keV Current :10 A Plasma density ~ 5 x 10^{12} cm^{-3} Current density ~ 30 mA/cm^2	5	-	11	50 ms
RF Generator	100 kW, 1 MHz	4	27	8	10 ms
Auxiliary power supply	Filament bias: 90V, 1A Heater: 16V, 10A Grid heating : 33V, 80A Grid bias : 60V, 10A	8	8	2	2 ms
Hydraulics	130 lpm @ 5.4 bar	35	12	35	200 ms
HVDC supply	(1) Extractor: 11kV, 35A (2) Accelerator: 35kV,15A	4	14	4	10ms
Gas system	0.3 – 1.5 Pa	3	35	1	10ms
Vacuum system	Speed =5000 l/s	7	18	4	50ms
Cs Oven system	---	5	5	1	200ms
Calorimeter	---	-	-	49	200ms
Spectrometer & other diagnostic	Plasus, 200 nm to 1100 nm, < 1.5 nm resolution	24	17	4	10 ms

Various schemes (PLC based, PXI based, VME or other system based, mix of PLC-PXI, Distributed vs. separated DACS and control) have been explored, based on the available technologies, their compatibility with existing system at IPP and operational requirements. A PLC (Siemens) based control system and a PXI (PCI eXtensions for Instrumentation) based data acquisition system (DAS) have been chosen for the ROBIN DACS because the PLC is industry proven and a widely used reliable and robust system; and PXI have capabilities to handle high speed data acquisition because of its backplane PCI bus having a bandwidth of 132 Mbytes/second, advanced timing and synchronization feature and PC based advanced hardware.

DATA ACQUISITION AND CONTROL SYSTEM (DACS)

Structure Of The DACS

The DACS of ROBIN is constructed using (1) PLC (Siemens) control system, (2) PXI (National Instruments) data acquisition system (DAS) and (3) Front end signal conditioning electronics. The structure of the DACS system is shown in Fig. 3 and a picture of the ROBIN control room having workstation 1 & 2, DVR system and various monitors, is shown in Fig. 4. The system is developed to provide an integrated control of each subsystem, which performs various functions. These functions involve automatic initialization and operation of various subsystems depending upon the various operating modes and timing sequence, setting of experimental values, providing safety protection for devices such as the ion source, TMP and Cryo pumps, RF generator and water cooling system; data acquisition, online monitoring, saving, transfer and analysis; providing proper front end-signal conditioning electronics and communication interface to each subsystem.

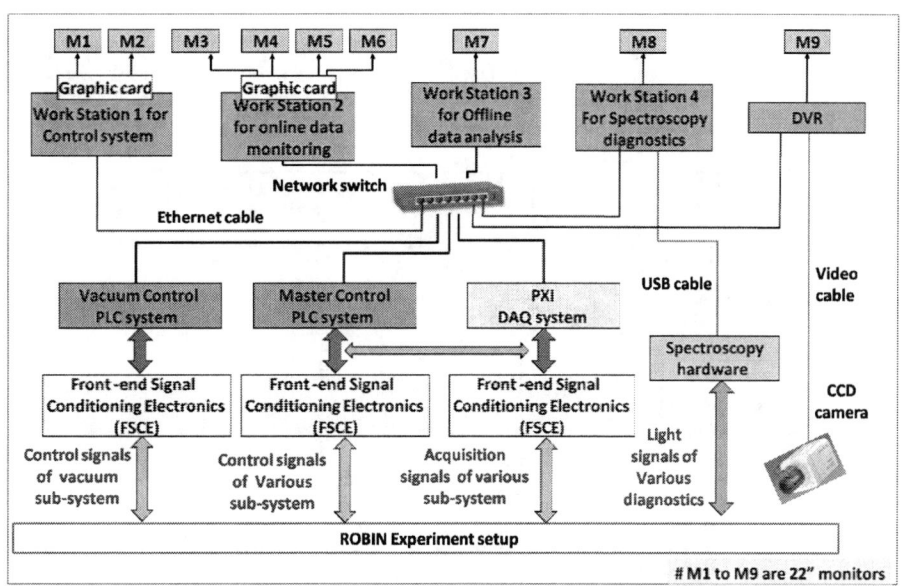

FIGURE 3. Structure of DACS.

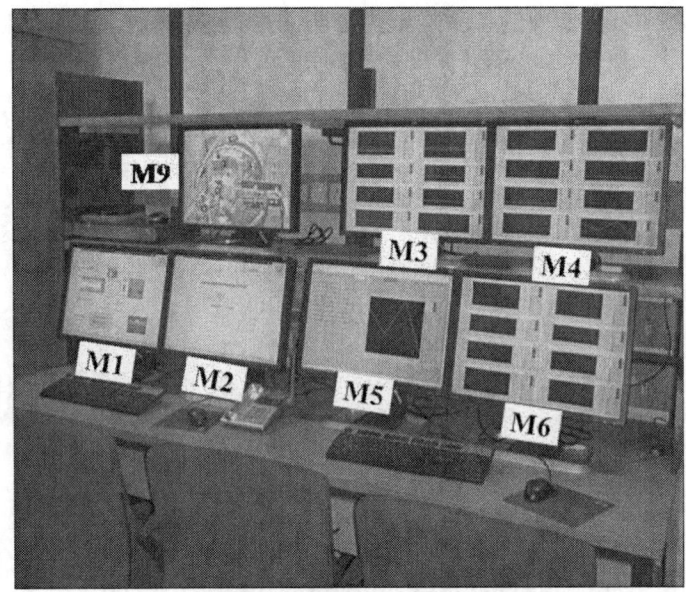

FIGURE 4. ROBIN control room.

Control System

The control system is employed using three PLC (Siemens) systems (1) S-7 400 system, (2) S-7 300 PLC system and (3) C-7 PLC system. The Structure of the control system is shown in Fig 5. The S-7 400 PLC system is constructed by CPU 414-2 DP, communication process CP- 443-1, Interface modules for two tier rack configuration, power supply modules and different analog and digital input/output modules. This

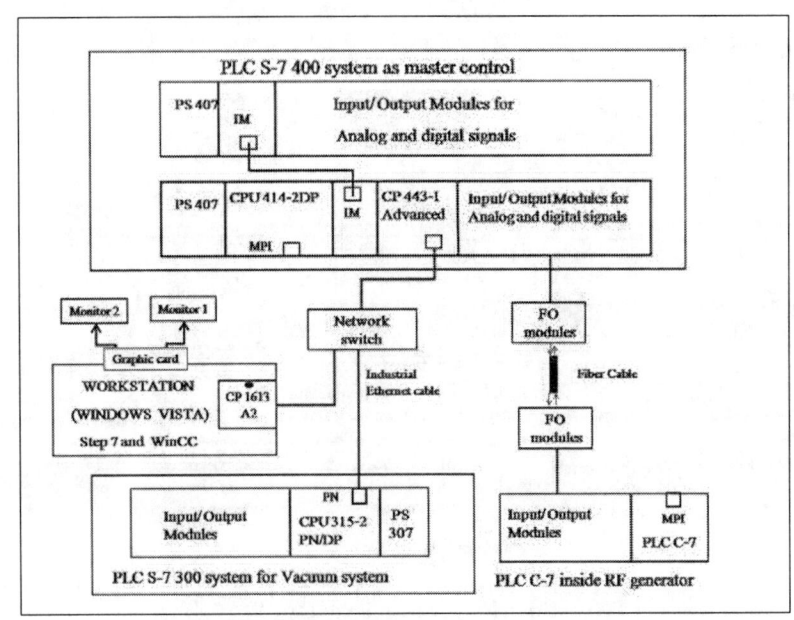

FIGURE 5. Structure of control system.

PLC system works as a master control for all the sub-systems and other two PLC systems. For controlling the vacuum sub-system, separate S-7 300 PLC system is employed. This system is constructed by CPU 315-2 PN/DP, power supply modules and various analog and digital input/output modules. RF generator is equipped with its own C-7 PLC system (supplied along with the generator) for the self-control and protection. The C-7 PLC communicates with master PLC through various control signals via fiber optics links to avoid ground loops and better noise immunity. Communication of the master control PLC, vacuum PLC system and central control room workstation is through a reliable and rugged Industrial Ethernet network. The control system workstation is playing a major role in the operation of ROBIN. It provides interface between operator/physicists and control system. In DACS the role of the control system is to provide sequential ON/OFF control and protection of the ion source and the various sub–systems. PLC Control program is developed using FBD (Functional Block Diagram) language in Siemens Step-7 professional tool that brings fast and error free programming facility and reduces malfunctions. User-friendly control GUI (Graphical User interface) is developed in Siemens Wincc SCADA (Supervisory Control and Data Acquisition) software. There are four different operating modes for ROBIN as detailed below:

a) **Standby mode:** In this mode all the sub-systems except vacuum system can be switched automatically from ON state to OFF state. When the automatic and proper shutdown of all the sub-systems is required, then this mode is used

b) **Single operating mode:** In this mode all the sub systems can be individually control and monitored.

c) **Pulse without beam mode:** This mode is used for conditioning of source before beam operation. Pulse timer is activated. For mode there are three different sub modes:
   (i) **Only filament:** In this mode only filaments power supply is turned ON during the Pulse. Only Filament light is observed in this mode.
   (ii) **Only Gas:** In this mode only gas feeding into the source during the pulse can be tested.
   (iii) **Only RF**: In this mode Plasma is generated but beam cannot be extracted and accelerated. All the sub-systems (except HV power supply) can be automatically switched ON and OFF for particular time during the shot.

d) **Beam Mode:** In this mode plasma can be generated. Beam can also be extracted and accelerated. All the sub-systems are controlled automatically during the shot as per the timing sequence.

Control GUI of only RF (plasma production) mode with timing details is shown in Fig.6. During operation if an exceptional condition or an interrupted input occurs, the PLC can stop the shot immediately and switch off all the Power supplies and other required sub-systems. The system is also providing start and stop triggers to the DAS system. The system can handle approximately 150 analog and 200 digital control and monitoring signals, required to perform complete control of the system.

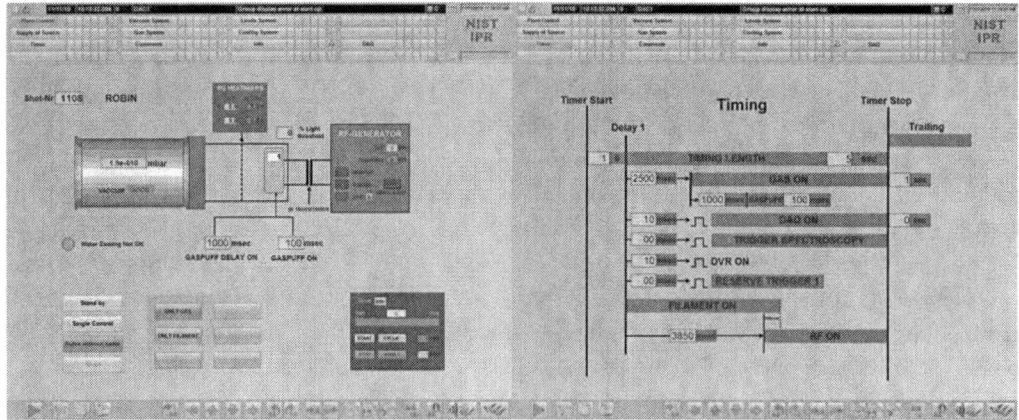

FIGURE 6. Control GUI for only RF (Plasma production) mode.

Data Acquisition System (DAS)

The Data acquisition system of the ROBIN test bed consists of PXI (National Instruments) based acquisition Hardware; LabVIEW based data acquisition (DAQ) software and workstation computers. The DAS is used for real time data acquisition, online monitoring, storage, backup and post analysis of 130 essential parameters like Voltage, Current, Frequency, RF Power, Temperature, Pressure, Flow, Plasma and beam density etc of the ROBIN Source and various sub-systems in order to verify and judge the system performance; and ensure its healthy operation. DAS system is dividing into two parts as mention below.

Hardware Architecture

The DAS hardware mainly employs **(1) PXI RT system:** having a NI PXI 1042 Chassis, NI PXI 8108 RT controller, NI PXI 6225 and NI PXI 6254 M-Series digitizer cards; and **(2) Host PC (Workstation):** having a Intel Core 2 Duo E 8500 Processor and NVIDIA Quadro NVS 450 graphic card for setting up multiple (4 nos.) displays for simultaneous online monitoring of all the acquisition channels. Connection of the RT system and host PC is through Ethernet. Data exchange between DAS and control system is done through Ethernet and NI OPC server.

Software Architecture

The DACS software developed in LabVIEW 2009 language have following features:
- Modular in design, flexible, reliable, robust and capable for future expansion.
- Perform complex algorithm on the wide range of incoming data.
- Reduce human interference in data management and handling.

Data acquisition software is mainly divided in two parts;
a. Real Time data acquisition module
b. PC based host application module

Both the modules run on the two different operating system platforms, LabVIEW RTOS (Real Time Operating System) and Windows Vista respectively. The real time data acquisition module demands required field data, processes as per the test requirements and configuration; and temporary store in RT system during the shot. After completion of the shot, RT system sends the stored file to the Host PC through TCP/IP. This method provides lossless acquisition of the data in the host PC. Host software provides graphical user interface to the system. The data acquisition software has following main facility:

(A) Configuration: The software has very good flexibility of user configuration for future requirements. In the configuration, operator can configure the number of channels to be acquired (out of 130 nos.) and data logged interval of the every channel.

(B) Data acquisition of 130 channels: The data acquisition has made on Real time operating system of PXI RT and it transferred data file to the host PC after the shot. Host application stores the data in binary file format. In addition, to that various calculations can be taken place in host application after data acquisition.

(C) Graphical User Interface (GUI): The software has very excellent and user friendly GUI. It's divided into two parts:
  (a) Online Graphs: The host PC has four different display screens (22" monitor). Four different screens show online and simultaneous channel graphs data and some numeric values indications of 130 channels. Number of graphs and List of channels that are plot in graphs is user selected.
  (b) Configuration, data management and backup GUI: Through this GUI user can generate configuration file and shot wise report. It also takes data backup in binary or ASCII format in various media.

(D) Data storage and Report generation: The host application stores the acquired data in such a manner that it can be utilize later on with minimum complexity. After completion of every shot, average means values (during the shot) of all the selected channels are stored in one master excel file (one file per day) and it is appended after every shot. The Software can generate 4 different types of test report in pre-defined formats in MS-Word.

(E) User group: Software has three different types of user group as mention below depending upon the access rights.
  (a) Administrator: Administrator has all rights to select/change channels as well as sampling rate and data logged interval in configuration file.
  (b) Supervisor: Supervisor can select channel to view the data acquisition but cannot select/change sampling rate and data logged interval.
  (c) User: User can only view the acquired data

Front-end Signal Conditioning Electronics

The ROBIN and its sub-systems are divided into two categories (1) Category 1: Sub-systems floated at high voltage (~ 35 KV) potential and (2) Category 2: Sub-systems connected at ground level. For control and data acquisition of the ROBIN system, various types of signals as mention below in table 2, are required to be generated and communicated between DACS and the sub-systems.

TABLE 2. Various signals and their level in ROBIN system.

Parameter	Types of the signal	
	Analog	Digital
Voltage level	0-10V, 0-20mV	0 & 24 V
Current level	4-20 mA, 0-20 mA	N.A
Signal frequency	(1) Fast signal > 1KHz	
	(2) Slow signal ≤ 1KHz	
Isolation required	(1) Category 1 sub systems : **35 kV**	
	(2) Category 2 sub systems: **2.5 kV**	
Signals	Temperatures, Pressure, Hydraulic flows, Voltage, Current, RF power etc.	Solenoid valves, RF and Power supplies ON/OFF signals; Various interlock signals etc.

Two types of signal isolation scheme for the front-end signal conditioning electronics, are implemented which are - (1) Fiber optics isolation for category 1 sub-systems and (2) Galvanic isolation for category 2 sub-systems.

EXPERIENCES DURING INTEGRATION AND COMMISSIONING OF THE DACS

Integration and commissioning of the indigenously developed DACS has been done successfully in the ROBIN test bed. During the commissioning various tests like functionality, stability, reliability and consistency of the system has been checked first with the dummy signals. After this the system was integrated in the ROBIN test bed. During the testing various software bugs like TCP/IP communication errors between Host PC and PXI RT, interlock, coding errors etc. were found and rectified. Campaign of RF Generator (RFG) commissioning in ROBIN tested bed has been done successfully with the DACS in October 2009. During that campaign the entire acceptance test of the RFG was done with required operation configuration i.e Modulation (3 Second ON and 20 second OFF) of RF power with 5 Hz notch, with the DACS in very reliable, fast and stable way. The data plots of the modulated RF power are shown in Fig. 7.

Further in the recent campaign of commissioning and plasma production phase of ROBIN (without grid assembly) has been completed successfully with the DACS in March 2010. In this campaign, complete integrated operation of the source with all the required protection, interlock and diagnostic signals has been realized through DACS. Noise pick-ups due to RF radiation were noticed in some signals during the plasma shot, in periphery of source. It was rectified by putting ferrite beads in the signal paths and by improving upon the shielding.

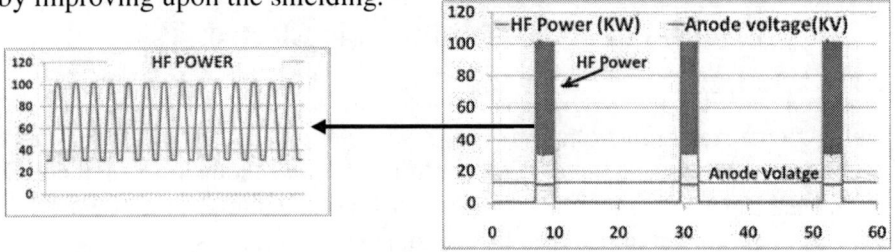

FIGURE 7. RF power modulation (3 Second ON, 20S OFF) with 5 HZ notch.

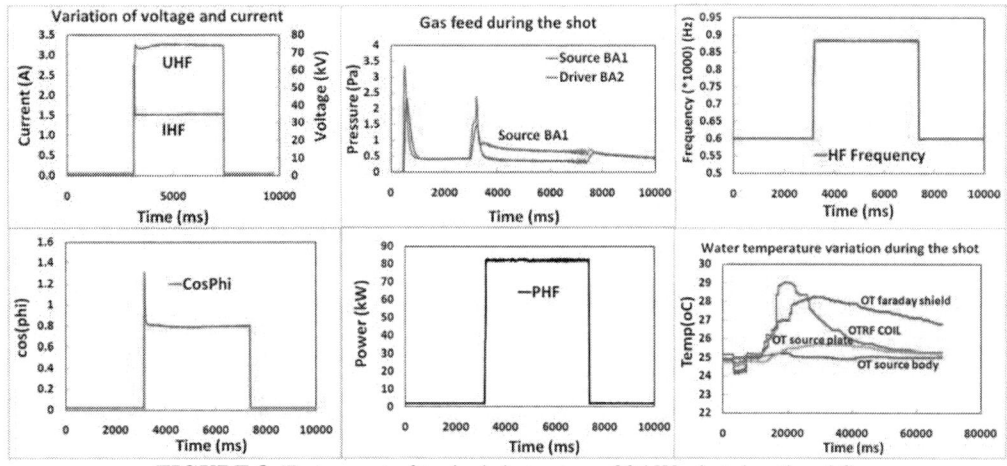

FIGURE 8. Test report of typical shot; power 80 kW; shot duration 4 S.

Loop time of the Control program has been measured and it was found to be maximum 2 ms only which prove that the control is very fast as required. Various experimental data has been acquired and stored in the DACS. A test report of the typical plasma shot of the ROBIN is shown in Fig. 8.

SUMMARY

The DACS has been conceptualized, implemented and commissioned in the ROBIN test bed at Institute for Plasma Research for first plasma production phase of the experiments. The system has worked reliably and successfully during acceptance tests, commissioning and operation of the RFG and ROBIN source (without grid assembly). The control algorithm and interlocks were tested. Real time data has been acquired, monitored online and stored in binary format. Proper front end signal conditioning electronics has been conceptualized and implemented in the DACS. Further expansion of the DACS will be implemented for second phase operation of the ROBIN for extraction and acceleration of the negative ion beam. Web-based DAS GUI will be developed for remote access of the ROBIN database for post analysis.

ACKNOWLEDGMENTS

The authors of the paper acknowledge the help and guidance received from the Negative NBI team at IPP, Garching, for conceptual the design of DACS and M/s Optimize solutions, Ahmedabad, India for contribution in software development in LabVIEW for the DAS.

REFERENCES

1. M.J. Singh et.al., *this conference* (2010).
2. G. Bansal et. al., *this conference* (2010).
3. G Bansal et. al, *J. Phys.: Conf. Ser.* 208 012060 (2010).
4. Agrajit Gahlaut et. al, *J. Phys.: Conf. Ser.* 208 012030 (2010).
5. Wang Yongjun,et. al, *Plasma Sci. Technol.* 7 2822 (2005).

Computer System for Unattended Control of Negative Ion Source

P.V.Zubarev, A.D.Khilchenko, A.N.Kvashnin, D.V.Moiseev, E.A.Puriga, A.L.Sanin and V.Ya.Savkin

Budker Institute of Nuclear Physics, Novosibirsk, Russia

Abstract. The computer system for control of cw surface-plasma source of negative ions is described. The system provides an automatic handling of source parameters by the specified scenario. It includes the automatic source start and long-term operation with switching and control of the power supplies blocks, setting and reading of source parameters like hydrogen feed, cesium seed, electrodes' temperature, checking of the protection and blockings elements like vacuum degradation, absence of cooling water, etc. The semi-automatic mode of control is also available, where the order of steps and magnitude of parameters, included to scenario, is corrected *in situ* by the operator. Control system includes the main controller and a set of peripheral local controllers. Commands execution is carried out by the main controller. Each peripheral controller is driven by the stand-alone program, stored in its ROM. Control system is handled from PC via Ethernet. The PC and controllers are connected by fiber optic lines, which provide the high voltage insulation and the stable system operation in spite the breakdowns and electromagnetic noise of cross-field discharge. The PC program for data setting and information display is developed under the LabView.

Keywords: Control System, Controller, Ion source
PACS: 52.70.Ds

INTRODUCTION

A Surface-Plasma negative ion Source (SPS) delivering cw H- beam is in operation on the tandem accelerator of BNCT experimental complex at Budker Institute. An advanced source version with beam current 15 mA was recently developed [1]. A surface-plasma source of negative ions is a plasma device, based on the cross-field discharge. An improved source performance – a high negative ion yield and low co-extracted electron current is provided by cesium addition to discharge. Optimal source activation by cesium requires the heating and temperature control of electrodes.

Several SPS features complicate the source systems design. Firstly, power and control units of the source operate at four different potentials. Secondly, repetitive breakdowns are typical for the plasma sources with ions extraction and high-voltage post-acceleration. Accompanying flux of hydrogen, of cesium atoms, of co-extracted electrons to the ion-optical system complicates the high-voltage holding in the source, so breakdowns could occur during initial conditioning of high-voltage electrodes and during regular source work. In addition, the SPS discharge in magnetic field is characterized by an increased level of fluctuations and noise. The breakdowns and noise pickup could lead to malfunction of electronics and make it difficult to use the standard power supply and control systems or to adapt the serially produced modules. The special control system was developed to supply the source operation with possibility of an unattended management. The peculiarities of control system design for intense negative ion sources with cesium are described below.

CONTROL SYSTEM STRUCTURE

Multiple power supplies and elements of negative ion source are biased at different potentials (cathode and anode potentials, extractor potential and ground) and they need about 50 channels of control. List of source units and required amount of control and measuring channels is presented in Table 1.

TABLE 1. Amount of measuring and control channels of the ion source.

Unit	Bias	Analog Inputs	Analog Outputs	Digital Inputs	Digital Outputs
Discharge PS, Heaters, Cs system	-40 kV	8	3	3	3
Extractor PS	-32 kV	2	1	4	1
Accelerator PS	ground	2	1	4	1
Magnetic Field, Gas System	ground	2	2	2	2
Interlocks and Protections Unit	ground	-	-	4	1

The distributed system for intense negative ion source control and data acquisition has been developed. An important advantage of the distributed control system is the reduction of required connections between remote units. The components of the distributed system are located at different potentials and can be easily connected with the help of the optical lines. The block diagram of the distributed system of the cw source is shown in Fig. 1.

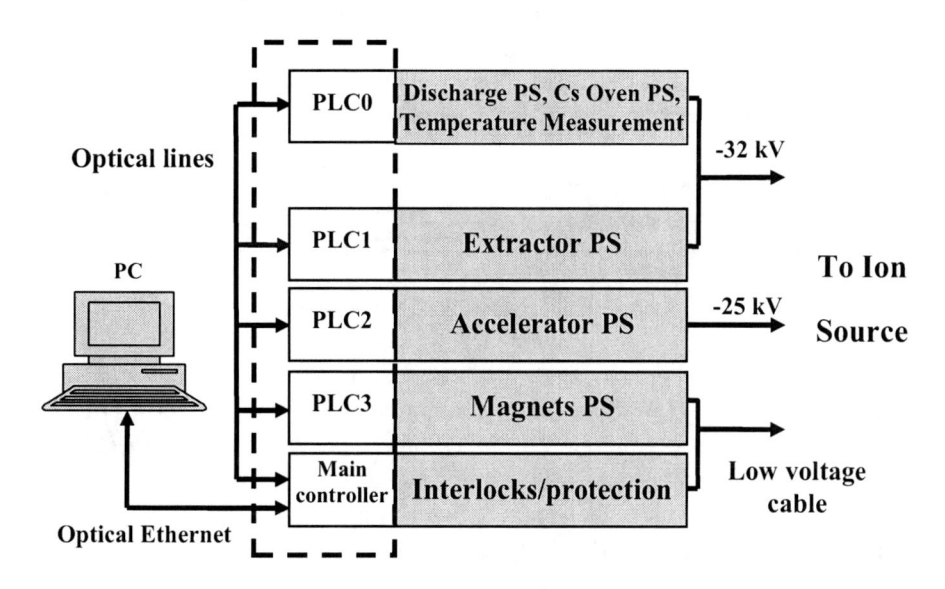

FIGURE 1. The block diagram of the control system.

The developed control system consists of a main controller, connected with PC, and a set of supplementary Peripheral Local Controllers (PLC in Fig. 1) embedded to each manageable unit of the source. The fiber-optic lines are used for communication between PLC units and main controller, while the optical Ethernet line is used for exchange between main controller and PC.

CONTROL SYSTEM MODULES

The basic element of the control system is the main controller. Its architecture is shown in Fig. 2. Main controller consists of a CPU board and of a 3 dual-channel communication microchips (UART in Fig. 2), providing the connection with the PLC via fiber optics. CPU board is based on the ARM7 processor and uses µLinux operating system.

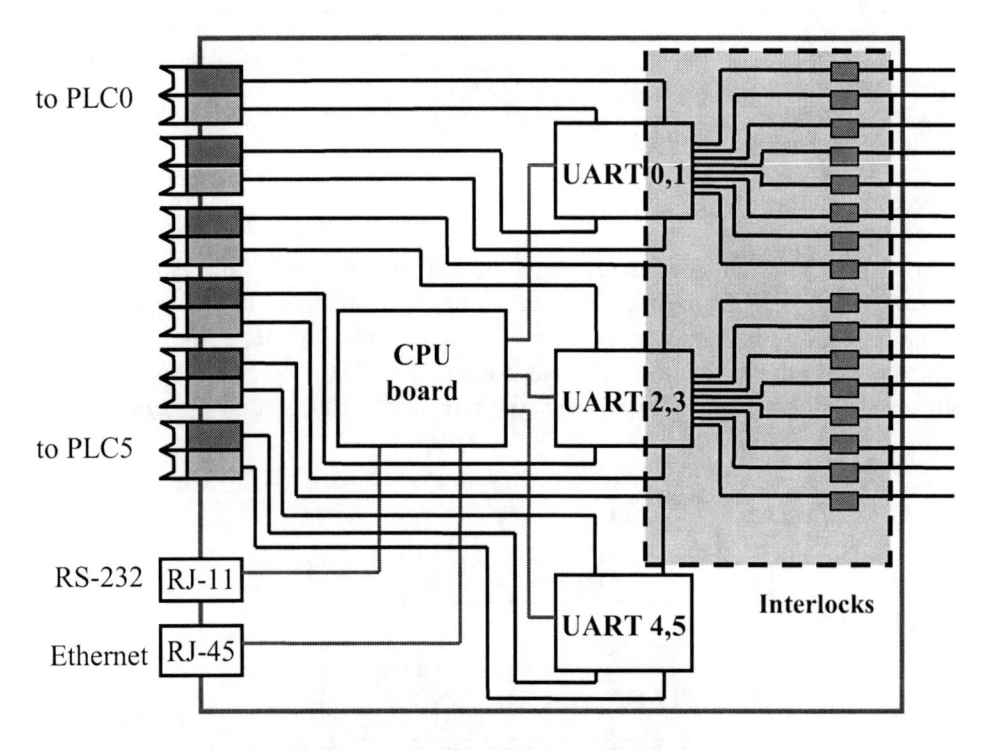

FIGURE 2. Architecture of the main controller.

The architecture of peripheral local controllers is shown in Fig. 3. The PLC core is a microcontroller. Each PLC module has an 8-channel ADC for measuring of analog signals, a 4-channel DAC to generate output voltages, an In/Out Register (IR/OR in Fig. 3) with 16 digital channels. All PLC modules are identical and interchangeable. Every PLC is equipped with the main power supply and with an additional autonomous power source (battery) for protection during power failure. PLC, located under high potential (PLC0 in Fig.1), is equipped with an additional shielding against the electromagnetic noise. Some parameters of the PLC are listed in Table 2.

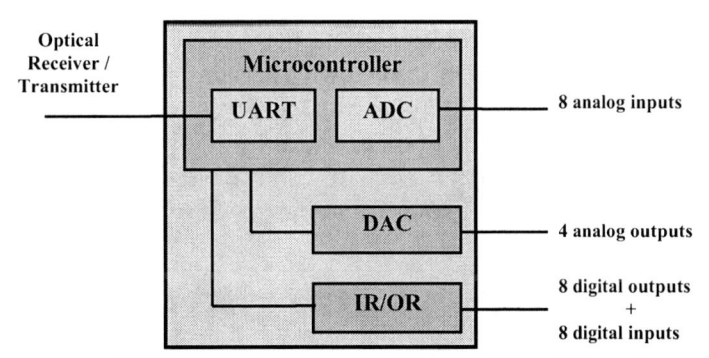

FIGURE 3. Architecture of Peripheral Local Controller.

TABLE 2. Some Parameters of every PLC

Unit	Amount of channels	Accuracy	Range
ADC	8	16 bit	-5÷5 V
DAC	4	16 Bit	0÷10 V
In/Out Register	8/8	-	-

THREE LEVELS OF SOFTWARE

Software of the source control system is divided into three levels: the lower level - PLC program, the central one – program of main controller, the upper – a PC program.

The lower level – software of PLC – provides the cyclic serial scanning of analog input signals with data acquisition, the collected data assembling into the packets and the following transmission to main controller. Also PLC software processes the control commands, received from the main controller. The protocol of PLC and main controller exchange provides the transfer of data packets of arbitrary length with the check of data integrity.

The central level – software of main controller – is responsible for data exchanges via Ethernet and UARTs interface channels and for running of algorithms of unattended control (scenarios). The main controller performs three separate program tasks. The first task is the cyclic (4 times per second) reading of data, measured by PLCs. The obtained data are sequentially stored into the table, containing the current data of all PLCs.

Second task of main controller is the Ethernet exchange with PC. Main controller interacts with PC via Ethernet-10/100 channel using client-server model, where the server is the main controller, while the client is the PC program. The second task processes the following operations: the request for the data measurement, the load, start and stop of scenario, and the transmission of control commands to PLCs.

The third task of main controller is scenario processing. This task uses three different types of scenario commands. The fastest type of scenario commands is a continuous check of emergency and assured interlocks. Two other types of commands are the "quick" commands with the run time 1 ms and the "slow" ones with the run time of 1s. Software of main controller could provide the source running and handling by scenario in the emergency cases independently without PC.

The upper level – program at PC – performs the request of measured data from main controller, the conversion of measured data to physical values, the visualization of data into graphical plots and the interaction with operator. The PC program has the window interface developed under the LabView. The configuration window of PC program permits to set the conversion coefficients of binary values to physical quantities and vice versa. The scenario window allows to create and to edit the scenario of unattended work. With the help of control window (shown in Fig. 4) operator can easily track the source parameters and to correct manually the source parameters during the scenario run.

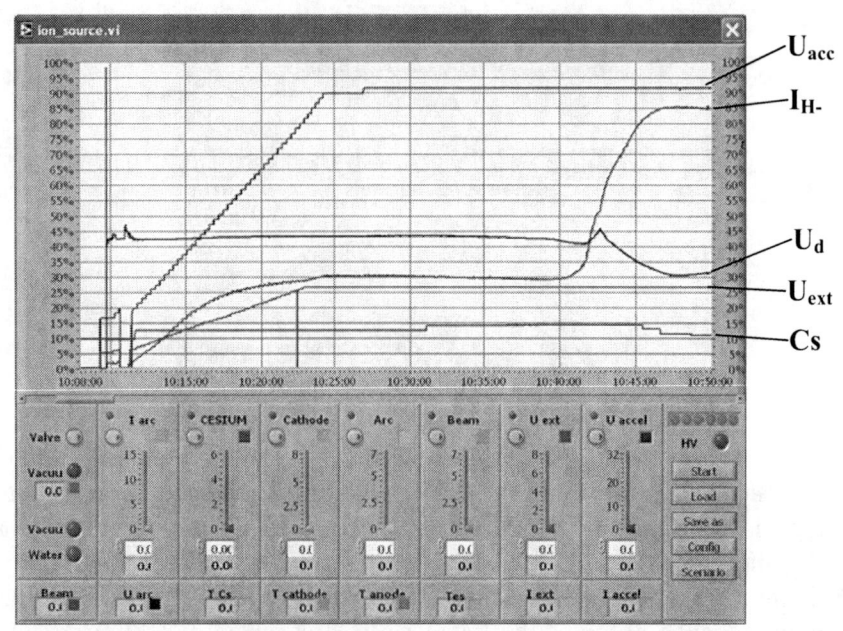

FIGURE 4. Control window of PC program. Operation by scenario during discharge ignition and HV conditioning is shown.

The view of control window of the PC program is shown in Fig. 4. Top part of the window contains the scan of source current parameters. Buttons and slides for parameters setting are located at the window bottom part. The current data of the source are also indicated on the bottom line of control window. Measured data are stored to PC and can be analyzed later (button "Load" at the right column of window).

Execution time of control command is 1 ms. Data packets are processed with repetition rate 4 times per second, the resulted graphics is updated 1 time per second.

SCENARIO FOR UNATTENDED CONTROL

Ion source control scenario, loaded to the main controller, could handle the following stages of source operation: discharge start; rise of high voltage; steady-state operation; processing of emergency cases. Source shutdown is usually produced manually, but it can be included to scenario as well.

Each listed stage of source operation consists of a certain sequence of actions, prescribed by scenario file. For example, the preliminary heating of electrodes and of cesium oven is executed during source initial phase for discharge start. The control of extraction/acceleration voltages and currents is applied on the stage of HV rise. An important task of scenario during the source conditioning and gradual rise of high voltage is to decrease the extraction or acceleration voltages after breakdown in the extraction or acceleration gap. The working range of discharge voltage, prescribed in scenario, is sustained during long-term runs by cesium oven temperature control. An example of source operation by scenario during discharge ignition and HV conditioning is shown in the scan of Fig. 4.

Special interpretive language has been developed. The interpretive language allows preparing scenarios by operators without change and compilation of PC program. It includes 30 basic commands, which allow checking and setting the source operation parameters, polling the state of interlocks, switching the components, displaying the error report or the request for manual action. Two independent software timers are used for implementing of programmed time delays. Commands of scenario use the clear names and physical values of source parameters. Scenario can prescribe the emergency actions under alert conditions, like absence of cooling water or poor vacuum condition. The necessary branching of the control algorithm steps is produced by the checking of source operation parameters and corresponding "go to" commands. The semi-automatic mode of control is also available, where the order of steps and magnitude of parameters, included to scenario, is corrected *in situ* by the operator.

The scenario editor is a part of the PC software. It allows creating and editing of scenario files. All parameters of the scenario commands are entered as physical quantities to the scenario file. The translation of the scenario parameters to appropriate binary codes for ADC and DAC is performed at the moment of transmission to the main controller. The editor uses the list of channels and conversion coefficients from the configuration file, containing information on used ADCs and DACs, their names, etc. Change in the conversion coefficients for ADCs and DACs does not require the modification of scenario. The emergency conditions and corresponding error messages could be defined by the operator.

CONCLUSION

Control and data acquisition system for unattended control of cw surface-plasma negative ion source has been developed. System has a distributed structure. Its units are properly protected against breakdowns, electromagnetic noise and power failures.

The multilevel software for source control system has been written. It contains the codes for peripheral local controllers, for the main controller and for PC. Software for PC has windowed interface. The scenario editor allows operator to write and execute different scenarios for source unattended or semi-automatic run. Various scenarios for source operation with fine-tuned parameters have been written and explored. Source work with unattended control by scenario is routinely produced at Budker Institute facilities. Described control and data acquisition system can be adopted for automation of different cw ion sources and other physical devices.

REFERENCES

1. Yu. Belchenko, A. Sanin, and A. Ivanov, in *Negative Ions, Beams and Sources*, edited by E. Surrey and A. Simonin, AIP Conference Proceedings 1097, American Institute of Physics, Melville, NY, 2009, pp. 214-222.

Status of NIO1 construction

M. Cavenago[a], E. Fagotti[a], M. Poggi[a], G. Serianni[b], V. Antoni[b], M. Bigi[b], F. Fellin[b], E. Gazza[b], M. Recchia[b], P. Veltri[b], S. Petrenko[c] and T. Kulevoy[c]

[a] *INFN-LNL, v.le dell'Universita' 2, 35020 Legnaro (PD), Italy*
[b] *Consorzio RFX, Associazione Euratom-ENEA sulla fusione, c.so S. Uniti 4, 35127 Padova, Italy*
[c] *INFN-LNL, Legnaro, Italy and ITEP, B. Cheremushkinskaya 25, 117218 Moscow, Russia*

Abstract. The NIO1 (Negative Ion Optimization phase 1) project consists of a multiaperture negative ion source mounted on a 60 kV accelerating column; up to 9 beamlets (15 mA H$^-$ each one) arranged in 3 x 3 matrix with 14 mm spacing can be extracted. The moderate size and the modular concept make some relative rotation of the source magnetic filter and the electrodes possible, so that the effect of crossed and aligned field can be easily compared. Other goals of source experimental program are emittance (and beam profile) measurement at several distances (for simulation code validation), testing of diagnostic components and of radiofrequency coupling. A full set of construction drawing was completed; also the Fast Emittance Scanner (FES) and its vacuum chambers were built (four mounting positions are reserved to FES). Some low power rf matching boxes were developed for a test plasma, approximately half the source size. A cesium oven compatible with NIO1 is being also developed; by using some industrial standard 100 W heaters (with a proper driver) a careful control of temperature is planned.

Keywords: ITER, ion, source, extraction
PACS: 41.75.Cn, 29.25.Ni, 52.50.Gj

INTRODUCTION

In this paper we describe a versatile and compact ion source (NIO1, Negative Ion Optimization phase 1, see Fig. 1), now under construction, useful for code validation and for testing of advanced beam diagnostic, and a fast emittance scanner (FES). The development of Neutral Beam Injectors (NBI) for the ITER project [1] and the large beam test facilities planned to be built at Consorzio RFX (Padova, Italy) described elsewhere [2] are a strong motivation to further improve the negative ion sources (NIS[3, 4, 5, 6]) and the understanding of their beam extraction, also in the perspective of fusion reactors beyond ITER.

The scheme of a typical NIS shown in Fig. 2 includes several stages in the accelerating column, as envisioned in a real 1 MV accelerator. Anyway NIO1 envisions only a 60 kV extraction voltage, with three major electrodes: the PG (Plasma Grid) held at -60 kV, which is the voltage reference for the source; the EG (extraction grid) held at -51 kV, to regulate the extracted ion beam current and to stop the coextracted electron (deflected by permanent magnets PM placed inside the EG itself); and the PA (post acceleration grid) which is held near the ground potential. After the PA, we add a repeller electrode REP that can be biased up to +150 V to optimize the space charge compensation in that region. The protection of PA and EG and their supplies against voltage transient and surges(possibly induced by electrode discharges) is based on passive elements (resis-

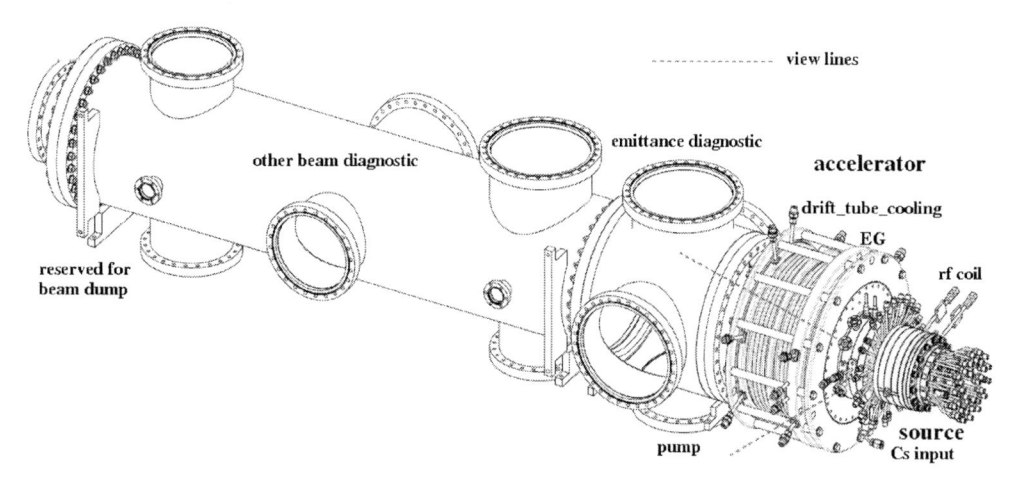

FIGURE 1. NIO1 overall design, final version. Total length 2.557 m (without cooling tubes).

tors, varistors and spark gaps) as discussed in a separated paper[7]. Accelerator column is compact and held together by compression bars as visible in Fig. 1. Let z be the beam axis and y be the dominant direction of the EG field.

Following current ITER choice, we use a radiofrequency coupling scheme to heat plasma (as opposite to the injection of 100 eV electrons by filaments) for the practical advantage of robustness; this implies the use of cesium inside the source to obtain the specified current (after reduction for expected stripping losses extracted current of D^- is specified as 200 A/m^2, roughly equivalent to 280 A/m^2 for H^-). Cesium regulation is obtained in NIO1 with a carefully stabilized external oven (see Fig. 1).

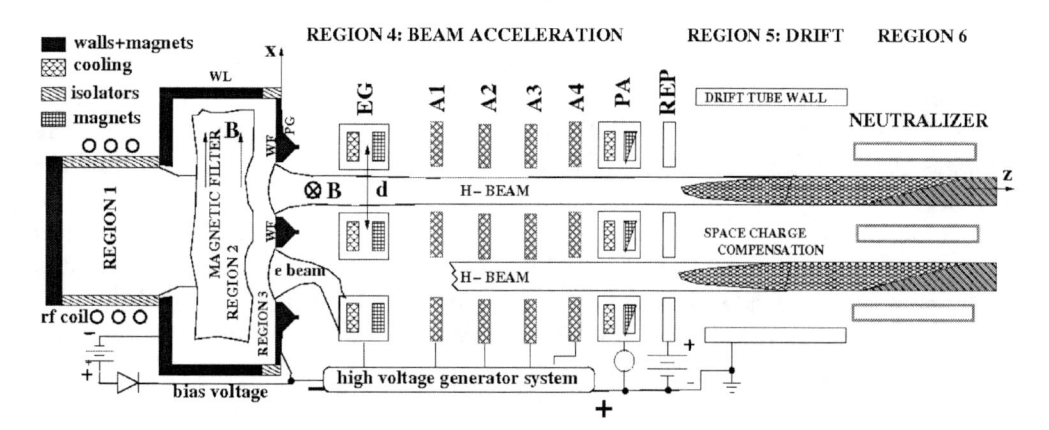

FIGURE 2. Scheme of an rf negative ion source and its accelerating column; for the sake of generality, we show A1, A2, A3 and A4 (cross-hatched rectangles), which are the additional grids envisioned in the MAMUG style accelerator (where the grid PA is called GG), but not in NIO1. Note that NIO1 design includes a low voltage repeller REP between the grounded drift tube and the nearly grounded PA. In the drift tube the H^- beam is space charge compensated (shaded area) by the slow ions H_2^+ produced. Conversion (neutralization) from H^- to fast H^0 (hatched beam area) happens later.

An electron temperature T_e about 4 eV is enough to produce the ionization required for plasma global balance [8, 9], while a lower temperature plasma $T_e \leq 1$ eV is optimal for H$^-$ propagation; let λ_a the mean free path traveled before destruction by a H$^-$ ion, usually longer than the elastic collision mean free path λ_c; the total mean free path is $\lambda = 1/(\lambda_c^{-1} + \lambda_a^{-1})$. So electrons heated to about $T_e \geq 4$ eV by radiofrequency are confined by a magnetic filter field $\mathbf{B}^f$ in a 1st region of the source (see Fig. 2), while a lower temperature plasma $T_e = 1$ eV diffuses in regions 2 and 3 towards the source exit, named plasma grid electrode (PG). An intermediate and optional electrode named bias plate (BP) returns the large flow of electrons, lost on PG, to the plasma. Ions H$^-$ are formed on the PG rear wall (cesiated and bombarded by fast H^0), or in the region 3 volume. The filter field $\mathbf{B}^f$ may be in the x or the y direction.

The ratio $R_j = j_e/j_{H^-}$ between the current density of electrons emitted from the plasma j_e and the ion current density j_{H^-} is to be minimized; qualitatively, it also depends on the filter magnetic field within a λ_a distance from PG, the bias voltages applied to source box, the Cesium both in plasma and on PG walls. Investigation of these effects is one of the goals of NIO1 design. In other words[10] the ratio $\alpha_R = n_{H^-}/n_e$ of particle densities n_{H^-} and n_e is to be maximized; these ratios can be roughly related at the plasma border by $R_j = R_m/\alpha_R$ with the constant $R_m = (m_{H^-}/m_e)^{1/2} \cong 43$.

Among other innovations to be tested at NIO1 we list: wall material effects; extended bias plate in the ion source; and many beam diagnostic systems, including calorimetric beam profile monitors (BPM). The following sections will discuss the detail of NIO1 design and a fast Alison scanners to measure emittance, under completion. Emittance measure will be a much more informative test for beam code validation that any BPM is. Beam simulations are described elsewhere [11, 12, 13, 14, 15].

THE NIO1 DESIGN

The compact dimension of ion source (and the installation in air) make substitution and rotation of parts easily feasible, which is clearly an advantage in experimental campaign. To take full advantage of this, the NIO1 design (Negative Ion Optimization 1, see Fig. 3) emphasizes modularity and symmetry (where possible); the source is a tower of disk assemblies (connected by O-rings). We have 9 beam holes in the PG (maximum diameter 8 mm) in a square pattern with $L_x = L_y = 14$ mm, where L_x is the spacing between centers in the x direction and L_y similarly for the y direction. To test the better direction of the source magnetic filter (crossed or parallel to the EG field) rotation of 90^0 is still possible; initial position corresponds to crossed fields. On the other side, integration of standard size components (feedthroughs, thermal insulators) in drawings is somewhat more difficult than in larger sources; see the Cs input pipe for example.

Among the several function of bias plate and PG assemblies [15, 16], we point out: 1) they provide a closed path for a current I_y (up to 500 A), to add a tunable B_x term to the filter field; 2) their central parts (90 mm diameter) can be heated up to 400 K to optimize Cs coverage, and the resulting thermal dilation are adsorbed by sliding seals (cycle frequency should be anyway kept as low as possible, ideally one cycle a day); 3) several voltage bias schemes are possible (see Fig. 4); 4) view lines are provided for actinometry and H$^-$ density measurement[17].

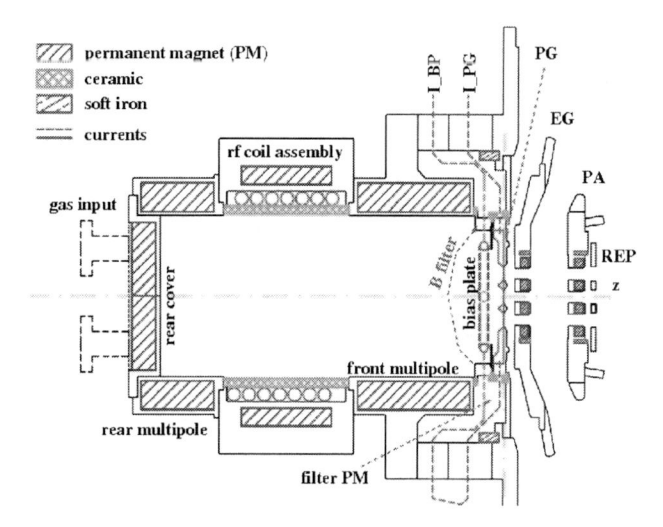

FIGURE 3. Scheme of NIO1 source: horizontal section zy (solid lines) and major connections not in section plane (dashed). Approximate position of filter magnets (not in this plane, but near) indicated by arrows. Only the central part of the PA and EG assemblies are shown.

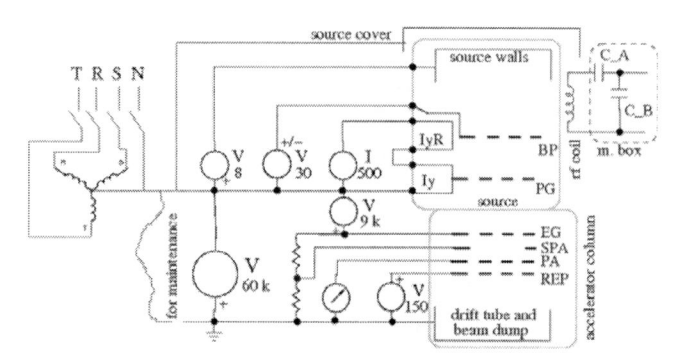

FIGURE 4. NIO1 major power supplies; some ratings are indicated. Matching box (m. box) scheme may be adjusted according to experience.

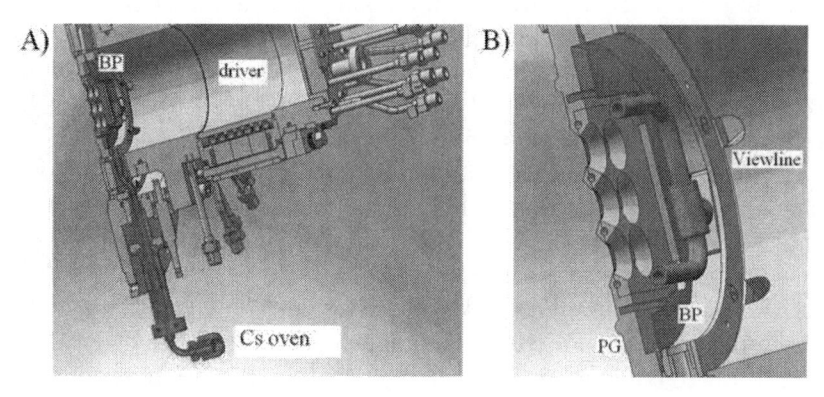

FIGURE 5. A) A vertical section of NIO1 source, with the Cs pipe put ion evidence (green color), as well as the Mo bias plate and return conductor of the bias plate assembly (BP); B) a closer view of BP and the plasma grid (PG).

The source walls are adequately cooled (note several external connections in Fig. 1) and are covered with magnets (see Fig. 3 in Ref [16]); the front multipole bars (72 mm long) can be replaced by shorter bars (36 mm long + 36 mm filler) to simulate the plasma expansion in larger sources. An iron ring encircle the filter magnet and the coupled conductors in the BP and PG assembly, with purpose of localizing (and enhancing) the filter field within 3 cm from the PG.

The $m = 7$ multipole field (14 poles) can merge smoothly with the dipole ($m = 1$) of the filter field. Multipole magnet bars allow space for cooling channels in between; also two lines of view pass between these bars (see Fig. 5), at a $d = 19$ mm distance from the PG. For economy and simplicity we used 14 bolts to keep the source together, even if a configuration with 16 bolts would have been more versatile.

The accelerator column design (including PM both in the EG and the PA) is practically unchanged from the conceptual design[15], with few minor improvements: 1) a 2 mm thick Vespel (TM, a polyimide-based plastics) insulator aligns the PA to the column base (at ground potential); 2) the six cooling circuits were better specialized; two cool the EG beam holes and two the PA beam holes, with flexible hoses to adjust beam electrode position (copper minimum thickness is 1.1 mm). The EG support is made of aluminum and cooled externally; PA assembly incorporate a fixed drift tube with the last cooling circuit, so to withstand beam losses during beam transients (source on/off, or discharge). 3) some external connection are available for the repeller electrode. In Fig. 1 we may note the external rims of the source baseplate, of the EG assembly and of the column base (500 mm diameter) which for economy and simplicity are the alignment references; the rims of the PA and of the intermediate spacer electrode SPA are somewhat smaller.

The column base is supported by a 6-way cross, where at least one 2000 l/s pump is mounted on a CF200 lateral port. This cross and the diagnostic chamber are joined together by 16.5 inches Conflat flanges (total mass 220 kg); enlarging 4 lateral ports to CF250 allowed to fit the fast emittance scanner (which seems the major mechanical vibration source).

The rf coil is designed to operate without an rf faraday shield (FS) inside, so that we can clarify its role fully; final drawing also envisions a removable rear cover, so that any FS can be retrofitted if needed. Rf window is a 80 mm long alumina tube. with $d_c = 100$ mm inner diameter and 110 mm outer diameter. Possible thermal loads are: the plasma impact on the rf window; the heating due to rf currents inside the coil (or inside the metal source chamber) easily computed and water cooled; the rf heating of the ferrite PM behind the coil, largely depending from material properties. To mitigate heating, air is forcedly circulated in the gaps between alumina and rf coil (required air speed order of 10 m/s); another air circuit cools the PM; all channels are carved in the polycarbonate disks that hold the coil and the PM and the rf window in place. Plasma impact on rf window is also mitigated by the ferrite PM field.

Resonance frequency f_r is determined by the coupled external capacitance C_c and the coil inductance L_c (now a 7 turn coil). Operation will be limited to clean H_2, with the purpose of verifying the minimum filling pressure p_f for rf plasma ignition, at several power levels P_c, and if possible, at several f_r; note that changing the matching box and the amplifier is a major cost. We expect the scaling $p_f \propto 1/d_c$ and $f_r \propto 1/d_c$, so from the ITER specification with $d_c = 250$ mm and $p_f = 0.3$ Pa, we expect $p_f = 0.75$ Pa in this smaller source and choose $f_r = 2 \pm 0.2$ MHz.

A benchmark experiment was conducted with a glass jar ($d_c = 60$ mm) and a 6 turn coil, using air as support gas. Two low power matching boxes with solid dielectric capacitances were produced; C_c regulation is obtained by changing capacitance number (or dielectric thickness). Within available rf power (total 350 W, more than half lost in cables, and so on), the plasma ignition pressure resulted extremely large ($p_f > 10^1$ Pa); the expected frequency jump at plasma switching on was not measurable; the effect of inserting a Faraday cage was not measurable. It is clear that more work (both theoretical and experimental) is necessary to clarify optimal rf coupling conditions.

Cesium oven

A quickly removable Cs oven (Fig. 6) is also under construction (at LNL mechanical workshop). Basic component is an economical all metal valve, which is left open during oven operation. At source and oven off, this valve can be closed, so that Cs reservoir can be vented with Argon, before removal. Still for economy and space constraints, Cs oven removal implies source air (or nitrogen) venting. The regulation of the Cs inflow inside source is in principle straightforward: the Cs reservoir temperature T_{cs} is adjusted within 1 K by a thermo resistance monitor and a PID (proportional integral differential) controller, moderating a 100 W heater with a SCR regulator. Oven may be heated for all the experiment duration or for a given period of time t_{cs}, according to the experiment schedule. The valve and the line are heated (by 4 distributed heaters) so that their minimal temperature T_p safely exceeds T_{cs}, accounting for the thermocouple reading errors (5 K) and the lack of temperature uniformity, so that all cesium evaporated is transferred to source (perhaps with some delay). According to previous experience[5, 17], oven may be switched off if source is used for short pulses. Aiming at a temperature

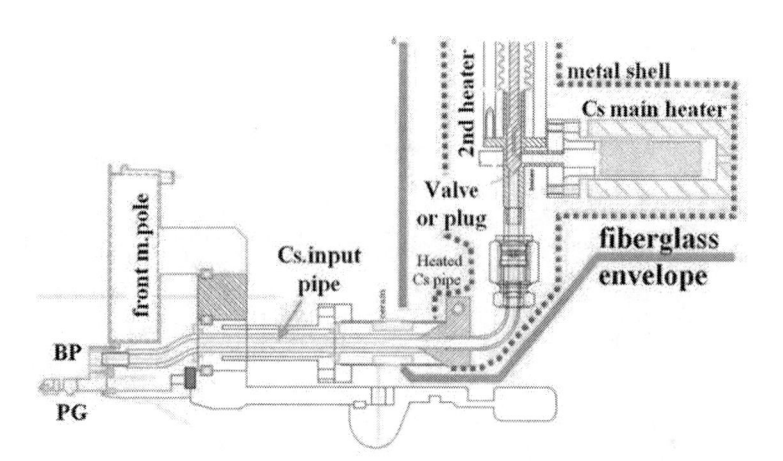

FIGURE 6. Concept of the removable Cs oven: a CF16 flange on the PG assembly allows a thermally isolated copper pipe for cesium transport (ID 4 mm, OD 8 mm) to reach the bias plate. Cesium oven is placed in a fiberglass pocket; by carefully regulating its temperature, cesium injection is controlled; on the contrary the copper pipe is overheated (at 470 K or more) to avoid Cs sticking.

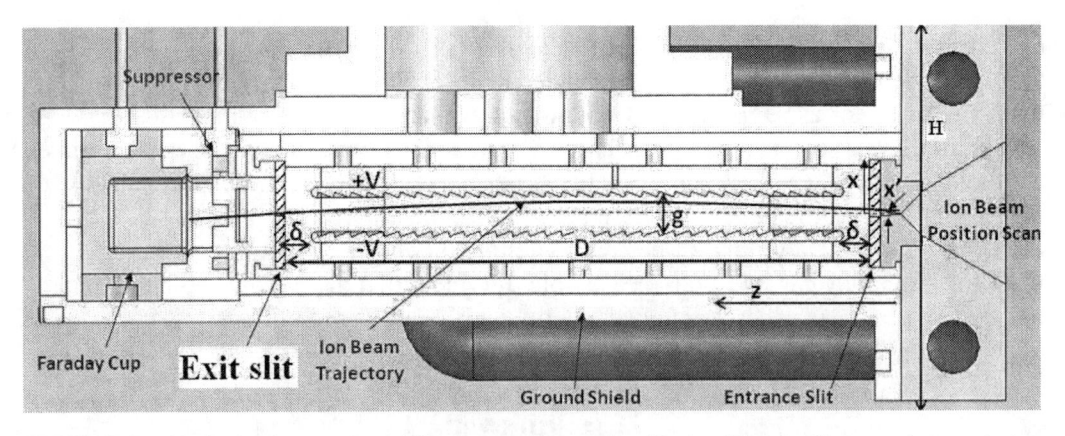

FIGURE 7. Typical scheme of an Allison meter measuring head, of height H; secondary emission suppression is used. Deflecting plate parameters V, D, δ and g are described in the text; the so called exit slit separates the deflecting plate region from the Faraday cup.

uniformity within 10 K, large shells of copper or Al encloses the oven parts, covered by suitable thermal insulator (and enclosures). The first goal of the prototype is to verify that the specified thermal stability and uniformity can be achieved, at least without a Cs load of the oven.

THE FAST EMITTANCE SCANNER

Allison emittance scanners[18, 19] are being developed in several laboratories to measure low energy ion beams, with the major advantage of requiring only one motor. As shown in Fig. 7 the measuring head has two selection slits (named entrance slit and exit slit) rigidly mounted and aligned on the same base, with a distance D; the first slit

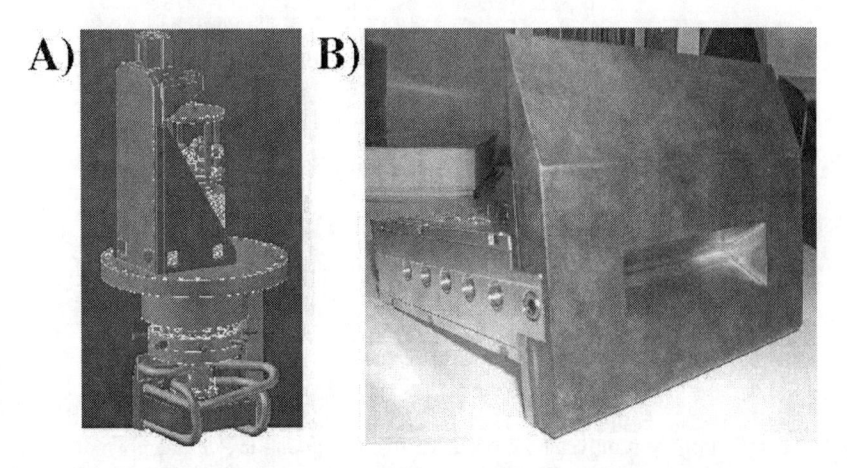

FIGURE 8. A) FES 3D view; note the water cooling manifold and the actuator B) The compact FES measure box in a closer view, showing the copper beam shield in the front.

TABLE 1. Beam parameters for FES

	Units	NIO1	TRIPS
Beam ion		H$^-$	H$^+$
Energy eV_b	keV	60	80
Current I_b	mA	130	50
Power density	MW/m^2	20.3	22.6
Beam dimension	mm	±40	±35
Beam divergence	mrad	±40	±130
Power (CW)	kW	7.8	4.0

selects particle position $x(t)$, where x is here the axis of insertion of the scanner head, visible in Fig. 8, and $x(t)$ is the instantaneous position of the first slit. Particles with different x' pass the first slit, where $x' = dx/dz$ is the tangent of the angle α_x between a particle and the beam axis z. Let its origin $z = 0$ be here at the first slit. In usual emittance plots, we represent the particle density in the $x - x'$ plane, or better in the phase space plane $x - \alpha_x$.

A fast voltage ramp $V(t)$ (formed by a function generator and buffered, with opposite signs, by two high voltage amplifiers) is applied between two deflecting plates placed between these slits, so ions are deflected by an amount $x_d(z) \propto V(t)$. Only the ions with $x_d(D) + x'D \cong 0$ pass the second slit and reach the Faraday cup FC, which has a suppression of the secondary particle emission (say electrons). The current $I(t)$ flowing into the Faraday cup thus measures the incoming ion current only (not the electron term), and is thus proportional to the ion phase space density at $(x(t), x'(t))$.

The water cooling of the head is designed to withstand beam power for considerable time without major damages. Anyway to avoid thermal deformation of the slits, we plan to insert and remove the scanner rapidly, typical times being: acceleration phase 200 ms, beam crossing 500 ms, motion reversal 400 ms, second beam crossing 500 ms, deceleration phase 200 ms; $I(t)$ recording is necessary only during beam crossing. The motor may be programmed, for adjusting the actual movement to the beam size; maximum head speed is 0.3 m/s, moving mass is 10 kg, plus vacuum static load (1500 N) and cables rigidity. In the event of motor failure, source is switched off.

The voltage ramp $V(t)$ is a symmetrical triangular wave (plus distortion), typical period being 66 μs ; it is convenient to measure $V(t)$ from the voltage monitor of a voltage amplifier and sample it at 1MS/s (or more), to measure amplifier distortion (load distortion/reflections must be modeled). The signal $I(t)$ from FC contains the information on the beam structure, so we aim at achieving a bandwidth as large as possible with small distortions; previous experience with matched FC make us confident in a 0 to 35 Mhz bandwidth with passive circuitry. Anyway, to make scanner usable with weaker beams, we provide a preamplification stage just inside the vacuum; properly speaking, this is a current to voltage converter matched to the following 50 ohm line.

The scanner is designed to work both with NIO1 and TRIPS[21] sources, with parameters shown in Table 1 and 2, where eV_b is ion energy and I_b the total beam current. TRPIS is a single beamlet H$^+$ source. Note that particle density is integrated along y, so that NIO1 beamlets are summed in groups of 3. Both for economy and for scanner

TABLE 2. FES design parameters

	Units	NIO1	TRIPS	FES
g	mm	8	8	8
D	mm	100	100	100
δ	mm	5	5	5
s	mm	0.05	0.05	0.05
V_M	kV	±0.43	±1.85	±2
x'_M	mrad	±40	±130	±145
θ_d	mrad	±0.5	±0.5	±0.5
Slew rate	V/µs	300	300	300
v_h	m/s	0.3	0.3	0.3
f bandwidth	MHz	31	23.3	35

prototype simplicity we decide not to include removable mask inside the scanner head; when a single beamlet measure is necessary, beamlet should be masked externally.

The measurable beam size ±40 mm is determined by NIO1 whole beam envelope, while cooling requirements are given by minimal beam size. The beam divergence range is determined by TRIPS, whose use is planned also in strongly convergent optics; the NIO1 optic is necessarily (as any other NBI optics) based on nearly parallel beams. Moreover, FES may measure single beamlet halos when mounted in the NIO1 crossing pump (closer position at 0.46 m from PG). Other positions at 0.97 m and 1.47 m from PG are available, both with vertical and horizontal insertion axis.

For the detailed geometry of Fig. 7 and analytical trajectories, the voltage V_d necessary to sample at a given x' deflection is $x' = V_d D_\delta/(4gV_b)$ where g is the average gap between plates and $D_\delta = D - 2\delta$, with δ the small spacing between ends of deflecting plates and slit. Moreover there is an upper bound $x'_M = \pm 2g/(D+2\delta)$ for $|x'|$: this is the instrument acceptance. Combining these relation, we found the maximum voltage V_d^M that needs to be supplied

$$V_d^M = \pm \frac{8g^2 V_b}{D_\delta(D+2\delta)} = \pm 2 \frac{(D+2\delta)}{D_\delta} V_b {x'_M}^2 \qquad (1)$$

The slit width s (here chosen equal) determines not only the resolution of the position x, but also the angular resolution $\propto s/D$. Thus, after the voltage scan period $2T$ is chosen, the necessary bandwidth is $f = 12^{1/2} x'_M D/(\pi s T)$. As apparent from table 2, the TRIPS case requires a nontrivial 2 kV voltage amplifier; deflecting plate are 90 mm long with a 93 mm effective length.

Scanner head is covered by aluminum panel with pumping holes. Insulator are in AlN, for better cooling. Note that the input slit is made by two bodies, a wider front slit which sustains most of beam thermal load, thank to a strong water cooling, followed by a block with a precisely machined slit (width $2s$ as said before). As usual, the deflecting plates have sawtooth surfaces, to trap most of the secondary emission of colliding ions.

Control and data acquisition electronics is an expensive and nonstandard item of the project, mostly for the memory depth required by the $I(t)$ fast storage. A sampling rate of 240 MS/s (about 8 points per shortest period in $I(t)$) is now specified. A minimal project envisions to use analog comparators on the signal $x(t)$ to form a trigger signal when beam is crossed, which enable data acquisition; the corresponding data acquisition system for $I(t)$, $V(t)$ and $x(t)$ is under procurement. Electronics may be upgraded and improved by experience. Preliminary tests are planned in 1st semester of 2011.

CONCLUSIONS

Construction of the NIO1 source has been begun, and drawings are completed. At the same time, the construction of the FES is nearly finished, and its electronics is being procured. The flexibility of NIO1 promises to be extremely valuable for a full validation of codes and for experimental campaigns.

ACKNOWLEDGMENTS

We thank R. Baruzzo (Cinel s.r.l.) and F. Boato for several technical discussions on the NIO1 drawings optimization.

REFERENCES

1. R. S. Hemsworth, J. H. Feist, M. Hanada, B. Heinemann, T. Inoue, E. Kussel, A. Krylov, P. Lotte, K. Miyamoto, N. Miyamoto, D. Murdoch, A. Nagase, Y. Ohara, Y. Okumura, J. Paméla, A. Panasenkov, K. Shibata, M. Tanii, and M. Watson. *Rev. Sci. Instrum.*, **67**, 1120 (1996).
2. Agostinetti et al., "Design of a Low Voltage, High Current Extraction System for the ITER Ion Source", *NIBS: Proceedings of the 1st International Symposium* (ed E. Surrey, A. Simonin, AIP-CP 1097, 2009), p 325
3. M. Bacal et al., Rev. Sci. Instrum., **71**, 1082 (2000).
4. J. Peters, *Rev. Sci. Instrum.*, **79**, 02A515 (2008).
5. W. Kraus et al., Rev. Sci. Instrum. 75 , 1832 (2004).
6. R. F. Welton et al., *Rev. Sci. Instrum.*, **75**, 1789 (2004).
7. M. Recchia, M. Bigi, M. Cavenago, submitted to Fusion Eng. Des. (2010).
8. P. N. Wainman et al., *J. Vac. Sci. Technol.*, A **13**, 2464 (1995).
9. M. A. Lieberman and A. J. Lichtenberg, *Principles of Plasma Discharges and Material Processing*, John Wiley, New York, 1994.
10. N. St J. Braithwaite and J. E. Allen, *J. Phys. D*, **21**, 1733 (2003).
11. J. H. Whealton et al. , *J. Appl. Phys.*, **64**, 6210(1988).
12. H.P.L. de Esch et al., *Rev. Sci. Instrum.*, **73**, 1045 (2002).
13. A. T. Forrester, *Large Ion Beams*, John Wiley, NY, 1996.
14. M. Cavenago, P. Veltri, F. Sattin, G. Serianni, V. Antoni, *IEEE Trans. on Plasma Science*, **36**, pp 1581-1588 (2008).
15. M. Cavenago et al., "Development of Small Multiaperture Negative Ion Beam Sources and Related Simulation Tools" in *NIBS: Proceedings of the 1st International Symposium* (ed E. Surrey, A. Simonin, AIP-CP 1097, 2009), p 149
16. M. Cavenago, T. Kulevoy, S. Petrenko, V. Antoni, M. Bigi, E. Gazza, M. Recchia, G. Serianni, and P. Veltri, *Rev. Sci. Instrum.*, **81**, 02A713 (2010).
17. D. Wunderlich, R. Gutser, U. Fantz, *AIP Conf. Proc.*, **925**, 46 (2007).
18. P. W. Allison, J. D. Sherman, and D. B. Holtkamp, *IEEE Trans. Nucl. Sci.*, NS-30, 2204-2206 (1983).
19. R. McAdams et al., *Rev. Sci. Instrum.*, **59** , 895 (1988).
20. M. P. Stockli et al., in *Proc. of the 16th Int. Work. on ECRIS*, (edt. M. Leitner, AIP Conference Proceedings AIP-CP 749, Melville, New York) p 108 (2008).
21. L. Celona et al. *Rev. Sci. Instrum.*, **75** , 1423 (2004).

Simulation And Design Of A Reflection Magnet For The EAST Neutral Beam System

Liang Li Zhen, Hu Chun Dong

Institute of Plasma Physics, Chinese Academy of Sciences, Hefei, 230031, China

Abstract. The simulation and design of a reflection magnet to be installed in the Experimental Advanced Superconducting Tokamak (EAST) neutral beam injection system are reported. A parametric design and simulation for the reflection magnet was carried out. For a deuterium beam with 42 cm as the bending radius, the intensity of reflection magnet field is about 1376 Gs at the energy of 80keV. In order to determine position of the ion dump and the surface power load, a particle simulation with Monte Carlo was developed to study ion trajectories. In addition, the louver design is introduced.

Keywords: Neutral Beam, Experimental Advanced Superconducting Tokamak, Reflection Magnet, Monte Carlo.
PACS: 52. 50. Gj 02.70.Ss

INTRODUCTION

In order to fully exploit capabilities of Experimental Advanced Superconducting Tokamak (EAST) and to study the physical issues involved in steady-state advanced tokamak devices. A neutral beam (NB) injection system is being constructed with two injectors at the energy of 80 keV and every injector will supply 4 MW neutral beam power at the dimension of 12 cm×48cm [1].

Heating with neutral beam injection has been successful in various magnetic confinement devices [2,3]. Once the neutral particles entering the plasma have become ionized, the resulting fast ions are slowed down by coulomb collisions. As the slowing down occurs energy is passed to the particles of the plasma, causing heating of both electrons and ions. The production of neutral beams for magnetic confining device involves several stages, a powerful ion beam generation, neutralization of ion beam, residual ions separation and treatment [4], as shown in Fig. 1.

It is the function of a separation magnet that dumps the residual ions and produces a neutral beam. Since the residual ions would hit some surface of the neutral beam injector and cause significant damage due to thermal overload and give rise to an additional source of a gas due to desorption, outgassing and recombination. A parametric design and simulation for the reflection magnet was carried out.

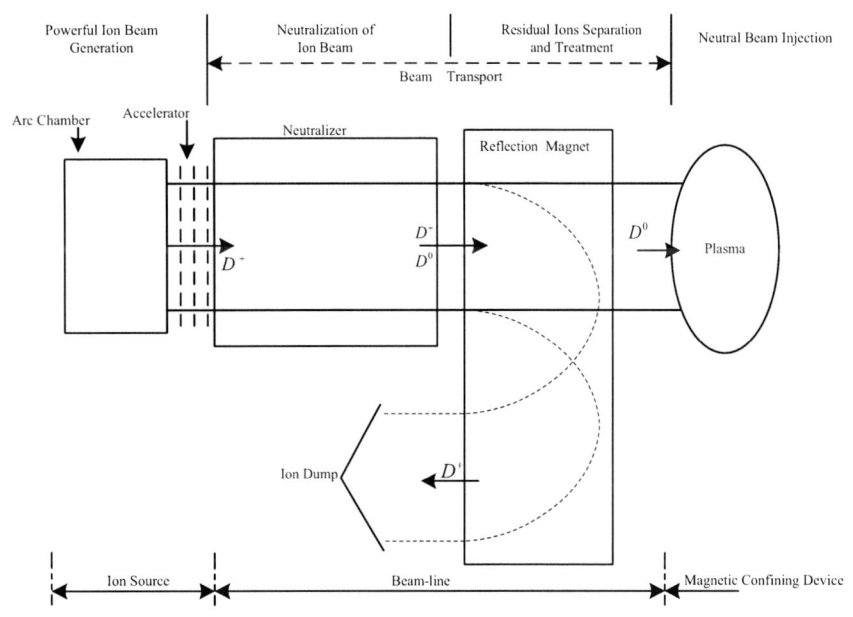

FIGURE 1. the Block Diagram of the Neutral Beam Injector

PARAMETRIC DESIGN FOR REFLECTION MAGNET

Reflection magnet make the residual ions deviated from the beam passage, using the ions' circular motion in uniform magnetic field by the Lorentz force. Considering the cost of device for residual ions separation, the vacuum distribution in the beam-line and additional source of gas due to residual ions deposition, a 180 degree reflection magnet with 90 degree entry and 90 degree exit is chosen for EAST neutral beam injector. In this paper, the parametric design includes magnetic field intensity design, the design of parameter for coil, the estimate for magnetic field region and protection louver design.

Magnetic Field Intensity Design

The archetypical motion of a charged particle in a magnetic field is circular, with the magnetic force providing the centripetal acceleration. Thus, the relationship of the ion energy and the bending radius is expressed as:

$$R = \frac{\sqrt{2mE}}{qB} \tag{1}$$

Where, the charge of the ion is q; the speed for the movement is v, the kinetic energy of the ion is E and B is the magnetic field intensity. The formula shows that the reflection radius is proportional to the square root of the ion's energy, withing the same charge/mass ratio. For the energetic beam from the neutralizer, there are three energy components, the full energy, the half energy and the third energy. For the beam at the dimension of 12 cm×48cm, in order to separate the residual ion from the beam passage and make full use of the space in the vacuum tank, the radius of the third energy ion is not less than 24 cm. Fig. 2 shows the process of the reflection magnet.

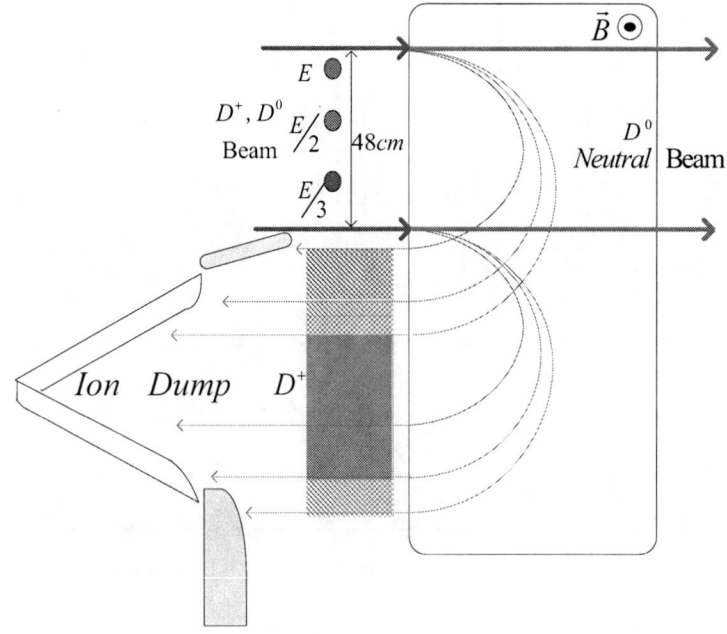

FIGURE 2. the Principle of the Reflection Magnet

When the third energy ion has the radius at 24 cm, the radius for the half energy and the full energy ion are 29.4 cm and 41.5 cm. Consequently, the reflection radius for the full energy ion is chosen as 42 cm. As shown in Fig. 3, the ions at the energy of 80 keV, a 1376 Gs magnetic field is essential for a 42 cm reflection radius. Therefore, an H type dipole magnet should be designed for the 1900 Gs uniform magnetic field, B=1900Gs.

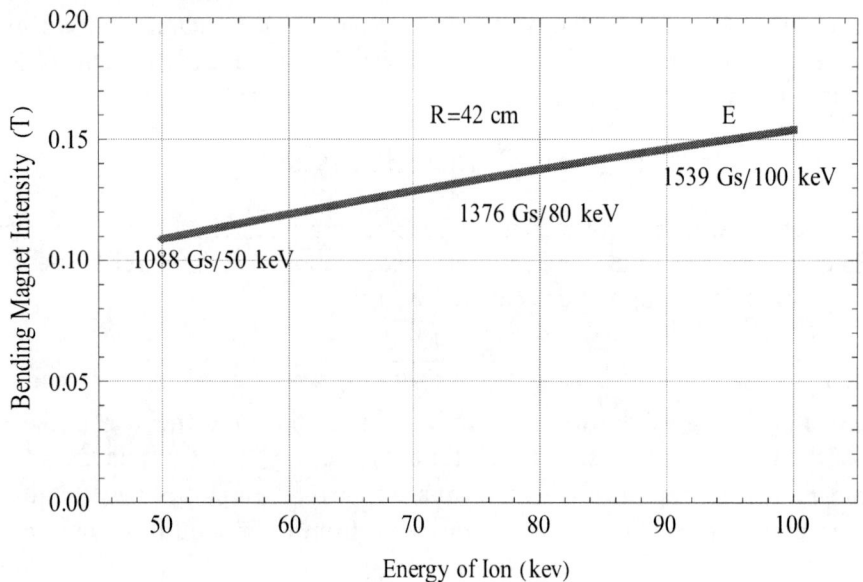

FIGURE 3. the Relationship between Magnetic Field Strength and Deuterium Ion Energy

Design of Parameter for Coil

According to Ampere's law, the excitation ampere turns can be expressed as:

$$NI = \oint H \cdot dl = \oint_{Lair} \frac{B}{\mu_0} \cdot dl + \oint_{Liron} \frac{B}{\mu} \cdot dl = \frac{B}{\mu_0} L_{air} + \frac{B}{\mu_0} L_{iron} \left(\frac{\mu_0}{\mu_{iron}} \right) \quad (2)$$

Where, μ_0 is the permeability of vacuum; μ_{iron} is the permeability of the pole and the magnet core. When the magnet core is not saturated, μ_{iron} is approximately constant and is much large than the permeability of the vacuum, $\mu_{iron} \gg \mu_0$. So, the above equation can be simplified as:

$$NI \approx L_{air} B / \mu_0 \quad (3)$$

The unit for magnetic field intensity B is Wb/m^2. The unit for L_{air} is meter. For an H type solenoid diode, L_{air} is the distance of the poles. For the EAST neutral beam, considering the beam divergence, the gap of the reflection magnet is not less than 20cm, L_{air}=20cm. Then, the design ampere-turn is about 33,500. According to electrical engineering handbook, the current density for external water cooling wire is about $3A/mm^2$, the current density for inner water cooling wire is about $10A/mm^2$. Sectional area of the coil is 11,166mm^2, 3,350 mm^2 for external and inner water cooling, respectively.

Estimation for Magnetic Field Region

In the case of the reflection radius r_m at 42cm, according to the experiential formula: $\xi = r_m/R$, where, ξ is available coefficient for the poles, $\xi = 0.78 \sim 0.91$; R is the radius for magnet pole. For the 180 degree reflection magnet with 90 degree entry and 90 degree exit, a rectangular section is employed for magnet pole. Considering the need to adjust the radius of ion deflection, the dimension for the magnet pole section is about 140 cm×50cm.

Protection Louver Design

During the bending process of the residual ion, few ions will be neutralized by collision with the background molecule in the gap of the magnet. Thus, the resulting energetic atom will move along the tangent direction of the bending track. However, at behind of the reflection magnet, there are located the cryopanels. At the same time, in order to forming a conductance for pumping at the magnet gap region, enough clearance must be reserved. Consequently, a louver is employed for protection the cryopanels. Fig. 4 shows the track for the particle in the magnet gap region. Thus, the injection angle of the energetic atom in the louver region can be expressed as:

$$\theta_1 = ArcSin\left[\frac{R}{\sqrt{w_p^2 + (w-R+r)^2}}\right] + ArcTan\left[\frac{w-R+r}{w_p}\right] \quad (4)$$

$$\theta_2 = ArcSin\left[\frac{R}{\sqrt{w_p^2 + (w-R+2R)^2}}\right] + ArcTan\left[\frac{w-R+2R}{w_p}\right] \quad (5)$$

$$\theta_3 = ArcSin[\frac{R}{\sqrt{w_p^2+(R-r)^2}}] - ArcTan[\frac{R-r}{w_p}] \qquad (6)$$

$$\theta_4 = ArcSin[\frac{R}{\sqrt{w_p^2+(R)^2}}] + ArcTan[\frac{R}{w_p}] \qquad (7)$$

FIGURE 4. the Track for Particle in Magnet Gap

Where, meaning of the parameters in formula is as shown in Fig. 4. The principle of the louver is shown in Fig. 5, the critical angle for louver protection can be expressed as:

$$\beta = ArcTan[\frac{s}{l \times \cos\alpha}] \qquad (8)$$

Where, meaning of the parameters in formula as shown in Fig. 5. Taking $R=42$ cm, $w_p=64$ cm, $r=18$ cm and $w_b=48$ cm as a reference, the magnet louver for EAST neutral beam injector is as shown in Fig. 5 (b). Thus, the energetic atom resulted from the neutralization in reflection magnet gap is kept from colliding with the cryopanels.

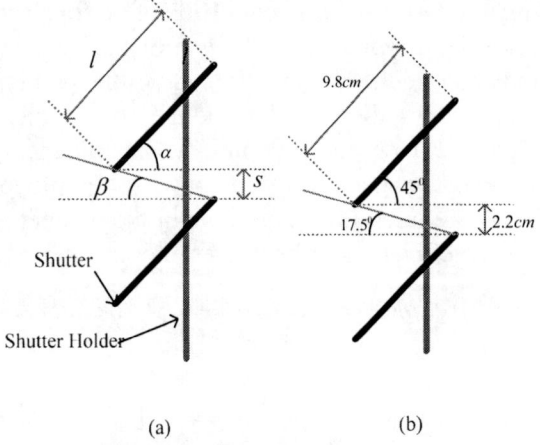

FIGURE 5. the Principle for Louver

MONTE CARLO SIMULATION FOR PARTICLE TRACK

The beam intensity change is core issue for the reflection magnet and the design of the inner elements. The ion orbits and the deposition profile of the power density must be designed carefully in order to ensure that the beam power intensity nowhere exceeds the limits given by the high heat flux elements. Therefore, a 2D bending magnet simulation code (BMS-2D) with Monte Carlo is developed.

Generally speaking, the field variation at the edge of a magnet on the median plane is shown in Fig. 6. The solid curve is the actual field pattern, the dash line is the equivalent hard edged field. At the condition of no saturation at the corners, the magnetic equi-potentials can be derived by using conformal transformation [5]. The result is

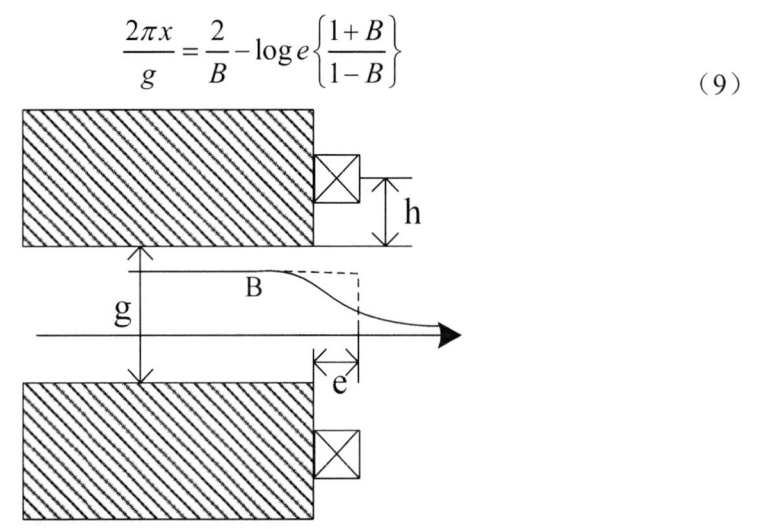

$$\frac{2\pi x}{g} = \frac{2}{B} - \log e\left\{\frac{1+B}{1-B}\right\} \tag{9}$$

FIGURE 6. the Fringing Field of a Magnet

Where, B is field strength; x is measured outward from the pole edge. This formula gives an infinitely distant equivalent hard edge. Considering the presence of exciting coils on the poles, the field distribution formula is modified. The equivalent hard edge occurs at a distance e from the pole face given by

$$e = \frac{g\left[2 - \log e\left(4/q\right)\right]}{2\pi} \tag{10}$$

Where, q is a parameter involving the coil position and is given by

$$\frac{\pi h}{g} = \sqrt{(q-1)} - Arc\tan\sqrt{(q-1)} \tag{11}$$

For the simulation, assuming the field in the equivalent hard edge is uniformity distribution. The beam intensity distribution formula is derived in reference 6. Firstly, using this formula and uniformly distributed random number, the injection position for the energetic particle at the entrance of the magnet can be obtained. Then, two random numbers are employed to determine the charge state and the energy state for the incident particles. In the neutralizer, the molecular ions will excitation by collision and dissociation by itself. Generally, after the neutralizer, there are just D^0 and D^+ with the full energy (E), half energy (E/2) and third energy (E/3) [7]. Consequently, if there is a half energy (E/2) particle injection, the code will have two same particles at this position. For the third energy particle, there will be three particles, using the same injection position, the charge state and the energy state. Because the pimping background pressure and collision cross section, the collisions between particles are ignored. Thus, the track of the energetic particle is determined by equations of motion.

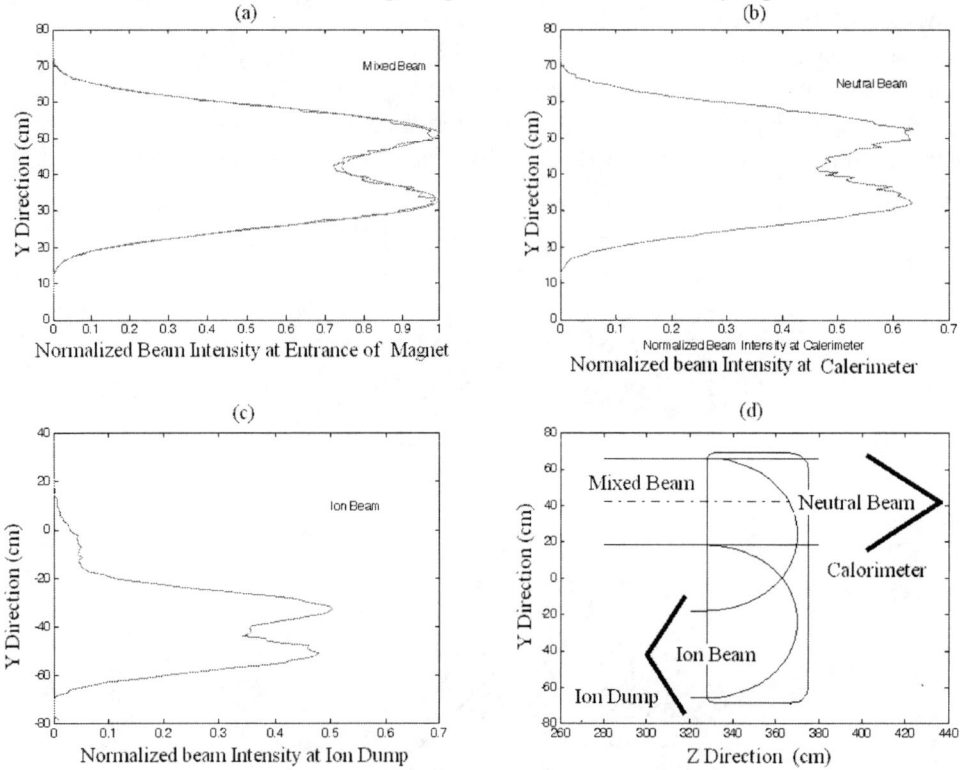

FIGURE 7. the Simulation Result.
(The beam intensity is normalized with the maximal intensity of mixed beam.)

Figure 7 shows the result with 107856 simulated particles, the beam profiles at the entrance of the magnet, calorimeter and ion dump are drawn. All of the beam intensity is normalized with the maximal intensity of mixed beam. In Fig. 7(a), the red dashed curve is the beam intensity given by the formula in reference 6. The blue solid curve is the simulated result by the code. The figure shows that two peaks, resulted from the mechanical convergence of the accelerator grids, will still be there for neutral beam and the residual ion beam. For the ion dump in Fig. 7(c), the width of the ion beam is expanded due to the different bending radius for full energy particles, half energy particles and third energy particles, resulting in the decrease of the beam intensity. The peak value of the ion beam intensity just 44% of that of the mixed beam.

In order to decrease the residual ion beam intensity incident on the ion dump, a magnet with an inclined entry and exit faces are simulated, as shown in Fig. 8. Generally, the inclined entry face makes the ion track to tend to the beam axis. The inclined exit face makes the ion track to away from the beam axis. However, this cut magnet will need more vacuum vessel space and a larger ion dump. At the same time, it will take more gas source at magnet region. So, the cut magnet is not chosen for EAST neutral beam injector.

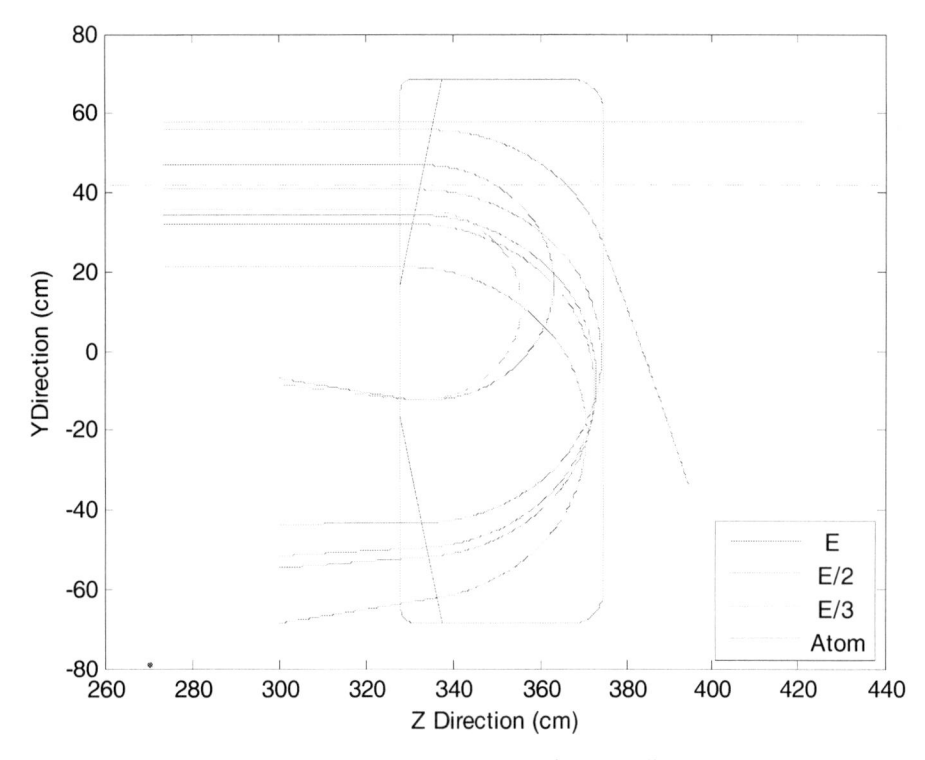

FIGURE 8. the Simulated Track for the Magnet with an Inclined Entry and Exit Faces

CONCLUSIONS AND DISCUSSION

A physical design of the reflection magnet for EAST neutral beam injector was carried out. For a deuterium beam with 42 cm as the bending radius, the intensity of reflection magnet field is about 1376Gs at the energy of 80 keV. In order to obtain this uniform magnetic field, a dipole magnet was designed. The design ampere-turn is about 33,500. The protection louver was introduced for keeping neutralized energetic particles from colliding with the cryopanels. The preliminary result of BMS code showed decrease of the beam intensity at the ion dump and calorimeter.

The BMS code is a 2D simulation code with Monte Carlo. The assumption on ignoring of collisions is not very reasonable. At the same time, the neutralized process in magnet gap region is not considered. In future, above deficiency will be consummated. The development of 3D simulation code is in the plan.

ACKNOWLEDGMENTS

This work is supported by National Natural Science Foundation of China under grant No. 10875146 and the Knowledge Innovation Program of the Chinese Academy of Sciences: the study and simulation on beam interaction with background particles in neutralization area for NBI.

REFERENCES

1. C.D.Hu, Y.L.Xie, Y.H.Xie, "Development of Long Pulse Neutral Beam Injector System for the EAST Tokamak", Chinese Nuclear Society Fall Meeting, Beijing(2009).
2. J. Wesson,.Tokamaks. second ed., Oxford: Oxford Clarendon Press, 1997,pp
3. G. Gibson, W. Lamb, E. Lauer, *Physical Review*, **14**(4):937(1959). pp. 249-281
4. E. Speth, *Rep. Prog. Phys.*. **52**:57 (1989).
5. A. P. Banford. the transport of charged particle beams. Great Britain:E.&F.N. Spon Ltd. 1966. pp. 77-81
6. Y.J.Xu,C.D.Hu,Y.L.Xie, et al., *J Fusin Energ*, (2010)in process.
7. L.Z.Liang, C.D.Hu, Z.M.Liu, et al., *High Power Laser and Particle Beams*. **20**,849(2008).(in Chinese)

NEW APPROACHES AND APPLICATIONS

Negative Ions for Emerging Interdisciplinary Applications

Samar K. Guharay

The MITRE Corporation, 7515 Colshire Dr., MS H407, McLean, VA 22102, USA

Abstract. In many applications related to ion beam-materials interactions negative ions are particularly desirable due to its merit to yield a very low surface charge-up voltage, ~ a few volts, for both electrically isolated surfaces and insulators. Some important applications pertaining to ion beam-material interactions include surface analysis by secondary ion mass spectrometry (SIMS), voltage-contrast microscopy for semiconductor device inspection, materials processing, and ion beam lithography. These applications primarily require vacuum environments. On the other hand, a distinct area of activities constitutes formation of ions and ion transport in ambient environmental conditions, i.e., at atmospheric pressures. In this context, ion mobility spectrometry (IMS) is an important analytical device that uses negative ions and operates at ambient conditions. IMS is widely used in both physical and biological sciences including monitoring environmental conditions, security screening and disease detection. This article highlights several critical issues related to the ionization sources and ion transport in IMS. Additionally, the critical issues related to ion sources, transport and focusing are discussed in the context of SIMS with sub-micrometer spatial resolution.

Keywords: Negative ions, interdisciplinary applications, ion mobility spectrometry, secondary ion mass spectrometry.
PACS: 07.77.Ka; 07.81.+a; 41.75.Cn; 34.50.Gb; 68.49.Sf.

INTRODUCTION

Significant progress has been made in the science and technology of negative ion sources, beam extraction, transport and focusing, especially in the context of high-energy accelerators and magnetic fusion [1]. Another area of major activities using negative ions covers ion beam-materials interactions. Due to its capability for self-regulation of charge-up voltage negative ions are desirable in the applications involving interactions with electrically insulating materials [2, 3]. Some important applications of ion beam-material interactions include new materials development or device fabrication by different processes, for example, ion implantation and ion beam lithography. Many analytical devices utilize ions to identify material characteristics. Secondary ion mass spectrometry (SIMS) is an important example in this pursuit [4, 5]. All of these applications require vacuum environments. On the other hand, ion mobility spectrometry (IMS) operates at ambient environments. The formation of ions and ion transport occurs at ambient conditions, i.e., at atmospheric pressures. This analytical device has the capability to detect and identify widely different materials in part per trillion quantities. IMS can be applied to analytical measurements, environment monitoring, biological and medical applications for diagnosing diseases, and above all, forensic examination and security screening [6, 7]. The capability of IMS to measure trace materials has engaged the scientific community over the years, and renewed interests have developed to enhance the scientific understanding of IMS and advance the state-of-the art.

This article first discusses the critical problems on ion sources and ion transport in the context of IMS. Next, the ion source, beam transport and beam focusing problems are discussed for SIMS with sub-micrometer spatial resolution.

DISTINCT ION SOURCES FOR DIFFERENT APPLICATIONS

As illustrated in Fig. 1 ion sources are, in general, coupled to a beam transport system, which is often assisted by beam acceleration and beam focusing.

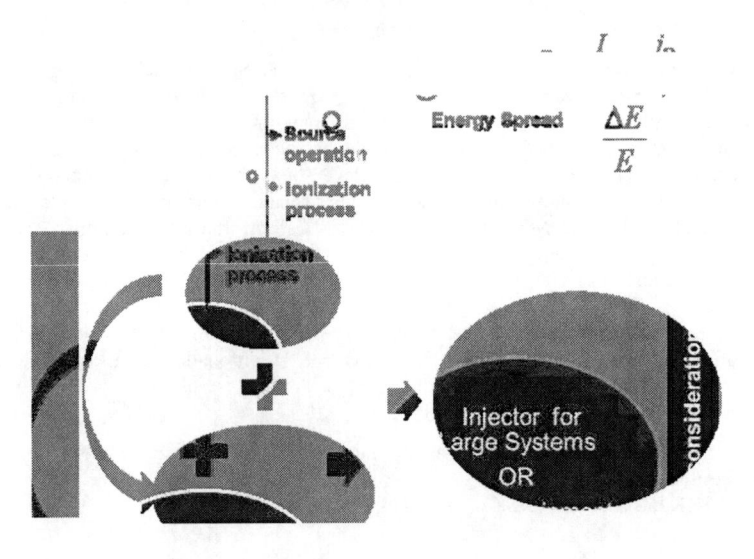

FIGURE 1. Ionization source and associated components

In addition to highlighting the coupling of the ion source to an appropriate downstream system this figure illustrates the key metrics defining the source performance and the quality of the ions extracted from a source. One of the metrics is source brightness (B) that includes intensity (I) of ions extracted from a source as well as the phase-space area of the ions. Often intensity of the ions is expressed in terms of angular intensity, i.e., current per unit solid angle (j_Ω). While intense sources delivering high current density are, by and large, desirable, the emittance (ε), virtual source size (r_v) and energy spread (ΔE) are critical when an ion source is coupled either to charged-particle optics for beam focusing or to an acceleration system that requires very stringent matching criteria at the input.

Figure 2 below shows three distinct areas of applications. These application areas capture both broad and point sources and operations at both ambient and under high vacuum conditions.

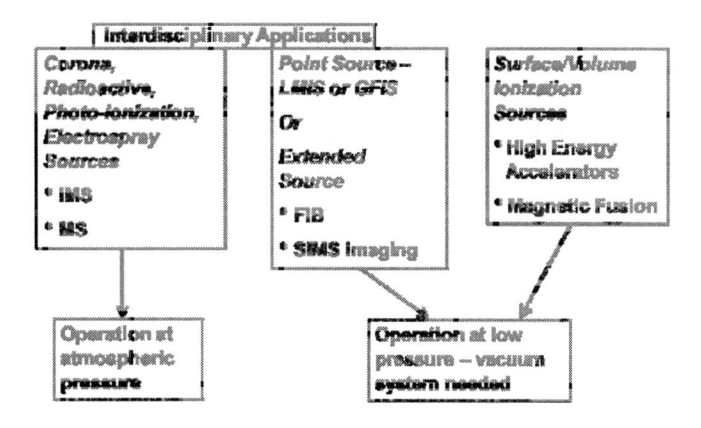

FIGURE 2. Sources for different applications; MS: Mass Spectrometry; LMIS: Liquid Metal Ion Source; GFIS: Gas Field Ionization Source; FIB: Focused Ion Beam

CRITICAL ION SOURCE ISSUES IN TWO DISTINCT INTERDISCIPLINARY APPLICATIONS

Ionization process, ion transport and focusing form the critical basic steps in applications. These issues are discussed below in the context of two distinct interdisciplinary applications: (a) ion mobility spectrometry (IMS) and (b) high spatial resolution secondary ion mass spectrometry (SIMS).

Ion Mobility Spectrometry (IMS)

The schematic diagram in Fig. 3 shows the essential components of an IMS. Ions from an atmospheric pressure ionization source in an IMS diffuse through a counter-streaming neutral background in a drift cell under the action of a uniform electric field E. The ionization source and the drift cell are usually separated by a gate that controls the time of injection of ions from the ionization source into the drift cell. The drift cell has a set of electrodes, typically in a cylindrically-symmetric ring configuration. An important criterion in designing the electrode system is that the axial electric field across the drift cell is uniform. A Faraday plate, accompanied by an aperture gate in front of it, registers the arrival of ions at the end of the drift cell. Typically, dry air is used as a counter-streaming gas through the drift tube.

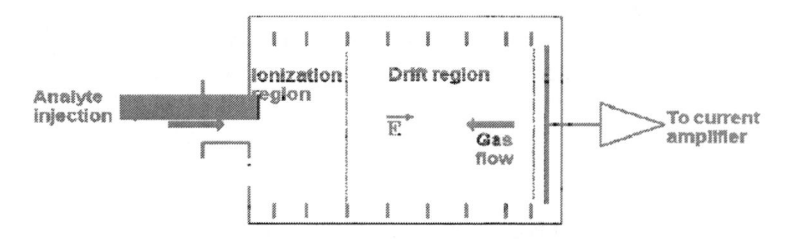

FIGURE 3. Schematics of an IMS system

The time of arrival of the ions at the Faraday plate is governed by the characteristic mobility of ions and the electric field in the drift tube, and it follows the simple expression: $t_d = L/(K\,E)$, where L is the drift cell length, K is the characteristic mobility of ions through the gas medium at the operating pressure and temperature of the gas, and E is the electric field in the drift cell. The spread in drift time (Δt_d) is related to the resolving power of IMS as $R = t_d/\Delta t_d$.

As indicated in Fig. 2, a variety of ion sources is used in IMS [7], and the state-of-the art of the ion sources is still evolving. A better understanding of the formation of the reactant ions and their charge-transfer process including possible pathway for selective ionization forms some of the areas of current investigation. The state-of-the art of IMS can be significantly advanced if

  (a) compact and intense ionization sources can be developed,
  (b) efficient analyte ion formation process is developed,
  (c) analyte ions are efficiently transported from the source to the IMS drift cell without any loss at the gate and through the drift cell.

The performance of IMS has been studied using different sources, and this is described elsewhere [8]. The use of carbon nanotubes [9, 10], nano-electromechanical system (NEMS) based soft-ionization [11], miniaturized electrospray array [12], and photo-ionization forms the crux of some recent developments of compact ionization sources for IMS.

Along with the development of intense ionization sources the other two critical issues are: (a) efficient injection of ions into the IMS drift cell and (b) transport of ions through the drift cell. A recent study on the effect of space-charge interactions [13] shows how this effect governs the key performance metrics of IMS, especially, ion loss and resolving power R. The following two expressions [13] give an insight into the coulomb interactions problem indicating the governing parameters.

$$\frac{1}{R} = \frac{a_2 - 1/R_0}{1 + e^{-a_1(\ln\sigma - a_3)}} + 1/R_0 = \frac{a_2 - 1/R_0}{1 + e^{a_1 a_3}\sigma^{-a_1}} + 1/R_0 \tag{1}$$

$$\ln\sigma = C_1 A + C_2 Q_L + C_3 \tag{2}$$

Here, the effective resolving power R is related to the resolving power R_0 without any consideration of the space charge effects. The notation σ refers to space-charge density, Q_L denotes loss fraction over a range of 0.05 to 0.6, the parameters a_1, a_2 and a_3 depend linearly on the aspect ratio A (= L/d, d being the inner diameter of the electrode rings in the drift tube), and the coefficients C_1, C_2, and C_3 are obtained from the fitting functions for ion loss fraction versus input charge density curves for different aspect ratios as described in detail in ref. 13. The goal is to maximize R and minimize Q_L for the highest input charge density. This is determined by finding out the optimum value of σ up to which these parameters are insensitive to the variation of σ.

The IMS systems show a current of about 1 nA at the Faraday plate for a typical pulse length of 0.2 ms when the diameter of the IMS tube is a few cm. The estimated charge density is about 10 fc/cm^2. It is shown in ref. 13 that the space-charge effect may not be serious until about 1 pc/cm^2. This indicates that there is ample room for injecting more charges into the IMS drift tube. The unresolved questions are:

(a) Is the IMS performance, especially the sensitivity, limited by the ionization source?

(b) Are ions generated in the ionization source efficiently transported through the drift tube? Or in other words, is the IMS performance limited by ion transport through the drift cell?

If the item (a) governs the issue, then developing intense ionization sources will enhance the IMS performance. On the other hand, if there is a significant loss of ions during ion transport through the drift cell, then it is important to develop a good understanding of the reason for ion loss and mitigate the problem. This is an unresolved issue, and further systematic studies using an axially moveable current monitor, for example, a Faraday cup, can be performed to identify the source of ion loss, if any.

High Spatial Resolution Secondary Ion Mass Spectrometry

Secondary ion mass spectrometry (SIMS) has captured interests in widely diverse disciplines. One of the recent activities is related to imaging with sub-100 nanometer spatial resolution. Understanding many basic biochemical processes requires study of mapping of chemicals in cells and tissues with nanometer or better spatial resolution. The critical issue in obtaining enough signals due to secondary ions from a surface under examination is the number of primary ions available within the sampling area under observation and the corresponding secondary ion yield. Without high-brightness ions from a source and a good coupling optics to converge these ions into a spot size of ~ nanometer scale, it is difficult to obtain sufficient secondary ions from the sample.

The primary ion species in SIMS are determined on the basis of the electron affinity or ionization potential of the materials that need to be liberated as secondary ions from a surface or matrix. Either Xe^+ or O^- is desirable from the considerations of efficient secondary ion yield from materials.

Coupling a high-brightness dc source with an appropriate focusing optics in a manner, as shown in Fig. 4, can deliver a spot size of ~ sub-100 nanometer at the target with a current of 10 pA [4]. The spot size at the target is primarily determined from the following expression – here the input beam is assumed to be almost paraxial so the contribution of geometric or spherical aberration can be neglected.

$$\delta_{total} = (\delta^2_{geometric} + \delta^2_{chromatic} + \delta^2_{stochastic})^{1/2} \qquad (3)$$

Here, the term $\delta_{geometric}$ is determined from the virtual source size r_v as $\delta_{geometric} = M\, r_v$ where M is the magnification factor of the lens system that focuses ions from the source to the final spot. The other parameters, namely, $\delta_{chromatic}$ and $\delta_{stochastic}$ represent the contribution to the spot size due to chromatic aberration and stochastic space-charge interactions. Note that while the distortion due to the global space-charge effect can be compensated or corrected by adjusting the position of the target, the role of stochastic space-charge effects is rather complex. The following expressions give an idea about the parameters that govern $\delta_{chromatic}$ and $\delta_{stochastic}$:

$$\delta_{chromatic} = M\, C_c\, \alpha_o\, \frac{\Delta E}{E} \qquad (4)$$

$$\delta_{stochastic} = const. \frac{I^{1/2} f_2 m^{1/4}}{E^{5/4} \alpha_c D^{1/2}} \tag{5}$$

Here, C_c is the chromatic aberration coefficient of the focusing lens, α_0 denotes the beam angle at the object side of the lens, ΔE is the energy spread of the ions and E is the axial energy of the ions. Additionally, the terms in the expression for $\delta_{stochastic}$ include I for the current, f_2 for the focal length for the final lens, m the mass of the ion species, E the energy of the beam, α_c the beam angle at cross over and D the diameter of the cross over. The article by Hammel [14] describes the significance of the terms for $\delta_{stochastic}$.

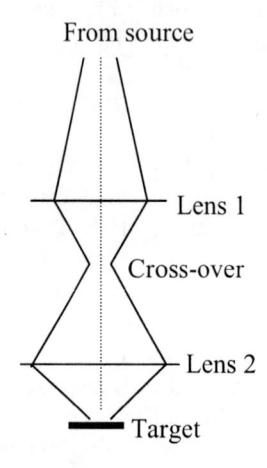

FIGURE 4. Schematics of a focused ion beam showing cross-over

The key issues for an overall high-performance SIMS imaging device are to:
(a) achieve a high-brightness and low energy spread dc O^- source,
(b) minimize ion-optical column length to reduce trajectory displacement effect which causes lateral displacement of the beam and reduces beam brightness,
(c) minimize number of cross-over to reduce Boersch effect which reduces the edge sharpness and increases the final probe size,
(d) use the lightest element suitable for applications,
(e) evaluate the optimum combination of source brightness and effective opening angle to get the maximum current within a probe.

CONCLUSIONS

Negative ions play important roles in many areas including basic and applied sciences. The needs for intense and compact negative ionization sources and associated beam transport and beam focusing systems are stressed upon in the context of ion mobility spectrometry and high spatial resolution secondary ion mass spectrometry. These two analytical devices have relevance to applications covering diverse disciplines.

ACKNOWLEDGMENTS

The author acknowledges the support of the MITRE Innovation Program. Approved for public release: 10-4688, Distribution Unlimited.

REFERENCES

1. Special issue on Ion Sources, Fundamentals and Applications, ed. L. R. Grisham, M. Bacal and S. K. Guharay, *IEEE Transactions on Plasma Science*, vol. 36, No.4, 2008.
2. J. Ishikawa, in *"The Physics and Technology of Ion Sources"*, ed. Ian G. Brown, Wiley, 2nd Ed., 2004; J. Ishikawa, Review Scientific Instruments, vol. 67, 1410, 1996.
3. P. N. Guzdar, A. S. Sharma, and S. K. Guharay, *Applied Physics Letters*, vol. 71, 3302, 1997.
4. S. K. Guharay, S. Douglass, and J. Orloff, *Applied Surface Science*, vol. 231-232, 926, 2004.
5. S. K. Guharay, J. Orloff and M. Wada, *IEEE Transactions on Plasma Science*, vol. 33, 1911, 2005.
6. G. A. Eiceman and Z. Karpas, *Ion Mobility Spectrometry*, Taylor & Francis, 2nd Ed., New York, 2005.
7. S. K. Guharay, P. Dwivedi and H. H. Hill, Jr., *IEEE Transactions on Plasma Science*, vol. 36, 1458, 2008.
8. V. Matsaev, M. Gumerov, L.K. Krasnobaev, et al., *International J. of Ion Mobility Spectrometry*, vol. 3, 112 (2002).
9. R.L. Fink, N. Jiang, K.N. Leung, A.J. Antolak, *Nanotech Conference & Expo*, Houston, May 3-7, 2009.
10. P.J. Traynor and R.G. Wright, US Patent 6,885,010, 2005.
11. Frank T. Hartley, and Isik Kanik, "A nanoscale soft-ionization membrane: a novel ionizer for ion mobility spectrometers for space applications", *Proc. SPIE, Nano- and Microtechnology: Materials, Processes, Packaging and Systems*, ed. D.K. Sood, A.P. Malshe, R. Maeda, vol. 4936, p.43, 2002.
12. X. Sun, R.T. Kelly, K. Tang, R.D. Smith, *Analyst*, vol. 135, 2296, 2010.
13. A. V. Mariano, W. Su, and S. K. Guharay, *Analytical Chemistry*, vol. 81, 3385, 2009.
14. E. Hammel, A. Chalupka, J. Fegeri, et al., *Journal of Vacuum Science and Technology B*, vol. 12, 3533, 1994.

Extraction and Acceleration of Ions from an Ion-Ion Plasma

Lara Popelier, Ane Aanesland, Pascal Chabert

Laboratoire de Physique des Plasmas – Ecole Polytechnique, Route de Saclay, 91128 Palaiseau, France

Abstract. Extraction and acceleration of positive and negative ions from a strong electronegative plasma and from an ion-ion plasma is investigated in the PEGASES thruster, working with SF_6. The plasma is generated in a cylindrical quartz tube terminated by metallic endplates. The electrons are confined by a static magnetic field along the axis of the cylinder. The electron mobility along the field is high and the electrons are determining the sheaths in front of the endplates. The core plasma potential can therefore be controlled by the bias applied to the endplates. An ion-ion plasma forms at the periphery as a result of electron confinement and ions can freely diffuse along the perpendicular direction or extraction axis. Langmuir probe and RFEA measurements are carried out along this axis. The measured ion energy distributions shows a single peak centered around a potential consistent with the plasma potential and the peak position could be controlled with a positive voltage applied to the endplates. When the endplates are biased negatively, the plasma potential saturates and remained close to 15 V. A beam of negatively charged particles can be observed under certain conditions when the endplates were biased negatively.

Keywords: Electronegative plasma, ion-ion plasma, magnetic filtering, extraction, acceleration, ion energy analyzer, thruster.
PACS: 52.27.Cm, 52.59.Bi

INTRODUCTION

Classical electric propulsion is based on the extraction and the acceleration of positive ions from electropositive plasmas, for example using xenon due to its combination of low ionization potential and high atomic mass [1,2]. The simplest concept for ion acceleration is to apply an electrostatic field between the source and an electrode. Gridded thrusters have isolating walls allowing the plasma to float and uses two or more grids for the extraction and acceleration: The first grid in contact with the plasma is highly positive and the plasma floats on top of this high potential. The sheath in front of the plasma grid is therefore of low potential and determined by the plasma density and electron temperature. The second one is negative with respect to the plasma grid, to create the potential drop through which ions fall. Thus ions fall through the sheath in front of the first grid with a current given by Child-Langmuir law, and then accelerated with the second grid to generate thrust.

Extracting positive charges induces a built up of a negative polarization of the whole system [1]. An electron emitting hollow cathod placed at the thruster exit is added to neutralize the expelled ions, this produces a beam of zero net charge and avoids thrust cancellation. But the ion-electron recombination is a rather slow process: a plasma in the downstream region of the engine, called plasma plume, forms and may damage the spacecraft and in particular the cathode. Thus the cathode has a short lifetime and this issue limits the lifetime and use of classical thrusters. New concepts for electric propulsion are intensively studied in order to increase cathode lifetime or even to avoid its use.

It has been reported that ion-ion plasmas can offer useful advantages in a variety of applications where neutral or quasi-neutral beams are used. Some examples are charge-free etching in the semiconductor industry, neutral beam injections for fusion, or for electric propulsion in space applications. Ideal ion-ion plasmas (or electron-free plasmas) that contain only positive and negative ions, or in practice, plasmas in which the main negative charge carriers are negative ions (a small amount of electrons can exist but do not rule plasma dynamics).

Knowledge of thruster principles and limitations together with a growing interest in ion-ion plasmas led to the development of the PEGASES thruster concept [1,2], meaning Plasma propulsion with Electronegative Gases. An electronegative plasma containing positive ions, negative ions and electrons, is used as an ion source. Electrons are magnetically filtered and an ion-ion plasma is obtained in the extraction stage. Positive and negative ions are then accelerated to generate the thrust. Ion-ion recombination is often more efficient than electron-ion recombination, so a beam with few charged particles in the downstream of the engine is expected and a hollow cathode is no longer necessary.

The PEGASES thruster is a gridded ion thruster, and also here the design of the grids depends on the sheath formation in front of the grid. Due to the lower temperature and greater mass of negative ions compared to electrons, the sheath structure in ion-ion plasmas differs significantly from that of conventional electron-ion plasmas. The PEGASES thruster concept does not rely on the way the power is coupled to the plasma. However, the first prototype, presented here, operates in helicon or inductive mode, depending on the power and the magnetic field.

In classical systems, the plasma is allowed to float and the plasma grid therefore sets the acceleration potential. This is due to the presence of electrons so that an electropositive space charge sheath forms and the plasma is always above the most positive potential [13]. Simulations of the sheath structure in ion-ion plasmas have shown that both positive and negative space charge sheaths can form, and the plasma does not have to be at the most positive potential [3-5]. Draghici et al. [6] showed that plasma potential in an ion-ion plasma could not be controlled with a positively biased electrode in the ion-ion region. We investigate here, with Langmuir probes (LP) and Retarding Field Energy Analyzer (RFEA) in an SF_6 plasma, how ion-ion plasma potential can be affected by a dc bias electrode in the core region which still contains mobile electrons. Results are compared to the case of an Argon plasma in the truster.

EXPERIMENTAL SET-UP

The geometry of the first PEGASES thruster prototype is cylindrical (Fig.1). The gas is introduced in the center of a 35cm long and 6cm diameter quartz tube. The plasma is created with a three-turn loop antenna wrapped around the tube and continuously excited by a power generator at 13.56MHz through a matching network using a Π-circuit; working power is 300W. A set of four coils with the same DC current flowing is placed symmetrically around the cylinder to create the magnetic field; working current is 1.5 A. The plasma is operated at low powers to avoid extensive heating and the RF coupling in these experiments are therefore most likely inductive. Two extraction zones are placed perpendicularly to the cylinder axis and the magnetic field lines (along the X axis). The magnetic field strength is chosen so that electrons are magnetized but ions are unmagnetized as their Larmor radius is larger than the cylinder radius. A set of two magnets yielding a 750G field can be arranged along one extraction tube to enhance the magnetic filtering in the extraction region where the ion-ion plasma is obtained.

Aluminium plates with a surface of 20 cm^2 are placed in contact with the plasma at the ends of the cylinder and are perpendicular to the magnetic field lines. Those plates can be either grounded or dc biased; endplate bias is addressed as V_{ep} and is referenced to the grounded vacuum chamber.

As there were no access to pressure gauges in the thruster, the value of entering gas flow in standard cubic centimeter per minute (sccm) was chosen as entry parameter. Experiments in SF_6 were carried out for 28 sccm; Argon flow was 40 sccm. These flows correspond to a pressure of 1-5 mTorr in the thruster.

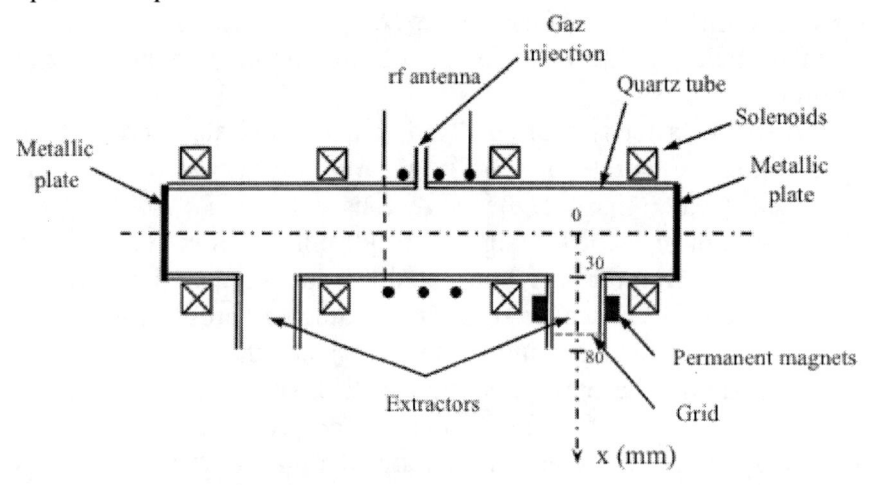

FIGURE 1. First prototype of the PEGASES thruster.

DIAGNOSTICS

Plasma potential V_p is the most interesting parameter in this study as it is essential in the creation of the potential drop through which ions fall. How V_p is deduced from the different diagnostic methods is detailed here.

A cylindrical Langmuir probe is used to access current-voltage characteristics from which local plasma parameters are deduced. The probe is rf-compensated for 13.56 MHz and 27 MHz with rf chokes. The probe tip, 6 mm long and 0.25 mm diameter, was made of platinum to avoid etching by SF_6. Langmuir probe measurements were spatially resolved along the x-axis as the probe was introduced in the thruster via the extractor. Time-averaged plasma potential at a given position along the x-axis was determined by using the crossing point given by drawing straight lines through the IV curve in the transition and electron saturation regions.

In electronegative plasmas, the current collected by a Langmuir probe biased positively is very large compared to the positive ion current collected when the probe is biased negatively. This is due to the presence of mobile electrons in addition to negative ions: electrons still dominate the negative current. In ion-ion plasmas the characteristics shows a negative particle saturation current of a magnitude comparable to the positive ion saturation current is therefore a good indication of an ion-ion plasma in which electrons are negligible.

In addition to plasma potential and density, IEDF is an important parameter for space propulsion application. Langmuir probe measurements cannot give this kind of information. Consequently, the use of an ion energy analyzer (or RFEA for Retarding Field Energy Analyzer) is necessary. The analyzer has been described previously [9], it has four plane grid, a plane collection plate and is differentially pumped. The entrance grid G, facing the plasma, is fixed on a grounded diaphragm with an aperture 3mm in diameter to reduce the open area seen by the plasma. The total system length is 1.02 mm. The Semion RFEA system application from Impedans was used as data acquisition unit. Because of the comparable dimensions of the extractor and the RFEA, all measurements with RFEA were done at x = 85 mm and could not be done inside the extraction zone.

A schematic layout of the ion energy analyzer is given on Fig. 2 for positive ions energy measurement. Positive ions that enter the analyzer through the aperture in G are accelerated towards a negatively biased grid while negative charges are repelled: this grid is called repeller and referred as R. After having passed grid R, ions are decelerated and filtered depending on their kinetic energy in the x direction by the retarding field between R et the discrimination grid D: ions having kinetic energies higher than the potential barrier put up by D are allowed to pass. The fourth grid is called the secondary grid S and fulfills two roles: the collected species is accelerated again, and a potential difference higher than the maximum energy of secondary electrons is set with the collector, ensuring that emitted electrons are collected. Collector C is maintained at a constant potential, here at -30V.

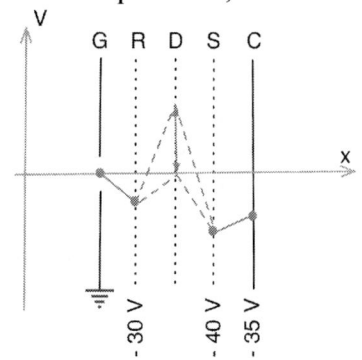

FIGURE 2. Layout of the ion energy analyzer or RFEA. Voltages are set to measure IEDF in the case of positive ions.

The ion current to the collector was recorded by the data acquisition unite as a function of the dc voltage applied to the discriminator: for each value of grid D, the collected current corresponds to all particles with a kinetic energy higher than the potential difference and the corresponding I-V_D curve decreases as V_D is increased. Its first derivative is proportional to the IEDF from which plasma parameters are deduced. If the derivative presents only one narrow peak, it can be approximated with a gaussian centered around V_1; it means a monoenergetic particle population of energy eV_1 and V_1 is usually taken as the plasma potential in the reactor as most of ions fall through the collisionless sheath in front of the grounded grid of the analyzer [11].

RESULTS AND DISCUSSION

It has been shown previously that there are different methods to generate an ion-ion plasma in the extractor of PEGASES [8,12]. In all cases, the region in the plasma cylinder (core) is dominated by the electron dynamics, where the electrons are mobile along the fieldlines, while in the region of the extractor the dynamics is dominated by heavy ions. Fig. 3 shows three Langmuir probe characteristics obtained for various potentials applied to the endplates: -40 V, 0V and +40V. In all cases an ion-ion plasma exists, seen by the symmetrical shape of the IV curve. However, the plasma potential is shifted according to the biased endplate voltages.

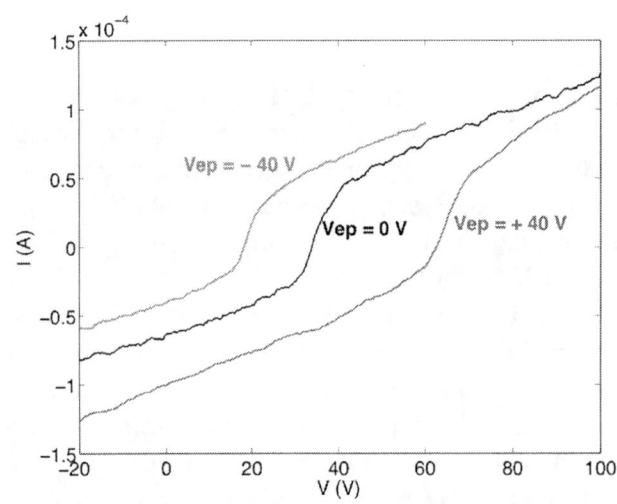

FIGURE 3. Langmuir characteristics in a SF_6 plasma at x = 82 mm for various endplates bias.

The IEDF measured in Argon plasmas and in ion-ion plasmas obtained using SF_6 with magnetic filtering shows narrow gaussian distributions indicating monoenergetic positive ion population in both cases. Fig. 3 shows the IEDF in the case of an ion-ion plasma for various biases on the endplates. It can be seen that the endplate voltage affects the peak position. The plasma potential, deduced from the IEDF, is shown in Fig. 5 as a function of the endplates bias, where diamond and square correspond to SF_6 and Ar, respectively. The linear increase in plasma potential is consistent with the linear increase in V_{ep} for positive voltages on the endplates. Indeed, the plasma potential floats up to the most positive potential [13] and equals the sum of the plasma potential at 0V and the applied voltage. For negative applied bias, a saturation at 15V is obtained. This saturation can be explained by a positive sheath created between the plasma and the grounded grid of the analyzer: ions acquired energy by falling through this sheath wich correspond to 15eV. Thus biasing endplates in the core of the electronegative plasma (where electrons rule the particles dynamics) did not affect ion-ion plasma formation, but enabled control of plasma potential.

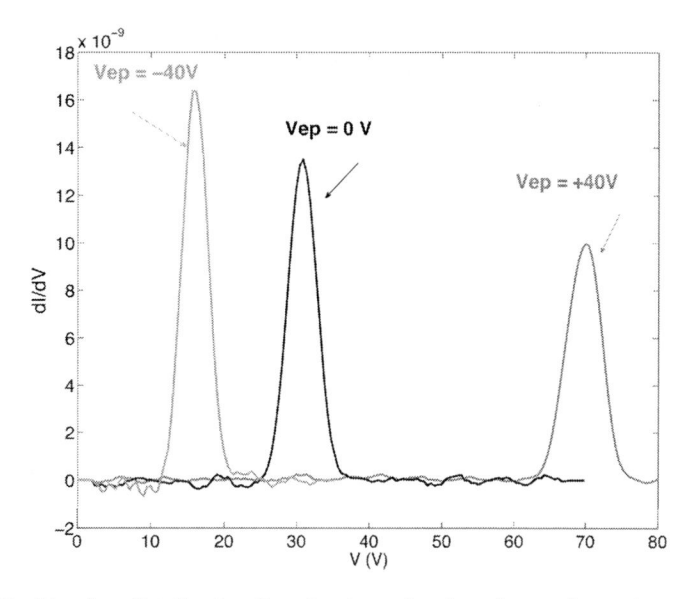

FIGURE 4. Positive ion distribution function in an ion-ion plasma for various endplates biases measured with RFEA at x = 82 mm.

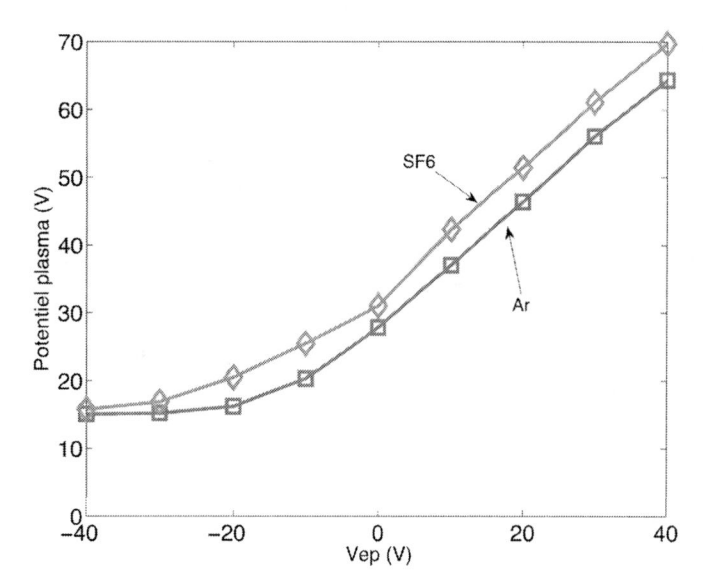

FIGURE 5. Plasma potential as a function of dc bias on endplates for Argon plasma and SF6 plasma measured with RFEA at x = 82 mm.

CONCLUSION

The plasma potential can be controlled in the ion-ion plasma region of the PEGASES thruster by dc-biased endplates placed in the electronegative core of the discharge. Although the measured ion energy distribution functions are not measuring accelerated ions, but ions falling through the sheath in front of the analyser, one can expect that by replasing the analyser by a grounded grid would allow the ions to be accelerated throught the grid and form an ion beam.

Negative ion energy distribution functions remains to be measured as the entrance of the analyser was grounded, and therefore very sensitive to any residue electrons. However, we expect a similar contol of the energies of negative ions, as in ion-ion plasmas negative space charge sheaths can form in front of the grid.

ACKNOWLEDGMENTS

This work is financially supported by EADS Astrium.

REFERENCES

1. R.G. Jahn, *Physics of Electric Propulsion*, New York: McGraw-Hill, chapter 7 (1968).
2. D.M. Goebel and I. Katz, *Fundamentals of Electric Propulsion: Ion and Hall Thrusters*, John Wiley & Sons (2008).
3. A. Meige, G. Leray, J.L. Raimbault, P. Chabert, Appl. Phys. Lett. **92**, 061501 (2008).
4. V. Midha and D. J. Economou, J. Appl. Phys. **90**, 1102 (2001).
5. V. Midha, B. Ramamurthi, and D. J. Economou, J. Appl. Phys. **91**, 6282 (2002).
6. M. Draghici, E. Stamate, J.phys. D: Appl. Phys. **43**, 155205 (2010) .
7. P. Chabert, 2007 WO 2007/065915 A1 (patent).
8. A. Aanesland, A, Meige, P. Chabert, Journal of Physics: Conference Series **162** (2009).
9. A. Perret, P. Chabert, J. Jolly, J.-P. Booth, Appl. Phys. Lett. **86**, 021501 (2005).
10. M.A. Lieberman and A.J. Lichtenberg, *Principles of Radiofrequency Discharges and Material Processing*, 2nd edition, New York: Wiley-Interscience (2005).
11. D. Gahan, B. Dolinaj B, M.B. Hopkins, Plasma Sources Sci. Technol. **17,** 5026 (2008).
12. A. Aanesland, L.Popelier, G. Leray, P. Chabert, S. Mazouffre, D. Gerst, *Plasma propulsion with Electronegative gases*, 31th International Electric Propulsion Conference, Ann Arbor, Michigan, USA, September 20-24, 2009.
13. F.F. Chen and J.P. Chang, *Lecture notes on Principles of Plasma Processing* (New York: Kluwer/Plenum, p34), 2003.

Development of a Negative Hydrogen Ion Source for Spatial Beam Profile Measurement of a High Intensity Positive Ion Beam

Katsuhiro Shinto[a], Motoi Wada[b], Tomoaki Nishida[b], Yasuhiro Demura[b], Daichi Sasaki[b], Katsuyoshi Tsumori[c], Masaki Nishiura[c], Osamu Kaneko[c], Masashi Kisaki[d] and Mamiko Sasao[d]

[a]*Japan Atomic Energy Agency (JAEA), Rokkasho, Aomori 039-3212, Japan*
[b]*Doshisha University, Kyotanabe, Kyoto 610-0394, Japan*
[c]*National Institute for Fusion Science (NIFS), Toki, Gifu 509-5292, Japan*
[d]*Tohoku University, Aoba, Sendai, Miyagi 980-8579, Japan*

Abstract. We have been developing a negative hydrogen ion (H$^-$ ion) source for a spatial beam profile monitor of a high intensity positive ion beam as a new diagnostic tool. In case of a high intensity continuous-wave (CW) deuteron (D$^+$) beam for the International Fusion Materials Irradiation Facility (IFMIF), it is difficult to measure the beam qualities in the severe high radiation environment during about one-year cyclic operation period. Conventional techniques are next to unusable for diagnostics in the operation period of about eleven months and for maintenance in the one-month shutdown period. Therefore, we have proposed an active beam probe system by using a negative ion beam and started an experimental study for the proof-of-principle (PoP) of the new spatial beam profile monitoring tool. In this paper, we present the status of development of the H$^-$ ion source as a probe beam source for the PoP experiment.

Keywords: Negative ion beam, Beam profile monitor, Active beam probe technique
PACS: 29.25.Ni, 41.75.Cn, 41.85.Qg

INTRODUCTION

Ion accelerators with their beam power of MW class, such as J-PARC (Japan Proton Accelerator Research Complex) at Tokai in Japan [1] and SNS (Spallation Neutron Source) at Oak Ridge in the United States [2], appear at the beginning of the 21st century. Other accelerators such as ESS (European Spallation Source) [3] in Europe and IFMIF (International Fusion Materials Irradiation Facility) [4], which deliver the beams of the power one order higher than the existing systems, are planned to be constructed. Figure 1 shows beam powers of the proton/deuteron accelerators in the world. As shown in the figure, most of the accelerators serve the beam with high energy (GeV region for nuclear and particle physics and several hundred MeV region for life and material sciences) and low average current, while the IFMIF accelerator and its prototype will produce beams with the energies of several tens of MeV and the average current intensities of several hundred mA.

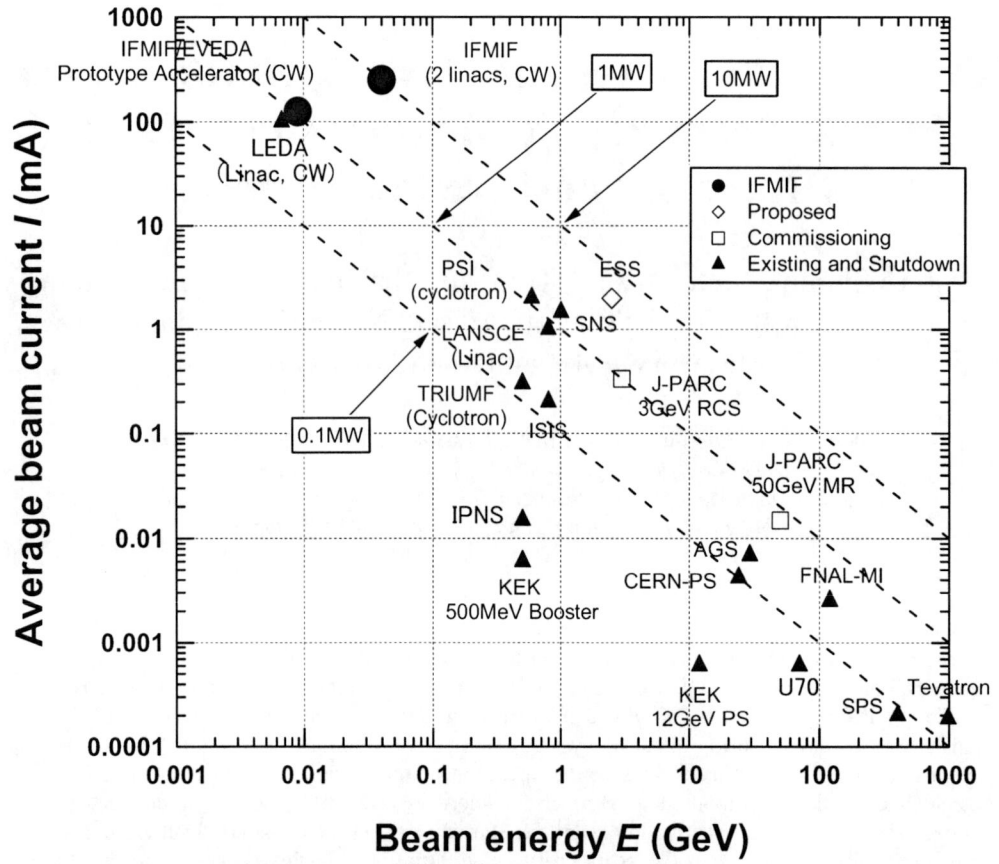

FIGURE 1. Beam power chart of proton/deuteron accelerators in the world.

The IFMIF is an accelerator driven neutron source to simulate the irradiation conditions of the intense neutron flux of 10^{18} n/m^2/s with the energy spectrum produced by D-T nuclear reactions. Though D-T nuclear experiments will be performed in ITER, the amount of neutrons produced in the ITER will be two orders of magnitude smaller than that produced by the next stage fusion demonstrator called DEMO. The IFMIF should be in charge of materials performance assessment under a high flux of energetic neutrons corresponding to the DEMO conditions.

The IFMIF accelerator consists of two linacs each of which provides a continuous-wave (CW) positive deuterium ion (D$^+$) beam. One single linac delivers a beam current of 125 mA (total 250 mA by two linacs) at the beam energy of 40 MeV. During the CW operation of the IFMIF linacs, the extremely high intensity D$^+$ beams and severe radiation environment make the beam diagnostics by conventional techniques in the transport lines next to impossible. A widely used wire-scanning type beam profile monitor can cause severe scattering of D$^+$ to damage beam line components. Components including the wire monitor itself become radioactive after a certain period of operation. A gas cell type beam profile monitor deteriorates vacuum of the beam line as well as the optics of the main D$^+$ beam. Other existing methods also have defects in the usage of the IFMIF beam line conditions.

We have proposed an active beam probe system by using a negative ion beam as a new tool to diagnose beam profiles of high-intensity positive ion beams such as the IFMIF D^+ beam [5]. To monitor beam profile of the positive ion beam from a remote position, the negative ion beam is injected into the positive ion beam perpendicularly, and the negative-ion-beam attenuation due to the beam-beam interaction is measured at each point. We have started an experimental study for the proof-of-principle (PoP) of the new spatial beam profile monitoring tool. This paper presents the concept of the system, and then progress of developing an H^- ion source as the probe beam source for the PoP experiment.

PRINCIPLE OF A NEGATIVE ION PROBE SYSTEM FOR HIGH INTENSITY BEAM PROFILE MONITORING

Figure 2 shows a conceptual drawing of the negative ion beam probe system to diagnose the beam profile of the high-intensity positive ion beams. A negative ion source is placed where the radiation level is low enough for hands-on maintenance. A negative ion beam extracted from the source must not affect the positive ion beam optics, while it must have a high brightness in order to fly a long distance from the probe beam source to detectors of positive/negative ion and neutral beams. To satisfy these constraints, the probe negative ion beam can have the cross section of a long rectangular shape. In case of the IFMIF accelerator, a rectangular H^- beam crosses the D^+ beam perpendicularly as shown in Fig.2. The long side of the rectangle can cover the entire cross section of the D^+ beam, while the short side should be thin so as not to disturb the target beam. Thin thickness of the H^- beam is also essential for measurement of the spatial distribution of the target beam with enough precision.

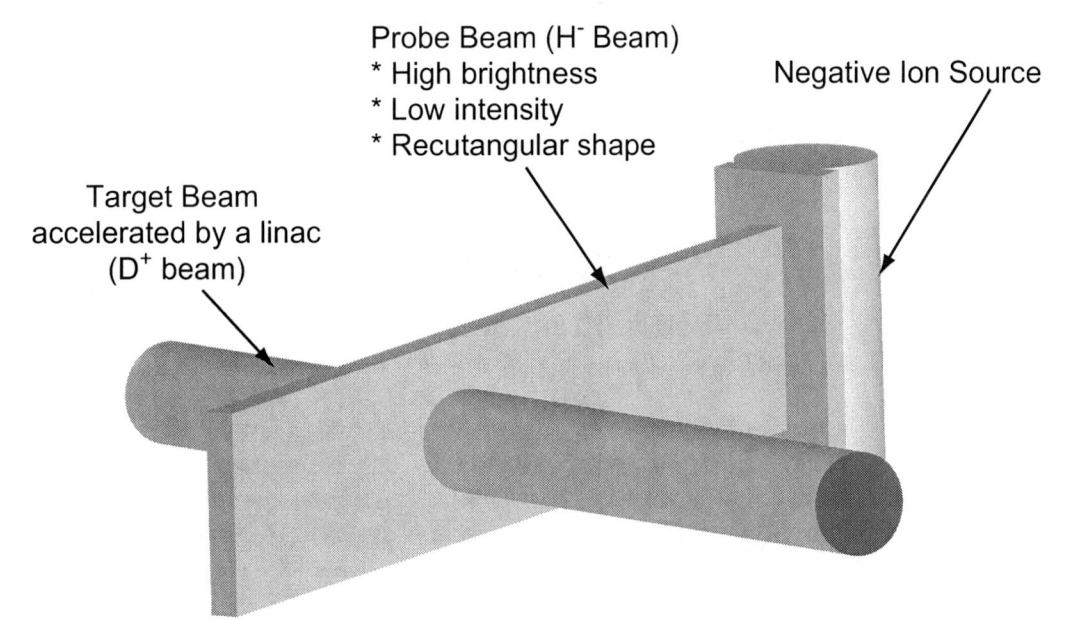

FIGURE 2. A conceptual drawing of the negative ion beam probe system.

The H⁻ beam is attenuated by the collisions with D⁺ ions in the beam. The produced neutrals are further converted to positive ions by the following processes.

$$H^- + D^+ \rightarrow H^0 + e^- + D^+$$

$$H^0 + D^+ \rightarrow H^+ + e^- + D^+$$

The relative velocity between H⁻ and D⁺ is so high that the cross section of the electron capture of D⁺ is negligible. At low energy regions, the cross section of electron detachment from H⁻ by H⁻+H⁺ collisions has been investigated and the results are reported [6, 7]. These are shown in Figure 3. The cross section data in the high barycentral energy regions are not available and the cross section is estimated by extrapolating the data in low energy region assuming classical energy dependence. The results show that the beam attenuation distance is larger, and a very small portion (~10^{-5}) of H⁻ is converted to neutrals due to collisions with D⁺ in IFMIF conditions.

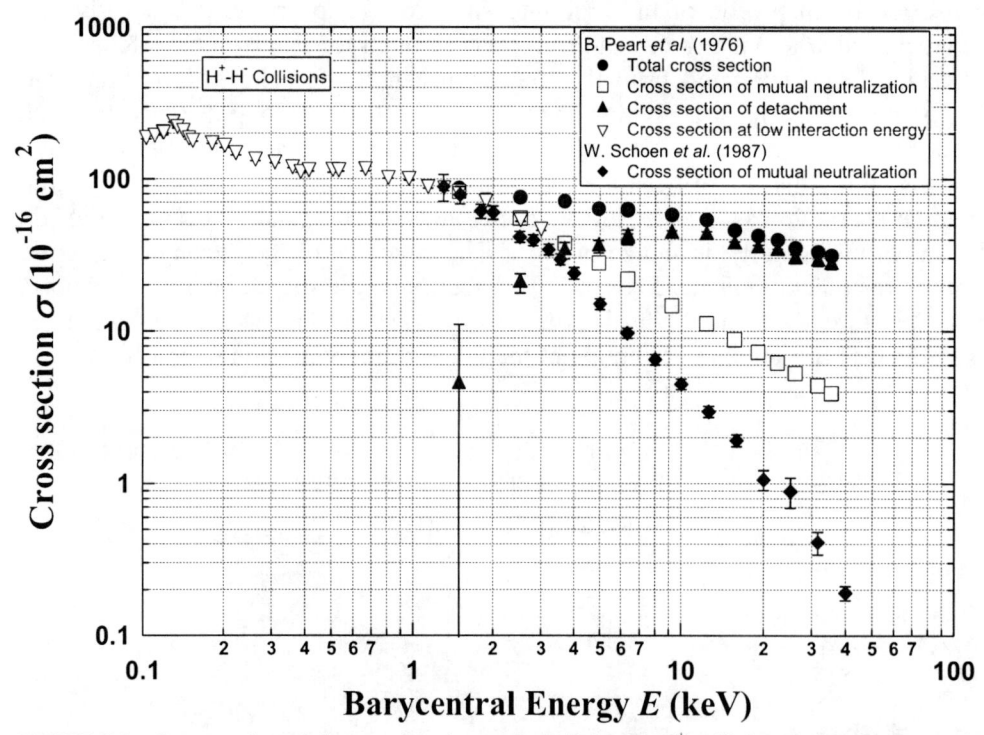

FIGURE 3. Cross section of electron detachment from H⁻ by H⁻+H⁺ collisions in low energy regions.

The beam trajectories of the mixed hydrogen ions and neutrals after passing through the D⁺ beam can be separated properly by an electromagnetic field. The spatial distribution of these beams will form mutually correlated signals which can enlarge the signal-to-noise ratio of the measurement system. The low signal level should be further enhanced by employing phase sensitive detection.

POP EXPERIMENT USING A HIGH-INTENSITY HE$^+$ BEAM

We have started an experimental study with a low energy intense ion beam system being tested at the National Institute for Fusion Science (NIFS) to validate the capability of the negative ion beam probe system. A strongly focusing He$^+$ ion source [8] is developed to measure the spatial profile and velocity distribution of alpha particles produced by D-T nuclear reactions in fusion plasmas [9]. The He$^+$ ion source has three concaved extraction electrodes and produces a 20 mm diameter He$^+$ beam with the current density of 500 mA/cm^2 at the focal point [10, 11]. We utilize the He$^+$ beam as a target beam for the PoP experiment. For the probe beam source, we have designed and assembled an H$^-$ ion source. The specifications of the H$^-$ ion source and experimental results of H$^-$ beam extraction test on a small test bench are described in the following section.

Figure 4 shows a schematic drawing of the experimental setup of the H$^-$ beam probe system to monitor the beam profile of the high current density He$^+$ beam formed by concaved extraction electrodes. The H$^-$ beam crosses the He$^+$ beam around the focal point. An electrostatic ion/neutral beam separator is coupled to an einzel lens system to focus/defocus the H$^+$/H$^-$ beam. At the end of the beam separator, a scintillation plate is placed to observe the beam spatial profile optically. We will validate this new technique through beam profile data comparing with those obtained by the existing IR imaging monitor [10].

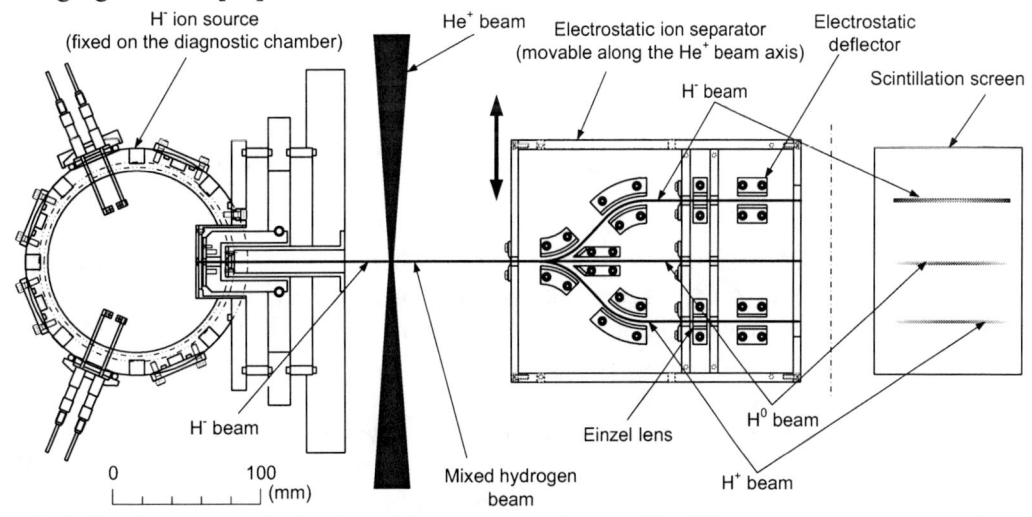

FIGURE 4. A schematic drawing of the experimental setup of the H$^-$ beam probe system to monitor the beam profile of the high current density He$^+$ beam extracted from an extraction system composed of three concaved electrodes.

H⁻ ION SOURCE FOR BEAM PROBE SYSTEM

Experimental Setup

We have designed and assembled the H⁻ ion source to produce a probe beam for the PoP experiment. Figure 5 shows a schematic drawing of a compact KAMABOKO type H⁻ ion source. This source has tungsten-filament hot cathodes and permanent magnets forming a tent-shaped magnetic filter [12] to confine the hydrogen plasma.

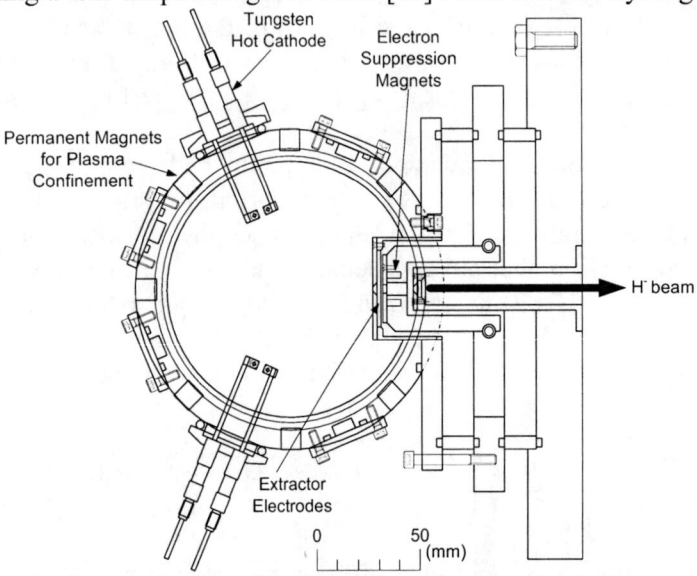

FIGURE 5. A schematic drawing of the H⁻ ion source for a negative ion beam probe system.

Figure 6 shows the measured distribution of the magnetic field strength and the direction of the magnetic flux in the ion source chamber. The magnetic field strengths at the center and that on the inner wall of the chamber are 20 G and 1500 G, respectively. The source chamber of 120 mm in inner diameter and 160 mm in length is made from an aluminum alloy. The H⁻ source is designed to produce the H⁻ beam with a rectangular shape of 70 mm×2 mm.

The H⁻ source was installed on a test bench to measure the plasma parameters in the source chamber, the efficiency of H⁻ production and brightness of the thin sheet H⁻ beam. Figure 7 is a photograph of the H⁻ beam test bench. Before the beam test, we measured plasma parameters of the hydrogen plasma in the source chamber. A tungsten-made Langmuir probe with the 3.0 mm length and 0.8 mm diameter electrode tip was located at 21 mm off-position from the plasma electrode in the source chamber. The measured electron temperature and the electron density for the 80W discharge power were $T_e = 2.0$ eV and $n_e = 1.0 \times 10^{11}$ cm^{-3}, respectively.

In order to measure the total H⁻ beam current extracted from the H⁻ ion source, a Faraday cup with the rectangular aperture of 100 mm×10 mm was installed in the beam diagnostic chamber.

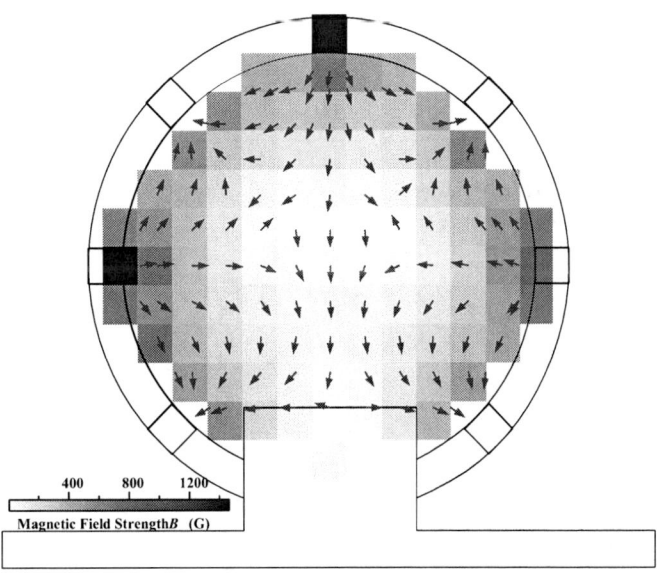

FIGURE 6. Measured result of the magnetic field strength distribution and direction of the magnetic flux.

FIGURE 7. A photograph of the H⁻ beam test bench.

Preliminary results of H⁻ beam extraction

The H⁻ beam was extracted from the compact KAMABOKO type H⁻ ion source. Figure 8 shows preliminary results of the H⁻ beam extraction study. Due to the limitation of the power supplies for both the hydrogen discharge and the beam extraction, measurements were made for the H⁻ beam current less than several tens µA.

The optimum ion source operation condition to obtain enough H⁻ beam current with the proper beam spatial homogeneity to utilize it as the probe beam is being investigated. In case of the PoP experiment, the discharge current will be increased so as to adjust the beam optics to 5~7 kV extraction potential. The estimated H⁻ beam current density will be of the order of mA/cm². By installing a beam profile monitor into the beam diagnostic chamber, the spatial beam profiles of the H⁻ beam will be monitored. After the measurement, the H⁻ ion source will be transported to NIFS to attach on the diagnostic chamber for the He⁺ beam and to start the PoP experiment.

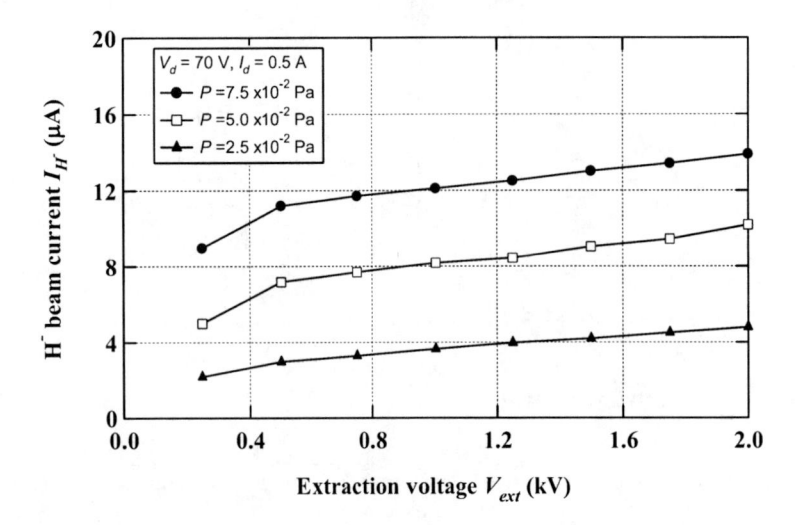

(a) Discharge voltage $V_d = 70$ V, Discharge current $I_d = 0.5$ A.

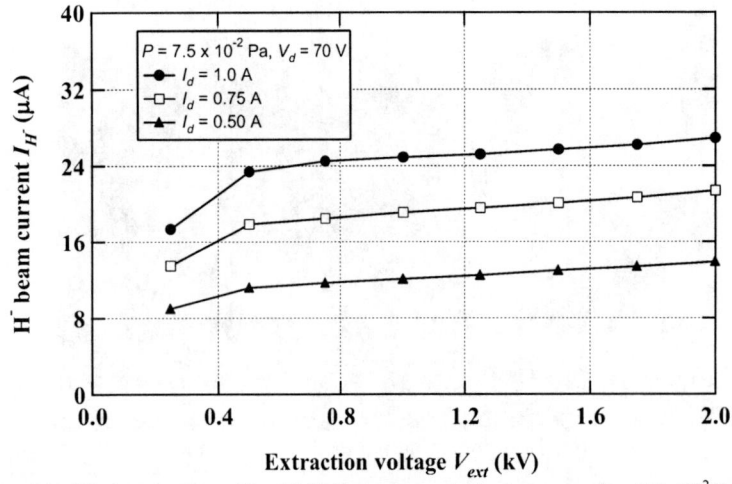

(b) Discharge voltage $V_d = 70$ V, Gas pressure in the ion source $P = 7.5 \times 10^{-2}$ Pa.

FIGURE 8. H⁻ beam current, I_{H^-}, extracted from the compact KAMABOKO type H⁻ ion source as a function of the extraction voltage V_{ext}.

SUMMARY AND FUTURE PROSPECTS

We have been developing a compact KAMABOKO type H⁻ ion source for the validation of the active beam probe technique as a new tool for the beam profile diagnostics of high intensity positive ion beams. We have built a small test bench to study the quality of the H⁻ beam extracted from the source and preliminary results are being obtained. The spatial H⁻ beam profile which is one of the most essential requirements for the active beam probe technique will be measured before the installation of the strongly focusing He$^+$ beam diagnostic chamber at NIFS NBI test stand.

ACKNOWLEDGMENTS

This work was supported by Japan Society for the Promotion of Science (JSPS) Grant-in-Aid for Scientific Research (KAKENHI) (C) 21560864. Also, it was supported partly by NIFS collaborative program NIFS10KCBR008. Authors appreciate the continuous encouragement of Dr. Yoshikazu Okumura, Deputy Director General of Fusion Research and Development Directorate in Japan Atomic Energy Agency (JAEA).

REFERENCES

1. N. Ouchi, *Proceedings of the 14th International Conference on RF Superconductivity (SRF'09)*, Berlin, Germany, 2009, pp. 934-940.
2. S. Henderson, *Proceedings of the 22nd Particle Accelerator Conference (PAC'07)*, Albuquerque, New Mexico, USA, 2007, pp.7-11.
3. M. Eshraqi, M. Brandin, I. Bustinduy, C. Carlile, H. Hahn, M. Lindroos, C. Oyon, S. Peggs, A. Ponton, K. Rathsman, R. Calaga, T. Satogata and A. Jansson, *Proceedings of the 1st International Particle Accelerator Conference (IPAC'10)*, Kyoto, Japan, 2010, pp. 804-806.
4. *IFMIF Comprehensive Design Report*, by the IFMIF International Team, an activity of the International Energy Agency (IEA), Implementing Agreement for a Program of Research and Development on Fusion Materials, January 2004.
5. K. Shinto, M. Wada, O. Kaneko, K. Tsumori, M. Nishiura, M. Sasao and M. Kisaki, *Proceedings of the 1st International Particle Accelerator Conference (IPAC'10)*, Kyoto, Japan, 2010, pp. 999-1001.
6. B. Peart, R. Grey and K. T. Dolder, *Journal of Physics B: Atomic and Molecular Physics*, **9**, 3047-3053 (1976).
7. W. Schön, S. Krüdener, F. Melchert, K. Rinn, M. Wagner and E. Salzborn, *Journal of Physics B: Atomic and Molecular Physics*, **20**, L759-L764 (1987).
8. K. Shinto, A. Okamoto, S. Kitajima, M. Sasao, M. Nishiura, O. Kaneko, S. Kiyama, H. Sakakita, Y. Hirano and M. Wada, *Proceedings of the 10th European Particle Accelerator Conference (EPAC'06)*, Edinburgh, Scotland, 2006, pp.1726-1728.
9. K. Shinto, H. Sugawara, S. Takeuchi, S. Kitajima, M. Takenaga, M. Sasao, M. Nishiura, O. Kaneko, S. Kiyama and M. Wada, *Proceedings of the 2005 Particle Accelerator Conference (PAC'05)*, Knoxville, Tennessee, USA, 2005, pp. 2630-2632.
10. M. Kisaki, K. Shinto, T. Kobuchi, A. Okamoto, S. Kitajima, M. Sasao, K. Tsumori, M. Nishiura, O. Kaneko, Y. Matsuda, M. Wada, H. Sakakita, S. Kiyama and Y. Hirano, *Review of Scientific Instruments*, **79**, 02C113 (2008).
11. M. Sasao, T. Kobuchi, M. Kisaki, H. Takahashi, A. Okamoto, S. Kitajima, O. Kaneko, K. Tsumori, K. Shinto and M. Wada, *Review of Scientific Instruments*, **81**, 02B701 (2010).
12. H. Tobari, M. Hanada, M. Kashiwagi, M. Taniguchi, N. Umeda, K. Watanabe, T. Inoue, K. Sakamoto and N. Takado, *Review of Scientific Instruments*, **79**, 02C111 (2008).

Electron Strippers for Compact Neutron Generators

K. Terai[a], N. Tanaka[a], M. Kisaki[a], K. Tsugawa[a], A. Okamoto[a], S. Kitajima[a], M. Sasao[a], T. Takeno[b], A. J. Antolak[c], K. N. Leung[d], M. Wada[e]

[a]*Graduate school of Engineering, Tohoku University, Aoba, Sendai, Miyagi 980-8579, Japan.*
[b]*Tohoku University International Advanced Research and Education Organization, Aoba, Sendai, Miyagi 980-8578, Japan*
[c]*Sandia National Laboratories, Livermore, California 94550, USA*
[d]*Lawrence Berkeley National Laboratory and Nuclear Engineering Dept., University of California, Berkeley, USA*
[e]*Doshisha University, Kyotanabe, Kyoto, 610-0321, Japan*

Abstract. The next generation of compact tandem-type DD or DT neutron generators requires a robust electron stripper with high charge exchange efficiency. In this study, stripping foils of various types were tested, and the H$^-$ to H$^+$ conversion efficiency, endurance to the heat load, and durability were investigated in terms of suitability in the tandem-type neutron generator. In the experiments, a H$^-$ beam was accelerated to about 180 keV, passes through a stripping foil, and produces a mixed beam of H$^-$, H^0, and H$^+$. These ions were separated by an electric field, and detected by a movable Faraday cup to determine the conversion efficiency. The experimental results using thin foils of diamond-like carbon, gold, and carbon nano-tubes revealed issues on the robustness. As a new concept, a H$^-$ beam was injected onto a metal surface with an oblique angle, and reflected H$^+$ ions are detected. It was found that the conversion efficiency, H$^+$ fraction in the reflected particles, depends on the surface condition, with the maximum value of about 90%.

Keywords: neutron generator, tandem accelerator, electron stripper, stripping foils
PACS: 29.25.D , 29.20.B , 28.52.S, 28.20.V

INTRODUCTION

There has been growing demand for compact and high intensity neutron generators in various fields, such as application to security systems, land mining, and structural evaluation of tankers or nuclear reactors. A tandem acceleration concept has been proposed for a DD or DT neutron generator [1-2]. In this approach, a D$^-$ beam is accelerated up to several-hundreds keV toward a high voltage stage for electron stripping, and the D$^+$ beam produced is re-accelerated and injected to a D- or T-target with almost doubled energy. This type of accelerator requires a compact, mechanically robust electron stripping system, which can endure high heat load. There are generally two types of electron stripping system, a gas stripping cell and a thin stripping foil. A stripping foil is preferable, because it is simple and the generator would be more compact. In order to produce neutrons by DD or DT reactions, D$^-$ ions need to be accelerated to the few hundred kilovolt range, but there has been almost no research of stripping foils applicable in this energy range.

In this study, various types of stripping foil, including diamond-like carbon, gold, and foils made from carbon nano-tubes, and from graphene were tested experimentally in the energy range of 60 – 150 keV. The foils were evaluated for their application to a tandem-type neutron generator in terms of H$^-$ to H$^+$ conversion efficiency, heat load, and durability. As a new concept, a solid plate electron stripper was tested, and the experimental results, and comparison with the theoretical prediction were reported.

EXPERIMENTAL SETUP

A test stand previously developed for an alpha particle diagnostics method, Advanced Beam Source 103 (ABS103) [4-7], was used as a H$^-$ beam source. Figure 1 shows the schematic diagram of the ABS103. An H$^+$ beam extracted from a compact bucket type source was focused at the center of a lithium charge exchange cell, and a H$^-$ beam produced was charge-separated and bent toward the accelerator column.

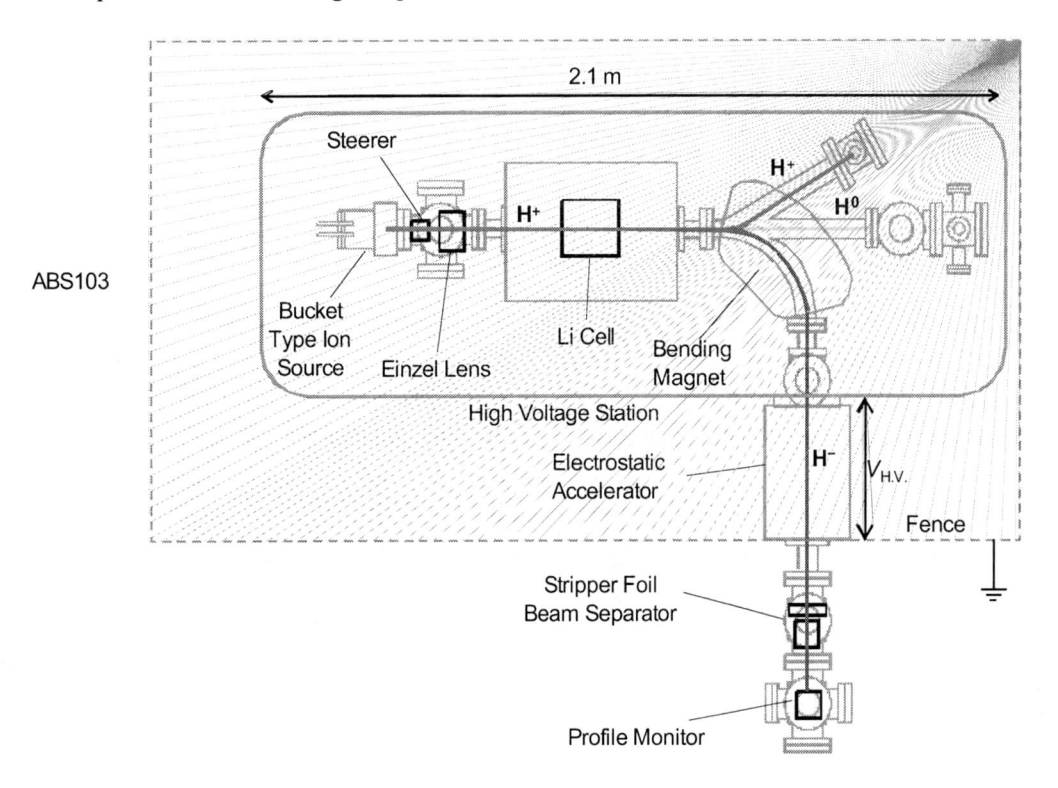

FIGURE 1. Schematic drawing of ABS103. An H$^-$ beam is produced through a lithium charge exchange cell. The details of the foil test setup, stripper foil, the beam separator, and the profile monitor are shown in Fig. 2.

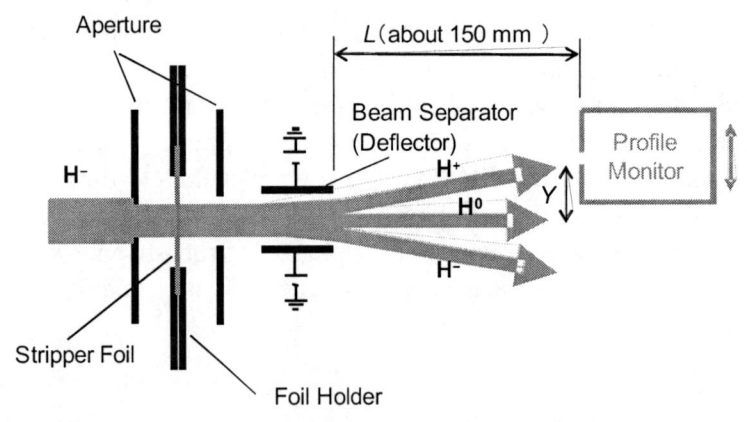

FIGURE 2. Schematic drawing of the experimental setup for the stripping foil test. The outgoing beam fractions from a foil are charge-separated by an electrostatic deflector in the beam separator, and detected by a movable Faraday cup (profile monitor).

Figure 2 shows the experimental setup for the stripping foil test. The H⁻ beam was extracted through a 4 mm-aperture is injected onto and passed through a test foil. The outgoing beam was charge-separated by an electrostatic deflector (36 mm in length) and detected by a movable Faraday cup (1 mm entrance slit). The distance between the center of the deflector and the entrance slit was 163 mm.

EXRERIMENTAL AND CALCULATION RESULTS

Stripping Foils

By using this apparatus, 5 types of stripping foils were tested, including a foil of evaporated amorphous carbon [8], a gold foil, a carbon nano-tube sheet [9], a double wall carbon nano-tube (DWCNT) on copper mesh and a graphene sheet. The thickness of each foil, the beam energy tested and results are summarized in Table 1. The H⁺ beam current was detected only from the amorphous carbon foil of 55 µg/cm² and the gold foil of 480 µg/cm². With other foils neither H⁺ beam current nor H⁻ beam current was detected because the beam energy was too low to penetrate them.

TABLE 1. Electron stripping foils tested.

Foil	Thickness	Beam Energy	Electron Stripping
amorphous carbon foil (Arizona Carbon Foil)	55 µg/cm²	105 – 147 keV	50 – 90 %
Gold	480 µg/cm²	127 keV	94 %, but beam divergence is large.
CNT sheet (specially provided by ANI)	unknown(~ 10 µm)	127 keV	not detected
DWCNT on Cu mesh	unknown(~ 10 µm)	147 keV	not detected
Graphene	unknown(~ 10 µm)	150 keV	not detected

The conversion efficiency from H⁻ to H⁺ was evaluated, by comparing theoretical prediction of outgoing beam profiles calculated using the TRIM code [3], where charge exchange probability was not taken into account. The beam profile at the faraday cup of the present geometry was calculated by using the energy loss, energy and angular straggling obtained from TRIM calculation. In Fig. 3 is shown the calculated beam profile from an amorphous carbon foil assuming all particles from the foil are H⁺, together with the measurement results. The conversion efficiency from H⁻ to H⁺ was calculated by integrating the measured beam currents profiles divided by that of theoretical prediction. The results are shown in Fig. 4.

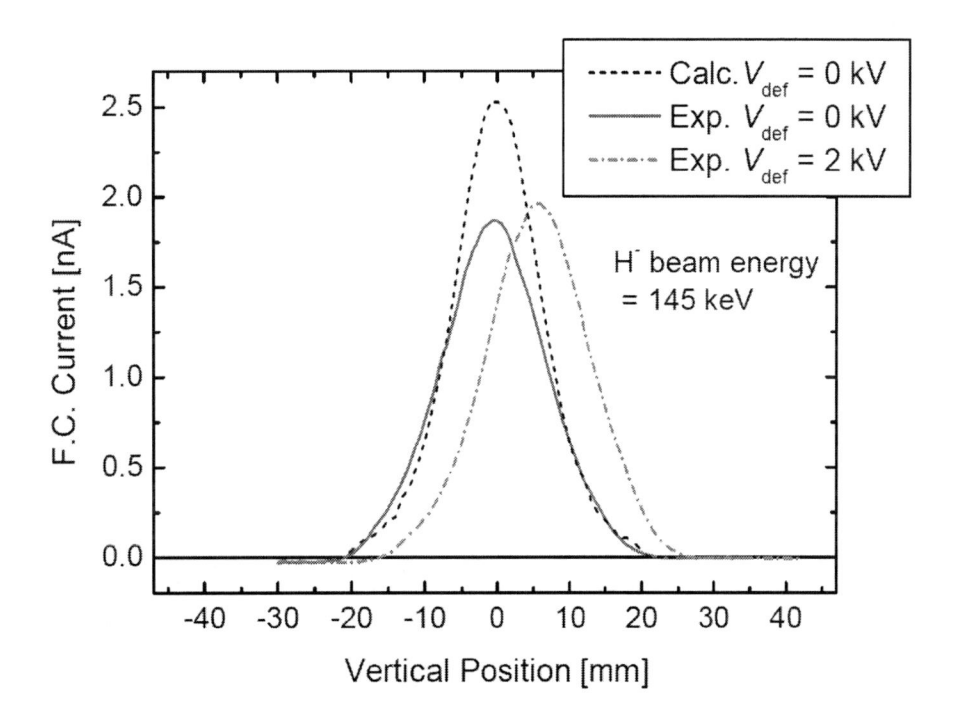

FIGURE 3. The H+ beam profiles from an amorphous carbon foil. Dashed line shows the theoretical profiles obtained by TRIM calculation assuming the same geometrical conditions, and solid lines show the measurement results. The peak shifted by 7 mm when the deflector voltage, V_{def}, of 2 kV was applied at the charge separator. It is consistent with the geometrical prediction.

Although the amorphous carbon foil showed high conversion efficiency, it is not suitable for the application to a compact neutron generator, because it is fragile, and is not able to withstand high heat load. For this reason, a gold foil was tested. The thinnest foil available was 480 µg/cm² in thickness. The measured spectrum is shown in Fig. 5. The output beam was dispersed due to the multiple scattering. The measured conversion efficiency for the gold foil was 94 %.

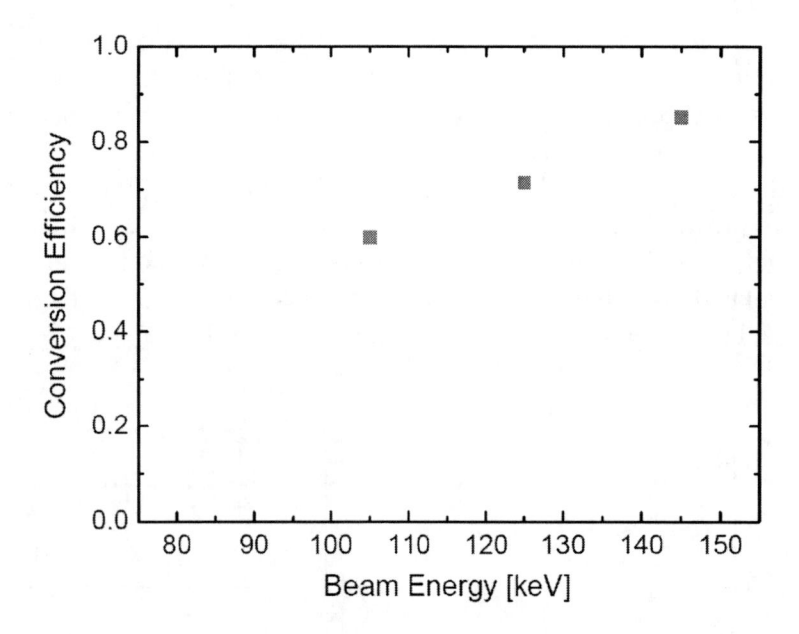

FIGURE 4. The H$^-$ to H$^+$ conversion efficiency from an amorphous carbon foil as a function of beam energy.

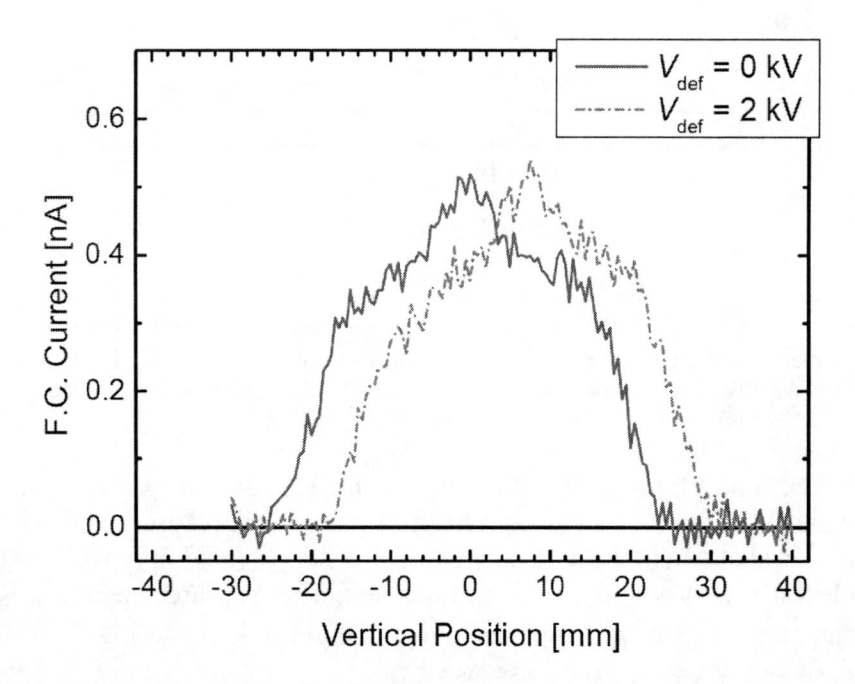

FIGURE 5. The beam profile from an gold foil. The 7 mm shift at the deflector voltage of 2 kV is consistent with the geometrical prediction.

Stripping Plates

As a new concept, a layered stripping plate was proposed by Terai et al. [10][#]. In this system, a H⁻ beam is injected onto a metal surface with an oblique angle. The schematic view of the arrangement is shown in Fig. 6.

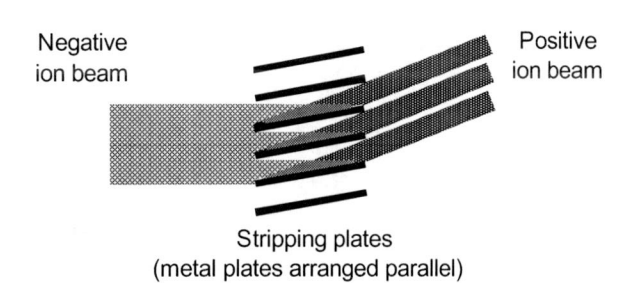

FIGURE 6. Schematic view of the stripping plate concept.

In order to test this concept, an experiment was performed using the present setup. The plates were highly polished molybdenum of 0.4 mm in thickness, 31mm wide and 25 in length to the beam direction. A set of 25 parallel plates separated by 0.2 mm in gap was tilted to the beam direction so that the beam incident angle to the plates was 89 degrees. The reflected H^+ ions were detected by the movable faraday cup. . Figure 7 shows the measured H^+ beam profile. The negative currents in the experimental profiles were caused by negative ions that passed through the small gap of plate. The conversion efficiency from H^- to H^+ was evaluated, in a similar manner to the carbon foil above. The beam profile at the faraday cup of the present geometry was calculated by using the energy loss, energy and angular straggling from a TRIM calculation. The calculated H^+ beam profile is shown in Fig. 7, by a dashed line, indicating the good agreement with the measured results. The conversion efficiency from H^- to H^+ was calculated by integrating the measured beam currents profiles divided by that of theoretical prediction. The results are shown in Fig. 8, as a function of beam energy. Here experimental results of two cases are shown. One was obtained with a factory supplied molybdenum plate (triangles), and the other was obtained with polished molybdenum plate (circles) [10], indicating that the reflected H^+ current depends on the surface condition. With the polished molybdenum plate, the conversion efficiency was about 90% at145 keV.

[#] Patent pending number: 2010-254663

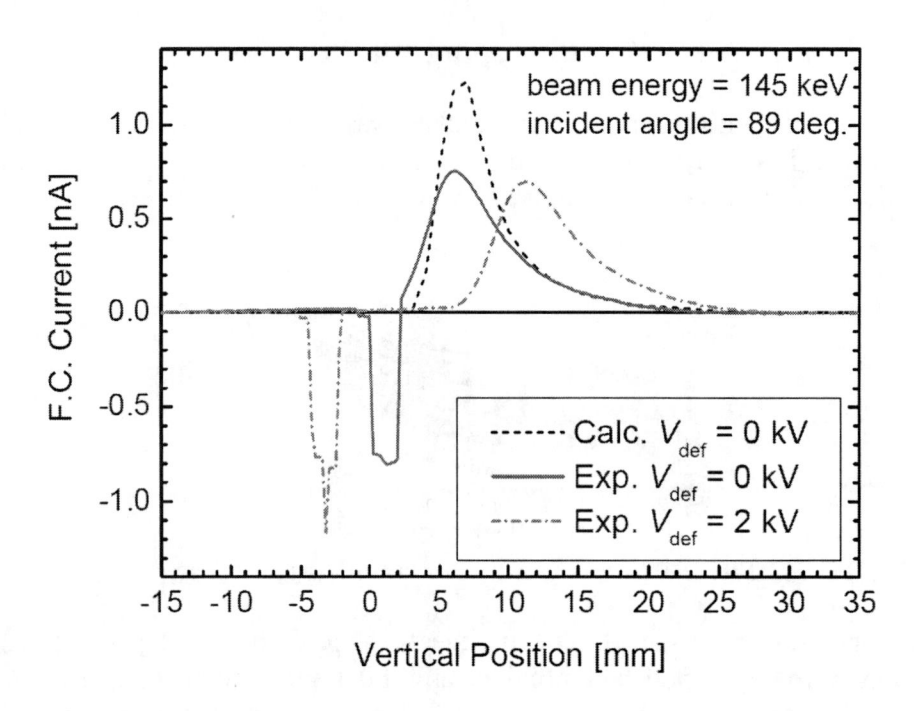

FIGURE 7. The H$^+$ beam profile from the stripping plate. Dashed line shows the theoretical profiles obtained by TRIM calculation assuming the same geometrical conditions, and solid lines show the measurement results. The peak shifted by 7 mm when the deflector voltage, V_{def}, of 2 kV was applied in the charge separator. It is consistent with the geometrical prediction. The minus signals of the experimental profiles were caused by negative ions that passed through the small gap of plates.

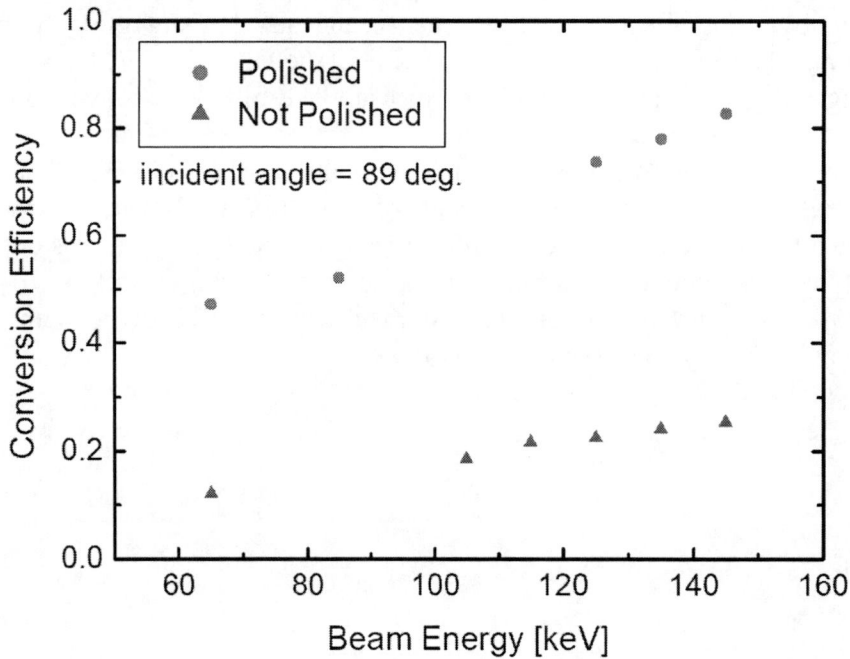

FIGURE 8. The H$^-$ to H$^+$ conversion efficiency from stripping plates as a function of beam energy. Triangles are obtained with a factory supplied molybdenum plate, and solid circles are obtained with polished molybdenum plate [10].

CONCLUSION

Stripping foils of various types including a foil of evaporated amorphous carbon, a gold foil, a carbon nano-tube sheet, a double wall carbon nano-tube (DWCNT) on copper mesh and a graphene sheet were tested in the energy range up to 150 keV for the application of a tandem type compact DD or DT neutron source. The H^+ beam current could only be detected from an amorphous carbon foil of 55 $\mu g/cm^2$ or a gold foil of 480 $\mu g/cm^2$. The experimental results using these foils revealed issues on the robustness.

Measurements for a solid plate electron stripper consisted of bombarding a H^- beam onto a stripping plate at an oblique angle. The reflected H^+ ions were detected and the reflected H^+ current was found to depend on the surface condition of the metal plates. The conversion efficiency of H^- to H^+ is nearly linear with the incident beam energy and, at 145 keV, has a value close to 90%.

ACKNOWLEDGMENTS

This work was supported by Grant-in-Aid for Priority Area 442-16082101, Ministry Education, Science and Culture Japan, and in part by Joint research program of the National Institute for Fusion Science. Sandia National Laboratories, the affiliation of one of authors (A.J.A) is a multi-program laboratory managed and operated by Sandia Corporation, a wholly owned subsidiary of Lockheed Martin Corporation for the U.S. Department of Energy National Nuclear Security Administration under contract DE-AC04-94AL85000.

REFERENCES

1. K. N. Leung, et al., *AIP conf. Proc.* 287, 368 (1994)
2. J. Reijonen, F. Gicquel, S. K. Hahto, M. King, T- P Lou and K-N Leung, *D-D Neutron Generator Development at LBNL*, Applied Radiation and Isotopes, 63, 757 (2005), LBNL-56673.
3. SRIM & TRIM, http://www.srim.org/.
4. K. Shinto, H. Sugawara, M. Takenaga, S. Takeuchi, N. Tanaka, A. Okamoto, S. Kitajima, M. Sasao, M. Nishiura and M. Wada, Optimization of a compact multicusp He^+ ion source for double charge exchanged He^- beam. [*Review of Scientific Instruments*,77(3),(2006)]
5. H. Sugawara, K. Shinto, N. Tanaka, S. Takeuchi, M. Kikuchi, A. Okamoto, S. Kitajima, M. Sasao, and M. Wada, Diagnostics of a He^+ beam extracted from a compact magnetic bucket-type ion source. *Rev. Sci. Instrum*,79,(2008),02B708.
6. M. Kikuchi, N. Tanaka, T. Nagamura, T. Kobuchi, A. Okamoto, S. Kitajima, M. Sasao ,H. Ymaoka , Study of the Beam Transport in a High-Energy Neutral Helium Beam System with Double Charge Exchange Cell. *Journal of Plasma and Fusion Research SERIES*,8,(2009),1539-1541]
7. N. Tanaka, T. Nagamura, M. Kikuchi, A. Okamoto , T. Kobuchi, S. Kitajima, M. Sasao, H. Yamaoka , M. Wada, Characteristics of a He^- Beam Produced in Lithium Vapor. AIP Conference Proceedings, 1097,(2009),443-448
8. Acf-Metals Arizona Carbon Inc.
9. Applied Nano Tech. Inc. (http://www.appliednanotech.net/)
10. K. Terai et al., "A stripping plate for a compact tandem-type neutron generator", will be submitted to Journal of Plasma and Fusion Research.

Extraction of low-energy negative oxygen ions for thin film formation

M. Vasquez Jr.*, D. Sasaki*, T. Kasuya*, S. Maeno† and M. Wada*

*Graduate School of Engineering, Doshisha University, Kyotanabe, Kyoto 610-0321 Japan
†Novelion Systems Co. Ltd., Kyotanabe, Kyoto 610-0332 Japan

Abstract. Coextraction of low-energy positive and negative ions were performed using a plasma sputter-type ion source system driven by a 13.56 MHz radio frequency (rf) power. Titanium (Ti) atoms were sputtered out from a target and the sputtered neutrals were postionized in oxygen/argon (O_2/Ar) plasma prior to extraction. The negative O ions were surface-produced and self-extracted. Mass spectral analyses of the extracted ion beams revealed the dependence of the ion current on the incident rf power, induced target bias and O_2/Ar partial pressure ratio. Ti^+ current was found to be dependent on Ar^+ current and reached a saturation value with increasing O_2 partial pressure while the O^- current showed a peak current at around 1:9 O_2/Ar partial pressure ratio. Ti^+ current was several orders of magnitude higher than that of the O^- current.

Keywords: ion beams, negative ions, positive ions
PACS: 52.27.Cm, 41.75.Cn, 52.50.Dg

INTRODUCTION

Utilization of ion beams with the energy below 100 eV has been considered to open wider area of application in semiconductor material processing. However, a severe space-charge problem arises on the surface where the beam forms a thin film on an insulating material. When the local electrical potential exceeds the beam energy, the beam never touches down on the surface with their original kinetic energy. Thus, negative ion beams have been preferred in surface modification and thin film growth due to the nature of reduced surface charge-up upon insulating materials [1]. Incoming positive ions as well as energetic neutral atoms cause the emission of secondary electrons upon their impact, and thus, the surface is positively charged up. When negative ion beam strikes the surface, a loosely bound electron at the affinity level is immediately liberated without knocking off an extra electron from the surface to keep the charge neutrality of the surface. Further adjustment of the surface charge neutrality can be attained by irradiating the surface with beams of both electrical polarities. Thus, a device that realizes simultaneous transport of positive and negative ion beams incident on a surface has been developed to form thin film materials by neutralizing the charge build-up [2].

The transport of a low-energy charged particle beam is extremely difficult as space-charge produced by beams makes the beam diverge and reduces the current density. Merging low-energy positive and negative ion beams can cancel out if not minimize the space-charge effect and allow the beams to travel as a neutralized beam. The merged beam with the electric charge neutralized condition, can also keep the charge neutrality as they are incident upon the insulating surface which can be an effective method to achieve charge-up-free ion beam based material synthesis.

Growth of titanium dioxide (TiO$_2$) thin films can be easily achieved using different methods such as reactive sputtering of a Ti target material in oxygen/argon (O$_2$/Ar) plasma [3, 4]. However, as devices shrink in size, more stringent control over the growth process and quality of the produced film is necessary. In order to achieve this, direct ion beam deposition method may be preferred over sputtering because of its superior characteristics in depositing high purity metal films at an isotope level [5] and the ability to control the growth process by independently controlling the ion flux and ion energy [6]. When a negative bias (ϕ_{bias}) is induced on a Ti target, energetic O/Ar ions liberate Ti neutrals from the surface while the O atoms gain electrons from the target surface [7]. The sputtered Ti neutrals are then postionized and extracted by a positive potential (ϕ_{ext}) induced on the extractors. On the other hand, the surface-produced O$^-$ gains an energy equivalent to ϕ_{bias}. The energetic O$^-$ are self-extracted as long as the ion energy is greater than ϕ_{ext}. Hence, low ϕ_{ext} are favorable. With this mechanism, extraction of both positive and negative ions can be realized. In this paper, extraction of low-energy positive and negative ions is evaluated.

EXPERIMENTAL DETAILS

Apparatus

The ion source was made of an 80 mm diameter 80 mm long cylindrical stainless steel chamber. Capacitively coupled plasma was produced via a 13.56 MHz radio frequency (rf) power source. Tuning the system to resonance was achieved via a simple L-C matching system. A 38 mm diameter Ti target with a planar surface was inserted to

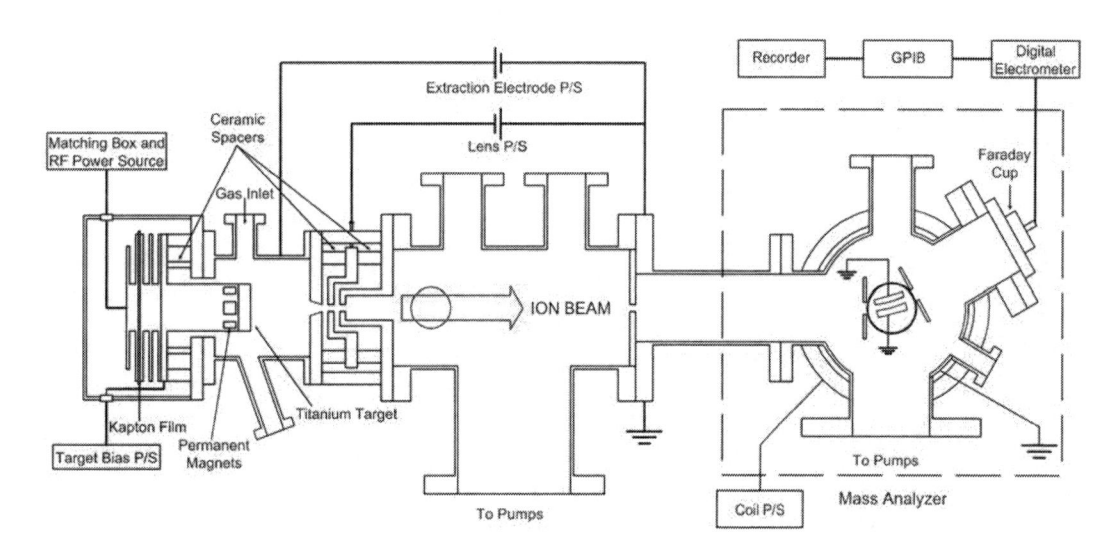

FIGURE 1. Schematic diagram of the sputter-type ion source system and the mass analyzer.

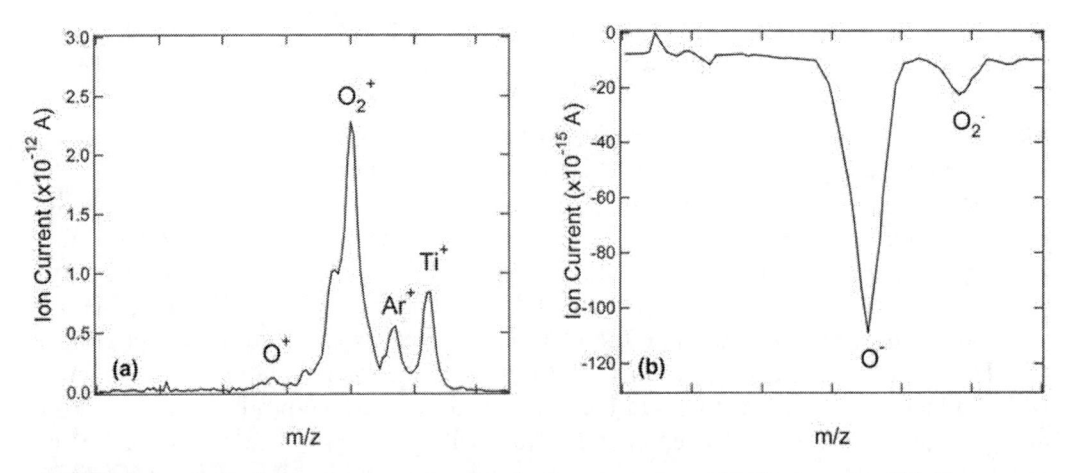

FIGURE 2. Typical mass spectra of an ion beam extracted at 100 W rf power, 100 V extraction potential, 0.11 O$_2$/Ar partial pressure ratio. Self-bias on the target was -210 V. (a) Positive ions, and; (b) Negative ions.

the ion source chamber. Mounted at the back side of the target were cylindrical and toroidal magnets which realized a planar magnetron magnetic field geometry to enhance ionization near the target surface. Aside from the self-bias induced on the target, a negative bias can be applied on the target through a dc power supply. A thin insulating film was inserted between the rf power feed and the target to realize a dc break. The extractor system was composed of three electrodes with each electrode having a single 6 mm diameter aperture. The extractors were mounted directly opposite to the target surface. Under positive ion extraction operation, the first electrode (plasma electrode) facing the target was biased positively. The second electrode (lens electrode) can be biased at any potential negatively or grounded while the third (ground electrode) was connected to ground. Gases were introduced through a gas inlet attached to the ion source chamber. Two turbomolecular pumps coupled to rotary pumps were installed in the downstream region. Gas pressure was monitored downstream of the extractor electrodes by an ionization gauge. Figure 1 shows the schematic of the experimental system.

Procedure

To measure species composition of the extracted ion beam, a momentum analyzer was positioned 30 cm from the ground electrode of the ion source. Between the ground electrode and the analyzer was a 2 mm diameter orifice used to collimate the ion beam. The differentially pumped analyzer was capable of determining positive and negative ionic species by changing the polarity of the magnetic coils. Process gas used were O$_2$ and Ar with varying partial pressures. Incident rf power was changed from 60 W to 110 W while the dc target bias was up to -300 V. Extraction potential was limited to 120 V. Figure 2 show a typical mass spectra of the positive (Fig. 2(a)) and that of the negative

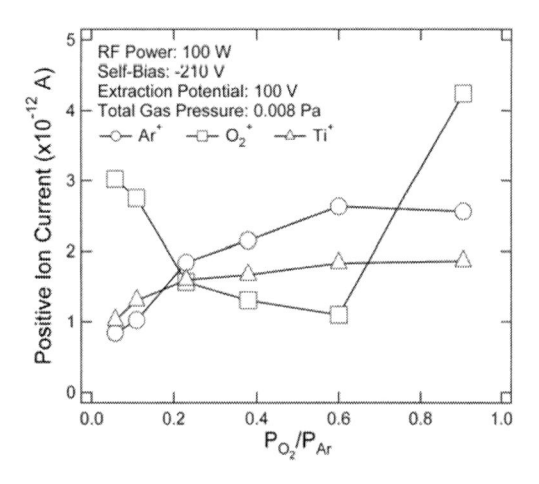

FIGURE 3. Variation in the positive ion currents with respect to O$_2$/Ar partial pressure ratio. The currents were determined from the current intensity of each ion from the mass spectra.

(Fig. 2(b)) scans of an ion beam extracted from the source. At 100 W rf power, the ϕ_{bias} on the target was -210 V. With $|\phi_{ext}| < |\phi_{bias}|$, self-extraction of negative ions could be achieved. Thus, species of O, Ar and Ti ions can be seen from the positive scan, while O ions can be identified from the negative scan. Due to the low electron affinity of Ti (0.08 eV), Ti$^-$ was not observed from the negative scan. From the mass spectra, the ion current (peak height) of the different species were compared and analyzed.

RESULTS

The positive ion currents were found to be several orders of magnitude higher than that of the negative ion currents. When $|\phi_{ext}| < |\phi_{bias}|$, negative ions were extracted and detected by the mass analyzer. When $|\phi_{ext}| > |\phi_{bias}|$, negative ions were not observed. This suggested that the negative ion energy was equivalent to or about the same as that of the target bias times the electron charge because the ions formed at the surface were accelerated away from the target.

From Fig. 3, comparison of the positive ion currents revealed the increasing trend of Ti$^+$ current with the increase in Ar$^+$ current and increase in O$_2$/Ar partial pressure ratio (P_{O_2}/P_{Ar}). However, as the pressure ratio increased to 1.0, the Ti$^+$ and Ar$^+$ currents reached their saturation values. In addition, Ti$^+$ current was found to be independent on the target bias. As for the O$_2^+$ current, at low partial pressure ratios, high current was observed and decreased with increasing partial pressure ratio. However, the trend reversed when the partial pressure ratio was greater than 0.6.

With regards to the negative ions, Fig. 4(a) shows the variation of the O$^-$ current with respect to partial pressure ratio of O$_2$ and Ar with a target self-bias (-210 V) and an induced target bias (-300 V). An increase in negative ion current was observed as the ϕ_{bias} was increased signifying the strong dependence of surface-produced negative ions on the ϕ_{bias}. An optimum negative ion current was observed when the partial pressure

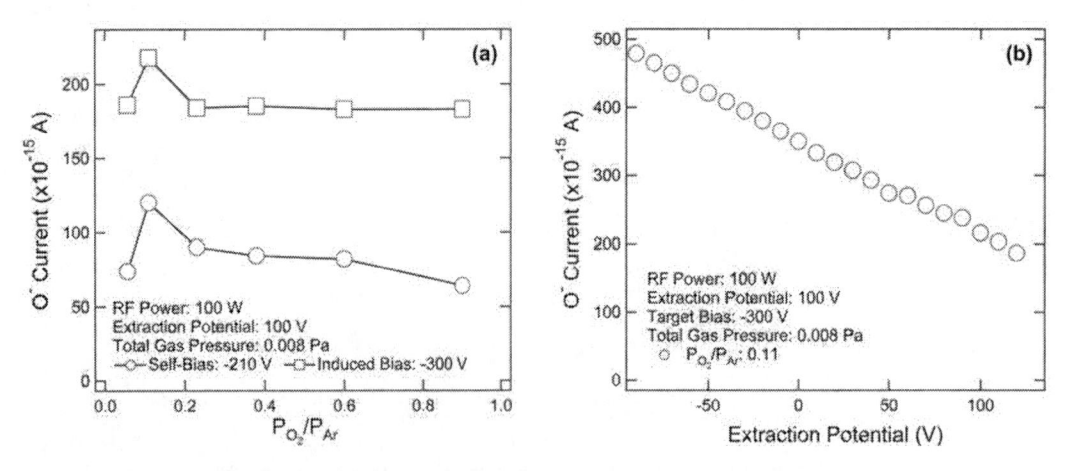

FIGURE 4. (a) Comparison of the O^- current with respect to the partial pressure ratio. (b) Dependence of the O^- current on the extraction potential.

ratio was around 0.10 for both the self-bias and induced bias conditions. Figure 4(b) shows the dependence of the O^- current on the extraction potential. Positive extraction potentials retarded the negative ions while negative extraction potentials accelerated the ions. As the lens electrode was grounded, the extraction potential could have created a weak lens effect to converge the beam of acceleration polarity.

DISCUSSION

The O_2^+ current tends to dominate the O^+ current at larger or smaller O_2 partial pressures (Fig. 2(a)). This behavior may be partially due to the reaction of O atoms with the target to form an oxide layer. Namely, atomic oxygen are gettered out swiftly from the ion source due to fresh coating of Ti. Furthermore, atomic O ions might be produced from sputtered oxide molecules which require deposition of O in Ti target and efficient sputtering of these O by plasma ions. Thus, low O^+ current was observed compared to O_2^+ current while O^- current was larger compared to O_2^- current.

As seen in Fig. 2(b), O^- was the only predominant negative ion species detected by the analyzer. As suggested above, formation of O^- may have occured from the dissociation of sputtered oxide molecule from the negatively-biased target. Afterwhich, the negative O ions were accelerated across the sheath from the surface with an energy in the range of the target bias potential. In sputter-based material synthesis, these energetic negative ions may severely damage sputter-deposited oxide thin films [7]. Thus, a decelerating mechanism must be employed in order to slow down these energetic ions which was achieved by inducing a positive ϕ_{ext} on the plasma electrode.

The increasing trend of the Ti^+ current with the increasing Ar^+ current might allude to the sole dependence of the Ti sputtering yield on the Ar^+ flux towards the target surface (Fig. 3). The density of the ions (n_i^+) in the discharge (where i can be Ar^+, O_2^+, Ti^+, etc.) is dependent on its neutral density (n_n), electron density (n_e), ionization cross-section (σ_i) and electron velocity (v_e),

$$n_i^+ = n_n n_e <\sigma_i v_e>. \tag{1}$$

The angular brackets enclosing $\sigma_i v_e$ indicate the average over velocity space. The flux (J_i^+) of each ion species is then

$$J_i^+ = n_i^+ v_i \tag{2}$$

where v_i is the ion velocity. As for the neutral Ti flux (J_{Ti}) from the target surface,

$$J_{Ti} = \sum_i (Y_{i \to Ti} J_i^+) \tag{3}$$

where $Y_{i \to Ti}$ is the sputtering yield of Ti from each ion component i. The neutral Ti density sputtered from the target is then J_{Ti}/v_{Ti}. The neutral Ti velocity (v_{Ti}) can be as large as $\sqrt{2E_B/m_{Ti}}$ where E_B is the sublimation energy of Ti. Due to the directionality of the Ti ions, the contribution of Ti^+ on the sputtering yield should be negligible. At 200 eV incident ion energy, the sputtering yield of Ti due to Ar^+ was more than twice to that due to O_2^+ using the Yamamura sputtering yield formula [8]. Hence from Eq. (3), Ti sputtering is expected to be dependent on the Ar ion flux towards the target ($J_{Ti} \propto Y_{Ar \to Ti} J_{Ar}^+$) especially at low partial pressures of O_2. As Ti atoms were liberated from the target, most of them would be deposited on the chamber wall near the plasma electrode.

As the O_2 partial pressure was increased, a decrease in molecular O ion current was observed. The decrease might be due to the reaction of O_2 with the target atoms to form the thin oxide layer and due to the adsorption of the molecules on the Ti layer formed on the chamber wall since Ti is known to be an efficient O_2 getter [9]. At higher O_2 partial pressure, the O_2 molecules were simply ionized in the discharge due to the increase in O_2 density. In addition, as the oxide thickness was increased, the rate of formation of a new oxide layer decreased hence more O_2 were available for ionization. Further increase in O_2 content would lead to target poisoning [10] which severely affected the sputtering yield of Ti.

The saturation of Ti^+ current implied that most of the sputtered Ti atoms came from the oxide layer on the target as the oxide thickness increased. A thick layer inhibited the penetration of Ar ions into the Ti target. In addition, it was found that Ti^+ current was independent on the target bias which suggested that the ions were formed from sputtered neutrals in the discharge and the sputtering yield of Ti should be reduced by some mechanism at higher bias voltage. Formation of TiO_2 layer on Ti surface can be the possible reason for observing nearly constant Ti yield against the target bias as will be described later.

The optimum current observed in Fig. 4(a) implied that negative ion production might depend on the oxide thickness. In turn, the oxide growth would be affected by the O_2 partial pressure. This may be similar to the observation of Tominaga, et al. [4] where a certain threshold pressure of O_2 existed which enhanced the negative ion production due to the presence of an oxide layer. Hence, high O^- current could be achieved at a critical O_2 partial pressure and oxide thickness.

Calculations using the stopping and range of ions in matter (SRIM) [11] have shown the dependence of O sputtering yield on the oxide thickness. In these simulations, Ar and

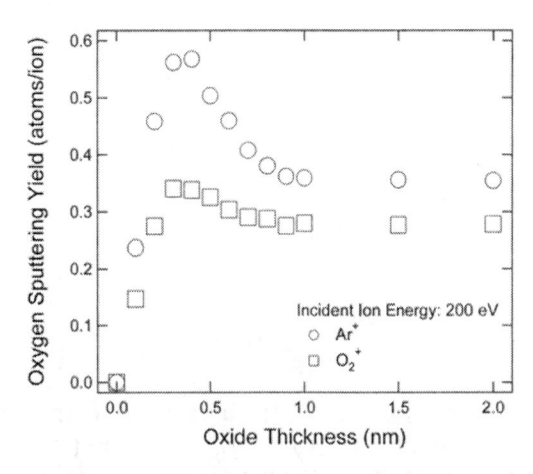

FIGURE 5. Simulated sputtering yield of O with respect to incident Ar and O_2 ions.

O_2 ions were incident upon a TiO_2 layer on top of a Ti target. By varying the thickness of the oxide, the sputtering yield of O was determined once an oxide molecule was sputtered and dissociated into Ti and O. Figure 5 shows the variation of the sputtering yield of O with respect to the oxide thickness at 200 eV incident ion energy. From the figure, an optimum sputtering yield existed when the oxide thickness was around 0.4 nm. This result was similar to the observations in Fig. 4(a). Thus, a critical oxide thickness exists which may correspond to a critical O_2/Ar partial pressure ratio that enhances O^- production.

To further increase the negative ion current, sputtering yield of the oxide layer should be increased. This can be achieved by increasing the supplied rf power which subsequently increase the target self-bias or by increasing the induced bias on the target as seen in Fig. 6. However, increase in rf power would increase the upstream plasma po-

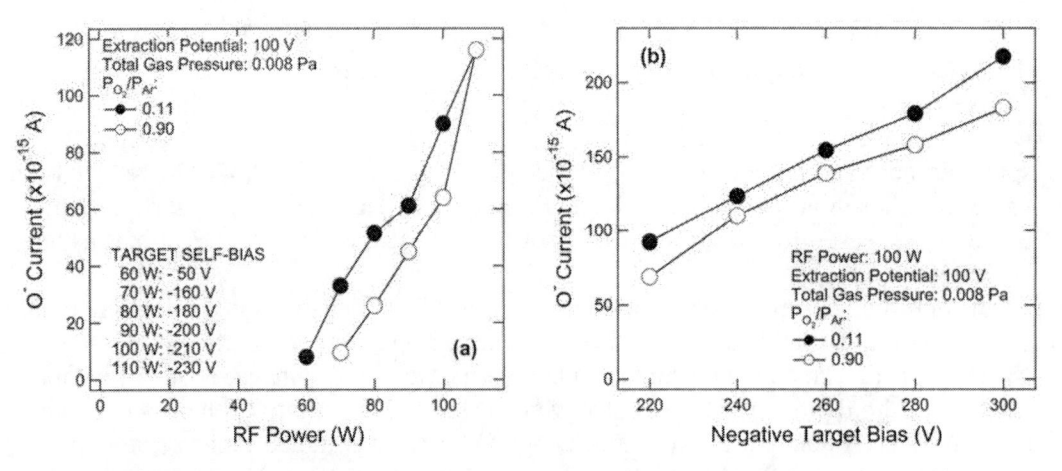

FIGURE 6. Dependence of the O^- current on (a) rf power, and; (b) induced target bias.

tential (ϕ_p) which would affect the positive ion energy (E_i^+) especially at low ϕ_{ext},

$$E_i^+ = e(\phi_p + \phi_{ext}). \tag{4}$$

In addition, the dc target bias affected the discharge which limited the allowable bias that can be induced on the target. Beyond -300 V, the discharge became unstable and was eventually extinguished. Hence, at 100 W rf power, up to -300 V bias could be supplied to the target in the present set-up. Target bias could also be increased by decreasing the system pressure to prevent arcing.

CONCLUSIONS

Coextraction of positive and negative ions without cesium seeding from a plasma sputter-type ion source has been demonstrated. Positive ions of O_2 and Ar were volume-produced in the bulk plasma and extracted through a positive potential induced on the plasma electrode of the extractors. Neutral Ti atoms were sputtered by energetic Ar ions impinging on a target surface and postionized to form Ti^+ in the discharge prior to extraction. Sputtering of the target material was found to be dependent on the flux of incoming Ar ions but was independent on the amount of O_2 especially at higher partial pressures. Presence of O_2 in the discharge led the the formation of a thin oxide layer on the target. The growth of the oxide layer on the target eventually affected the sputtering efficiency. The energy of the surface-produced negative ions was in the same order as that of the target bias potential. Self-extraction of negative ions out of the source was possible provided that the bias potential was greater than the positive extraction potential. The thickness of the oxide layer might have contributed in the O^- emission. The oxide thickness could be controlled by varying the partial pressure of O_2. Optimum O^- current was achieved at 1:9 O_2/Ar partial pressure ratio in the present system. As seen from the mass spectral analyses, good amount of ion current was measured with the mass analyzer located 30 cm from the extractors as positive ions were extracted with negative ions. Thus, the low-energy ion beam was able to traverse from the extractors to the mass analyzer while minimizing the space-charge effect due to the presence of positive and negative ions. The amount of negative ions was small but increase in the negative ion current can be realized by increasing the target bias. Increase in target bias can be achieved by increasing the rf power, inducing a separate dc bias and by decreasing the system pressure. The system is now being optimized for growing thin oxide films.

ACKNOWLEDGMENTS

The authors would like to acknowledge the High Tech Research Center of Doshisha University. One of the authors, M. Vasquez, would like to thank the Graduate Research Enhancement Grant of Doshisha University.

REFERENCES

1. J. Ishikawa, *Rev. Sci. Instrum.* **63** 2368–2073 (1992).

2. Y. Horino, *Mat. Chem. Phys.* **54** 224–228 (1998).
3. K. Okimura, A. Shibata, N. Maeda, K. Tachibana, Y. Noguchi and K. Tsuchida, *Jpn. J. Appl. Phys.* **34** 4950–4955 (1996).
4. K. Tominaga, D. Ito and Y. Miyamoto, *Vacuum* **80** 654–657 (2006).
5. K. Miyake and K. Ohashi, *Mater. Chem. Phys.* **54** 321–324 (1998).
6. N. Sasaki, S. Shimizu and S. Ogata, *Thin Solid Films* **281–282** 175–178 (1996).
7. T. Ishijima, K. Goto, N. Ohshima, K. Kinoshita and H. Toyoda, *Jpn. J. Appl. Phys.* **48** 116004 (2009).
8. Y. Yamamura and H. Tawara, *Atomic Data and Nuclear Tables* **62** 149–253 (1996).
9. J. Müller, B. Singh and N. Surplice, *J. Phys. D: Appl. Phys.* **5** 1177–1184 (1972).
10. M. Zeuner, H. Neumann, J. Zalman and H. Biederman, *J. Appl. Phys.* **83** 5083–5086 (1998).
11. J. Ziegler, J. Biersack and U. Littmark, *The Stopping and Range of Ions in Matter*, Pergamon Press, New York, 1985.

AUTHOR INDEX

A

Aanesland, Ane 668
Agostinetti, P. 381, 526, 574, 594
Akino, N. 536
Ando, A. 322
Antolak, A. J. 684
Antoni, V. 526, 574, 594, 640
Arias, J. Sanchez 245
Ärje, J. 439
Asano, E. 374, 517
Asano, Eiji 359

B

Bacal, M. 13, 58
Bandyopadhyay, M. 555, 604, 614, 624
Bansal, G. 604, 614, 624
Belchenko, Yuri 401
Bertolo, S. 245
Bigi, M. 545, 640
Blondel, C. 494
Boilson, D. 545
Bollinger, D. S. 284
Bonicelli, T. 545
Brillet, A. 494
Brinkley, J. 439

C

Capitelli, Mario 88
Carmichael, J. 113
Castel, A. 245
Cavenago, M. 97, 526, 574, 594, 640
Chabert, Pascal 668
Chaibi, W. 494
Chakraborty, A. 614, 624
Chakraborty, A. K. 555, 604
Chang, D. H. 567
Chareyre, J. 555
Chaudet, E. 245

Chaudet, Elodie 255
Chitarin, G. 381, 526, 545, 574, 594
Christin, L. 494
Chun Dong, Hu 650
Clark, H. 439
Czarnetzki, Uwe 140

D

Dairaku, M. 449
Dairaku, Masayuki 457
Dalla Palma, M. 584
Davydenko, V. 393
de Esch, H. 494
De Esch, H. 545
De Muri, M. 584
Decamps, H. 545
Delsart, C. 494
Demura, Yasuhiro 675
Desai, N. J. 226
Drag, C. 494
Dudnikov, Vadim 411

E

Ecarnot, J.-F. 245
Endo, Y. 536

F

Fagotti, E. 640
Faircloth, D. C. 205
Fantz, U. 303, 310, 348, 367, 604
Fantz, Ursel 359
Favre, G. 245
Favre, Gilles 255
Fayet, F. 245
Felden, O. 272
Fellin, F. 640
Franzen, P. 310, 545, 604
Fukano, A. 68, 339

Fukuyama, T. 58
Funaoi, T. 322

G

Gabor, C. 205
Gahlaut, A. 604, 614, 624
Gallo, A. 381
Garibaldi, P. 494
Gazza, E. 640
Gebel, R. 272
Geisser, J.-M. 245
Geros, E. 113
Graceffa, J. 545
Grand, C. 494
Grisham, L. R. 466, 476, 536
Grisham, Larry R. 457
Guharay, Samar K. 661

H

Haase, M. 245, 265
Habert, A. 245
Han, B. 411
Han, B. X. 123, 216, 226
Hanada, M. 449, 466, 536
Hansen, J. 245
Hardek, T. 216
Hatayama, A. 22, 39, 58, 68, 339
Hayami, T. 339
Hayashi, H. 374
Heinemann, B. 545, 604
Hemsworth, R. 545, 555, 567
Higuchi, T. 430
Holmes, A. J. T. 78

I

Ikeda, K. 367, 374, 517
Ikeda, Katsunori 359
Ikegami, K. 235, 292
Inoue, T. 22, 175, 449, 466, 536, 545
Inoue, Takashi 457
Ito, Y. 374
Ivanov, A. 393

J

Joffe, S. 245
Johnson, K. F. 113
Johnson, Rolland P. 411
JT-60 NBI Group 466

K

Kalvas, T. 113, 439
Kameyama, N. 39
Kaneko, O. 367, 374, 517
Kaneko, Osamu 359, 675
Kang, Y. 216
Kashiwagi, M. 449, 466, 536, 545
Kashiwagi, Mieko 457
Kasuya, T. 107, 186, 692
Kawai, M. 536
Kazawa, M. 536
Kenik, E. A. 226
Kenmotsu, T. 107
Kenmotsu, Takahiro 134
Khilchenko, A. D. 634
Kikuchi, K. 536
Kimura, Y. 186
King, D. B. 78
Kisaki, M. 367, 374, 684
Kisaki, Masashi 359, 675
Kitajima, S. 684
Kobayashi, S. 466
Koga, S. 39
Koivisto, H. 113
Kojima, A. 449, 466, 536
Kolmogorov, A. 393
Komada, S. 374, 517
Komada, Seiji 359
Komata, M. 536
Komppula, J. 113
Kondo, T. 374, 517
Kondo, Tomoki 359
Kraus, W. 303, 545, 604
Kronberger, M. 245
Kronberger, Matthias 255
Kubo, N. 536
Küchler, Detlef 255
Kulevoy, T. 640

Kumar, Ajeet   150
Kumar, Sunil   604
Kuppel, S.   58
Kuriyama, M.   545
Kvashnin, A. N.   634

L

Lang, B. R.   226
Lawrie, S. R.   205
Lee, S.   567
Lepetit, B.   13
Letchford, A. P.   205
Lettry, J.   245
Lettry, Jacques   255
Leung, K. N.   684
Li Zhen, Liang   650
Lifschitz, A. F.   30
Liu, Y.   505
Lombard, D.   245
Longo, Savino   88
Luchetta, A.   545

M

Maeda, T.   430
Maeno, S.   692
Maier, R.   272
Marconato, N.   381, 594
Marcuzzi, D.   545, 574
Marmillon, A.   245
Marques Balula, J.   245, 265
Martens, C.   604
Masiello, A.   545
Mathot, S.   245
Matsumoto, Y.   175
Matsuno, T.   322
Matsushita, D.   39
McAdams, R.   13, 78
Midttun, O.   245
Minea, T.   30
Minelli, Pierpaolo   88
Miyamoto, K.   58
Miyamoto, N.   422
Mochalskyy, S.   30
Mogaki, K.   536

Moiseev, D. V.   634
Moyret, P.   245
Moyret, Pierre   255
Murray, S. N.   123, 226, 411
Murray, Jr., S. N.   216

N

Nagaoka, K.   367, 374, 517
Nagaoka, Kenichi   359
Nakano, H.   367, 374, 381, 517, 526
Nakano, Haruhisa   359
Namekawa, Y.   235, 292
NBI Group   374
Nemoto, S.   536
NIFS-NBI Group   329
Nisbet, D.   245, 265
Nishad, S.   555
Nishida, Tomoaki   675
Nishiura, M.   175, 186
Nishiura, Masaki   675
NNBI Team   303, 310, 348

O

Oasa, K.   536
Obermayer, S.   604
Ogino, K.   186
Oguri, H.   235, 292
Ohkoshi, K.   235, 292
Ohshima, K.   536
Oka, Y.   329
Okamoto, A.   684
Okayama, K.   567
O'Neil, M.   245
Oohara, W.   430
Osakabe, M.   367, 374, 517
Osakabe, Masaki   359

P

Pandya, K.   604, 614, 624
Paoluzzi, M.   245
Paoluzzi, M. M.   265
Paoluzzi, Mauro   255
Parmar, K. G.   604, 614, 624

Pasqualotto, R. 545, 584
Paunska, Tsvetelina V. 165
Pennisi, T. R. 123, 216, 226, 411
Perkins, M. 205
Petrenko, S. 640
Pilan, N. 526, 574, 594
Piller, C. 216
Poggi, M. 640
Pomaro, N. 545
Popelier, Lara 668
Porton, Michael 482
Potter, K. G. 226
Prever-Loiri, L. 245
Prever-Loiri, Laurent 255
Puriga, E. A. 634

R

Rajesh, S. 555
Recchia, M. 640
Riedl, R. 604
Rizzolo, A. 584
Roopesh, G. 555
Rotti, C. 555
Rouleau, G. 113

S

Sagot, F. 567
Sakamoto, K. 449
Sakamoto, Keishi 457
Sanin, A. L. 634
Sanin, Andrey 401
Santana, M. 123, 216, 226, 411
Sartori, E. 594
Sasaki, D. 692
Sasaki, Daichi 675
Sasaki, S. 536
Sasao, M. 684
Sasao, Mamiko 675
Sato, M. 374, 517
Sato, Mamoru 359
Savkin, V. Ya. 634
Savkin, Valery 401
Schmitzer, C. 245
Schmitzer, Claus 255

Schunke, B. 545, 555, 567
Scrivens, R. 245
Scrivens, Richard 255
Sekiguchi, H. 517
Sekiguchi, Haruo 359
Senecha, V. K. 150
Serianni, G. 381, 526, 545, 574, 584, 594, 640
Shah, Sejal 555
Shibata, T. 22
Shibuya, M. 374, 517
Shibuya, Masayuki 359
Shimamoto, S. 186
Shinto, Katsuhiro 675
Shivarova, Antonia P. 48, 165, 192
Shoji, T. 329
Simizu, T. 536
Simonin, A. 494
Singh, M. J. 555, 604, 614, 624
Sonara, J. 604
Sonato, P. 545
Soni, J. 604, 614
Soni, Jignesh 624
Srusti, B. 555
St. Lishev, Stiliyan 48, 192
Stelzer, J. 113
Steyaert, D. 245
Steyaert, Didier 255
Stockli, M. P. 216, 226
Stockli, Martin P. 123, 411
Surrey, E. 78
Surrey, Elizabeth 482
Svensson, L. 545, 555, 567

T

Taccogna, Francesco 88
Takeiri, Y. 322, 367, 374, 381, 517, 526
Takeiri, Yasuhiko 1, 359
Takemoto, Jumpei 457
Takeno, T. 684
Tanaka, M. 545
Tanaka, N. 322, 684
Tanaka, Y. 466
Taniguchi, M. 449, 466, 536, 545
Taniguchi, Masaki 457

Tarnev, Khristo Ts. 48, 165
Tarvainen, O. 113, 439
Terai, K. 684
Terasaki, R. 22, 39, 339
Tobari, H. 449, 466, 536, 545
Tobari, Hiroyuki 457
Toigo, V. 545
Tokushige, S. 107
Tokuzawa, T. 374
Tsankov, Tsanko 140
Tsuchida, K. 449
Tsuchida, Kazuki 457
Tsugawa, K. 684
Tsumori, K. 175, 322, 367, 374, 381, 517, 526
Tsumori, Katsuyoshi 359, 675

U

Ueno, A. 235, 292
Umeda, N. 449, 466, 536
Umeda, Naotaka 457
Usui, K. 536

V

Vandevraye, M. 494
van Houtte, D. 567
Vasquez, Jr., M. 692
Veltri, P. 526, 574, 584, 594, 640
Vestergard, H. 245
Villecroze, F. 494

W

Wada, M. 107, 175, 186, 422, 684, 692
Wada, Motoi 134, 675
Wada, S. 58
Watanabe, K. 449, 466, 536, 545
Watanabe, Kazuhiro 457
Welton, R. F. 123, 216, 226, 411
Whitehead, M. 205
Wilhelmsson, M. 245
Wimmer, Ch. 348
Wood, T. 205
Wünderlich, D. 303

Y

Yadav, Ratnakar 624
Yadava, Ratnakar 604
Yamanaka, H. 449
Yamanaka, Haruhiko 457
Yamano, Y. 466
Yoshinari, S. 339

Z

Zaccaria, P. 545
Zelenski, A. 393
Zubarev, P. V. 634